MATHEMATICAL METHODS FOR OPTICAL PHYSICS AND ENGINEERING

The first textbook on mathematical methods focusing on techniques for optical science and engineering, this textbook is ideal for advanced undergraduates and graduate students in optical physics.

Containing detailed sections on the basic theory, the textbook places strong emphasis on connecting the abstract mathematical concepts to the optical systems to which they are applied. It covers many topics which usually only appear in more specialized books, such as Zernike polynomials, wavelet and fractional Fourier transforms, vector spherical harmonics, the z-transform, and the angular spectrum representation.

Most chapters end by showing how the techniques covered can be used to solve an optical problem. Essay problems in each chapter based on research publications, together with numerous exercises, help to further strengthen the connection between the theory and its application.

GREGORY J. GBUR is an Associate Professor of Physics and Optical Science at the University of North Carolina at Charlotte, where he has taught a graduate course on mathematical methods for optics for the past five years and a course on advanced physical optics for two years.

MATHEMATICAL METHODS FOR OPTICAL PHYSICS AND ENGINEERING

GREGORY J. GBUR
University of North Carolina

CAMBRIDGE
UNIVERSITY PRESS

Shaftesbury Road, Cambridge CB2 8EA, United Kingdom

One Liberty Plaza, 20th Floor, New York, NY 10006, USA

477 Williamstown Road, Port Melbourne, VIC 3207, Australia

314–321, 3rd Floor, Plot 3, Splendor Forum, Jasola District Centre, New Delhi – 110025, India

103 Penang Road, #05–06/07, Visioncrest Commercial, Singapore 238467

Cambridge University Press is part of Cambridge University Press & Assessment,
a department of the University of Cambridge.

We share the University's mission to contribute to society through the pursuit of
education, learning and research at the highest international levels of excellence.

www.cambridge.org
Information on this title: www.cambridge.org/9780521516105

© G. J. Gbur 2011

First published 2011

A catalogue record for this publication is available from the British Library

Library of Congress Cataloging-in-Publication data
Gbur, Greg.
Mathematical methods for optical physics and engineering / Greg Gbur.
p. cm.
Includes bibliographical references and index.
ISBN 978-0-521-51610-5 (hardback)
1. Optics–Mathematics. I. Title.
QC355.3.G38 2011
535.01´51–dc22   2010036195

ISBN   978-0-521-51610-5   Hardback

Dedicated to my wife Beth,
and my parents

Contents

Preface   *page* xv

1   Vector algebra   1
     1.1   Preliminaries   1
     1.2   Coordinate system invariance   4
     1.3   Vector multiplication   9
     1.4   Useful products of vectors   12
     1.5   Linear vector spaces   13
     1.6   Focus: periodic media and reciprocal lattice vectors   17
     1.7   Additional reading   24
     1.8   Exercises   24
2   Vector calculus   28
     2.1   Introduction   28
     2.2   Vector integration   29
     2.3   The gradient, ∇ 35
     2.4   Divergence, $\nabla\cdot$ 37
     2.5   The curl, $\nabla\times$ 41
     2.6   Further applications of ∇ 43
     2.7   Gauss' theorem (divergence theorem)   45
     2.8   Stokes' theorem   47
     2.9   Potential theory   48
     2.10   Focus: Maxwell's equations in integral and differential form   51
     2.11   Focus: gauge freedom in Maxwell's equations   57
     2.12   Additional reading   60
     2.13   Exercises   60
3   Vector calculus in curvilinear coordinate systems   64
     3.1   Introduction: systems with different symmetries   64
     3.2   General orthogonal coordinate systems   65
     3.3   Vector operators in curvilinear coordinates   69
     3.4   Cylindrical coordinates   73

	3.5	Spherical coordinates	76
	3.6	Exercises	79
4		Matrices and linear algebra	83
	4.1	Introduction: Polarization and Jones vectors	83
	4.2	Matrix algebra	88
	4.3	Systems of equations, determinants, and inverses	93
	4.4	Orthogonal matrices	102
	4.5	Hermitian matrices and unitary matrices	105
	4.6	Diagonalization of matrices, eigenvectors, and eigenvalues	107
	4.7	Gram–Schmidt orthonormalization	115
	4.8	Orthonormal vectors and basis vectors	118
	4.9	Functions of matrices	120
	4.10	Focus: matrix methods for geometrical optics	120
	4.11	Additional reading	133
	4.12	Exercises	133
5		Advanced matrix techniques and tensors	139
	5.1	Introduction: Foldy–Lax scattering theory	139
	5.2	Advanced matrix terminology	142
	5.3	Left–right eigenvalues and biorthogonality	143
	5.4	Singular value decomposition	146
	5.5	Other matrix manipulations	153
	5.6	Tensors	159
	5.7	Additional reading	174
	5.8	Exercises	174
6		Distributions	177
	6.1	Introduction: Gauss' law and the Poisson equation	177
	6.2	Introduction to delta functions	181
	6.3	Calculus of delta functions	184
	6.4	Other representations of the delta function	185
	6.5	Heaviside step function	187
	6.6	Delta functions of more than one variable	188
	6.7	Additional reading	192
	6.8	Exercises	192
7		Infinite series	195
	7.1	Introduction: the Fabry–Perot interferometer	195
	7.2	Sequences and series	198
	7.3	Series convergence	201
	7.4	Series of functions	210
	7.5	Taylor series	213
	7.6	Taylor series in more than one variable	218
	7.7	Power series	220
	7.8	Focus: convergence of the Born series	221

	7.9	Additional reading	226
	7.10	Exercises	226
8	Fourier series		230
	8.1	Introduction: diffraction gratings	230
	8.2	Real-valued Fourier series	233
	8.3	Examples	236
	8.4	Integration range of the Fourier series	239
	8.5	Complex-valued Fourier series	239
	8.6	Properties of Fourier series	240
	8.7	Gibbs phenomenon and convergence in the mean	243
	8.8	Focus: X-ray diffraction from crystals	246
	8.9	Additional reading	249
	8.10	Exercises	249
9	Complex analysis		252
	9.1	Introduction: electric potential in an infinite cylinder	252
	9.2	Complex algebra	254
	9.3	Functions of a complex variable	258
	9.4	Complex derivatives and analyticity	261
	9.5	Complex integration and Cauchy's integral theorem	265
	9.6	Cauchy's integral formula	269
	9.7	Taylor series	271
	9.8	Laurent series	273
	9.9	Classification of isolated singularities	276
	9.10	Branch points and Riemann surfaces	278
	9.11	Residue theorem	285
	9.12	Evaluation of definite integrals	288
	9.13	Cauchy principal value	297
	9.14	Focus: Kramers–Kronig relations	299
	9.15	Focus: optical vortices	302
	9.16	Additional reading	308
	9.17	Exercises	308
10	Advanced complex analysis		312
	10.1	Introduction	312
	10.2	Analytic continuation	312
	10.3	Stereographic projection	316
	10.4	Conformal mapping	325
	10.5	Significant theorems in complex analysis	332
	10.6	Focus: analytic properties of wavefields	340
	10.7	Focus: optical cloaking and transformation optics	345
	10.8	Exercises	348
11	Fourier transforms		350
	11.1	Introduction: Fraunhofer diffraction	350

11.2	The Fourier transform and its inverse	352
11.3	Examples of Fourier transforms	354
11.4	Mathematical properties of the Fourier transform	358
11.5	Physical properties of the Fourier transform	365
11.6	Eigenfunctions of the Fourier operator	372
11.7	Higher-dimensional transforms	373
11.8	Focus: spatial filtering	375
11.9	Focus: angular spectrum representation	377
11.10	Additional reading	382
11.11	Exercises	383
12	Other integral transforms	386
12.1	Introduction: the Fresnel transform	386
12.2	Linear canonical transforms	391
12.3	The Laplace transform	395
12.4	Fractional Fourier transform	400
12.5	Mixed domain transforms	402
12.6	The wavelet transform	406
12.7	The Wigner transform	409
12.8	Focus: the Radon transform and computed axial tomography (CAT)	410
12.9	Additional reading	416
12.10	Exercises	416
13	Discrete transforms	419
13.1	Introduction: the sampling theorem	419
13.2	Sampling and the Poisson sum formula	423
13.3	The discrete Fourier transform	427
13.4	Properties of the DFT	430
13.5	Convolution	432
13.6	Fast Fourier transform	433
13.7	The z-transform	437
13.8	Focus: z-transforms in the numerical solution of Maxwell's equations	445
13.9	Focus: the Talbot effect	449
13.10	Exercises	456
14	Ordinary differential equations	458
14.1	Introduction: the classic ODEs	458
14.2	Classification of ODEs	459
14.3	Ordinary differential equations and phase space	460
14.4	First-order ODEs	469
14.5	Second-order ODEs with constant coefficients	474
14.6	The Wronskian and associated strategies	476
14.7	Variation of parameters	478
14.8	Series solutions	480
14.9	Singularities, complex analysis, and general Frobenius solutions	481

14.10	Integral transform solutions	485
14.11	Systems of differential equations	486
14.12	Numerical analysis of differential equations	488
14.13	Additional reading	501
14.14	Exercises	501
15	**Partial differential equations**	**505**
15.1	Introduction: propagation in a rectangular waveguide	505
15.2	Classification of second-order linear PDEs	508
15.3	Separation of variables	517
15.4	Hyperbolic equations	519
15.5	Elliptic equations	525
15.6	Parabolic equations	530
15.7	Solutions by integral transforms	534
15.8	Inhomogeneous problems and eigenfunction solutions	538
15.9	Infinite domains; the d'Alembert solution	539
15.10	Method of images	544
15.11	Additional reading	545
15.12	Exercises	545
16	**Bessel functions**	**550**
16.1	Introduction: propagation in a circular waveguide	550
16.2	Bessel's equation and series solutions	552
16.3	The generating function	555
16.4	Recurrence relations	557
16.5	Integral representations	560
16.6	Hankel functions	564
16.7	Modified Bessel functions	565
16.8	Asymptotic behavior of Bessel functions	566
16.9	Zeros of Bessel functions	567
16.10	Orthogonality relations	569
16.11	Bessel functions of fractional order	572
16.12	Addition theorems, sum theorems, and product relations	576
16.13	Focus: nondiffracting beams	579
16.14	Additional reading	582
16.15	Exercises	582
17	**Legendre functions and spherical harmonics**	**585**
17.1	Introduction: Laplace's equation in spherical coordinates	585
17.2	Series solution of the Legendre equation	587
17.3	Generating function	589
17.4	Recurrence relations	590
17.5	Integral formulas	592
17.6	Orthogonality	594
17.7	Associated Legendre functions	597

17.8	Spherical harmonics	602
17.9	Spherical harmonic addition theorem	605
17.10	Solution of PDEs in spherical coordinates	608
17.11	Gegenbauer polynomials	610
17.12	Focus: multipole expansion for static electric fields	611
17.13	Focus: vector spherical harmonics and radiation fields	614
17.14	Exercises	618
18	**Orthogonal functions**	622
18.1	Introduction: Sturm–Liouville equations	622
18.2	Hermite polynomials	627
18.3	Laguerre functions	641
18.4	Chebyshev polynomials	650
18.5	Jacobi polynomials	654
18.6	Focus: Zernike polynomials	655
18.7	Additional reading	662
18.8	Exercises	662
19	**Green's functions**	665
19.1	Introduction: the Huygens–Fresnel integral	665
19.2	Inhomogeneous Sturm–Liouville equations	669
19.3	Properties of Green's functions	674
19.4	Green's functions of second-order PDEs	676
19.5	Method of images	685
19.6	Modal expansion of Green's functions	689
19.7	Integral equations	693
19.8	Focus: Rayleigh–Sommerfeld diffraction	701
19.9	Focus: dyadic Green's function for Maxwell's equations	704
19.10	Focus: scattering theory and the Born series	709
19.11	Exercises	712
20	**The calculus of variations**	715
20.1	Introduction: principle of Fermat	715
20.2	Extrema of functions and functionals	718
20.3	Euler's equation	721
20.4	Second form of Euler's equation	727
20.5	Calculus of variations with several dependent variables	730
20.6	Calculus of variations with several independent variables	732
20.7	Euler's equation with auxiliary conditions: Lagrange multipliers	734
20.8	Hamiltonian dynamics	739
20.9	Focus: aperture apodization	742
20.10	Additional reading	745
20.11	Exercises	745
21	**Asymptotic techniques**	748
21.1	Introduction: foundations of geometrical optics	748

21.2   Definition of an asymptotic series   753

21.3   Asymptotic behavior of integrals   756

21.4   Method of stationary phase   763

21.5   Method of steepest descents   766

21.6   Method of stationary phase for double integrals   771

21.7   Additional reading   772

21.8   Exercises   773

Appendix A The gamma function   775

A.1   Definition   775

A.2   Basic properties   776

A.3   Stirling's formula   778

A.4   Beta function   779

A.5   Useful integrals   780

Appendix B Hypergeometric functions   783

B.1   Hypergeometric function   784

B.2   Confluent hypergeometric function   785

B.3   Integral representations   785

References   787

Index   793

Preface

Why another textbook on Mathematical Methods for Scientists? Certainly there are quite a few good, indeed classic texts on the subject. What can another text add that these others have not already done?

I began to ponder these questions, and my answers to them, over the past several years while teaching a graduate course on Mathematical Methods for Physics and Optical Science at the University of North Carolina at Charlotte. Although every student has his or her own difficulties in learning mathematical techniques, a few problems amongst the students have remained common and constant. The foremost among these is the "wall" between the mathematics the students learn in math class and the applications they study in other classes. The Fourier transform learned in math class is internally treated differently than the Fourier transform used in, say, Fraunhofer diffraction. The end result is that the student effectively learns the same topic twice, and is unable to use the intuition learned in a physics class to help aid in mathematical understanding, or to use the techniques learned in math class to formulate and solve physical problems.

To try and correct for this, I began to devote special lectures to the consequences of the math the students were studying. Lectures on complex analysis would be followed by discussions of the analytic properties of wavefields and the Kramers–Kronig relations. Lectures on infinite series could be highlighted by the discussion of the Fabry–Perot interferometer.

Students in my classes were uniformly dissatisfied with the standard textbooks. Part of this dissatisfaction arises from the broad topics from which examples are drawn: quantum physics, field theory, general relativity, optics, mechanics, and thermodynamics, to name a few. Even the most dedicated theoretical physics students do not have a great physical intuition about all these subfields, and consequently many of the examples are no more useful in their minds than problems in abstract mathematics.

Given that students in my class are studying optics, I have focused most of my attention on methods directly related to optical science. Here again the standard texts became a problem, as there is not a perfect overlap between important methods for general physics and important methods for optics. For example, group theory is not commonly used among most optics researchers, and Fourier transforms, essential to the optics researcher, are not used as much by the rest of the general physics community. Teaching to an optics crowd would require that the emphasis on material be refocused. It was in view of these various observations that I decided that a new mathematical methods book, with an emphasis on optics, would be useful.

Optics as both an industry and a field of study in its own right has grown dramatically over the past two decades. Optics programs at universities have grown in size and number in recent years. The University of Rochester and the University of Arizona are schools which have had degrees for some time, while the University of Central Florida and the University of North Carolina at Charlotte have started programs within recent years. With countless electrical engineering programs emphasizing studies in optics, it seems likely that more optics degrees will follow in the years to come. A textbook which serves such programs and optical researchers in general seems to have the potential to be a popular resource.

My goal, then, was to write a textbook on mathematical methods for physics and optical science, with an emphasis on those techniques relevant to optical scientists and engineers. The level of the book is intended for an advanced undergraduate or beginning graduate level class on math methods. One of my main objectives was to write a "leaner" book than many of the 1000+ page math books currently available, and do so by pushing much of the abstract mathematical subtlety into references. Instead, the emphasis is placed on making the connection between the mathematical techniques and the optics problems they are intimately related to. To make this connection, most chapters begin with a short introduction which illustrates the relevance of the mathematical technique being considered, and ends with one or more applications, in which the technique is used to solve a problem. Physical examples within the chapters are drawn predominantly from optics, though examples from other fields will be used when appropriate.

A book of this type will address a number of mathematical techniques which are normally not compiled into a single volume. It is hoped that this book will therefore serve not only as a textbook but also potentially as a reference book for researchers in optics.

Another "wall" in students' understanding is making the connection between the topics learned in class and research results in the literature. A number of exercises in each chapter are essay-style questions, in which a journal article must be read and its relevance to the mathematical method discussed. I have also endeavored to provide an appreciable number of exercises in each chapter, with some similar problems to facilitate teaching a class multiple times. Some more advanced chapters have fewer exercises, mainly because it is difficult to find exercises that are simultaneously solvable and enlightening.

Early chapters cover the basics that are essential for any student of the physical sciences, including vectors, curvilinear coordinate systems, differential equations, sequences and series, matrices, and this part of the book might be used for any math methods for physics course. Later chapters concentrate on techniques significant to optics, including Fourier analysis, asymptotic methods, Green's functions, and more general types of integral transform.

A book of this sort requires a lot of help, and I have sought plenty of insight from colleagues. I would like to thank Professor John Schotland, Professor Daniel James, Professor Tom Suleski, Dr Choon How Gan, Mike Fairchild and Casey Rhodes for helpful suggestions during the course of writing. I give special thanks to Professor Taco Visser and Dr Damon

Diehl, each of whom read significant sections of the manuscript and provided corrections, and to Professor Emil Wolf, who gave me encouragement and inspiration during the writing process. I am grateful to Professor John Foley who some time ago gave me access to his collection of math methods exercises, which were useful in developing my own problems. Professor Daniel S. Jones generously provided a photograph of X-ray diffraction, and Professor Visser provided a figure on the Poincaré sphere. I would also like to express my appreciation to the very helpful people at Cambridge University Press, including Simon Capelin, John Fowler, Megan Waddington, and Lindsay Barnes. Special thanks goes to Frances Nex for her careful editing of the text.

I also have to thank a number of people for their help in keeping me sane during the writing process! Among them, let me thank my guitar instructor Toby Watson, my skating coach Tappie Dellinger, and my friends at Skydive Carolina, particularly my regular jump buddies Nancy Newman, Mickey Turner, John Solomon, Robyn Porter, Mike Reinhart, and Heiko Lotz! I would also like to give a "shout out" to Eric Smith and Mahy El-Kouedi for their friendship and support.

Finally, let me thank my wife Beth Szabo for her support, understanding and patience during this rather strenuous writing process.

1
Vector algebra

1.1 Preliminaries

In introductory physics, we often deal with physical quantities that can be described by a single number. The temperature of a heated body, the mass of an object, and the electric potential of an insulated metal sphere are all examples of such *scalar* quantities.

Descriptions of physical phenomena are not always (indeed, rarely) that simple, however, and often we must use multiple, but related, numbers to offer a complete description of an effect. The next level of complexity is the introduction of *vector* quantities.

A vector may be described as a conceptual object having both *magnitude* and *direction*. Graphically, vectors can be represented by an arrow:

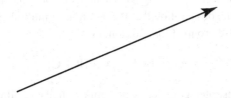

The length of the arrow is the *magnitude* of the vector, and the direction of the arrow indicates the *direction* of the vector.

Examples of vectors in elementary physics include displacement, velocity, force, momentum, and angular momentum, though the concept can be extended to more complicated and abstract systems. Algebraically, we will usually represent vectors by boldface characters, i.e. $\mathbf{F}$ for force, $\mathbf{v}$ for velocity, and so on.

It is worth noting at this point that the word "vector" is used in mathematics with somewhat broader meaning. In mathematics, a *vector space* is defined quite generally as a set of elements (called vectors) together with rules relating to their addition and scalar multiplication of vectors. In this sense, the set of real numbers form a vector space, as does any ordered set of numbers, including matrices, to be discussed in Chapter 4, and complex numbers, to be discussed in Chapter 9. For most of this chapter we reserve the term "vector" for quantities which possess magnitude and direction in three-dimensional space, and are independent of the specific choice of coordinate system in a manner to be discussed in

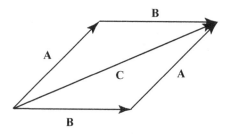

Figure 1.1 The parallelogram law of vector addition. Adding **B** to **A** (the addition above the **C**-line) is equivalent to adding **A** to **B** (the addition below the **C**-line).

Section 1.2. We briefly describe vector spaces at the end of this chapter, in Section 1.5. The interested reader can also consult Ref. [Kre78, Sec. 2.1].

Vector addition is *commutative* and *associative*; commutativity refers to the observation that the addition of vectors is order independent, i.e.

$$\mathbf{A} + \mathbf{B} = \mathbf{B} + \mathbf{A} = \mathbf{C}. \tag{1.1}$$

This can be depicted graphically by the parallelogram law of vector addition, illustrated in Fig. 1.1. A pair of vectors are added "tip-to-tail"; that is, the second vector is added to the first by putting its tail at the end of the tip of the first vector. The resultant vector is found by drawing an arrow from the origin of the first vector to the tip of the second vector. Associativity refers to the observation that the addition of multiple vectors is independent of the way the vectors are grouped for addition, i.e.

$$(\mathbf{A} + \mathbf{B}) + \mathbf{C} = \mathbf{A} + (\mathbf{B} + \mathbf{C}). \tag{1.2}$$

This may also be demonstrated graphically if we first define the following vector additions:

$$\mathbf{E} \equiv \mathbf{A} + \mathbf{B}, \tag{1.3}$$

$$\mathbf{D} \equiv \mathbf{E} + \mathbf{C}, \tag{1.4}$$

$$\mathbf{F} \equiv \mathbf{B} + \mathbf{C}. \tag{1.5}$$

The vectors and their additions are illustrated in Fig. 1.2. It can be immediately seen that

$$\mathbf{E} + \mathbf{C} = \mathbf{A} + \mathbf{F}. \tag{1.6}$$

So far, we have introduced vectors as purely geometrical objects which are independent of any specific coordinate system. Intuitively, this is an obvious requirement: where I am standing in a room (my "position vector") is independent of whether I choose to describe it by measuring it from the rear left corner of the room or the front right corner. In other words, the vector has a physical significance which does not change when I change my method of describing it.

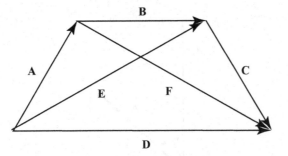

Figure 1.2 The trapezoid rule of vector addition. It makes no difference if we first add **A** and **B**, and then **C**, or first add **B** and **C**, and then **A**.

By choosing a coordinate system, however, we may create a *representation* of the vector in terms of these coordinates. We start by considering a Cartesian coordinate system with coordinates x, y, z which are all mutually perpendicular and form a right-handed coordinate system.[1] For a given Cartesian coordinate system, the vector **A**, which starts at the origin and ends at the point with coordinates (A_x, A_y, A_z), is completely described by the coordinates of the end point.

It is highly convenient to express a vector in terms of these components by use of unit vectors $\hat{\mathbf{x}}, \hat{\mathbf{y}}, \hat{\mathbf{z}}$, vectors of unit magnitude pointing in the directions of the positive coordinate axes,

$$\mathbf{A} = A_x\hat{\mathbf{x}} + A_y\hat{\mathbf{y}} + A_z\hat{\mathbf{z}}. \tag{1.7}$$

This equation indicates that a vector equals the vector sum of its components. In three dimensions, the *position vector* **r** which measures the distance from a chosen origin is written as

$$\mathbf{r} = x\hat{\mathbf{x}} + y\hat{\mathbf{y}} + z\hat{\mathbf{z}}, \tag{1.8}$$

where x, y, and z are the lengths along the different coordinate axes.

The sum of two vectors can be found by taking the sum of their individual components. This means that the sum of two vectors **A** and **B** can be written as

$$\mathbf{A} + \mathbf{B} = (A_x + B_x)\hat{\mathbf{x}} + (A_y + B_y)\hat{\mathbf{y}} + (A_z + B_z)\hat{\mathbf{z}}. \tag{1.9}$$

The magnitude (length) of a vector in terms of its components can be found by two successive applications of the Pythagorean theorem. The magnitude A of the complete vector, also written as $|\mathbf{A}|$, is found to be

$$A = \sqrt{A_x^2 + A_y^2 + A_z^2}. \tag{1.10}$$

Another way to represent the vector in a particular coordinate system is by its magnitude A and the angles α, β, γ that the vector makes with each of the positive coordinate axes.

[1] If x is the outward-pointing index finger of the right hand, y is the folded-in ring finger and z is the thumb, pointing straight up.

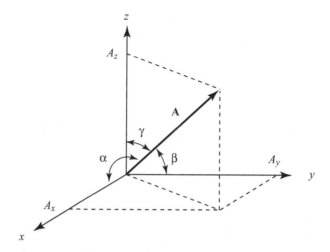

Figure 1.3 Illustration of the vector **A**, its components (A_x, A_y, A_z), and the angles α, β, γ.

These angles and their relationship to the vector and its components are illustrated in Fig. 1.3. The quantities $\cos\alpha$, $\cos\beta$, and $\cos\gamma$ are called *direction cosines*. It might seem that there is an inconsistency with this representation, since we now evidently need four numbers $(A, \alpha, \beta, \gamma)$ to describe the vector, where we needed only three (A_x, A_y, A_z) before. This seeming contradiction is resolved by the observation that α, β, and γ are not independent quantities; they are related by the equation,

$$\cos^2\alpha + \cos^2\beta + \cos^2\gamma = 1. \tag{1.11}$$

In the spherical coordinate system to be discussed in Chapter 3, we will see that we may completely specify the position vector by its magnitude r and two angles θ and ϕ.

It is to be noted that we usually see vectors in physics in two distinct classes:

1. Vectors associated with the property of a single, localized object, such as the velocity of a car, or the force of gravity acting on a moving projectile.
2. Vectors associated with the property of a nonlocalized "object" or system, such as the electric field of a light wave, or the velocity of a fluid. In such a case, the vector quantity is a function of position and possibly time and we may do calculus with respect to the position and time variables. This vector quantity is usually referred to as a *vector field*.

Vector fields are extremely important quantities in physics and we will return to them often.

1.2 Coordinate system invariance

We have said that a vector is independent of any specific coordinate system – in other words, that a vector is independent of how we choose to characterize it. This seems like an obvious criterion, but there are physical quantities which have magnitude and direction but are not vectors; an example of this in optics is the set of principle indices of refraction of an anisotropic crystal. Thus, to define a vector properly, we need to formulate mathematically this concept of *coordinate system invariance*. Furthermore, it is not uncommon to require, in the solution of a physical problem, the transformation from one coordinate system to another. We therefore take some time to study the mathematics relating to the behavior of a vector under a change of coordinates.

The simplest coordinate transformation is a change of origin, leaving the orientation of the axes unchanged. The only vector that depends explicitly upon the origin is the position vector $\mathbf{r}$, which is a measure of the vector distance from the origin. If the new origin of a new coordinate system, described by position vector $\mathbf{r}'$, is located at the position $\mathbf{r}_0$ from the old origin, the coordinates are related by the formula

$$\mathbf{r}' = \mathbf{r} - \mathbf{r}_0. \tag{1.12}$$

Most other basic vectors depend upon the *displacement vector* $\mathbf{R} = \mathbf{r}_2 - \mathbf{r}_1$, i.e. the change in position, and therefore are unaffected by a change in origin. Examples include the velocity, momentum, and force upon an object.

A less trivial example of a change of coordinate system is a change of the orientation of coordinate axes, and its effect on a position vector $\mathbf{r}$. For simplicity, we first consider the two dimensional case. The vector $\mathbf{r}$ may be written in one coordinate system as $\mathbf{r} = x\hat{\mathbf{x}} + y\hat{\mathbf{y}}$, while in a second coordinate system this vector may be written as $\mathbf{r}' = x'\hat{\mathbf{x}}' + y'\hat{\mathbf{y}}'$. The (x, y) coordinate axes are rotated to a new location to become the (x', y') axes, while leaving the vector $\mathbf{r}$ (in particular, the location of the tip of $\mathbf{r}$) fixed. The question we ask: what are the components of the vector $\mathbf{r}$ in the new coordinate system, which makes an angle ϕ with the old system? The relation between the two systems is illustrated in Fig. 1.4.

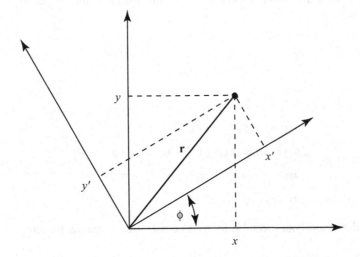

Figure 1.4 Illustration of the position vector $\mathbf{r}$ and its components in the two coordinate systems.

By straightforward trigonometry, one can readily find that the new coordinates of the vector (x', y') may be written in terms of the old coordinates as

$$x' = x\cos\phi + y\sin\phi, \tag{1.13}$$

$$y' = -x\sin\phi + y\cos\phi. \tag{1.14}$$

These equations are based on the assumption that the magnitude and direction of the vector is independent of the coordinate system, and this assumption should hold for anything we refer to as a vector. We therefore *define* a vector as a quantity whose components transform under rotations just as the position vector **r** does, i.e. a vector **A** with components A_x and A_y in the unprimed system should have components

$$A'_x = A_x\cos\phi + A_y\sin\phi, \tag{1.15}$$

$$A'_y = -A_x\sin\phi + A_y\cos\phi \tag{1.16}$$

in the primed system.

It is important to emphasize again that we are only rotating the coordinate axes, and that the vector **A** does not change: (A_x, A_y) and (A'_x, A'_y) are *representations* of the vector, different ways of describing the same physical property. Indeed, another way to define a vector is that it is a quantity with magnitude and direction that is independent of the coordinate system.

We can also interpret Eqs. (1.15) and (1.16) in an entirely different manner: if we were to physically rotate the vector **A** over an angle $-\phi$ about the origin of the coordinate system, the new direction of the vector in the *same* coordinate system would be given by (A'_x, A'_y). A rotation of the coordinate system in the direction ϕ is mathematically equivalent to a rotation of the vector by an angle $-\phi$.

To generalize the discussion to three (or more) dimensions, it helps to modify the notation somewhat. We write

$$x \to x_1, \tag{1.17}$$

$$y \to x_2, \tag{1.18}$$

and define

$$a_{11} = \cos\phi,$$

$$a_{12} = \sin\phi = \cos(\pi/2 - \phi) = \cos(\phi - \pi/2),$$

$$a_{21} = -\sin\phi = -a_{12} = \cos(\phi + \pi/2),$$

$$a_{22} = \cos\phi. \tag{1.19}$$

With these definitions, our formulas for a coordinate transformation become

$$x'_1 = a_{11}x_1 + a_{12}x_2, \tag{1.20}$$

$$x'_2 = a_{21}x_1 + a_{22}x_2. \tag{1.21}$$

These transformations may be written in a summation format,

$$x_i' = \sum_{j=1}^{2} a_{ij} x_j, \quad i = 1, 2,$$ (1.22)

where x_i' is the ith component of the vector in the primed frame, x_j is the jth component of the vector in the unprimed frame, and the notation $\sum_{j=n}^{m}$ indicates summation over all terms with index j ranging from n to m. The quantity a_{ij} can be seen from Eqs. (1.19) to be the *direction cosine* with the respect to the ith primed coordinate and the jth unprimed coordinate.

What happens if we run the rotation in reverse? We can still use Eq. (1.21), but we replace ϕ by $-\phi$ and switch the primed and unprimed coordinates, i.e. the primed coordinates are now the start of the rotation and the unprimed coordinates are now the end of the rotation. With these changes, Eq. (1.21) becomes

$$x_1 = a_{11} x_1' - a_{12} x_2',$$ (1.23)

$$x_2 = -a_{21} x_1' + a_{22} x_2'.$$ (1.24)

Noting that $a_{12} = -a_{21}$, these formulas can be rewritten in the compact summation form,

$$x_j = \sum_{i=1}^{2} a_{ij} x_i', \quad j = 1, 2.$$ (1.25)

Generalizing to N dimensions may now be done by analogy, simply introducing higher-order direction cosines a_{ij} and an N-dimensional position vector $\mathbf{r} = (x_1, \dots, x_N)$, which will satisfy the relations

$$x_i' = \sum_{j=1}^{N} a_{ij} x_j, \quad i = 1, \dots, N.$$ (1.26)

The a_{ij}s may be written in a differential form with respect to the two coordinate systems as

$$a_{ij} = \frac{\partial x_i'}{\partial x_j}.$$ (1.27)

This formula can be derived by taking partial derivatives of the transformation equations for $\mathbf{r}$, namely Eq. (1.26). The quantity a_{ij} has more components than a vector and will be seen to be a *tensor*, which can be represented in matrix form; we will discuss such beasts in Chapter 5.

The reverse rotation may also be written by analogy,

$$x_j = \sum_{i=1}^{N} a_{ij} x_i',$$ (1.28)

and from this expression we may also write

$$a_{ij} = \frac{\partial x_j}{\partial x_i'}. \tag{1.29}$$

It is evident that the coordinates of a vector must in the end be unchanged if the axes are first rotated and then rotated back to their original positions; from this we can derive an *orthogonality condition* for the coefficients a_{ij}. We begin with the transformation of the vector **V** and its reverse,

$$V_k = \sum_{i=1}^{N} a_{ik} V_i', \tag{1.30}$$

$$V_i' = \sum_{j=1}^{N} a_{ij} V_j. \tag{1.31}$$

On substitution of the latter equation into the former, we have

$$V_k = \sum_{i=1}^{N} a_{ik} \left[\sum_{j=1}^{N} a_{ij} V_j \right] = \sum_{j=1}^{N} \left[\sum_{i=1}^{N} a_{ik} a_{ij} \right] V_j. \tag{1.32}$$

The left-hand side of this equation is the kth component of the vector in the unprimed frame. The right-hand side of the equation is a weighted sum of all components of the vector in the unprimed frame. By the use of Eqs. (1.27) and (1.29), we may write the quantity in square brackets as

$$\sum_{i=1}^{N} a_{ik} a_{ij} = \sum_{i=1}^{N} \frac{\partial x_k}{\partial x_i'} \frac{\partial x_i'}{\partial x_j} = \frac{\partial x_k}{\partial x_j}, \tag{1.33}$$

where the final step is the result of application of the chain rule of calculus. Because the variables x_j, for $j = 1, \ldots, N$, are independent of one another, we readily find that

$$\sum_{i=1}^{N} a_{ik} a_{ij} = \delta_{jk}, \tag{1.34}$$

where δ_{jk} is the *Kronecker delta*, defined as

$$\delta_{jk} = \begin{cases} 1 & \text{for} \quad j = k, \\ 0 & \text{for} \quad j \neq k. \end{cases} \tag{1.35}$$

We will see a lot of the Kronecker delta in the future – remember it! It is another example of a *tensor*, like the rotation tensor a_{ij}.

1.3 Vector multiplication

In Section 1.1, we looked at the addition of vectors, which may be considered a generalization of the addition of ordinary numbers. One can envision that there exists a generalized form of multiplication for vectors, as well; with three components for each vector, however, there are a large number of possibilities for what we might call "vector multiplication". Just as vectors themselves are invariant under a change of coordinates, any vector multiplication should also be invariant under a change of coordinates. It turns out that for three-dimensional vectors there exist four possibilities, three of which we discuss here.[2]

1.3.1 Multiplication by a scalar

The simplest form of multiplication involving vectors is the multiplication of a vector by a scalar. The effect of such a multiplication is the "scaling" of each component of the vector equally by the scalar α, i.e.

$$\alpha \mathbf{V} = \alpha (V_x \hat{\mathbf{x}} + V_y \hat{\mathbf{y}} + V_z \hat{\mathbf{z}}) = (\alpha V_x)\hat{\mathbf{x}} + (\alpha V_y)\hat{\mathbf{y}} + (\alpha V_z)\hat{\mathbf{z}}. \tag{1.36}$$

It is clear from the above that the act of multiplying by a scalar does not change the direction of the vector, but only scales its length by the factor α; we may formally write $|\alpha \mathbf{V}| = |\alpha||\mathbf{V}|$. The result of the multiplication is also a vector, as it is clear that this product is invariant under a rotation of the coordinate axes, which does not affect the vector length.

It is to be noted that we may also consider scalar multiplication "backwards", i.e. that it represents the multiplication of a scalar by a vector, with the end result being a vector. This interpretation will be employed in Chapter 2 to help categorize the different types of vector differentiation.

1.3.2 Scalar or dot product

The *scalar product* (or "dot product") between two vectors is represented by a dot and is defined as

$$\mathbf{A} \cdot \mathbf{B} \equiv AB \cos \theta, \tag{1.37}$$

where A and B are the magnitudes of vectors $\mathbf{A}$ and $\mathbf{B}$, respectively, and θ is the angle between the two vectors.

The rotational invariance of this quantity is almost obvious from its definition, for we know that the magnitudes of vectors and the angles between any two vectors are all unchanged under rotations. We will confirm this more rigorously in a moment.

In terms of components in a particular Cartesian coordinate system, the scalar product is given by[3]

$$\mathbf{A} \cdot \mathbf{B} = \sum_i A_i B_i = \sum_i B_i A_i = \mathbf{B} \cdot \mathbf{A}. \tag{1.38}$$

---

[2] A discussion of the fourth, the direct product, will be deferred until Section 5.6.
[3] From now on, we no longer write the upper and lower ranges of the summations.

We can use this representation of the scalar product to rigorously prove that it is invariant under rotations. We start with the scalar product in the primed coordinate system, and substitute into it the representation of the primed vectors in terms of the unprimed coordinates, Eq. (1.31),

$$\sum_i A_i' B_i' = \sum_i \left[\sum_j a_{ij} A_j \sum_k a_{ik} B_k \right] = \sum_j \sum_k \left[\sum_i a_{ij} a_{ik} \right] A_j B_k. \qquad (1.39)$$

The expression in the last brackets is simply the orthogonality relation between the direction cosines, Eq. (1.34), and may be set to δ_{jk}. We thus have

$$\sum_i A_i' B_i' = \sum_i A_i B_i \qquad (1.40)$$

and we have proven that the scalar product is invariant under rotations.

The dot product may be used to demonstrate another familiar geometrical formula, the *law of cosines*. Defining

$$\mathbf{C} = \mathbf{A} + \mathbf{B}, \qquad (1.41)$$

it follows from the parallelogram law of vector addition that $\mathbf{A}$, $\mathbf{B}$, and $\mathbf{C}$ form the sides of a triangle. If we take the dot product of $\mathbf{C}$ with itself,

$$\mathbf{C} \cdot \mathbf{C} = (\mathbf{A} + \mathbf{B}) \cdot (\mathbf{A} + \mathbf{B}) = \mathbf{A} \cdot \mathbf{A} + \mathbf{B} \cdot \mathbf{B} + 2\mathbf{A} \cdot \mathbf{B}. \qquad (1.42)$$

The dot product of a vector with itself is simply the squared magnitude of the vector, and the dot product of $\mathbf{A}$ and $\mathbf{B}$ is defined by Eq. (1.37). We thus arrive at

$$C^2 = A^2 + B^2 + 2AB\cos\theta, \qquad (1.43)$$

the law of cosines.

1.3.3 Vector or cross product

The third form of rotationally invariant product involving vectors is the *vector product* or "cross product". Just as the scalar product is named such because the result of the product is a scalar, the result of the vector product is another vector. It is represented by a cross between vectors,

$$\mathbf{C} = \mathbf{A} \times \mathbf{B}. \qquad (1.44)$$

In evident contrast with the scalar product, the magnitude of the vector product is defined as

$$C = AB\sin\theta, \qquad (1.45)$$

where θ is again the angle between the vectors $\mathbf{A}$ and $\mathbf{B}$. From this definition, it is to be noted that the magnitude of $\mathbf{C}$ is the area of the parallelogram formed by $\mathbf{A}$ and $\mathbf{B}$, and that

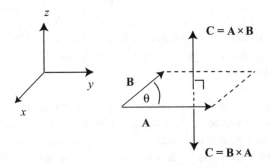

Figure 1.5 Illustration of the cross product. Vector $\mathbf{A}$ points in the $+\hat{\mathbf{y}}$ direction, while $\mathbf{B}$ points in the $-\hat{\mathbf{x}}$ direction. Then $\mathbf{C} = \mathbf{A} \times \mathbf{B}$ points in the $+\hat{\mathbf{z}}$ direction, while $\mathbf{C} = \mathbf{B} \times \mathbf{A}$ points in the $-\hat{\mathbf{z}}$ direction.

$\mathbf{A} \times \mathbf{A} = 0$ for any vector $\mathbf{A}$. The parallelogram formed by $\mathbf{A}$ and $\mathbf{B}$ defines a plane, and we assign to the vector $\mathbf{C}$ a direction such that it is perpendicular to the plane of $\mathbf{A}$ and $\mathbf{B}$ in a manner such that $\mathbf{A}$, $\mathbf{B}$, and $\mathbf{C}$ form a right-handed system. This is illustrated in Fig. 1.5.

It immediately follows from this definition of direction that

$$\mathbf{A} \times \mathbf{B} = -\mathbf{B} \times \mathbf{A}. \tag{1.46}$$

We may also ascertain the effect of the cross product on the unit vectors of the Cartesian coordinate system,

$$\hat{\mathbf{x}} \times \hat{\mathbf{x}} = \hat{\mathbf{y}} \times \hat{\mathbf{y}} = \hat{\mathbf{z}} \times \hat{\mathbf{z}} = 0, \tag{1.47}$$

and

$$\begin{aligned} \hat{\mathbf{x}} \times \hat{\mathbf{y}} = \hat{\mathbf{z}}, \quad & \hat{\mathbf{y}} \times \hat{\mathbf{z}} = \hat{\mathbf{x}}, \quad \hat{\mathbf{z}} \times \hat{\mathbf{x}} = \hat{\mathbf{y}} \\ \hat{\mathbf{y}} \times \hat{\mathbf{x}} = -\hat{\mathbf{z}}, \quad & \hat{\mathbf{z}} \times \hat{\mathbf{y}} = -\hat{\mathbf{x}}, \quad \hat{\mathbf{x}} \times \hat{\mathbf{z}} = -\hat{\mathbf{y}}. \end{aligned} \tag{1.48}$$

From these results, we may determine the Cartesian components of $\mathbf{C}$ in terms of the components of $\mathbf{A}$ and $\mathbf{B}$,

$$C_x = A_y B_z - A_z B_y, \quad C_y = A_z B_x - A_x B_z, \quad C_z = A_x B_y - A_y B_x. \tag{1.49}$$

These rules can be put in a convenient determinant form,[4]

$$\mathbf{C} = \begin{vmatrix} \hat{\mathbf{x}} & \hat{\mathbf{y}} & \hat{\mathbf{z}} \\ A_x & A_y & A_z \\ B_x & B_y & B_z \end{vmatrix}. \tag{1.50}$$

We may also write the cross product in a summation form, though we need to introduce an additional tensor, the *Levi-Civita tensor* ϵ_{ijk},

$$C_i = \sum_j \sum_k \epsilon_{ijk} A_j B_k, \tag{1.51}$$

---

[4] Determinants will be discussed in Section 4.3. They are not necessary to understand this chapter.

where

$$\epsilon_{ijk} = \begin{cases} 1 & \text{if } ijk \text{ are cyclic (counting upwards), i.e. } ijk = 123,231,312, \\ -1 & \text{if } ijk \text{ are anti-cyclic (counting downwards), i.e. } ijk = 321,213,132, \\ 0 & \text{otherwise.} \end{cases}$$

(1.52)

The Levi-Civita tensor is noteworthy because it depends on *three* indices, i,j,k, whereas the previous tensors introduced, such as δ_{ij}, depended only on two. The Levi-Civita tensor is a more complicated type of tensor than the Kronecker delta.

One other property of the cross product will be found to be consistently useful. It follows immediately from the definition of the direction of the cross product that

$$\mathbf{A} \cdot \mathbf{C} = \mathbf{A} \cdot (\mathbf{A} \times \mathbf{B}) = 0.$$

(1.53)

In other words, the cross product of two vectors is orthogonal (perpendicular) to each of the vectors individually. This can also be readily shown by using the component definitions of the dot and cross products.

1.4 Useful products of vectors

Two types of multiple product of vectors appear quite regularly in physical problems; we briefly discuss each of them.

1.4.1 Triple scalar product

The triple scalar product is given by

$$\mathbf{A} \cdot (\mathbf{B} \times \mathbf{C}).$$

(1.54)

It can readily be shown that the triple scalar product may be written in terms of components as

$$\mathbf{A} \cdot (\mathbf{B} \times \mathbf{C}) = A_x(B_yC_z - B_zC_y) + A_y(B_zC_x - B_xC_z) + A_z(B_xC_y - B_yC_x).$$

(1.55)

From this result, the following symmetry properties of the triple scalar product can be deduced:

$$\mathbf{A} \cdot (\mathbf{B} \times \mathbf{C}) = \mathbf{B} \cdot (\mathbf{C} \times \mathbf{A}) = \mathbf{C} \cdot (\mathbf{A} \times \mathbf{B}).$$

(1.56)

In other words, the triple scalar product is invariant under cyclic permutations of the vectors, as illustrated in Fig. 1.6.

Just as the cross product may be conveniently represented using determinants, the triple scalar product may also be represented by determinants,

$$\mathbf{A} \cdot \mathbf{B} \times \mathbf{C} = \begin{vmatrix} A_x & A_y & A_z \\ B_x & B_y & B_z \\ C_x & C_y & C_z \end{vmatrix}.$$

(1.57)

$$\overbrace{\mathbf{A} \cdot (\mathbf{B} \times \mathbf{C})}} = \overbrace{\mathbf{C} \cdot (\mathbf{A} \times \mathbf{B})}} = \mathbf{B} \cdot (\mathbf{C} \times \mathbf{A})$$

Figure 1.6 Illustrating the cyclic nature of the triple scalar product. The vectors may all be "moved" to the left or right without changing the value of the product.

Geometrically, one can show that the triple scalar product represents the volume of the parallepiped formed by the vectors $\mathbf{A}$, $\mathbf{B}$, and $\mathbf{C}$.

1.4.2 Triple vector product

The triple vector product is the familiar rule, usually known as the BAC-CAB rule,

$$\mathbf{A} \times (\mathbf{B} \times \mathbf{C}) = \mathbf{B}(\mathbf{A} \cdot \mathbf{C}) - \mathbf{C}(\mathbf{A} \cdot \mathbf{B}). \tag{1.58}$$

The BAC-CAB rule, along with its vector calculus cousin (to be discussed in Section 2.6.3), are perhaps the most commonly useful vector formulas, and they should be burned into memory!

The triple vector product may be confirmed by a straightforward representation of the vectors $\mathbf{A}$, $\mathbf{B}$, and $\mathbf{C}$ in a Cartesian coordinate system.

1.5 Linear vector spaces

In this chapter, we have so far looked at the properties of individual vectors. We may also, however, consider the set of all possible vectors of a certain class *as a whole*, in what is referred to as a *linear vector space*. In anticipation of future discussion, we take some time to formally define a linear vector space and some other related spaces.

Definition 1.1 (Linear vector space) *A linear vector space S is defined as a set of elements called vectors, which satisfy the following ten properties related to elements* $|x\rangle$, $|y\rangle$ *and* $|z\rangle$:

1. $|x\rangle + |y\rangle \in S$ *(completeness with respect to addition)*,
2. $(|x\rangle + |y\rangle) + |z\rangle = |x\rangle + (|y\rangle + |z\rangle)$ *(associativity)*,
3. $|x\rangle + |0\rangle = |x\rangle$ *(existence of zero)*,
4. $\exists |y\rangle$ *such that* $|x\rangle + |y\rangle = |0\rangle$ *(existence of negative element)*,
5. $|x\rangle + |y\rangle = |y\rangle + |x\rangle$ *(commutativity)*,
6. $a|x\rangle \in S$ *(completeness with respect to scalar multiplication)*,
7. $a(b|x\rangle) = (ab)|x\rangle$ *(associativity of scalar multiplication)*,
8. $(a+b)|x\rangle = a|x\rangle + b|x\rangle$ *(first distribution rule)*,
9. $a(|x\rangle + |y\rangle) = a|x\rangle + a|y\rangle$ *(second distribution rule)*,
10. $1|x\rangle = |x\rangle$ *(existence of unit element)*.

To emphasize the general nature of this definition, we have adopted the so-called "bra-ket" notation for vectors, where $|x\rangle$ is an element of the vector space, and referred to as a "ket". We will discuss the "bra" form of a vector, $\langle x|$, momentarily. The symbol $\in$ indicates that the given vector belongs to the given set, i.e. $|x\rangle \in S$ means that $|x\rangle$ is a member of the set S. The symbol $\exists$ is mathematical shorthand for "there exists".

The first five items of the definition describe the behavior of vectors under addition; the last five items describe the behavior of vectors under scalar multiplication. Most of these items (namely, 2–5, 7–10) are clearly satisfied by the three-dimensional vectors discussed throughout this chapter. The first item is a statement that the object resulting from the addition of two vectors is itself a vector, and the sixth item is a statement that the object resulting from scalar multiplication of a vector is itself a vector. The first four items above classify the vector space as a mathematical *group*.[5]

This definition is broad enough that we may also refer to the complete set of numbers x on the real line $-\infty < x < \infty$ as forming a vector space, with the numbers x serving the role of "vectors". Elements of a vector space may also consist of an array of numbers: the set of all real-valued matrices also satisfy the definition of a linear vector space. Elements of a vector space may have complex values: the set of all complex numbers $z = x + iy$ in the plane satisfy the definition of a linear vector space. We will assume throughout the rest of this section that the vectors are in general complex-valued.

With such a broad definition, it is not clear what we have gained, other than listing properties of vectors that are obvious or seemingly trivial. When we begin to study more general classes of vectors, however, we will see that there are important properties which have to do with the set of vectors (the space) as a whole, and this general formalism will be useful. We briefly consider some of these concepts to prepare for future discussions.

Definition 1.2 (Linear independence) *A set of vectors* $X = \{|x_1\rangle, |x_2\rangle, \ldots, |x_N\rangle\}$ *are said to be* linearly independent *if the only solution to the equation*

$$a_1 |x_1\rangle + a_2 |x_2\rangle + \cdots + a_N |x_N\rangle = |0\rangle \tag{1.59}$$

is $a_i = 0$ for all i, where a_i are scalars.

Linear independence implies that no vector in the set X may be written as the sum of any combination of the others. A set X of vectors which is not linearly independent is *linearly dependent*.

Example 1.1 A trivial example of linearly independent vectors in three-dimensional space are the unit vectors $\hat{\mathbf{x}}$, $\hat{\mathbf{y}}$, $\hat{\mathbf{z}}$. The only way to construct a **0**-vector from these is by setting their coefficients all equal to zero, i.e. $x\hat{\mathbf{x}} + y\hat{\mathbf{y}} + z\hat{\mathbf{z}} = \mathbf{0}$ only for $x = y = z = 0$. A less trivial example is the trio of vectors in three dimensions listed below,

$$\mathbf{x}_1 = \hat{\mathbf{x}} + 2\hat{\mathbf{y}}, \quad \mathbf{x}_2 = 2\hat{\mathbf{x}} + 4\hat{\mathbf{y}} + \hat{\mathbf{z}}, \quad \mathbf{x}_3 = 5\hat{\mathbf{z}}. \tag{1.60}$$

---

[5] Group theory is a beautiful and general mathematical theory which is especially useful in atomic, nuclear and particle physics. See [Ham64, Cor97] for a detailed discussion.

If we take the addition of these vectors, with coefficients $a_1 = 1$, $a_2 = -1/2$ and $a_3 = 1/10$, the result is $\mathbf{0}$; the vectors are therefore linearly dependent.

◊

We see that *all* three-dimensional vectors may be written as some combination of the unit vectors $\hat{\mathbf{x}}$, $\hat{\mathbf{y}}$, $\hat{\mathbf{z}}$. The trio of unit vectors is then said to *span* the three-dimensional space. In general, a set of vectors is said to *span* a vector space if all vectors in the space may be written as a linear combination of that set. Returning to three-dimensional space, it can be seen quite clearly that one can find *at most* three linearly independent vectors at once, i.e. any collection of four or more vectors is necessarily linearly dependent. We may use this observation as the definition of the *dimension* of a vector space: the dimension of a vector space is the maximum number of linearly independent vectors which may be found. We will, however, find many important cases where the dimension of the vector space is *infinite*.

A linearly independent set of vectors which span a vector space are referred to as a *basis* for that space. There are always many possible choices of basis for any particular vector space. For instance, we may use the unit vectors $(\hat{\mathbf{x}}, \hat{\mathbf{y}}, \hat{\mathbf{z}})$ as a basis in three dimensions, but we may also use $(\hat{\mathbf{x}} + \hat{\mathbf{y}}, \hat{\mathbf{x}} - \hat{\mathbf{y}}, \hat{\mathbf{z}})$, or even $(\hat{\mathbf{x}}, \hat{\mathbf{x}} + \hat{\mathbf{y}}, \hat{\mathbf{z}})$. It is to be noted that in the last case the basis vectors are not even perpendicular to one another. A basis need not consist of mutually perpendicular vectors; the important characteristic is linear independence.

For a finite-dimensional space, finding a basis is usually a straightforward process. When the space is infinite-dimensional, however, things become much less clear. An important problem in dealing with infinite-dimensional vector spaces is proving that a set of vectors forms a complete basis.

There are several other types of space which should be mentioned. The first of these is an *inner product space*, also known as a *scalar product space*. To define such a space, we must first define an inner product.

Definition 1.3 (Inner product) *Suppose we have a linear vector space S. An inner product is a rule that associates with any pair of vectors $|x\rangle$ and $|y\rangle$ a complex number z, which is written as $z = \langle y| x\rangle$ and satisfies the following properties:*

1. *$\langle y| x\rangle = (\langle x| y\rangle)^*$, where $*$ refers to the complex conjugate,*
2. *if $|w\rangle = a|x\rangle + b|y\rangle$, then $\langle v| w\rangle = a\langle v| x\rangle + b\langle v| y\rangle$,*
3. *$\langle x| x\rangle \geq 0$, with equality occurring only for $|x\rangle = 0$.*

The inner product is also known as the *scalar product*, as the scalar or dot product of three-dimensional vectors satisfies the conditions for an inner product. Other examples of inner products will appear in future chapters.

We may now define an inner product space in a straightforward manner.

Definition 1.4 (Inner product space) *A linear vector space which has a scalar product associated with it is known as an inner product space.*

The way we have written the scalar product suggests that we may associate with every "ket" $|x\rangle$ a corresponding "bra" vector $\langle x|$, and the inner product is the formation of a "bracket" from these types of vectors. In general, we cannot say much about the nature of the "bra" vector, other than the fact that every "ket" has an associated "bra". For a three-dimensional, real-valued vector, the "bra" and "ket" vectors are the same.

We may define the orthogonality of vectors quite generally through their inner product.

Definition 1.5 (Orthogonality) *A pair of vectors* $|x\rangle$, $|y\rangle$ *are* orthogonal *if their inner product vanishes, i.e.*

$$\langle y|\,x\rangle = \langle x|\,y\rangle = 0. \tag{1.61}$$

As yet, we have not introduced a notion of length or "distance" between our general vectors $|x\rangle$, $|y\rangle$. In three-dimensional space, we known that the length of a vector can be defined in terms of its scalar product,

$$\text{length}(\mathbf{x}) = \sqrt{\mathbf{x} \cdot \mathbf{x}}. \tag{1.62}$$

Also, the separation between two vectors may be defined using the scalar product in a similar manner,

$$\text{distance}(\mathbf{x}_1, \mathbf{x}_2) = \sqrt{(\mathbf{x}_1 - \mathbf{x}_2) \cdot (\mathbf{x}_1 - \mathbf{x}_2)}. \tag{1.63}$$

In a general inner product space, we may associate a distance between two vectors by the inner product,

$$\text{distance}(|x\rangle, |y\rangle) = \sqrt{((\langle x| - \langle y|)(|x\rangle - |y\rangle))} = \sqrt{\langle x|\,x\rangle + \langle y|\,y\rangle - 2\text{Re}\{\langle x|\,y\rangle\}}. \tag{1.64}$$

It is possible to introduce a more general concept of length without reference to an inner product. This results in a distinct type of space known as a *metric space*.

Definition 1.6 (Metric space) *A metric space is defined as a set of elements which has a real, positive number* $\rho(x,y)$, *called the metric, associated with any pair of elements x, y. The metric must satisfy the conditions*

1. $\rho(x,y) = \rho(y,x)$,
2. $\rho(x,y) = 0$ *only for* $x = y$,
3. $\rho(x,y) + \rho(y,z) \geq \rho(x,z)$.

It is to be noted that a metric space is a different beast than a linear vector space, as a metric space is defined without any reference to the properties of addition and scalar multiplication. Therefore there can exist linear vector spaces which are not metric spaces, and vice versa. An inner product space, however, with a "distance" defined as in Eq. (1.64), is automatically a metric space. One can show with some effort that the "distance" satisfies the three properties of a metric space. An inner product space is automatically a metric space, but a metric space is not necessarily an inner product space.

The third property of the metric listed above deserves a bit of attention. It is the generalization of the familiar triangle inequality of geometry: the sum of the lengths of any sides of the triangle is greater than the length of the third side.

This section contains a lot of definition without much application! The purpose, however, is to demonstrate that the familiar three-dimensional vector space is part of a larger "family" of vector spaces, all of which have the same basic properties. This big picture shows us which properties of vectors are important, and what properties to look for when defining new vector spaces. We will get much more practice in the study and analysis of linear vector spaces in Chapter 4, on linear algebra.

1.6 Focus: periodic media and reciprocal lattice vectors

The properties of waves propagating in a periodic medium have been the basis of a number of important physical discoveries and applications. In the early 1900s, Max von Laue first demonstrated that X-rays could be diffracted by crystals [vL13], and soon afterwards William Henry Bragg and William Lawrence Bragg applied X-ray diffraction to the analysis of crystal structure [Bra14]. X-ray diffraction occurs because the wavelengths of X-rays are comparable to the spacing of atoms in a crystal; visible light, with much longer wavelength, is not sensitive to this atomic lattice and its propagation can be treated as though the medium were homogeneous. In recent decades, however, researchers have engineered materials with periodic structures on the order of the wavelength of visible light. These materials, known as *photonic crystals*, have optical properties analogous to the X-ray properties of ordinary crystals [JMW95].

The basic theory of crystals and other periodic media involves the application of a number of concepts of vector algebra from this chapter. We restrict ourselves for the moment to discussions of X-ray diffraction, though the mathematics also applies broadly to discussions of photonic crystals and even to electron propagation in crystals.

The basic structure used to model crystals is the *Bravais lattice*, which is defined as an infinite array of discrete units arranged such that the system looks exactly the same from any of the lattice points. Some examples of two-dimensional lattices are illustrated in Fig. 1.7.

The units of the Bravais lattice may represent individual atoms, collections of atoms, or molecules; the units are identical throughout the lattice, however.

As the lattice is of infinite spatial extent, it looks the same under translations from one unit to another. For instance, if we jump vertically from one unit to another in the square lattice, its appearance is unchanged. We can characterize this mathematically in a second, equivalent, definition of a Bravais lattice, as the set of all points which have position vectors $\mathbf{R}$ of the form

$$\mathbf{R} = m_1\mathbf{a}_1 + m_2\mathbf{a}_2 + m_3\mathbf{a}_3, \tag{1.65}$$

where $-\infty < m_i < \infty$ are integers, with $i = 1, 2, 3$, and $\mathbf{a}_i$ are known as *primitive vectors* which characterize the lattice. Different sets of integers m_i characterize different points in the lattice. The primitive vectors are a linearly independent, but not necessarily orthogonal,

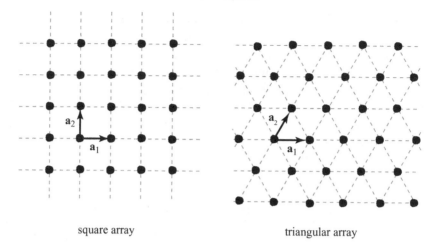

square array   triangular array

Figure 1.7 A pair of two-dimensional Bravais lattices: a square array of units and a triangular array of units. The units are represented as black dots. Examples of primitive vectors $\mathbf{a}_1$ and $\mathbf{a}_2$ are shown.

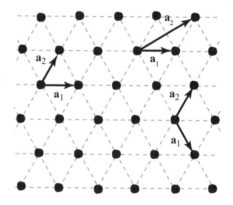

Figure 1.8 Illustration of multiple choices of primitive vectors for a Bravais lattice. Any unit of the lattice can be contructed by an appropriate combination of any pair of vectors.

set of vectors. These vectors have dimensions of length and their lengths characterize the separation between the units of the lattice; examples are shown in Fig. 1.8. Any unit of a Bravais lattice can be characterized by specifying a set of primitive vectors and the integers m_i; in such a case, the primitive vectors are said to *span* the lattice.

One immediate consequence of our vector definition of the Bravais lattice is that we can define the volume of a unit of the lattice. The volume of the parallelepiped with all $m_i = 1$ is simply $V = |\mathbf{a}_1 \cdot (\mathbf{a}_2 \times \mathbf{a}_3)|$.

It is important to note that the choice of primitive vectors is not unique. Figure 1.8 shows several different pairs of primitive vectors which span the lattice. A little thought

Figure 1.9 A precession photograph of a zero-level of the X-ray diffraction pattern. The photo was made on Polaroid film with unfiltered copper radiation. (Courtesy of Professor Daniel S. Jones of UNC Charlotte.)

will convince the reader that any pair of these primitive vectors can be constructed out of any other pair, thus demonstrating their equivalence.

We now turn to the diffraction of X-rays, and see how the tools of vector algebra aid in the understanding of this phenomenon. When a crystal is illuminated by a broadband (multifrequency) X-ray beam, the X-rays are scattered only in isolated directions, forming a pattern of spots at the detector; an example is shown in Fig. 1.9. There are two ways to interpret the experimental results, due to the Braggs and von Laue, and we consider each interpretation and then show them to be equivalent.

The Braggs' approach involves the introduction of *lattice planes* in the crystal. Any three noncollinear points of a Bravais lattice define a planar surface which intersects an infinite number of lattice points. Because of the periodicity of the lattice, we may then construct an infinite family of parallel lattice planes for a given choice of noncollinear points. There are an infinite number of families of lattice planes, several of which are illustrated for a two-dimensional square lattice in Fig. 1.10.

Bragg assumed that these planes of atoms act essentially as planar surfaces from which X-rays reflect in a specular manner. Rays that are reflected from adjacent planes travel different distances, with the ray reflected from the lower plane traveling farther by a distance $\Delta = 2d\cos\theta$, where θ is the angle of incidence with respect to the normal of the plane[6] and d is the distance between planes; this is illustrated in Fig. 1.11.

---

[6] In crystallography, the angle of incidence is typically measured from the lattice plane rather than the normal to the plane, as is done in optics; we stick to the optics definition for consistency with future discussions.

Vector algebra

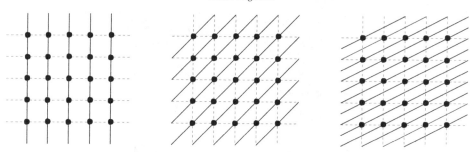

Figure 1.10 Illustration of different families of lattice planes of a square lattice.

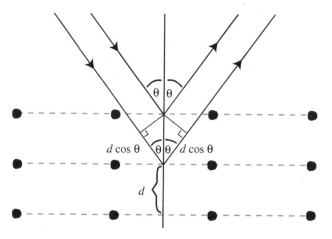

Figure 1.11 Derivation of the Bragg condition.

If we assume that the illuminating rays are monochromatic plane waves, any difference Δ in path between the two rays introduces a phase difference $k\Delta$ between them, where $k = 2\pi/\lambda$ is the wavenumber of the rays and λ is the wavelength. When the two reflected rays are in phase, i.e. their phase is a multiple of 2π, there is constructive interference between them and a strong reflected signal. The Bragg condition therefore states that X-ray diffraction peaks appear for angles of illumination such that

$$2kd\cos\theta = 2\pi N, \tag{1.66}$$

where N is an integer.

The Bragg condition is the easiest way to understand X-ray diffraction phenomena, but it has complexity hidden within it, namely in the plane spacing d and the angle to the plane θ, both of which depend upon the specific crystal structure. An arguably more elegant formulation was produced by von Laue, and this formulation will require some tools of vector algebra.

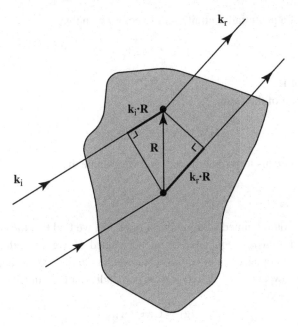

Figure 1.12 Derivation of the von Laue condition.

Instead of considering the crystal as various families of planes, von Laue looked at the X-rays scattered by individual units. Let us suppose that an X-ray is incident upon the crystal with wavevector $\mathbf{k}_i$ and is scattered with wavevector $\mathbf{k}_r$, where $|\mathbf{k}_i| = |\mathbf{k}_r| = k$. The phase of a plane wave with wavevector $\mathbf{k}$ may be written as $\mathbf{k} \cdot \mathbf{r}$. For two atoms separated by a lattice vector $\mathbf{R}$, the relative phase difference between them is $(\mathbf{k}_i - \mathbf{k}_r) \cdot \mathbf{R}$, as illustrated in Fig. 1.12. For constructive interference to occur, this phase difference must be equal to a multiple of 2π, and furthermore this must hold for every pair of units in the lattice, i.e. every $\mathbf{R}$, for if it did not, there would inevitably be some pairs of units out of phase and the total contribution from all pairs would tend to cancel out.

We may write this requirement as

$$e^{i(\mathbf{k}_i - \mathbf{k}_r) \cdot \mathbf{R}} = 1, \tag{1.67}$$

or, equivalently, as

$$(\mathbf{k}_i - \mathbf{k}_r) \cdot \mathbf{R} = 2\pi N, \tag{1.68}$$

where N is any integer. More concisely, we may state that scattering only occurs when

$$\mathbf{k}_i - \mathbf{k}_r = \mathbf{G}, \tag{1.69}$$

where **G** is one of a potentially infinite set of vectors such that

$$e^{i\mathbf{G}\cdot\mathbf{R}} = 1, \tag{1.70}$$

for every value of **R**.

We try a solution of the form

$$\mathbf{G} = [n_1\mathbf{g}_1 + n_2\mathbf{g}_2 + n_3\mathbf{g}_3], \tag{1.71}$$

where n_i are coefficients to be determined, and we assume that

$$\mathbf{g}_i \cdot \mathbf{a}_j = 2\pi \delta_{ij}. \tag{1.72}$$

On substitution of our assumed solution into Eq. (1.70), we find that the condition is automatically satisfied if the n_i are taken to be integer-valued. However, finding vectors which satisfy Eq. (1.72) is not necessarily trivial, as the vectors $\mathbf{a}_j$ themselves are not necessarily orthogonal. We know, however, that $\mathbf{a}_2 \times \mathbf{a}_3$ is perpendicular to both $\mathbf{a}_2$ and $\mathbf{a}_3$, so $\mathbf{g}_1$ must be of the form

$$\mathbf{g}_1 = C\mathbf{a}_2 \times \mathbf{a}_3. \tag{1.73}$$

The constant of normalization C is found by requiring that $\mathbf{a}_1 \cdot \mathbf{g}_1 = 2\pi$. Applying similar reasoning to $\mathbf{g}_2$ and $\mathbf{g}_3$, we find that

$$\mathbf{g}_1 = 2\pi \frac{\mathbf{a}_2 \times \mathbf{a}_3}{\mathbf{a}_1 \cdot (\mathbf{a}_2 \times \mathbf{a}_3)}, \tag{1.74}$$

$$\mathbf{g}_2 = 2\pi \frac{\mathbf{a}_3 \times \mathbf{a}_1}{\mathbf{a}_1 \cdot (\mathbf{a}_2 \times \mathbf{a}_3)}, \tag{1.75}$$

$$\mathbf{g}_3 = 2\pi \frac{\mathbf{a}_1 \times \mathbf{a}_2}{\mathbf{a}_1 \cdot (\mathbf{a}_2 \times \mathbf{a}_3)}. \tag{1.76}$$

The normalizations of each of the vectors can be shown to be the same using the permutation property of the triple scalar product, discussed in Section 1.4.

Equation (1.71) shows that the set of all vectors **G** form a Bravais lattice themselves, called the *reciprocal lattice*. The X-ray diffraction condition of von Laue reduces to the statement that *an X-ray is diffracted if the difference between the wavenumber of the incident and reflected ray lies on the reciprocal lattice.*

It can be readily shown that the reciprocal of the reciprocal lattice is the original Bravais lattice; for instance, if we define

$$\mathbf{b}_1 = 2\pi \frac{\mathbf{g}_2 \times \mathbf{g}_3}{\mathbf{g}_1 \cdot (\mathbf{g}_2 \times \mathbf{g}_3)}, \tag{1.77}$$

we may use vector algebra relations to show that $\mathbf{b}_1 = \mathbf{a}_1$, and so forth.

The von Laue condition for X-ray diffraction can be readily shown to be equivalent to the Bragg condition. To do so, we first demonstrate that every family of lattice planes is perpendicular to a set of parallel reciprocal lattice vectors, and conversely that every reciprocal lattice vector is perpendicular to a family of lattice planes.

As we have noted, a lattice plane may be defined by three noncollinear points of the Bravais lattice. Choosing the origin as one of these points, we write the other two as

$$\mathbf{R}_1 = A_1\mathbf{a}_1 + B_1\mathbf{a}_2 + C_1\mathbf{a}_3, \qquad (1.78)$$

$$\mathbf{R}_2 = A_2\mathbf{a}_1 + B_2\mathbf{a}_2 + C_2\mathbf{a}_3, \qquad (1.79)$$

where A_i, etc., are integer valued. The direction normal to this surface, denoted $\mathbf{n}$, can be found by the cross product of $\mathbf{R}_1$ and $\mathbf{R}_2$, so that

$$\mathbf{n} \equiv \mathbf{R}_1 \times \mathbf{R}_2 = (A_1B_2 - B_1A_2)\mathbf{a}_1 \times \mathbf{a}_2 + (A_2C_1 - C_2A_1)\mathbf{a}_3 \times \mathbf{a}_1 + (B_1C_2 - B_2C_1)\mathbf{a}_2 \times \mathbf{a}_3. \qquad (1.80)$$

The quantities in parenthesis are all integer-valued. Dividing this expression by the normalization $\mathbf{a}_1 \cdot (\mathbf{a}_2 \times \mathbf{a}_3)$, we immediately find that this normal satisfies our definition (1.71) for a reciprocal lattice vector.

One of the lattice planes includes the origin, and that plane is therefore defined by $\mathbf{G} \cdot \mathbf{r} = 0$. Because of the definition of the reciprocal lattice, we know that the next nearest plane must satisfy $\mathbf{G} \cdot \mathbf{r} = 2\pi$. If we denote the perpendicular distance between the planes (direction parallel to $\mathbf{G}$) as d, we may then note that the shortest reciprocal lattice vector associated with this family of lattice planes has length $|\mathbf{G}| = 2\pi/d$.

To prove the converse, we consider the set of all planes in space that satisfy the expression $\exp[i\mathbf{G} \cdot \mathbf{r}] = 1$. These planes must be perpendicular to $\mathbf{G}$ and are separated by a distance $d = 2\pi/|\mathbf{G}|$. Because we define the reciprocal lattice by the expression $\exp[i\mathbf{G} \cdot \mathbf{R}] = 1$, it follows that our set of spatial planes must include all points of the Bravais lattice, and therefore our spatial planes contain the lattice planes within them.

We are now in a position to relate the Bragg condition to the von Laue condition. A given reciprocal lattice vector must be of a length which is an integer multiple of $2\pi/d$, where d is the separation of the family of lattice planes associated with $\mathbf{G}$. We note that $|\mathbf{k}_i| = |\mathbf{k}_r| = k$, and that Eq. (1.69) therefore implies that the component $k_\perp$ of the incident wave normal to the family of planes must satisfy

$$2k_\perp = N2\pi/d. \qquad (1.81)$$

Referring to Fig. 1.12, $k_\perp = k\cos\theta$, which results in the expression

$$2k\cos\theta = 2\pi N/d. \qquad (1.82)$$

This is, of course, just the Bragg condition (1.66).

Reciprocal lattice vectors play a role not only in scattering from periodic media but also in understanding the propagation of waves in periodic media. We will return to them in later chapters.

1.7 Additional reading

Further information about the relationship between X-ray scattering and crystal structure can be found in the following references.

- N. W. Ashcroft and N. D. Mermin, *Solid State Physics* (London, Thomson Learning, 1976).
- P. M. Chaikin and T. C. Lubensky, *Principles of Condensed Matter Physics* (Cambridge, Cambridge University Press, 1995).
- A. Guiner, *X-ray Diffraction in Crystals, Imperfect Crystals and Amorphous Bodies* (San Francisco, W. H. Freeman, 1963). Also published by Dover.

1.8 Exercises

1. A vector $\mathbf{A} = 3\hat{\mathbf{x}} + 5\hat{\mathbf{y}} + 2\hat{\mathbf{z}}$. Determine the magnitude of this vector, and its direction cosines.
2. A vector $\mathbf{A} = -2\hat{\mathbf{x}} + 3\hat{\mathbf{y}} + \hat{\mathbf{z}}$. Determine the magnitude of this vector, and its direction cosines.
3. Consider the vectors $\mathbf{A} = 3\hat{\mathbf{x}} + 2\hat{\mathbf{y}}$ and $\mathbf{B} = \hat{\mathbf{x}} - \hat{\mathbf{y}}$. Calculate the scalar product of these vectors using Eq. (1.38). Also, determine the magnitudes of the vectors and the angle between them, and confirm that the scalar product satisfies Eq. (1.37).
4. Consider the vectors $\mathbf{A} = 5\hat{\mathbf{x}} - 3\hat{\mathbf{z}}$ and $\mathbf{B} = 2\hat{\mathbf{x}} + \hat{\mathbf{z}}$. Calculate the scalar product of these vectors using Eq. (1.38). Also, determine the magnitudes of the vectors and the angle between them, and confirm that the scalar product satisfies Eq. (1.37).
5. Calculate the cross product $\mathbf{A} \times \mathbf{B}$ of the vectors $\mathbf{A} = 3\hat{\mathbf{x}} + 2\hat{\mathbf{y}} - \hat{\mathbf{z}}$ and $\mathbf{B} = \hat{\mathbf{x}} - \hat{\mathbf{y}} + 4\hat{\mathbf{z}}$.
6. Calculate the cross product $\mathbf{A} \times \mathbf{B}$ of the vectors $\mathbf{A} = \hat{\mathbf{x}} + 7\hat{\mathbf{z}}$ and $\mathbf{B} = 2\hat{\mathbf{x}} + \hat{\mathbf{y}} + 4\hat{\mathbf{z}}$.
7. Consider the vectors $\mathbf{A} = \hat{\mathbf{x}} + 2\hat{\mathbf{y}} + \hat{\mathbf{z}}$, $\mathbf{B} = \hat{\mathbf{y}} + \hat{\mathbf{z}}$, and $\mathbf{C} = 2\hat{\mathbf{x}} - 3\hat{\mathbf{z}}$. Calculate the triple scalar product $\mathbf{A} \cdot (\mathbf{B} \times \mathbf{C})$ and the triple vector product $\mathbf{A} \times (\mathbf{B} \times \mathbf{C})$ of these vectors.
8. Consider the vectors $\mathbf{A} = 3\hat{\mathbf{x}} - \hat{\mathbf{y}} + 2\hat{\mathbf{z}}$, $\mathbf{B} = \hat{\mathbf{x}} + 2\hat{\mathbf{y}} + \hat{\mathbf{z}}$, and $\mathbf{C} = 2\hat{\mathbf{x}} - \hat{\mathbf{y}}$. Calculate the triple scalar product $\mathbf{A} \cdot (\mathbf{B} \times \mathbf{C})$ and the triple vector product $\mathbf{A} \times (\mathbf{B} \times \mathbf{C})$ of these vectors.
9. Let $\mathbf{A}$, $\mathbf{B}$, and $\mathbf{C}$ be vectors in three-dimensional space. By writing all vectors in Cartesian coordinates, explicitly demonstrate the BAC-CAB rule,

$$\mathbf{A} \times (\mathbf{B} \times \mathbf{C}) = \mathbf{B}(\mathbf{A} \cdot \mathbf{C}) - \mathbf{C}(\mathbf{A} \cdot \mathbf{B}).$$

10. Let $\mathbf{A}$, $\mathbf{B}$ be vectors in three-dimensional space. By writing both vectors in Cartesian coordinates, explicitly show that

$$\mathbf{A} \cdot (\mathbf{A} \times \mathbf{B}) = 0$$

for any choice of vectors $\mathbf{A}$ and $\mathbf{B}$.

11. Using the properties of the triple vector product, show that

$$\mathbf{a} \times [\mathbf{b} \times \mathbf{c}] + \mathbf{b} \times [\mathbf{c} \times \mathbf{a}] + \mathbf{c} \times [\mathbf{a} \times \mathbf{b}] = 0.$$

This is known as the *Jacobi identity*.

12. For vectors **a**, **b**, **c** and **d**, show that

$$(\mathbf{a} \times \mathbf{b}) \times (\mathbf{c} \times \mathbf{d}) = [\mathbf{a} \cdot (\mathbf{b} \times \mathbf{d})]\mathbf{c} - [\mathbf{a} \cdot (\mathbf{b} \times \mathbf{c})]\mathbf{d}.$$

13. Show that

$$(\mathbf{a} \times \mathbf{b}) \cdot (\mathbf{c} \times \mathbf{d}) = (\mathbf{a} \cdot \mathbf{c})(\mathbf{b} \cdot \mathbf{d}) - (\mathbf{a} \cdot \mathbf{d})(\mathbf{b} \cdot \mathbf{c}).$$

14. For the following collections of vectors in three-dimensional space, determine whether each collection is linearly independent:
   (a) $\mathbf{v}_1 = \hat{\mathbf{x}} + \hat{\mathbf{y}} + \hat{\mathbf{z}}$, $\mathbf{v}_2 = \hat{\mathbf{x}} + \hat{\mathbf{y}}$, $\mathbf{v}_3 = \hat{\mathbf{z}}$,
   (b) $\mathbf{v}_1 = 2\hat{\mathbf{x}} + \hat{\mathbf{y}}$, $\mathbf{v}_2 = 2\hat{\mathbf{y}} + \hat{\mathbf{z}}$, $\mathbf{v}_3 = 2\hat{\mathbf{z}} + \hat{\mathbf{y}}$,
   (c) $\mathbf{v}_1 = 2\hat{\mathbf{x}} + \hat{\mathbf{y}}$, $\mathbf{v}_2 = 2\hat{\mathbf{y}} + \hat{\mathbf{z}}$, $\mathbf{v}_3 = -2\hat{\mathbf{x}} + \hat{\mathbf{y}} + \hat{\mathbf{z}}$.
15. For the following collections of vectors in three-dimensional space, determine whether each collection is linearly independent:
   (a) $\mathbf{v}_1 = \hat{\mathbf{x}} + \hat{\mathbf{z}}$, $\mathbf{v}_2 = 2\hat{\mathbf{y}} + \hat{\mathbf{z}}$, $\mathbf{v}_3 = \hat{\mathbf{x}} - \hat{\mathbf{y}}$,
   (b) $\mathbf{v}_1 = 2\hat{\mathbf{x}} + \hat{\mathbf{y}} + \hat{\mathbf{z}}$, $\mathbf{v}_2 = 2\hat{\mathbf{x}} + \hat{\mathbf{y}} + \hat{\mathbf{z}}$, $\mathbf{v}_3 = 2\hat{\mathbf{x}} + 2\hat{\mathbf{y}} + \hat{\mathbf{z}}$,
   (c) $\mathbf{v}_1 = 2\hat{\mathbf{x}}$, $\mathbf{v}_2 = \hat{\mathbf{x}} + 2\hat{\mathbf{y}}$, $\mathbf{v}_3 = \hat{\mathbf{x}} + 2\hat{\mathbf{y}} + 3\hat{\mathbf{z}}$.
16. Consider the set of complex numbers $z = x + iy$, where $i^2 = -1$ and $-\infty < x < \infty$ and $-\infty < y < \infty$. Justify that this set forms a linear vector space.
17. Consider the set of 2×2 matrices **A** given by

$$\mathbf{A} = \begin{bmatrix} a_{11} & a_{12} \\ a_{21} & a_{22} \end{bmatrix},$$

where the coefficients a_{ij} are real-valued and $-\infty < a_{ij} < \infty$. Scalar multiplication of a matrix is defined as

$$\alpha\mathbf{A} = \begin{bmatrix} \alpha a_{11} & \alpha a_{12} \\ \alpha a_{21} & \alpha a_{22} \end{bmatrix},$$

and addition of matrices is defined as

$$\mathbf{A} + \mathbf{B} = \begin{bmatrix} a_{11} + b_{11} & a_{12} + b_{12} \\ a_{21} + b_{21} & a_{22} + b_{22} \end{bmatrix}.$$

Justify that this set forms a linear vector space.

18. A *corner reflector* consists of three mutually perpendicular, intersecting flat reflecting surfaces, such as the interior of the corner of a cube:

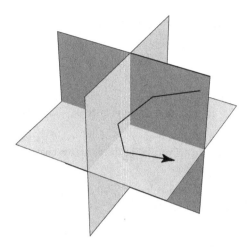

    Describe the behavior of a ray of light which strikes all three surfaces of the reflector.

19. In a simple model of spontaneous parametric down-conversion, a high-intensity "pump" photon of frequency ω_p and momentum $\mathbf{k}_p$ is converted into two photons of frequency ω_0 and momenta $\mathbf{k}_1$ and $\mathbf{k}_2$ through a nonlinear interaction with a material. By energy conservation, $\omega_p = 2\omega_0$. Also, by momentum conservation, $\mathbf{k}_p = \mathbf{k}_1 + \mathbf{k}_2$, and $|\mathbf{k}_i| = n_i \omega_i / c$, $i = p, 0$, with n_i the index of refraction of the medium and c the speed of light in vacuum. Assuming the pump photon propagates along the z-direction, and that the two converted photons propagate in the xz-plane,

    (a) What restriction does energy and momentum conservation place on the possible values of n_p/n_0?
    (b) When condition (a) is satisfied, find the angle θ between the converted and the pump photons.

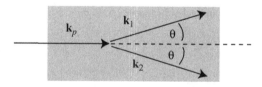

20. Using the stated properties of the Levi-Civita tensor ϵ_{ijk}, show that

$$\epsilon_{kji} = -\epsilon_{ijk},$$

$$\epsilon_{ikj} = -\epsilon_{ijk},$$

$$\epsilon_{jik} = -\epsilon_{ijk},$$

    for all possible values of i, j, and k.

21. Using the stated properties of the Levi-Civita tensor ϵ_{ijk}, show that

$$\sum_k \epsilon_{ijk}\epsilon_{klm} = \delta_{il}\delta_{jm} - \delta_{im}\delta_{jl},$$

    and use this result to prove Eq. (1.58).

22. A hexagonal lattice has primitive vectors

$$\mathbf{a}_1 = \frac{\sqrt{3}a}{2}\hat{\mathbf{x}} + \frac{a}{2}\hat{\mathbf{y}}, \quad \mathbf{a}_2 = -\frac{\sqrt{3}a}{2}\hat{\mathbf{x}} + \frac{a}{2}\hat{\mathbf{y}}, \quad \mathbf{a}_3 = c\hat{\mathbf{z}}.$$

    Show that the reciprocal lattice is also a hexagonal lattice, but with rotated axes.

23. A body-centered cubic lattice has primitive vectors

$$\mathbf{a}_1 = \frac{1}{2}a(-\hat{\mathbf{x}} + \hat{\mathbf{y}} + \hat{\mathbf{z}}), \quad \mathbf{a}_2 = \frac{1}{2}a(\hat{\mathbf{x}} - \hat{\mathbf{y}} + \hat{\mathbf{z}}), \quad \mathbf{a}_3 = \frac{1}{2}a(\hat{\mathbf{x}} + \hat{\mathbf{y}} - \hat{\mathbf{z}}).$$

    Show that the reciprocal lattice is a face-centered cubic lattice, with primitive vectors

$$\mathbf{a}_1 = \frac{1}{2}\alpha(\hat{\mathbf{y}} + \hat{\mathbf{z}}), \quad \mathbf{a}_2 = \frac{1}{2}\alpha(\hat{\mathbf{x}} + \hat{\mathbf{z}}), \quad \mathbf{a}_3 = \frac{1}{2}\alpha(\hat{\mathbf{x}} + \hat{\mathbf{y}}).$$

24. Let $\mathbf{a}_1$, $\mathbf{a}_2$, and $\mathbf{a}_3$ represent a set of primitive vectors, and $\mathbf{g}_1$, $\mathbf{g}_2$, and $\mathbf{g}_3$ represent the reciprocal lattice. Demonstrate using vector algebra identities that the reciprocal of the reciprocal lattice vectors are equal to the original vectors $\mathbf{a}_1$, $\mathbf{a}_2$, and $\mathbf{a}_3$.

25. The position vector of a straight line passing through a vector position $\mathbf{a}$ and running parallel to a unit vector $\mathbf{e}$ may be written as a parametric equation of the form $\mathbf{r}(\lambda) = \mathbf{a} + \lambda\mathbf{e}$, where λ is a real-valued parameter.
    (a) By using vector multiplication, rewrite this equation in the form $f(\mathbf{r}) = 0$, i.e. without a parameter.
    (b) Write an equation for $\mathbf{r}(\lambda)$ for a line which passes through two points $\mathbf{a}$ and $\mathbf{b}$, and determine a parameterless form of this equation.
    (c) Write an equation for $\mathbf{r}(\lambda)$ for a line which passes through $\mathbf{a}$ and is perpendicular to $\mathbf{e}_1$ and $\mathbf{e}_2$, and determine a parameterless form of this equation.

26. The position vector of a plane passing through three vector positions $\mathbf{a}$, $\mathbf{b}$ and $\mathbf{c}$ may be written as a parametric equation of the form $\mathbf{r}(\lambda, \mu) = \mathbf{a} + \lambda(\mathbf{b} - \mathbf{a}) + \mu(\mathbf{c} - \mathbf{a})$, where λ and μ are real-valued parameters.
    (a) By using vector multiplication, rewrite this equation in the form $f(\mathbf{r}) = 0$, i.e. without parameters.
    (b) Write an equation for $\mathbf{r}(\lambda, \mu)$ for a plane which passes through points $\mathbf{a}$ and $\mathbf{b}$ and is parallel to $\mathbf{e}$, and determine a parameterless form of this equation.
    (c) Write an equation for $\mathbf{r}(\lambda, \mu)$ for a plane which passes through point $\mathbf{a}$ and is parallel to $\mathbf{e}_1$ and $\mathbf{e}_2$, and determine a parameterless form of this equation.

2

Vector calculus

2.1 Introduction

So far, we have discussed the basic concepts of vector algebra. In particular, we described several different ways of performing multiplication with vectors: the scalar product, the dot product, and the cross product.

When we turn to dealing with *fields*, i.e. vectors $\mathbf{V}(\mathbf{r}) = \mathbf{V}(x,y,z)$ and scalars $\phi(\mathbf{r}) = \phi(x,y,z)$ which are functions of position, we now have the option of taking derivatives of such fields and integrating them with respect to their position variables. Though there are in principle many different ways to combine vector multiplication and differentiation, we are restricted by the requirement that any physically relevant form of differentiation must be invariant under a change of coordinate system. It can be shown that there are three basic types of spatial derivative that involve vector fields and satisfy coordinate invariance, one corresponding to each of the types of vector multiplication:

vector multiplication	**vector operation**
$\mathbf{A}\alpha$ (scalar multiplication),	$\nabla\phi(x,y,z)$ (gradient),
$\mathbf{A}\cdot\mathbf{B}$ (dot product),	$\nabla\cdot\mathbf{V}(x,y,z)$ (divergence),
$\mathbf{A}\times\mathbf{B}$ (cross product),	$\nabla\times\mathbf{V}(x,y,z)$ (curl),

where $\nabla = \frac{\partial}{\partial x}\hat{\mathbf{x}} + \frac{\partial}{\partial y}\hat{\mathbf{y}} + \frac{\partial}{\partial z}\hat{\mathbf{z}}$ is the "del" operator (called an operator because it "operates" on a field). Each of these operations is distinguished by its input/output products: the gradient takes a scalar function and relates a vector derivative to it, the divergence takes a vector function and relates a scalar derivative to it, and the curl takes a vector function and relates a vector derivative to it.

We will discuss each of these operators in turn, and consider the following properties of each:

1. their mathematical definition,
2. their physical (mathematical) meaning,
3. alternative definitions and mathematical properties.

A complete understanding of the physical meaning of vector derivatives, however, involves the application of vector integration. We begin this chapter with an introduction to the techniques of vector integration, and then use these to discuss vector derivatives.

2.2 Vector integration

Integration problems in three-dimensional space may be broadly divided into three classes: *path integrals*, in which we integrate over a particular path in space, *surface integrals*, in which we integrate over a two-dimensional surface, and *volume integrals*, in which we integrate through a three-dimensional volume in space. There are numerous sub-types of each class of integral that may appear, depending on its physical origin.

It is to be noted that there are a relatively small number of vector integrals that can be evaluated exactly. Often, however, we are more interested in the relationship between different classes of vector integrals, as we will see in discussing Gauss' theorem in Section 2.7 and Stokes' theorem in Section 2.8. For specific problems in which the integral must be evaluated, numerical integration is often the best approach.

2.2.1 Path integrals

Path integrals involve the integration of a function along a path in three-dimensional space. Depending on the problem, the integral may be taken over a *vector* path element $d\mathbf{r}$ (the direction of which points along the path) or a *scalar* path element dr. The basic idea is illustrated in Fig. 2.1(a).

There are numerous path integrals which may be encountered in physics; among them we have

$$\int_C \phi d\mathbf{r}, \tag{2.1}$$

$$\int_C \mathbf{V} \cdot d\mathbf{r}, \tag{2.2}$$

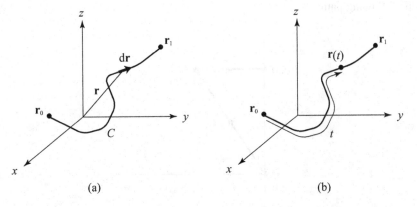

$$(a) \qquad\qquad (b)$$

Figure 2.1 Illustrating the idea of a path integral in three-dimensional space from point $\mathbf{r}_0$ to point $\mathbf{r}_1$. (a) The general notation, illustrating the vector $\mathbf{r}$ and vector element $d\mathbf{r}$. (b) A parameterization of the path, such that variable t is a measure of the distance along it.

$$\int_C \mathbf{V} \times d\mathbf{r}, \tag{2.3}$$

$$\int_C \phi d\mathbf{r}, \tag{2.4}$$

$$\int_C \mathbf{V} d\mathbf{r}, \tag{2.5}$$

where C is a contour which may be open or closed.

Evaluation of path integrals involves reducing them to the form of an ordinary integration of a function of a single variable. If the contour consists of a finite collection of straight line segments, a path integral may be broken up into a sum of line integrals over each of the segments. If the contour is curved, it is usually most useful to parameterize it by a dimensionless parameter t, defined on the range $0 \le t \le 1$, so that the position $\mathbf{r}$ on the curve is a function of t, i.e. $\mathbf{r} \equiv \mathbf{r}(t)$; this is illustrated in Fig. 2.1(b). Then $d\mathbf{r}$ becomes

$$d\mathbf{r} = \frac{d\mathbf{r}(t)}{dt} dt, \tag{2.6}$$

and the integral reduces to a single integral over the parameter t. In Cartesian coordinates, this amounts to defining the path as $\mathbf{r}(t) = x(t)\hat{\mathbf{x}} + y(t)\hat{\mathbf{y}} + z(t)\hat{\mathbf{z}}$. Equation (2.2) above, for instance, takes on the form

$$\int_C \mathbf{v} \cdot d\mathbf{r} = \int_0^1 \left\{ v_x[\mathbf{r}(t)] \frac{dx}{dt} + v_y[\mathbf{r}(t)] \frac{dy}{dt} + v_z[\mathbf{r}(t)] \frac{dy}{dt} \right\} dt. \tag{2.7}$$

This is now a scalar integral over a single variable t, and can be evaluated using ordinary techniques of single-variable integral calculus.

Example 2.1 Calculate the path integral $\int_C \mathbf{v} \cdot d\mathbf{r}$ of the vector function $\mathbf{v}(x,y) = xy\hat{\mathbf{x}} + 3y\hat{\mathbf{y}}$ over the following path:

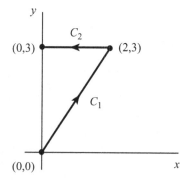

As this integral consists of two separate straight-line segments, it is helpful to consider the total integral as consisting of two "legs", the first leg C_1 being the diagonal and the second leg C_2 being the horizontal. For the first leg, we may parameterize the path as $x(t) = 2t$ and $y(t) = 3t$, where $t : 0 \rightarrow 1$. Equation (2.7) takes on the form

$$\int_{C_1} \mathbf{v} \cdot d\mathbf{r} = \int_0^1 [2t \cdot 3t \cdot 2 + 3 \cdot 3t \cdot 3] dt = \frac{35}{2}. \tag{2.8}$$

For the second leg, the path can be parameterized as $x(t) = 2t$, $y(t) = 3$, with $t : 1 \rightarrow 0$ for convenience. We then have

$$\int_{C_2} \mathbf{v} \cdot d\mathbf{r} = \int_1^0 2t \cdot 3 \cdot 2dt = -6. \tag{2.9}$$

The total result is therefore $-23/2$.

◇

If the integral is taken with respect to a scalar element dr, as for instance in Eqs. (2.4) and (2.5), it may still be parameterized by noting that $r = \sqrt{x^2 + y^2 + z^2}$; we then have

$$dr = \frac{dr(t)}{dt}dt = \frac{x(t) + y(t) + z(t)}{\sqrt{x^2(t) + y^2(t) + z^2(t)}}dt. \tag{2.10}$$

An integral such as Eq. (2.5) would have the parameterized form

$$\int_C \mathbf{V}dr = \int_0^1 \{\hat{\mathbf{x}}V_x[\mathbf{r}(t)] + \hat{\mathbf{y}}V_y[\mathbf{r}(t)] + \hat{\mathbf{z}}V_z[\mathbf{r}(t)]\}\frac{x(t) + y(t) + z(t)}{\sqrt{x^2(t) + y^2(t) + z^2(t)}}dt. \tag{2.11}$$

This expression is in the form of three independent single-variable integrals, one for each unit vector.

2.2.2 *Surface integrals*

Surface integrals involve an integration of a function over a two-dimensional manifold in three-dimensional space. As in the path integral case, the integration may be taken over an infinitesimal *vector* surface element $d\mathbf{a}$ or a scalar surface element da.

Area elements $d\mathbf{a}$ of vector form are defined as having a magnitude equal to the area da and a direction $\mathbf{n}$ defined as normal to the surface. This leaves some ambiguity in the definition of the direction, because there are two normals to any surface, such as the two sides of a sheet of paper. To remove the ambiguity, we use the following conventions. (1) If the surface is closed (like the surface of a sphere, with no openings) the area normal is defined as being the outward normal. (2) Open surfaces (like the area defined by a circle) usually only appear in connection with line integrals taken along the edge of the surface. The normal for an open surface is chosen using the "right-hand" rule based on the direction the line integral is taken around the circle: if the fingers of the right-hand curve in the direction of the path, the thumb points in the direction of the surface normal. These distinctions are illustrated in Fig. 2.2.

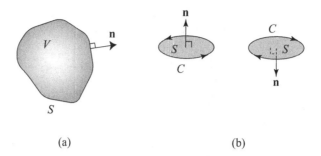

Figure 2.2 Illustrating the unit normal for (a) a closed surface S surrounding a volume V and (b) an open surface S bounded by a curve C.

The possible forms of surface integrals are analogous to those of the line integrals,

$$\int \phi d\mathbf{a}, \tag{2.12}$$

$$\int \mathbf{v} \cdot d\mathbf{a}, \tag{2.13}$$

$$\int \mathbf{v} \times d\mathbf{a}, \tag{2.14}$$

$$\int \phi da, \tag{2.15}$$

$$\int \mathbf{v} da. \tag{2.16}$$

Of these surface integrals, the dot product form, Eq. (2.13), is often encountered because of its association with flux through a surface, notably in Gauss' law, to be discussed in Section 6.1. Equations (2.15) and (2.16) arise in calculations of the electrostatic potential and electrostatic field of a surface charge distribution, respectively.

Surface integrals can also be parameterized; because the surface is two dimensional, however, it must be parameterized by two variables. For example, defining a location in the xy-plane requires one to specify both x and y. This sort of parameterization is commonly done in geometrical optics, in which the propagation of a field can be characterized by describing a surface of constant phase – the light rays are then perpendicular to this surface. It is to be noted that because surface integrals are integrals over two variables, they should formally be written with a double integral ($\int\int$); for simplicity, we only write the double integral when the limits of integration are given.

Example 2.2 Calculate a surface integral $\int \mathbf{v} \cdot d\mathbf{a}$ for a planar surface defined by the equation $z = by$, which is confined to the values $0 \leq y \leq 1$, $-1 \leq x \leq 1$, and for the function $\mathbf{v} = x^2 y \hat{\mathbf{z}}$.

The normal to the surface is defined as pointing towards the $+\hat{\mathbf{z}}$-direction. A cross-section of this surface is shown below:

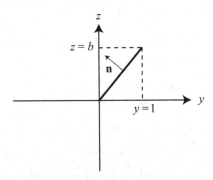

In order to evaluate this integral, we need to parameterize the surface. The variable x serves as a perfectly good parameterization of the x-position on the surface, so we need to find a parameter which properly describes the position along the yz-direction. We define s as the perpendicular distance from the x-axis to a point on the plane; is is clear that s ranges from 0 to $\sqrt{b^2+1}$. We may also write $y = s/\sqrt{b^2+1}$ and $z = bs/\sqrt{b^2+1}$. The normal to the surface must be perpendicular to the direction of increasing s, which is proportional to $\mathbf{s} = \hat{\mathbf{y}} + b\hat{\mathbf{z}}$. The normal may be taken to be $\mathbf{n} = \pm(-b\hat{\mathbf{y}} + \hat{\mathbf{z}})/\sqrt{b^2+1}$; if we choose the normal in the $+\hat{\mathbf{z}}$-direction, we have $\mathbf{n} = (-b\hat{\mathbf{y}} + \hat{\mathbf{z}})/\sqrt{b^2+1}$. Our surface integral then becomes

$$\int \mathbf{v} \cdot d\mathbf{a} = \int_{-1}^{1} \int_{0}^{\sqrt{b^2+1}} x^2 y \hat{\mathbf{z}} \cdot (-b\hat{\mathbf{y}} + \hat{\mathbf{z}})/\sqrt{b^2+1} \, dx \, ds. \tag{2.17}$$

Substituting for y and z, we have

$$\int \mathbf{v} \cdot d\mathbf{a} = \int_{-1}^{1} \int_{0}^{\sqrt{b^2+1}} \frac{x^2 s}{b^2+1} \, dx \, ds = \frac{1}{3}. \tag{2.18}$$

A good check on the parameterization of the surface is that the integral $\int da$ produces the expected surface area; in this case we find that $\int da = 2\sqrt{b^2+1}$.

◇

For planar surfaces, it is relatively easy to find a parameterization; for curved surfaces, however, the process is more subtle. Let us suppose we have a surface defined by the equation

$$z = f(x, y). \tag{2.19}$$

The most straightforward parameterization is one that uses the x and y variables; however, the surface element da is *not* simply the product $dx \, dy$. In the example above, we considered a small element of surface which is constant along the x-direction and linear in the y-direction, such that

$$z = f(y) = by. \tag{2.20}$$

By simple geometry, we may argue that the area of this surface is of the form

$$da = \sqrt{1+b^2}dx\,dy.$$

(2.21)

Noting that $b = \frac{\partial f}{\partial y}$, we may further write

$$da = \sqrt{1+\left(\frac{\partial f}{\partial y}\right)^2}\,dx\,dy.$$

(2.22)

For an arbitrary surface with a linear slope, the area element is

$$da = \sqrt{1+\left(\frac{\partial f}{\partial x}\right)^2+\left(\frac{\partial f}{\partial y}\right)^2}\,dx\,dy.$$

(2.23)

We may use this expression to calculate surface integrals of various forms. To determine the unit normal to the surface, if needed, we may use the gradient operation, discussed in Section 2.3.

Example 2.3 Let us calculate the area of a parabolic surface satisfying the equation

$$z = ax^2, \quad -1 \le x \le 1,$$

(2.24)

and $-1 \le y \le 1$. This surface is shown below.

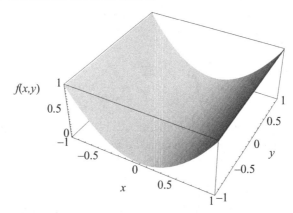

The area element is given by

$$da = \sqrt{1+4a^2x^2}dx\,dy.$$

(2.25)

The area of the surface may then be written in the form

$$A = \int_{-1}^{1}\int_{-1}^{1}\sqrt{1+4a^2x^2}dx\,dy = 2\int_{-1}^{1}\sqrt{1+4a^2x^2}dx.$$

(2.26)

This nontrivial integral may be looked up, and the result is

$$A = \sqrt{1+4a^2} + \frac{\sinh^{-1}(2a)}{2a}. \tag{2.27}$$

◇

2.2.3 Volume integrals

Volume integrals of a vector quantity are somewhat simpler, at least in principle, because the volume element is always a scalar quantity – in Cartesian coordinates, for instance, we may write the volume element as $d\tau = dx\,dy\,dz$. Therefore, the components of the vector $\mathbf{v}$ may be integrated individually and the vector nature typically presents no new challenges.

Some examples of this form include:

$$\int \phi d\tau, \tag{2.28}$$

$$\int \mathbf{v} d\tau. \tag{2.29}$$

With an understanding of the basics of vector integration, we are now better equipped to discuss the vector differential operators and their interpretation.

2.3 The gradient, ∇

The gradient is an operation upon a *scalar* field which results in a vector field: if $\phi(x,y,z)$ is a scalar field defined in three-dimensional space, $\nabla\phi(x,y,z)$ is a vector field. In Cartesian coordinates, it has the form

$$\nabla\phi(x,y,z) \equiv \frac{\partial\phi}{\partial x}\hat{\mathbf{x}} + \frac{\partial\phi}{\partial y}\hat{\mathbf{y}} + \frac{\partial\phi}{\partial z}\hat{\mathbf{z}}. \tag{2.30}$$

It is to be recalled from Section 1.2 that a vector must satisfy specific transformation properties under a rotation of coordinates; we will show that this is the case for $\nabla\phi$ at the end of this section.

To understand the significance of the gradient, we first consider a one-dimensional scalar function $f(t)$. We know from elementary calculus that the instantaneous rate of change of this function is given by its derivative, $df(t)/dt$. Furthermore, as t is changed by an infinitesimal amount dt, the function $f(t)$ changes by an amount $df = [df(t)/dt]dt$.

Let us now consider a scalar field $\phi(x,y,z)$ defined in three-dimensional space, and we consider a path through that three-dimensional space characterized by $\mathbf{s}(t)$, where t is a real parameter which varies from 0 to 1 – that is, as t increases from 0 to 1, our position on the path changes continuously from $\mathbf{s}_0$ to $\mathbf{s}_1$. This formulation is the same used in Section 2.2.1, and was illustrated in Fig. 2.1.

Let us investigate how the function ϕ changes as we move along the path $\mathbf{s}$, i.e. as we move an infinitesimal distance $d\mathbf{s}$ along the path. We can partly characterize this by determining the derivative of ϕ with respect to the parameter t, i.e. $d\phi[x(t), y(t), z(t)]/dt$. Using the chain rule, we can write this path derivative in terms of Cartesian coordinates as

$$\frac{d\phi}{dt} = \frac{\partial\phi}{\partial x}\frac{dx}{dt} + \frac{\partial\phi}{\partial y}\frac{dy}{dt} + \frac{\partial\phi}{\partial z}\frac{dz}{dt}. \tag{2.31}$$

It is to be noted that this derivative is taken with respect to the dimensionless parameter t; we now need to use this formula to find a true spatial derivative of the function ϕ, i.e. $d\phi/ds$. From Eq. (2.31), the total differential of ϕ may be written as

$$d\phi = \frac{\partial\phi}{\partial x}dx + \frac{\partial\phi}{\partial y}dy + \frac{\partial\phi}{\partial z}dz. \tag{2.32}$$

Similarly, writing $\mathbf{s}(t) = x(t)\hat{\mathbf{x}} + y(t)\hat{\mathbf{y}} + z(t)\hat{\mathbf{z}}$, the infinitesimal vector change in position along the path is given by

$$d\mathbf{s} = \hat{\mathbf{x}}\frac{\partial x}{\partial t}dt + \hat{\mathbf{y}}\frac{\partial y}{\partial t}dt + \hat{\mathbf{z}}\frac{\partial z}{\partial t}dt = \hat{\mathbf{x}}dx + \hat{\mathbf{y}}dy + \hat{\mathbf{z}}dz. \tag{2.33}$$

If we define a "vector" ∇ such that

$$\nabla \equiv \frac{\partial}{\partial x}\hat{\mathbf{x}} + \frac{\partial}{\partial y}\hat{\mathbf{y}} + \frac{\partial}{\partial z}\hat{\mathbf{z}}, \tag{2.34}$$

we may formally write

$$d\phi = \nabla\phi \cdot d\mathbf{s}. \tag{2.35}$$

Furthermore, the unit vector in the direction of $d\mathbf{s}$, denoted $\hat{\mathbf{s}}$, is simply the normalized form of $d\mathbf{s}$,

$$\hat{\mathbf{s}} = \frac{d\mathbf{s}}{ds} = \hat{\mathbf{x}}\frac{dx}{ds} + \hat{\mathbf{y}}\frac{dy}{ds} + \hat{\mathbf{z}}\frac{dz}{ds}. \tag{2.36}$$

Applying this to Eq. (2.35), we have

$$\frac{d\phi}{ds} = \nabla\phi \cdot \hat{\mathbf{s}}. \tag{2.37}$$

We may then state, from Eq. (2.35), that the infinitesimal change in ϕ as one moves along the path $\mathbf{s}$ is given by the gradient of ϕ at that point dotted with the infinitesimal path element $d\mathbf{s}$. Also, from Eq. (2.37), the spatial derivative of ϕ along the path is given by the gradient of ϕ dotted with the local direction of the path $\hat{\mathbf{s}}$.

From this discussion we may conclude that $\nabla\phi$ is a three-dimensional generalization of an ordinary (single-variable) derivative. To understand further the meaning of the gradient, suppose we consider the field ϕ at a given point in space and examine how the field changes as we move an infinitesimal distance away from that point. The quantity $\nabla\phi$ defines a direction in three-dimensional space. Looking at Eq. (2.35), and assuming that the magnitude of $d\mathbf{s}$ is fixed and only the direction is changed, the maximum change in the value of ϕ will occur when d**s** is *parallel* to $\nabla\phi$. Thus, the direction of the gradient indicates the direction of the greatest rate of change of the field ϕ.

One question we have not addressed is whether or not $\nabla\phi$ is actually a vector: does it transform under rotations of the coordinate system in the same way as a vector? We first note that ϕ is assumed to be a physical quantity; that is, its value is independent of the coordinate system,

$$\phi'(x_1',x_2',x_3') = \phi(x_1,x_2,x_3). \tag{2.38}$$

If we look at the derivative of ϕ' with respect to x_i', we find that

$$\frac{\partial\phi'(x_1',x_2',x_3')}{\partial x_i'} = \frac{\partial\phi(x_1,x_2,x_3)}{\partial x_i'} = \sum_j \frac{\partial x_j}{\partial x_i'}\frac{\partial\phi}{\partial x_j}. \tag{2.39}$$

Referring back to Eq. (1.29), we have

$$a_{ij} = \frac{\partial x_j}{\partial x_i'}, \tag{2.40}$$

and therefore

$$\frac{\partial\phi'(x_1',x_2',x_3')}{\partial x_i'} = \sum_j a_{ij}\frac{\partial\phi}{\partial x_j}. \tag{2.41}$$

On comparison of this equation with Eq. (1.26), it follows that $\nabla\phi$ is a vector.

The gradient operator appears frequently in physics and optics; for example, the potential U of a conservative force field $\mathbf{F}$ is related by $\mathbf{F} = -\nabla U$. We will return to this topic in Section 2.9.

One can also write an *integral* definition of the gradient. The derivation follows very closely the integral definition of the divergence, to be covered next, so we will refer to it at the end of that section.

2.4 Divergence, ∇·

Because the "del" (∇) operator is a vector operator, one may apply it not only directly to a scalar field but may take the scalar product of it with a vector field, producing a scalar quantity. This quantity $\nabla\cdot\mathbf{v}$, is known as the *divergence* of the vector $\mathbf{v}$. In Cartesian coordinates it takes on the form

$$\nabla\cdot\mathbf{v} = \frac{\partial v_x}{\partial x} + \frac{\partial v_y}{\partial y} + \frac{\partial v_z}{\partial z}. \tag{2.42}$$

While the gradient is analogous to a simple derivative of a single variable function, the divergence (and the curl to follow) has no such simple analogy. It is usually described as a measure of the net "flow" of a vector field away from a point. We analyze this description by considering a physical problem. Let $\mathbf{p}(x,y,z)$ be the momentum density (momentum

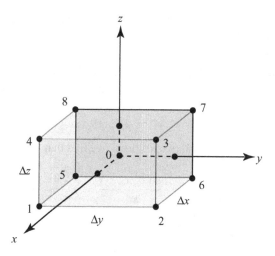

Figure 2.3 An infinitesimal box in three-dimensional space, used to understand the divergence of a vector.

per unit volume, or mass per unit area per unit time) of a compressible fluid. In Cartesian coordinates we will write this function as

$$\mathbf{p}(x,y,z) = p_x(x,y,z)\hat{\mathbf{x}} + p_y(x,y,z)\hat{\mathbf{y}} + p_z(x,y,z)\hat{\mathbf{z}}. \qquad (2.43)$$

We consider a small volume $\Delta\tau = \Delta x \Delta y \Delta z$ of space, and examine the mass of fluid flowing out of this volume per unit time. The volume can be assumed to be centered on the origin of our coordinate system without loss of generality, and is illustrated in Fig. 2.3.

We make one assumption concerning the behavior of the function $\mathbf{p}$ and the size of the box $\Delta\tau$. It is assumed that the function $\mathbf{p}$ has continuous first partial derivatives, and that the volume is sufficiently small so that the functions p_x, p_y, and p_z can all be well-represented by their first-order Taylor series expansions, i.e.

$$\mathbf{p}(x,y,z) = \mathbf{p}(0) + x\partial_x\mathbf{p}(0) + y\partial_y\mathbf{p}(0) + z\partial_z\mathbf{p}(0) + O(\mathbf{r}^2). \qquad (2.44)$$

Here ∂_x represents the partial derivative with respect to x, with similar identifications for ∂_y and ∂_z; all quantities are evaluated at the origin. The O in $O(\mathbf{r}^2)$ is known as an *order parameter* and indicates that the remainder of the Taylor series has a quadratic dependence on the variables x, y, and z. Taylor series will be discussed in Chapter 7 and Taylor series in more than one variable will be discussed in detail in Section 7.6; for now, we make this an *assumption* of our investigation.

Let us look at the flow of fluid Q through each face of the parallelepiped separately. The net flow *out of* the volume through the face 1234 is given by the integral

$$Q_{1234} = \int_{1234} \mathbf{p} \cdot \mathbf{da}. \qquad (2.45)$$

In this case, $\mathbf{da} = dy\,dz\hat{\mathbf{x}}$, and the integration is over $-\Delta y/2 \leq y \leq \Delta y/2$, $-\Delta z/2 \leq z \leq \Delta z/2$. We then have

$$Q_{1234} = \int_{1234} \mathbf{p} \cdot \mathbf{da} \approx p_x(0)\Delta y\Delta z + \frac{\Delta x}{2}\partial_x p_x(0)\Delta y\Delta z. \quad (2.46)$$

The use of the "approximately equals" sign ($\approx$) indicates that we have, for the moment, neglected the remainder. Similarly, the net flow out of the volume through the face 5678 is given by the integral

$$Q_{5678} = \int_{5678} \mathbf{p} \cdot \mathbf{da}, \quad (2.47)$$

where now $\mathbf{da} = -dy\,dz\hat{\mathbf{x}}$. We then have

$$Q_{5678} = \int_{5678} \mathbf{p} \cdot \mathbf{da} \approx -p_x(0)\Delta y\Delta z + \frac{\Delta x}{2}\partial_x p_x(0)\Delta y\Delta z. \quad (2.48)$$

The net flow out through the x-faces of the cube is therefore

$$Q_x = Q_{1234} + Q_{5678} \approx \Delta x\Delta y\Delta z\partial_x p_x(0). \quad (2.49)$$

We may make similar arguments for the net flow out through the y- and z-faces of the cube. The total flow out of the cube may be represented as

$$Q_{net} \approx \left[\partial_x p_x(0) + \partial_y p_y(0) + \partial_z p_z(0)\right]\Delta x\Delta y\Delta z = \nabla \cdot \mathbf{p}(0)\Delta\tau. \quad (2.50)$$

The flow out of the volume is approximately equal to the divergence evaluated at the origin multiplied by the small volume $\Delta\tau$. What about the contribution of the remainder term of Eq. (2.44)? The remainder consists of terms which behave as x^2, y^2, xy, and so forth. So while the terms in Eq. (2.50) are proportional to the volume, or a length to the third power, the integral of the remainder terms will be proportional to a length to the *fourth power*. In the limit as $\Delta\tau \to 0$, the remainder terms will become negligible in comparison with the divergence term listed above, and may be neglected.

The divergence of a vector thus represents the net "flow" of that vector out of a unit volume. We can use this to formulate an integral/limit definition of the divergence,

$$\nabla \cdot \mathbf{p} \equiv \lim_{\Delta\tau \to 0} \frac{\int_{\Delta S} \mathbf{p} \cdot \mathbf{da}}{\Delta\tau}, \quad (2.51)$$

where $\Delta\tau$ is a small volume element, the integral is taken over the small surface ΔS of that volume, and $\mathbf{da}$ is an infinitesimal vector surface element of that volume, with direction normal to the surface.

Furthermore, within the validity of our Taylor series representation (2.44), we may also prove an important infinitesimal vector identity. The integral of the divergence of $\mathbf{p}$ throughout the volume $\Delta\tau$ can be readily shown to be of the form

$$\int_{\Delta\tau} \nabla \cdot \mathbf{p}\,dx\,dy\,dz = \nabla \cdot \mathbf{p}(0)\Delta\tau. \quad (2.52)$$

Using this equation and Eq. (2.50), we may write the following infinitesimal vector identity,

$$\int_{\Delta S} \mathbf{p} \cdot d\mathbf{a} = \int_{\Delta \tau} \nabla \cdot \mathbf{p} d\tau. \tag{2.53}$$

We have demonstrated this relationship only for infinitesimally small volumes; in Section 2.7 we will show that it can be extended to volumes of finite size.

The integral definition of divergence, Eq. (2.51), can also be used to determine an integral definition of the gradient. Let $\mathbf{v} \equiv \hat{\mathbf{x}}\phi$, where ϕ is a scalar field. Because the unit vectors are constant, the del operator does not affect them, and we have

$$\frac{\partial \phi}{\partial x} = \lim_{\Delta \tau \to 0} \frac{\int_{\Delta S} \phi da_x}{\Delta \tau}, \tag{2.54}$$

where da_x is the infinitesimal surface element(s) of the x-faces of the cube. Letting $\mathbf{v} = \hat{\mathbf{y}}\phi$ and then $\hat{\mathbf{z}}\phi$, we may combine the three results to find that

$$\nabla \phi = \lim_{\Delta \tau \to 0} \frac{\int_{\Delta S} \phi d\mathbf{a}}{\Delta \tau}, \tag{2.55}$$

where $d\mathbf{a}$ is a vector area element which points in the direction normal to the surface.

Returning briefly to our physical interpretation of the divergence, it is to be noted that the momentum density of a fluid may be written as $\mathbf{p} = \rho \mathbf{v}$, where ρ is the density of the fluid and $\mathbf{v}$ is the velocity field. The divergence of $\mathbf{p}$ is therefore the divergence of the product of a scalar and vector function. Such divergences are quite common in physics, and it is to be noted that one can simplify it into the sum of a gradient term and a divergence term,

$$\nabla \cdot (f\mathbf{v}) = (\nabla f) \cdot \mathbf{v} + f \nabla \cdot \mathbf{v}. \tag{2.56}$$

This expression may be interpreted as a vector form of the usual product rule,

$$\frac{d}{dx}(uv) = v\frac{du}{dx} + u\frac{dv}{dx}, \tag{2.57}$$

and can in fact be demonstrated by writing out the divergence in Cartesian components and applying the product rule to each component of the resulting expression. One can also use the product rule directly on the expression, with some thought, by noting that the dot product can only act on $\mathbf{v}$, while the differential operator must act on both f and $\mathbf{v}$. Therefore there should be two terms, one with ∇ acting on f (as a gradient), one with it acting on $\mathbf{v}$ (as a divergence). One should always use some care in working with ∇ acting on complicated products of vector and scalar functions.

2.5 The curl, ∇×

The last of the basic operations involving vector fields is the curl, denoted by $\nabla\times$. It is expressed in Cartesian coordinates as

$$\nabla \times \mathbf{v} = \hat{\mathbf{x}}\left(\frac{\partial}{\partial y}v_z - \frac{\partial}{\partial z}v_y\right) - \hat{\mathbf{y}}\left(\frac{\partial}{\partial x}v_z - \frac{\partial}{\partial z}v_x\right) + \hat{\mathbf{z}}\left(\frac{\partial}{\partial x}v_y - \frac{\partial}{\partial y}v_x\right). \tag{2.58}$$

Just like its relative, the cross product, the curl can be represented in a convenient determinant form:

$$\nabla \times \mathbf{v} = \begin{vmatrix} \hat{\mathbf{x}} & \hat{\mathbf{y}} & \hat{\mathbf{z}} \\ \frac{\partial}{\partial x} & \frac{\partial}{\partial y} & \frac{\partial}{\partial z} \\ v_x & v_y & v_z \end{vmatrix}. \tag{2.59}$$

The curl is usually described as a measure of the "circulation" of a vector field, or how much the vector field "goes in circles". To quantify this, we define the circulation of a field as an integral around a differential path ΔC in the xy-plane, as illustrated in Fig. 2.4. The sides of the loop are taken to be of length Δx and Δy, and the loop has a total area Δa. The origin is taken to lie in the center of the loop, without loss of generality.

Formally, circulation C_{1234} is written as the following line integral of a vector around a closed path,

$$C_{1234} \equiv \oint_{\Delta C} \mathbf{v} \cdot \mathrm{d}\mathbf{r} = \int_1 v_x \mathrm{d}r_x + \int_2 v_y \mathrm{d}r_y + \int_3 v_x \mathrm{d}r_x + \int_4 v_y \mathrm{d}r_y. \tag{2.60}$$

It is to be noted that the components of $\mathrm{d}\mathbf{r}$ have been written as $(\mathrm{d}r_x, \mathrm{d}r_y)$ instead of $(\mathrm{d}x, \mathrm{d}y)$; this is due to the vector nature of the path element. For instance, if we always write our path integrals along x such that the lower limit is the lower value of x and the upper

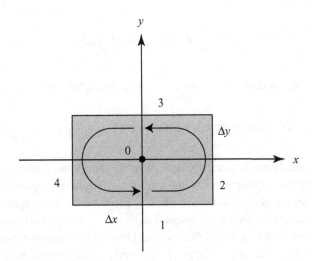

Figure 2.4 Illustrating the loop used to understand the curl.

limit is the higher value of x, $dr_x = dx$ for a path in the positive x-direction and $dr_x = -dx$ for a path along the negative x-direction.

One can see from Eq. (2.60) that C_{1234} does in fact measure how much a vector field "goes in circles": if the vector $\mathbf{v}$ is always parallel to the path element $d\mathbf{r}$, the dot product is always positive and the integral will have a nonzero value. If $\mathbf{v}$ is not always parallel to $d\mathbf{r}$, the dot product will have positive and negative values and the "circulation" will be less.

Provided Δa is sufficiently small, and $\mathbf{v}$ has continuous first partial derivatives, we may again represent our vector function $\mathbf{v}$ by its first-order Taylor series representation,

$$\mathbf{v}(x,y,z) = \mathbf{v}(0) + x\partial_x\mathbf{v}(0) + y\partial_y\mathbf{v}(0) + z\partial_z\mathbf{v}(0) + O(\mathbf{r}^2). \tag{2.61}$$

Each of the integrals in Eq. (2.60) may then be determined as follows:

$$\int_1 v_x dr_x = \int_{-\Delta x/2}^{\Delta x/2} v_x(x, -\Delta y/2, 0)dx = v_x(0) - \Delta y \Delta x \partial_y v_x(0)/2, \tag{2.62}$$

$$\int_2 v_y dr_y = \int_{-\Delta y/2}^{\Delta y/2} v_y(\Delta x/2, y, 0)dy = v_y(0) + \Delta x \Delta y \partial_x v_y(0)/2, \tag{2.63}$$

$$\int_3 v_x dr_x = \int_{-\Delta x/2}^{\Delta x/2} v_x(x, \Delta y/2, 0)(-dx) = -v_x(0) - \Delta y \Delta x \partial_y v_x(0)/2, \tag{2.64}$$

$$\int_4 v_y dr_y = \int_{-\Delta y/2}^{\Delta y/2} v_y(-\Delta x/2, y, 0)(-dy) = -v_y(0) + \Delta x \Delta y \partial_x v_y(0)/2. \tag{2.65}$$

With these results, the total value for the circulation may be written as

$$C_{1234} = \left(\frac{\partial}{\partial x} v_y(0) - \frac{\partial}{\partial y} v_x(0) \right) \Delta x \Delta y. \tag{2.66}$$

On comparison with our definition of the curl, Eq. (2.58), it follows that

$$\text{circulation in the x-y plane} = \hat{\mathbf{z}} \cdot [\nabla \times \mathbf{v}](0) \Delta x \Delta y. \tag{2.67}$$

In other words, circulation in the x-y plane about an infinitesimal loop centered on the origin is given by the z-component of the curl at the origin, multiplied by the area of the loop. Likewise, circulation in the y-z plane is given by the x-component of the curl, with a similar statement for the y-component. The association of the curl with circulation is therefore confirmed. An inspection of the remainder term of the Taylor series approximation demonstrates that it is of the order of length to the third power, whereas the terms above are of length to the second power. In the limit $\Delta S \to 0$, the remainder terms become negligible.

This demonstration has also, in fact, given us an integral definition of the components of the curl; for instance

$$\nabla \times \mathbf{v}|_x = \lim_{\Delta S \to 0} \frac{\oint_{yz} \mathbf{v} \cdot d\mathbf{r}}{\Delta S} \tag{2.68}$$

for the x-component, where the path is taken to be a loop in the yz-plane.

Furthermore, we can prove an infinitesimal vector identity related to the curl. By evaluating the surface integral of the curl under the approximation of Eq. (2.61), we can show that

$$\oint \mathbf{v} \cdot d\mathbf{r} = \int_{\Delta S} \nabla \times \mathbf{v} \cdot d\mathbf{a}. \tag{2.69}$$

That is, the path integral of $\mathbf{v}$ over an infinitesimal closed loop is equal to the integral of the curl of $\mathbf{v}$ over the area ΔS of the loop. We will use this in proving Stokes' theorem in Section 2.8.

2.6 Further applications of ∇

One can further apply the ∇ operator to the divergence, curl, and gradient of functions. For instance, the divergence of a vector is a scalar function, of which we may then take the gradient, i.e. $\nabla(\nabla \cdot \mathbf{v})$. Three successive applications of ∇ are of particular physical interest, and we consider them briefly.

2.6.1 ∇ × ∇φ

Writing this formula out explicitly in Cartesian coordinates, we find that

$$\nabla \times \nabla \phi = \hat{\mathbf{x}} \left(\frac{\partial^2 \phi}{\partial y \partial z} - \frac{\partial^2 \phi}{\partial z \partial y} \right) + \hat{\mathbf{y}} \left(\frac{\partial^2 \phi}{\partial z \partial x} - \frac{\partial^2 \phi}{\partial x \partial z} \right) + \hat{\mathbf{z}} \left(\frac{\partial^2 \phi}{\partial x \partial y} - \frac{\partial^2 \phi}{\partial y \partial x} \right) = 0, \tag{2.70}$$

where we have used the fact that $\partial^2/\partial x \partial y = \partial^2/\partial y \partial x$. Thus the curl of a gradient is always zero. In fact, any vector field $\mathbf{v}$ for which the curl vanishes is referred to as *irrotational*, and it can be shown that it can always be expressed as the gradient of a scalar function. Physical examples of this are gravitational and electrostatic forces: for instance, the gravitational force $\mathbf{F}$ can be written as the gradient of the potential energy U, $\mathbf{F} = -\nabla U$.

2.6.2 ∇ · (∇ × v)

It can be shown be writing this out explicitly in Cartesian coordinates that

$$\nabla \cdot (\nabla \times \mathbf{v}) = 0. \tag{2.71}$$

Thus the divergence of a curl always vanishes. A vector field $\mathbf{v}$ for which the divergence vanishes is referred to as *solenoidal*, and it can be shown that it can always be expressed as the curl of a vector function. A physical example of this is the magnetic field $\mathbf{B}$, which satisfies the Maxwell's equation,

$$\nabla \cdot \mathbf{B} = 0. \tag{2.72}$$

We will see that one can always express the magnetic field in terms of a *vector potential* $\mathbf{A}$,

$$\mathbf{B} = \nabla \times \mathbf{A}. \tag{2.73}$$

2.6.3 $\nabla \times (\nabla \times \mathbf{v})$

This triple product has the form of a vector triple product, and it is tempting to simply use the "BAC-CAB" rule to immediately simplify it. One must take care, however, because a naïve application of the rule results in the formula

$$\nabla \times (\nabla \times \mathbf{v}) = \nabla(\nabla \cdot \mathbf{v}) - \mathbf{v}(\nabla \cdot \nabla), \tag{2.74}$$

which is clearly flawed, as the "del" operator must act upon the vector $\mathbf{v}$. By direct expansion of the triple product, one finds that the correct answer is very close to this one,

$$\nabla \times (\nabla \times \mathbf{v}) = \nabla(\nabla \cdot \mathbf{v}) - (\nabla \cdot \nabla)\mathbf{v} = \nabla(\nabla \cdot \mathbf{v}) - \nabla^2 \mathbf{v}, \tag{2.75}$$

where ∇^2 is the Laplacian, discussed in the next section.

2.6.4 Laplacian, $\nabla \cdot \nabla\phi$

The Laplacian appears often in theoretical physics and partial differential equations; it can be shown on substitution that it takes on the form

$$\nabla \cdot \nabla\phi \equiv \frac{\partial^2 \phi}{\partial x^2} + \frac{\partial^2 \phi}{\partial y^2} + \frac{\partial^2 \phi}{\partial z^2}. \tag{2.76}$$

The Laplacian $\nabla \cdot \nabla$ is usually abbreviated as ∇^2, and often as $\triangle$. When $\nabla^2\phi = 0$, we have *Laplace's equation*, which describes, among other things, the electric potential of a static electric field in a region free of charges. It also can be used to describe the steady-state distribution of temperature in a region with no sources of heat, the steady-state flow of an incompressible fluid, and even a static magnetic potential in a region which contains no currents. From the above examples, it can be seen that the Laplace's equation is generally connected with solutions to problems that are (1) static, steady-state or otherwise time-independent, and (2) solved in regions with no sources. We will come back to this equation numerous times throughout this text.

In optics, Laplace's equation is often used to study the interaction of optical fields in regions much smaller than the wavelength of light. In such a region, the electric field looks essentially like an electrostatic field with a time-harmonically varying amplitude.

2.7 Gauss' theorem (divergence theorem)

We have already seen in Section 2.4 that, on an infinitesimal scale, the divergence of a vector times a unit volume is associated with the flow through the surface of that unit volume. It is possible to extend this observation to finite size surfaces and volumes, in which form it is commonly known as *Gauss' theorem* or the *divergence theorem*. This theorem may be stated as follows.

Theorem 2.1 (Gauss' theorem) *Let us assume we have a vector field* **v** *which is continuous and has continuous first partial derivatives in a volume V and on the surface S of that volume. Then the vector field* **v** *satisfies the relation*

$$\oint_S \mathbf{v} \cdot d\mathbf{a} = \int_V \nabla \cdot \mathbf{v} d\tau. \tag{2.77}$$

We can prove this by dividing up the volume V of interest into a large number of infinitesimal parallelepipeds, as illustrated in Fig. 2.5.

For each parallelepiped, the infinitesimal rule, Eq. (2.53), is satisfied,

$$\int_{\Delta S} \mathbf{v} \cdot d\mathbf{a} = \int_{\Delta \tau} \nabla \cdot \mathbf{v} d\tau. \tag{2.78}$$

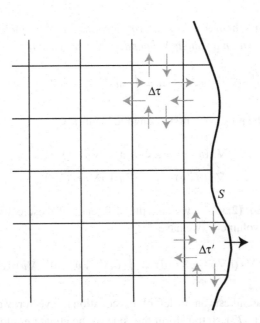

Figure 2.5 The technique used to demonstrate Gauss' theorem. A volume V is divided into a small number of infinitesimal volumes $\Delta \tau$. The flux through the surface of the inner volume $(\Delta \tau)$ is canceled by the flux from adjoining volumes, while the boundary volume $(\Delta \tau')$ has one contribution which is not canceled.

Let us consider the flow between two parallelepipeds with a common surface. Because the vector field **v** is continuous, the flow out of the common surface for one parallelepiped is equal to the flow into the surface for the other parallelepiped. Therefore, if we sum Eq. (2.78) over all infinitesimal volumes, the contributions to the left of the equation will cancel out for all interior surfaces, leaving only the contribution which comes from the exterior surface S of the total volume. On the right-hand side of the equation, the sum over all parallelepipeds becomes an integral over the total volume V. Thus we have

$$\sum_{\text{exterior surfaces}} \int_{\Delta S} \mathbf{v} \cdot \mathbf{da} \rightarrow \int_S \mathbf{v} \cdot \mathbf{da}, \qquad (2.79)$$

$$\sum_{\text{volumes}} \int_{\Delta \tau} \nabla \cdot \mathbf{v} \mathrm{d}\tau \rightarrow \int_V \nabla \cdot \mathbf{v} \mathrm{d}\tau, \qquad (2.80)$$

which leads us directly to Gauss' theorem.

Gauss' theorem requires continuity of the first partial derivatives of the field so that the infinitesimal rule (2.78) is satisfied. This theorem suggests that for any vector field **v** with such continuous derivatives and any volume V within that field, the flow of **v** across the surface of the volume is equal to the integrated value of the divergence within the volume.

An extension of Gauss' theorem which plays an important role in diffraction theory is *Green's theorem*.

Theorem 2.2 (Green's theorem) *Two scalar functions u and v which are continuous and differentiable on and within a volume V bounded by a closed surface S satisfy the relation*

$$\int_S (u\nabla v - v\nabla u) \cdot \mathbf{da} = \int_V (u\nabla \cdot \nabla v - v\nabla \cdot \nabla u) \mathrm{d}\tau. \qquad (2.81)$$

This theorem can be proven by first noting the identities

$$\nabla \cdot (u\nabla v) = u\nabla \cdot \nabla v + (\nabla u) \cdot (\nabla v), \qquad (2.82)$$

$$\nabla \cdot (v\nabla u) = v\nabla \cdot \nabla u + (\nabla v) \cdot (\nabla u), \qquad (2.83)$$

which follow from Eq. (2.56). If we take the difference of these equations and integrate what remains over a volume V, we have

$$\int_V \nabla \cdot (u\nabla v - v\nabla u) \mathrm{d}\tau = \int_V (u\nabla \cdot \nabla v - v\nabla \cdot \nabla u) \mathrm{d}\tau. \qquad (2.84)$$

The quantity in the parenthesis on the left of this equation is a vector; we may therefore use Gauss' theorem, Eq. (2.77), to transform this side of the equation into a surface integral, and we arrive at the result of Green's theorem.

When we have a little more mathematics in our toolkit we will return to Green's theorem and see how it can be used to derive diffraction formulas, in Section 19.1.

2.8 Stokes' theorem

In the previous section, we extended our earlier physical interpretation of the divergence into a general theorem concerning the relation of a vector field on a surface to its divergence in the volume enclosed by the surface; in this section, we extend our physical interpretation of the curl into a general theorem concerning the relation of a vector field on a closed path to its curl in the area bounded by the path. This theorem is generally known as Stokes' theorem. We saw from our analysis of the curl that, for an infinitesimal open surface, relation (2.69) holds, i.e.

$$\oint_{\Delta C} \mathbf{v} \cdot d\mathbf{r} = \int_{\Delta S} \nabla \times \mathbf{v} \cdot d\mathbf{a}. \qquad (2.85)$$

We wish to extend this theorem to a finite-sized open surface, so we follow an analogous procedure to that of the previous section. We consider a vector field $\mathbf{v}$ which is continuous and has a continuous first derivative on an open surface S, bounded by a closed curve C. We then subdivide this surface into a large number of infinitesimal surfaces ΔS bounded by curves ΔC. Similar to the case of Gauss' theorem, because $\mathbf{v}$ is continuous we find that the contributions of those line segments which are shared between two surfaces ΔS are of equal magnitude and of opposite sign, because the paths are in opposite directions, as illustrated in Fig. 2.6.

The only contributions which survive when we sum over all the infinitesimal path elements are the line segments on the boundary. On the right-hand side of the equation, the sum over all surface elements results in an integral over the total surface. We are thus left with Stokes' theorem.

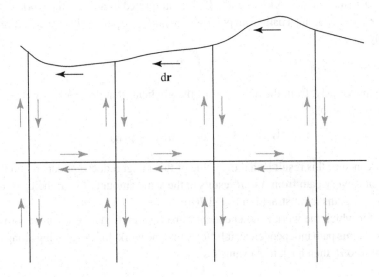

Figure 2.6 The technique used to demonstrate Stokes' theorem.

Theorem 2.3 (Stokes' theorem) *Let us assume we have a vector field* **v** *which is continuous and has continuous first partial derivatives on an open surface S and on the boundary C of that surface. Then the vector field* **v** *satisfies the relation*

$$\oint_C \mathbf{v} \cdot d\mathbf{r} = \int_S \nabla \times \mathbf{v} \cdot d\mathbf{a}. \tag{2.86}$$

Again, the requirement of the continuity of the derivatives follows from the use of the infinitesimal rule, Eq. (2.85).

2.9 Potential theory

One of the most significant classes of vector fields is the gradient field, defined by

$$\mathbf{F} = -\nabla\phi. \tag{2.87}$$

As mentioned earlier, we refer to such a vector field as "irrotational". Obvious examples of this are the gravitational force, the electric field produced by a static electric charge and the electric force derived from that field. When **F** represents a force, ϕ can be interpreted as a potential energy – hence the name of this section, "potential theory". For convenience, we will typically refer to **F** as a force and ϕ as a potential.

From the incredibly simple equation (2.87) follows a large collection of mathematical consequences, and we now consider some of the most significant of these – more will follow when we discuss differential equations and complex variables.

One of the most important consequences of the force being represented by a gradient is the path independence of work integrals. If **F** is interpreted as a force, the work W done by the force in moving along a path from point A to point B (the path is illustrated in Fig. 2.7(a)) is given by

$$W = \int_A^B \mathbf{F} \cdot d\mathbf{r}. \tag{2.88}$$

We saw previously, from the definition of the gradient, that $d\phi = \nabla\phi \cdot d\mathbf{r} = -\mathbf{F} \cdot d\mathbf{r}$, so that we may write that

$$W = -\int_{\phi(A)}^{\phi(B)} d\phi = \phi(A) - \phi(B). \tag{2.89}$$

The significance of this result is that the value of this integral does not depend on the exact path chosen – every path from A to B results in the same amount of work being done. Such a collection of paths is illustrated in Fig. 2.7(b).

A force for which the work is independent of path is referred to as a *conservative force*.

Because of this path independence, it follows that the work done in moving along a closed loop (from A to B, then back to A) vanishes,

$$\oint \mathbf{F} \cdot d\mathbf{r} = 0. \tag{2.90}$$

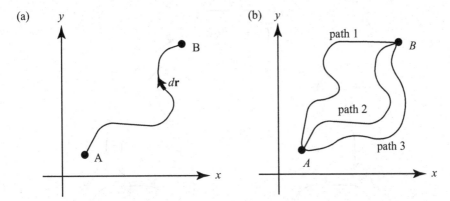

Figure 2.7 Illustrating (a) a path from point A to point B, and (b) the path independence of gradient fields. The work done in following any of the paths from A to B is the same.

Recalling Stokes' theorem, we know that

$$\oint \mathbf{F} \cdot \mathbf{dr} = \int_S \nabla \times \mathbf{F} \cdot \mathbf{da}, \tag{2.91}$$

where S is a surface defined by the closed path and $\mathbf{da}$ is an infinitesimal element on that surface. We may therefore write that

$$\int_S \nabla \times \mathbf{F} \cdot \mathbf{da} = 0. \tag{2.92}$$

If path independence is satisfied over a region of space, we may choose any infinitesimal loop in that region as a closed path; Eq. (2.92) must then be satisfied for every loop in this region. This can only happen if the integrand vanishes identically through that region, i.e.

$$\nabla \times \mathbf{F} = 0. \tag{2.93}$$

We have thus seen that, if a vector field is (1) a gradient field, then (2) all path integrals of that vector field are path independent. From this path independence, we determined that (3) the curl of $\mathbf{F}$ vanishes. We may in fact show that these three conditions are equivalent. To do this, we must demonstrate that if any of the equations are satisfied, all are satisfied. We have so far shown that (1) leads to (3); all we need to do now is show that (3) leads back to (1). Starting with Eq. (2.93), we can integrate both sides of the equation over an arbitrary open surface and apply Stokes' theorem. We immediately arrive back at Eq. (2.90); by choosing points A and B on the path encircling the area, we divide the closed loop into two pieces and demonstrate that the integral along each path has the same value:

$$\int_A^B \mathbf{F} \cdot \mathbf{dr} \bigg|_{\text{path 1}} = -\int_B^A \mathbf{F} \cdot \mathbf{dr} \bigg|_{\text{path 2}} = \int_A^B \mathbf{F} \cdot \mathbf{dr} \bigg|_{\text{path 2}}. \tag{2.94}$$

Vector calculus

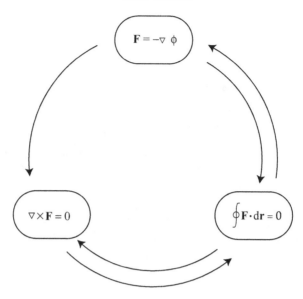

Figure 2.8 The three equivalent formulations of a conservative force field, and the paths which lead between them.

Thus path independence is confirmed! To show again that $\mathbf{F} = -\nabla\phi$, we *define* $\phi(x,y,z)$ as the path integral of $\mathbf{F}$ from a fixed point M to the point (x,y,z), i.e.

$$\phi(x,y,z) = \int_M^{(x,y,z)} \mathbf{F} \cdot d\mathbf{r}. \qquad (2.95)$$

Using this equation, we may write that

$$\phi(A) - \phi(B) = \int_M^A \mathbf{F} \cdot d\mathbf{r} - \int_M^B \mathbf{F} \cdot d\mathbf{r} = -\int_A^B \mathbf{F} \cdot d\mathbf{r}. \qquad (2.96)$$

Therefore the path integral of $\mathbf{F}$ is a *total differential* and it follows that $\mathbf{F} \cdot d\mathbf{r} = -d\phi = -\nabla\phi \cdot d\mathbf{r}$. From this we have $\mathbf{F} = -\nabla\phi$, and we have shown that (1) follows from (2).

The three equivalent formulations of a conservative force are summarized in Fig. 2.8.

Just as we have seen that a scalar potential may be used to represent a vector field which is irrotational, a *vector potential* may be used to represent a vector field for which the divergence is zero (a solenoidal field) – that is, if

$$\nabla \cdot \mathbf{F} = 0, \qquad (2.97)$$

we may write

$$\mathbf{F} = \nabla \times \mathbf{A}, \qquad (2.98)$$

where $\mathbf{A}$ is the vector potential of the field. The vector potential is typically more difficult to determine for a particular given vector field, due to its vector nature.

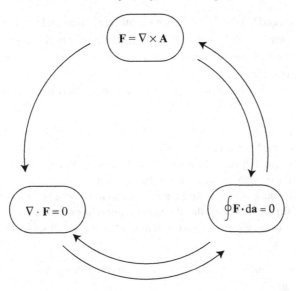

Figure 2.9 The three equivalent formulations of a field defined by a vector potential.

A set of relationships directly analogous to Fig. 2.8 can be made for the vector potential, and are illustrated in Fig. 2.9. If $\mathbf{F} = \nabla \times \mathbf{A}$, for instance, it follows immediately from Eq. (2.71) that $\nabla \cdot \mathbf{F} = 0$. We can then use Gauss' theorem to demonstrate that $\oint \mathbf{F} \cdot d\mathbf{a} = 0$.

One important observation concerning both the scalar and vector potentials is their *nonuniqueness*. A given irrotational or solenoidal vector field has an infinite number of potentials that can be used to represent it. This nonuniqueness arises from the fact that the fields are determined from the potentials by derivative relations.

For instance, for the scalar potential ϕ, we may add any constant to the potential without affecting the vector field which is derived from it. This should be familiar from elementary mechanics or electromagnetics as the freedom to choose the zero of potential energy.

For vector fields, the situation is more complicated; because of the curl operator, different components of $\mathbf{A}$ are mixed together to determine the vector field $\mathbf{F}$. However, we have already seen in Section 2.6 that $\nabla \times \nabla \Lambda = 0$, where Λ is an arbitrary (continuous) scalar function. Thus for any vector potential for which $\mathbf{F} = \nabla \times \mathbf{A}$, we may add to this vector potential the vector $\nabla \Lambda$ without affecting the value of the field $\mathbf{F}$ which is derived from it. In electromagnetic theory, this freedom to choose the form of the vector potential is called *gauge freedom* and is useful in finding solutions to Maxwell's equations; we discuss this in more detail in Section 2.11.

2.10 Focus: Maxwell's equations in integral and differential form

It might be said that there are two broad categories of problems in vector calculus: (1) the direct calculation of vector quantities (e.g. the curl of a given vector field, the path integral of a given vector field) and (2) the use of vector calculus relations and theorems to simplify the form of physical formulas. A famous and important example of the latter is the conversion of experimentally determined electromagnetic relations into the concise set of equations now known as *Maxwell's equations*.

We will work with the following four experimentally-determined physical phenomena.

- Gauss' law: the flux of the electric field over a closed surface S is proportional to the total charge enclosed in the volume bounded by that surface.
- Ampère's law: the line integral of the magnetic field around a closed path C is proportional to the total current passing through the area bounded by that path.
- Faraday's law: the induced electromotive force around a closed circuit C is proportional to the time rate of change of the magnetic field flux through the area of the circuit.
- No magnetic monopoles: the flux of the magnetic field over a closed surface S is always equal to zero.

We can reduce all of these formulas to differential equations by the use of Gauss' theorem and Stokes' theorem.

2.10.1 Gauss' law and "no magnetic monopoles"

In Gaussian units, Gauss' law has the mathematical form

$$\oint_S \mathbf{E}(\mathbf{r},t) \cdot \mathrm{d}\mathbf{a} = 4\pi Q_{\text{enc}}, \tag{2.99}$$

where $\mathbf{E}(\mathbf{r},t)$ is the electric field at position $\mathbf{r}$ and time t, S is a surface enclosing a volume V, and $Q_{\text{enc}}(t)$ is the net charge enclosed by that surface. We can use Gauss' theorem to convert the left-hand side of this equation to a volume integral, i.e.

$$\oint_S \mathbf{E}(\mathbf{r},t) \cdot \mathrm{d}\mathbf{a} = \int_V \nabla \cdot \mathbf{E}(\mathbf{r},t)\mathrm{d}\tau. \tag{2.100}$$

Furthermore, the total charge $Q_{\text{enc}}(t)$ may be written as the volume integral of the charge density $\rho(\mathbf{r},t)$ throughout the volume V,

$$Q_{\text{enc}}(t) = \int_V \rho(\mathbf{r},t)\mathrm{d}\tau. \tag{2.101}$$

On substitution from this pair of equations into Gauss' law, we find that

$$\int_V \nabla \cdot \mathbf{E}(\mathbf{r},t)\mathrm{d}\tau = 4\pi \int_V \rho(\mathbf{r},t)\mathrm{d}\tau. \tag{2.102}$$

But this integral relationship must hold for any choice of surface S and volume V. This directly leads to the conclusion that the integrands must be identical, or that

$$\nabla \cdot \mathbf{E}(\mathbf{r},t) = 4\pi\rho(\mathbf{r},t). \tag{2.103}$$

Similarly, the "no magnetic monopoles rule" has the mathematical form

$$\oint_S \mathbf{B}(\mathbf{r},t) \cdot \mathbf{da} = 0. \tag{2.104}$$

Using Gauss' theorem again, we readily find that

$$\nabla \cdot \mathbf{B}(\mathbf{r},t) = 0. \tag{2.105}$$

2.10.2 Faraday's law and Ampère's law

Faraday's law states that *the induced electromotive force around a circuit is proportional (and opposite) to the time rate of change of the magnetic field flux through the circuit*. To be more precise, let C be a closed loop bounded by an (open) surface S with a unit normal $\mathbf{n}$, as shown in Fig. 2.10, and let $\mathbf{B}(\mathbf{r},t)$ be the magnetic induction in the neighborhood of the circuit. The magnetic flux passing through the circuit is defined as (Gaussian units assumed)

$$F(t) = \int_S \mathbf{B}(\mathbf{r},t) \cdot \mathbf{da}, \tag{2.106}$$

where $\mathbf{da}$ is an infinitesimal area element. The electromotive force around the circuit is given by

$$\mathcal{E}(t) = \oint_C \mathbf{E}(\mathbf{r},t) \cdot \mathbf{dr}, \tag{2.107}$$

where $\mathbf{E}(\mathbf{r},t)$ is the field at the location of the path element $\mathbf{dr}$ at time t. Faraday's law states that

$$\mathcal{E}(t) = -\frac{1}{c}\frac{dF(t)}{dt}. \tag{2.108}$$

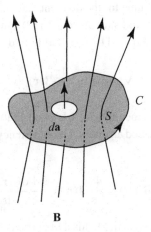

B

Figure 2.10 The open surface and loop used to demonstrate Faraday's law.

We may use Stokes' theorem to rewrite the electromotive force in terms of the curl of $\mathbf{E}$,

$$\oint_C \mathbf{E}(\mathbf{r},t) \cdot d\mathbf{r} = \int_S \nabla \times \mathbf{E}(\mathbf{r},t) \cdot d\mathbf{a}. \tag{2.109}$$

Faraday's law then becomes,

$$\int_S \nabla \times \mathbf{E}(\mathbf{r},t) \cdot d\mathbf{a} = -\frac{1}{c}\frac{d}{dt}\int_S \mathbf{B}(\mathbf{r},t) \cdot d\mathbf{a}. \tag{2.110}$$

This law is found to be independent of the exact choice of loop C and area S, so it follows that the integrands must be equal. We arrive at

$$\nabla \times \mathbf{E}(\mathbf{r},t) = -\frac{1}{c}\frac{d\mathbf{B}(\mathbf{r},t)}{dt}, \tag{2.111}$$

which is one of Maxwell's equations and the differential form of Faraday's law.

One can do a similar derivation to derive a differential form of Ampère's law: *the path integral of the magnetic field around a closed loop is proportional to the total current passing through the area bounded by that loop.* The law may be written as

$$\oint_C \mathbf{B}(\mathbf{r}) \cdot d\mathbf{r} = \frac{4\pi}{c}I_{enc}, \tag{2.112}$$

where I_{enc} is the total current passing through the area S of the closed path C, and may be written in terms of the current density $\mathbf{J}$ as

$$I_{enc} = \int_S \mathbf{J}(\mathbf{r}) \cdot d\mathbf{a}. \tag{2.113}$$

It is important to note that we assume for the moment that the current and magnetic field are independent of time t. The left-hand side of Eq. (2.112) can be transformed using Stokes' theorem, and on substitution of Eq. (2.113) we readily find

$$\nabla \times \mathbf{B}(\mathbf{r}) = \frac{4\pi}{c}\mathbf{J}(\mathbf{r}). \tag{2.114}$$

This last equation is, however, incomplete; Maxwell realized that the four differential equations as derived so far possessed an internal inconsistency, which he fixed by modifying Ampère's law to the form

$$\nabla \times \mathbf{B}(\mathbf{r},t) = \frac{4\pi}{c}\mathbf{J}(\mathbf{r},t) + \frac{1}{c}\frac{d\mathbf{E}(\mathbf{r},t)}{dt}, \tag{2.115}$$

where we now can include time-varying fields and currents. The latter term of Eq. (2.115) was referred to by Maxwell as the *displacement current*. We can understand its physical significance by converting the modified Ampère's law *back* into integral form, but with $\mathbf{J} = 0$ for simplicity. Integrating Eq. (2.115) over an open surface S, we find that

$$\int_V \nabla \times \mathbf{B}(\mathbf{r},t) \cdot d\mathbf{a} = \int_S \frac{1}{c} \frac{d\mathbf{E}(\mathbf{r},t)}{dt} \cdot d\mathbf{a}. \qquad (2.116)$$

The left-hand side of this equation may be written, using Stokes' theorem again, as a path integral,

$$\int_V \nabla \times \mathbf{B}(\mathbf{r},t) \cdot d\mathbf{a} = \oint_C \mathbf{B}(\mathbf{r},t) \cdot d\mathbf{r}. \qquad (2.117)$$

Let us refer to the path integral of $\mathbf{B}$ around a closed loop as the "magnetic circulation" $\mathcal{C}$ of the field. We note that the right-hand side of Eq. (2.116) is simply the time derivative of the electric flux F_e passing through the open surface S,

$$F_e(t) = \int_S \mathbf{E}(\mathbf{r},t) \cdot d\mathbf{a}. \qquad (2.118)$$

We may therefore write the modified Ampère's law in integral form as

$$\mathcal{C}(t) = \frac{1}{c} \frac{dF_e(t)}{dt}. \qquad (2.119)$$

This equation should be compared to Eq. (2.108). With the exception of a minus sign, the two equations are complementary. Faraday's law states, in short, that a *time-varying magnetic field induces a circulating electric field*; the modified Ampère's law states in turn that a *time-varying electric field induces a circulating magnetic field*. The two laws are therefore quite complementary to one another;[1] the effect is illustrated in Fig. 2.11.

The complete set of Maxwell's equations are therefore

$$\nabla \cdot \mathbf{E}(\mathbf{r},t) = 4\pi\rho(\mathbf{r},t) \quad \text{(Gauss' law)}, \qquad (2.120)$$

$$\nabla \cdot \mathbf{B}(\mathbf{r},t) = 0 \quad \text{("no magnetic monopoles")}, \qquad (2.121)$$

$$\nabla \times \mathbf{E}(\mathbf{r},t) = -\frac{1}{c} \frac{d\mathbf{B}(\mathbf{r},t)}{dt} \quad \text{(Faraday's law)}, \qquad (2.122)$$

$$\nabla \times \mathbf{B}(\mathbf{r},t) = \frac{4\pi}{c} \mathbf{J}(\mathbf{r},t) + \frac{1}{c} \frac{d\mathbf{E}(\mathbf{r},t)}{dt} \quad \text{(modified Ampère's law)}. \qquad (2.123)$$

Even though many readers are no doubt familiar with the calculation, it would be remiss to discuss vector calculus and Maxwell's equations without describing the most famous and important result to arise from their application: the discovery of electromagnetic wave

---

[1] It is interesting to note that Faraday himself, inspired by the relation between electricity and magnetism, attempted to experimentally demonstrate a relation between gravity and electricity in which a changing *gravitational* flux would induce an electric current. His experiments failed, but make for fascinating reading [Far51].

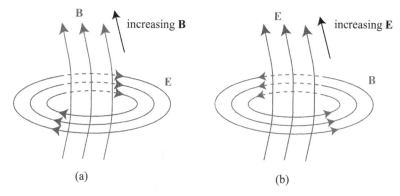

Figure 2.11 (a) Illustration of Faraday's law: time-varying magnetic fields induce electric fields. (b) Illustration of the modified Ampère's law: time-varying electric fields induce magnetic fields.

equations. Let us consider Maxwell's equations in a region free of electric charges and currents; in this case the equations take on the form

$$\nabla \cdot \mathbf{E}(\mathbf{r}, t) = 0, \tag{2.124}$$

$$\nabla \cdot \mathbf{B}(\mathbf{r}, t) = 0, \tag{2.125}$$

$$\nabla \times \mathbf{E}(\mathbf{r}, t) = -\frac{1}{c}\frac{d\mathbf{B}(\mathbf{r}, t)}{dt}, \tag{2.126}$$

$$\nabla \times \mathbf{B}(\mathbf{r}, t) = \frac{1}{c}\frac{d\mathbf{E}(\mathbf{r}, t)}{dt}. \tag{2.127}$$

Let us take the curl of the third of these equations (Faraday's law). We then have the formula

$$\nabla \times [\nabla \times \mathbf{E}(\mathbf{r}, t)] = -\frac{1}{c}\nabla \times \frac{d\mathbf{B}(\mathbf{r}, t)}{dt}. \tag{2.128}$$

On the right-hand side of this equation, we may exchange the order of differentiation, taking the curl of $\mathbf{B}$ first and then the time derivative. Once we have done so, we may use the modified Ampère's law to rewrite the result in terms of the electric field,

$$-\frac{1}{c}\nabla \times \frac{d\mathbf{B}(\mathbf{r}, t)}{dt} = -\frac{1}{c}\frac{d}{dt}[\nabla \times \mathbf{B}(\mathbf{r}, t)] = -\frac{1}{c}\frac{d}{dt}\left[\frac{1}{c}\frac{d\mathbf{E}(\mathbf{r}, t)}{dt}\right] = -\frac{1}{c^2}\frac{d^2\mathbf{E}(\mathbf{r}, t)}{dt^2}. \tag{2.129}$$

The left-hand side of Eq. (2.128) can be simplified by the use of Eq. (2.75); we then have

$$\nabla[\nabla \cdot \mathbf{E}(\mathbf{r}, t)] - \nabla^2\mathbf{E}(\mathbf{r}, t) = -\frac{1}{c^2}\frac{d^2\mathbf{E}(\mathbf{r}, t)}{dt^2}. \tag{2.130}$$

In the absence of sources, however, we have $\nabla \cdot \mathbf{E} = 0$, i.e. Eq. (2.124). Equation (2.130) reduces to the form

$$\nabla^2 \mathbf{E}(\mathbf{r},t) - \frac{1}{c^2} \frac{d^2 \mathbf{E}(\mathbf{r},t)}{dt^2} = 0. \tag{2.131}$$

This is a wave equation for the electric field $\mathbf{E}$. Similarly, beginning with the curl of the modified Ampère's law, one can show that the magnetic field $\mathbf{B}$ also satisfies a wave equation. This discovery by Maxwell was an important first step towards recognizing light as an electromagnetic wave.

2.11 Focus: gauge freedom in Maxwell's equations

We noted at the end of Section 2.9 that there is a significant amount of freedom in choosing a scalar potential $\phi(\mathbf{r})$ which represents an irrotational field $\mathbf{E} = -\nabla\phi$ and in choosing a vector potential $\mathbf{A}(\mathbf{r})$ which represents a solenoidal field $\mathbf{B} = \nabla \times \mathbf{A}$. In electrostatics and magnetostatics, the electric and magnetic field (and hence the scalar and vector potential) are independent quantities, and may be derived independently of one another. In general time-varying electrodynamics, however, the electric and magnetic fields are interrelated and the vector and scalar potentials must be developed simultaneously. There is still significant freedom in choosing the potentials, however, and an appropriate choice of potentials can simplify the derivation of the electromagnetic fields. We now discuss such *gauge freedom* in more detail.

The second of Maxwell's equations ("no magnetic monopoles"), Eq. (2.121), tells us that the divergence of the magnetic field is always zero. Regardless of the behavior of the electric field, then, we may always express the magnetic field as the curl of a vector potential $\mathbf{A}$,

$$\mathbf{B}(\mathbf{r},t) = \nabla \times \mathbf{A}(\mathbf{r},t). \tag{2.132}$$

The scalar potential is a little more difficult to come by. We have seen that a vector field $\mathbf{V}(\mathbf{r},t)$ can only be represented by a scalar potential if its curl vanishes, i.e.

$$\nabla \times \mathbf{V}(\mathbf{r},t) = 0. \tag{2.133}$$

It follows from Faraday's law, Eq. (2.122), that the electric field in general has a non-vanishing curl. However, if we substitute from Eq. (2.132) into Eq. (2.122), we find that

$$\nabla \times \left[\mathbf{E}(\mathbf{r},t) + \frac{1}{c} \frac{\partial \mathbf{A}(\mathbf{r},t)}{\partial t} \right] = 0. \tag{2.134}$$

Evidently the quantity $\mathbf{E} + \frac{1}{c} \frac{\partial \mathbf{A}}{\partial t}$ can be expressed as the gradient of a scalar potential, as its curl vanishes. We write this quantity in the form

$$\mathbf{E}(\mathbf{r},t) + \frac{1}{c} \frac{\partial \mathbf{A}(\mathbf{r},t)}{\partial t} = -\nabla\phi(\mathbf{r},t). \tag{2.135}$$

We may rearrange this formula to write the electric field in terms of a scalar and vector potential,

$$\mathbf{E}(\mathbf{r},t) = -\nabla\phi(\mathbf{r},t) - \frac{1}{c}\frac{\partial\mathbf{A}(\mathbf{r},t)}{\partial t}. \tag{2.136}$$

Differential equations for these potentials may be written by substituting from Eqs. (2.132) and (2.136) into Maxwell's equations; Maxwell's equations then take on the form

$$\nabla^2\phi(\mathbf{r},t) + \frac{1}{c}\frac{\partial}{\partial t}[\nabla\cdot\mathbf{A}(\mathbf{r},t)] = -4\pi\rho(\mathbf{r},t) \quad \text{(Gauss' law)}, \tag{2.137}$$

$$0 = 0 \quad \text{("no magnetic monopoles")}, \tag{2.138}$$

$$-\frac{1}{c}\nabla\times\frac{\partial\mathbf{A}(\mathbf{r},t)}{\partial t} = -\frac{1}{c}\nabla\times\frac{d\mathbf{A}(\mathbf{r},t)}{dt} \quad \text{(Faraday's law)}, \tag{2.139}$$

$$\nabla^2\mathbf{A}(\mathbf{r},t) - \frac{1}{c^2}\frac{\partial^2\mathbf{A}(\mathbf{r},t)}{\partial t^2}$$

$$-\nabla\left[\nabla\cdot\mathbf{A}(\mathbf{r},t) + \frac{1}{c}\frac{\partial\phi(\mathbf{r},t)}{\partial t}\right] = -\frac{4\pi}{c}\mathbf{J}(\mathbf{r},t) \quad \text{(modified Ampère's law)}. \tag{2.140}$$

Two of these equations ("no magnetic monopoles" and Faraday's law) have reduced to trivial identities. By using the potentials, we have therefore made a substantial simplification in the solution of Maxwell's equations: instead of solving four vector equations, with six unknowns (the components of $\mathbf{E}$ and $\mathbf{B}$), we now only need solve two vector equations, Eqs. (2.137) and (2.140), with four unknowns (ϕ and the components of $\mathbf{A}$). It can be seen from these equations that this is still a nontrivial task: the equations are coupled (both contain ϕ and $\mathbf{A}$).

Now we may exploit the freedom we have in choosing our potentials, though we must take care because both $\mathbf{B}$ and $\mathbf{E}$ depend on the vector potential $\mathbf{A}$. First, we note that we may add the vector $\nabla\Lambda(\mathbf{r},t)$ to the vector potential without changing the value of the magnetic field. This addition, however, will also change the electric field by an amount $\frac{1}{c}\nabla\frac{\partial\Lambda(\mathbf{r},t)}{\partial t}$, so we must also subtract $\frac{1}{c}\frac{\partial\Lambda(\mathbf{r},t)}{\partial t}$ from the scalar potential simultaneously to leave the electric field unchanged. The following pair of gauge transformations, then, will result in the same value of the electric and magnetic fields,

$$\mathbf{A} \rightarrow \mathbf{A} + \nabla\Lambda, \quad \phi \rightarrow \phi - \frac{1}{c}\frac{\partial\Lambda}{\partial t}. \tag{2.141}$$

By an appropriate choice of the function Λ, we can simplify the form of Eqs. (2.137) and (2.140) and make them much more amenable to solution. There are two commonly-used and extremely helpful choices, which are referred to as working in the *Lorenz gauge*[2] and the *Coulomb gauge*.

---

[2] Though typically referred to as the "Lorentz gauge", this result is evidently due to Ludwig *Lorenz* [Lor67], not Hendrik Antoon Lorentz.

In the Lorenz gauge, we assume that we can find a function Λ such that

$$\nabla \cdot \mathbf{A}(\mathbf{r},t) + \frac{1}{c}\frac{\partial \phi(\mathbf{r},t)}{\partial t} = 0. \tag{2.142}$$

It is to be noted that this quantity appears in the square brackets of Eq. (2.140). Assuming that this can be done, we readily reduce our two equations for the potentials to the form

$$\nabla^2 \phi(\mathbf{r},t) - \frac{1}{c^2}\frac{\partial^2 \phi(\mathbf{r},t)}{\partial t^2} = -4\pi\rho(\mathbf{r},t), \tag{2.143}$$

$$\nabla^2 \mathbf{A}(\mathbf{r},t) - \frac{1}{c^2}\frac{\partial^2 \mathbf{A}(\mathbf{r},t)}{\partial t^2} = -\frac{4\pi}{c}\mathbf{J}(\mathbf{r},t), \tag{2.144}$$

which are simply a pair of (uncoupled) wave equations for ϕ and $\mathbf{A}$!

What sort of function Λ will result in ϕ, $\mathbf{A}$ satisfying Eq. (2.142)? Let us suppose that $\mathbf{A}$ and ϕ *do not* satisfy the Lorenz condition, and look at a transformed pair of potentials

$$\mathbf{A}' = \mathbf{A} + \nabla\Lambda, \quad \phi' = \phi - \frac{1}{c}\frac{\partial \Lambda}{\partial t}. \tag{2.145}$$

The new potentials will satisfy the Lorenz condition if

$$\nabla^2\Lambda - \frac{1}{c^2}\frac{\partial^2\Lambda}{\partial t^2} = -\left(\nabla \cdot \mathbf{A} + \frac{1}{c}\frac{\partial \phi}{\partial t}\right). \tag{2.146}$$

This is simply an inhomogeneous wave equation for the function Λ, with a source term determined by $\mathbf{A}$ and ϕ. It can in principle be solved, so there is in general a choice of potential which satisfies the Lorenz condition.

One almost never needs to find the actual function Λ. Instead, one directly solves Eqs. (2.143) and (2.144) for the potential, while *assuming* that the gauge condition (2.142) is satisfied.

In the Coulomb gauge, we make a simpler assumption that

$$\nabla \cdot \mathbf{A} = 0. \tag{2.147}$$

With this assumption, our equations for the potentials reduce to the form

$$\nabla^2 \phi(\mathbf{r},t) = -4\pi\rho(\mathbf{r},t), \tag{2.148}$$

$$\nabla^2 \mathbf{A}(\mathbf{r},t) - \frac{1}{c^2}\frac{\partial^2 \mathbf{A}(\mathbf{r},t)}{\partial t^2} = -\frac{4\pi}{c}\mathbf{J}(\mathbf{r},t) + \frac{1}{c}\nabla\frac{\partial \phi(\mathbf{r},t)}{\partial t}. \tag{2.149}$$

The equation for the scalar potential is simply *Poisson's equation*, which may be solved quite readily using methods from electrostatics. The equation for the vector potential is then an inhomogeneous wave equation which can be solved once the scalar potential has been determined.

What sort of function Λ will satisfy Eq. (2.147)? Again assuming that **A** *does not* satisfy the Coulomb condition, we look for a function $\mathbf{A}' = \mathbf{A} + \nabla\Lambda$. It will satisfy the Coulomb condition if

$$\nabla^2\Lambda = -\nabla \cdot \mathbf{A}. \tag{2.150}$$

The function Λ must satisfy Poisson's equation, with a source term dependent upon the starting potential **A**. Poisson's equation can be solved, so in general there is a choice of potential which results in the Coulomb gauge.

2.12 Additional reading

- H. M. Schey, *Div Grad Curl and All That* (New York, W. W. Norton, 2005, 4th edition). A great introductory discussion of vector calculus, with many examples of vector integration.
- J. D. Jackson, From Lorenz to Coulomb and other explicit gauge transformations, *Am. J. Phys.* **70** (2002), 917–928. A discussion of a variety of gauges in electromagnetic theory.
- J. D. Jackson and L. B. Okun, Historical roots of gauge invariance, *Rev. Mod. Phys.* **73** (2001), 663–680. A detailed historical overview of the use of gauge theory in physics.
- O. D. Kellogg, *Foundations of Potential Theory* (NY, Dover, 1953). A rigorous and detailed mathematical text on potential theory.

2.13 Exercises

1. Determine the gradient of the function $\phi(x,y) = x^2 + 2xy + y^2$, and find all points where the $\nabla\phi = 0$. What does $\nabla\phi = 0$ imply about the behavior of the function ϕ?
2. Determine the gradient of the function $\phi(x,y) = x^3 + 2x^2y + 3xy^2$, and find all points where the $\nabla\phi = 0$. What does $\nabla\phi = 0$ imply about the behavior of the function ϕ?
3. A vector function **v** is defined as $\mathbf{v} = x^2\hat{\mathbf{x}} + y^2\hat{\mathbf{y}} + z^2\hat{\mathbf{z}}$. Find $\nabla \cdot \mathbf{v}$ and $\nabla \times \mathbf{v}$.
4. A vector function **v** is defined as $\mathbf{v} = y^2\hat{\mathbf{x}} + x^2\hat{\mathbf{y}} + xz\hat{\mathbf{z}}$. Find $\nabla \cdot \mathbf{v}$ and $\nabla \times \mathbf{v}$.
5. A surface is defined by the equation $x^2 + y^2 = 2z$. Find an expression for a unit vector perpendicular to this surface.
6. A surface is defined by the equation $x^2/4 + y^2/4 + z^2 = 1$. Find an expression for a unit vector perpendicular to this surface.
7. We consider the path integral of the vector function $\mathbf{v}(x,y) = x^2\hat{\mathbf{x}}$, where the path C is the unit circle in the xy-plane centered on the origin. Determine the value of
   (a) $\oint_C \mathbf{v}dr$, $\oint_C \mathbf{v} \cdot d\mathbf{r}$.
   (Parameterize the curve by choosing $x = \cos\theta$, $y = \sin\theta$.)
8. We consider the surface integral of the vector function $\mathbf{v}(x,y,z) = y\hat{\mathbf{x}} + xz\hat{\mathbf{z}}$, where the surface S lies entirely in the first octant of three-dimensional space and satisfies $x + y + z = 1$. Determine the value of
   (a) $\int_S \mathbf{v} \cdot d\mathbf{a}$, $\int_S \mathbf{v}da$.
   (Parameterize the surface by x and y.)

9. We consider the path integral of the vector function $v(x,y) = x^2 y\hat{x} + y\hat{y}$, where the path is the isosceles triangle shown below:

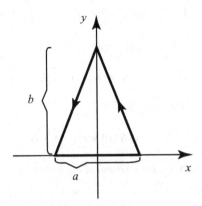

Determine the value of

(a) $\oint_C \mathbf{v} \cdot d\mathbf{r}$, $\oint_C \mathbf{v} dr$.

10. We consider the surface integral of the vector function $v(x,y,z) = x^2 z\hat{x} + 2z^2\hat{y} + yz\hat{z}$, where the path satisfies the equation $z = ax^2$, $-1 \leq x \leq 1$, $-1 \leq y \leq 1$. Determine the value of

(a) $\int_S \mathbf{v} \cdot d\mathbf{a}$, $\int_S \mathbf{v} da$.

11. It is somewhat surprising to note that almost any parameterization of a path integral will work, provided it accurately characterized the path. Repeat Example 2.1 with the parameterization $\mathbf{r}(t) = 2t^2\hat{x} + 3t^2\hat{y}$ along C_1, with $t : 0 \rightarrow 1$, and $\mathbf{r}(\theta) = 2\cos\theta\hat{x} + 3\hat{y}$ along C_2, with $\theta : 0 \rightarrow \pi/2$, and show that the result is the same.

12. Prove that $\nabla \cdot (\mathbf{a} \times \mathbf{b}) = \mathbf{b} \cdot (\nabla \times \mathbf{a}) - \mathbf{a} \cdot (\nabla \times \mathbf{b})$.

13. Prove that $\mathbf{a} \times (\nabla \times \mathbf{a}) = \nabla\left(a^2/2\right) - (\mathbf{a} \cdot \nabla)\mathbf{a}$, where $a = |\mathbf{a}|$.

14. Prove that

$$\nabla(\mathbf{a} \cdot \mathbf{b}) = (\mathbf{a} \times \nabla) \times \mathbf{b} + (\mathbf{b} \times \nabla) \times \mathbf{a} + \mathbf{a}(\nabla \cdot \mathbf{b}) + \mathbf{b}(\nabla \cdot \mathbf{a}).$$

15. Prove that

$$\nabla \times (\mathbf{a} \times \mathbf{b}) = (\mathbf{b} \cdot \nabla)\mathbf{a} - (\mathbf{a} \cdot \nabla)\mathbf{b} - \mathbf{b}(\nabla \cdot \mathbf{a}) + \mathbf{a}(\nabla \cdot \mathbf{b}).$$

16. By use of the vector function $\mathbf{v} = \mathbf{c}\phi$, where $\mathbf{c}$ is an *arbitrary* constant vector, prove the following alternative versions of Gauss' theorem and Stokes' theorem:

(a) $\oint_S \phi d\mathbf{a} = \int_V \nabla\phi d\tau$,

(b) $\oint_C \phi d\mathbf{r} = \int_S d\mathbf{a} \times \nabla\phi$.

17. By use of the vector function $\mathbf{v} = \mathbf{c} \times \mathbf{A}$, where $\mathbf{c}$ is an *arbitrary* constant vector, prove the following alternative versions of Gauss' theorem and Stokes' theorem:

(a) $\oint_S d\mathbf{a} \times \mathbf{A} = \int_V (\nabla \times \mathbf{A})d\tau$,

(b) $\oint_C d\mathbf{r} \times \mathbf{A} = \int_S (d\mathbf{a} \times \nabla) \times \mathbf{A}$.

18. By making the substitution $\mathbf{v} = \nabla \times \mathbf{u}$ in Stokes' theorem, show that

$$\int_S \nabla^2 \mathbf{u} \cdot d\mathbf{a} = \int_S \nabla(\nabla \cdot \mathbf{u}) \cdot d\mathbf{a} - \oint_C \nabla \times \mathbf{u} \cdot d\mathbf{r}.$$

19. By making the substitution $\mathbf{v} = u\nabla w$ in Stokes' theorem, show that

$$\int_S \nabla u \times \nabla w \cdot d\mathbf{a} = \oint_C u\nabla \cdot d\mathbf{r} - \oint_C w\nabla u \cdot d\mathbf{r}.$$

20. For the position vector $\mathbf{r}$ and a constant vector $\mathbf{A}$, show that

$$\nabla \times (\mathbf{A} \times \mathbf{r}) = 2\mathbf{A}.$$

21. For constant vectors $\mathbf{A}$ and $\mathbf{B}$, show that

$$\nabla[\mathbf{A} \cdot (\mathbf{B} \times \mathbf{r})] = \mathbf{A} \times \mathbf{B}.$$

22. An electric point dipole has a scalar potential of the form

$$\phi(\mathbf{r}) = \frac{\mathbf{p} \cdot \mathbf{r}}{r^3},$$

where $\mathbf{p}$ is the dipole moment, a constant vector. Calculate the electric field using $\mathbf{E} = -\nabla\phi$.

23. A magnetic point dipole has a vector potential of the form

$$\mathbf{A}(\mathbf{r}) = \frac{\mathbf{m} \times \mathbf{r}}{r^3},$$

where $\mathbf{m}$ is the magnetic dipole moment, a constant vector. Calculate the magnetic field using $\mathbf{B} = \nabla \times \mathbf{A}$.

24. Demonstrate that the divergence theorem is satisfied in a cube of unit side with edges along the coordinate axes for the function

$$\mathbf{v} = x^2\hat{\mathbf{x}} + y^2\hat{\mathbf{y}} + z^2\hat{\mathbf{z}}.$$

25. Demonstrate that the divergence theorem is satisfied in a cube of unit side with edges along the coordinate axes for the function

$$\mathbf{v} = xy\hat{\mathbf{x}} + z^2\hat{\mathbf{y}} + x^2 z\hat{\mathbf{z}}.$$

26. Demonstrate that Stokes' theorem is satisfied for a square path of unit side with edges along the x and y coordinate axes, for the function

$$\mathbf{v} = x^2 y\hat{\mathbf{x}} + yx^2\hat{\mathbf{y}}.$$

27. Demonstrate that Stokes' theorem is satisfied for a square path of unit side with edges along the x and y coordinate axes, for the function

$$\mathbf{v} = x^2 y^2\hat{\mathbf{x}} + y^2 x^3\hat{\mathbf{y}}.$$

28. Read the paper by J. D. Jackson, From Lorenz to Coulomb and other explicit gauge transformations, *Am. J. Phys.* **70** (2002), 917–928, which discusses uncommon gauge transformations. Name the additional gauges described and how they compare to the familiar Lorenz and Coulomb gauges, and give the gauge conditions for each.

29. Vector potentials are generally considered to be "fictitious" quantities that provides ease of calculation, but in quantum mechanics they can play a very significant role. Read the paper by Y. Aharonov and D. Bohm, Significance of electromagnetic potentials in the quantum theory, *Phys. Rev.* **115** (1959), 485–491. Explain the experiment described in the paper, and explain why Aharonov and Bohm come to the conclusion that the electrons are being influenced by the vector potential. What mathematical quantity characterizes the effect of the vector potential?

3

Vector calculus in curvilinear coordinate systems

3.1 Introduction: systems with different symmetries

It is obvious that the Cartesian coordinate system which we have been using up to this point is ideally suited for those situations in which one is trying to solve problems with a rectangular, or "box"-type geometry. Determining the electric potential inside of a cube, for instance, would be an ideal problem to use Cartesian coordinates, for one can choose coordinates such that faces of the cube each correspond to a coordinate being constant. However, it is also clear that there exist problems in which Cartesian coordinates are poorly suited. Two such situations are illustrated in Fig. 3.1. A monochromatic laser beam has a preferred direction of propagation z, but is often rotationally symmetric about this axis. Instead of using (x,y) to describe the transverse characteristics of the beam, it is natural to instead use polar coordinates (ρ,ϕ), which indicate the distance from the axis and the angle with respect to the x-axis, respectively. A point in three-dimensional space can be represented in such *cylindrical coordinates* by (ρ,ϕ,z). In the scattering of light from a localized scattering object, the field behaves far from the scatterer as a distorted spherical wave whose amplitude decays as $1/r$, r being the distance from the origin of the scatterer. In this case, instead of expressing the scattered field in terms of coordinates (x,y,z), one typically uses *spherical coordinates*, in which a point in space is represented by (r,θ,ϕ), θ being the angle between the position vector and the z-axis and ϕ defined as in cylindrical coordinates.

Though one can envision many possible coordinate systems which could be used in the solution of various problems, the most significant systems are those which simplify the solution of partial differential equations, in particular Laplace's equation. There are in fact exactly 11 coordinate systems which can be defined which always produce *separable solutions* for Laplace's equation, $\nabla^2 f = 0$, a separable solution being one which may be expressed as a product of three functions, each dependent on only one coordinate, e.g. $f(x,y,z) = A(x)B(y)C(z)$. (There are also several more systems which can be separated only under certain circumstances.) A list of these separable systems looks quite daunting: Cartesian, circular cylindrical, conical, ellipsoidal, elliptic cylindrical, oblate spheroidal, parabolic, parabolic cylindrical, paraboloidal, prolate spheroidal, and spherical. Of these, though, scientists commonly find use for just three: spherical, (circular) cylindrical, and

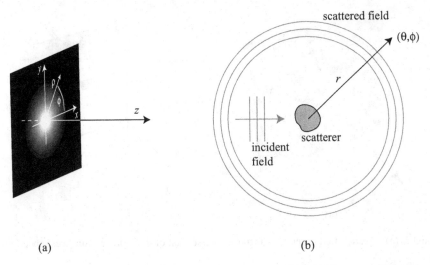

Figure 3.1 Two systems in which Cartesian coordinates are not ideal. (a) The propagation of a rotationally symmetric laser beam along the z-axis. (b) Measurement of the field scattered by a localized scattering object.

Cartesian. It might seem easiest, then, to study in detail the spherical and cylindrical systems (Cartesian, of course, was already done in Chapter 2) and be on our way, without trying to solve the problem generally. However, it is extremely illuminating, even for our two systems of interest, to solve the general case first, for then the specific solutions practically fall out of the mathematics. We will see that the differential operators have quite different-looking and nontrivial forms in different coordinate systems, but may all be related to the theory of general coordinate systems.

3.2 General orthogonal coordinate systems

We begin our discussion by reexamining the meaning of a particular component of a given coordinate. What does, for instance, the x-component in Cartesian coordinates specify? It is clear that specifying x restricts the possible positions of the coordinate to a plane, in particular the (y,z)-plane, in three-dimensional space; this is illustrated in Fig. 3.2(a).

This plane also has a unit vector related to it, defined by the direction *normal* to the plane. As noted in Section 2.2.2, there are two possible normals; we define the unit normal to be in the direction of *increasing* coordinate. Similar planes (and unit vectors) exist for the y-component and the z-component. In Cartesian coordinates, then, a point is specified by the intersection of three planes, as illustrated in Fig. 3.2(b).

As noted in the introduction, for problems defined in a sphere or in a cylinder, it is more useful to define one or more of the coordinate surfaces as a *curved* surface. For instance, in a cylindrical system, the coordinates are $\rho = \sqrt{x^2+y^2}$, $\phi =$ "angle between the x-axis and a vector $\rho = (x,y)$", and z. The coordinate ρ specifies a cylindrical surface, as illustrated

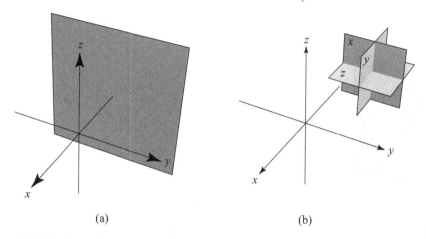

(a)   (b)

Figure 3.2 (a) A plane of constant x, and (b) the intersection of three planes (surfaces) specifying a point.

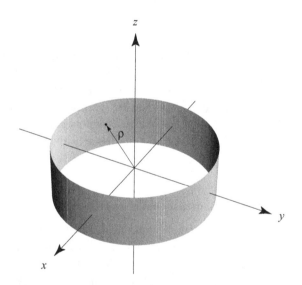

Figure 3.3 A surface of constant ρ.

in Fig. 3.3. A system which includes one or more curved coordinates is referred to as a *curvilinear coordinate system*.

An important observation is that the direction of the unit vector associated with a curved surface now depends upon the specific location on this surface; in other words, the unit vectors of a curvilinear coordinate system are *position dependent*. This will have a tremendous influence upon the definition of vector differential quantities such as the gradient, divergence and curl. It is to be noted that the coordinates we use, such as ϕ in cylindrical coordinates, need not be physical distances, but may also represent angles or other dimensionless quantities.

In general we specify a coordinate system in three dimensions by three coordinates q_1, q_2, q_3 which define three surfaces. We will find it convenient to write these coordinates in shorthand as $\mathbf{q} \equiv (q_1, q_2, q_3)$, though $\mathbf{q}$ is not typically a vector and does not transform as one under rotations. We may relate these coordinates to the Cartesian coordinates, and vice versa, i.e.

$$
\begin{array}{ll}
\text{general coordinates} & \text{cylindrical coordinates} \\[4pt]
q_1, q_2, q_3 & \rho, \phi, z \\[4pt]
x = x(q_1, q_2, q_3) & x = \rho \cos \phi \\[4pt]
y = y(q_1, q_2, q_3) & y = \rho \sin \phi \\[4pt]
z = z(q_1, q_2, q_3) & z = z
\end{array}
\tag{3.1}
$$

and

$$
\begin{array}{ll}
q_1 = q_1(x, y, z) & \rho = \sqrt{x^2 + y^2} \\[4pt]
q_2 = q_2(x, y, z) & \phi = \arctan(y/x) \\[4pt]
q_3 = q_3(x, y, z) & z = z.
\end{array}
\tag{3.2}
$$

It is to be noted that the specification of ϕ in Eq. (3.2) above is not quite complete: there is a π ambiguity because of the π-periodicity of arctan.

The unit vectors $\mathbf{e}_1$, $\mathbf{e}_2$, $\mathbf{e}_3$ of the system coordinates are defined as the positive normal of each of the coordinate surfaces. A vector $\mathbf{v}$ specified at a point (q_1, q_2, q_3) may then be written in terms of these coordinates and unit vectors as

$$
\mathbf{v} = \mathbf{e}_1 v_1 + \mathbf{e}_2 v_2 + \mathbf{e}_3 v_3.
\tag{3.3}
$$

In order to perform calculus in a general coordinate system, we need to be able to specify an infinitesimal measure of distance. In Cartesian coordinates, we know that the distance $\mathrm{d}s$ between two neighboring points is given by

$$
\mathrm{d}s^2 = \mathrm{d}x^2 + \mathrm{d}y^2 + \mathrm{d}z^2,
\tag{3.4}
$$

which follows from the Pythagorean theorem. We may use the chain rule to determine how $\mathrm{d}x$, $\mathrm{d}y$ and $\mathrm{d}z$ are related to the curvilinear coordinate system,

$$
\mathrm{d}x = \frac{\partial x}{\partial q_1} \mathrm{d}q_1 + \frac{\partial x}{\partial q_2} \mathrm{d}q_2 + \frac{\partial x}{\partial q_3} \mathrm{d}q_3,
\tag{3.5}
$$

with similar results for dy, dz. On substituting back into the expression (3.4) for the distance element, we find that the differential for distance is given in curvilinear coordinates as

$$ds^2 = \sum_{i,j=1}^{3} g_{ij} dq_i dq_j, \tag{3.6}$$

where the coefficients g_{ij} (nine in total) are referred to collectively as the *metric* of the coordinate system,

$$g_{ij} = \frac{\partial x}{\partial q_i}\frac{\partial x}{\partial q_j} + \frac{\partial y}{\partial q_i}\frac{\partial y}{\partial q_j} + \frac{\partial z}{\partial q_i}\frac{\partial z}{\partial q_j} = \sum_{l=1}^{3} \frac{\partial x_l}{\partial q_i}\frac{\partial x_l}{\partial q_j}. \tag{3.7}$$

This metric quantifies how the length element ds depends on the structure of the curvilinear coordinate system. It is another example of a tensor, to be discussed in Section 5.6. A similar metric is extremely important in general relativity, where it describes the curvature not of the coordinate system but of space-time itself.

Equation (3.6) simplifies greatly if we limit ourselves to orthogonal coordinate systems, in which the coordinate surfaces are mutually perpendicular and hence $\mathbf{e}_i \cdot \mathbf{e}_j = \delta_{ij}$; in this case it is found that

$$g_{ij} = 0, \quad i \neq j. \tag{3.8}$$

We may then write the distance element ds^2 in the simplified form,

$$ds^2 = \sum_i (h_i dq_i)^2, \tag{3.9}$$

where $h_i^2 = g_{ii}$ are the diagonal elements of the metric and will be referred to as the *scale factors* of the coordinates. Now ds has dimensions of length; because, as we noted before, the curved coordinates q_i may represent angles or other dimensionless variables, the scale factors may or may not have dimensions of length, depending on what type of coordinate they represent.

To give the reader a feel for what these scale factors are like, we list them below for Cartesian, cylindrical, and spherical coordinates (we will discuss these specific curvilinear systems in more detail momentarily):

system	q_1	h_1	q_2	h_2	q_3	h_3
Cartesian	x	1	y	1	z	1
cylindrical	ρ	1	ϕ	ρ	z	1
spherical	r	1	θ	r	ϕ	$r\sin\theta$

$$(3.10)$$

Using these scale factors, the differential distance vector $d\mathbf{s}$ is

$$d\mathbf{s} = h_1 dq_1 \mathbf{e}_1 + h_2 dq_2 \mathbf{e}_2 + h_2 dq_2 \mathbf{e}_2 = \sum_i h_i dq_i \mathbf{e}_i. \tag{3.11}$$

From Eq. (3.11), it can be seen that the infinitesimal length element in the ith direction is $h_i dq_i$; from this we may immediately see that the infinitesimal volume element is

$$d\tau = h_1 h_2 h_3 dq_1 dq_2 dq_3. \tag{3.12}$$

An infinitesimal area element which points in the e_1 direction consists of an area which is orthogonal to this direction, i.e. an area in the q_2, q_3 coordinates. We thus have $da_1 = h_2 h_3 dq_2 dq_3$, with analogous results for da_2 and da_3. Putting these results together, the vector area element is generally given by

$$d\mathbf{a} = h_2 h_3 dq_2 dq_3 \mathbf{e}_1 + h_1 h_3 dq_1 dq_3 \mathbf{e}_2 + h_1 h_2 dq_1 dq_2 \mathbf{e}_3. \tag{3.13}$$

The infinitesimal elements described by Eqs. (3.11), (3.12), and (3.13) can be used to calculate path, volume, and surface integrals in curvilinear coordinate systems, respectively. One must take care, however, and remember that the direction of infinitesimal vector elements depend upon the position in the coordinate system.

3.3 Vector operators in curvilinear coordinates

We have mentioned earlier that vector operators such as the gradient, divergence, and curl have significantly different forms in curvilinear systems because of the position dependence of the unit vectors $\mathbf{e}_i$. In this section we derive the forms of these operators for a general curvilinear coordinate system.

3.3.1 The gradient

The best way to determine the gradient in a curvilinear coordinate system is to return to the geometric interpretation we developed in Section 2.3. Let us consider an infinitesimal displacement along the direction $\mathbf{e}_1$. Referring to the relation $d\phi = \nabla\phi \cdot d\mathbf{s}$, Eq. (2.35), the infinitesimal distance element in the direction $\mathbf{e}_1$ is $h_1 dq_1$, so that $d\mathbf{s} = h_1 q_1 \mathbf{e}_1$. On substitution into Eq. (2.35), we find that

$$d\phi = \nabla\phi|_1 h_1 dq_1, \tag{3.14}$$

which implies that

$$\nabla\phi|_1 = \frac{1}{h_1}\frac{\partial\phi}{\partial q_1}. \tag{3.15}$$

Performing the same steps for q_2 and q_3, we find that

$$\nabla\phi = \sum_i \mathbf{e}_i \frac{1}{h_i}\frac{\partial\phi}{\partial q_i}, \tag{3.16}$$

This form is mathematically similar to the basic expression for the gradient in Cartesian coordinates; however, it is to be noted that now the scale factors h_i appear in the denominators.

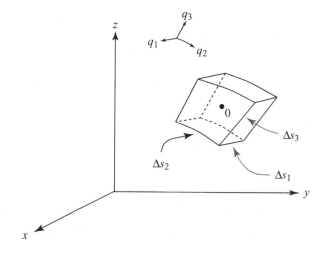

Figure 3.4 The volume used to define divergence in a curvilinear coordinate system.

3.3.2 The divergence

To determine the divergence, we use the integral definition of the divergence which we derived in Section 2.4,

$$\nabla \cdot \mathbf{v} \equiv \lim_{\Delta\tau \to 0} \frac{\int_{\Delta S} \mathbf{v} \cdot \mathbf{da}}{\Delta\tau}, \tag{3.17}$$

where we remind the reader that $\Delta\tau$ is an infinitesimal volume element and $\mathbf{da}$ is an infinitesimal area element. To calculate the surface integral, we construct a six-sided region whose sides are chosen such that one of the coordinates is constant on each side, as shown in Fig. 3.4. For convenience, the origin is defined as the center of the volume.

We must take some care in our analysis here, because the direction of our coordinate unit vectors as well as the size of the scale factors are dependent upon position. The volume of our six-sided region may formally be written as $\Delta\tau = \Delta s_1 \Delta s_2 \Delta s_3$, but only in the limit $\Delta\tau \to 0$ may we write $\Delta s_i = h_i(0)\Delta q_i$.

Let us focus upon the integrals over the faces of constant q_1 first. The integral over the positive-q_1 face (at $q_1 = +\Delta q_1/2$) may be written as

$$\int_{+\Delta q_1/2} \mathbf{v} \cdot \mathbf{da} = \int_{q_2} \int_{q_3} v_1(\mathbf{q}) h_2(\mathbf{q}) h_3(\mathbf{q}) \mathrm{d}q_2 \mathrm{d}q_3, \tag{3.18}$$

where we have used the infinitesimal area element given by Eq. (3.13). We can get an estimate of this integral by using a Taylor series expansion, but this expansion must be of all functional dependencies, i.e. $f(\mathbf{q}) \equiv v_1(\mathbf{q}) h_2(\mathbf{q}) h_3(\mathbf{q})$, so that

$$f(\mathbf{q}) = f(0) + q_1 \frac{\partial}{\partial q_1} f(0) + q_2 \frac{\partial}{\partial q_2} f(0) + q_3 \frac{\partial}{\partial q_3} f(0) + O(\mathbf{q}^2). \tag{3.19}$$

With this expression, we may write

$$\int_{+\Delta q_1/2} \mathbf{v} \cdot d\mathbf{a} = (v_1 h_2 h_3)(0)\Delta q_2 \Delta q_3 + \frac{\partial}{\partial q_1}(v_1 h_2 h_3)(0)\Delta q_1 \Delta q_2 \Delta q_3/2. \qquad (3.20)$$

Similarly, we have

$$\int_{-\Delta q_1/2} \mathbf{v} \cdot d\mathbf{a} = -(v_1 h_2 h_3)(0)\Delta q_2 \Delta q_3 + \frac{\partial}{\partial q_1}(v_1 h_2 h_3)(0)\Delta q_1 \Delta q_2 \Delta q_3/2. \qquad (3.21)$$

Adding these together, the q_1-components give the value

$$\int_{q_1} \mathbf{v} \cdot d\mathbf{a} = \frac{\partial}{\partial q_1}(v_1 h_2 h_3)(0)\Delta q_1 \Delta q_2 \Delta q_3. \qquad (3.22)$$

Performing a similar analysis for the q_2- and q_3-faces, we have

$$\int_{\Delta S} \mathbf{v} \cdot d\mathbf{a} = \left[\frac{\partial}{\partial q_1}(v_1 h_2 h_3)(0) + \frac{\partial}{\partial q_2}(v_2 h_1 h_3)(0) + \frac{\partial}{\partial q_3}(v_3 h_1 h_2)(0)\right]\Delta q_1 \Delta q_2 \Delta q_3. \qquad (3.23)$$

Dividing by $\Delta\tau$, and taking the limit $\Delta\tau \to 0$, we find that the divergence in curvilinear coordinates is given by

$$\nabla \cdot \mathbf{v} = \frac{1}{h_1 h_2 h_3}\left[\frac{\partial}{\partial q_1}(v_1 h_2 h_3) + \frac{\partial}{\partial q_2}(v_2 h_3 h_1) + \frac{\partial}{\partial q_3}(v_3 h_1 h_2)\right]. \qquad (3.24)$$

The Laplacian of a scalar ($\nabla^2\phi$) can be readily found by combining Eqs. (3.16) and (3.24). We find that

$$\nabla^2\phi = \frac{1}{h_1 h_2 h_3}\left[\frac{\partial}{\partial q_1}\left(\frac{h_2 h_3}{h_1}\frac{\partial\phi}{\partial q_1}\right) + \frac{\partial}{\partial q_2}\left(\frac{h_3 h_1}{h_2}\frac{\partial\phi}{\partial q_2}\right) + \frac{\partial}{\partial q_3}\left(\frac{h_1 h_2}{h_3}\frac{\partial\phi}{\partial q_3}\right)\right]. \qquad (3.25)$$

3.3.3 The curl

To determine the curl in curvilinear coordinates we use its integral definition, developed in Section 2.5,

$$(\nabla \times \mathbf{v}) \cdot \mathbf{n} = \lim_{\Delta S \to 0} \frac{\oint_{\Delta S} \mathbf{v} \cdot d\mathbf{r}}{\Delta S}. \qquad (3.26)$$

This formula gives the component of the curl in the direction of the unit vector $\mathbf{n}$ by performing a path integral over an infinitesimal loop with area element pointing in the $\mathbf{n}$-direction.

We consider first a loop with surface normal pointing in the $\mathbf{e}_1$ direction, as illustrated in Fig. 3.5. As in the case of the divergence, the area of the loop may be written as $\Delta s_2 \Delta s_3 = h_2(0)h_3(0)\Delta q_2 \Delta q_3$ only in the limit as $\Delta S \to 0$.

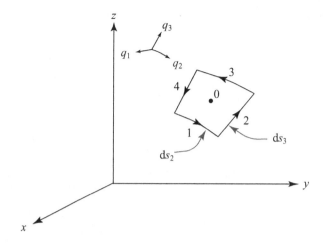

Figure 3.5 The surface and closed path used to define the curl in curved coordinates.

Looking at the path integral, we see that the first leg may be written as

$$\int_1 \mathbf{v} \cdot d\mathbf{r} = \int_{-\Delta q_2/2}^{+\Delta q_2/2} v_2 h_2 dq_2,$$ (3.27)

where we have used a path element $ds_2 = h_2 dq_2$. Similarly, the path integral along the third leg is given by

$$\int_3 \mathbf{v} \cdot d\mathbf{r} = \int_{\Delta q_2/2}^{-\Delta q_2/2} v_2 h_2 dq_2.$$ (3.28)

These integrals can be evaluated by considering a Taylor series expansion of the function $g(\mathbf{q}) \equiv v_2(\mathbf{q})h_2(\mathbf{q})$, i.e.

$$g(\mathbf{q}) = g(0) + q_1 \frac{\partial}{\partial q_1} g(0) + q_2 \frac{\partial}{\partial q_2} g(0) + q_3 \frac{\partial}{\partial q_3} g(0) + O(\mathbf{q}^2).$$ (3.29)

On substitution and adding the results for legs 1 and 3, it is readily found that the only surviving components are of the form

$$\int_{1+3} \mathbf{v} \cdot d\mathbf{r} = -\frac{\partial}{\partial q_3}(h_2 v_2)(0)\Delta q_2 \Delta q_3.$$ (3.30)

If we perform a similar analysis for legs 2 and 4, we find the total path integral,

$$\oint \mathbf{v} \cdot d\mathbf{r} = \left[\frac{\partial}{\partial q_2}(h_3 v_3) - \frac{\partial}{\partial q_3}(h_2 v_2)\right]\Delta q_2 \Delta q_3.$$ (3.31)

On substituting from this equation back into Eq. (3.26), we readily find that

$$(\nabla \times \mathbf{v}) \cdot \mathbf{e}_1 = \frac{1}{h_2 h_3}\left[\frac{\partial}{\partial q_2}(h_3 v_3) - \frac{\partial}{\partial q_3}(h_2 v_2)\right].$$ (3.32)

The other components can be found by performing a similar calculation with an e_2, e_3 loop, and the total curl is then found to be

$$\nabla \times \mathbf{v} = \frac{1}{h_1 h_2 h_3} \left\{ \left[\frac{\partial}{\partial q_2}(h_3 v_3) - \frac{\partial}{\partial q_3}(h_2 v_2) \right] h_1 \mathbf{e}_1 \right.$$

$$- \left[\frac{\partial}{\partial q_1}(h_3 v_3) - \frac{\partial}{\partial q_3}(h_1 v_1) \right] h_2 \mathbf{e}_2$$

$$\left. + \left[\frac{\partial}{\partial q_1}(h_2 v_2) - \frac{\partial}{\partial q_2}(h_1 v_1) \right] h_3 \mathbf{e}_3 \right\}. \tag{3.33}$$

As in Cartesian coordinates, the curl for curved coordinates can be written in a compact determinant form,

$$\nabla \times \mathbf{v} = \frac{1}{h_1 h_2 h_3} \begin{vmatrix} \mathbf{e}_1 h_1 & \mathbf{e}_2 h_2 & \mathbf{e}_3 h_3 \\ \frac{\partial}{\partial q_1} & \frac{\partial}{\partial q_2} & \frac{\partial}{\partial q_2} \\ h_1 v_1 & h_2 v_2 & h_3 v_3 \end{vmatrix}. \tag{3.34}$$

Where are we now, in terms of vector calculus in curvilinear coordinates? Though it may have seemed initially unpleasant, we have determined quite generally the form for various vector operators in a curvilinear system. Now the application to a specific system simply involves finding the scale factors h_i.

An alternative method for deriving the various vector operators involves returning to their Cartesian definitions and considering the appropriate transformation of both coordinates and unit vectors.

It is to be noted that in Cartesian coordinates, for which $h_x = h_y = h_z = 1$, all the expressions we have developed reduce to the forms we derived in Chapter 1.

3.4 Cylindrical coordinates

The circular cylindrical coordinate system (as distinguished from the more daunting elliptical cylindrical or parabolic cylindrical coordinate systems) uses as coordinate surfaces the following choices.

1. Right circular cylinders with the z-axis as the center axis, and coordinate $\rho = \sqrt{x^2 + y^2}$.
2. Half-planes through the z-axis, defining an angle ϕ, $0 \leq \phi < 2\pi$ such that $\phi = \arctan(y/x)$. There is an ambiguity in defining the angle with the arctangent function, which is resolved by adding that $\phi = 0$ when the coordinate lies on the positive x-axis.
3. Planes normal to the z-direction, as in Cartesian coordinates, with coordinate z.

These surfaces are shown in Fig. 3.6.

The inverse relations between (ρ, ϕ, z) and (x, y, z) are given by

$$x = \rho \cos \phi, \quad y = \rho \sin \phi, \quad z = z. \tag{3.35}$$

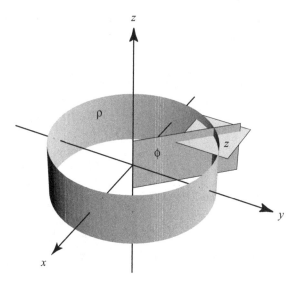

Figure 3.6 The coordinate surfaces for cylindrical coordinates.

We can determine the scale factors by using the formula

$$h_i^2 = g_{ii} = \sum_{l=1}^{3} \left(\frac{\partial x_l}{\partial q_i} \right)^2.$$ (3.36)

We may take the appropriate derivatives of Eq. (3.35),

$$\frac{\partial x}{\partial \rho} = \cos\phi, \qquad \frac{\partial y}{\partial \rho} = \sin\phi, \qquad \frac{\partial z}{\partial \rho} = 0,$$

$$\frac{\partial x}{\partial \phi} = -\rho\sin\phi, \qquad \frac{\partial y}{\partial \phi} = \rho\cos\phi, \qquad \frac{\partial z}{\partial \phi} = 0,$$ (3.37)

$$\frac{\partial x}{\partial z} = 0, \qquad \frac{\partial y}{\partial z} = 0, \qquad \frac{\partial z}{\partial z} = 1,$$

from which we readily find the values of h_i to be

$$h_\rho = 1, \quad h_\phi = \rho, \quad h_z = 1.$$ (3.38)

As we have mentioned before, the unit vectors $\hat{\rho}$, $\hat{\phi}$, $\hat{z}$ are defined as being normal to the coordinate surfaces, and in the direction of increasing coordinate. Thus $\hat{z}$ points in the positive z direction, $\hat{\rho}$ points in the direction of increasing ρ, and $\hat{\phi}$ points in the direction of increasing ϕ. This is shown in Fig. 3.7 for two different position vectors $\mathbf{r}$.

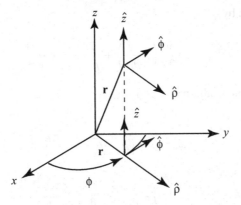

Figure 3.7 The unit vectors in cylindrical coordinates.

It is often helpful to have these unit vectors expressed in terms of the familiar Cartesian unit vectors. One can find, via some careful geometry, that

$$\hat{\rho} = \hat{x}\cos\phi + \hat{y}\sin\phi,$$
$$\hat{\phi} = -\hat{x}\sin\phi + \hat{y}\cos\phi, \qquad (3.39)$$
$$\hat{z} = \hat{z}.$$

Problems in cylindrical coordinates often involve the calculation of volume integrals throughout a cylindrical region or surface integrals on a surface of constant ρ. The expressions for these infinitesimal elements are

$$d\tau = \rho\, d\rho\, d\phi\, dz, \qquad (3.40)$$

and

$$da_\rho = \rho\, d\phi\, dz, \qquad (3.41)$$

as can be determined using the general expressions of Section 3.2.

The gradient, divergence, curl, and Laplacian may be found by substituting the scale factors into the Eqs. (3.16), (3.24), (3.33), and (3.25), respectively. The gradient is given by

$$\nabla\varphi = \frac{\partial\varphi}{\partial\rho}\hat{\rho} + \frac{1}{\rho}\frac{\partial\varphi}{\partial\phi}\hat{\phi} + \frac{\partial\varphi}{\partial z}\hat{z}, \qquad (3.42)$$

and the divergence is given by

$$\nabla\cdot\mathbf{v} = \frac{1}{\rho}\frac{\partial}{\partial\rho}(\rho v_\rho) + \frac{1}{\rho}\frac{\partial}{\partial\phi}v_\phi + \frac{\partial}{\partial z}v_z. \qquad (3.43)$$

The Laplacian is given by

$$\nabla^2 \varphi = \frac{1}{\rho}\frac{\partial}{\partial \rho}\left(\rho\frac{\partial \varphi}{\partial \rho}\right) + \frac{1}{\rho^2}\frac{\partial^2 \varphi}{\partial \phi^2} + \frac{\partial^2 \varphi}{\partial z^2}, \tag{3.44}$$

which may be written in an alternative form as

$$\nabla^2 \varphi = \frac{\partial^2 \varphi}{\partial \rho^2} + \frac{1}{\rho}\frac{\partial \varphi}{\partial \rho} + \frac{1}{\rho^2}\frac{\partial^2 \varphi}{\partial \phi^2} + \frac{\partial^2 \varphi}{\partial z^2}. \tag{3.45}$$

The curl may be written as

$$\nabla \times \mathbf{v} = \frac{1}{\rho}\left\{\left[\frac{\partial}{\partial \phi}v_z - \frac{\partial}{\partial z}(\rho v_\phi)\right]\hat{\boldsymbol{\rho}}\right.$$
$$- \left[\frac{\partial}{\partial \rho}v_z - \frac{\partial}{\partial z}v_\rho\right]\rho\hat{\boldsymbol{\phi}}$$
$$\left.+ \left[\frac{\partial}{\partial \rho}(\rho v_\phi) - \frac{\partial}{\partial \phi}v_\rho\right]\hat{\mathbf{z}}\right\}. \tag{3.46}$$

3.5 Spherical coordinates

The spherical coordinate system uses as coordinate surfaces the following choices.

1. Concentric spheres centered at the origin, coordinate $r = \sqrt{x^2 + y^2 + z^2}$.
2. Right circular cones centered on the z-axis, with vertices at the origin, $\theta = \arccos(z/r)$, where $0 \le \theta \le \pi$. $\theta = 0$ when the point is on the positive z-axis.
3. Half planes through the z-axis, defining an angle ϕ such that $\phi = \arctan(y/x)$, as in the cylindrical coordinate case.

These surfaces and their intersection are shown in Fig. 3.8.

The expression of (x, y, z) in terms of (ρ, θ, ϕ) is

$$x = r\sin\theta\cos\phi, \quad y = r\sin\theta\sin\phi, \quad z = r\cos\theta. \tag{3.47}$$

In a manner similar to that employed in cylindrical coordinates, we can readily find that

$$h_r = 1, \quad h_\theta = r, \quad h_\phi = r\sin\theta. \tag{3.48}$$

The forms of the gradient, divergence, curl, and Laplacian can be found at the end of the section in all their awful glory. We first focus on a few other useful relations in this coordinate system: the area elements and the behavior of the unit vectors.

The volume element in spherical coordinates can be found using Eq. (3.12) to be

$$d\tau = r^2\sin\theta\, dr\, d\theta\, d\phi. \tag{3.49}$$

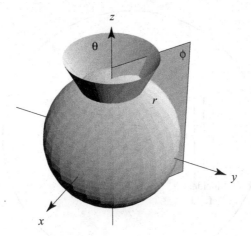

Figure 3.8 The coordinate surfaces for spherical coordinates.

What is the area element on a spherical surface centered on the origin (i.e. for $r =$ constant)? It follows from the general formulation that $da_r = h_\theta h_\phi d\theta d\phi$; on substituting in for the values of the scale factors, we find that

$$da_r = r^2 \sin\theta d\theta d\phi. \tag{3.50}$$

In scattering theory, we are often interested in the flux of a certain quantity (such as light, or high-energy particles) through a sphere of large size. The geometry of the problem is illustrated in Fig. 3.9.

Because physical laws generally require flux conservation (the amount of "stuff" leaving a sphere does not change if we increase the size of the sphere), it is helpful to consider a radius-free measure of area element, usually referred to as an element of *solid angle* $d\Omega$, where

$$d\Omega = da_r/r^2 = \sin\theta d\theta d\phi. \tag{3.51}$$

The total solid angle on a sphere can be readily found by integrating ϕ over its range from 0 to 2π, and θ over its range of 0 to π. We find that

$$\int_{\text{sphere}} d\Omega = 4\pi. \tag{3.52}$$

In terms of solid angle, the volume element may be written as

$$d\tau = r^2 dr d\Omega. \tag{3.53}$$

Such integrals are common in not only scattering theory, but also radiation and antenna theory.

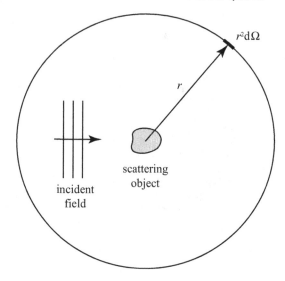

Figure 3.9 The geometry of scattering problems.

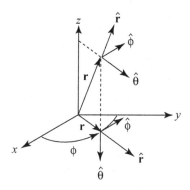

Figure 3.10 The unit vectors in spherical coordinates.

The directions of the unit vectors are shown in Fig. 3.10, for two different position vectors **r**.

The directions of all three unit vectors depend on position and hence the coordinate values, and change as the coordinate values are changed. As in the cylindrical case, it is often helpful to have the unit vectors expressed in terms of Cartesian unit vectors,

$$\hat{\mathbf{r}} = \hat{\mathbf{x}}\sin\theta\cos\phi + \hat{\mathbf{y}}\sin\theta\sin\phi + \hat{\mathbf{z}}\cos\theta,$$

$$\hat{\boldsymbol{\theta}} = \hat{\mathbf{x}}\cos\theta\cos\phi + \hat{\mathbf{y}}\cos\theta\sin\phi - \hat{\mathbf{z}}\sin\theta,$$

$$\hat{\boldsymbol{\phi}} = -\hat{\mathbf{x}}\sin\phi + \hat{\mathbf{y}}\cos\phi. \tag{3.54}$$

The vector operators can be readily determined by use of the general formulas with the appropriate scale factors. The gradient may be written as

$$\nabla\varphi = \frac{\partial\varphi}{\partial r}\hat{\mathbf{r}} + \frac{1}{r}\frac{\partial\varphi}{\partial\theta}\hat{\boldsymbol{\theta}} + \frac{1}{r\sin\theta}\frac{\partial\varphi}{\partial\phi}\hat{\boldsymbol{\phi}}. \tag{3.55}$$

The divergence takes on the form

$$\nabla\cdot\mathbf{v} = \frac{1}{r^2\sin\theta}\left[\frac{\partial}{\partial r}(r^2\sin\theta\, v_r) + \frac{\partial}{\partial\theta}(r\sin\theta\, v_\theta) + \frac{\partial}{\partial\phi}(rv_\phi)\right]. \tag{3.56}$$

The curl may be written as

$$\begin{aligned}
\nabla\times\mathbf{v} = \frac{1}{r^2\sin\theta}&\left\{\left[\frac{\partial}{\partial\theta}(r\sin\theta\, v_\phi) - \frac{\partial}{\partial\phi}(rv_\theta)\right]\hat{\mathbf{r}}\right.\\
&- \left[\frac{\partial}{\partial r}(r\sin\theta\, v_\phi) - \frac{\partial}{\partial\phi}(v_r)\right]r\hat{\boldsymbol{\theta}}\\
&\left.+ \left[\frac{\partial}{\partial r}(rv_\theta) - \frac{\partial}{\partial\theta}(v_r)\right]r\sin\theta\,\hat{\boldsymbol{\phi}}\right\}.
\end{aligned} \tag{3.57}$$

The Laplacian can be determined by combining the gradient and divergence formulas,

$$\nabla^2\varphi = \frac{1}{r^2\sin\theta}\left[\sin\theta\frac{\partial}{\partial r}\left(r^2\frac{\partial\varphi}{\partial r}\right) + \frac{\partial}{\partial\theta}\left(\sin\theta\frac{\partial\varphi}{\partial\theta}\right) + \frac{1}{\sin\theta}\frac{\partial^2\varphi}{\partial\phi^2}\right]. \tag{3.58}$$

It also has an alternative form,

$$\nabla^2\varphi = \frac{1}{r}\frac{\partial^2}{\partial r^2}(r\varphi) + \frac{1}{r^2\sin\theta}\left[\frac{\partial}{\partial\theta}\left(\sin\theta\frac{\partial\varphi}{\partial\theta}\right) + \frac{1}{\sin\theta}\frac{\partial^2\varphi}{\partial\phi^2}\right]. \tag{3.59}$$

3.6 Exercises

1. Parabolic cylindrical coordinates are described by (σ,τ,z), where

$$x = \sigma\tau,$$
$$y = \frac{1}{2}(\tau^2 - \sigma^2),$$
$$z = z.$$

   Determine equations for the coordinate surfaces of σ, τ, and z and calculate the scale factors for these coordinates.

2. Prolate spheroidal coordinates are described by (μ,ν,ϕ), with $\mu \geq 0, 0 \leq \nu \leq \pi$, and $0 \leq \phi < 2\pi$, where

$$x = a\sinh\mu\sin\nu\cos\phi,$$
$$y = a\sinh\mu\sin\nu\sin\phi,$$
$$z = a\cosh\mu\cos\nu,$$

and a is a positive real constant. Determine equations for the coordinate surfaces of μ, v, and ϕ and calculate the scale factors for these coordinates.

3. The elliptic cylindrical coordinate system uses coordinates u, v, and z, described by the equations

$$x = a \cosh u \cos v,$$
$$y = a \sinh u \sin v,$$
$$z = z,$$

with a a constant. Determine equations for the coordinate surfaces of u and v, and determine the unit vectors for these coordinates in terms of u and v and $\hat{\mathbf{x}}$ and $\hat{\mathbf{y}}$. Show that the unit vectors are orthogonal.

4. The oblate spheroidal coordinate system uses coordinates μ, v, and ϕ, described by the equations

$$x = a \cosh \mu \cos v \cos \phi,$$
$$y = a \cosh \mu \cos v \sin \phi,$$
$$z = a \sinh \mu \sin v,$$

with a a constant and ϕ defined as in spherical coordinates. Determine equations for the coordinate surfaces μ and v, and determine the unit vectors for these coordinates in terms of μ, v, ϕ, and $\hat{\mathbf{x}}$, $\hat{\mathbf{y}}$, and $\hat{\mathbf{z}}$. Show that the unit vectors are orthogonal.

5. For a unit vector $\mathbf{e}_1$ in a curved coordinate system, show that

(a) $\nabla \cdot \mathbf{e}_1 = \frac{1}{h_1 h_2 h_3} \frac{\partial (h_2 h_3)}{\partial q_1}$,

(b) $\nabla \times \mathbf{e}_1 = \frac{1}{h_1} \left[\mathbf{e}_2 \frac{1}{h_3} \frac{\partial h_1}{\partial q_3} - \mathbf{e}_3 \frac{1}{h_2} \frac{\partial h_1}{\partial q_2} \right]$.

This result illustrates that unit vectors are not constant in curved coordinate systems.

6. Show that unit vectors $\mathbf{e}_i$ of a general orthogonal coordinate system satisfy

(a) $\mathbf{e}_i = \frac{1}{h_i} \frac{\partial \mathbf{r}}{\partial q_i}$,

(b) $\frac{\partial \mathbf{e}_i}{\partial q_j} = \mathbf{e}_j \frac{1}{h_i} \frac{\partial h_j}{\partial q_i}$,

(c) $\frac{\partial \mathbf{e}_i}{\partial q_i} = -\sum_{j \neq i} \mathbf{e}_j \frac{1}{h_j} \frac{\partial h_i}{\partial q_j}$.

(Hint: Compare $\mathbf{e}_i \cdot \mathbf{e}_j$ with the definition of the metric.)

7. Calculate the Laplacian of the function

$$\phi(x,y,z) = \frac{x+z}{\sqrt{x^2+y^2}}$$

in both Cartesian coordinates and cylindrical coordinates.

8. Calculate the Laplacian of the function

$$\phi(x,y,z) = z\sqrt{x^2+y^2+z^2}$$

in both Cartesian coordinates and spherical coordinates.

9. A cylindrical surface is defined by the expression

$$f(x,y) = \sqrt{x^2+y^2} - \alpha = 0,$$

where α is a constant. Show that the normalized gradient of $f(x,y)$ is equal to the unit vector $\hat{\rho}$,

$$\hat{\rho} = \hat{x}\cos\phi + \hat{y}\sin\phi.$$

10. A spherical surface is defined by the expression

$$f(x,y,z) = \sqrt{x^2 + y^2 + z^2} - \alpha = 0,$$

where α is a constant. Show that the normalized gradient of $f(x,y,z)$ is equal to the unit vector $\hat{r}$,

$$\hat{r} = \hat{x}\sin\theta\cos\phi + \hat{y}\sin\theta\sin\phi + \hat{z}\cos\theta.$$

11. Determine expressions for $\partial/\partial x$, $\partial/\partial y$, and $\partial/\partial z$ in cylindrical coordinates.
12. Determine expressions for $\partial/\partial x$, $\partial/\partial y$, and $\partial/\partial z$ in spherical coordinates.
13. Conical coordinates are defined by r, μ, and ν, described by the equations

$$x = \frac{r\mu\nu}{bc},$$

$$y = \frac{r}{b}\sqrt{\frac{(\mu^2 - b^2)(\nu^2 - b^2)}{(b^2 - c^2)}},$$

$$z = \frac{r}{c}\sqrt{\frac{(\mu^2 - c^2)(\nu^2 - c^2)}{(c^2 - b^2)}}, \tag{3.60}$$

where b and c are constants and the coordinates are constrained by $\nu^2 < c^2 < \mu^2 < b^2$. Determine the scale factors h_r, h_μ, and h_ν, and write an expression for the gradient in conical coordinates.
14. Paraboloidal coordinates are defined by λ, μ, and ν, described by the equations

$$x^2 = \frac{(a-\lambda)(a-\mu)(a-\nu)}{b-a},$$

$$y^2 = \frac{(b-\lambda)(b-\mu)(b-\nu)}{a-b},$$

$$z = \frac{1}{2}(a+b-\lambda-\mu-\nu), \tag{3.61}$$

where b and c are constants and the coordinates are constrained by $\lambda < b < \mu < a < \nu$. Determine the scale factors h_λ, h_μ, and h_ν, and write an expression for the divergence in conical coordinates. Note that the expressions for x and y involve the *square* of the coordinates.
15. Derivatives in curved coordinates often produce unexpected and seemingly nonintuitive results. Calculate the divergence of $v = \hat{r}$ and $v = \frac{1}{r^2}\hat{r}$ in spherical coordinates.
16. Derivatives in curved coordinates often produce unexpected and seemingly nonintuitive results. Calculate the divergence of $v = \hat{\rho}$ and $v = \frac{1}{\rho}\hat{\rho}$ in cylindrical coordinates.
17. Orthogonality of the unit vectors allows us to determine one of them from the other two. Calculate the expression for $\hat{\phi}$ by use of the expressions (3.54) for $\hat{r}$ and $\hat{\theta}$.
18. Orthogonality of the unit vectors allows us to determine one of them from the other two. Calculate the expression for $\hat{\phi}$ by use of the expressions (3.39) for $\hat{\rho}$ and $\hat{z}$.

19. We consider a vector function **v** defined by

$$\mathbf{v} = -\hat{\mathbf{x}}\frac{y}{x^2+y^2} + \hat{\mathbf{y}}\frac{x}{x^2+y^2}.$$

In cylindrical coordinates, determine

(a) $\int_S \nabla \times \mathbf{v} \cdot d\mathbf{a}$, where S is the area of the unit circle;

(b) $\oint_C \mathbf{v} \cdot d\mathbf{r}$, where the path C is the unit circle traversed counterclockwise.

(c) Compare the two results, and show that they fail to satisfy Stokes' theorem. What has gone wrong?

20. We consider a vector function **v** defined by

$$\mathbf{v} = \hat{\mathbf{x}}\frac{zx}{x^2+y^2} + \hat{\mathbf{y}}\frac{zy}{x^2+y^2}.$$

In cylindrical coordinates, determine

(a) $\int_V \nabla \cdot \mathbf{v} d\tau$, where V is the volume of a cylinder of radius a and height L, centered on the z-axis with its base in the xy-plane;

(b) $\oint_S \mathbf{v} \cdot d\mathbf{a}$, where the path S is the surface of the cylinder described in (a).

(c) Compare the two results, and show that they fail to satisfy Gauss' theorem. What has gone wrong?

21. Read the paper by M. Kerker, Invisible bodies, *J. Opt. Soc. Am. A* **65** (1975), 376–379. Describe the nature (size, shape, interior structure) of the objects being studied, and name the coordinate system being used to characterize their behavior. What is the surprising result of Kerker's paper?

22. Read the paper by M. A. Bandres, J. C. Gutiérrez-Vega and S. Chávez-Cerda, Parabolic non-diffracting optical wave fields, *Opt. Lett.* **29** (2004), 44–46. Explain the unusual physical property of the wavefields sought in the paper, and explain how the use of a curvilinear coordinate system helps find these wavefields.

4

Matrices and linear algebra

4.1 Introduction: Polarization and Jones vectors

A light wave traveling in free space, far away from any material boundaries, is generally a *transverse wave*; that is, the electric field **E** is perpendicular to the direction of propagation. Let us suppose we have a monochromatic electromagnetic plane wave traveling in the positive z-direction, described by the complex expression,

$$\mathbf{E}(z,t) = \mathbf{E}_0 e^{ikz - i\omega t}, \tag{4.1}$$

where k is the wavenumber of the wave, ω is the angular frequency, and $\mathbf{E}_0$ is the *polarization* of the light wave, a generally complex vector which lies in the x–y plane, i.e. $\mathbf{E}_0 \cdot \hat{\mathbf{z}} = 0$. The physical wave is simply the real part of Eq. (4.1). The polarization of the field can be further decomposed into its x and y-components, in the form

$$\mathbf{E}_0 = E_x \hat{\mathbf{x}} + E_y \hat{\mathbf{y}} \equiv \epsilon_x e^{i\phi_x} \hat{\mathbf{x}} + \epsilon_y e^{i\phi_y} \hat{\mathbf{y}}, \tag{4.2}$$

where ϵ_x and ϵ_y are non-negative, real-valued numbers and ϕ_x, ϕ_y are real-valued. The behavior of the polarization in a plane of fixed z depends on the relative values of these four parameters. The possibilities may be summarized as follows.

- Linear polarization: $\phi_y - \phi_x = 0, \pm \pi$. In this case, the tip of the electric field oscillates along a straight line.
- Circular polarization: $\phi_y = \phi_x \pm \pi/2$, $\epsilon_x = \epsilon_y$. In this case, the tip of the electric field follows a circular path.
- Elliptical polarization: all cases not covered above. For elliptical polarization, the tip of the electric field follows an elliptical path.

Examples of these cases are illustrated in Fig. 4.1.

The vector defined by Eq. (4.2) is referred to as the *Jones vector* of the field, after the scientist who introduced it [Jon41]. It can be seen from Eq. (4.1) that the Jones vector of a plane wave is not changed on propagation in free space: a circularly polarized field remains a circularly polarized field on propagation, and so forth. Devices exist, however, which can transform the state of polarization, such as linear polarizers, quarter-wave plates, half-wave plates and optical rotators. We briefly consider the effect of each of these.

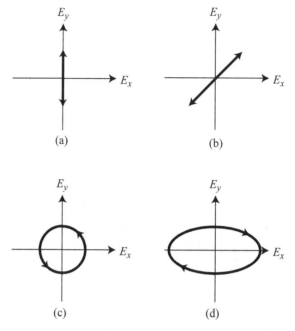

Figure 4.1 The different possible polarization states of a monochromatic wave: (a) linear polarization, with $\epsilon_x = 1$, $\epsilon_y = 0$, (b) linear polarization, with $\epsilon_x = \epsilon_y$, (c) circular polarization, with $\phi_y = \phi_x + \pi/2$, (d) elliptical polarization, with $\epsilon_x = 2\epsilon_y$ and $\phi_y = \phi_x - \pi/2$. It is to be noted that, because of the phase, the electric field goes counterclockwise in case (c) and clockwise in case (d).

Linear polarizers. An ideal linear polarizer completely blocks light polarized along one axis, while completely transmitting light along the perpendicular axis. This can be represented quite readily using the dot product: for a polarizer which transmits along the axis defined by the unit vector $\mathbf{w}$, the transmitted field is given by

$$\mathbf{E}' = \mathbf{w}(\mathbf{E} \cdot \mathbf{w}). \tag{4.3}$$

If we define $\mathbf{w} = \hat{\mathbf{x}} \cos\theta + \hat{\mathbf{y}} \sin\theta$, the Jones vector is transformed by a polarizer as follows:

$$E'_x = \cos^2\theta E_x + \cos\theta \sin\theta E_y, \tag{4.4}$$

$$E'_y = \sin\theta \cos\theta E_x + \sin^2\theta E_y. \tag{4.5}$$

Quarter-wave plate. All phase retarding devices use the birefringence property of an anisotropic material to selectively change the phase of one component of the electric field. In the case of a quarter-wave plate, one component of the field has its phase advanced by a factor of $\pi/2$ relative to the other. We can write this effect in compact vector notation as follows, assuming $\mathbf{w}$ represents the "fast" axis of the plate,

$$\mathbf{E}' = -i\mathbf{w}(\mathbf{E} \cdot \mathbf{w}) + (\mathbf{w} \times \mathbf{z})[\mathbf{E} \cdot (\mathbf{w} \times \mathbf{z})]. \tag{4.6}$$

Using $\mathbf{w} = \hat{\mathbf{x}}\cos\theta + \hat{\mathbf{y}}\sin\theta$, the Jones vector is in general transformed as follows,

$$E_x' = (-i\cos^2\theta + \sin^2\theta)E_x - (i+1)\sin\theta\cos\theta E_y, \tag{4.7}$$

$$E_y' = -(i+1)\sin\theta\cos\theta E_x + (-i\sin^2\theta + \cos^2\theta)E_y. \tag{4.8}$$

For example, if the y-component of the field is the fast axis ($\theta = \pi/2$), the transformation of the Jones vector reduces to the form

$$E_x' = 1E_x + 0E_y = E_x, \tag{4.9}$$

$$E_y' = 0E_x - iE_y = -iE_y. \tag{4.10}$$

Half-wave plate. For a half-wave plate, one component of the field has its phase advanced by a factor of π. In vector form, this transformation may be written as

$$\mathbf{E}' = -\mathbf{w}(\mathbf{E} \cdot \mathbf{w}) + (\mathbf{w} \times \mathbf{z})[\mathbf{E} \cdot (\mathbf{w} \times \mathbf{z})]. \tag{4.11}$$

In terms of components of the Jones vector, this takes on the form

$$E_x' = -\cos(2\theta)E_x - \sin(2\theta)E_y, \tag{4.12}$$

$$E_y' = -\sin(2\theta)E_x + \cos(2\theta)E_y. \tag{4.13}$$

If the half-wave plate retards the y-component of the field ($\theta = \pi/2$), the Jones vector is transformed as

$$E_x' = 1E_x + 0E_y = E_x, \tag{4.14}$$

$$E_y' = 0E_x - 1E_y = -E_y. \tag{4.15}$$

Faraday rotator. This device is based on the Faraday effect [Far46], in which a light wave traveling through a material parallel to a static magnetic field has its polarization rotated by an angle θ_0, regardless of its initial direction of polarization. We may use Eqs. (1.15) and (1.16) to describe the rotation of the electric field vector,

$$E_x' = E_x\cos\theta_0 - E_y\sin\theta_0, \tag{4.16}$$

$$E_y' = E_x\sin\theta_0 + E_y\cos\theta_0. \tag{4.17}$$

The angle θ_0 is positive or negative depending on whether the magnetic field is parallel or antiparallel to the direction of propagation, respectively.

Looking at the transformation equations for each of the devices described, it is clear that the general behavior of all them can be summarized in a single pair of equations,

$$E_x' = a_{xx}E_x + a_{xy}E_y, \tag{4.18}$$

$$E_y' = a_{yx}E_x + a_{yy}E_y. \tag{4.19}$$

The effect of these devices upon the Jones vector follows a regular, mathematical structure which we would do well to exploit. We introduce a two-dimensional *column vector* for the Jones vector, in the form

$$|E\rangle = \begin{bmatrix} E_x \\ E_y \end{bmatrix}. \qquad (4.20)$$

At the same time, we introduce a two-dimensional *matrix* of the form,

$$\mathbf{A} \equiv \begin{bmatrix} a_{xx} & a_{xy} \\ a_{yx} & a_{yy} \end{bmatrix}. \qquad (4.21)$$

The effect of a device upon the Jones vector is simply a multiplication of the vector by the matrix, producing another vector $|E'\rangle$,

$$|E'\rangle = \mathbf{A}\,|E\rangle. \qquad (4.22)$$

The multiplication of a matrix and a vector results in another vector, and the determination of the *i*th element of the resulting vector comes from the dot product of the *i*th row of the matrix with the multiplied vector. We may treat the multiplication of two matrices together as multiplication of the first matrix by a pair of column vectors constructed from the second.

The matrices for each of our individual components may therefore be written as follows. For a linear polarizer,

$$\mathbf{M}_{\text{linear}}(\theta) = \begin{bmatrix} \cos^2\theta & \cos\theta\sin\theta \\ \sin\theta\cos\theta & \sin^2\theta \end{bmatrix}. \qquad (4.23)$$

A quarter-wave plate may be written as

$$\mathbf{M}_{\text{quarter}}(\theta) = \begin{bmatrix} (-i\cos^2\theta + \sin^2\theta) & -(i+1)\sin\theta\cos\theta \\ -(i+1)\sin\theta\cos\theta & (-i\sin^2\theta + \cos^2\theta) \end{bmatrix}, \qquad (4.24)$$

while a half-wave plate may be written as

$$\mathbf{M}_{\text{half}}(\theta) = \begin{bmatrix} -\cos(2\theta) & -\sin(2\theta) \\ -\sin(2\theta) & \cos(2\theta) \end{bmatrix}. \qquad (4.25)$$

The Faraday rotator is described by

$$\mathbf{M}_{\text{rot}}(\theta_0) = \begin{bmatrix} \cos(\theta_0) & -\sin(\theta_0) \\ \sin(\theta_0) & \cos(\theta_0) \end{bmatrix}. \qquad (4.26)$$

What is the advantage of such a formalism? The effects of all polarization-modifying devices have been unified by recognizing their similar mathematical effects on the state of polarization. Furthermore, we can easily determine the effect of one or more devices upon a polarization state by simple matrix multiplication. We briefly demonstrate a few of these below.

4.1.1 Linear polarization to circular polarization

Linear polarization can be converted to circular polarization by the use of a quarter-wave plate alone. Let us consider a quarter-wave plate aligned along the y-axis ($\theta = \pi/2$); the matrix of this component is therefore

$$\mathbf{M}_{\text{quarter}} = \begin{bmatrix} 1 & 0 \\ 0 & -i \end{bmatrix}. \tag{4.27}$$

We consider the effect of this component on linearly polarized light, oriented at 45° from the x-axis, which has a polarization state of the form

$$|E\rangle = \frac{1}{\sqrt{2}} \begin{bmatrix} E_0 \\ E_0 \end{bmatrix}. \tag{4.28}$$

The effect of the quarter-wave plate on the polarization state is

$$|E'\rangle = \frac{1}{\sqrt{2}} \begin{bmatrix} 1 & 0 \\ 0 & -i \end{bmatrix} \begin{bmatrix} E_0 \\ E_0 \end{bmatrix} = \frac{1}{\sqrt{2}} \begin{bmatrix} E_0 \\ -iE_0 \end{bmatrix}, \tag{4.29}$$

which is circularly polarized light.

4.1.2 Rotation of polarization

A state of linear polarization may be rotated by the use of a half-wave plate. Let us imagine a half-wave plate, oriented at an arbitrary angle θ, illuminated by light polarized along the x-axis, i.e.

$$|E\rangle = \begin{bmatrix} E_0 \\ 0 \end{bmatrix}. \tag{4.30}$$

The polarization of the light after passing through the plate is given by

$$|E'\rangle = \begin{bmatrix} -\cos(2\theta) & -\sin(2\theta) \\ -\sin(2\theta) & \cos(2\theta) \end{bmatrix} \begin{bmatrix} E_0 \\ 0 \end{bmatrix} = \begin{bmatrix} -E_0\cos(2\theta) \\ -E_0\sin(2\theta) \end{bmatrix}. \tag{4.31}$$

Neglecting the overall phase factor (the minus sign), the polarization has been rotated over an angle 2θ.

4.1.3 Optical attenuator

We consider a system which consists of four elements: a linear polarizer polarized along the x-axis (with matrix $\mathbf{P}_1$), a half-wave plate oriented at an angle θ_1 from the x-axis (with matrix $\mathbf{H}_1$), a linear polarizer polarized along the y-axis (with matrix $\mathbf{P}_2$), and then a half-wave plate oriented at an angle θ_2 from the x-axis (with matrix $\mathbf{H}_2$). The effect of this system on a general state of polarization may be found by performing matrix multiplication of the different elements, i.e.

$$\mathbf{H_2 P_2 H_1 P_1} = \begin{bmatrix} -\cos(2\theta_2) & -\sin(2\theta_2) \\ -\sin(2\theta_2) & \cos(2\theta_2) \end{bmatrix} \begin{bmatrix} 0 & 0 \\ 0 & 1 \end{bmatrix} \begin{bmatrix} -\cos(2\theta_1) & -\sin(2\theta_1) \\ -\sin(2\theta_1) & \cos(2\theta_1) \end{bmatrix} \begin{bmatrix} 1 & 0 \\ 0 & 0 \end{bmatrix}.$$

(4.32)

It is important to note that the first element of the optical system is the *last* matrix in Eq. (4.32), and hence the first one which will influence the polarization state. We simplify this formula by performing matrix multiplication between $\mathbf{H_2}$ and $\mathbf{P_2}$, and $\mathbf{H_1}$ and $\mathbf{P_1}$, and then multiplying the two resulting matrices, which yields the result

$$\mathbf{H_2 P_2 H_1 P_1} = \begin{bmatrix} \sin(2\theta_2)\sin(2\theta_1) & 0 \\ -\cos(2\theta_2)\sin(2\theta_1) & 0 \end{bmatrix}.$$

(4.33)

What is the effect of this system on a field polarized along the x-axis? We may multiply by the appropriate vector to find that

$$\mathbf{H_2 P_2 H_1 P_1} |E\rangle = \begin{bmatrix} \sin(2\theta_2)\sin(2\theta_1) & 0 \\ -\cos(2\theta_2)\sin(2\theta_1) & 0 \end{bmatrix} \begin{bmatrix} E_0 \\ 0 \end{bmatrix} = \begin{bmatrix} E_0\sin(2\theta_2)\sin(2\theta_1) \\ -E_0\cos(2\theta_2)\sin(2\theta_1) \end{bmatrix}.$$

(4.34)

Let us now take $\theta_2 = \pi/4$. The formula reduces to the form

$$\mathbf{H_2 P_2 H_1 P_1} |E\rangle = \begin{bmatrix} E_0\sin(2\theta_1) \\ 0 \end{bmatrix}.$$

(4.35)

The field remains polarized along the x-direction, but now has its amplitude attenuated by a factor of $\sin(2\theta_1)$. By varying the angle of the half-wave plate which lies between the two polarizers, we are able to adjust the amount of light which is transmitted, i.e. we have made an *optical attenuator*.

From these examples, it should be clear that a good knowledge of the mathematics of matrices simplifies the calculation of polarization effects in optical systems. We will see that matrix manipulation is extremely useful in numerous other applications. In this chapter we discuss the fundamentals of matrices and linear algebra.

4.2 Matrix algebra

A matrix $\mathbf{A}$ is, at first glance, simply an array of numbers, usually written with brackets around it. A 3×3 square matrix, for example, would have the form

$$\mathbf{A} = \begin{bmatrix} a_{11} & a_{12} & a_{13} \\ a_{21} & a_{22} & a_{23} \\ a_{31} & a_{32} & a_{33} \end{bmatrix}.$$

(4.36)

The elements of a matrix are usually distinguished by a pair of subscripts: the first indicates the row of the element, the second the column; a general element of the matrix might be written as a_{ij}. We assume at first that the a_{ij} are real-valued.

Matrices may have an equal or unequal number of rows and columns; when the numbers are equal, we have a *square* matrix. In elementary physics problems, square matrices are most common, but are by no means the only ones which are important. Matrices which have only one column or one row, i.e.

$$\mathbf{v} = \begin{bmatrix} v_1 \\ v_2 \\ v_3 \end{bmatrix}, \quad \mathbf{u} = \begin{bmatrix} u_1 & u_2 & u_3 \end{bmatrix}, \tag{4.37}$$

are referred to as *column vectors* or *row vectors*, respectively. Depending on the interpretation of the vectors, they may or may not actually be "physical" vectors of the sort discussed in Chapter 1; an example of a "nonphysical" vector is the vector representation of a geometric light ray to be given in Section 4.10. When proving general results relating to column and row vectors, the "bra-ket" notation of Section 1.5 is extremely convenient; for real-valued vectors, we write $|\mathbf{v}\rangle$ as the column vector version of $\mathbf{v}$, and $\langle\mathbf{v}|$ as the row vector version. When using the "bra-ket" form, we will write matrices as capital, bold-faced letters, e.g. $\mathbf{A}$. We will adapt this notation in future operations involving vectors.

Matrices, as a logically grouped set of numbers, are an extension of the theory of ordinary numbers. The set of all real-valued matrices of fixed size form a linear vector space, as can be readily shown. Specific rules for addition and multiplication of matrices form a new algebra called *linear algebra*. These specific rules are listed below.

1. Equality. Matrix $\mathbf{A} = \mathbf{B}$ if and only if $a_{ij} = b_{ij}$ for all i, j. This implies that both matrices have the same number of rows and columns.
2. Addition. $\mathbf{A} + \mathbf{B} = \mathbf{C}$ if and only if $a_{ij} + b_{ij} = c_{ij}$. That is, to add two matrices together one simply adds the corresponding elements of each matrix together. From this definition, because the a_{ij}s, etc., are ordinary numbers, it follows that matrix addition is commutative, $\mathbf{A} + \mathbf{B} = \mathbf{B} + \mathbf{A}$, and associative, $(\mathbf{A} + \mathbf{B}) + \mathbf{C} = \mathbf{A} + (\mathbf{B} + \mathbf{C})$.
3. Null matrix. There exists a zero, or null matrix, for which all elements are zero, e.g.:

$$\mathbf{0} = \begin{bmatrix} 0 & 0 & 0 \\ 0 & 0 & 0 \\ 0 & 0 & 0 \end{bmatrix}. \tag{4.38}$$

The null matrix satisfies the relation, $\mathbf{A} + \mathbf{0} = \mathbf{0} + \mathbf{A} = \mathbf{A}$.
4. Scalar multiplication. The multiplication of a matrix by a scalar α is defined as

$$\alpha\mathbf{A} = (\alpha\mathbf{A}). \tag{4.39}$$

The parenthesis here indicates that the elements of the new matrix $\alpha\mathbf{A}$ are given by: $(\alpha\mathbf{A})_{ij} = \alpha a_{ij}$; that is, each element of the matrix is multiplied by α.

5. Matrix multiplication. To define matrix multiplication, we "go physical" and return to the rotations discussed in Section 1.2:

$$x_i' = \sum_{j=1}^{N} b_{ij} x_j, \quad i = 1, \ldots, N. \tag{4.40}$$

We now interpret b_{ij} as the elements of a matrix **B**. After the first rotation of coordinates, let us consider a second rotation, defined by a matrix **A** (elements a_{ij}) which converts the elements x_i' into a double-primed system, x_i'':

$$x_i'' = \sum_{j=1}^{N} a_{ij} x_j', \quad i = 1, \ldots, N. \tag{4.41}$$

How do we express x_i'' in terms of the original system coordinates, x_i? We may substitute from Eq. (4.40) into Eq. (4.41) to arrive at the result

$$x_i'' = \sum_{k=1}^{N} a_{ik} \left[\sum_{j=1}^{N} b_{kj} x_j \right] = \sum_{j=1}^{N} \left[\sum_{k=1}^{N} a_{ik} b_{kj} \right] x_j. \tag{4.42}$$

The last part of this equation looks like another coordinate transformation **C**, with elements c_{ij} given by

$$c_{ij} = \sum_{k=1}^{N} a_{ik} b_{kj}. \tag{4.43}$$

This appears to be a "multiplication" of the effects of two rotations **A**, **B** to produce a combined rotation **C**. We *define* matrix multiplication by Eq. (4.43); in other words,

$$\mathbf{AB} = \mathbf{C} \quad \text{if and only if} \quad c_{ij} = \sum_{k=1}^{N} a_{ik} b_{kj}. \tag{4.44}$$

What does this equation mean? It indicates that if we want the element of **C** which is in the ith row and jth column, we take the elements of the ith row of **A** and the jth row of **B**, multiply them together *in sequence* and add the results.

Example 4.1 We consider the multiplication of two 3×3 matrices, given by

$$\mathbf{A} = \begin{bmatrix} 1 & 3 & 2 \\ 1 & 0 & 1 \\ 2 & 2 & 4 \end{bmatrix} \tag{4.45}$$

and

$$\mathbf{B} = \begin{bmatrix} 1 & 1 & 0 \\ 0 & 1 & 0 \\ 2 & 0 & 1 \end{bmatrix}. \tag{4.46}$$

The multiplication gives, for instance,

$$(\mathbf{AB})_{11} = \begin{bmatrix} 1 & 3 & 2 \\ 1 & 0 & 1 \\ 2 & 2 & 4 \end{bmatrix} \begin{bmatrix} 1 & 1 & 0 \\ 0 & 1 & 0 \\ 2 & 0 & 1 \end{bmatrix} = 1 \times 1 + 3 \times 0 + 2 \times 2 = 5. \tag{4.47}$$

$$(\mathbf{AB})_{21} = \begin{bmatrix} 1 & 3 & 2 \\ 1 & 0 & 1 \\ 2 & 2 & 4 \end{bmatrix} \begin{bmatrix} 1 & 1 & 0 \\ 0 & 1 & 0 \\ 2 & 0 & 1 \end{bmatrix} = 1 \times 1 + 0 \times 0 + 1 \times 2 = 3. \tag{4.48}$$

$$(\mathbf{AB})_{13} = \begin{bmatrix} 1 & 3 & 2 \\ 1 & 0 & 1 \\ 2 & 2 & 4 \end{bmatrix} \begin{bmatrix} 1 & 1 & 0 \\ 0 & 1 & 0 \\ 2 & 0 & 1 \end{bmatrix} = 1 \times 0 + 3 \times 0 + 2 \times 1 = 2. \tag{4.49}$$

The solution for the whole multiplication is

$$\mathbf{AB} = \begin{bmatrix} 1 & 3 & 2 \\ 1 & 0 & 1 \\ 2 & 2 & 4 \end{bmatrix} \begin{bmatrix} 1 & 1 & 0 \\ 0 & 1 & 0 \\ 2 & 0 & 1 \end{bmatrix} = \begin{bmatrix} 5 & 4 & 2 \\ 3 & 1 & 1 \\ 10 & 4 & 4 \end{bmatrix}. \tag{4.50}$$

◇

One important observation about this definition of multiplication is that it is *not* in general commutative, i.e.

$$\mathbf{AB} \neq \mathbf{BA}. \tag{4.51}$$

For instance, consider

$$\begin{bmatrix} 1 & 2 \\ 3 & 1 \end{bmatrix} \begin{bmatrix} 2 & 2 \\ 0 & 1 \end{bmatrix} = \begin{bmatrix} 2 & 4 \\ 6 & 7 \end{bmatrix} \neq \begin{bmatrix} 2 & 2 \\ 0 & 1 \end{bmatrix} \begin{bmatrix} 1 & 2 \\ 3 & 1 \end{bmatrix} = \begin{bmatrix} 8 & 6 \\ 3 & 1 \end{bmatrix}. \tag{4.52}$$

Let us introduce a matrix known as the *commutator* of $\mathbf{A}$ and $\mathbf{B}$ as

$$[\mathbf{A}, \mathbf{B}] \equiv \mathbf{AB} - \mathbf{BA}. \tag{4.53}$$

In general, the commutator of a pair of matrices is non-zero. An important exception of this non-commutativity is when both $\mathbf{A}$ and $\mathbf{B}$ are square *diagonal matrices*, in which only the elements on diagonal are nonzero. For example,

$$\mathbf{A} = \begin{bmatrix} 1 & 0 & 0 \\ 0 & 4 & 0 \\ 0 & 0 & 2 \end{bmatrix} \tag{4.54}$$

is a diagonal matrix. Commuting matrices and diagonal matrices are significant for a number of reasons, and we will discuss them in more detail in Section 4.6.

It is to be noted that irregular matrices (those not square) may be multiplied together, as long as the number of *columns* of **A** matches the number of *rows* of **B**; the result is a matrix with a number of rows equal to the rows of **A** and columns equal to the columns of **B**. For example,

$$\begin{bmatrix} 3 & 3 & 1 \\ 1 & 0 & 0 \end{bmatrix} \begin{bmatrix} 1 & 2 \\ 0 & 0 \\ 3 & 1 \end{bmatrix} = \begin{bmatrix} 6 & 7 \\ 1 & 2 \end{bmatrix}. \tag{4.55}$$

This is particularly useful for now we have a description of a matrix multiplying a vector,

$$\mathbf{A}\,|\mathbf{x}\rangle = \begin{bmatrix} a_{11} & a_{12} \\ a_{21} & a_{22} \end{bmatrix} \begin{bmatrix} x_1 \\ x_2 \end{bmatrix} = \begin{bmatrix} a_{11}x_1 + a_{12}x_2 \\ a_{21}x_1 + a_{22}x_2 \end{bmatrix} = \begin{bmatrix} x_1' \\ x_2' \end{bmatrix}. \tag{4.56}$$

Equating terms in the latter two pieces reproduces the results of Section 1.2, in particular Eqs. (1.21).

This form of matrix multiplication also allows a compact way of writing the dot product, as a product of a row vector with a column vector. Thus

$$\mathbf{x} \cdot \mathbf{y} = \langle \mathbf{x} | \mathbf{y} \rangle = \begin{bmatrix} x_1 & x_2 & x_3 \end{bmatrix} \begin{bmatrix} y_1 \\ y_2 \\ y_3 \end{bmatrix} = x_1y_1 + x_2y_2 + x_3y_3. \tag{4.57}$$

As we have mentioned previously, this form of multiplication is referred to as the inner product. It is to be noted that the inner product may be written as the product of a "bra" and a "ket".

We may also construct an *outer product* of vectors, of the form $|x\rangle\langle y|$. It is not difficult to show that this product results in a square matrix; for instance,

$$|\mathbf{x}\rangle\langle\mathbf{y}| = \begin{bmatrix} x_1 \\ x_2 \\ x_3 \end{bmatrix} \begin{bmatrix} y_1 & y_2 & y_3 \end{bmatrix} = \begin{bmatrix} x_1y_1 & x_1y_2 & x_1y_3 \\ x_2y_1 & x_2y_2 & x_2y_3 \\ x_3y_1 & x_3y_2 & x_3y_3 \end{bmatrix}. \tag{4.58}$$

6. Unit matrix. With multiplication defined, it is worth asking if there exists a unit matrix, i.e. a square matrix **I** such that, for any square matrix **A** of the same size,

$$\mathbf{IA} = \mathbf{AI} = \mathbf{A}. \tag{4.59}$$

It can readily be shown that such a matrix exists, and its components are given by the Kronecker delta, δ_{ij}. For a 3×3 matrix, this takes the form

$$\mathbf{I} = \begin{bmatrix} 1 & 0 & 0 \\ 0 & 1 & 0 \\ 0 & 0 & 1 \end{bmatrix}. \tag{4.60}$$

7. Inverse matrix. There is no direct analogue of division in linear algebra, but a related concept of great importance is the possible existence of an *inverse* of a square ($n \times n$) matrix. An inverse of a matrix $\mathbf{A}$ is denoted as $\mathbf{A}^{-1}$, and it satisfies the following equation,

$$\mathbf{A}^{-1}\mathbf{A} = \mathbf{A}\mathbf{A}^{-1} = \mathbf{I}. \tag{4.61}$$

Not every square matrix possesses an inverse. We will begin to explore the circumstances under which a matrix possesses an inverse in the next section.

One important and useful relation involves the inverse of a product of matrices. Assuming $\mathbf{C} = \mathbf{AB}$, and that $\mathbf{A}^{-1}$ and $\mathbf{B}^{-1}$ exist, how does the inverse of $\mathbf{C}$ relate to the inverses of $\mathbf{A}$ and $\mathbf{B}$? We postulate that

$$\mathbf{C}^{-1} = (\mathbf{AB})^{-1} = \mathbf{B}^{-1}\mathbf{A}^{-1}. \tag{4.62}$$

To prove this, we simply pre- and post- multiply $\mathbf{AB}$ by $\mathbf{B}^{-1}\mathbf{A}^{-1}$:

$$\begin{aligned} \mathbf{B}^{-1}\mathbf{A}^{-1}\mathbf{AB} &= \mathbf{B}^{-1}\mathbf{IB} = \mathbf{B}^{-1}\mathbf{B} = \mathbf{I}, \\ \mathbf{ABB}^{-1}\mathbf{A}^{-1} &= \mathbf{AIA}^{-1} = \mathbf{AA}^{-1} = \mathbf{I}. \end{aligned} \tag{4.63}$$

4.3 Systems of equations, determinants, and inverses

Let us step away briefly from matrix algebra and consider the following problem: we are interested in solving a system of three homogeneous equations of the general form,

$$\begin{aligned} a_{11}x_1 + a_{12}x_2 + a_{13}x_3 &= 0, \\ a_{21}x_1 + a_{22}x_2 + a_{23}x_3 &= 0, \\ a_{31}x_1 + a_{32}x_2 + a_{33}x_3 &= 0, \end{aligned} \tag{4.64}$$

where we are given the quantities a_{ij} and wish to solve for the unknowns x_i, neglecting the trivial solution $x_1 = x_2 = x_3 = 0$. We present this as a system of three equations with three unknowns, but it could easily be generalized to n equations with n unknowns. Under what conditions does this system of equations possess a nontrivial solution?

If we write our unknowns and our coefficients in a vector form, i.e. $\mathbf{x} \equiv (x_1, x_2, x_3)$, $\mathbf{a}_1 \equiv (a_{11}, a_{12}, a_{13})$, $\mathbf{a}_2 \equiv (a_{21}, a_{22}, a_{23})$, etc., Eqs. (4.64) take on the abbreviated form

$$\mathbf{a}_1 \cdot \mathbf{x} = 0, \quad \mathbf{a}_2 \cdot \mathbf{x} = 0, \quad \mathbf{a}_3 \cdot \mathbf{x} = 0. \tag{4.65}$$

Geometrically, it may be said that $\mathbf{x}$ must be perpendicular to $\mathbf{a}_1$, $\mathbf{a}_2$, and $\mathbf{a}_3$. This itself is only possible if $\mathbf{a}_1$, $\mathbf{a}_2$, and $\mathbf{a}_3$ all lie in the same plane, because only in such a case can one find a vector perpendicular to the plane. In other words, the three vectors must be *linearly dependent* (as discussed in Section 1.5). This can be expressed using vector algebra by the requirement that the vector triple product must vanish,

$$\mathbf{a}_1 \cdot (\mathbf{a}_2 \times \mathbf{a}_3) = 0. \tag{4.66}$$

This follows from that fact that the cross product of $\mathbf{a}_2$ and $\mathbf{a}_3$ is perpendicular to both; if $\mathbf{a}_1$ is in the same plane as $\mathbf{a}_2$ and $\mathbf{a}_3$, the triple product vanishes. In three dimensions, this triple product is what is called the *determinant* of the system of equations.

In the case that the determinant vanishes, one of the equations may be written as a linear combination of the other two, and we can solve any two of the equations. In terms of x_3, the solutions are given by the expressions,

$$\frac{x_1}{x_3} = \frac{a_{12}a_{23} - a_{13}a_{22}}{a_{11}a_{22} - a_{12}a_{21}},$$

$$\frac{x_2}{x_3} = -\frac{a_{11}a_{23} - a_{13}a_{21}}{a_{11}a_{22} - a_{12}a_{21}}. \tag{4.67}$$

The original system of equations may also be written compactly in matrix form, as follows:

$$\begin{bmatrix} a_{11} & a_{12} & a_{13} \\ a_{21} & a_{22} & a_{23} \\ a_{31} & a_{32} & a_{33} \end{bmatrix} \begin{bmatrix} x_1 \\ x_2 \\ x_3 \end{bmatrix} = \begin{bmatrix} 0 \\ 0 \\ 0 \end{bmatrix}, \tag{4.68}$$

or in the simple form

$$\mathbf{A}\,|\mathbf{x}\rangle = |\mathbf{0}\rangle, \tag{4.69}$$

where $|\mathbf{0}\rangle$ is the null vector of length 0. It can be seen from Eq. (4.66) that the determinant is a characteristic of the matrix $\mathbf{A}$, and in fact tells us quite a bit about the behavior of the matrix. For instance, suppose that there exists an inverse matrix $\mathbf{A}^{-1}$; we could then pre-multiply Eq. (4.69) by this inverse, to find that

$$\mathbf{A}^{-1}\mathbf{A}\,|\mathbf{x}\rangle = \mathbf{A}^{-1}\,|\mathbf{0}\rangle,$$

$$\mathbf{I}\,|\mathbf{x}\rangle = |\mathbf{0}\rangle,$$

$$|\mathbf{x}\rangle = |\mathbf{0}\rangle. \tag{4.70}$$

In other words, if the inverse of the matrix exists, the only solution to the system of equations is the null vector $|\mathbf{0}\rangle$. Conversely, if a nontrivial solution to the system of equations exists, the inverse cannot exist. Since a nontrivial solution exists only when the determinant vanishes, we can conclude that *the inverse of a matrix does not exist if the determinant of the matrix vanishes*.

We come to similar conclusions by looking at a system of two homogeneous equations,

$$a_{11}x_1 + a_{12}x_2 = 0, \tag{4.71}$$

$$a_{21}x_1 + a_{22}x_2 = 0. \tag{4.72}$$

We may express this in terms of vectors in a two-dimensional vector space, and our vectors must satisfy the pair of equations

$$\mathbf{a}_1 \cdot \mathbf{x} = 0, \quad \mathbf{a}_2 \cdot \mathbf{x} = 0. \tag{4.73}$$

This pair of equations can be satisfied only if $\mathbf{a}_1$ and $\mathbf{a}_2$ point in the same direction; we may quantify this mathematically by requiring that the dot product of the two vectors is equal to the product of their magnitudes,

$$\mathbf{a}_1 \cdot \mathbf{a}_2 = a_1 a_2. \tag{4.74}$$

Squaring both sides of this equation, and writing each vector in component form, the equation reduces to

$$(a_{11}a_{22} - a_{21}a_{12})^2 = 0. \tag{4.75}$$

The quantity $a_{11}a_{22} - a_{21}a_{12}$ is the determinant of the 2×2 matrix

$$\mathbf{A} = \begin{bmatrix} a_{11} & a_{12} \\ a_{21} & a_{22} \end{bmatrix}. \tag{4.76}$$

The determinant also plays an important role in the solution of an inhomogeneous set of equations. We consider first the system of three equations shown below,

$$
\begin{aligned}
a_{11}x_1 + a_{12}x_2 + a_{13}x_3 &= \alpha_1, \\
a_{21}x_1 + a_{22}x_2 + a_{23}x_3 &= \alpha_2, \\
a_{31}x_1 + a_{32}x_2 + a_{33}x_3 &= \alpha_3,
\end{aligned}
\tag{4.77}
$$

which may be written in matrix form as

$$\mathbf{A}\,|\mathbf{x}\rangle = |\boldsymbol{\alpha}\rangle, \tag{4.78}$$

where $|\boldsymbol{\alpha}\rangle \equiv (\alpha_1, \alpha_2, \alpha_3)$. In contrast to the homogeneous case, if the inverse of the matrix $\mathbf{A}$ exists, we may write

$$
\begin{aligned}
\mathbf{A}^{-1}\mathbf{A}\,|\mathbf{x}\rangle &= \mathbf{A}^{-1}\,|\boldsymbol{\alpha}\rangle, \\
\mathbf{I}\,|\mathbf{x}\rangle &= \mathbf{A}^{-1}\,|\boldsymbol{\alpha}\rangle, \\
|\mathbf{x}\rangle &= \mathbf{A}^{-1}\,|\boldsymbol{\alpha}\rangle.
\end{aligned}
\tag{4.79}
$$

We therefore may make the argument that the solution to an inhomogeneous set of equations exists only if the inverse of the matrix exists. Since the existence of the inverse depends on the determinant, we may state that *the solution to an inhomogeneous set of equations exists if and only if the determinant of the matrix is nonzero.*

For the 3×3 case considered here, this can be shown explicitly. With some effort, the solution to the system of equations (4.77) can be shown to be given by

$$
x_1 = \frac{\det(\mathbf{a}_1^{(1)}, \mathbf{a}_2^{(1)}, \mathbf{a}_3^{(1)})}{\det(\mathbf{A})},
$$

$$
x_2 = \frac{\det(\mathbf{a}_1^{(2)}, \mathbf{a}_2^{(2)}, \mathbf{a}_3^{(2)})}{\det(\mathbf{A})}, \tag{4.80}
$$

$$
x_3 = \frac{\det(\mathbf{a}_1^{(3)}, \mathbf{a}_2^{(3)}, \mathbf{a}_3^{(3)})}{\det(\mathbf{A})}.
$$

In this equation, $\det(\mathbf{a}, \mathbf{b}, \mathbf{c})$ refers to the determinant of a trio of vectors, while $\det(\mathbf{A})$ is the determinant of a matrix. We define $\mathbf{a}_1^{(1)} = (\alpha_1, a_{12}, a_{13})$, $\mathbf{a}_1^{(2)} = (a_{11}, \alpha_2, a_{13})$, and $\mathbf{a}_3^{(3)} = (a_{31}, a_{32}, \alpha_3)$, with similar expressions for the other vectors: for example, $\mathbf{a}_i^{(j)}$ is the vector $\mathbf{a}_i$ with the jth component replaced by α_j. It can be seen that this solution does not exist unless $\det(\mathbf{A}) \neq 0$.

The determinant is thus an important property of a system of equations, both homogeneous and inhomogeneous. It is conveniently expressed in a matrix form, with vertical lines replacing square brackets,

$$
\det(\mathbf{a}_1, \mathbf{a}_2, \mathbf{a}_3) = \det(\mathbf{A}) = \begin{vmatrix} a_{11} & a_{12} & a_{13} \\ a_{21} & a_{22} & a_{23} \\ a_{31} & a_{32} & a_{33} \end{vmatrix}. \tag{4.81}
$$

For a general $n \times n$ matrix, the determinant D_n can be written in summation form,

$$
D_n = \sum_{i,j,k...} \epsilon_{ijk...} a_{1i} a_{2j} a_{3k} \dots, \tag{4.82}
$$

where $\epsilon_{ijk...}$ is an n-index generalization of the Levi-Civita symbol (1.52), taking on value $+1$ when $ijk \dots$ are an even permutation, -1 when $ijk \dots$ are an odd permutation, and zero otherwise.

For a 3×3 matrix, we have already seen that the determinant takes on the value of the scalar triple product,

$$
D_3 = a_{11}(a_{22}a_{33} - a_{23}a_{32}) + a_{12}(a_{23}a_{31} - a_{21}a_{33}) + a_{13}(a_{21}a_{32} - a_{22}a_{31}). \tag{4.83}
$$

We have also seen that, for a 2×2 matrix, the determinant has the simple form

$$
D_2 = \begin{vmatrix} a_{11} & a_{12} \\ a_{21} & a_{22} \end{vmatrix} = a_{11}a_{22} - a_{12}a_{21}, \tag{4.84}
$$

and for a 1×1 matrix (a matrix with one element, i.e. a scalar), the determinant is simply the value of the scalar.

Determinants have many useful mathematical properties, all of which can be shown using Eq. (4.88). The first of these is the expression of the determinant of an $N \times N$ matrix $\mathbf{A}$ by sub-determinants of that matrix,

$$\det(\mathbf{A}) = \sum_{j=1}^{N}(-1)^{i+j} a_{ij}\det(\mathbf{M}_{ij}) = \sum_{i=1}^{N}(-1)^{i+j} a_{ij}\det(\mathbf{M}_{ij}). \qquad (4.85)$$

The quantity $\mathbf{M}_{ij}$ is the matrix formed by removing the ith row and jth column of the original matrix. The determinant of this smaller matrix is known as a *minor* of the determinant. The product

$$C_{ij} \equiv \det(\mathbf{M}_{ij})(-1)^{i+j} \qquad (4.86)$$

is referred to as the *cofactor* of the matrix. There are two forms of this equation: the first one involves calculating the determinant using an arbitrary row i of the matrix, while the second one involves calculating the determinant using an arbitrary column j of the matrix. Both methods give the same result.

This definition of a determinant specifies a calculation procedure for an arbitrary matrix: For each element in a particular row (or column) of the matrix, draw a line through the row and column that the element appears in. Multiply that element by the determinant of the matrix which remains, and also by the factor $(-1)^{i+j}$. Add together the contributions for all elements of the row; this gives the determinant.

Example 4.2 Here we calculate the determinant of a 3×3 matrix first by using its top row,

$$\begin{vmatrix} 1 & 1 & 0 \\ 0 & 1 & 1 \\ 1 & 0 & 1 \end{vmatrix} = 1 \times \begin{vmatrix} \cancel{1} & \cancel{1} & \cancel{0} \\ \cancel{0} & 1 & 1 \\ \cancel{1} & 0 & 1 \end{vmatrix} + (-1) \times \begin{vmatrix} \cancel{1} & \cancel{1} & \cancel{0} \\ 0 & \cancel{1} & 1 \\ 1 & \cancel{0} & 1 \end{vmatrix} + 0 \times \begin{vmatrix} \cancel{1} & \cancel{1} & \cancel{0} \\ 0 & 1 & \cancel{1} \\ 1 & 0 & \cancel{1} \end{vmatrix} = 2. \qquad (4.87)$$

We arrive at the same result if we instead calculate the determinant by using its first column,

$$\begin{vmatrix} 1 & 1 & 0 \\ 0 & 1 & 1 \\ 1 & 0 & 1 \end{vmatrix} = 1 \times \begin{vmatrix} \cancel{1} & \cancel{1} & \cancel{0} \\ \cancel{0} & 1 & 1 \\ \cancel{1} & 0 & 1 \end{vmatrix} + (0) \times \begin{vmatrix} \cancel{1} & 1 & 0 \\ \cancel{0} & \cancel{1} & \cancel{1} \\ \cancel{1} & 0 & 1 \end{vmatrix} + 1 \times \begin{vmatrix} \cancel{1} & 1 & 0 \\ \cancel{0} & 1 & 1 \\ \cancel{1} & \cancel{0} & \cancel{1} \end{vmatrix} = 2. \qquad (4.88)$$

◊

Other properties of determinants follow from the general definition (4.82). An interchange of any two rows or columns of a matrix changes the sign of the determinant. Sometimes this can be a useful technique for finding the determinant of a matrix rapidly; for instance

$$\begin{vmatrix} 0 & 1 & 0 \\ 0 & 0 & 1 \\ 1 & 0 & 0 \end{vmatrix} = - \begin{vmatrix} 1 & 0 & 0 \\ 0 & 0 & 1 \\ 0 & 1 & 0 \end{vmatrix} = + \begin{vmatrix} 1 & 0 & 0 \\ 0 & 1 & 0 \\ 0 & 0 & 1 \end{vmatrix} = 1. \qquad (4.89)$$

From this antisymmetry property, it follows that if any two rows or columns of a determinant are equal, the determinant is zero. This can also be seen from our original discussion regarding linear equations; the vanishing of the determinant indicates that at least two rows are linearly dependent.

If each element of a row or column is multiplied by a constant, the determinant is multiplied by a constant.

Also, if a multiple of one row is added (column by column) to another row, the value of the determinant is unchanged.

The determinant of the product of two matrices, **AB**, is equal to the product of the determinants, i.e. $\det(\mathbf{AB}) = \det(\mathbf{A})\det(\mathbf{B})$.

As we have already noted, the determinant of a square matrix **A** is intimately related to the existence (or nonexistence) of its inverse matrix, $\mathbf{A}^{-1}$. One can show, for a square matrix which possesses an inverse, that the elements of the inverse matrix are given by the formula

$$a_{ij}^{-1} = \frac{C_{ji}}{\det(\mathbf{A})}, \tag{4.90}$$

where C_{kl} is the k, lth cofactor of the determinant of **A** [see Eq. (4.86)]. This representation of the inverse of a matrix is not terribly useful in practice, however, because its solution requires the calculation of many determinants: the determinant of **A**, plus the sub-determinants which are buried in the cofactors. We will shortly consider an easier and more reliable method of calculating the inverse of a matrix.

One of the most significant properties of the determinant of a matrix is its invariance under coordinate transformations. In Section 4.4, we will discuss the transformation of matrices into different coordinate systems by a rotation of the coordinate axes. The determinant is unchanged by such a rotation of the axes. There is only one other property of a matrix which is also invariant under rotations: the *trace*, defined as the sum of the diagonal elements of the matrix **A**,

$$\text{Tr}(\mathbf{A}) = \sum_{i=1}^{N} a_{ii}. \tag{4.91}$$

The solution of inhomogeneous systems of equations, and the calculation of inverse matrices, are both extremely important operations, especially in computational physics. Though we formally have a solution for both problems – Eq. (4.80) for an inhomogeneous system of equations and Eq. (4.90) for the inverse of a matrix – both solutions involve the calculation of one or more determinants of the matrix.

These methods, though straightforward in principle, are problematic in practice, mainly for two reasons: (1) the calculation of a determinant is time consuming, and for large matrices, can be extremely slow even for fast computers! (2) the determinant, for large matrices, involves many, many multiplications and additions, which typically introduces large errors in the final result, associated with the finite precision of the computations.

Fortunately, other methods exist for determining the solution of a system of equations and the inverse of a matrix that do not involve calculation of the determinant. We discuss two of these methods below; more advanced techniques will be considered in Chapter 5.

4.3.1 Gaussian elimination

Gaussian elimination[1] is a technique for solving a system of equations without the direct use of determinants. We illustrate it below for a simple system of three equations with three unknowns.

We want to solve the system of equations,

$$2x + y + 2z = 10,$$
$$x + 3y + 2z = 16, \tag{4.92}$$
$$3x - y + 4z = 6.$$

The goal is to use the first equation to eliminate the first unknown (x) from the latter two equations. Then we may use the modified second equation to eliminate y from the last equation, giving us the value of z. We may then substitute back up into the second equation to determine y, and the first equation to determine x.

First, we divide each equation by its x coefficient, to arrive at

$$x + \frac{1}{2}y + z = 5,$$
$$x + 3y + 2z = 16, \tag{4.93}$$
$$x - \frac{1}{3}y + \frac{4}{3}z = 2.$$

Now, subtract the first equation from the second and third:

$$x + \frac{1}{2}y + z = 5,$$
$$\frac{5}{2}y + z = 11, \tag{4.94}$$
$$-\frac{5}{6}y + \frac{1}{3}z = -3.$$

---

[1] The method in fact originally appeared long before Gauss, in the Chinese text *The Nine Chapters on the Mathematical Art*, circa 150 BCE.

Dividing the latter two equations by their new y coefficients, we have

$$x + \frac{1}{2}y + z = 5,$$

$$y + \frac{2}{5}z = \frac{22}{5}, \tag{4.95}$$

$$y - \frac{2}{5}z = \frac{18}{5}.$$

We now subtract the second equation from the third:

$$x + \frac{1}{2}y + z = 5,$$

$$y + \frac{2}{5}z = \frac{22}{5}, \tag{4.96}$$

$$-\frac{4}{5}z = -\frac{4}{5}.$$

We have found $z = 1$; substituting this result back into the second equation, we have $y = 4$. substituting both y and z into the first equation, we find that $x = 2$.

This method works quite well and is used often in computer solutions. There are other methods to arrange equations so that they can be readily solved; we will discuss some of them after we learn a bit more about matrices.

4.3.2 Gauss–Jordan matrix inversion

Matrix inversion is hugely important because a large number of computational problems reduce, in essence, to the solution of a large system of equations of the form

$$\mathbf{A}\,|\mathbf{x}\rangle = |\boldsymbol{\alpha}\rangle, \tag{4.97}$$

where the matrix $\mathbf{A}$ is $n \times n$, with n typically very large (anywhere from 100 to a million). If one is solving this system of equations for a number of different vectors $\boldsymbol{\alpha}$ and fixed $\mathbf{A}$, the most efficient strategy is to determine $\mathbf{A}^{-1}$, so that the solution of the problem reduces to a simple matrix multiplication: $|\mathbf{x}\rangle = \mathbf{A}^{-1}|\boldsymbol{\alpha}\rangle$. A large variety of methods have been introduced to perform matrix inversion; we focus here on one of the most straightforward: Gauss–Jordan matrix inversion.

This method involves performing simultaneous operations on a matrix and the unit matrix, gradually transforming the unit matrix into the inverse and the original matrix into the unit matrix. The technique is best described by an example: we start by writing the initial matrix $\mathbf{A}$ and the unit matrix $\mathbf{I}$ side by side,

$$\begin{bmatrix} 1 & 2 & 1 \\ 2 & 2 & 3 \\ 1 & 3 & 1 \end{bmatrix} \quad \begin{bmatrix} 1 & 0 & 0 \\ 0 & 1 & 0 \\ 0 & 0 & 1 \end{bmatrix}. \tag{4.98}$$

We now multiply each row by a number so that the first column is replaced by unity,

$$
\begin{bmatrix} 1 & 2 & 1 \\ 1 & 1 & 3/2 \\ 1 & 3 & 1 \end{bmatrix}
\begin{bmatrix} 1 & 0 & 0 \\ 0 & 1/2 & 0 \\ 0 & 0 & 1 \end{bmatrix}.
\tag{4.99}
$$

We now subtract the first row from the second and third,

$$
\begin{bmatrix} 1 & 2 & 1 \\ 0 & -1 & 1/2 \\ 0 & 1 & 0 \end{bmatrix}
\begin{bmatrix} 1 & 0 & 0 \\ -1 & 1/2 & 0 \\ -1 & 0 & 1 \end{bmatrix}.
\tag{4.100}
$$

The first column of the left matrix is now the same as that of the identity matrix! We now wish to do the same thing to the middle column. First, we normalize the second and third rows to unity,

$$
\begin{bmatrix} 1 & 2 & 1 \\ 0 & 1 & -1/2 \\ 0 & 1 & 0 \end{bmatrix}
\begin{bmatrix} 1 & 0 & 0 \\ 1 & -1/2 & 0 \\ -1 & 0 & 1 \end{bmatrix},
\tag{4.101}
$$

which in this case is a fairly trivial operation. We next use the middle row to eliminate the terms at the top and bottom of the center column,

$$
\begin{bmatrix} 1 & 0 & 2 \\ 0 & 1 & -1/2 \\ 0 & 0 & 1/2 \end{bmatrix}
\begin{bmatrix} -1 & 1 & 0 \\ 1 & -1/2 & 0 \\ -2 & 1/2 & 1 \end{bmatrix}.
\tag{4.102}
$$

It is to be noted that the left two columns are now equal to the left two columns of the identity matrix. We now normalize the third row term to unity,

$$
\begin{bmatrix} 1 & 0 & 2 \\ 0 & 1 & -1/2 \\ 0 & 0 & 1 \end{bmatrix}
\begin{bmatrix} -1 & 1 & 0 \\ 1 & -1/2 & 0 \\ -4 & 1 & 2 \end{bmatrix}.
\tag{4.103}
$$

To finish, we use the third row of the left matrix to eliminate the contributions above and below it; we are left with

$$
\begin{bmatrix} 1 & 0 & 0 \\ 0 & 1 & 0 \\ 0 & 0 & 1 \end{bmatrix}
\begin{bmatrix} 7 & -1 & 4 \\ -1 & 0 & 1 \\ -4 & 1 & 2 \end{bmatrix}.
\tag{4.104}
$$

The matrix on the right hand side is now equal to $\mathbf{A}^{-1}$, which can be checked by a simple matrix multiplication. What, in fact, have we just done? In effect, the Gauss–Jordan method is equivalent to performing a series of matrix operations on the equation:

$$
\mathbf{A}\mathbf{A}^{-1} = \mathbf{I}.
\tag{4.105}
$$

Each step we performed was equivalent to a matrix operation $\mathbf{M}_i$ acting on both sides of this equation, leaving us after three steps with an equation of the form

$$(\mathbf{M}_3\mathbf{M}_2\mathbf{M}_1\mathbf{A})\mathbf{A}^{-1} = (\mathbf{M}_3\mathbf{M}_2\mathbf{M}_1\mathbf{I}). \tag{4.106}$$

When we have performed enough steps so that $\mathbf{M}_N \cdots \mathbf{M}_3\mathbf{M}_2\mathbf{M}_1\mathbf{A} = \mathbf{I}$, it follows that $\mathbf{A}^{-1} = \mathbf{M}_N \cdots \mathbf{M}_3\mathbf{M}_2\mathbf{M}_1\mathbf{I}$. The Gauss–Jordan method is a way of informally performing this matrix algebra.

4.4 Orthogonal matrices

Though much can be done using general matrix theory, certain classes of matrices have special properties which make them both physically relevant and mathematically useful. We now consider several of these classes, starting with so-called *orthogonal matrices*.

In Section 1.2, when we dealt with rotations of coordinate systems, we found that the matrix $\mathbf{A}$, with components a_{ij}, satisfied the following property, derived by looking at the effect of a rotation and its reverse,

$$\sum_{i=1}^{N} a_{ik} a_{ij} = \delta_{jk}, \tag{4.107}$$

where δ_{jk} is again the Kronecker delta (1.35). We will call a matrix which satisfies the condition (4.107) an *orthogonal matrix*. This looks very much like the multiplication of matrix $\mathbf{A}$ by itself,

$$(\mathbf{A}\mathbf{A})_{jk} = \sum_{i=1}^{N} a_{ji} a_{ik} = \sum_{i=1}^{N} a_{ik} a_{ji}, \tag{4.108}$$

except that the summation of Eq. (4.107) is over the *first* rows of each matrix, while the summation in multiplication is over the columns of one matrix and the rows of the other. We may, however, define a new matrix $\tilde{\mathbf{A}}$ such that the components satisfy the relation,

$$\tilde{a}_{ij} = a_{ji}. \tag{4.109}$$

This new matrix, directly related to $\mathbf{A}$, is called the *transpose* of $\mathbf{A}$, and it is created by interchanging the row and column of every element of $\mathbf{A}$; for instance,

$$\text{if} \quad \mathbf{A} = \begin{bmatrix} 1 & 2 & 3 \\ 4 & 5 & 6 \\ 7 & 8 & 9 \end{bmatrix}, \quad \text{then} \quad \tilde{\mathbf{A}} = \begin{bmatrix} 1 & 4 & 7 \\ 2 & 5 & 8 \\ 3 & 6 & 9 \end{bmatrix}. \tag{4.110}$$

It is to be noted that the diagonal elements of $\mathbf{A}$ and $\tilde{\mathbf{A}}$ are the same. For future reference, we also note that it follows that the transpose of a real-valued column vector is the corresponding row vector; i.e.

$$\text{for } |\mathbf{v}\rangle = \begin{bmatrix} 1 \\ 2 \\ 3 \end{bmatrix}, \quad |\tilde{\mathbf{v}}\rangle = \begin{bmatrix} 1 & 2 & 3 \end{bmatrix} = \langle\mathbf{v}|. \tag{4.111}$$

With this definition of $\tilde{\mathbf{A}}$, the orthogonality condition (4.107) may be written as

$$\sum_{i=1}^{N} \tilde{a}_{ji}a_{ik} = \delta_{jk}, \tag{4.112}$$

or in matrix form as

$$\tilde{\mathbf{A}}\mathbf{A} = \mathbf{A}\tilde{\mathbf{A}} = \mathbf{I}. \tag{4.113}$$

Since the inverse of a matrix satisfies the equation,

$$\mathbf{A}^{-1}\mathbf{A} = \mathbf{A}\mathbf{A}^{-1} = \mathbf{I}, \tag{4.114}$$

it follows that $\tilde{\mathbf{A}} = \mathbf{A}^{-1}$, or that *an orthogonal matrix is a matrix whose transpose is its inverse.*

As an example, we return to rotations in two dimensions, and consider the matrix for a rotation by an angle ϕ,

$$\mathbf{A}(\phi) = \begin{bmatrix} \cos\phi & \sin\phi \\ -\sin\phi & \cos\phi \end{bmatrix}. \tag{4.115}$$

The transpose of this matrix is

$$\tilde{\mathbf{A}}(\phi) = \begin{bmatrix} \cos\phi & -\sin\phi \\ \sin\phi & \cos\phi \end{bmatrix}, \tag{4.116}$$

which is equal to $\mathbf{A}(-\phi)$, i.e. to a rotation over an angle $-\phi$. Multiplying them together gives

$$\begin{bmatrix} \cos\phi & \sin\phi \\ -\sin\phi & \cos\phi \end{bmatrix} \begin{bmatrix} \cos\phi & -\sin\phi \\ \sin\phi & \cos\phi \end{bmatrix}$$

$$= \begin{bmatrix} \cos^2\phi + \sin^2\phi & -\sin\phi\cos\phi + \sin\phi\cos\phi \\ -\sin\phi\cos\phi + \sin\phi\cos\phi & \cos^2\phi + \sin^2\phi \end{bmatrix}$$

$$= \begin{bmatrix} 1 & 0 \\ 0 & 1 \end{bmatrix} = \mathbf{I}. \tag{4.117}$$

Therefore the rotation by $-\phi$, followed by a rotation by ϕ, gives no rotation at all. It is to be noted that we initially derived the orthogonality condition by assuming such a result, that

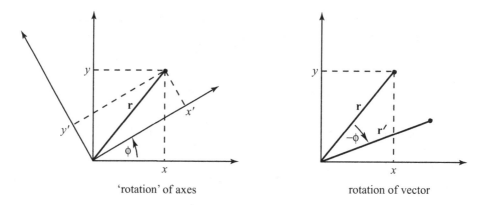

'rotation' of axes   rotation of vector

Figure 4.2 The double-duty of orthogonal matrices: rotation of coordinates or rotation of vector.

two opposite rotations must cancel. One can also derive this condition by the requirement that the length of a vector must be unchanged under coordinate rotations,

$$\sum_i x_i^2 = \sum_i x_i'^2 = \sum_i \left(\sum_j a_{ij}x_j\right)\left(\sum_k a_{ik}x_k\right) = \sum_{j,k} x_j x_k \sum_i a_{ij}a_{ik}. \qquad (4.118)$$

This relation will only be correct if the orthogonality condition is satisfied.

In fact, orthogonal matrices can be used to perform a double-duty of sorts: such a matrix may represent, as we have already seen, a rotation of the coordinate axes to a new set, but it may also represent the effect of rotating a vector **r** in the opposite direction, as illustrated in Fig. 4.2.

It is to be emphasized that the former case is a pseudo-rotation: nothing physical rotates, only the representation of the vector. In the latter case, we may use the rotation matrices to describe the new position of a vector which is rotated over a given angle in a *fixed* coordinate system.

An interesting result can be found if we consider mixing the two uses of the rotation matrices together. Let us interpret matrix **A** as rotating the vector $|\mathbf{r}\rangle$ to a new position $|\mathbf{r}_1\rangle = \mathbf{A}|\mathbf{r}\rangle$: the vector is now pointing in a different direction. Now let us consider a rotation of the coordinate system by a matrix **B**. The new position of the vector $|\mathbf{r}_1\rangle$ in the new coordinate system is given by

$$\left|\mathbf{r}_1'\right\rangle = \mathbf{B}\,|\mathbf{r}_1\rangle = \mathbf{BA}\,|\mathbf{r}\rangle. \qquad (4.119)$$

This expression mixes the two coordinate systems, as there is a primed vector on the left side, and an unprimed vector on the right. We would like to express both sides in terms of the primed coordinate system, i.e. we would like to determine the effect of **A** on vectors represented in the primed system. To do so, we note that $\mathbf{B}^{-1}\mathbf{B} = \mathbf{I}$, and insert this into our expression,

$$|\mathbf{r}_1'\rangle = \mathbf{BAI}\,|\mathbf{r}\rangle = \mathbf{BA}(\mathbf{B}^{-1}\mathbf{B})\,|\mathbf{r}\rangle = \mathbf{BAB}^{-1}\,|\mathbf{r}'\rangle, \qquad (4.120)$$

where we have used in the last step the relation $|\mathbf{r}'\rangle = \mathbf{B}\,|\mathbf{r}\rangle$.

Thus the matrix $\mathbf{A}'$ which performs the rotation of $|\mathbf{r}'\rangle$ into $|\mathbf{r}_1'\rangle$ in the new coordinate system can be written in the form

$$\mathbf{A}' = \mathbf{BAB}^{-1}. \qquad (4.121)$$

Thus not only the vectors $|\mathbf{r}\rangle$ but also the matrix $\mathbf{A}$ have their representations changed in the new coordinate system. Such a transformation of a matrix is known as a *similarity transformation*.

For the quantum-mechanical-minded: the two interpretations of the rotation matrix as (1) a rotation of system coordinates and (2) a rotation of the vectors of the system are analogous to the Heisenberg and Schrödinger pictures of quantum mechanics; see, for instance [Sha80, Section 4.3] and [CTDL77, Chapter 3].

4.5 Hermitian matrices and unitary matrices

In dealing with vectors and matrices up to this point we have assumed that their elements are strictly real-valued. In numerous important physical problems, however, vectors and matrices with complex elements are unavoidable; examples include quantum mechanics, in which the state of a particle is represented by a complex wavevector, and electromagnetics, in which a monochromatic wavefield is typically represented by a complex function.

In extending our discussion of matrices to those with complex elements, a number of new definitions are necessary. From these definitions come several classes of matrices which, like orthogonal matrices, have special significance. First, the definitions.

1. Conjugate. We define $\mathbf{A}^*$ as the complex conjugate of the matrix $\mathbf{A}$, such that the elements of $\mathbf{A}^*$ are simply the complex conjugates of the elements of $\mathbf{A}$, i.e. $a_{ij}^* \equiv (a_{ij})^*$. The complex conjugate is formed by replacing i by $-$i everywhere it appears.
2. Adjoint. We define the *adjoint* $\mathbf{A}^\dagger$ of $\mathbf{A}$ as the matrix formed by taking the transpose of $\mathbf{A}^*$, i.e.

$$\mathbf{A}^\dagger \equiv \widetilde{\mathbf{A}^*} = \tilde{\mathbf{A}}^*. \qquad (4.122)$$

The elements of $\mathbf{A}^\dagger$ are given by $a_{ij}^\dagger = (a_{ji})^*$.
We also redefine the row vector $\langle\mathbf{v}|$ to be the adjoint of the corresponding column vector, $|\mathbf{v}\rangle$, i.e. $|\mathbf{v}\rangle^\dagger = \langle\mathbf{v}|$. For real-valued vectors, this reduces to the familiar relationship $|\tilde{\mathbf{v}}\rangle = \langle\mathbf{v}|$. One important reason for doing this is that the inner product of a vector with itself becomes a real-valued quantity, even if its elements are complex, and we can therefore use this inner product as a metric (recall Section 1.5).
3. Hermitian matrix. A matrix $\mathbf{A}$ is called *Hermitian* (or self-adjoint) if it satisfies

$$\mathbf{A} = \mathbf{A}^\dagger. \qquad (4.123)$$

In other words, a matrix is Hermitian if it is equal to its adjoint. These matrices have special significance in quantum mechanics, where they are used to characterize observable quantities. Note that if a matrix is real and symmetric ($\tilde{\mathbf{A}} = \mathbf{A}$) then it is automatically Hermitian, i.e. Hermitian matrices are a generalization of real symmetric matrices.

4. Unitary matrix. A matrix $\mathbf{U}$ is called *unitary* if

$$\mathbf{U}^\dagger = \mathbf{U}^{-1}. \tag{4.124}$$

In other words, a matrix is unitary if its adjoint is its inverse. If $\mathbf{U}$ is unitary and real, then $\tilde{\mathbf{U}} = \mathbf{U}^{-1}$ and the matrix is orthogonal. Unitary matrices are a generalization of orthogonal matrices and can be considered a complex-valued generalization of rotations.

5. The adjoint of a product. An important property of adjoint matrices is that the adjoint of a product of matrices is equal to the reverse product of the adjoints of the matrices,

$$(\mathbf{AB})^\dagger = \mathbf{B}^\dagger \mathbf{A}^\dagger. \tag{4.125}$$

To prove this, we may return to the definition of the product of matrices and employ the adjoint property,

$$(ab)_{ij} = \sum_k a_{ik} b_{kj}. \tag{4.126}$$

The adjoint of this matrix is

$$(ab)_{ij}^\dagger = \left(\sum_k a_{ik} b_{kj} \right)^\dagger = \sum_k a_{jk}^* b_{ki}^* = \sum_k b_{ki}^* a_{jk}^* = \sum_k b_{ik}^\dagger a_{kj}^\dagger = (b^\dagger a^\dagger)_{ij}. \tag{4.127}$$

This is similar to a result previously derived for the inverse of a product of matrices, Eq. (4.62). This property may be expanded to the adjoint of any product of matrices,

$$(\mathbf{A}_1 \mathbf{A}_2 \cdots \mathbf{A}_N)^\dagger = \mathbf{A}_N^\dagger \mathbf{A}_{N-1}^\dagger \cdots \mathbf{A}_1^\dagger. \tag{4.128}$$

An important special case of this is the adjoint of the inner product,

$$(\langle \mathbf{u} | \mathbf{v} \rangle)^\dagger = |\mathbf{v}\rangle^\dagger \langle \mathbf{u}|^\dagger = \langle \mathbf{v} | \mathbf{u} \rangle. \tag{4.129}$$

We will have need of these results shortly.

6. Normal matrix. A *normal matrix* $\mathbf{A}$ is one which commutes with its Hermitian conjugate, i.e. one such that

$$\mathbf{A}\mathbf{A}^\dagger = \mathbf{A}^\dagger \mathbf{A}, \tag{4.130}$$

or $[\mathbf{A}, \mathbf{A}^\dagger] = 0$. Normal matrices include, as special cases, both Hermitian matrices and unitary matrices. They will play an important role in our discussion of diagonalization in the next section.

Example 4.3 (Pauli matrices) An important collection of matrices which are both Hermitian and unitary are the Pauli matrices, usually denoted by $\boldsymbol{\sigma}$,

$$\sigma_1 = \begin{bmatrix} 0 & 1 \\ 1 & 0 \end{bmatrix}, \quad \sigma_2 = \begin{bmatrix} 0 & -i \\ i & 0 \end{bmatrix}, \quad \sigma_3 = \begin{bmatrix} 1 & 0 \\ 0 & -1 \end{bmatrix}. \tag{4.131}$$

The Hermitian property can be demonstrated by direct calculation. For instance,

$$\sigma_2^\dagger = \tilde{\sigma}_2^* = \begin{bmatrix} 0 & i \\ -i & 0 \end{bmatrix}^* = \begin{bmatrix} 0 & -i \\ i & 0 \end{bmatrix} = \sigma_2. \tag{4.132}$$

Likewise, unitarity can be demonstrated, using the Hermiticity property for simplicity,

$$\sigma_2^\dagger \sigma_2 = \sigma_2 \sigma_2 = \begin{bmatrix} 0 & -i \\ i & 0 \end{bmatrix}\begin{bmatrix} 0 & -i \\ i & 0 \end{bmatrix} = \begin{bmatrix} 1 & 0 \\ 0 & 1 \end{bmatrix}. \tag{4.133}$$

The Pauli matrices' most familiar use is in quantum mechanics, in which the matrices are interpreted as the operators representing the components of a particle of spin $1/2$. If the quantum state of the spin particle is represented by the vector $|s\rangle$ (a column vector), the observable value s_2 of the y-component of the spin, for instance, is given by

$$s_2 = \langle s | \sigma_2 | s \rangle, \tag{4.134}$$

where $\langle s |$ is the row vector corresponding to $|s\rangle$. These matrices have a number of interesting mathematical properties, among them:

1. $\sigma_i \sigma_j + \sigma_j \sigma_i = 2\delta_{ij}\mathbf{I}$ (anticommutation);
2. $\sigma_i \sigma_j = i\epsilon_{ijk}\sigma_k$, where ϵ_{ijk} is the Levi-Civita tensor;
3. $(\sigma_i)^2 = \mathbf{I}$;
4. $\sigma_i \sigma_j - \sigma_j \sigma_i = 2i\epsilon_{ijk}\sigma_k$ (commutation);
5. along with the unit matrix $\mathbf{I}$, the three Pauli matrices form a complete set of 2×2 matrices; in other words, any 2×2 matrix $\mathbf{M}$ may be expanded as a (complex) sum of these matrices:

$$\mathbf{M} = m_0\mathbf{I} + m_1\sigma_1 + m_2\sigma_2 + m_3\sigma_3, \tag{4.135}$$

   where the m_i are complex numbers;
6. these same four matrices form a *group*, and may be used as generators of the so-called SU(2) group.

◇

4.6 Diagonalization of matrices, eigenvectors, and eigenvalues

We discussed in Section 4.2 some of the advantages of working with diagonal matrices, e.g. their commutativity and their simplicity. In this section we will demonstrate that a broad class of matrices can be converted, by a similarity transformation, into a diagonal form. For matrices which represent a real-valued quantity in three-dimensional space, this transformation could represent a rotation of the coordinate axes. Often much insight can be gained by changing the coordinate system of a problem to one in which matrices describing a physical quantity of interest are diagonal.

Example 4.4 Let us consider the angular momentum vector **L** of a complicated rotating object, which is given by the matrix equation[2]

$$\mathbf{L} = \mathcal{I}\,|\boldsymbol{\omega}\rangle, \tag{4.136}$$

where $|\boldsymbol{\omega}\rangle$ is the angular velocity vector and $\mathcal{I}$ is the moment of inertia matrix, whose components are defined by

$$\mathcal{I}_{ij} = \int \rho(\mathbf{r}) \left[\delta_{ij} \sum_k x_k^2 - x_i x_j \right] d\tau, \tag{4.137}$$

where $\rho(\mathbf{r})$ is the mass density of the rotating object and $\mathbf{r} = (x_1, x_2, x_3)$. It can be seen by an interchange of i and j that this matrix is real symmetric.

Suppose we have a specific moment of inertia matrix given by

$$\mathcal{I} = \begin{bmatrix} 11/24 & 0 & 7/24 \\ 0 & 18/24 & 0 \\ 7/24 & 0 & 11/24 \end{bmatrix}. \tag{4.138}$$

It is not immediately obvious what sort of object might have a moment of inertia of this form. Suppose we consider a change of coordinates by rotating the system by $\theta = -\pi/4$ about the y-axis; we saw from Eq. (4.121) that the form of the matrix in the new coordinate system is

$$\mathcal{I}' = \mathbf{R}_y \mathcal{I} \mathbf{R}_y^{-1}, \tag{4.139}$$

where the matrix $\mathbf{R}_y$ is given by

$$\mathbf{R}_y = \begin{bmatrix} \cos\theta & 0 & -\sin\theta \\ 0 & 1 & 0 \\ \sin\theta & 0 & \cos\theta \end{bmatrix} = \begin{bmatrix} \sqrt{2}/2 & 0 & \sqrt{2}/2 \\ 0 & 1 & 0 \\ -\sqrt{2}/2 & 0 & \sqrt{2}/2 \end{bmatrix}. \tag{4.140}$$

On substitution, we find that

$$\mathcal{I}' = \begin{bmatrix} \sqrt{2}/2 & 0 & \sqrt{2}/2 \\ 0 & 1 & 0 \\ -\sqrt{2}/2 & 0 & \sqrt{2}/2 \end{bmatrix} \begin{bmatrix} 11/24 & 0 & 7/24 \\ 0 & 18/24 & 0 \\ 7/24 & 0 & 11/24 \end{bmatrix} \begin{bmatrix} \sqrt{2}/2 & 0 & -\sqrt{2}/2 \\ 0 & 1 & 0 \\ \sqrt{2}/2 & 0 & \sqrt{2}/2 \end{bmatrix}$$

$$= \begin{bmatrix} 9/12 & 0 & 0 \\ 0 & 9/12 & 0 \\ 0 & 0 & 2/12 \end{bmatrix}. \tag{4.141}$$

In this coordinate system, it is clear that the moment of inertia is smaller when the object is rotating about the z-axis than when it is rotating about x or y; it is also clear that the moment

---

[2] A more optics-specific example would be the permittivity of an anisotropic optical material, but such an example does not have the intuitive appeal of mechanical rotation problems.

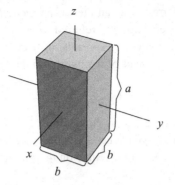

Figure 4.3 The shape of the object used in Example 4.4.

of inertia is the same rotating about x and y. In fact, the object described is a rectangular object of uniform density and of length $a = 2$ in the z direction, and $b = 1$ in the x and y directions, as illustrated in Fig. 4.3.

◇

We now determine a general procedure to diagonalize a matrix and, ideally, to determine the coordinate system in which the matrix is diagonal. To do this, let us assume that a matrix $\mathbf{A}$ can be converted into a diagonal form $\mathbf{A}'$ by a similarity transformation $\mathbf{R}$, such that

$$\mathbf{A}' = \mathbf{R}\mathbf{A}\mathbf{R}^{-1}. \tag{4.142}$$

We write this equation in a slightly different form by pre-multiplying it by the matrix $\mathbf{R}^{-1}$,

$$\mathbf{R}^{-1}\mathbf{A}' = \mathbf{A}\mathbf{R}^{-1}. \tag{4.143}$$

If we write the matrix $\mathbf{R}^{-1}$ as the collection of three column vectors $|\mathbf{v}_i\rangle$, $i = 1, 2, 3$, i.e. $\mathbf{R}^{-1} = (|\mathbf{v}_1\rangle \quad |\mathbf{v}_2\rangle \quad |\mathbf{v}_3\rangle)$, this equation becomes a collection of three vector–matrix equations,

$$\mathbf{A}|\mathbf{v}_i\rangle = A_i'|\mathbf{v}_i\rangle, \tag{4.144}$$

where A_i' are the diagonal elements of the diagonal matrix $\mathbf{A}'$. This is a homogeneous set of equations, which can be seen by moving the right-hand side of Eq. (4.144) to the left:

$$(\mathbf{A} - A_i'\mathbf{I})|\mathbf{v}_i\rangle = \begin{bmatrix} A_{11} - A_i' & A_{12} & A_{13} \\ A_{21} & A_{22} - A_i' & A_{23} \\ A_{31} & A_{32} & A_{33} - A_i' \end{bmatrix} \begin{bmatrix} v_{ix} \\ v_{iy} \\ v_{iz} \end{bmatrix} = 0, \tag{4.145}$$

where v_{ix} is the x-component of the vector $|\mathbf{v}_i\rangle$; likewise for y, z. For notational simplicity, we replace the number A_i' by λ and the vector $|\mathbf{v}_i\rangle$ by $|\mathbf{v}\rangle$. The matrix–vector equation becomes

$$(\mathbf{A} - \lambda\mathbf{I})|\mathbf{v}\rangle = 0. \tag{4.146}$$

This equation is usually referred to as an *eigenvalue equation*, and a value of λ and a vector $|\mathbf{v}\rangle$ which satisfy this equation are referred to as an *eigenvalue* and *eigenvector*, respectively. From the discussion of determinants in Section 4.3, we know this equation only has a solution if the determinant of the matrix on the left-hand side of the equation vanishes:

$$|\mathbf{A} - \lambda\mathbf{I}| = 0. \tag{4.147}$$

For a 3×3 matrix, this results in a cubic equation for the values of λ (called the *secular equation*), which implies three eigenvalues λ_i, with $i = 1, 2, 3$. Once we know the eigenvalues, we may input them back into the equation (4.146) and solve this equation for the corresponding eigenvector $|\mathbf{v}_i\rangle$.

It can be shown that eigenvalues and eigenvectors exist for *any* square matrix. However, *not every square matrix can be diagonalized*. It can be shown that diagonalization is only possible if the eigenvectors of the matrix $\mathbf{A}$ are linearly independent, which will in general only be true if the eigenvalues of the matrix $\mathbf{A}$ are distinct.

Fortunately, we will see that most matrices of broad physical interest, namely Hermitian matrices (and their real symmetric counterparts) and normal matrices, have eigenvectors which can be made orthonormal to one another; we will prove this explicitly in Section 4.6.1. A variety of similarity matrices $\mathbf{R}$ can be constructed from orthonormal vectors by using different arrangements of the eigenvectors $|\mathbf{v}_i\rangle$. For instance, with a set of three orthonormal eigenvectors $|\mathbf{v}_i\rangle$, $i = 1, 2, 3$, the similarity matrix $\mathbf{R}^{-1}$ may be written as $(|\mathbf{v}_1\rangle \quad |\mathbf{v}_2\rangle \quad |\mathbf{v}_3\rangle)$, or $(|\mathbf{v}_2\rangle \quad |\mathbf{v}_1\rangle \quad |\mathbf{v}_3\rangle)$, and so on. The elements of the resulting diagonal matrix $\mathbf{A}'$ are simply the eigenvectors λ_i.

Example 4.5 We consider the diagonalization of the matrix

$$\mathbf{A} = \begin{bmatrix} 1 & 0 & 0 \\ 0 & 1 & 1 \\ 0 & 1 & 1 \end{bmatrix}. \tag{4.148}$$

We construct the secular equation by taking the determinant of the matrix given by Eq. (4.147):

$$|\mathbf{A} - \lambda\mathbf{I}| = \begin{vmatrix} 1-\lambda & 0 & 0 \\ 0 & 1-\lambda & 1 \\ 0 & 1 & 1-\lambda \end{vmatrix} = (1-\lambda)[(1-\lambda)^2 - 1] = 0. \tag{4.149}$$

One obvious root of this cubic equation is $\lambda = 1$ (the first term in parenthesis). The other roots can be found by solution of the quadratic equation,

$$(1-\lambda)^2 - 1 = 0. \tag{4.150}$$

The roots of this equation are $\lambda = 0$ and $\lambda = 2$; our complete set of eigenvalues is thus $\lambda = 0, 1, 2$.

Now we need to find the eigenvectors corresponding to each of the eigenvalues. Starting with $\lambda = 0$, we have the matrix equation,

$$\begin{bmatrix} 1-0 & 0 & 0 \\ 0 & 1-0 & 1 \\ 0 & 1 & 1-0 \end{bmatrix} \begin{bmatrix} v_{1x} \\ v_{1y} \\ v_{1z} \end{bmatrix} = \begin{bmatrix} 1 & 0 & 0 \\ 0 & 1 & 1 \\ 0 & 1 & 1 \end{bmatrix} \begin{bmatrix} v_{1x} \\ v_{1y} \\ v_{1z} \end{bmatrix} = 0. \tag{4.151}$$

We can solve this set of equations for v_{1x}, v_{1y}, v_{1z}; we find that

$$v_{1x} = 0, \quad v_{1y} = -v_{1z}. \tag{4.152}$$

We can normalize this choice and make it a unit vector by requiring

$$\langle \mathbf{v}_1 | \mathbf{v}_1 \rangle = 1 = 2v_{1y}^2. \tag{4.153}$$

This results in the vector

$$|\mathbf{v}_1\rangle = \begin{bmatrix} 0 \\ 1/\sqrt{2} \\ -1/\sqrt{2} \end{bmatrix}. \tag{4.154}$$

Similarly, for $\lambda = 1$,

$$\begin{bmatrix} 1-1 & 0 & 0 \\ 0 & 1-1 & 1 \\ 0 & 1 & 1-1 \end{bmatrix} \begin{bmatrix} v_{2x} \\ v_{2y} \\ v_{2z} \end{bmatrix} = \begin{bmatrix} 0 & 0 & 0 \\ 0 & 0 & 1 \\ 0 & 1 & 0 \end{bmatrix} \begin{bmatrix} v_{2x} \\ v_{2y} \\ v_{2z} \end{bmatrix} = 0. \tag{4.155}$$

We thus have that $v_{2y} = v_{2z} = 0$ and v_{2x} is unspecified; requiring $\langle \mathbf{v}_2 | \mathbf{v}_2 \rangle = 1$, we have

$$|\mathbf{v}_2\rangle = \begin{bmatrix} 1 \\ 0 \\ 0 \end{bmatrix}. \tag{4.156}$$

Finally, we have $\lambda = 2$; for this we have the matrix equation

$$\begin{bmatrix} 1-2 & 0 & 0 \\ 0 & 1-2 & 1 \\ 0 & 1 & 1-2 \end{bmatrix} \begin{bmatrix} v_{3x} \\ v_{3y} \\ v_{3z} \end{bmatrix} = \begin{bmatrix} -1 & 0 & 0 \\ 0 & -1 & 1 \\ 0 & 1 & -1 \end{bmatrix} \begin{bmatrix} v_{3x} \\ v_{3y} \\ v_{3z} \end{bmatrix} = 0. \tag{4.157}$$

We have $v_{3x} = 0$, $v_{3y} = v_{3z}$. The normalized vector is

$$|\mathbf{v}_3\rangle = \begin{bmatrix} 0 \\ 1/\sqrt{2} \\ 1/\sqrt{2} \end{bmatrix}. \tag{4.158}$$

These three vectors are orthonormal, i.e. $\langle \mathbf{v}_i | \mathbf{v}_j \rangle = \delta_{ij}$, which can be readily verified. Furthermore, the matrix $\mathbf{A}$ is real symmetric, so we expect that we can diagonalize the matrix by constructing an orthogonal matrix from the eigenvectors. We choose the arrangement given by

$$\mathbf{R}^{-1} = \begin{bmatrix} |\mathbf{v}_3\rangle & |\mathbf{v}_1\rangle & |\mathbf{v}_2\rangle \end{bmatrix} = \begin{bmatrix} 1 & 0 & 0 \\ 0 & 1/\sqrt{2} & 1/\sqrt{2} \\ 0 & -1/\sqrt{2} & 1/\sqrt{2} \end{bmatrix}. \tag{4.159}$$

The diagonal form of the matrix $\mathbf{A}$ is therefore given by

$$\mathbf{A}' = \begin{bmatrix} 1 & 0 & 0 \\ 0 & 0 & 0 \\ 0 & 0 & 2 \end{bmatrix}. \tag{4.160}$$

◇

4.6.1 Hermitian matrices, normal matrices, and eigenvalues

It is to be noted that in the example of Section 4.5, the three eigenvectors ($|\mathbf{v}_1\rangle$, $|\mathbf{v}_2\rangle$, $|\mathbf{v}_3\rangle$) are orthogonal – that is, $\langle \mathbf{v}_i | \mathbf{v}_j \rangle = \delta_{ij}$. We have stated that this follows from the fact that the matrix is real and symmetric. Such a result can be generalized to Hermitian matrices, of which the real symmetric matrix is a special case.

Theorem 4.1 *The eigenvalues of a Hermitian matrix are real-valued, and the eigenvectors corresponding to distinct eigenvalues are orthogonal.*

To prove this, we start with the eigenvalue equation for several distinct eigenvalues and eigenvectors such that $\lambda_i \neq \lambda_j$ for $i \neq j$,

$$\begin{aligned} \mathbf{A}|\mathbf{v}_i\rangle &= \lambda_i|\mathbf{v}_i\rangle, \\ \mathbf{A}|\mathbf{v}_j\rangle &= \lambda_j|\mathbf{v}_j\rangle. \end{aligned} \tag{4.161}$$

Let us multiply the first of these equations by the corresponding row vector of the second equation,

$$\langle \mathbf{v}_j |\mathbf{A}|\mathbf{v}_i\rangle = \lambda_i \langle \mathbf{v}_j |\mathbf{v}_i\rangle, \tag{4.162}$$

and the second of the equations by the corresponding row vector of the first,

$$\langle \mathbf{v}_i |\mathbf{A}|\mathbf{v}_j\rangle = \lambda_j \langle \mathbf{v}_i |\mathbf{v}_j\rangle. \tag{4.163}$$

We then take the adjoint of this latter equation,

$$\left(\langle \mathbf{v}_i |\mathbf{A}|\mathbf{v}_j\rangle\right)^\dagger = \langle \mathbf{v}_j |\mathbf{A}^\dagger|\mathbf{v}_i\rangle = \left(\lambda_j \langle \mathbf{v}_i |\mathbf{v}_j\rangle\right)^\dagger = \lambda_j^* \langle \mathbf{v}_j |\mathbf{v}_i\rangle. \tag{4.164}$$

Because $\mathbf{A}$ is Hermitian, it follows that

$$\langle \mathbf{v}_j |\mathbf{A}|\mathbf{v}_i\rangle = \lambda_j^* \langle \mathbf{v}_j |\mathbf{v}_i\rangle. \tag{4.165}$$

Subtracting Eq. (4.165) from Eq. (4.162), we have the result that

$$\left(\lambda_i - \lambda_j^*\right)\langle \mathbf{v}_j |\mathbf{v}_i\rangle = 0. \tag{4.166}$$

If we first consider $i = j$, it follows that if $\langle v_i | v_i \rangle \neq 0$,

$$\lambda_i = \lambda_i^*, \tag{4.167}$$

or that the eigenvalues are real. For $i \neq j$, we return to Eq. (4.166), and note that if $\lambda_i \neq \lambda_j$, it follows that

$$\langle v_j | v_i \rangle = 0, \tag{4.168}$$

or that the eigenvectors are orthogonal.

Therefore the eigenvalues of a Hermitian (or real symmetric) matrix are all real-valued, and the eigenvectors of a Hermitian (or real symmetric) matrix are all orthogonal.

What happens if a pair (or more) of the eigenvalues are the same (in this situation, the eigenvalues are called *degenerate*)? Then the preceding argument about orthogonality does not hold. If we solve for the eigenvectors with respect to the degenerate eigenvalue, we find that they are not completely specified; in fact, the equations which result only specify that the degenerate eigenvectors must be orthogonal to all of the nondegenerate eigenvectors. We thus have extra freedom to choose the degenerate eigenvectors, and may readily choose them to be orthonormal.

In quantum mechanics, matrices (and operators) which represent physically observable quantities are always Hermitian. The real-valuedness of the eigenvalues can be given a physical interpretation: physically observable quantities are always real-valued.

Example 4.6 We consider the matrix

$$\begin{bmatrix} 1 & 0 & 0 \\ 0 & 0 & 1 \\ 0 & 1 & 0 \end{bmatrix}. \tag{4.169}$$

The secular equation for this matrix is given by

$$(1 - \lambda)(\lambda^2 - 1) = 0, \tag{4.170}$$

or $\lambda = -1, 1, 1$; the eigenvalue $\lambda = 1$ is therefore degenerate. Solving for a suitably normalized eigenvector corresponding to $\lambda = -1$, we have

$$|v_1\rangle = \begin{bmatrix} 0 \\ 1/\sqrt{2} \\ -1/\sqrt{2} \end{bmatrix}. \tag{4.171}$$

For $\lambda = 1$ the solution for the eigenvector results only in the single equation

$$-v_y + v_z = 0, \tag{4.172}$$

which is not enough to uniquely specify a vector, let alone two! We have an infinite number of choices for the pair of degenerate eigenvectors; one choice is to let $v_y = v_z$ and $v_x = 0$:

$$|\mathbf{v}_2\rangle = \begin{bmatrix} 0 \\ 1/\sqrt{2} \\ 1/\sqrt{2} \end{bmatrix}, \tag{4.173}$$

which can be seen to be orthogonal to $|\mathbf{v}_1\rangle$. Another choice can be found by using the curl of vectors $\mathbf{v}_1$ and $\mathbf{v}_2$, $\mathbf{v}_3 \propto \mathbf{v}_1 \times \mathbf{v}_2$, or we may simply select $v_y = v_z = 0$, which amounts to the same thing,

$$|\mathbf{v}_3\rangle = \begin{bmatrix} 1 \\ 0 \\ 0 \end{bmatrix}. \tag{4.174}$$

We may thus always find eigenvectors for a Hermitian matrix which are orthonormal, regardless of whether there exist degenerate eigenvalues or not.

◇

We may also describe the properties of normal matrices quite generally, as stated in the following theorem.

Theorem 4.2 *The eigenvalues of a normal matrix $\mathbf{A}$ are the complex conjugates of the eigenvalues of $\mathbf{A}^\dagger$, and the eigenvectors of $\mathbf{A}$ and $\mathbf{A}^\dagger$ are the same. The eigenvectors corresponding to distinct eigenvalues are orthogonal.*

Let us consider the secular equation of a normal matrix $\mathbf{A}$, i.e.

$$(\mathbf{A} - \lambda_i \mathbf{I}) |\mathbf{v}_i\rangle = 0. \tag{4.175}$$

The Hermitian conjugate of this equation is given by

$$\langle \mathbf{v}_i| (\mathbf{A}^\dagger - \lambda_i^* \mathbf{I}) = 0. \tag{4.176}$$

Considering the product of Eqs. (4.175) and (4.176), we have

$$\langle \mathbf{v}_i| (\mathbf{A}^\dagger - \lambda_i^* \mathbf{I})(\mathbf{A} - \lambda_i \mathbf{I}) |\mathbf{v}_i\rangle = 0. \tag{4.177}$$

The product between brackets may be expanded to give,

$$\langle \mathbf{v}_i| (\mathbf{A}^\dagger \mathbf{A} - \lambda_i^* \mathbf{A} - \lambda_i \mathbf{A}^\dagger - |\lambda_i|^2) |\mathbf{v}_i\rangle = 0. \tag{4.178}$$

Because $\mathbf{A}$ is a normal matrix, however, we may reverse the order of the product in the first term, $\mathbf{A}^\dagger \mathbf{A} = \mathbf{A}\mathbf{A}^\dagger$. With this change, the above equation may be rewritten as

$$\langle \mathbf{v}_i| (\mathbf{A} - \lambda_i \mathbf{I})(\mathbf{A}^\dagger - \lambda_i^* \mathbf{I}) |\mathbf{v}_i\rangle = 0. \tag{4.179}$$

This in turn may be written as

$$[(\mathbf{A}^\dagger - \lambda_i^* \mathbf{I}) \, |\mathbf{v}_i\rangle]^\dagger (\mathbf{A}^\dagger - \lambda_i^* \mathbf{I}) \, |\mathbf{v}_i\rangle = 0. \tag{4.180}$$

This equation will only be satisfied if the secular equation for $\mathbf{A}^\dagger$ satisfies

$$(\mathbf{A}^\dagger - \lambda_i^* \mathbf{I}) \, |\mathbf{v}_i\rangle = 0. \tag{4.181}$$

We therefore find that the eigenvectors of $\mathbf{A}^\dagger$ for a normal matrix are the same as for $\mathbf{A}$, and the eigenvalues are the complex conjugates of the eigenvalues of $\mathbf{A}$.

To demonstrate orthogonality, we consider two eigenvectors with two distinct eigenvalues, which satisfy

$$\begin{aligned}
\mathbf{A} \, |\mathbf{v}_i\rangle &= \lambda_i \, |\mathbf{v}_i\rangle , \\
\mathbf{A} \, |\mathbf{v}_j\rangle &= \lambda_j \, |\mathbf{v}_j\rangle .
\end{aligned} \tag{4.182}$$

We multiply the first equation by $\langle \mathbf{v}_j |$,

$$\langle \mathbf{v}_j | \, \mathbf{A} \, |\mathbf{v}_i\rangle = \lambda_i \langle \mathbf{v}_j | \, \mathbf{v}_i\rangle . \tag{4.183}$$

We may write the left-hand side of this equation as

$$(\mathbf{A}^\dagger \, |\mathbf{v}_j\rangle)^\dagger \, |\mathbf{v}_i\rangle = \lambda_i \langle \mathbf{v}_j | \, \mathbf{v}_i\rangle . \tag{4.184}$$

However, we have seen that the eigenvalue of $\mathbf{A}^\dagger$ is λ_j^*, so we may write

$$(\mathbf{A}^\dagger \, |\mathbf{v}_j\rangle)^\dagger = (\lambda_j^* \, |\mathbf{v}_j\rangle)^\dagger = \langle \mathbf{v}_j | \lambda_j. \tag{4.185}$$

On substitution into Eq. (4.184), we find that

$$(\lambda_j - \lambda_i) \langle \mathbf{v}_j | \, \mathbf{v}_i\rangle = 0. \tag{4.186}$$

Because the eigenvalues were assumed to be distinct, the eigenvectors must be orthogonal.

For non-normal matrices, much less can be said about the eigensystem. The eigenvectors are not necessarily orthogonal, though they are typically linearly independent if the eigenvalues are distinct. In such a case, the matrix can be diagonalized, though the diagonal elements of the matrix are no longer the eigenvalues. If a non-normal matrix has degenerate eigenvalues, a linearly independent set of eigenvectors cannot be found and the matrix cannot be diagonalized. Fortunately, most physical problems involve matrices which are at least normal, if not Hermitian.

4.7 Gram–Schmidt orthonormalization

When a Hermitian matrix has degenerate eigenvalues, the corresponding eigenvectors are not automatically orthogonal. It is helpful to have a process by which an orthonormal set of vectors can be constructed out of a given linearly independent set; one such process is the so-called *Gram–Schmidt orthonormalization*.

Let us suppose we have a set of N linearly independent vectors $\{|\mathbf{v}_1\rangle,\ldots,|\mathbf{v}_N\rangle\}$, and we wish to construct an orthogonal set $\{|\mathbf{u}_1\rangle,\ldots,|\mathbf{u}_N\rangle\}$. Starting with vector $|\mathbf{v}_1\rangle$, we let $|\mathbf{u}_1\rangle = |\mathbf{v}_1\rangle$ and seek a vector $|\mathbf{u}_2\rangle$ of the form,

$$|\mathbf{u}_2\rangle = |\mathbf{v}_2\rangle + c\,|\mathbf{v}_1\rangle, \tag{4.187}$$

where c is a constant to be determined. Requiring $|\mathbf{u}_2\rangle$ to be orthogonal to $|\mathbf{u}_1\rangle$, we have

$$\langle \mathbf{u}_1|\,\mathbf{u}_2\rangle = \langle \mathbf{v}_1|\,\mathbf{v}_2\rangle + c\,\langle \mathbf{v}_1|\,\mathbf{v}_1\rangle = 0. \tag{4.188}$$

Solving for c, we have

$$c = -\frac{\langle \mathbf{v}_1|\,\mathbf{v}_2\rangle}{\langle \mathbf{v}_1|\,\mathbf{v}_1\rangle}, \tag{4.189}$$

or that

$$|\mathbf{u}_2\rangle = |\mathbf{v}_2\rangle - \frac{\langle \mathbf{v}_1|\,\mathbf{v}_2\rangle}{\langle \mathbf{v}_1|\,\mathbf{v}_1\rangle}\,|\mathbf{v}_1\rangle. \tag{4.190}$$

This equation has a simple geometric interpretation: to construct a vector orthogonal to $|\mathbf{v}_1\rangle$, we subtract that component of $|\mathbf{v}_2\rangle$ which points in the direction of $|\mathbf{v}_1\rangle$; this is illustrated in Fig. 4.4.

To construct an third orthogonal vector, we start with $|\mathbf{v}_3\rangle$ and subtract off those components which point in the direction of vectors $|\mathbf{u}_1\rangle$ and $|\mathbf{u}_2\rangle$. This process may be continued

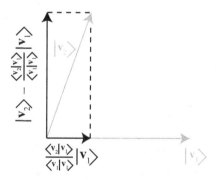

Figure 4.4 Illustration of the Gram–Schmidt process for two vectors.

for N vectors, in the form

$$|\mathbf{u}_1\rangle = |\mathbf{v}_1\rangle, \tag{4.191}$$

$$|\mathbf{u}_2\rangle = |\mathbf{v}_2\rangle - \frac{\langle\mathbf{u}_1|\,\mathbf{v}_2\rangle}{\langle\mathbf{u}_1|\,\mathbf{u}_1\rangle}\,|\mathbf{u}_1\rangle, \tag{4.192}$$

$$|\mathbf{u}_3\rangle = |\mathbf{v}_3\rangle - \frac{\langle\mathbf{u}_1|\,\mathbf{v}_3\rangle}{\langle\mathbf{u}_1|\,\mathbf{u}_1\rangle}\,|\mathbf{u}_1\rangle - \frac{\langle\mathbf{u}_2|\,\mathbf{v}_3\rangle}{\langle\mathbf{u}_2|\,\mathbf{u}_2\rangle}\,|\mathbf{u}_2\rangle, \tag{4.193}$$

$$\cdots$$

$$|\mathbf{u}_N\rangle = |\mathbf{v}_N\rangle - \sum_{n=1}^{N-1} \frac{\langle\mathbf{u}_n|\,\mathbf{v}_N\rangle}{\langle\mathbf{u}_n|\,\mathbf{u}_n\rangle}\,|\mathbf{u}_n\rangle. \tag{4.194}$$

A set of orthonormal vectors $\{|\mathbf{e}_i\rangle\}$ may be derived by simply normalizing the set $\{|\mathbf{u}_i\rangle\}$, i.e.

$$|\mathbf{e}_i\rangle \equiv \frac{|\mathbf{u}_i\rangle}{\sqrt{\langle\mathbf{u}_i|\,\mathbf{u}_i\rangle}}. \tag{4.195}$$

Example 4.7 We consider the set of vectors

$$|\mathbf{v}_1\rangle = \begin{bmatrix} 1 \\ 1 \\ 0 \end{bmatrix}, \quad |\mathbf{v}_2\rangle = \begin{bmatrix} 1 \\ 0 \\ 1 \end{bmatrix}, \quad |\mathbf{v}_3\rangle = \begin{bmatrix} 0 \\ 1 \\ 1 \end{bmatrix}. \tag{4.196}$$

We let $|\mathbf{u}_1\rangle = |\mathbf{v}_1\rangle$, and we may find $|\mathbf{u}_2\rangle$ from Eq. (4.192),

$$|\mathbf{u}_2\rangle = \begin{bmatrix} 1 \\ 0 \\ 1 \end{bmatrix} - \frac{1}{2}\begin{bmatrix} 1 \\ 1 \\ 0 \end{bmatrix} = \begin{bmatrix} 1/2 \\ -1/2 \\ 1 \end{bmatrix}. \tag{4.197}$$

We may find $|\mathbf{u}_3\rangle$ from Eq. (4.194), in the form

$$|\mathbf{u}_3\rangle = \begin{bmatrix} 0 \\ 1 \\ 1 \end{bmatrix} - \frac{1}{2}\begin{bmatrix} 1 \\ 1 \\ 0 \end{bmatrix} - \frac{1}{3}\begin{bmatrix} 1/2 \\ -1/2 \\ 1 \end{bmatrix} = \begin{bmatrix} -2/3 \\ 2/3 \\ 2/3 \end{bmatrix}. \tag{4.198}$$

It can be shown by direct calculation that these vectors are orthogonal.

◇

It is to be noted that a variety of different orthogonal sets may be constructed from any starting set of linearly independent vectors, depending on the order in which the vectors are orthogonalized.

The standard Gram–Schmidt technique is not entirely suitable for computational work, as round-off errors can result in the final set of vectors being very far from orthogonal. A much more stable modified version breaks up the calculation of the jth orthogonal vector into a series of $j - 1$ successive approximations, i.e.

$$\left|\mathbf{u}_j^{(1)}\right\rangle = \left|\mathbf{v}_j\right\rangle - \frac{\langle\mathbf{u}_1|\mathbf{v}_j\rangle}{\langle\mathbf{u}_1|\mathbf{u}_1\rangle}\left|\mathbf{u}_1\right\rangle, \tag{4.199}$$

$$\left|\mathbf{u}_j^{(2)}\right\rangle = \left|\mathbf{u}_j^{(1)}\right\rangle - \frac{\langle\mathbf{u}_2|\mathbf{u}_j^{(1)}\rangle}{\langle\mathbf{u}_2|\mathbf{u}_2\rangle}\left|\mathbf{u}_2\right\rangle, \tag{4.200}$$

$$\cdots$$

$$\left|\mathbf{u}_j^{(j-2)}\right\rangle = \left|\mathbf{u}_j^{(j-3)}\right\rangle - \frac{\langle\mathbf{u}_{j-2}|\mathbf{u}_j^{(j-3)}\rangle}{\langle\mathbf{u}_{j-2}|\mathbf{u}_{j-2}\rangle}\left|\mathbf{u}_{j-2}\right\rangle, \tag{4.201}$$

$$\left|\mathbf{u}_j\right\rangle = \left|\mathbf{u}_j^{(j-2)}\right\rangle - \frac{\langle\mathbf{u}_{j-1}|\mathbf{u}_j^{(j-2)}\rangle}{\langle\mathbf{u}_{j-1}|\mathbf{u}_{j-1}\rangle}\left|\mathbf{u}_{j-1}\right\rangle. \tag{4.202}$$

Further details can be found in [GL96, Chapter 5].

4.8 Orthonormal vectors and basis vectors

We have noted previously, in Section 1.5, that a collection of vectors forms a basis for the vector space if they are all linearly independent and span the space. If we form a square matrix from the collection of vectors, they form a basis if the determinant of the matrix is non zero.

If we have an orthonormal set of basis vectors for a finite-dimensional vector space, we may make one more relevant construction: we may construct the identity matrix out of the basis vectors by use of the *direct product* (or *outer product*). Suppose we have an N-dimensional vector space with orthonormal basis vectors $|\mathbf{n}\rangle$; it can be shown that the identity matrix is equal to

$$\mathbf{I} = \sum_{n=1}^{N} |\mathbf{n}\rangle\langle\mathbf{n}|, \tag{4.203}$$

where $|\mathbf{n}\rangle\langle\mathbf{n}|$ represents the direct product of the two vectors. An expression of the form of Eq. (4.203) is generally referred to as a *closure relation*.

We can prove that Eq. (4.203) is true by noting, if $\{|\mathbf{n}\rangle\}$ form a basis of the N-dimensional space, that an arbitrary vector $|\mathbf{q}\rangle$ may be written as

$$|\mathbf{q}\rangle = \sum_{n=1}^{N} q_n |\mathbf{n}\rangle. \tag{4.204}$$

On multiplying this equation by Eq. (4.203), we find that

$$\mathbf{I}|\mathbf{q}\rangle = \left(\sum_{n=1}^{N} |\mathbf{n}\rangle\langle\mathbf{n}|\right)\left(\sum_{n'=1}^{N} q'_n |\mathbf{n}'\rangle\right). \tag{4.205}$$

By using $\langle \mathbf{n} | \mathbf{n}' \rangle = \delta_{nn'}$, we readily find that

$$\mathbf{I} | \mathbf{q} \rangle = | \mathbf{q} \rangle, \tag{4.206}$$

which is the first part of the definition of the identity matrix, Eq. (4.59); a similar argument can be made using row vectors $\langle \mathbf{q} |$.

Example 4.8 We consider a set of orthonormal basis vectors for a four-dimensional space, defined as follows,

$$|1\rangle = \frac{1}{2} \begin{bmatrix} 1 \\ 1 \\ 1 \\ 1 \end{bmatrix}, \quad |2\rangle = \frac{1}{\sqrt{6}} \begin{bmatrix} -1 \\ -1 \\ 0 \\ 2 \end{bmatrix}, \quad |3\rangle = \frac{1}{\sqrt{2}} \begin{bmatrix} -1 \\ 1 \\ 0 \\ 0 \end{bmatrix}, \quad |4\rangle = \frac{1}{\sqrt{12}} \begin{bmatrix} 1 \\ 1 \\ -3 \\ 1 \end{bmatrix}. \tag{4.207}$$

We can immediately show that this set of vectors is linearly independent by calculating the determinant of a matrix constructed from them,

$$\det(|1\rangle, |2\rangle, |3\rangle, |4\rangle) = \begin{vmatrix} 1/2 & -1/\sqrt{6} & -1/\sqrt{2} & 1/\sqrt{12} \\ 1/2 & -1/\sqrt{6} & 1/\sqrt{2} & 1/\sqrt{12} \\ 1/2 & 0 & 0 & -3/\sqrt{12} \\ 1/2 & 2/\sqrt{6} & 0 & 1/\sqrt{12} \end{vmatrix} = -1. \tag{4.208}$$

Orthogonality of the vectors can be shown by taking the various dot products between the vectors; for instance,

$$\langle 1 | 2 \rangle = \frac{1}{\sqrt{12}} \begin{bmatrix} 1 & 1 & 1 & 1 \end{bmatrix} \begin{bmatrix} -1 \\ -1 \\ 0 \\ 2 \end{bmatrix} = 0. \tag{4.209}$$

If we attempt to construct an identity matrix from the vectors, we find that

$$\mathbf{I} = |1\rangle \langle 1| + |2\rangle \langle 2| + |3\rangle \langle 3| + |4\rangle \langle 4|$$

$$= \frac{1}{4} \begin{bmatrix} 1 & 1 & 1 & 1 \\ 1 & 1 & 1 & 1 \\ 1 & 1 & 1 & 1 \\ 1 & 1 & 1 & 1 \end{bmatrix} + \frac{1}{6} \begin{bmatrix} 1 & 1 & 0 & -2 \\ 1 & 1 & 0 & -2 \\ 0 & 0 & 0 & 0 \\ -2 & -2 & 0 & 4 \end{bmatrix}$$

$$+\frac{1}{2}\begin{bmatrix} 1 & -1 & 0 & 0 \\ -1 & 1 & 0 & 0 \\ 0 & 0 & 0 & 0 \\ 0 & 0 & 0 & 0 \end{bmatrix} + \frac{1}{12}\begin{bmatrix} 1 & 1 & -3 & 1 \\ 1 & 1 & -3 & 1 \\ -3 & -3 & 9 & -3 \\ 1 & 1 & -3 & 1 \end{bmatrix}$$

$$= \begin{bmatrix} 1 & 0 & 0 & 0 \\ 0 & 1 & 0 & 0 \\ 0 & 0 & 1 & 0 \\ 0 & 0 & 0 & 1 \end{bmatrix}. \tag{4.210}$$

◇

Why is this helpful? For finite-dimensional vector spaces, this construction is not essential: it is almost obvious, from a given set of vectors, whether they form an orthonormal basis or not. When we discuss infinite-dimensional vector spaces [where a "vector" is defined as a function $f(x)$], we will be interested in finding a set of basis *functions* which can be used to represent any arbitrary function. We can test a set of functions to see if they form a basis by determining whether or not they can be used to construct the Dirac delta function $\delta(x-x')$, which represents, in essence, the identity matrix in an infinite-dimensional vector space. We will see that the sines and cosines of the Fourier series (Chapter 8) and the complex exponentials of the Fourier transform (Chapter 11) form a complete set of basis functions over their respective domains.

4.9 Functions of matrices

Occasionally in dealing with matrices we encounter *functions* of matrices; for instance, if $\mathbf{A}$ is a matrix, we may see $\exp(\mathbf{A})$, or $\sin(\mathbf{A})$. Such formulas can be made meaningful by using the Taylor series expansion of the functions, to be discussed in Section 7.5. For $\exp(\mathbf{A})$, for instance, this means

$$\exp(\mathbf{A}) = \sum_{n=0}^{\infty} \frac{1}{n!}\mathbf{A}^n. \tag{4.211}$$

In fact, such a Taylor series representation is the way we *define* functions of matrices, because we may evaluate any finite product of matrices.

Such functions of matrices occur quite often in quantum mechanical theory; they also play a significant role in group theory. One might consider this a final generalization of ordinary numbers to matrices – just as one can have functions of ordinary numbers, one can also have functions of matrices.

4.10 Focus: matrix methods for geometrical optics

4.10.1 Introduction

In the design of optical imaging systems and laser resonators, the propagation of light is typically described in terms of rays. Light is assumed to follow straight line trajectories

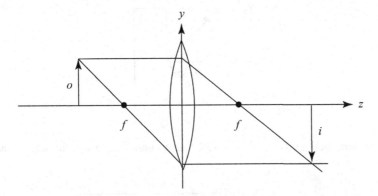

Figure 4.5 The two rays which can be readily used to determine image formation in a thin lens. It should be noted that the figure does not accurately portray refraction inside the lens.

through space, only deviating in direction upon reflection and refraction. As an example of this paradigm of light propagation, we consider image formation by a thin lens, illustrated in Fig. 4.5. The image can be found by tracing the paths of two rays through the system: the ray which is parallel to the z-axis will be refracted to pass through the rear focal point of the lens, while the ray which passes through the front focal point will be refracted to be parallel to the z-axis.

The focal point can be related to the refractive index of the lens and the radii of curvature of the front surface, r_1, and the rear surface, r_2, by the lensmaker's formula,

$$\frac{1}{f} = (n-1)\left(\frac{1}{r_1} - \frac{1}{r_2}\right), \tag{4.212}$$

where f is the focal distance of the lens. A *positive lens* has a positive focal length, while a *negative lens* has a negative focal length.

This formula, and others like it, can be derived by using the laws of reflection and refraction. The law of reflection at a flat surface is given by

$$\theta_i = \theta_r, \tag{4.213}$$

where θ_i is the angle of incidence and θ_r is the angle of reflection; this is illustrated in Fig. 4.6.

The law of refraction at a flat surface is given by Snell's law,

$$n_1 \sin\theta_1 = n_2 \sin\theta_2, \tag{4.214}$$

where n_1 and n_2 are the indices of refraction in the two media, θ_1 is the angle of incidence of ray propagating in medium 1, and θ_2 is the angle of transmission of the ray propagating in medium 2. This is illustrated in Fig. 4.7.

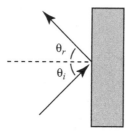

Figure 4.6 Illustration of the law of reflection: the angle of reflection equals the angle of incidence.

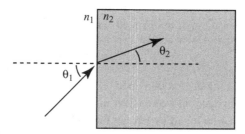

Figure 4.7 Illustration of Snell's law.

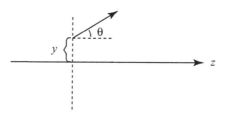

Figure 4.8 A meridional ray is uniquely specified by its angle with respect to the system axis and its height above the axis.

How do we specify the behavior of a given ray? Let us restrict ourselves to *meridional rays*, which are those rays which intersect the optical axis and remain in a fixed plane as they propagate through the system. Furthermore, we consider only systems which are rotationally symmetric about the z-axis. In a fixed z-plane of an optical system, a meridional ray is specified by its y-position and the angle that it makes with the z-axis, shown in Fig. 4.8.

For a single lens, or a very simple system of lenses, one can perform ray-tracing to determine the path of a ray. But what if we are dealing with a more complicated system, with multiple lenses of different types and focal lengths?

It is useful to have a method to easily evaluate the effect of a general optical system on a ray. We will see that the way to do this is through matrix methods.

4.10.2 Translation and refraction in matrix form

We begin by making a simplifying, yet quite general assumption: that the rays are all paraxial, i.e. they are only propagating in a direction nearly parallel to the z-axis. This means that we may make small angle approximations for trigonometric functions, i.e.

$$\sin\theta \approx \theta, \quad \tan\theta \approx \theta. \tag{4.215}$$

This in turn means that Snell's law can be written in the simplified form

$$n_1\theta_1 = n_2\theta_2. \tag{4.216}$$

We may divide ray propagation into two cases (three, if we count reflection, which we will not consider here, but is readily accounted for): propagation of a ray through a uniform medium, and refraction of a ray at the boundary between two media. We consider each in turn.

1. Propagation through a uniform medium. The behavior of the ray as it propagates between two planes of constant z can be determined from Fig. 4.9.
   We denote by θ the angle of the ray at the first plane, and θ' the angle of the ray at the second plane; similarly, we denote by y the y-position of the ray at the first plane, and y' the position of the ray at the second plane. Using elementary geometry, and the small angle approximation, $\tan\theta \approx \theta$, it follows from the figure that
   $$\theta' = \theta, \quad y' = y + l\theta. \tag{4.217}$$
   This is a linear system of equations which can be rewritten in a matrix form,
   $$|y'\rangle = \mathcal{T}|y\rangle, \tag{4.218}$$

where

$$\mathcal{T} = \begin{bmatrix} 1 & l/n \\ 0 & 1 \end{bmatrix}, \tag{4.219}$$

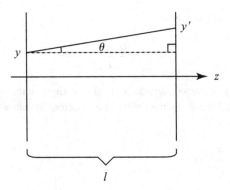

Figure 4.9 The effect of free-space propagation over a distance l on the ray parameters.

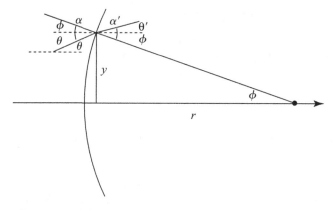

Figure 4.10 Notation used to determine the effect of a refractive index boundary on ray behavior.

and

$$|\mathbf{y}\rangle = \begin{bmatrix} y \\ n\theta \end{bmatrix}, \tag{4.220}$$

with a similar expression for $|\mathbf{y}'\rangle$. It is to be noted that we have chosen as our second variable $n\theta$, instead of θ; this allows us to incorporate Snell's law in a very natural way, as we will see in the next item.

2. Refraction at the boundary between two media. We consider the effect of refraction at a curved surface on the behavior of a ray incident upon it; the notation is illustrated in Fig. 4.10.
   The angle of incidence on the surface is denoted by α, and the angle of refraction is denoted by α'; from elementary geometrical considerations, it follows that

$$\alpha = \theta + \phi, \quad \alpha' = \theta' + \phi, \tag{4.221}$$

where $\phi \approx y/r$, r being the radius of curvature of the surface. Snell's law at the surface therefore becomes

$$n(\theta + \phi) = n'(\theta' + \phi). \tag{4.222}$$

In terms of $n'\theta'$, this can be written as

$$n'\theta' = \frac{(n-n')}{r}y + n\theta. \tag{4.223}$$

A second equation $y = y'$ is immediately determined by noting that the position of the ray does not change upon refraction. We may then write the two equations in matrix form as

$$|\mathbf{y}'\rangle = \mathcal{R}\,|\mathbf{y}\rangle, \tag{4.224}$$

where

$$\mathcal{R} = \begin{bmatrix} 1 & 0 \\ (n-n')/r & 1 \end{bmatrix}, \tag{4.225}$$

with $|y\rangle$ and $|y'\rangle$ the same as in item 1. Typically, the quantity

$$P \equiv \frac{n' - n}{r} \tag{4.226}$$

is called the *power* of the refracting surface, and the matrix for refraction can be written as

$$\mathcal{R} = \begin{bmatrix} 1 & 0 \\ -P & 1 \end{bmatrix}. \tag{4.227}$$

It is to be noted that, if the refracting surface is planar, $r \to \infty$, and we have

$$\mathcal{R}_{\mathrm{plane}} = \begin{bmatrix} 1 & 0 \\ 0 & 1 \end{bmatrix}. \tag{4.228}$$

A planar surface does not affect the ray vector at all; this is why we defined the second component of the ray vector as $n\theta$, instead of simply θ.

It is very important to note that our calculation does not take into account the fact that the z-position at which refraction occurs depends on the height of the ray; a ray arriving with larger y is refracted at a different distance than a ray arriving with smaller y, due to the curvature of the lens. The matrix method described here treats all refraction as occurring at a single z-plane. In paraxial optical systems, lens curvature is typically not very large and this approximation is justified.

The behavior of any paraxial optical system may therefore be described by a system matrix $\mathcal{S}$, consisting of the ordered product of all refractions and translations in the system. It is to be noted that the determinants of $\mathcal{R}$ and $\mathcal{T}$ are equal to unity. Because $\det(\mathbf{AB}) = \det(\mathbf{A})\det(\mathbf{B})$, as discussed in Section 4.3, we see that *the determinant of any system matrix is equal to unity*. This observation can be used as a simple check of a calculation of a system matrix.

4.10.3 Lenses

Using the approach outlined above, the construction of a lens, and an entire lens system, has been reduced to the multiplication of a number of matrices. For example, a general lens may be considered to be the product of three matrices: $\mathcal{R}_1$, the refraction at the first surface, $\mathcal{T}$, the propagation inside the lens, and $\mathcal{R}_2$, the refraction at the second surface. If we denote the effect of the total lens as $\mathcal{L}$, we have that

$$\mathcal{L} = \mathcal{R}_2 \mathcal{T} \mathcal{R}_1. \tag{4.229}$$

It is to be noted that the first refraction is on the right of this equation, since this is the first surface that the ray encounters.

If the lens is exceedingly thin, we may use the "thin lens" approximation; in matrix terms, this amounts to assuming that the lens is thin enough that the propagation within it has a negligible effect,

$$\mathcal{L}_{\mathrm{thin}} \approx \mathcal{R}_2 \mathbf{I} \mathcal{R}_1 = \mathcal{R}_2 \mathcal{R}_1. \tag{4.230}$$

Multiplying these out explicitly, we have

$$\mathcal{L}_{\text{thin}} = \begin{bmatrix} 1 & 0 \\ -P_2 & 1 \end{bmatrix} \begin{bmatrix} 1 & 0 \\ -P_1 & 1 \end{bmatrix} = \begin{bmatrix} 1 & 0 \\ -P_1 - P_2 & 1 \end{bmatrix}. \tag{4.231}$$

However, $P_1 + P_2 = (n-1)(1/r_1 - 1/r_2) = 1/f$, so we have

$$\mathcal{L}_{\text{thin}} = \begin{bmatrix} 1 & 0 \\ -1/f & 1 \end{bmatrix}. \tag{4.232}$$

A more general, non-thin lens will be described by a general matrix $\mathcal{L}_{\text{gen}}$ such that

$$\mathcal{L}_{\text{gen}} = \begin{bmatrix} a_{11} & a_{12} \\ a_{21} & a_{22} \end{bmatrix}. \tag{4.233}$$

Such lenses are typically described by their cardinal points, namely their front and rear *focal planes*, and their front and rear principal planes. The rear focal plane is defined as the distance behind the lens at which all incident rays parallel to the optical axis intersect the optical axis. In matrix notation, we may write

$$\begin{bmatrix} 0 \\ n\theta_o \end{bmatrix} = \begin{bmatrix} 1 & f_r \\ 0 & 1 \end{bmatrix} \begin{bmatrix} a_{11} & a_{12} \\ a_{21} & a_{22} \end{bmatrix} \begin{bmatrix} y_i \\ 0 \end{bmatrix}. \tag{4.234}$$

An arbitrary parallel ray of height y_1 intersects the optical axis at a distance f_r past the end of the optical system. On comparing sides of this matrix equation, we find that we require

$$0 = (a_{11} + f_r a_{21})y_1. \tag{4.235}$$

To form an imaging system, this must be true for all y_1, which implies that

$$f_r = -\frac{a_{11}}{a_{21}}. \tag{4.236}$$

Similarly, the front focal plane of the system is taken to be the distance f_f at which a ray crossing the optical axis in front of the lens will exit parallel to the axis at height y_2, i.e.

$$\begin{bmatrix} y_2 \\ 0 \end{bmatrix} = \begin{bmatrix} a_{11} & a_{12} \\ a_{21} & a_{22} \end{bmatrix} \begin{bmatrix} 1 & f_f \\ 0 & 1 \end{bmatrix} \begin{bmatrix} 0 \\ n\theta_i \end{bmatrix}. \tag{4.237}$$

From this equation, we may deduce the formula,

$$f_f = -\frac{a_{22}}{a_{21}}. \tag{4.238}$$

The front and back focal planes are illustrated in Fig. 4.11.

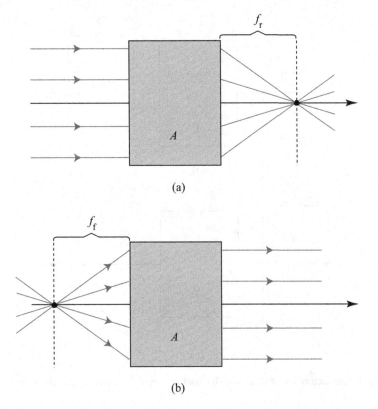

Figure 4.11 Illustration of the (a) rear and (b) front focal planes of a general imaging system A.

The principal planes are defined by looking for a set of translation matrices which make the system matrix equivalent to a thin lens, i.e.

$$\begin{bmatrix} 1 & p_r \\ 0 & 1 \end{bmatrix} \begin{bmatrix} a_{11} & a_{12} \\ a_{21} & a_{22} \end{bmatrix} \begin{bmatrix} 1 & p_f \\ 0 & 1 \end{bmatrix} = \begin{bmatrix} 1 & 0 \\ -1/f & 1 \end{bmatrix}, \tag{4.239}$$

where we have assumed that $n = 1$. The quantities p_r and p_f are the positions of the rear and front principal planes, respectively. The matrices may be multiplied directly to find

$$\begin{bmatrix} a_{11} + a_{21}p_r & a_{12} + a_{22}p_r + p_f(a_{11} + a_{21}p_r) \\ a_{21} & a_{22} + a_{21}p_f \end{bmatrix} = \begin{bmatrix} 1 & 0 \\ -1/f & 1 \end{bmatrix}. \tag{4.240}$$

Matching the diagonal elements, we have

$$p_r = \frac{1 - a_{11}}{a_{21}}, \tag{4.241}$$

$$p_f = \frac{1 - a_{22}}{a_{21}}. \tag{4.242}$$

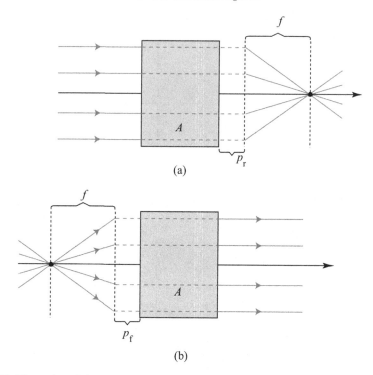

Figure 4.12 Illustration of the (a) rear and (b) front principal planes of a general imaging system A.

The upper-right element of the matrix is automatically zero using $a_{11}a_{22} - a_{21}a_{12} = 1$, and the effective thin lens focal length (measured from the principal planes) is $f = -1/a_{21}$. The principal planes are illustrated in Fig. 4.12. As can be seen from the figure, all refraction may be treated as occurring at the principal planes. The use of principal planes can be a great simplification in matrix calculations; the region between them can in a sense be "ignored" and the lens can be treated as an effective thin lens.

4.10.4 Conjugate matrix

We have seen that it is relatively easy to construct a matrix which defines a lens system; the question then becomes how do we use this matrix to get results of interest? For problems in which one is constructing an imaging system, an important quantity is the so-called *conjugate matrix* or image matrix, defined as

$$\mathcal{C} = \mathcal{T}_i S \mathcal{T}_o, \tag{4.243}$$

where $\mathcal{T}_o$ is the translation matrix which describes the propagation of rays from the object plane to the system input, S is the system matrix which describes the propagation of rays through the system, and $\mathcal{T}_i$ describes the propagation of rays from the output of the system

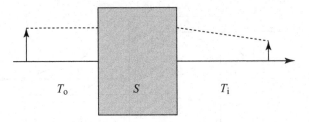

Figure 4.13 The notation associated with the conjugate matrix.

to the image plane (the position of which is probably an unknown). The conjugate matrix describes the total effect of the system on a ray as it propagates from the object plane to the image plane; the geometry is illustrated in Fig. 4.13.

We will write the conjugate matrix as

$$\mathcal{C} = \begin{bmatrix} c_{11} & c_{12} \\ c_{21} & c_{22} \end{bmatrix}. \tag{4.244}$$

It is then fair to ask what the coefficients c_{ij} might physically represent. If we let this matrix operate on a ray, we find that

$$y' = c_{11}y + c_{12}n\theta. \tag{4.245}$$

If this system is an imaging system, however, every ray which emanates from a given point on the object must arrive at the same point on the image, regardless of angle. This means we have the requirement that $c_{12} = 0$; this condition can be used to determine the location of the image, given the object position.

If $c_{12} = 0$, our equation above becomes

$$y' = c_{11}y, \tag{4.246}$$

which implies that the height of the image is proportional to the height of the object, and the proportionality is c_{11}. If we define the magnification of the system as m, we have $c_{11} = m$. A negative magnification represents an inverted image.

We now consider a parallel ray which is passing through the system. We know that the focal point of the system is the point through which all parallel rays pass; if we consider the effect of our imaging system on an object ray with $\theta = 0$, the output angle of this ray will be given by

$$n'\theta' = c_{21}y. \tag{4.247}$$

Geometrically, we may write that $-y/f = \tan\theta'$, where f is the focal length of the system from the front principal plane; this is illustrated in Fig. 4.14.

Using the small angle approximation, we therefore can see that $c_{21} = -n'/f$, which we call the *power* of the system. The quantity c_{22} describes how the angle of the ray is magnified on passing through the system, and may be referred to as the *angular magnification*. The

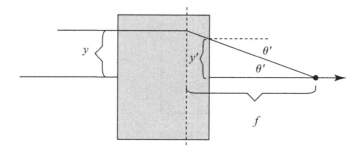

Figure 4.14 The effect of a system on a parallel ray.

conjugate matrix itself can give us most of the important imaging properties of the optical system.

We elaborate on the preceding arguments with a few simple examples.

4.10.5 Imaging by a thin lens

We can use the conjugate matrix to determine some of the familiar properties of thin lens imaging. We assume that an object is present at the plane x_0 and that the image will be found at x_i, as illustrated in Fig. 4.15.

The conjugate matrix for this system is given by

$$C = \begin{bmatrix} 1 & x_i \\ 0 & 1 \end{bmatrix} \begin{bmatrix} 1 & 0 \\ -1/f & 1 \end{bmatrix} \begin{bmatrix} 1 & x_0 \\ 0 & 1 \end{bmatrix} = \begin{bmatrix} 1 - x_i/f & x_0 - x_i x_0/f + x_i \\ -1/f & -x_0/f + 1 \end{bmatrix}. \quad (4.248)$$

Recalling the earlier discussion, the image location is specified by $x_0 - x_i x_0/f + x_i = 0$ ($c_{12} = 0$), or

$$x_i = \frac{x_0}{x_0/f - 1}. \quad (4.249)$$

If $x_0 < f$, x_i is negative and the image is virtual; otherwise, the image is real. This expression can be reorganized by taking its inverse,

$$\frac{1}{x_i} + \frac{1}{x_0} = \frac{1}{f}. \quad (4.250)$$

This equation is the *thin lens formula* of geometrical optics, which relates the object and image position.

We can find the magnification by evaluating $c_{11} = m$,

$$c_{11} = m = 1 - x_i/f = 1 - \frac{x_0/f}{x_0/f - 1} = \frac{1}{1 - x_0/f}. \quad (4.251)$$

If $x_0 > f$, the magnification is negative, i.e. the image is inverted.

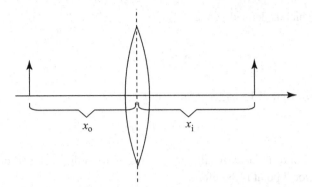

Figure 4.15 The notation for thin lens imaging.

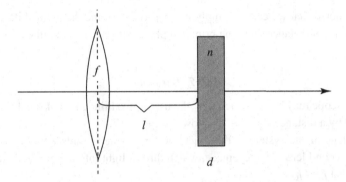

Figure 4.16 The notation for a thin lens focusing through a dielectric plate.

4.10.6 Lens with a dielectric plate

Here we consider a system which consists of a thin lens followed after a distance l by a dielectric plate of thickness d and index of refraction n, as illustrated in Fig. 4.16.

The system matrix is given by:

$$\mathcal{S} = \begin{bmatrix} 1 & d/n \\ 0 & 1 \end{bmatrix} \begin{bmatrix} 1 & l \\ 0 & 1 \end{bmatrix} \begin{bmatrix} 1 & 0 \\ -1/f & 1 \end{bmatrix} = \begin{bmatrix} 1 - l/f - d/nf & l + d/n \\ -1/f & 1 \end{bmatrix}. \quad (4.252)$$

It is to be noted that the surfaces of the plate do not contribute to the system matrix, because they are planar.

A quantity of some interest is the shift in the focal point of the lens due to the dielectric plate. The focal point is defined as the point at which a parallel ray which enters the system crosses the axis. We consider, then, an "almost" conjugate matrix,

$$\mathcal{S}' = \begin{bmatrix} 1 & x_i \\ 0 & 1 \end{bmatrix} \begin{bmatrix} 1 - l/f - d/nf & l + d/n \\ -1/f & 1 \end{bmatrix} = \begin{bmatrix} 1 - l/f - d/nf - x_i/f & l + d/n + x_i \\ -1/f & 1 \end{bmatrix}. \quad (4.253)$$

A parallel ray, which is defined by the vector

$$\mathbf{y} = \begin{bmatrix} y \\ 0 \end{bmatrix},$$ (4.254)

will intersect the origin if $1 - l/f - d/nf - x_i/f = 0$, or

$$x_i = f - l - d/n.$$ (4.255)

This is the position of the new focal point, measured from the back end of the dielectric. The shift in the focal point is therefore

$$\Delta f = d - d/n.$$ (4.256)

It is to be noted that we cannot simply use c_{21} to determine the focal shift, because c_{21} measures the focal distance from the principal plane which lies within the system.

4.10.7 Telescope

A simple telescope may be constructed by the use of two thin lenses of focal lengths f_o and f_e, separated by a distance $d = f_o + f_e$, shown in Fig. 4.17.

The first lens of the system is the objective lens, which points towards the object of interest; the second lens is the eyepiece, which directs light into the eye of the observer. It is assumed that $f_o > f_e$.

The system matrix is readily found to be given by the formula

$$S_{\text{telescope}} = \begin{bmatrix} 1 & 0 \\ -1/f_e & 1 \end{bmatrix} \begin{bmatrix} 1 & d \\ 0 & 1 \end{bmatrix} \begin{bmatrix} 1 & 0 \\ -1/f_o & 1 \end{bmatrix},$$

$$= \begin{bmatrix} 1 - d/f_o & d \\ -1/f_e - 1/f_o + d/f_e f_o & -d/f_e + 1 \end{bmatrix}.$$ (4.257)

This equation simplifies greatly if we use the fact that $d = f_e + f_o$; we then may write that

$$S_{\text{telescope}} = \begin{bmatrix} -f_e/f_o & d \\ 0 & -f_o/f_e \end{bmatrix}.$$ (4.258)

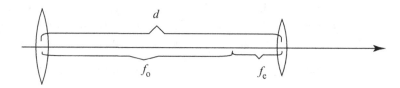

Figure 4.17 The notation for a telescope.

If we introduce an object position x_o and an image position x_i, we can find the conjugate matrix,

$$
C_{\text{telescope}} = \begin{bmatrix} 1 & x_i \\ 0 & 1 \end{bmatrix} \begin{bmatrix} -f_e/f_o & d \\ 0 & -f_o/f_e \end{bmatrix} \begin{bmatrix} 1 & x_o \\ 0 & 1 \end{bmatrix},
$$
$$
= \begin{bmatrix} -f_e/f_o & -(f_e/f_o)x_o - (f_o/f_e)x_i + d \\ 0 & -f_o/f_e \end{bmatrix}. \tag{4.259}
$$

From this equation we can determine the properties of the telescope. Noting that the image is located at the position for which $c_{12} = 0$, we have

$$
x_i = \frac{df_e}{f_o} - \left(\frac{f_e}{f_o}\right)^2 x_o. \tag{4.260}
$$

A telescope is typically used to image objects which are far away. If we consider $x_o \to \infty$, we find that $x_i \to -\infty$, or that the image of the object is a virtual image located at infinity.

The absolute value of the magnification (c_{11}) is less than unity, suggesting that the image is smaller than the original object. However, the magnitude of the angle of the ray is increased by a factor $c_{22} = -f_o/f_e$; this means that the object appears to be much closer, and can be resolved much better by the eye.

4.11 Additional reading

A detailed description of matrix methods for polarization and geometrical optics may be found in the following references.

- A. Gerard and J. M. Birch, *Introduction to Matrix Methods in Optics* (New York, Wiley, 1975). This book also discusses the use of matrices in polarization optics.
- G. Kloos, *Matrix Methods for Optical Layout* (Bellingham, WA, SPIE Press, 2007).
- A. Nussbaum, *Geometrical Optics: An Introduction* (Reading, MA, Addison-Wesley, 1968).
- C. Brosseau, *Fundamentals of Polarized Light* (New York, Wiley, 1998). A detailed book on polarization effects that includes much discussion of the various matrix representations.

4.12 Exercises

1. Determine the product **AB** for the following pairs of matrices:

   (a) $\mathbf{A} = \begin{bmatrix} 1 & 2 \\ 1 & 1 \end{bmatrix}$, $\mathbf{B} = \begin{bmatrix} 3 & 1 \\ 2 & 0 \end{bmatrix}$,

   (b) $\mathbf{A} = \begin{bmatrix} 1 & 2 & 4 \\ 3 & 1 & 0 \\ 1 & 1 & 1 \end{bmatrix}$, $\mathbf{B} = \begin{bmatrix} 4 & 4 & 1 \\ 2 & 2 & 3 \\ 3 & 1 & 5 \end{bmatrix}$,

(c) $\mathbf{A} = \begin{bmatrix} 1 & 1/2 & 1/3 \\ 2 & 1/3 & 2/3 \\ 1/3 & 1/2 & 4 \end{bmatrix}$, $\mathbf{B} = \begin{bmatrix} 1/2 & 0 & 1 \\ 0 & 0 & 3 \\ 1 & 1/3 & 2 \end{bmatrix}$.

2. Determine the product **AB** for the following pairs of matrices:

(a) $\mathbf{A} = \begin{bmatrix} 1 & 1 \\ 3 & 0 \end{bmatrix}$, $\mathbf{B} = \begin{bmatrix} 4 & 2 \\ 2 & 2 \end{bmatrix}$,

(b) $\mathbf{A} = \begin{bmatrix} 1 & 1 & 1 \\ 3 & 2 & 4 \\ 1 & 0 & 1 \end{bmatrix}$, $\mathbf{B} = \begin{bmatrix} 3 & 5 & 2 \\ 2 & 1 & 2 \\ 3 & 2 & 4 \end{bmatrix}$,

(c) $\mathbf{A} = \begin{bmatrix} 1/4 & 1/2 & 3/4 \\ 1/2 & 2/4 & 5/4 \\ 1 & 1/2 & 2 \end{bmatrix}$, $\mathbf{B} = \begin{bmatrix} 1/2 & 1/4 & 1 \\ 3/4 & 1 & 2 \\ 1 & 1/2 & 1 \end{bmatrix}$.

3. Calculate the determinant of the following matrices:

(a) $\begin{bmatrix} 1 & 3 \\ 2 & 1 \end{bmatrix}$, (b) $\begin{bmatrix} 1 & 1 & 4 \\ 1 & 2 & 1 \\ 2 & 2 & 2 \end{bmatrix}$, (c) $\begin{bmatrix} 1 & 3 & 1 & 1 \\ 2 & 1 & 0 & 4 \\ 0 & 0 & 0 & 1 \\ 1 & 2 & 3 & 4 \end{bmatrix}$.

4. Calculate the determinant of the following matrices:

(a) $\begin{bmatrix} 1/2 & 2 \\ 1 & 1 \end{bmatrix}$, (b) $\begin{bmatrix} 3 & 1 & 3 \\ 1 & 3 & 1 \\ 1 & 0 & 2 \end{bmatrix}$, (c) $\begin{bmatrix} 0 & 2 & 1 & 0 \\ 1 & 1 & 2 & 2 \\ 0 & 3 & 2 & 1 \\ 0 & 2 & 2 & 0 \end{bmatrix}$.

5. Determine whether the following matrices have inverses and, if so, find them using Gauss–Jordan inversion:

(a) $\begin{bmatrix} 1 & 1 \\ 2 & 3 \end{bmatrix}$, (b) $\begin{bmatrix} 1 & 1 & 0 \\ -2 & 3 & 1 \\ 1 & -2 & 1 \end{bmatrix}$, (c) $\begin{bmatrix} 1 & 0 & 0 \\ 2 & 2 & 1 \\ 1 & 2 & 1 \end{bmatrix}$.

6. Determine whether the following matrices have inverses and, if so, find them using Gauss–Jordan inversion:

(a) $\begin{bmatrix} 1 & 2 \\ 3 & 2 \end{bmatrix}$, (b) $\begin{bmatrix} 1 & 1 & 1 \\ 2 & 1 & 4 \\ 1 & 0 & 1 \end{bmatrix}$, (c) $\begin{bmatrix} 5 & 3 & 3 \\ 1 & 5 & 1 \\ 2 & -1 & 1 \end{bmatrix}$.

7. Find solutions for the following systems of equations:

(a) $2x - y = 0,$
$x + y + z = 6,$
$2x - 2y + 2z = 0.$

(b) $x + 2y - z = 0,$
$2x + y + 2z = 6,$
$3x + y + 2z = 10.$

(c) $2x + 2y + 2z = 0,$
$x - y + 3z = 0,$
$x + 3y - z = 0.$

8. Find solutions for the following systems of equations:
   (a) $x - y = 0$,
      $3x + y + z = 0$,
      $2x + 2y + z = 0$.
   (b) $x - 2y + z = -1$,
      $2x + y - 3z = 8$,
      $x + 2y + 2z = 12$.
   (c) $-x + 2y + 2z = 3$,
      $2x + y + z = 4$,
      $3x - y + 3z = 9$.

9. Find eigenvalues and eigenvectors of the following matrices:

   (a) $\begin{bmatrix} 1 & 0 & 0 \\ -2 & 3 & 0 \\ -3 & 3 & 1 \end{bmatrix}$, (b) $\begin{bmatrix} \sqrt{2} & 0 & -1 \\ 0 & \sqrt{2} & -1 \\ -1 & -1 & \sqrt{2} \end{bmatrix}$.

10. Find eigenvalues and eigenvectors of the following matrices:

    (a) $\begin{bmatrix} 0 & 1 & 0 \\ -2 & 3 & 0 \\ -3 & 3 & 1 \end{bmatrix}$, (b) $\begin{bmatrix} 3 & 1 & 0 \\ 1 & 3 & 0 \\ 0 & 0 & 6 \end{bmatrix}$.

11. One model of a periscope consists of six identical thin lenses, separated by a distance equal to the focal length of the lenses:

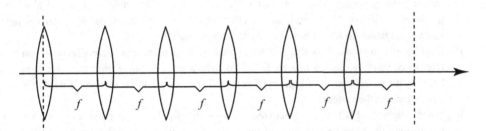

    Calculate the system matrix for the periscope, assuming that the entrance to the system is at the first lens and the exit from the system is at distance f from the last lens. Where is the image formed by this system?

12. Show that the distance between an object and the real image produced by a thin positive lens cannot be less than four times the focal length.

13. The polarization $\mathbf{P}$ induced in an anisotropic crystal by the electric field $\mathbf{E}$ of an illuminating light wave is given by the matrix equation

    $$\mathbf{P} = \boldsymbol{\epsilon} \cdot \mathbf{E},$$

    where $\boldsymbol{\epsilon}$ is the electric permittivity tensor of the crystal, which may be written as a 3×3 matrix. Suppose, in a particular coordinate system, this matrix has the form

    $$\boldsymbol{\epsilon} = \epsilon_0 \begin{bmatrix} 1 & 2 & 1 \\ 2 & 2 & 2 \\ 1 & 2 & 1 \end{bmatrix},$$

where ϵ_0 is the permittivity of free space. Find a rotation matrix which converts ϵ into a diagonal form.

14. The coherency matrix, which describes the statistical and polarization properties of an electromagnetic field at a point in a paraxial beam propagating in the z-direction, may be written as

$$\mathbf{W} = \begin{bmatrix} W_{xx} & W_{xy} \\ W_{xy}^* & W_{yy} \end{bmatrix},$$

where $W_{ij} \equiv \langle E_i^* E_j \rangle$, and $\langle \cdots \rangle$ denotes averaging. Assuming a general coherency matrix as given above, find a rotation which reduces it to a diagonal form.

15. In the explanation of Gauss–Jordan inversion, we described the method as a series of operations of matrices $\mathbf{M}_i$ which converted $\mathbf{A}$ to $\mathbf{A}^{-1}$. For 3×3 matrices, write the matrices $\mathbf{M}$ which, when acting on $\mathbf{A}$, perform separately the following operations:
    (a) divides the second row by a factor a, while leaving the other rows unchanged;
    (b) adds b times the first row to the third row.

16. Prove the following statements.
    (a) If a row of a matrix is multiplied by a constant, the determinant is multiplied by a constant.
    (b) If a multiple of one row of a matrix is added (column by column) to another row, the value of the determinant is unchanged.
    (Hint: derive matrices $\mathbf{M}$ which perform the above actions on a matrix $\mathbf{A}$, and use the property $\det(\mathbf{MA}) = \det(\mathbf{M})\det(\mathbf{A})$.)

17. A polarization-dependent optical isolator consists of three elements: a vertically-oriented polarizer, a Faraday rotator which rotates the electric field over an angle of $45°$, and a second polarizer oriented at $45°$. Determine the effect of this isolator on a vertically polarized light wave propagating (a) forwards and (b) backwards through the device.

18. Using matrix methods, show that two orthogonally-arranged linear polarizers (vertical and horizontal) completely block a linearly polarized light beam. Assume a third polarizer is placed between the others two at an angle θ from the vertical; calculate the amount of light that now passes through the system.

19. When a light field is not fully polarized, i.e. its state of polarization possesses some random fluctuations, the Jones vectors are inadequate to describe the polarization properties of the field. One may then use a more general vector known as a *Stokes column*, written as

$$|\mathbf{S}\rangle = \begin{bmatrix} \langle E_x^* E_x + E_y^* E_y \rangle \\ \langle E_x^* E_x - E_y^* E_y \rangle \\ \langle E_x^* E_y + E_y^* E_x \rangle \\ -i \langle E_x^* E_y - E_y^* E_x \rangle \end{bmatrix},$$

where the angle brackets ($\langle \cdots \rangle$) represent time averaging. The effects of optical devices are described by the use of 4×4 matrices known as *Mueller matrices*. Determine the expressions for the Mueller matrices of:
    (a) a quarter-wave plate with its fast axis along the x-direction,
    (b) a linear polarizer aligned at an angle θ from the x-direction.
    (Hint: the statistical properties of the light are irrelevant for these devices. Determine the effect of the devices on E_x and E_y, and write expressions for the transformation of the quantities described in the Stokes column.)

20. Use the Gram–Schmidt process to orthogonalize the following set of vectors:

$$|\mathbf{v}_1\rangle = \begin{bmatrix} 1 \\ 1 \\ 0 \\ 1 \end{bmatrix}, \quad |\mathbf{v}_2\rangle = \begin{bmatrix} 0 \\ 1 \\ 1 \\ 0 \end{bmatrix}, \quad |\mathbf{v}_3\rangle = \begin{bmatrix} 1 \\ 0 \\ 1 \\ 0 \end{bmatrix}, \quad |\mathbf{v}_4\rangle = \begin{bmatrix} 0 \\ 0 \\ 1 \\ 1 \end{bmatrix}.$$

Orthogonalize the vectors both in the given order $(1,2,3,4)$ and the reverse order $(4,3,2,1)$.

21. Use the Gram–Schmidt process to orthogonalize the following set of vectors:

$$|\mathbf{v}_1\rangle = \begin{bmatrix} 2 \\ 1 \\ 0 \\ 1 \end{bmatrix}, \quad |\mathbf{v}_2\rangle = \begin{bmatrix} 1 \\ 1 \\ 1 \\ 1 \end{bmatrix}, \quad |\mathbf{v}_3\rangle = \begin{bmatrix} 0 \\ 0 \\ 1 \\ 0 \end{bmatrix}, \quad |\mathbf{v}_4\rangle = \begin{bmatrix} 1 \\ 2 \\ 0 \\ 0 \end{bmatrix}.$$

Orthogonalize the vectors both in the given order $(1,2,3,4)$ and the reverse order $(4,3,2,1)$.

22. Read the paper by H. Kogelnik and T. Li, Laser beams and resonators, *Appl. Opt.* **5** (1966), 1550–1567, and discuss how matrix methods for geometrical optics are used to describe the stability properties of laser resonators.

23. Read the paper by E. Wolf, Coherence properties of partially polarized electromagnetic radiation, *Nuovo Cimento* **13** (1959), 1165–1181. Explain the meaning of the *coherency matrix* of an electromagnetic beam. What sorts of coherency matrix result in degrees of polarization $P = 1$ and $P = 0$?

24. It is often helpful to work with a Jones vector defined in a circular polarization basis, such that

$$|E\rangle = \begin{bmatrix} E_+ \\ E_- \end{bmatrix},$$

where E_+ and E_- represent the right- and left-handed components of the field. Determine the matrices which describe a linear polarizer of angle θ and a rotator of angle θ_0 in this basis. (Hint: determine the unitary matrix which converts from the linear basis to the circular basis.)

25. The Sohncke theory of optical activity states that a Faraday rotator can be constructed by an appropriate collection of phase retarding plates. The simplest system of this form consists of a quarter-wave plate, a wave plate with "fast" axis phase α and orientation θ, followed by another quarter-wave plate perpendicular to the first. Show that, for an appropriate choice of α and θ, this system works as a Faraday rotator.

26. Prove that the product of two Hermitian matrices $\mathbf{A}$ and $\mathbf{B}$ is itself Hermitian if and only if $\mathbf{A}$ and $\mathbf{B}$ commute.

27. For a general square matrix $\mathbf{C}$, show that the matrices $\mathbf{C} + \mathbf{C}^\dagger$ and $\mathrm{i}(\mathbf{C} - \mathbf{C}^\dagger)$ are Hermitian. From this, we may argue that any square matrix may be decomposed into the sum of a Hermitian and an anti-Hermitian matrix, in the form

$$\mathbf{C} = \frac{1}{2}(\mathbf{C} + \mathbf{C}^\dagger) - \frac{\mathrm{i}}{2}[\mathrm{i}(\mathbf{C} - \mathbf{C}^\dagger)].$$

28. A plano-convex lens of index n, illustrated below, consists of one surface of curvature R, a slab of thickness d, and a planar output surface.

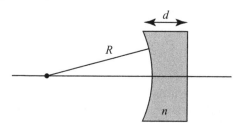

Find the location of the focal planes and the principal planes, and describe the imaging properties of the lens.

5

Advanced matrix techniques and tensors

5.1 Introduction: Foldy–Lax scattering theory

As problems in optical science become more complicated, it has become increasingly common to solve them using computational techniques. In the process of setting up the problem in a form which can be solved on a computer, it is often reduced to a matrix equation.

An example of this is the scattering of a wave by a system of discrete particles, as illustrated in Fig. 5.1. We consider the case of a scalar wave for simplicity, though the method is readily extended to electromagnetic waves. A monochromatic incident field of spatial dependence $U^{(i)}(\mathbf{r})$ is assumed to be illuminating a system of N particles, each of which is labeled by an index j. The illuminating field is scattered off each of the particles, producing secondary spherical waves $U_j(\mathbf{r})$. The total field $U(\mathbf{r})$ in the region of the system of particles may be written as the sum of the illuminating field and the field scattered by the collection of particles, i.e.

$$U(\mathbf{r}) = U^{(i)}(\mathbf{r}) + U^{(s)}(\mathbf{r}). \tag{5.1}$$

The scattered field may be written as the total field scattered by all of the particles in the system,

$$U^{(s)}(\mathbf{r}) = \sum_{j=1}^{N} U_j(\mathbf{r}). \tag{5.2}$$

The field scattered by the jth particle is assumed to be a spherical wave $G(\mathbf{r} - \mathbf{r}_j)$, centered on the jth particle, whose strength is proportional to the "polarizability" α_j of the particle and the *total* field at the position $\mathbf{r}_j$ of the particle,

$$U_j(\mathbf{r}) = \alpha_j G(\mathbf{r} - \mathbf{r}_j) U(\mathbf{r}_j). \tag{5.3}$$

The latter point is crucial; an individual particle interacts not only with the illuminating field, but the scattered field produced by all other particles in the system. In three dimensions,

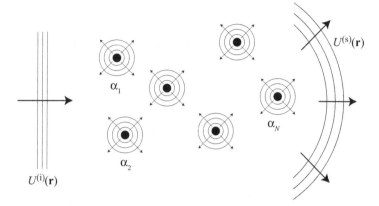

Figure 5.1 A scalar wave $U^{(i)}(\mathbf{r})$ interacts with a collection of particles of polarizability α_j.

the spherical wave $G(\mathbf{r} - \mathbf{r}_j)$ has the form

$$G(\mathbf{r} - \mathbf{r}_j) = \frac{\exp[ik|\mathbf{r} - \mathbf{r}_j|]}{|\mathbf{r} - \mathbf{r}_j|}. \tag{5.4}$$

We may write an expression for the total field $U(\mathbf{r})$ as

$$U(\mathbf{r}) = U^{(i)}(\mathbf{r}) + \sum_{j=1}^{N} \alpha_j G(\mathbf{r} - \mathbf{r}_j) U(\mathbf{r}_j). \tag{5.5}$$

This equation is a simple version of what is generally referred to as a Foldy–Lax equation, after the researchers who first developed it [Fol45, Lax52]. The solution to such an equation is nontrivial, because of the presence of the total field $U(\mathbf{r})$ on both sides of the equation.

We can attempt to solve the equation self-consistently by considering the total field only at the points of the scatterers $\mathbf{r}_i$. Because our scatterers are considered to be point-like, however, we must remove the singular behavior of the scattered field by excluding the "self-energy" contribution to the scattered field, i.e. exclude $i = j$. We then have a set of equations,

$$U(\mathbf{r}_i) = U^{(i)}(\mathbf{r}_i) + \sum_{j \neq i} \alpha_j G(\mathbf{r}_i - \mathbf{r}_j) U(\mathbf{r}_j). \tag{5.6}$$

This looks very much like a matrix equation. Defining $U_j \equiv U(\mathbf{r}_j)$, $U^{(i)}(\mathbf{r}_j) \equiv U_j^{(i)}$, and letting

$$\mathbf{M}_{ij} \equiv \begin{cases} \alpha_j G(\mathbf{r}_i - \mathbf{r}_j), & i \neq j, \\ 0, & i = j, \end{cases} \tag{5.7}$$

we may write

$$U_i = U_i^{(i)} + \sum_{j=1}^{N} \mathbf{M}_{ij} U_j. \tag{5.8}$$

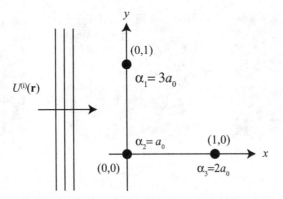

Figure 5.2 A simple system of three point scatterers.

Introducing vectors $\mathbf{U} = \{U_j\}$, $\mathbf{U}^{(i)} = \{U_j^{(i)}\}$, and the identity matrix $\mathbf{I}$, we may write this in the vector–matrix equation

$$\mathbf{IU} = \mathbf{U}^{(i)} + \mathbf{MU}. \tag{5.9}$$

This equation may be rewritten in the form

$$(\mathbf{I} - \mathbf{M})\mathbf{U} = \mathbf{U}^{(i)}, \tag{5.10}$$

which is a matrix equation of the form solved in Section 4.3, with a formal solution

$$\mathbf{U} = (\mathbf{I} - \mathbf{M})^{-1}\mathbf{U}^{(i)}. \tag{5.11}$$

Here $(\mathbf{I} - \mathbf{M})^{-1}$ represents the inverse of the matrix $(\mathbf{I} - \mathbf{M})$. Once the field $U(\mathbf{r}_j)$ is known at the positions of all point scatterers, the field everywhere else in space may be calculated directly from Eq. (5.5).

As an example, we consider three point scatterers in a two-dimensional geometry is shown in Fig. 5.2. The scatterers are located at positions $\mathbf{r}_1 = (0, 1)$, $\mathbf{r}_2 = (0, 0)$, and $\mathbf{r}_3 = (0, 1)$ (in dimensionless units), with polarizabilities taken to be $\alpha_1 = 3a_0$, $\alpha_1 = a_0$, and $\alpha_2 = 2a_0$. The wavelength is taken to be $\lambda = 0.11$, and the illuminating field is taken to be a plane wave coming from the left.

Figure 5.3 shows the intensity $|U^{(s)}(\mathbf{r})|^2$ of the scattered field for two different strengths of the polarizability. In the weak scattering case (a_0 very small), we expect that the scattered field will be very small compared to the illuminating field, and that $U(\mathbf{r})$ on the right-hand side of Eq. (5.5) may to a good approximation be replaced by $U^{(i)}(\mathbf{r})$; this is known as the *Born approximation*, and will be discussed further in Section 7.8. We see that the Born approximation for the scattered field in (c) agrees well with the weak scattering case.

We have seen that the problem of wave scattering from a system of point particles may be formulated as a problem of matrix inversion. Similarly, the scattering of light from continuous media can be put in matrix form by an appropriate discretization of the medium properties; see, for instance, Refs. [MP98] and [VBL99].

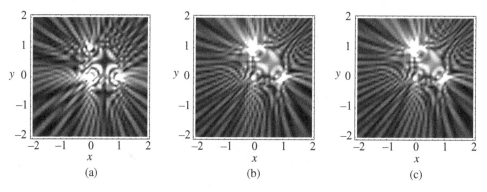

Figure 5.3 The scattered intensity for (a) $a_0 = 50$, (b) $a_0 = 1$, and (c) the Born approximation result for $a_0 = 1$.

In such situations, solution of matrix equations and the inversion of matrices is essential. The simple methods described in the previous chapter, however, provide unstable and inaccurate results for the large matrices required. In this chapter, we discuss a number of more advanced methods for manipulating matrices and finding solutions to matrix equations.

We also discuss several generalizations of concepts described in the previous chapter. When a matrix is non-Hermitian, it may not be written in terms of single set of eigenvectors, but a more general expansion known as a *biorthogonal expansion* may still be applied. Just as matrices may be considered a "generalization" of the idea of vectors, we may introduce a broader generalization in the form of *tensors*. We end the chapter with a basic discussion of tensor properties.

5.2 Advanced matrix terminology

In this section we introduce some more advanced terminology that will be useful in characterizing the properties of matrices. To begin, however, it is to be noted that many applications require the use of non-square, or rectangular, matrices with m rows and n columns. Such a matrix may appear in an *underspecified* set of equations, such as

$$a_{11}x_1 + a_{12}x_2 + a_{13}x_3 = \alpha_1,$$

$$a_{21}x_1 + a_{22}x_2 + a_{23}x_3 = \alpha_2,$$

where the number of unknowns (three in this case) is greater than the number of equations (two in this case). The matrix **A** characterizing this equation has $m = 2$ and $n = 3$. A rectangular matrix may also appear in an *overspecified* set of equations, such as

$$a_{11}x_1 + a_{12}x_2 + a_{13}x_3 = \alpha_1,$$

$$a_{21}x_1 + a_{22}x_2 + a_{23}x_3 = \alpha_2,$$

$$a_{31}x_1 + a_{32}x_2 + a_{33}x_3 = \alpha_3,$$

$$a_{41}x_1 + a_{42}x_2 + a_{43}x_3 = \alpha_4,$$

where the number of unknowns (three in this case) is less than the number of equations (four in this case). Now the matrix $\mathbf{A}$ has $m = 4$ and $n = 3$.

Both situations can occur in the analysis of linear optical imaging systems. Such systems have a finite number of detectors m, and that data may be linearly related to n features of the object. If one measures more data than needed for a unique reconstruction of the desired object features ($m > n$), the system is overspecified. Often this is not possible, and typically the amount of data is less than the number of desired object features ($m < n$), and the system is underspecified. In an underspecified case, one cannot expect a unique solution, but the solution may be restricted to a subspace of the complete space of possible solutions. In any case, the features of the object are typically reconstructed by the use of some sort of matrix inversion technique.

We now introduce a number of definitions that are important in the existence and characterization of a matrix inverse.

Definition 5.1 (Range) *The range of an $m \times n$ matrix $\mathbf{A}$ is the set of all m-dimensional vectors $|\mathbf{y}\rangle$ that can be constructed from the matrix equation $|\mathbf{y}\rangle = \mathbf{A}|\mathbf{x}\rangle$, where $|\mathbf{x}\rangle$ is any n-dimensional vector.*

Definition 5.2 (Null space) *The null space of an $m \times n$ matrix $\mathbf{A}$ is the set of n-dimensional vectors $|\mathbf{x}\rangle$ such that $\mathbf{A}|\mathbf{x}\rangle = 0$.*

Definition 5.3 (Rank) *The rank of an $m \times n$ matrix is the dimension of the space spanned by the vectors in the range of $\mathbf{A}$. A matrix is said to be rank deficient if its rank is less than the minimum of m and n.*

These concepts are most easily understood by considering the special case of an $n \times n$ orthogonal square matrix; such a matrix can be diagonalized to find n orthogonal eigenvectors $|v_i\rangle$ and n corresponding eigenvalues λ_i. Suppose that $p \leq n$ of the eigenvalues vanish, i.e. $\lambda_i = 0$ for $i = 1, \ldots, p$. The corresponding eigenvectors $|v_i\rangle$ satisfy the equation $\mathbf{A}|v_i\rangle = 0$, and so these eigenvectors lie in the null space of the matrix. The set of eigenvectors $|\mathbf{v}_i\rangle$ such that $i = p+1, \ldots, n$ satisfy the equation $\lambda_i |\mathbf{v}_i\rangle = \mathbf{A}|\mathbf{v}_i\rangle$, and the set of all vectors that can be constructed from $|\mathbf{v}_i\rangle$ is the range of the matrix. It is clear that the null space of the matrix is a p-dimensional space, while the range of the matrix is an $n - p$-dimensional space; the rank of the matrix is therefore $n - p$. The sum of the null space dimension and the rank of the matrix is equal to n; this can be generally shown to be true, even for rectangular $m \times n$ matrices.

5.3 Left–right eigenvalues and biorthogonality

We noted in Section 4.6.1 that normal matrices are the broadest class of square matrices whose eigenvectors are guaranteed to be orthogonal to one another. The eigenvectors are generally linearly independent of one another, but only provided the eigenvalues are distinct. We may, however, introduce an additional set of eigenvectors that are orthogonal to the original set, and from this we may find a "diagonal" representation of such a matrix in terms of these two sets of eigenvectors.

We introduce some new terminology to incorporate our new eigenvectors. The traditional eigenvectors of a square matrix satisfy the equation

$$\mathbf{A}\,|\mathbf{y}_i\rangle = \lambda_i\,|\mathbf{y}_i\rangle. \tag{5.12}$$

These eigenvectors are column vectors that right-multiply the matrix $\mathbf{A}$; we now refer to them as *right eigenvectors*.

We may also introduce a set of eigenvectors that left-multiply the matrix $\mathbf{A}$ and satisfy the equation

$$\langle\mathbf{x}_i|\,\mathbf{A} = \lambda_i\,\langle\mathbf{x}_i|. \tag{5.13}$$

These are row vectors, and we refer to them as *left eigenvectors*. We may attempt to relate the two sets of eigenvectors by taking the adjoint of Eq. (5.13),

$$\mathbf{A}^\dagger\,|\mathbf{x}_i\rangle = \lambda_i^*\,|\mathbf{x}_i\rangle. \tag{5.14}$$

If the matrix $\mathbf{A}$ is Hermitian, then $\mathbf{A}^\dagger = \mathbf{A}$ and $\lambda_i^* = \lambda_i$, and we may say that the left and right eigenvectors are equal to one another. In general, however, we find that the left eigenvector $\langle\mathbf{x}_i|$ of $\mathbf{A}$ is the adjoint of the right eigenvector of $\mathbf{A}^\dagger$, and vice versa.

We have seen that Eq. (5.12) may be rewritten in the form

$$(\mathbf{A} - \lambda_i\mathbf{I})\,|\mathbf{y}_i\rangle = 0, \tag{5.15}$$

and this equation can be solved only if $\det(\mathbf{A} - \lambda\mathbf{I}) = 0$. Similarly, Eq. (5.13) may be written as

$$\langle\mathbf{x}_i|\,(\mathbf{A} - \lambda_i\mathbf{I}) = 0, \tag{5.16}$$

which also can only be solved if $\det(\mathbf{A} - \lambda\mathbf{I}) = 0$. We therefore find that *the left and right eigenvalues are equal to one another*.

The most important property of the left and right eigenvectors is their mutual orthogonality, which we now prove. Starting with Eq. (5.15), we premultiply the expression by $\langle\mathbf{x}_j|$, to get

$$\langle\mathbf{x}_j|(\mathbf{A} - \lambda_i\mathbf{I})\,|\mathbf{y}_i\rangle = 0. \tag{5.17}$$

By use of Eq. (5.13), we may write this expression in the form

$$(\lambda_j - \lambda_i)\langle\mathbf{x}_j|\mathbf{y}_i\rangle = 0. \tag{5.18}$$

If $\lambda_i \neq \lambda_j$, we necessarily find that $\langle\mathbf{x}_j|\,\mathbf{y}_i\rangle = 0$, or *the left and right eigenvectors with distinct eigenvalues are orthogonal to one another*. If the eigenvalues are degenerate, we may take appropriate linear combinations of the left and right eigenvectors to make them orthogonal. The left and right eigenvectors therefore independently span the space of all possible vectors.

With this in mind, let us assume that we have furthermore taken the left and right eigenvectors to be orthonormal, i.e. $\langle \mathbf{x}_j | \mathbf{y}_i \rangle = \delta_{ij}$. It follows that we can expand out matrix $\mathbf{A}$ in the form

$$\mathbf{A} = \sum_{i=1}^{n} \lambda_i |\mathbf{y}_i\rangle \langle \mathbf{x}_i|, \tag{5.19}$$

where n is the number of columns of the square matrix. This is as close to a "diagonal" representation of a non-normal matrix that we can achieve.

Example 5.1 We consider the decomposition of the matrix

$$\mathbf{A} = \begin{bmatrix} 1 & 2 \\ 1 & 1 \end{bmatrix} \tag{5.20}$$

into left and right eigenvectors. We may find the eigenvalues from the secular equation, $\det(\mathbf{A} - \lambda \mathbf{I}) = 0$, which leads to $\lambda_+ = 1 + \sqrt{2}$ and $\lambda_- = 1 - \sqrt{2}$. The right eigenvectors are readily found by solving the equation

$$\begin{bmatrix} 1 - \lambda_i & 2 \\ 1 & 1 - \lambda_i \end{bmatrix} \begin{bmatrix} u_i \\ v_i \end{bmatrix} = \begin{bmatrix} 0 \\ 0 \end{bmatrix}, \tag{5.21}$$

with $i = +, -$, which leads to

$$|\mathbf{y}_+\rangle = \sqrt{\frac{1}{2\sqrt{2}}} \begin{bmatrix} \sqrt{2} \\ 1 \end{bmatrix}, \quad |\mathbf{y}_-\rangle = \sqrt{\frac{1}{2\sqrt{2}}} \begin{bmatrix} \sqrt{2} \\ -1 \end{bmatrix}. \tag{5.22}$$

The left eigenvectors can be found from the equation

$$\begin{bmatrix} u_i & v_i \end{bmatrix} \begin{bmatrix} 1 - \lambda_i & 2 \\ 1 & 1 - \lambda_i \end{bmatrix} = \begin{bmatrix} 0 & 0 \end{bmatrix}. \tag{5.23}$$

This leads to the solutions

$$\langle \mathbf{x}_+| = \sqrt{\frac{1}{2\sqrt{2}}} \begin{bmatrix} 1 & \sqrt{2} \end{bmatrix}, \quad \langle \mathbf{x}_-| = \sqrt{\frac{1}{2\sqrt{2}}} \begin{bmatrix} 1 & -\sqrt{2} \end{bmatrix}. \tag{5.24}$$

By direct calculation one can show that $\langle \mathbf{x}_+ | \mathbf{y}_- \rangle = \langle \mathbf{x}_- | \mathbf{y}_+ \rangle = 0$. We may also show that

$$(1 + \sqrt{2}) |\mathbf{y}_+\rangle \langle \mathbf{x}_+| + (1 - \sqrt{2}) |\mathbf{y}_-\rangle \langle \mathbf{x}_-| = \mathbf{A}, \tag{5.25}$$

as expected.

◇

The system of left and right eigenvectors is often referred to as a *biorthogonal system*.

Definition 5.4 (Biorthogonal system) *The sets of vectors* $\{\langle \mathbf{x}_i|\}$ *and* $\{|\mathbf{y}_i\rangle\}$, *for* $i = 1, \ldots, N$, *is said to be* biorthogonal *if a one-to-one correspondence can be established between the two sets and* $\langle \mathbf{x}_i | \mathbf{y}_j \rangle = 0$ *for* $i \neq j$, *and* $\langle \mathbf{x}_i | \mathbf{y}_j \rangle \neq 0$ *for* $i = j$.

The concept of a biorthogonal system can be extended to systems of *functions*, as well; the general theory may be found in [Pel11].

5.4 Singular value decomposition

In the previous section, we showed that it is possible to write a general $n \times n$ square matrix in the form

$$\mathbf{A} = \sum_{i=1}^{n} \lambda_i |\mathbf{y}_i\rangle \langle \mathbf{x}_i|, \tag{5.26}$$

where $\langle \mathbf{x}_i|$ and $|\mathbf{y}_i\rangle$ are the left and right eigenvectors of the matrix, respectively, and λ_i are the eigenvalues. One might wonder whether it is possible to extend this result to a $m \times n$ rectangular matrix using an expansion of the form

$$\mathbf{A} = \sum_{i=1}^{\min(m,n)} \lambda_i |\mathbf{y}_i\rangle \langle \mathbf{x}_i|, \tag{5.27}$$

where $|\mathbf{y}_i\rangle$ are n-dimensional column vectors and $\langle \mathbf{x}_i|$ are m-dimensional row vectors, and the sum is over the smallest of m and n. In seeking such an expansion, we develop what is perhaps the most important technique in matrix computations, the *singular value decomposition*.

Let us assume that we start with a general $m \times n$ matrix $\mathbf{A}$, and assume for the moment that $m < n$. We cannot write an eigenvector equation for this matrix because $m \neq n$, but we can construct two square Hermitian matrices from it: $\mathbf{X} = \mathbf{A}^\dagger \mathbf{A}$, which is $n \times n$, and $\mathbf{Y} = \mathbf{A}\mathbf{A}^\dagger$, which is $m \times m$. Because each of these matrices is Hermitian, we may find a set of eigenvectors and eigenvalues for them; we label the eigenvectors and eigenvalues of $\mathbf{X}$ as $|\mathbf{x}_i\rangle$ and x_i, with $i = 1, \ldots, n$, and the eigenvectors and eigenvalues of $\mathbf{Y}$ as $|\mathbf{y}_j\rangle$ and y_j, with $j = 1, \ldots, m$.

The matrix $\mathbf{Y}$ may be written in the form

$$\mathbf{Y} = \mathbf{A}\mathbf{A}^\dagger = \sum_{i=1}^{m} y_i |\mathbf{y}_i\rangle \langle \mathbf{y}_i|. \tag{5.28}$$

Because the vectors $|\mathbf{x}_i\rangle$ satisfy $\langle \mathbf{x}_i | \mathbf{x}_i \rangle = 1$, we may also write this as

$$\mathbf{Y} = \sum_{i=1}^{m} \sqrt{y_i} |\mathbf{y}_i\rangle \langle \mathbf{x}_i| \sum_{j=1}^{m} \sqrt{y_j} |\mathbf{x}_j\rangle \langle \mathbf{y}_j|. \tag{5.29}$$

This implies that the matrix $\mathbf{A}$ can be written in the form

$$\mathbf{A} = \sum_{i=1}^{m} \sqrt{y_i}\,|\mathbf{y}_i\rangle\,\langle\mathbf{x}_i|. \tag{5.30}$$

Similarly, the matrix $\mathbf{X}$ may be written as

$$\mathbf{X} = \mathbf{A}^\dagger\mathbf{A} = \sum_{i=1}^{n} x_i\,|\mathbf{x}_i\rangle\,\langle\mathbf{x}_i|. \tag{5.31}$$

There is a bit of a conundrum in attempting to substitute $\langle\mathbf{y}_i|\,\mathbf{y}_i\rangle$ into this expression, however, as there are only m vectors $|\mathbf{y}_i\rangle$, and $n > m$. This is resolved by noting that the rank of the matrix $\mathbf{A}$ is at most m, because the most m-dimensional vectors that can be constructed in the range of $\mathbf{A}$ is m. Assuming that $x_i = 0$ for $i > m$, we may therefore write

$$\mathbf{X} = \mathbf{A}^\dagger\mathbf{A}\sum_{i=1}^{m} x_i\,|\mathbf{x}_i\rangle\,\langle\mathbf{y}_i|\,|\mathbf{y}_i\rangle\,\langle\mathbf{x}_i|, \tag{5.32}$$

which may then be written as

$$\mathbf{X} = \mathbf{A}^\dagger\mathbf{A} = \sum_{i=1}^{m} \sqrt{x_i}\,|\mathbf{x}_i\rangle\,\langle\mathbf{y}_i|\sum_{j=1}^{m} \sqrt{x_j}\,|\mathbf{y}_j\rangle\langle\mathbf{x}_j|. \tag{5.33}$$

We have the implication that

$$\mathbf{A} = \sum_{i=1}^{m} \sqrt{x_i}\,|\mathbf{y}_i\rangle\,\langle\mathbf{x}_i|. \tag{5.34}$$

For this to be consistent with Eq. (5.30), we must have $\sqrt{x_i} = \sqrt{y_i}$. The expression that we now have for $\mathbf{A}$ is of the desired form of Eq. (5.27); this form is what we refer to as the singular value decomposition.

Definition 5.5 *The* singular value decomposition *(SVD) of an $m \times n$ matrix $\mathbf{A}$ is an expansion of the matrix in the form*

$$\mathbf{A} = \sum_{i=1}^{r} s_i\,|\mathbf{y}_i\rangle\,\langle\mathbf{x}_i|, \tag{5.35}$$

where $|\mathbf{y}_i\rangle$ are the eigenvectors of the $m \times m$ matrix $\mathbf{Y} = \mathbf{A}\mathbf{A}^\dagger$, $|\mathbf{x}_i\rangle$ are the eigenvectors of the $n \times n$ matrix $\mathbf{X} = \mathbf{A}^\dagger\mathbf{A}$, and s_i are the singular values *of the matrix, such that $s_i = \sqrt{x_i}$ and are non-negative real-valued quantities. The sum is over $r = \min(m,n)$, and it is to be noted that it is possible that some $s_i = 0$.*

From Eq. (5.35), it can be seen that the vectors $|\mathbf{y}_i\rangle$ and $|\mathbf{x}_i\rangle$ satisfy the following complementary set of "eigenvalue equations",

$$\mathbf{A}\,|\mathbf{x}_i\rangle = s_i\,|\mathbf{y}_i\rangle, \tag{5.36}$$

$$\mathbf{A}^\dagger\,|\mathbf{y}_i\rangle = s_i\,|\mathbf{x}_i\rangle. \tag{5.37}$$

This provides a straightforward way to derive the vectors $|\mathbf{x}_i\rangle$ given the vectors $|\mathbf{y}_i\rangle$ and the singular values s_i, provided none of the $s_i = 0$.

In matrix form, the SVD may be written as

$$\mathbf{A} = \mathbf{USV}^\dagger, \tag{5.38}$$

where $\mathbf{U}$ is an $m \times m$ unitary matrix constructed from the eigenvectors of $\mathbf{Y}$, $\mathbf{V}$ is an $n \times n$ unitary matrix constructed from the eigenvectors of $\mathbf{X}$, and $\mathbf{S}$ is an $m \times n$ "diagonal" matrix[1] such that $S_{ij} = \delta_{ij}s_i$. Though our initial argument assumed $m < n$, we may construct an SVD for any values of m and n.

It is hopefully clear that we have not rigorously proven the existence of the SVD and its form; rather, we have provided a plausibility argument for it. A rigorous proof may be found in [GL96]. Fortunately, our plausibility argument provides a strategy for constructing the SVD of a given matrix.

Example 5.2 We will determine the SVD of the matrix

$$\mathbf{A} = \begin{bmatrix} 1 & 1 & 0 \\ 1 & 0 & 1 \end{bmatrix}. \tag{5.39}$$

To do so, we can construct the matrices $\mathbf{X}$ and $\mathbf{Y}$, determine their eigenvalues and eigenvectors, and add them in the appropriate manner.

The matrix $\mathbf{Y}$ is of the form

$$\mathbf{Y} = \mathbf{AA}^\dagger = \begin{bmatrix} 1 & 1 & 0 \\ 1 & 0 & 1 \end{bmatrix} \begin{bmatrix} 1 & 1 \\ 1 & 0 \\ 0 & 1 \end{bmatrix} = \begin{bmatrix} 2 & 1 \\ 1 & 2 \end{bmatrix}. \tag{5.40}$$

The secular equation is $\lambda^2 - 4\lambda + 3 = 0$, and so the eigenvalues of $\mathbf{Y}$ are $\lambda = 1, 3$. By the usual methods we find the eigenvectors to be of the form

$$|\mathbf{y}_1\rangle = \frac{1}{\sqrt{2}} \begin{bmatrix} 1 \\ -1 \end{bmatrix}, \quad |\mathbf{y}_3\rangle = \frac{1}{\sqrt{2}} \begin{bmatrix} 1 \\ 1 \end{bmatrix}. \tag{5.41}$$

---

[1] Some references use an alternative expansion in which $\mathbf{U}$ is an $m \times n$ matrix, $\mathbf{S}$ is an $n \times n$ matrix and $\mathbf{V}$ is an $n \times n$ matrix. Such an SVD provides the same information as the one described here.

We may perform a similar analysis for the matrix $\mathbf{X}$, which is of the form

$$\mathbf{X} = \mathbf{A}^\dagger \mathbf{A} = \begin{bmatrix} 1 & 1 \\ 1 & 0 \\ 0 & 1 \end{bmatrix} \begin{bmatrix} 1 & 1 & 0 \\ 1 & 0 & 1 \end{bmatrix} = \begin{bmatrix} 2 & 1 & 1 \\ 1 & 1 & 0 \\ 1 & 0 & 1 \end{bmatrix}. \tag{5.42}$$

The secular equation for $\mathbf{X}$ is of the form $(1 - \lambda)[\lambda^2 - 3\lambda] = 0$, which leads to eigenvectors $\lambda = 0, 1, 3$, as expected. The eigenvectors are

$$|\mathbf{x}_0\rangle = \frac{1}{\sqrt{3}} \begin{bmatrix} 1 \\ -1 \\ -1 \end{bmatrix}, \quad |\mathbf{x}_1\rangle = \frac{1}{\sqrt{2}} \begin{bmatrix} 0 \\ 1 \\ -1 \end{bmatrix}, \quad |\mathbf{x}_3\rangle = \frac{1}{\sqrt{6}} \begin{bmatrix} 2 \\ 1 \\ 1 \end{bmatrix}. \tag{5.43}$$

The singular values of the matrix $\mathbf{A}$ are $s_1 = 1$, $s_3 = \sqrt{3}$; the singular value decomposition takes the form

$$\mathbf{A} = |\mathbf{y}_1\rangle \langle \mathbf{x}_1| + \sqrt{3} |\mathbf{y}_3\rangle \langle \mathbf{x}_3|. \tag{5.44}$$

We may verify that this is equal to the matrix $\mathbf{A}$,

$$|\mathbf{y}_1\rangle \langle \mathbf{x}_1| + \sqrt{3} |\mathbf{y}_3\rangle \langle \mathbf{x}_3| = \frac{1}{2} \begin{bmatrix} 1 \\ -1 \end{bmatrix} \begin{bmatrix} 0 & 1 & -1 \end{bmatrix} + \frac{1}{2} \begin{bmatrix} 1 \\ 1 \end{bmatrix} \begin{bmatrix} 2 & 1 & 1 \end{bmatrix}$$

$$= \frac{1}{2} \begin{bmatrix} 0 & 1 & -1 \\ 0 & -1 & 1 \end{bmatrix} + \frac{1}{2} \begin{bmatrix} 2 & 1 & 1 \\ 2 & 1 & 1 \end{bmatrix} = \mathbf{A}. \tag{5.45}$$

We may write this in matrix form as

$$\mathbf{A} = \mathbf{U}\mathbf{S}\mathbf{V}^\dagger = \begin{bmatrix} |\mathbf{x}_1\rangle & |\mathbf{x}_3\rangle \end{bmatrix} \mathrm{diag}(1, \sqrt{3}) \begin{bmatrix} \langle \mathbf{y}_1| \\ \langle \mathbf{y}_3| \\ \langle \mathbf{y}_0| \end{bmatrix}, \tag{5.46}$$

where $\mathbf{U}$ is the matrix made up of column vectors $|\mathbf{x}_i\rangle$, while $\mathbf{V}$ is the matrix made up of column vectors $|\mathbf{y}_i\rangle$, and $\mathrm{diag}(\cdots)$ represents a diagonal matrix with elements $(\cdots)$. As shown, we have

$$\mathbf{U} = \begin{bmatrix} 1/\sqrt{2} & 1/\sqrt{2} \\ -1/\sqrt{2} & 1/\sqrt{2} \end{bmatrix}, \quad \mathbf{V}^\dagger = \begin{bmatrix} 0 & 1/\sqrt{2} & -1/\sqrt{2} \\ 2/\sqrt{6} & 1/\sqrt{6} & 1/\sqrt{6} \\ 1/\sqrt{3} & -1\sqrt{3} & -1\sqrt{3} \end{bmatrix},$$

$$\mathrm{diag}(1, \sqrt{3}) = \begin{bmatrix} 1 & 0 & 0 \\ 0 & \sqrt{3} & 0 \end{bmatrix}. \tag{5.47}$$

It is to be noted that the ordering of the diagonal elements in the decomposition can be changed by changing the order of the vectors $|\mathbf{x}_i\rangle$ and $\langle \mathbf{y}_i|$ in the above matrices. This is important, as by convention the singular values are arranged in order of decreasing size.

◇

It should be noted that the method of deriving the SVD used above is conceptually straightforward, but not necessarily the most computationally efficient or stable method. Plenty of pre-packaged SVD routines are available; see, for instance, [PTVF92].

The power of the SVD lies in its ability to characterize the behavior of a matrix and even diagnose problems associated with matrix inversion and the solution of systems of equations. From the SVD, we may immediately make a number of observations. First, the rank of the matrix $\mathbf{A}$ is equal to the total number of nonzero singular values. Any vector $|\mathbf{x}_i\rangle$ for which the singular value vanishes satisfies $\mathbf{A}|\mathbf{x}_i\rangle = 0$ and therefore lies in the null space. The range of the matrix $\mathbf{A}$ is therefore the domain spanned by the set of vectors $|\mathbf{y}_i\rangle$ which have nonzero singular values.

If we consider the solution of systems of equations, we may also make some general observations. The equation $\mathbf{A}|\mathbf{x}\rangle = 0$ has a nontrivial solution if and only if the matrix has a non-empty null space, i.e. if the matrix has singular values equal to zero. Furthermore, the number of zero singular values is equal to the number of undetermined parameters in the solution.

The inhomogeneous equation $\mathbf{A}|\mathbf{x}\rangle = |\mathbf{b}\rangle$ has a solution if and only if $|\mathbf{b}\rangle$ lies in the range of $\mathbf{A}$. This in turn suggests that the vector $|\mathbf{b}\rangle$ must be expressible as a linear combination of the vectors $|\mathbf{y}_i\rangle$ which possess nonzero singular values.

It often turns out in practice that a system of equations $\mathbf{A}|\mathbf{x}\rangle = |\mathbf{b}\rangle$ has no solution. For instance, imaging systems often result in an overspecified system of equations where the amount of data taken, characterized by $|\mathbf{b}\rangle$, is larger than the number of unknowns to be determined, characterized by $|\mathbf{x}\rangle$. Though the system of equations may be solvable under perfect conditions, even a small amount of noise, added to the vector $|\mathbf{b}\rangle$, may move it out of the range of $\mathbf{A}$ and make the system unsolvable.

Even with an unsolvable system, the SVD may be used to determine the *best possible* solution, namely a solution $|\hat{\mathbf{x}}\rangle$ that minimizes the length

$$L \equiv \left| \mathbf{A}|\hat{\mathbf{x}}\rangle - |\mathbf{b}\rangle \right|. \tag{5.48}$$

We postulate that this minimum solution is given by

$$|\hat{\mathbf{x}}\rangle = \mathbf{V}\hat{\mathbf{S}}\mathbf{U}^\dagger|\mathbf{b}\rangle. \tag{5.49}$$

If the matrix $\mathbf{S}$ is an $m \times n$ diagonal matrix with diagonal elements s_i, then $\hat{\mathbf{S}}$ is an $n \times m$ diagonal matrix with diagonal elements equal to $1/s_i$ for $s_i \neq 0$ and 0 for $s_i = 0$. This may be written in vector form as

$$|\hat{\mathbf{x}}\rangle = \sum_{j=1}^{r} \frac{1}{s_j} |\mathbf{x}_j\rangle\langle\mathbf{y}_j|\mathbf{b}\rangle. \tag{5.50}$$

We first consider the case for which $|\mathbf{b}\rangle$ lies in the range of $\mathbf{A}$. We may then write

$$\mathbf{A}|\hat{\mathbf{x}}\rangle = \sum_{i=1}^{r}\sum_{j=1}^{r} |\mathbf{y}_i\rangle\langle\mathbf{x}_i|\mathbf{x}_j\rangle\langle\mathbf{y}_j|\mathbf{b}\rangle. \tag{5.51}$$

Because the $\{|x_j\rangle\}$ are orthonormal, we have $\langle x_i | x_j \rangle = \delta_{ij}$, and we may write

$$A |\hat{x}\rangle = \sum_{i=1}^{r} \langle y_j | b \rangle |y_i\rangle . \tag{5.52}$$

If $|b\rangle$ lies in the range of $|y_j\rangle$, it may be completely written in terms of those vectors. The right-hand side of the above expression is then the eigenfunction expansion of $|b\rangle$, and the equation reduces to $A |\hat{x}\rangle - |b\rangle = 0$. If $|b\rangle$ lies in the range of A, the exact solution to the equation is given by $|\hat{x}\rangle$.

Let us now consider the case for which $|b\rangle$ does not lie in the range of A, and demonstrate that the solution $|\hat{x}\rangle$ is the solution that minimizes Eq. (5.48).

We introduce a new solution $|\hat{x}'\rangle$, defined as

$$|\hat{x}'\rangle = \sum_{j=1}^{r} \frac{1}{s_j} |x_j\rangle \langle y_j | b \rangle + \sum_{j=1}^{r} a_j |x_j\rangle, \tag{5.53}$$

where the a_j are constants. If we apply our matrix A to this vector, we have

$$A |\hat{x}'\rangle = \sum_{i=1}^{r} \langle y_j | b \rangle |y_i\rangle + \sum_{j=1}^{r} a_j |y_j\rangle . \tag{5.54}$$

Because $|b\rangle$ does not lies in the range of A, it may not be expressed in terms of the first r vectors of the set $\{|y_j\rangle\}$; however, it may be determined from the entire set, i.e.

$$|b\rangle = \sum_{i=1}^{m} \langle y_j | b \rangle |y_i\rangle . \tag{5.55}$$

We may therefore write

$$A |\hat{x}'\rangle - |b\rangle = \sum_{i=r+1}^{m} \langle y_j | b \rangle |y_i\rangle + \sum_{j=1}^{r} a_j |y_j\rangle, \tag{5.56}$$

where the difference is in the range of summation of the first term. Because the vectors $|y_j\rangle$ are orthogonal, it readily follows that

$$\left| A |\hat{x}'\rangle - |b\rangle \right|^2 = \sum_{j=r+1}^{m} |\langle y_i | b \rangle|^2 + \sum_{j=1}^{r} |a_j|^2. \tag{5.57}$$

The solution that minimizes this quantity has $a_j = 0$, i.e. $|\hat{x}\rangle$.

Example 5.3 (Underspecified system of equations) We consider the solution of the system of equations $\mathbf{A}\,|\mathbf{x}\rangle = |\mathbf{b}\rangle$, where

$$\mathbf{A} = \begin{bmatrix} 1 & 1 & 0 \\ 1 & 0 & 1 \end{bmatrix} \tag{5.58}$$

is the matrix from the previous example, and

$$|\mathbf{x}\rangle = \begin{bmatrix} x_1 \\ x_2 \\ x_3 \end{bmatrix}, \quad |\mathbf{b}\rangle = \begin{bmatrix} a \\ b \end{bmatrix}. \tag{5.59}$$

This is an underspecified system of equations (more unknowns than equations), and we expect to find a non unique solution. Using the results of the previous example, we have

$$\mathbf{U} = \begin{bmatrix} 1/\sqrt{2} & 1/\sqrt{2} \\ -1/\sqrt{2} & 1/\sqrt{2} \end{bmatrix}, \quad \mathbf{V}^\dagger = \begin{bmatrix} 0 & 1/\sqrt{2} & -1/\sqrt{2} \\ 2/\sqrt{6} & 1/\sqrt{6} & 1/\sqrt{6} \\ 1/\sqrt{3} & -1\sqrt{3} & -1\sqrt{3} \end{bmatrix},$$

$$\mathbf{S} = \begin{bmatrix} 1 & 0 & 0 \\ 0 & \sqrt{3} & 0 \end{bmatrix}. \tag{5.60}$$

The matrix $\hat{\mathbf{S}}$ is then

$$\hat{\mathbf{S}} = \begin{bmatrix} 1 & 0 \\ 0 & 1/\sqrt{3} \\ 0 & 0 \end{bmatrix}. \tag{5.61}$$

The solution $|\hat{\mathbf{x}}\rangle$ can be found by direct matrix multiplication. We get the result

$$|\hat{\mathbf{x}}\rangle = \mathbf{V}\hat{\mathbf{S}}\mathbf{U}^\dagger\,|\mathbf{b}\rangle = \begin{bmatrix} \frac{1}{3}(a+b) \\ \frac{1}{3}(2a-b) \\ \frac{1}{3}(2b-a) \end{bmatrix}. \tag{5.62}$$

This is not the most general solution, however. We may add any multiple of the null vector,

$$|\mathbf{x}_0\rangle = \frac{1}{\sqrt{3}} \begin{bmatrix} 1 \\ -1 \\ -1 \end{bmatrix}, \tag{5.63}$$

to the solution without changing it, because $\mathbf{A}\,|\mathbf{x}_0\rangle = 0$. The general solution is therefore

$$|\mathbf{x}\rangle = \begin{bmatrix} \frac{1}{3}(a+b) \\ \frac{1}{3}(2a-b) \\ \frac{1}{3}(2b-a) \end{bmatrix} + \frac{\alpha}{\sqrt{3}} \begin{bmatrix} 1 \\ -1 \\ -1 \end{bmatrix}, \tag{5.64}$$

where α is an arbitrary constant.

◇

The singular value decomposition is also useful in matrix computations. We have seen that the matrix $\mathbf{A}$ can be characterized by its singular values s_i. From Eq. (5.50), we see that the solution to $\mathbf{A}\,|\mathbf{x}\rangle = |\mathbf{b}\rangle$ involves scaling the component of $|\mathbf{b}\rangle$ in the $|\mathbf{y}_i\rangle$-direction by a factor $1/s_i$. The solution of the system of equations will therefore favor those vectors corresponding to small singular values. This is typically characterized by the *condition number*.

Definition 5.6 (Condition number) *The condition number $\kappa(\mathbf{A})$ of a matrix $\mathbf{A}$ is defined as the ratio of the largest singular value of the matrix to the smallest.*

When the condition number becomes comparable to the floating-point precision of the computer system, the solution to the system of equations becomes corrupted by contributions from vectors for the smallest singular values. Any noise in vectors corresponding to those vectors gets amplified to the point that it dominates the solution, resulting in incorrect solutions.

A solution to this problem is to simply set $1/s_i = 0$ in the matrix $\hat{\mathbf{S}}$ for those singular values that are anomalously low. The computational solution to the system of equations will not be exact, but it will typically be closer to the correct answer than a solution that includes these values, and will still minimize L, subject to the additional constraint on the singular values. For very large matrices with a broad spectrum of singular values, however, it somewhat of an art form to determine which values should be included in the solution and which should be set to zero.

5.5 Other matrix manipulations

There are a large number of different techniques that can be used to manipulate matrices, usually with the goal of solving a system of equations or finding the inverse of the matrix. We describe a few of these to highlight the broad variety of techniques available.

5.5.1 Gaussian elimination with pivoting

Gaussian elimination, described in Section 4.3.1, is a remarkably effective technique for solving systems of equations. Algorithmic implementations of it can run into a significant hiccup, however, as the following example shows.

Example 5.4 We consider the solution of a 4×4 system of equations of the form $\mathbf{A}\,|\mathbf{x}\rangle = |\mathbf{b}\rangle$, which we write in so-called *augmented matrix* format,

$$\left[\begin{array}{cccc|c} 1 & 2 & 1 & 1 & 1 \\ 2 & 4 & 5 & 3 & 2 \\ 1 & 4 & 2 & 5 & 2 \\ 3 & 7 & 4 & 4 & 4 \end{array}\right], \tag{5.65}$$

where the matrix and the vector $|\mathbf{b}\rangle$ are written in a single matrix, separated by a vertical bar, for convenience. Performing the first step in the elimination, we have

$$
\left[
\begin{array}{cccc|c}
1 & 2 & 1 & 1 & 1 \\
0 & 0 & 3 & 1 & 0 \\
0 & 2 & 1 & 4 & 1 \\
0 & 1 & 1 & 1 & 1
\end{array}
\right].
\tag{5.66}
$$

Here we have run into the hiccup: we cannot use the second row to perform elimination of the second column, because the element $a_{22} = 0$. A computational algorithm that automatically attempted to use this row to perform row reduction would fail.

The problem is easily surmountable, however. As each row represents a single equation of the system, we can interchange any two rows without changing the solution, a process called *pivoting*; the new element used to perform row reduction is referred to as the *pivot element*. In our example, we exchange the second and third rows, to get

$$
\left[
\begin{array}{cccc|c}
1 & 2 & 1 & 1 & 1 \\
0 & 2 & 1 & 4 & 1 \\
0 & 0 & 3 & 1 & 0 \\
0 & 1 & 1 & 1 & 1
\end{array}
\right].
\tag{5.67}
$$

Proceeding as usual from this point, we arrive at the final form

$$
\left[
\begin{array}{cccc|c}
1 & 2 & 1 & 1 & 1 \\
0 & 1 & 1/2 & 2 & 1/2 \\
0 & 0 & 1 & 1/3 & 0 \\
0 & 0 & 0 & -7/6 & 1/2
\end{array}
\right].
\tag{5.68}
$$

The system of equations is therefore solved by $x_1 = -9/7$, $x_2 = 9/7$, $x_3 = 1/7$, $x_4 = -3/7$.

$\diamond$

Pivoting is not just important when the pivot element is identically zero. If the pivot element is extremely small compared to other elements in the same column, it will need to be multiplied by a very large number. If such a scaling needs to be done many times for a very large matrix, it can introduce significant round-off error that can produce a nonsensical solution. Algorithms with pivoting therefore use the largest element in a particular column as the pivot element.

Good algorithms typically do not physically swap the pivoted rows in memory, which is an inefficient process, but rather merely keep track of the location and order of any pivoting done.

The process described here is known as *partial pivoting*. In some cases, *complete pivoting* is used, in which both rows and columns are interchanged to produce the most stable solution. Complete pivoting is typically not necessary, however, and in most applications partial pivoting is sufficient.

5.5.2 LU decomposition

While the singular value decomposition (SVD) of Section 5.4 is arguably the most important and useful matrix decomposition technique, numerous other decompositions exist that
can be used to simplify the solution of a system of equations or the determination of an
inverse matrix. Perhaps the most commonly encountered of these is the LU, or lower-upper,
decomposition, which we take some time to describe.

The LU decomposition is the factorization of a general square matrix $\mathbf{A}$ of the form

$$\mathbf{A} = \mathbf{LU}, \tag{5.69}$$

where $\mathbf{L}$ is a *lower triangular matrix* with all elements zero above the diagonal, i.e.

$$\mathbf{L} = \begin{bmatrix} l_{11} & 0 & 0 & \cdots & 0 \\ l_{21} & l_{22} & 0 & & 0 \\ l_{31} & l_{32} & l_{33} & & 0 \\ \vdots & & & \ddots & 0 \\ l_{n1} & l_{n2} & l_{n3} & \cdots & l_{nn} \end{bmatrix}, \tag{5.70}$$

and $\mathbf{U}$ is an *upper triangular matrix* with all elements zero below the diagonal, i.e.

$$\mathbf{U} = \begin{bmatrix} u_{11} & u_{12} & u_{13} & \cdots & u_{1n} \\ 0 & u_{22} & u_{23} & & u_{2n} \\ 0 & 0 & u_{33} & & u_{3n} \\ \vdots & & & \ddots & 0 \\ 0 & 0 & 0 & \cdots & u_{nn} \end{bmatrix}. \tag{5.71}$$

Such a decomposition is unique if the diagonal elements of the matrix $\mathbf{L}$ are taken to be
unity, i.e. $l_{ii} = 1$. Not every matrix can be factorized in this form, however, as we will
discuss below.

If the LU decomposition of a matrix can be found, the solution of the system of equations
$\mathbf{A}\,|\mathbf{x}\rangle = |\mathbf{b}\rangle$ may be readily found as follows. We start with $\mathbf{LU}\,|\mathbf{x}\rangle = |\mathbf{b}\rangle$, and define $|\mathbf{y}\rangle \equiv$
$\mathbf{U}\,|\mathbf{x}\rangle$. We first solve the system of equations $\mathbf{L}\,|\mathbf{y}\rangle = |\mathbf{b}\rangle$, which is straightforward due
to the lower triangular nature of $\mathbf{L}$. Starting at the top of the matrix, we have $y_1 = b_1$,
$l_{21}y_1 + y_2 = b_2$, and so on. Then we may solve for $|\mathbf{x}\rangle$ by using the equation $\mathbf{U}\,|\mathbf{x}\rangle = |\mathbf{y}\rangle$,
which is also straightforward due to the upper triangular nature of $\mathbf{U}$. We therefore are able
to solve the system of equations without explicitly calculating the inverse of the matrix.

The LU decomposition of a general matrix may be determined by the application of
Gaussian elimination to the matrix $\mathbf{A}$ and its adjoint, $\mathbf{A}^{\dagger}$. The process of Gaussian elimination
may be imagined as the application of a number of transformation matrices $\mathbf{M}_i$ to the
matrix $\mathbf{A}$, the result of which is a row-reduced matrix in upper diagonal form, i.e.

$$\mathbf{M}_N \cdots \mathbf{M}_3\mathbf{M}_2\mathbf{M}_1\mathbf{A} = \mathbf{U}. \tag{5.72}$$

The matrix $\mathbf{U}$ is simply the result of a Gaussian elimination of $\mathbf{A}$, *without normalizing each of the diagonal elements*, by convention. If we apply the inverse $\mathbf{M}_i^{-1}$ of each transformation matrix to this equation in reverse order, we have

$$\mathbf{A} = \mathbf{M}_1^{-1}\mathbf{M}_2^{-1}\mathbf{M}_3^{-1}\cdots\mathbf{M}_N^{-1}\mathbf{U}, \tag{5.73}$$

which implies that

$$\mathbf{L} = \mathbf{M}_1^{-1}\mathbf{M}_2^{-1}\mathbf{M}_3^{-1}\cdots\mathbf{M}_N^{-1}. \tag{5.74}$$

To prove that this is a lower diagonal matrix, we need the explicit form of the transformation matrices. A matrix that scales the kth row by a constant α is simply the identity matrix with the kth diagonal element replaced by α. A matrix that adds α times the kth row to the lth row, with $l > k$, is the identity matrix with the l,kth element replaced by α. Both types of transformation matrix are lower triangular, with $m_{ij} = 0$ for $j > i$.

We can readily show that the product of two lower triangular matrices is also lower triangular. Suppose we take the product of two matrices $\mathbf{M}$ and $\mathbf{N}$, such that $m_{ij} = 0$ for $j > i$ and $n_{ij} = 0$ for $j > i$. The coefficients of the product are given by

$$(\mathbf{MN})_{ij} = \sum_{k=1}^{N} m_{ik}n_{kj} = \sum_{k=1}^{i} m_{ik}n_{kj} + \sum_{k=i+1}^{N} m_{ik}n_{kj}. \tag{5.75}$$

Let us suppose that $j > i$. In the first summation, $n_{kj} = 0$ since $j > k$, and in the second summation, $m_{ik} = 0$ since $k > i$. Therefore the matrix $\mathbf{MN}$ is also lower triangular, and we have shown that the matrix $\mathbf{L}$ as given by Eq. (5.74) is also lower triangular.

We may also find the form of the matrix $\mathbf{L}$ by Gaussian elimination. If we take the adjoint of Eq. (5.69), we have

$$\mathbf{U}^\dagger\mathbf{L}^\dagger = \mathbf{A}^\dagger. \tag{5.76}$$

Because $\mathbf{L}$ is a lower triangular matrix, $\mathbf{L}^\dagger$ is an upper triangular matrix and the result of performing Gaussian elimination on $\mathbf{A}^\dagger$, with the diagonal elements normalized to unity.

Example 5.5 We attempt to find the LU decomposition of the matrix $\mathbf{A}$, given by

$$\mathbf{A} = \begin{bmatrix} 1 & 2 & 2 \\ 1 & 0 & 1 \\ 0 & 1 & 0 \end{bmatrix}. \tag{5.77}$$

If we perform Gaussian elimination on $\mathbf{A}$, we find that the matrix $\mathbf{U}$ is given by

$$\mathbf{U} = \begin{bmatrix} 1 & 2 & 2 \\ 0 & -2 & -1 \\ 0 & 0 & -1/2 \end{bmatrix}. \tag{5.78}$$

The adjoint of **A** is

$$\mathbf{A}^\dagger = \begin{bmatrix} 1 & 1 & 0 \\ 2 & 0 & 1 \\ 2 & 1 & 0 \end{bmatrix}. \tag{5.79}$$

Applying Gaussian elimination to this matrix, with normalization such that the diagonal elements are unity, we have

$$\mathbf{L}^\dagger = \begin{bmatrix} 1 & 1 & 0 \\ 0 & 1 & -1/2 \\ 0 & 0 & 1 \end{bmatrix}. \tag{5.80}$$

We may readily verify that the product **LU** is equal to **A**,

$$\begin{bmatrix} 1 & 0 & 0 \\ 1 & 1 & 0 \\ 0 & -1/2 & 1 \end{bmatrix} \begin{bmatrix} 1 & 2 & 2 \\ 0 & -2 & -1 \\ 0 & 0 & -1/2 \end{bmatrix} = \begin{bmatrix} 1 & 2 & 2 \\ 1 & 0 & 1 \\ 0 & 1 & 0 \end{bmatrix}. \tag{5.81}$$

◇

It is to be noted that this method of determining the LU decomposition, though conceptually simple, is not the most efficient technique. The most commonly used technique is known as *Crout's algorithm*, which is described in [PTVF92]. When efficiently implemented, the LU decomposition allows the solution of a system of equations without the extra effort of finding the inverse matrix $\mathbf{A}^{-1}$.

Not all matrices possess an LU decomposition, however. Consider a matrix which, in the course of Gaussian elimination, naturally results in a zero pivot element. A lower row can then be used to perform the elimination, but the matrix that results at the end of the calculation will not be upper triangular and hence cannot be part of an LU decomposition. It can be shown that the LU decomposition exists if and only if all of its *leading principal minors* are non zero. These minors are the determinants of all square matrices formed with the quadratic upper-left part of the original matrix.

When the LU decomposition of a matrix **A** does not exist, a more general decomposition known as the *LUP decomposition* can be found, of the form

$$\mathbf{PA} = \mathbf{LU}. \tag{5.82}$$

The matrix **P** is a permutation matrix that pivots the matrix to a form that does not have zero leading principal minors. For instance, the 3×3 matrix that interchanges the second and third rows of a matrix **A** is of the form

$$\mathbf{P} = \begin{bmatrix} 1 & 0 & 0 \\ 0 & 0 & 1 \\ 0 & 1 & 0 \end{bmatrix}. \tag{5.83}$$

An LUP decomposition of a matrix always exists; if the goal of the decomposition is to find the solution of a system of equations, that solution is unchanged by the operation of the matrix **P**.

Example 5.6 We consider the decomposition of the matrix

$$\mathbf{A} = \begin{bmatrix} 1 & 2 & 1 \\ 2 & 4 & 4 \\ 1 & 3 & 3 \end{bmatrix}. \tag{5.84}$$

The first step of Gaussian elimination reduces this matrix to the form

$$\begin{bmatrix} 1 & 2 & 1 \\ 0 & 0 & 2 \\ 0 & 1 & 2 \end{bmatrix}, \tag{5.85}$$

which has a zero pivot element at the $(2,2)$ location, implying that this matrix cannot be decomposed in a standard LU decomposition. Alternatively, we note that the 2×2 leading principal minor has determinant

$$\det \begin{bmatrix} 1 & 2 \\ 2 & 4 \end{bmatrix} = 0. \tag{5.86}$$

We seek an LUP decomposition by applying the permutation matrix of Eq. (5.83) to **A**,

$$\mathbf{PA} = \begin{bmatrix} 1 & 2 & 1 \\ 1 & 3 & 3 \\ 2 & 4 & 4 \end{bmatrix}. \tag{5.87}$$

Performing Gaussian elimination on this matrix, we have

$$\mathbf{U} = \begin{bmatrix} 1 & 2 & 1 \\ 0 & 1 & 2 \\ 0 & 0 & 2 \end{bmatrix}. \tag{5.88}$$

Performing Gaussian elimination on the matrix $(\mathbf{PA})^\dagger$, we have

$$\mathbf{L}^\dagger = \begin{bmatrix} 1 & 1 & 2 \\ 0 & 1 & 0 \\ 0 & 0 & 1 \end{bmatrix}. \tag{5.89}$$

We may readily confirm that the product **LU** results in the matrix **PA**,

$$\mathbf{LU} = \begin{bmatrix} 1 & 0 & 0 \\ 1 & 1 & 0 \\ 2 & 0 & 1 \end{bmatrix} \begin{bmatrix} 1 & 2 & 1 \\ 0 & 1 & 2 \\ 0 & 0 & 2 \end{bmatrix} = \begin{bmatrix} 1 & 2 & 1 \\ 1 & 3 & 3 \\ 2 & 4 & 4 \end{bmatrix}. \tag{5.90}$$

5.6 Tensors

With the introduction of matrices, we now have three very distinct types of physical quantity: scalars, vectors, and matrices. The components of vectors can be characterized by a single index i, i.e. the components of $|\mathbf{v}\rangle$ are v_i, while the elements of matrices can be characterized by two indices i and j, i.e. the components of $\mathbf{A}$ are a_{ij}. A scalar is characterized by a single number with no indices. We have also seen a quantity that possess three indices, however, namely the Levi-Civita symbol ϵ_{ijk}, defined by Eq. (1.52).

All of these are examples of a general class of objects referred to as *tensors*. The *rank* of a tensor refers to the number of indices needed to characterize it, and hence the number of elements needed to describe it. A scalar is a tensor of rank 0, a vector is a tensor of rank 1, and a matrix is a tensor of rank 2. The Levi-Civita symbol ϵ_{ijk} is a tensor of rank 3. In three-dimensional space, a general tensor of rank N will possess 3^N elements.

Not every collection of numbers represents a tensor. In Section 1.2 we noted that a vector must transform in a certain way under a change of coordinates. Similarly, a general object $T_{ijk...mnp}$ is a tensor only if it satisfies a more general transformation relation.

The advantage of tensor analysis is at least twofold. First, it allows for the construction of physical quantities more general than the matrices we have been discussing, and to deal with their transformation properties in an elegant manner. Second, we may use tensor analysis to characterize such physical quantities in curved coordinates and non-orthogonal (skew) coordinate systems.

It is relatively rare to require the full power of tensor analysis in optical problems, though this is quickly changing thanks to the introduction of the field of *transformation optics*, to be discussed in Section 10.7. In this section we sketch some of the basic ideas and results from the theory.[2]

5.6.1 Einstein summation convention

It will be convenient before we begin to introduce a summation convention first used by Einstein. It is generally observed that the only time a pair of equal indices appear on the same side of a tensor equation, they are summed over; for instance, the coordinate system rotations of Section 1.2 have the form

$$x'_i = \sum_{j=1}^{3} a_{ij}x_j, \tag{5.91}$$

where x'_i are the components in the new coordinate system, x_j are the components in the old system, and a_{ij} are the components of the rotation tensor. The summation convention is that repeated indices always imply summation unless explicitly stated otherwise; the previously

---

[2] The original paper of Levi-Civita and Ricci on tensor analysis is translated in [Her75]; also available freely through Google books.

stated rotation equation reduces to the compact form,

$$x'_i = a_{ij}x_j. \tag{5.92}$$

5.6.2 Covariant and contravariant vectors and tensors

When working in Cartesian coordinates, we have seen that we may expand an arbitrary vector into a set of orthogonal components, of the form

$$\mathbf{A} = A_x\hat{\mathbf{x}} + A_y\hat{\mathbf{y}} + A_z\hat{\mathbf{z}}. \tag{5.93}$$

The components can be determined using the orthonormality of the unit vectors, e.g. $A_x = \mathbf{A} \cdot \hat{\mathbf{x}}$.

When we consider the expansion of a vector in a basis where the basis vectors $(\mathbf{e}_1, \mathbf{e}_2, \mathbf{e}_3)$ are not necessarily orthogonal or of unit length, i.e. a skew coordinate system, the situation is more complicated. The vector should have a representation of the form

$$\mathbf{A} = A^1\mathbf{e}_1 + A^2\mathbf{e}_2 + A^3\mathbf{e}_3 = A^i\mathbf{e}_i. \tag{5.94}$$

In order to determine the components A^i, we need to find a complementary set of vectors $(\mathbf{e}^1, \mathbf{e}^2, \mathbf{e}^3)$ such that

$$\mathbf{e}_i \cdot \mathbf{e}^j = \delta_{ij}. \tag{5.95}$$

We have already found the solution to this problem in the context of crystal structure in Section 1.6; the set of vectors defined by

$$\mathbf{e}^1 = \frac{\mathbf{e}_2 \times \mathbf{e}_3}{\mathbf{e}_1 \cdot (\mathbf{e}_2 \times \mathbf{e}_3)}, \quad \mathbf{e}^2 = \frac{\mathbf{e}_3 \times \mathbf{e}_1}{\mathbf{e}_1 \cdot (\mathbf{e}_2 \times \mathbf{e}_3)}, \quad \mathbf{e}^3 = \frac{\mathbf{e}_1 \times \mathbf{e}_2}{\mathbf{e}_1 \cdot (\mathbf{e}_2 \times \mathbf{e}_3)}, \tag{5.96}$$

satisfy Eq. (5.95). With this set, we may write the components of $\mathbf{A}$ in the form $A^i = \mathbf{A} \cdot \mathbf{e}^i$.

However, the vectors $\mathbf{e}^i$ also span three-dimensional space, and we may also represent our vector $\mathbf{A}$ in the form

$$\mathbf{A} = A_1\mathbf{e}^1 + A_2\mathbf{e}^2 + A_3\mathbf{e}^3 = A_i\mathbf{e}^i, \tag{5.97}$$

where $A_i = \mathbf{A} \cdot \mathbf{e}_i$.

The vector components A_i are referred to as the *covariant* components of $\mathbf{A}$, while the vector components A^i are referred to as the *contravariant* components of $\mathbf{A}$. The set of vectors $\{\mathbf{e}_i\}$ are called the covariant basis vectors and the set $\{\mathbf{e}^i\}$ the contravariant basis vectors; the two sets form a biorthogonal system, as discussed in Section 5.3. The distinction between these two representations of vectors lies in the transformation properties of the vector components.

Let us suppose we have a coordinate transformation, represented by a tensor α^i_j, that can be used to represent the components of a contravariant vector in the new coordinate system, i.e.

$$\hat{A}^i = \alpha^i_j A^j. \tag{5.98}$$

We will address the placement of the indices momentarily. This transformation will also affect the form of the original basis vectors, which take on the form

$$\hat{\mathbf{e}}_i = \beta_i^l \mathbf{e}_l. \tag{5.99}$$

Because the coordinate transformation cannot change the physical meaning of a vector, we must have

$$\mathbf{A} = A^i \mathbf{e}_i = \hat{A}^i \hat{\mathbf{e}}_i = A^j \alpha_j^i \beta_i^l \mathbf{e}_l. \tag{5.100}$$

It is to be noted that the latter part of this expression has three summations, over i, j, and l. The invariance of the vector under transformations implies that

$$\alpha_j^i \beta_i^l = \delta_j^l, \tag{5.101}$$

where δ_j^i is the Kronecker delta. In terms of rotation matrices, this implies that $\beta_i^l = (\alpha_l^i)^{\mathrm{T}}$.

We also expect that the inner product of a contravariant and covariant vector must be invariant under rotations. Assuming that a contravariant vector transforms according to the rule $\hat{A}_i = \gamma_i^j A_j$, we may write

$$\hat{A}^i \hat{B}_i = \alpha_j^i \gamma_i^l A^j B_l. \tag{5.102}$$

For this equation to be satisfied, we must have

$$\alpha_j^i \gamma_i^l = \delta_j^l, \tag{5.103}$$

which implies that a contravariant vector transforms as $\gamma_i^l = (\alpha_l^i)^{\mathrm{T}}$.

We are led to the argument that the covariant components of a vector transform according to the transformation

$$\hat{A}_i = \alpha_i^k A_k, \tag{5.104}$$

while the contravariant components of a vector transform according to the transformation

$$\hat{A}^i = \alpha_k^i A^k. \tag{5.105}$$

We may now introduce the idea of contravariant, covariant, and mixed tensors. A *covariant tensor* $T_{lmn...pqr}$ (all indices down) is one whose components transform as

$$\hat{T}_{lmn...pqr} = \alpha_l^{l'} \alpha_m^{m'} \cdots \alpha_q^{q'} \alpha_r^{r'} T_{l'm'n'...p'q'r'}. \tag{5.106}$$

A *contravariant tensor* (all indices up) is one whose components transform as

$$\hat{T}^{lmn...pqr} = \alpha_{l'}^l \alpha_{m'}^m \cdots \alpha_{q'}^q \alpha_{r'}^r T^{l'm'n'...p'q'r'}. \tag{5.107}$$

Finally, a *mixed tensor* (some indices up, some down) is one whose components transform as

$$\hat{T}^{lmn...}_{...pqr} = \alpha_{l'}^l \alpha_{m'}^m \cdots \alpha_q^{q'} \alpha_r^{r'} T^{l'm'n'...}_{...p'q'r'}. \tag{5.108}$$

There is some subtlety in the definition of a mixed tensor because the order of the indices is important. Referring back to the coordinate rotations of Section 1.2, it is clear that the transformation $x_i' = a_{ij}x_j$ is distinct from $x_i'' = a_{ji}x_j$; in fact the two transformations are rotations in opposing directions. When necessary, we will indicate the order of indices with a placement of dots. For instance, the tensor $A^k_{\cdot lm}$ has first a contravariant, then two covariant, indices, while $A^{\cdot k}_{l \cdot m}$ has a covariant, then a contravariant, then a covariant index. Such distinctions typically only matter, however, when an index is raised or lowered by use of the metric tensor to be discussed in Section 5.6.4.

A natural question remains: what is the significance of contravariance versus covariance? We consider a rotation from a set of skew coordinates x_i to a different set x_i'. We first note, as in Section 1.2, that the position vector transforms under rotation as

$$x_i' = \frac{\partial x_i'}{\partial x_j} x_j. \tag{5.109}$$

Labeling $\alpha_j^i = \frac{\partial x_i'}{\partial x_j}$, we designate this a contravariant transformation; the coordinates of the position vector will now formally be written as x^i. However, in Section 2.3, the gradient of a scalar function transforms as

$$\frac{\partial \phi'(x_1', x_2', x_3')}{\partial x_i'} = \frac{\partial x_j}{\partial x_i'} \frac{\partial \phi}{\partial x_j} = \alpha_i^j \frac{\partial \phi}{\partial x_j}. \tag{5.110}$$

The gradient transforms as a covariant vector, while the position vector transforms as a contravariant vector. For Cartesian coordinates, the distinction between the types of vectors (and tensors) vanishes, but it proves significant in skew systems and curvilinear systems.

It is worth emphasizing once more that a contravariant vector and a covariant vector are different representations of the *same* physical vector; some authors refer to them instead as the "contravariant representation" and "covariant representation". The distinction between the representations is that a contravariant vector transforms in an ordinary manner, while the covariant vector transforms in such a way as to maintain the biorthogonal relationship between the two representations.

5.6.3 *Tensor addition and tensor products*

Tensor addition may be performed term by term, the only criterion being that the rank of the tensors must be the same and the number of covariant and contravariant indices must be equal. Therefore, $A^{ij}_{\cdot\cdot lmn} + B^{ij}_{\cdot\cdot lmn}$ is allowed, while $A^{ij}_{\cdot\cdot lmn} + B^i_{\cdot jlmn}$ is not.

There are two types of tensor product that may be introduced, an *inner product*, also known as a *contraction*, and an *outer product*, also known as the *direct product*. Contraction involves setting a pair of indices, one contravariant and one covariant, equal and summing over them. The contraction of the tensor A^j_i is A^i_i; the inner product of two vectors A^i and B_j, is given by $A^i B_i$. The act of contraction reduces the rank of a tensor by 2, removing one contravariant index and one covariant index.

The outer product is simply the direct product of two tensors of any rank, resulting in a tensor with a rank equal to the sum of the individual tensor ranks. The product of A_{ij} and B^{lm}, in that order, is $C_{ij\cdot\cdot}^{\cdot\cdot lm}$. It is important to note that tensor multiplication is non-commutative, i.e.

$$C_{ijkl} = A_{ij}B_{kl} \neq C_{klij} = A_{kl}B_{ij}. \tag{5.111}$$

There is an important theorem related to products of tensors, known as the *quotient theorem*. We define it for second-rank tensors; extensions to higher-order tensors can be readily shown.

Theorem 5.1 (Quotient theorem) *We consider an array of numbers X_i^j, and an arbitrary tensor A_{kl}. If the contraction $X_i^j A_{jl}$ results in a second-rank covariant tensor C_{il}, then the object X_i^j must be a second-rank mixed tensor. Similarly, if the outer product $X_i^j A_{kl}$ results in a second-rank mixed tensor C_{ikl}^j, then the object X_i^j must be a second-rank mixed tensor.*

In short, the quotient theorem suggests that only the product (inner or outer) of tensors can produce another tensor. If the product of a tensor with an array of numbers results in another tensor, then the array must be a tensor itself.

We can prove this theorem using the transformation properties of tensors. Starting with the equation

$$X_i^j A_{jl} = C_{il}, \tag{5.112}$$

we consider a rotation into a new coordinate system, in which the equation has the form

$$\hat{X}_i^j \hat{A}_{jl} = \hat{C}_{il}. \tag{5.113}$$

Because C_{il} transforms as a tensor, we may write

$$\hat{X}_i^j \hat{A}_{jl} = \alpha_i^{i'} \alpha_l^{l'} C_{i'l'}. \tag{5.114}$$

We may operate on both sides of this expression by the inverse coordinate transformation; noting that $\alpha_k^i \alpha_i^{i'} = \delta_k^{i'}$, we may write

$$\alpha_k^i \alpha_m^l \hat{X}_i^j \hat{A}_{jl} = \delta_k^{i'} \delta_m^{l'} C_{i'l'} = C_{km}. \tag{5.115}$$

Because A_{jl} also transforms as a tensor, this expression may be written as

$$\alpha_k^i \alpha_m^l \hat{X}_i^j \alpha_j^{j'} \alpha_l^{l'} A_{j'l'} = C_{km}. \tag{5.116}$$

We may then use $\alpha_m^l \alpha_l^{l'} = \delta_m^{l'}$; our equation reduces to the form

$$\left[\alpha_k^i \alpha_j^{j'} \hat{X}_i^j \right] A_{j'm} = C_{km}. \tag{5.117}$$

To be consistent with Eq. (5.112), we find that we must have

$$\alpha_k^i \alpha_j^{j'} \hat{X}_i^j = X_k^{j'}.$$ (5.118)

We may invert this expression by again applying rotations to find that

$$\hat{X}_i^j = \alpha_i^k \alpha_l^j X_k^l.$$ (5.119)

This equation represents the expected transformation properties of a second rank mixed tensor.

5.6.4 The metric tensor

One natural question that arises is the relationship between the contravariant and covariant representations of a vector. We address this by first considering a seemingly different problem, namely the infinitesimal length element in a skew coordinate system.

We know that the infinitesimal length element in Cartesian coordinates may be represented as

$$ds^2 = (dx^i)^2,$$ (5.120)

where the summation convention is applied and the infinitesimal components are contravariant. If we assume a transformation α_j^i to a non-orthogonal set of coordinates y^i such that $dx^i = \alpha_j^i dy^j$, we may write

$$ds^2 = \alpha_j^i \alpha_k^i dy^j dy^k.$$ (5.121)

The components of the transformation tensor α_j^i are readily found to be

$$\alpha_j^i = \frac{\partial x^i}{\partial y^j},$$ (5.122)

which implies that the length element in non-orthogonal coordinates may be written as

$$ds^2 = g_{jk} dy^j dy^k,$$ (5.123)

where g_{jk} is the *metric tensor*, given by

$$g_{jk} = \frac{\partial x^i}{\partial y^j} \frac{\partial x^i}{\partial y^k}.$$ (5.124)

On comparison with Eq. (5.110), it is clear that the metric tensor behaves as a covariant tensor. From Eq. (5.124), we can see that it is a symmetric tensor, i.e. $g_{ij} = g_{ji}$. This is the same definition of a metric as given in the discussion of curvilinear coordinates in Section 3.2.

As the length element is a contravariant vector, we expect that any contravariant vector
will have a length given by

$$|A|^2 = g_{ij}A^jA^k, \tag{5.125}$$

and this length is independent of the choice of coordinate system. However, the length of
a vector may also be written as $|A|^2 = A^iA_i$, which implies that we may write

$$A_i = g_{ij}A^j. \tag{5.126}$$

The metric tensor operating on a contravariant vector transforms it into a covariant vector.
More generally, the metric tensor may be used to lower a contravariant index on a tensor
to a covariant one, e.g.

$$g_{ij}A^{jk\cdots}_{\cdots lm} = A^{\cdot k\cdot}_{i\cdot lm}. \tag{5.127}$$

We may see this process more explicitly by use of the following argument. A contravariant
vector may be written as $\mathbf{A}_+ = A^i\mathbf{e}_i$, while a covariant vector may be written as $\mathbf{A}_- = A_i\mathbf{e}^i$;
it should be emphasized again that $\mathbf{A}_-$ and $\mathbf{A}_+$ are representations of the *same* vector $\mathbf{A}$,
i.e. $\mathbf{A}_+ = \mathbf{A}_- = \mathbf{A}$. We may introduce an operator that transforms a contravariant vector
into a covariant one of the form

$$\mathbf{P}_- \equiv \mathbf{e}^i\mathbf{e}_i, \tag{5.128}$$

which is a direct product, or *dyadic product*, of two vectors. Dyads will be discussed
much later, in Section 19.9; for now we note that $\mathbf{P}_-$ "operates" on a vector $\mathbf{A}$ by taking
the dot product of the latter unit vector with $\mathbf{A}$, with the former unit vector remaining,
i.e. $\mathbf{P}_-\mathbf{A} = \mathbf{e}^i(\mathbf{e}_i \cdot \mathbf{A})$.

It is clear that the following relation holds, since $\mathbf{e}_i \cdot \mathbf{e}^j = \delta^j_i$,

$$\mathbf{P}_-\mathbf{A} = [\mathbf{e}^i\mathbf{e}_i] \cdot \mathbf{A}_- = \mathbf{e}^iA_i = \mathbf{A}_-. \tag{5.129}$$

Alternatively, we may write

$$\mathbf{P}_-\mathbf{A} = [\mathbf{e}^i\mathbf{e}_i] \cdot \mathbf{A}_+ = \mathbf{e}^i[\mathbf{e}_i \cdot \mathbf{e}_j]A^j. \tag{5.130}$$

This shows that the *i*th component of the covariant vector may be written as

$$A_i = [\mathbf{e}_i \cdot \mathbf{e}_j]A^j. \tag{5.131}$$

If we consider the transformation of a contravariant vector from a Cartesian basis to a skew
basis, we may write

$$A^i = \frac{\partial y^i}{\partial x^j}A^j. \tag{5.132}$$

For the vector to remain invariant under this transformation, however, the covariant basis
vector $\mathbf{e}_i$ must transform as

$$\mathbf{e}_i = \frac{\partial x_i}{\partial y_j}\mathbf{x}_j. \tag{5.133}$$

This suggests that we may write

$$[\mathbf{e}_i \cdot \mathbf{e}_j]_{lm} = \frac{\partial x_i}{\partial y_l}\mathbf{x}_l \cdot \mathbf{x}_m\frac{\partial x_j}{\partial y_m} = \frac{\partial x_i}{\partial y_l}\frac{\partial x_i}{\partial y_m} = g_{lm}. \tag{5.134}$$

We have arrived at the metric tensor again, illustrating that it converts a contravariant vector into a covariant one.

We may also consider the tensor that converts a covariant vector into a contravariant one, which presumably must satisfy

$$A^i = g^{ij}A_j. \tag{5.135}$$

We may contract this vector with the metric g_{ki}, resulting in the expression

$$A_k = g_{ki}A^i = g_{ki}g^{ij}A_j. \tag{5.136}$$

Evidently the tensor g^{ij} is simply the inverse of the tensor g_{ij}, and must satisfy

$$g_{ki}g^{ij} = \delta_k^j. \tag{5.137}$$

Provided our skew coordinate system is invertible, which will be true if the basis vectors are linearly independent, the tensor g^{ij} can be determined from g_{ij} by the use of familiar matrix inversion techniques.

Example 5.7 The introduction of contravariant and covariant vectors and their associated metric tensors is enough to drive one quite mad;[3] however, these are generalizations of things we are already familiar with and not quite as mysterious as they first appear. A simple example, showing these concepts in action, will prove quite helpful here.

We consider a vector that has a Cartesian representation,

$$\mathbf{A} = 2\hat{\mathbf{x}} + \hat{\mathbf{y}} + 2\hat{\mathbf{z}}, \tag{5.138}$$

and we wish to represent it in a skew coordinate system with (covariant) basis vectors

$$\mathbf{e}_1 = \hat{\mathbf{x}} + \hat{\mathbf{y}}, \tag{5.139}$$

$$\mathbf{e}_2 = \hat{\mathbf{y}}, \tag{5.140}$$

$$\mathbf{e}_3 = \hat{\mathbf{z}}. \tag{5.141}$$

It is not difficult to figure out that the contravariant vector can be written as

$$\mathbf{A}_+ = 2\mathbf{e}_1 - \mathbf{e}_2 + 2\mathbf{e}_3. \tag{5.142}$$

[3] It has already had that effect on the author.

We may figure out the contravariant basis vectors by use of Eqs. (5.96); we then have

$$\mathbf{e}^1 = \frac{\mathbf{e}_2 \times \mathbf{e}_3}{\mathbf{e}_1 \cdot (\mathbf{e}_2 \times \mathbf{e}_3)} = \hat{\mathbf{x}}, \tag{5.143}$$

$$\mathbf{e}^2 = \frac{\mathbf{e}_3 \times \mathbf{e}_1}{\mathbf{e}_1 \cdot (\mathbf{e}_2 \times \mathbf{e}_3)} = \hat{\mathbf{y}} - \hat{\mathbf{x}}, \tag{5.144}$$

$$\mathbf{e}^3 = \frac{\mathbf{e}_1 \times \mathbf{e}_2}{\mathbf{e}_1 \cdot (\mathbf{e}_2 \times \mathbf{e}_3)} = \hat{\mathbf{z}}. \tag{5.145}$$

One can readily verify that $\mathbf{e}_i \cdot \mathbf{e}^j = \delta_i^j$. In this contravariant basis, we may write

$$\mathbf{A}_- = 3\mathbf{e}^1 + \mathbf{e}^2 + 2\mathbf{e}^3. \tag{5.146}$$

We can construct the metric tensor for our skew coordinate system by writing the position vector in both Cartesian and skew coordinates,

$$\mathbf{r} = x\hat{\mathbf{x}} + y\hat{\mathbf{y}} + z\hat{\mathbf{z}}, \tag{5.147}$$

$$\mathbf{r} = y_1\mathbf{e}_1 + y_2\mathbf{e}_2 + y_3\mathbf{e}_3, \tag{5.148}$$

or as

$$\mathbf{r} = y_1\hat{\mathbf{x}} + (y_1 + y_2)\hat{\mathbf{y}} + y_3\hat{\mathbf{z}}. \tag{5.149}$$

From here, we may calculate all partial derivatives of (x,y,z) with respect to (y_1,y_2,y_3); by the use of Eq. (5.124), we may write the matrix $\mathbf{G}$ of the metric g_{ij} as

$$\mathbf{G} = \begin{bmatrix} 2 & 1 & 0 \\ 1 & 1 & 0 \\ 0 & 0 & 1 \end{bmatrix}. \tag{5.150}$$

We may now use straightforward matrix multiplication to demonstrate

$$\mathbf{GA}_+ = \begin{bmatrix} 2 & 1 & 0 \\ 1 & 1 & 0 \\ 0 & 0 & 1 \end{bmatrix} \begin{bmatrix} 2 \\ -1 \\ 2 \end{bmatrix} = \begin{bmatrix} 3 \\ 1 \\ 2 \end{bmatrix} = \mathbf{A}_-. \tag{5.151}$$

We have confirmed that the metric converts the contravariant vector to a covariant one! We may also determine the tensor g^{ij}, to be written in matrix form as $\mathbf{H}$, by matrix inversion, and find that

$$\mathbf{H} = \begin{bmatrix} 1 & -1 & 0 \\ -1 & 2 & 0 \\ 0 & 0 & 1 \end{bmatrix}. \tag{5.152}$$

Straightforward matrix multiplication shows that $\mathbf{HA}_- = \mathbf{A}_+$.

◇

5.6.5 Pseudotensors

There is one small hiccup in the theory as yet described that should be addressed; for simplicity in this section we restrict ourselves to Cartesian coordinates and ignore the distinction between covariance and contravariance.

The coordinate transformations we have considered so far have been primarily rigid rotations (also known as *proper rotations*), which rotate the coordinate axes but leave the handedness of the coordinate system unchanged. We may also consider *improper rotations*, which include not only a rotation but an inversion of the coordinate axes through the origin. Such an inversion by itself would have the following effect on the coordinates:

$$x_i = -x_i. \tag{5.153}$$

An ordinary vector will also have its coordinates entirely reversed in the new system. However, let us consider a vector such as the angular momentum of a particle, $\mathbf{L} = \mathbf{r} \times \mathbf{p}$, with $\mathbf{r}$ the position vector and $\mathbf{p}$ the momentum. The reversal of axes will flip the sign of $\mathbf{r}$ and the sign of $\mathbf{p}$, but the sign of $\mathbf{L}$ will remain unchanged! In the new coordinate system, the vector $\mathbf{L}$ has in fact changed its direction, something that a physical vector would not do; this idea is illustrated in Fig. 5.4. The vector $\mathbf{L}$, and any vector that is defined by a cross-product of two physical vectors, is known as a *pseudovector*.

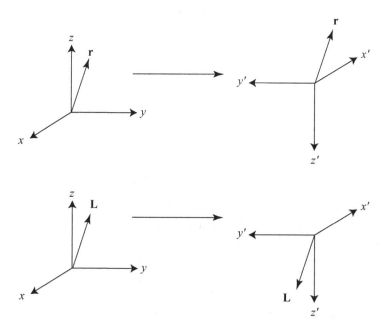

Figure 5.4 The transformation of a (a) vector $\mathbf{r}$ and a (b) pseudovector $\mathbf{L}$ under reflections.

In tensor notation, the cross product can be written with the help of the Levi-Civita tensor,

$$\mathbf{L} = \hat{\mathbf{r}}_i \epsilon_{ijk} x_j p_k, \tag{5.154}$$

where $\hat{\mathbf{r}}_i$ is the ith unit vector of the Cartesian system. Because x_j, p_k, and $\hat{\mathbf{r}}_i$ all change sign under reflection, it follows that the Levi-Civita tensor also changes sign under reflection. The quantity ϵ_{ijk} does not satisfy the definition of a tensor under reflection; it is known as a *pseudotensor*.

The behavior of pseudotensors, at least in the cross-product case, arises because the cross product implicitly includes the handedness of the coordinate system in its definition. When the handedness is changed by a reflection, the definition of the cross product must change as well.

Strictly speaking, the sign change of a pseudotensor under transformation is equal to the determinant of the transformation matrix. If one or three axes are reflected, the sign of a pseudotensor changes; if only two axes are reflected, however, the sign of the pseudotensor remains invariant.

One useful property of the Levi-Civita tensor is worth noting here. The contraction of two such tensors satisfies

$$\sum_k \epsilon_{ijk} \epsilon_{klm} = \delta_{il}\delta_{jm} - \delta_{im}\delta_{jl}. \tag{5.155}$$

5.6.6 Tensors in curvilinear coordinates

We now extend the mathematics of tensors to the case of curvilinear coordinate systems, which simultaneously increases the usefulness and complexity of the theory.

We begin by considering a general coordinate system (u^1, u^2, u^3). For reasons which will become clear momentarily, we avoid tying this system explicitly to a Cartesian coordinate system. An infinitesimal displacement ds may be written in the form

$$d\mathbf{s} = \frac{\partial \mathbf{r}}{\partial u^i} du^i. \tag{5.156}$$

The infinitesimal distance ds satisfies the expression

$$ds^2 = g_{ij} du^i du^j, \tag{5.157}$$

where g_{ij} is the metric tensor defined as

$$g_{ij} = \frac{\partial \mathbf{r}}{\partial u^i} \cdot \frac{\partial \mathbf{r}}{\partial u^j}. \tag{5.158}$$

If we write $\mathbf{r}$ in Cartesian coordinates, this definition is comparable to that of Eq. (5.124), except that the metric is no longer necessarily a dimensionless quantity. Following our alternative definition of the metric using unit vectors, Eq. (5.134), this suggests that the basis vectors (not necessarily unit) for our coordinate system are given by

$$\mathbf{e}_i = \frac{\partial \mathbf{r}}{\partial u^i}. \tag{5.159}$$

These are the covariant basis vectors; a contravariant set may be constructed of the form

$$\mathbf{e}^i = \nabla u^i. \tag{5.160}$$

This can be demonstrated from Eq. (2.35), which defines the gradient in terms of the differential of a scalar function ϕ,

$$d\phi = \nabla \phi \cdot d\mathbf{r}. \tag{5.161}$$

If we let $\phi = u^i$, and divide both sides of this expression by du^j, we have

$$\frac{du^i}{du^j} = (\nabla u^i) \cdot \frac{d\mathbf{r}}{du^j} = \delta^j_i, \tag{5.162}$$

in accordance with our definition of a biorthogonal expansion. The metric tensor may therefore also be written in the form

$$g_{ij} = \mathbf{e}_i \cdot \mathbf{e}_j. \tag{5.163}$$

If we are working in an orthogonal coordinate system, we expect that the vectors $\mathbf{e}_i$ will be mutually orthogonal, as will $\mathbf{e}^i$; however, they are not necessarily of unit length and do not necessarily have the same normalization.

Example 5.8 (Spherical coordinates) We briefly consider the metric and basis vectors in spherical coordinates. In spherical coordinates, the position vector may be written as

$$\mathbf{r} = r \sin\theta \cos\phi \hat{\mathbf{x}} + r \sin\theta \sin\phi \hat{\mathbf{y}} + r \cos\theta \hat{\mathbf{z}}. \tag{5.164}$$

This leads us to the set of contravariant basis vectors, using Eq. (5.159),

$$\mathbf{e}_1 = \hat{\mathbf{x}} \sin\theta \cos\phi + \hat{\mathbf{y}} \sin\theta \sin\phi + \hat{\mathbf{z}} \cos\theta = \hat{\mathbf{r}}, \tag{5.165}$$

$$\mathbf{e}_2 = r\hat{\mathbf{x}} \cos\theta \cos\phi + r\hat{\mathbf{y}} \cos\theta \sin\phi - r\hat{\mathbf{z}} \sin\theta = r\hat{\boldsymbol{\theta}}, \tag{5.166}$$

$$\mathbf{e}_3 = -r \sin\theta \hat{\mathbf{x}} \sin\phi + r \sin\theta \hat{\mathbf{y}} \cos\phi = r \sin\theta \hat{\boldsymbol{\phi}}. \tag{5.167}$$

On comparison with Eqs. (3.54), we see that these vectors are parallel to the ordinary unit vectors of spherical coordinates, $(\hat{\mathbf{r}}, \hat{\boldsymbol{\theta}}, \hat{\boldsymbol{\phi}})$, but have different normalization. The covariant basis vectors may be determined using Eq. (5.160) and the definition of the gradient in spherical coordinates, Eq. (3.55),

$$\mathbf{e}^1 = \nabla r = \hat{\mathbf{r}}, \tag{5.168}$$

$$\mathbf{e}^2 = \nabla \theta = \frac{1}{r}\hat{\boldsymbol{\theta}}, \tag{5.169}$$

$$\mathbf{e}^3 = \nabla \phi = \frac{1}{r\sin\theta}\hat{\boldsymbol{\phi}}. \tag{5.170}$$

These vectors are parallel to the contravariant basis vectors, but have different normalization. It readily follows that $\mathbf{e}_i \cdot \mathbf{e}^j = \delta_i^j$, but that $\mathbf{e}_i \cdot \mathbf{e}^j \neq \delta_i^j$.

$\diamond$

On a change of basis, the vectors $\mathbf{e}_i$ transform as follows

$$\mathbf{e}'_i = \frac{\partial \mathbf{r}}{\partial u'_i} = \frac{\partial u_j}{\partial u'_i}\frac{\partial \mathbf{r}}{\partial u_j} = \frac{\partial u_j}{\partial u'_i}\mathbf{e}_j. \tag{5.171}$$

This in turn implies that the components of the position vector, to be denoted r^i, must transform as

$$r'^i = \frac{\partial u'_i}{\partial u_j}r^j. \tag{5.172}$$

This is the definition of a contravariant vector. A covariant vector will have the complementary transformation

$$A'_i = \frac{\partial u_j}{\partial u'_i}A_i. \tag{5.173}$$

We may then define contravariant and covariant tensors for a general coordinate system as follows. A second-rank contravariant tensor satisfies the transformation law

$$A'^{ij} = \frac{\partial u'_i}{\partial u_k}\frac{\partial u'_j}{\partial u_l}A^{kl}, \tag{5.174}$$

while a second-rank covariant tensor satisfies the transformation law

$$A'_{ij} = \frac{\partial u_k}{\partial u'_i}\frac{\partial u_l}{\partial u'_j}A_{kl}. \tag{5.175}$$

A similar definition applies to mixed tensors.

We have meticulously avoided referring our curvilinear coordinates back to a Cartesian system in this section, and with good reason. It is possible to find curved spaces in which it is not possible to construct a metric which satisfies Eq. (5.124). Such *non-Euclidean spaces* may still be treated using the methods of this section, which make no assumptions about the shape of the underlying space.

Example 5.9 (Spherical space) The simplest example of a non-Euclidean space is the surface of a sphere of radius R. This surface is clearly two-dimensional, and location on the sphere can be specified by the angles (θ, ϕ). Infinitesimal displacement on the sphere is given by the expression

$$ds^2 = R^2(d\theta)^2 + R^2 \sin^2 \theta (d\phi)^2, \tag{5.176}$$

which follows from the scale factors of Section 3.5. The metric for this space, written in matrix form as **G**, is

$$\mathbf{G} = \begin{bmatrix} R^2 & 0 \\ 0 & R^2 \sin^2 \theta \end{bmatrix}. \tag{5.177}$$

We now ask if there exists a Cartesian coordinate system (x, y) lying on the surface of the sphere that can be used to describe this metric. If so, the functions $x(\theta, \phi)$ and $y(\theta, \phi)$ must satisfy the partial differential equations

$$\left(\frac{\partial x}{\partial \theta}\right)^2 + \left(\frac{\partial y}{\partial \theta}\right)^2 = R^2, \tag{5.178}$$

$$\left(\frac{\partial x}{\partial \phi}\right)^2 + \left(\frac{\partial y}{\partial \phi}\right)^2 = R^2 \sin^2 \theta, \tag{5.179}$$

$$\frac{\partial x}{\partial \theta}\frac{\partial x}{\partial \phi} + \frac{\partial y}{\partial \theta}\frac{\partial y}{\partial \phi} = 0. \tag{5.180}$$

It can be shown that this set of equations possesses no solution.

◇

5.6.7 The covariant derivative

In Chapter 3, we observed that special care must be taken when evaluating derivatives in curvilinear coordinate systems. We can see this quite readily by making a naïve attempt to construct a derivative of a vector v^i in such a system, of the form $\partial v^i / \partial u^j$. Under a general coordinate transformation, this derivative may be written as

$$\frac{\partial v'^i}{\partial u'^j} = \frac{\partial u^k}{\partial u'^j}\frac{\partial v'^i}{\partial u^k}. \tag{5.181}$$

This expression is not complete, however, because v'^i itself must be written in terms of the original coordinate system. We have

$$\frac{\partial v'^i}{\partial u'^j} = \frac{\partial u^k}{\partial u'^j}\frac{\partial}{\partial u^k}\left(\frac{\partial u'^i}{\partial u^l}v^l\right) = \frac{\partial u^k}{\partial u'^j}\frac{\partial u'^i}{\partial u^l}\frac{\partial v^l}{\partial u^k} + \frac{\partial u^k}{\partial u'^j}\frac{\partial^2 u'^i}{\partial u^k \partial u^l}v^l. \tag{5.182}$$

The first term on the right is what we would expect for the transformation of a second-rank mixed tensor, but the second term on the right is not of tensor form at all and our naïve form for the derivative is *not* a tensor. This problem arises because we are working in a curved coordinate system; the basis vectors change their behavior with position and this property is inadvertently being left out of the naïve choice $\partial v^i / \partial u^j$.

We try again and explicitly include the unit vector in our expression for the derivative,

$$\frac{\partial(v^i \mathbf{e}_i)}{\partial u^j} = \frac{\partial v^i}{\partial u^j}\mathbf{e}_i + v^i \frac{\partial \mathbf{e}_i}{\partial u^j}. \tag{5.183}$$

The latter term in this expression must itself be expressible in terms of the basis vectors $\mathbf{e}_j$. We define a new symbol, Γ_{ij}^k such that

$$\frac{\partial \mathbf{e}_i}{\partial u^j} = \Gamma_{ij}^k \mathbf{e}_k. \tag{5.184}$$

The quantity Γ_{ij}^k is known as the *Christoffel symbol*; it is to be noted that it is in general a function of the coordinates u^j. A more explicit form may be derived by the use of the covariant basis vectors $\mathbf{e}^i$,

$$\Gamma_{ij}^k = \mathbf{e}^k \cdot \frac{\partial \mathbf{e}_i}{\partial u^j}. \tag{5.185}$$

We may similarly introduce the derivative of the contravariant basis vectors in the form

$$\frac{\partial \mathbf{e}^i}{\partial u^j} = -\Gamma_{kj}^i \mathbf{e}^k. \tag{5.186}$$

With the use of the Christoffel symbol, Eq. (5.183) may be written as

$$\frac{\partial v^i \mathbf{e}_i}{\partial u^j} = \frac{\partial v^i}{\partial u^j}\mathbf{e}_i + v^i \Gamma_{ij}^k \mathbf{e}_k. \tag{5.187}$$

The indices i and k are contracted and therefore dummy indices; we exchange them and are then able to write the derivative in the form

$$\frac{\partial v^i \mathbf{e}_i}{\partial u^j} = \left(\frac{\partial v^i}{\partial u^j} + v^k \Gamma_{kj}^i \right) \mathbf{e}_i. \tag{5.188}$$

The expression in parenthesis is known as the *covariant derivative*; since we have essentially factored out the basis vectors, we may write this in the compact form

$$v_{;j}^i \equiv \frac{\partial v^i}{\partial u^j} + v^k \Gamma_{kj}^i. \tag{5.189}$$

The semicolon in the subscript is the shorthand notation for the covariant derivative. With expressions for the basis vectors in terms of the coordinates u^i, we can determine the value of the Christoffel symbol and therefore determine the value of the covariant derivative. Using the properties of the Christoffel symbol, we can also show that $v_{;j}^i$ acts as a mixed second-rank tensor.

It is important to note that the Christoffel symbol itself does not transform as a second-rank tensor; in essence, its presence cancels the "un-tensor-like" behavior that we found in Eq. (5.182).

With the introduction of the covariant derivative, it is possible to develop a more general calculus of tensors, which includes generalized forms of the familiar vector derivative operators and vector identities such as Stokes' theorem and Gauss' theorem.

5.7 Additional reading

Several books that are useful for computational matrix methods, also mentioned within this chapter:

- G. H. Golub and C. F. Van Loan, *Matrix Computations* (Baltimore, John Hopkins Press, 1996, 3rd edition).
- W. H. Press, S. A. Teukolsky, W. T. Vetterling and B. P. Flannery, *Numerical Recipes in C++* (Cambridge, Cambridge University Press, 2002, 2nd edition).

For those who would dare, most books on general relativity include a primer on tensors. I can also recommend the introductory book:

- A. I. Borisenko and I. E. Tarapov, *Vector and Tensor Analysis with Applications* New York, Dover, 1979.

5.8 Exercises

1. Find the eigenvalues and the left–right eigenvectors of the following matrix

$$\mathbf{A} = \begin{bmatrix} 1/2 & -1/2 & -1 \\ 1/2 & 3/2 & 1 \\ -1/2 & 1/2 & 1 \end{bmatrix}.$$

2. Find the eigenvalues and the left–right eigenvectors of the following matrix

$$\mathbf{A} = \begin{bmatrix} 1 & 0 & 0 \\ 0 & 2 & -2 \\ 0 & 3 & 3 \end{bmatrix}.$$

3. Consider the matrix

$$\mathbf{A} = \begin{bmatrix} 1 & 1 \\ \epsilon & 3 \end{bmatrix},$$

where $0 \leq \epsilon \leq 1$. Find the eigenvalues and left–right eigenvectors of the system. Demonstrate that, in the limit $\epsilon \to 1$, the left and right eigenvectors are the same.

4. Determine the SVD of the matrix such that the singular values are arranged in decreasing order,

$$\mathbf{A} = \begin{bmatrix} 2 & 0 \\ 0 & 1 \\ 2 & 0 \end{bmatrix}.$$

5. Determine the SVD of the matrix such that the singular values are arranged in decreasing order,

$$A = \begin{bmatrix} 1 & 0 & 0 \\ 1 & 0 & 0 \end{bmatrix}.$$

6. Determine the SVD of the matrix such that the singular values are arranged in decreasing order,

$$A = \begin{bmatrix} 1/\sqrt{2} & -1 & 1 & 0 \\ 0 & 0 & 0 & 0 \\ 1/\sqrt{2} & 1 & -1 & 0 \end{bmatrix}.$$

7. Determine the SVD of the matrix such that the singular values are arranged in decreasing order,

$$A = \begin{bmatrix} 1/\sqrt{2} & 0 & -\sqrt{2} \\ 0 & 3/\sqrt{2} & 0 \\ 1/\sqrt{2} & 0 & \sqrt{2} \\ 0 & 3/\sqrt{2} & 0 \end{bmatrix}.$$

8. Find the LU decomposition of the following matrix

$$A = \begin{bmatrix} 2 & 2 & 3 \\ 1 & 0 & 4 \\ 3 & 1 & 2 \end{bmatrix}.$$

9. Find the LU decomposition of the following matrix

$$A = \begin{bmatrix} 1 & 4 & 3 \\ 2 & 1 & 2 \\ 6 & 3 & 0 \end{bmatrix}.$$

10. Find the LUP decomposition of the following matrix

$$A = \begin{bmatrix} 0 & 1 & 3 \\ 2 & 1 & 0 \\ 0 & 4 & 1 \end{bmatrix}.$$

   Be sure to specify the form of the matrix $\mathbf{P}$.
11. Find the LUP decomposition of the following matrix

$$A = \begin{bmatrix} 1 & 1 & 1 \\ 1 & 1 & 2 \\ 3 & 1 & 0 \end{bmatrix}.$$

   Be sure to specify the form of the matrix $\mathbf{P}$.
12. Prove the quotient rule for the case where A_j^k is a mixed tensor, X_k^j is a set of numbers, and the product is a scalar C of the form $A_j^k X_k^j = C$. In other words, show that X_k^j is a second-rank mixed tensor.

13. Prove the quotient rule for a direct product, where A_{ij} is a tensor, X^{kl} is a set of numbers, and the product is a mixed tensor of the form $A_{ij}X^{kl} = C_{ij}^{kl}$. In other words, show that X^{kl} is a second-rank contravariant tensor.

14. Argue that the form of the Kronecker delta δ_i^j is the same under any coordinate transformation. Can the same be said for the forms δ_{ij}, δ^{ij}, derived using the metric tensor?

15. Verify that the components of the metric tensor g^{ij} can be written using the contravariant basis vectors in the form $g^{ij} = \mathbf{e}^i \cdot \mathbf{e}^j$.

16. Consider a skew coordinate system with basis vectors $\mathbf{e}_1 = \hat{\mathbf{x}} - \hat{\mathbf{z}}$, $\mathbf{e}_2 = \hat{\mathbf{y}} + \hat{\mathbf{x}}$, $\mathbf{e}_3 = \hat{\mathbf{z}} - \hat{\mathbf{y}}$, and a vector with the corresponding Cartesian representation $\mathbf{A} = \hat{\mathbf{x}} + 4\hat{\mathbf{y}} + 3\hat{\mathbf{z}}$. Determine the contravariant and covariant representation of the vector in the skew coordinate system, and determine the tensors g_{ij} and g^{ij}. Verify that the tensors transform a contravariant vector into a covariant one, and vice versa.

17. Consider a skew coordinate system with basis vectors $\mathbf{e}_1 = \hat{\mathbf{x}} + 3\hat{\mathbf{y}}$, $\mathbf{e}_2 = \hat{\mathbf{x}} + \hat{\mathbf{y}} + \hat{\mathbf{x}}$, $\mathbf{e}_3 = \hat{\mathbf{z}} + 2\hat{\mathbf{y}}$, and a vector with the corresponding Cartesian representation $\mathbf{A} = 2\hat{\mathbf{x}} + 3\hat{\mathbf{z}}$. Determine the contravariant and covariant representation of the vector in the skew coordinate system, and determine the tensors g_{ij} and g^{ij}. Verify that the tensors transform a contravariant vector into a covariant one, and vice versa.

18. Parabolic cylindrical coordinates are described by (σ, τ, z), where

$$x = \sigma\tau,$$
$$y = \frac{1}{2}(\tau^2 - \sigma^2),$$
$$z = z.$$

Determine the contravariant and covariant basis vectors for this coordinate system, verify their biorthogonality, and determine the metric tensor.

19. The elliptic cylindrical coordinate system uses coordinates u, v, and z, described by the equations

$$x = a\cosh u\cos v,$$
$$y = a\sinh u\sin v,$$
$$z = z,$$

with a a constant. Determine the contravariant and covariant basis vectors for this coordinate system, verify their biorthogonality, and determine the metric tensor.

20. Tensor notation can often make difficult vector identities easy to demonstrate. Prove the following identity using properties of the Kronecker and Levi-Civita tensors:

$$[\mathbf{A} \times \mathbf{B}] \cdot [\mathbf{C} \times \mathbf{D}] = (\mathbf{A} \cdot \mathbf{C})(\mathbf{B} \cdot \mathbf{D}) - (\mathbf{A} \cdot \mathbf{D}) - (\mathbf{A} \cdot \mathbf{D})(\mathbf{B} \cdot \mathbf{C}).$$

21. Tensor notation can often make difficult vector identities easy to demonstrate. Prove the following identity using properties of the Kronecker and Levi-Civita tensors:

$$[\mathbf{A} \times [\mathbf{B} \times \mathbf{C}]] + [\mathbf{B} \times [\mathbf{C} \times \mathbf{A}]] + [\mathbf{C} \times [\mathbf{A} \times \mathbf{B}]] = 0.$$

6

Distributions

6.1 Introduction: Gauss' law and the Poisson equation

The fundamental law of electrostatics is Coulomb's law, which states that the electric force on a point charge Q (the field charge) due to a point charge q at the origin (the source charge) is given by

$$\mathbf{F} = \frac{qQ\hat{\mathbf{r}}}{r^2},\tag{6.1}$$

where $\hat{\mathbf{r}} = \mathbf{r}/r$ is the unit vector in the $\mathbf{r}$ direction, pointing from the source to the field charge. Coulomb's law may also be expressed in terms of the electric field $\mathbf{E}$ generated by the charge q as $\mathbf{F} = Q\mathbf{E}$, such that

$$\mathbf{E} = \frac{q\hat{\mathbf{r}}}{r^2},\tag{6.2}$$

It may be demonstrated that this law is equivalent to Gauss' law, which states that the flux of the electric field satisfies the relation,

$$\int_S \mathbf{E}\cdot\mathbf{da} = \begin{cases} 0, & q \notin S, \\ 4\pi q, & q \in S, \end{cases}\tag{6.3}$$

where S is a closed surface which encloses ($\in$) the charge q or does not enclose ($\notin$) the charge q, as illustrated in Fig. 6.1. Gauss's law was mentioned in the context of Maxwell's equations in Chapter 2, but was not explicitly proven; we now do so.

We can prove it using a number of the techniques developed in Chapter 2. First, let us consider the case when q is not inside S. We may use the divergence theorem, Eq. (2.77), to turn the surface integral of $\mathbf{E}$ into a volume integral,

$$\int_S \mathbf{E}\cdot\mathbf{d\sigma} = \int_V \nabla\cdot\mathbf{E}d\tau,\tag{6.4}$$

177

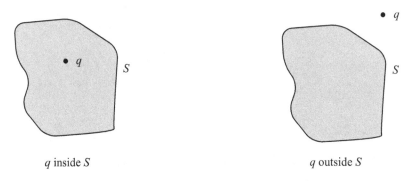

q inside *S*   *q* outside *S*

Figure 6.1 A charge inside and outside of the closed surface *S*.

where V is the volume enclosed by the surface S. The divergence of $\mathbf{E}$ can be readily found using Cartesian coordinates (we ignore the q factor):

$$\nabla \cdot \left(\frac{\hat{\mathbf{r}}}{r^2}\right) = \nabla \cdot \left(\frac{\mathbf{r}}{r^3}\right) = \left(\frac{\partial}{\partial x}\hat{\mathbf{x}} + \frac{\partial}{\partial y}\hat{\mathbf{y}} + \frac{\partial}{\partial z}\hat{\mathbf{z}}\right) \cdot \frac{x\hat{\mathbf{x}} + y\hat{\mathbf{y}} + z\hat{\mathbf{z}}}{(x^2 + y^2 + z^2)^{3/2}}$$

$$= \frac{3}{(x^2 + y^2 + z^2)^{3/2}} - 3\frac{x^2 + y^2 + z^2}{(x^2 + y^2 + z^2)^{5/2}} = 0. \tag{6.5}$$

Thus the upper part of Eq. (6.3) is proven.

Proving the second part of Gauss' law is more challenging, because the value of the electric field diverges at $r = 0$ (in addition, the direction of the vector $\mathbf{r}$ is unspecified at that point!). Because the divergence theorem only applies to functions with continuous first partial derivatives, we cannot directly use it on any volume which includes the origin. What we can do, however, is exclude a small spherical region around the origin from our volume integral, and exclude a little tunnel from that inner region to the outside of the sphere. This tunnel mathematically places the charge outside of the volume V, and allows us to apply the divergence theorem to this volume. The construction of the excluded region is illustrated in Fig. 6.2.

In the limit that the radius of the inner sphere and the width of the tunnel vanish, we should recover that part of Gauss' law in which the surface S includes the charge at the origin. The question then naturally arises: does this construction produce sensible results?

The total surface integral consists of three parts, the integral over the outer surface S, the integral over the inner spherical surface S', and the integral over the tunnel area, S_{tunnel}. This surface does *not* enclose any charge, thanks to the presence of the tunnel, so we may apply the first part of Gauss' law to this surface integral,

$$\int_S \frac{\hat{\mathbf{r}} \cdot d\mathbf{a}}{r^2} + \int_{S'} \frac{\hat{\mathbf{r}} \cdot d\mathbf{a}'}{\alpha^2} + \int_{S_{\text{tunnel}}} \frac{\hat{\mathbf{r}} \cdot d\mathbf{a}}{r^2} = 0. \tag{6.6}$$

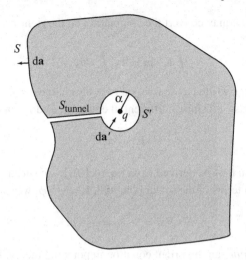

Figure 6.2 The notation relating to the excluded regions for the volume V.

Here $\mathbf{da}'$ is the area element of the inner sphere and α is the radius of that sphere. Because the field is continuous over the region which contains the tunnel, the tunnel's contribution to the integral vanishes as it is made arbitrarily narrow. The integral over the sphere can be evaluated exactly, because $\mathbf{da}' = -\hat{\mathbf{r}}\alpha^2 d\Omega$, where $d\Omega$ is an element of solid angle. The minus sign is present because we define area elements as pointing *outward* from a volume. Evaluating the integral over the surface S' (the solid angle integral gives 4π), we have

$$\int_{S'} \frac{\hat{\mathbf{r}} \cdot \mathbf{da}'}{\alpha^2} = -\int_{S'} \frac{\hat{\mathbf{r}} \cdot \hat{\mathbf{r}}\alpha^2}{\alpha^2} d\Omega = -4\pi, \tag{6.7}$$

which is *independent* of the radius α. On substituting this into Eq. (6.6), we have proven the second part of Gauss' law.

As derived so far, Gauss' law applies only to a single point charge, but we may readily generalize it to collections of point charges and continuous distributions of charge. First, let us suppose we have a large number of point charges q_i located at points $\mathbf{r}_i$ within the volume V, each producing an electric field $\mathbf{E}_i$. Gauss' law applies to each of these charges individually, and we may add together the relations for each of them as follows:

$$\int_S \left(\sum_i \mathbf{E}_i \right) \cdot \mathbf{da} = \int_S \mathbf{E} \cdot \mathbf{da} = 4\pi \sum_i q_i. \tag{6.8}$$

Here we have used the fact that the total field $\mathbf{E} = \sum_i \mathbf{E}_i$, i.e. that the principle of linear superposition holds for electric fields. This expression demonstrates that Gauss' law holds for any finite number of point charges. Introducing a continuous charge density $\rho(\mathbf{r})$, and letting $q_i \rightarrow \rho(\mathbf{r}_i)d\tau$, we may convert the summation on the right-hand side to an integral over the volume,

$$\int_S \mathbf{E} \cdot d\mathbf{a} = 4\pi \int_V \rho\, d\tau. \tag{6.9}$$

With the enclosed charge written as an integral over the charge density, we may use Gauss' *theorem* to convert Gauss' *law* into a differential equation, as done in Section 2.10,

$$\nabla \cdot \mathbf{E} = 4\pi\rho. \tag{6.10}$$

One additional equation may be derived from Eq. (6.10). If we substitute into this equation the expression for $\mathbf{E}$ in terms of the scalar potential, $\mathbf{E} = -\nabla\phi$, we find that

$$\nabla \cdot \nabla\phi = \nabla^2\phi = -4\pi\rho. \tag{6.11}$$

This is *Poisson's equation*, an important equation in potential theory. If the charge density is equal to zero in the region of interest, we have

$$\nabla^2\phi = 0, \tag{6.12}$$

which is Laplace's equation, which was discussed in Section 2.9 and is relevant to numerous diverse physical problems.

It should be noted that the method which we used to prove Gauss' law, i.e. determining the integral of a singular quantity by defining a new volume which excludes the singular point, is a technique which appears a number of times in mathematical physics. In particular, we will see it again in Chapter 9, in deriving results in complex analysis.

There is one curious oversight in our discussion of Gauss' law. We began by deriving that law for a point charge q, and then generalized our result to a *continuous* charge density $\rho(\mathbf{r})$. But what does this charge density $\rho(\mathbf{r})$ look like for the single point charge q we started with? This would be an important question to answer, for instance, if we wanted to use Eq. (6.10) to solve for the field of one or more point charges directly. The charge density $\rho_p(\mathbf{r})$ of a point charge would evidently be a strange beast: because the charge is a mathematical point, the its density must be zero everywhere except at a single point in space, while the integral of the charge density over an enclosing volume must be equal to a finite number q, in accordance with Eq. (6.9). The function $\rho_p(\mathbf{r})$ must be an unusual object indeed, zero at every point except at the location of the charge, where it must be infinite. A safe physical answer to this difficulty would be to state that there is no such thing as a true point charge, as every charge presumably has a finite size, and hence finite density, associated with it. Mathematically, however, point charges are extremely convenient entities, and we would like to avoid doing away with them if at all possible.

The resolution of this problem is to represent the charge density of a point charge located at $\mathbf{r}_0$ by a special mathematical entity known as a three-dimensional Dirac delta function,

$$\rho_p(\mathbf{r}) = q\delta^{(3)}(\mathbf{r} - \mathbf{r}_0). \tag{6.13}$$

A delta function is an example of a *distribution*, a generalization of the ordinary functions we are used to working with. It is to be noted that the function $\delta^{(3)}$ must have dimensions of $[\text{length}]^{-3}$, since the charge density has dimensions of $[\text{charge}]/[\text{length}]^3$. To produce agreement between the continuous and discrete forms of Gauss' law, Eqs. (6.9) and (6.3), we require the distribution $\delta^{(3)}(\mathbf{r} - \mathbf{r}_0)$ to satisfy the following property:

$$\int_V \delta^{(3)}(\mathbf{r} - \mathbf{r}_0)\mathrm{d}^3 r = \begin{cases} 0, & \mathbf{r}_0 \notin V, \\ 1, & \mathbf{r}_0 \in V. \end{cases} \tag{6.14}$$

In fact, an integral identity of this form is the only way we *can* define the delta function, as its behavior falls outside the bounds of what is acceptable according to the normal theory of functions. In spite of its strangeness, the delta function is an extremely useful mathematical tool which will continue to make an appearance throughout the rest of this book. This distribution, and others like it, form the subject of this chapter.

6.2 Introduction to delta functions

We restrict ourselves for the moment to functions of one variable x. We introduce a *Dirac delta function* $\delta(x)$ as a mathematical object which satisfies the following pair of properties.

1. It must vanish away from the origin:

$$\delta(x) = 0, \quad x \neq 0. \tag{6.15}$$

2. The integral of $\delta(x)$ times a smooth function $f(x)$ is equal to the value of the function at the origin:

$$\int_{-\infty}^{\infty} \delta(x)f(x)\mathrm{d}x = f(0). \tag{6.16}$$

This latter property is often referred to as the *sifting property* of the delta function, because the delta function "sifts" through the function and keeps only the value at the origin, much like a prospector panning for gold in a muddy river.

It follows from the sifting property of the delta function that it must appear as an infinitely high, infinitely thin spike at the origin. Qualitatively, one can appreciate that this object does not really satisfy the definition of a function, since its value at the origin is not clearly defined. From a rigorous mathematical point of view, it does not satisfy a basic theorem of Lebesgue integration.

Theorem 6.1 (Integral equivalence) *If $f(x)$ and $g(x)$ are functions such that $f = g$ almost everywhere (except on a set of measure zero), then f is Lebesgue integrable if and only if g is Lebesgue integrable and the integrals of f and g are the same.*

Let us assume $f(x) = \delta(x)$ is a well-behaved function. If we choose $g(x) = 0$, then f and g are equal everywhere except at the point $x = 0$ (which is a "set of measure zero"). However, the integrals of both functions are *not* equal: the integral of f is nonzero and the integral of g is identically zero. Our assumption is violated and $\delta(x)$ does not satisfy the conditions of a function with respect to Lebesgue integration.

When physicist P. A. M. Dirac championed[1] the use of delta functions in his quantum text in 1930 [Dir30], mathematicians initially disregarded them as improper mathematical shenanigans. However, the usefulness of Dirac's approach rapidly became evident, and a rigorous theory which describes such strange objects was developed, known as the *theory of distributions*.

Formally, a distribution is a mathematical operation which assigns to any function $f(x)$ a unique number f_0. In the case of the Dirac delta function $\delta(x)$, the operation assigns the number $f_0 = f(0)$. We write this operation as the integration of the delta function with the function $f(x)$ in the form of the sifting integral, Eq. (6.16).

In fact, we may say that a distribution such as the delta function is only rigorously defined by its integral with respect to another function; we may consider it to be *defined* by Eq. (6.16), and it is not generally healthy to think of it as an object that exists outside of an integration.

We can, however, establish the delta function as the *limit* of a sequence of well-behaved functions, known as a *delta-sequence*. Numerous delta-sequences exist; several of them are shown below:

$$\delta_n(x) = \begin{cases} 0, & |x| > 1/2n, \\ n, & |x| \le 1/2n, \end{cases} \tag{6.17}$$

$$\delta_n(x) = \frac{n}{\sqrt{\pi}} \exp\left(-n^2 x^2\right), \tag{6.18}$$

$$\delta_n(x) = \frac{n}{2} \exp[-n|x|], \tag{6.19}$$

$$\delta_n(x) = \frac{\sin(nx)}{\pi x} = \frac{n}{\pi} j_0(nx). \tag{6.20}$$

The symbol j_0 represents a spherical Bessel function of order 0, to be described in detail in Chapter 16.

It is to be noted that the integral of each of these functions is 1; that is,

$$\int_{-\infty}^{\infty} \delta_n(x) \mathrm{d}x = 1. \tag{6.21}$$

What happens to these delta-sequences in the limit that $n \to \infty$? As n increases, each of the delta-sequences becomes taller and narrower, becoming arbitrarily tall and narrow in the

---

[1] As is often the case, the history is much more complicated. Other researchers introduced delta functions and sequences long before Dirac popularized them. See [Jac08] for details.

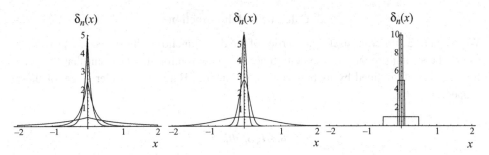

Figure 6.3 Delta sequences: (a) exponential, Eq. (6.19), (b) Gaussian, Eq. (6.18), (c) step function, Eq. (6.17). Plots show $n = 1$, $n = 5$, and $n = 10$.

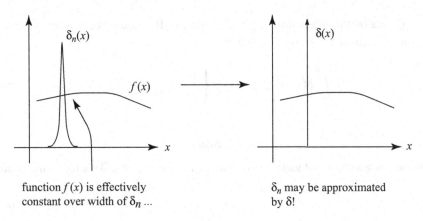

function $f(x)$ is effectively constant over width of δ_n ...

δ_n may be approximated by δ!

Figure 6.4 The process by which a delta sequence $\delta_n(x)$, for large n, behaves like the delta function $\delta(x)$.

limit. This is illustrated for several of these functions in Fig. 6.3. If the function $f(x)$ we are sifting is continuous and slowly varying near the origin, it follows that, for large n, we may approximate $f(x)$ by the lowest term of its Taylor series at the origin, $f(0)$, in the integral. Therefore we have that

$$\lim_{n\to\infty} \int_{-\infty}^{\infty} f(x)\delta_n(x)\mathrm{d}x \approx f(0) \lim_{n\to\infty} \int_{-\infty}^{\infty} \delta_n(x)\mathrm{d}x = f(0). \qquad (6.22)$$

This reproduces the sifting property of the Dirac delta function. The reasoning is illustrated in Fig. 6.4.

Delta sequences can often be useful for proving or testing results relating to delta functions. Provided the integrals of delta sequences can be evaluated for finite n, one can then take the limit $n \to \infty$ to determine the idealized behavior of the delta function.

6.3 Calculus of delta functions

We can derive numerous useful properties of the delta function. All of these are proven by use of the sifting property of the delta function, again reinforcing the point that the delta function is only defined by its behavior in an integral. Here we consider a few of these properties.

6.3.1 Shifting

A delta function need not be centered on the origin. By a simple change of variable, we may readily demonstrate that

$$\int_{-\infty}^{\infty} \delta(x-x')f(x')dx' = f(x). \tag{6.23}$$

We may generalize this to a finite domain of integration. Because the delta function vanishes at all points except at $x = x'$, we may write

$$\int_{a}^{b} \delta(x-x')f(x')dx' = \begin{cases} 0, & x \notin [a,b], \\ f(x), & x \in [a,b]. \end{cases} \tag{6.24}$$

6.3.2 Scaling

We may also use a change of variable to demonstrate from Eq. (6.16) the following relation:

$$\delta(ax) = \frac{1}{|a|}\delta(x). \tag{6.25}$$

The proof of this follows almost immediately. If $a > 0$, we may make a change to $y = ax$, so that

$$\int_{-\infty}^{\infty} f(x)\delta(ax)dx = \frac{1}{a}\int_{-\infty}^{\infty} f(y/a)\delta(y)dy = \frac{1}{a}f(0). \tag{6.26}$$

If $a < 0$, we make the same change of variable to $y = ax$, the only difference being the effect on the integration order,

$$\int_{-\infty}^{\infty} f(x)\delta(ax)dx = \frac{1}{a}\int_{\infty}^{-\infty} f(y/a)\delta(y)dy = -\frac{1}{a}f(0). \tag{6.27}$$

6.3.3 Composition

Suppose $g(x)$ is a smooth function with zeros at points a; it then follows that

$$\delta[g(x)] = \sum_{a=\text{zeros of } g} \frac{\delta(x-a)}{|g'(a)|}. \tag{6.28}$$

This can be proven by first noting that the delta function only has nonzero value in the neighborhood of the zeros of $g(x)$; we may therefore break up the integral into a sum of integrals over the immediate neighborhoods of the points a,

$$\int_{-\infty}^{\infty} f(x)\delta[g(x)]dx = \sum_a \int_{a-\epsilon}^{a+\epsilon} f(x)\delta[g(x)]dx, \tag{6.29}$$

where ϵ is a small positive number. For small enough ϵ, we may approximate $g(x)$ within each integral by the lowest term of its Taylor expansion (to be discussed in detail in Chapter 7) about the point $x = a$,

$$g(x) \approx g(a) + (x-a)g'(a) = (x-a)g'(a), \tag{6.30}$$

where we have used the fact that $g(a) = 0$. On substituting from this equation into Eq. (6.29), and using the scaling property, Eq. (6.25), we immediately prove Eq. (6.28).

6.3.4 Derivatives

One further series of useful relations can be proven using the sifting property: the behavior of the derivatives of the delta function.

$$\int_{-\infty}^{\infty} \frac{d^m \delta(x)}{dx^m} f(x)dx = (-1)^m \frac{d^m f(x)}{dx^m}\bigg|_{x=0}. \tag{6.31}$$

This can be proven using integration by parts; for example, with $m = 1$,

$$\int_{-\infty}^{\infty} \frac{d\delta(x)}{dx} f(x)dx = \delta(x)f(x)\big|_{-\infty}^{\infty} - \int_{-\infty}^{\infty} \delta(x)\frac{df(x)}{dx}dx = -\frac{df(x)}{dx}\bigg|_{x=0}. \tag{6.32}$$

It should seem very suspicious to use integration by parts on a non-function like the delta function. We may make this treatment rigorous by using a delta-sequence which possesses continuous mth derivatives, such as Eq. (6.18), and performing the integration by parts with respect to it. In the limit $m \to \infty$, we should recover Eq. (6.31).

6.4 Other representations of the delta function

Though historically the delta function might have started as a helpful little tool for a quantum physicist, it is now a fundamental component of the theory of linear vector spaces. We will use the delta function numerous times over the course of this book, and see it represented in a variety of forms. The delta-sequence is just one way of representing a delta function; there are in fact a potentially infinite number of different representations. Each of these methods has its own uses depending on the circumstances. Many of these, it should be noted, are eigenfunction expansions of the delta function. In Section 4.8 we described how an orthonormal basis for an N-dimensional vector space could be used to construct the identity matrix, in the form

$$\mathbf{I} = \sum_{n=1}^{N} |n\rangle \langle n|. \tag{6.33}$$

Such an expression is generally referred to as a *closure relation*, and can be generalized to infinite-dimensional vector spaces, provided the identity matrix $\mathbf{I}$ is replaced with the Dirac delta function.

When we discuss Fourier series in Chapter 8, we will expand an arbitrary piecewise continuous function $f(t)$ over the domain $t = 0$ to $t = T$ in terms of sine and cosine functions, in the form

$$f(t) = \frac{a_0}{2} + \sum_{n=1}^{\infty} [a_n \cos(n\omega_0 t) + b_n \sin(n\omega_0 t)], \tag{6.34}$$

where $\omega_0 = 2\pi/T$. In terms of a vector space, we may consider $f(t)$ to represent an infinite-dimensional vector whose components are given by the coefficients a_n and b_n. We have a set of orthonormal basis vectors of the form

$$|0,1\rangle = \sqrt{\frac{1}{T}}, \quad |n,1\rangle = \sqrt{\frac{2}{T}} \cos(n\omega_0 t), \quad |n,2\rangle = \sqrt{\frac{2}{T}} \sin(n\omega_0 t), \quad n > 1. \tag{6.35}$$

These basis vectors are orthogonal according to the "dot product",

$$\langle n,1| m,1\rangle = \frac{2}{T} \int_0^T \cos(n\omega_0 t)\cos(m\omega_0 t)dt = \delta_{nm}, \quad \langle n,1| m,2\rangle = 0. \tag{6.36}$$

It can be shown that the delta function can be constructed using these basis vectors in the form

$$\delta(t - t') = |0,1\rangle\langle 0,1| + \sum_{n=1}^{\infty} |n,1\rangle\langle n,1| + \sum_{n=1}^{\infty} |n,2\rangle\langle n,2|, \tag{6.37}$$

or, in terms of functions,

$$\delta(t - t') = \frac{1}{T} + \sum_{n=1}^{\infty} \frac{2}{T} \cos(n\omega_0 t')\cos(n\omega_0 t) + \sum_{n=1}^{\infty} \frac{2}{T} \sin(n\omega_0 t')\sin(n\omega_0 t). \tag{6.38}$$

We will discuss this representation in more detail in Chapter 8.

When we discuss Fourier transforms in Chapter 11, we will expand an arbitrary square-integrable function $f(t)$ defined on the domain $-\infty < t < \infty$ in terms of complex exponentials, in the form

$$f(t) = \int_{-\infty}^{\infty} F(\omega)e^{-i\omega t}d\omega. \tag{6.39}$$

Here the "basis vectors" are the complex exponentials,

$$|\omega\rangle = \sqrt{\frac{1}{2\pi}} e^{-i\omega t}, \tag{6.40}$$

and the delta function can be constructed by an integral over these basis vectors,

$$\delta(t-t') = \int_{-\infty}^{\infty} |\omega\rangle \langle\omega| \, d\omega = \frac{1}{2\pi} \int_{-\infty}^{\infty} e^{i\omega(t-t')} d\omega. \tag{6.41}$$

This representation is extremely useful in the derivation of many theorems relating to the Fourier transform. We will discuss it again when we discuss Fourier transforms, in Chapter 11.

Similar closure relations can be written for any set of orthonormal functions which form a basis over a given domain.

6.5 Heaviside step function

Closely related to the delta function is the *step function*, or Heaviside step function, defined as

$$S(x-a) = \begin{cases} 0, & x < a, \\ 1, & x \geq a. \end{cases} \tag{6.42}$$

Such an object is particularly useful in describing the beginning of temporal events, such as the arrival of an optical pulse at a detector, or the switching on of an electrical circuit. The step function is itself a function in the ordinary sense of the term, but it can be shown quite readily that its derivative is the delta function, i.e.

$$\frac{dS(x-a)}{dx} = \delta(x-a). \tag{6.43}$$

A straightforward way to prove this is to define a "step function-sequence" analogous to the delta-sequence,

$$S_n(x) = \begin{cases} 0, & x < -n, \\ (x+n)/2n, & |x| \leq n, \\ 1, & x > n. \end{cases} \tag{6.44}$$

The evolution of this sequence is illustrated in Fig. 6.5.

We can take the derivative of this function to find,

$$\frac{dS_n(x)}{dx} = \begin{cases} 0, & |x| > n, \\ 1/2n, & |x| \leq n, \end{cases} \tag{6.45}$$

which can be seen on comparison with Eq. (6.17) to be a delta-sequence! Thus the limit of the derivative of the step-sequence is equal to the limit of the delta-sequence.

It is worth noting how one progresses, via derivatives, from a continuous but non-smooth function to the step function and then to the delta function. This progression is illustrated in Fig. 6.6. We start with a function which is continuous but has a "kink" in it at $x = 0$, of the form

$$f(x) = \alpha|x|, \tag{6.46}$$

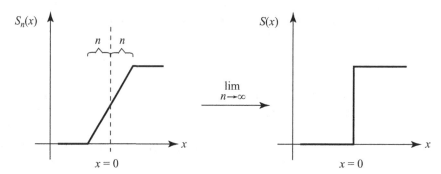

Figure 6.5 The evolution of the step function-sequence.

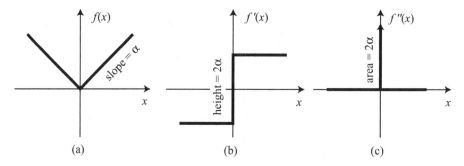

Figure 6.6 The progression of derivatives of distributions: a "kink" (a) becomes a "step" (b) which becomes a delta function (c).

with $\alpha > 0$. For $x < 0$, the slope of this function is $-\alpha$; for $x > 0$, the slope of this function is $+\alpha$. The derivative of $f(x)$ therefore looks like

$$f'(x) = 2\alpha S(x) - \alpha. \tag{6.47}$$

In turn, the derivative of this function will be a delta function,

$$f''(x) = 2\alpha \delta(x). \tag{6.48}$$

Though, in general, the exact functional progression is not as simple as given in the specific case of Eqs. (6.46), (6.47) and (6.48), we may postulate the general rule: the derivative of a "kink" leads to a "step", and the derivative of a "step" leads to a delta function. Higher derivatives follow the rule given by Eq. (6.31).

6.6 Delta functions of more than one variable

We began this chapter by introducing the three-dimensional Dirac delta function $\delta^{(3)}(\mathbf{r} - \mathbf{r}')$, which is a distribution infinitely localized in three-dimensional space whose volume integral is unity. In Cartesian coordinates, this can be readily expressed as the product of three one-variable delta functions, i.e.

$$\delta^{(3)}(\mathbf{r} - \mathbf{r}') = \delta(x - x')\delta(y - y')\delta(z - z'). \tag{6.49}$$

It is also useful, however, to also express this delta function in terms of curvilinear coordinate systems such as the cylindrical and spherical systems.

It is tempting to write the three-dimensional delta function as the product of delta functions with respect to each of the variables q_i, i.e. $\delta^{(3)}(\mathbf{r} - \mathbf{r}') = \delta(q_1 - q_1')\delta(q_2 - q_2')\delta(q_3 - q_3')$. In general, this can be immediately seen to be incorrect because the different sides of the equation will have different dimensionality. For instance, in spherical coordinates (r, θ, ϕ), the left-hand side has dimensions $[\text{length}]^{-3}$ while the right-hand side has only dimensions $[\text{length}]^{-1}$.

Let us try instead a delta function of the form

$$\delta^{(3)}(\mathbf{r} - \mathbf{r}') = D(q_1, q_2, q_3)\delta(q_1 - q_1')\delta(q_2 - q_2')\delta(q_3 - q_3'), \tag{6.50}$$

where $D(q_1, q_2, q_3)$ is a function to be determined. Let us integrate $\delta^{(3)}$ over a small volume $\Delta\tau$ centered on the point $\mathbf{r}$, i.e.

$$\int_{\Delta\tau} \delta^{(3)}(\mathbf{r} - \mathbf{r}')d\tau' = \int_{\Delta\tau} D(q_1, q_2, q_3)\delta(q_1 - q_1')\delta(q_2 - q_2')\delta(q_3 - q_3')d\tau'. \tag{6.51}$$

The infinitesimal volume element $d\tau'$ can be expressed using Eq. (3.12) in terms of the curvilinear coordinate differentials and scale factors as

$$d\tau' = h_1(q_1', q_2', q_3')h_2(q_1', q_2', q_3')h_3(q_1', q_2', q_3')dq_1'dq_2'dq_3'. \tag{6.52}$$

On substitution from this equation into Eq. (6.51), we readily find that

$$\int_{\Delta\tau} \delta^{(3)}(\mathbf{r} - \mathbf{r}')d\tau' = D(\mathbf{r})h_1(\mathbf{r})h_2(\mathbf{r})h_3(\mathbf{r}). \tag{6.53}$$

Because this integral should be equal to unity, we find that

$$D(\mathbf{r}) = \frac{1}{h_1(\mathbf{r})h_2(\mathbf{r})h_3(\mathbf{r})}, \tag{6.54}$$

and using Eq. (6.50), we may write the three-dimensional delta function in curvilinear coordinates as

$$\delta^{(3)}(\mathbf{r} - \mathbf{r}') = \frac{\delta(q_1 - q_1')\delta(q_2 - q_2')\delta(q_3 - q_3')}{h_1(\mathbf{r})h_2(\mathbf{r})h_3(\mathbf{r})}. \tag{6.55}$$

In spherical coordinates, we may therefore write

$$\delta^{(3)}(\mathbf{r} - \mathbf{r}') = \frac{\delta(r - r')\delta(\theta - \theta')\delta(\phi - \phi')}{r^2 \sin\theta}. \tag{6.56}$$

In terms of solid angle, this expression may be written as

$$\delta^{(3)}(\mathbf{r} - \mathbf{r}') = \delta(r - r')\delta(\Omega - \Omega').$$ (6.57)

In cylindrical coordinates, we may write

$$\delta^{(3)}(\mathbf{r} - \mathbf{r}') = \frac{\delta(\rho - \rho')\delta(\phi - \phi')\delta(z - z')}{\rho}.$$ (6.58)

It is worth noting that a curvilinear coordinate representation of the delta function poten-tially suffers from singular behavior in certain special regions. These special regions are locations where the position vector in three-dimensional space is uniquely determined by specifying only one or two coordinates, with the second and/or third coordinate left unde-fined. For instance, in both spherical and cylindrical coordinates, ϕ is undefined when the position vector $\mathbf{r}$ lies on the positive z-axis. In spherical coordinates, a position on the positive z-axis can be uniquely determined by specifying r and $\theta = 0$, while in cylindrical coordinates a position on the z-axis can be uniquely determined by specifying $\rho = 0$ and z. In spherical coordinates, the point at the origin is uniquely defined by the single equation $r = 0$.

Two options exist in this case. One can simply impose arbitrary values of the unspecified coordinate(s) for the integration in three-dimensional space; for instance, with a delta cen-tered at $r = 0$, one could choose any value of θ and ϕ and perform the integration normally. This solution forces us to specify one or more coordinates which possess no physical mean-ing, and for this reason it is unsatisfying. A more elegant solution follows by considering a reduced version of Eq. (6.51) in the special case when the position vector lies within one of the singular regions and the meaningless delta functions are left out. For instance, let us consider points on the positive z-axis in spherical coordinates, and take the delta function to be of the form

$$\delta_z^{(3)}(\mathbf{r} - \mathbf{r}') = D(r',\theta')\delta(r - r')\delta(\theta').$$ (6.59)

We integrate this over a small volume centered on the point $(r, \theta = 0)$; requiring that the integral be equal to unity, we readily find that

$$D(r',\theta') = \frac{1}{r'^2 \sin\theta'},$$ (6.60)

and that

$$\delta_z^{(3)}(\mathbf{r} - \mathbf{r}') = \frac{\delta(r - r')\delta(\theta')}{2\pi r'^2 \sin\theta'}.$$ (6.61)

For the case of a delta function centered on the origin in spherical coordinates, we assume

$$\delta_0^{(3)}(\mathbf{r} - \mathbf{r}') = D(r')\delta(r'),$$ (6.62)

which leads to the result

$$\delta_0^{(3)}(\mathbf{r} - \mathbf{r}') = \frac{\delta(r')}{4\pi r^2}.$$ (6.63)

Returning to Eq. (6.49) for the three-dimensional delta function in Cartesian coordinates, it is to be noted that it may be written as the product of three independent one-variable delta functions. When this function is integrated over a volume, each delta function fixes the value of a single coordinate. For the three-dimensional delta function, the result of the sifting integral is the value of a function at a single point. We may also consider the effect of a function of the form

$$\delta^{(p)}(\mathbf{r} - \mathbf{r}') = \delta(x - x')\delta(y - y'),$$ (6.64)

where the delta for z has been removed. If we use this delta to sift a function by integrating over all of three-dimensional space, we find that

$$\int f(x', y', z')\delta^{(p)}(\mathbf{r} - \mathbf{r}')d\tau' = \int_{-\infty}^{\infty} f(x, y, z')dz'.$$ (6.65)

The result is therefore an integration of the function $f(x, y, z)$ over the z-axis, i.e. a *path integral*. In Section 12.8, we will see that this relationship is the basis of the Radon transform and the technique of computed tomography.

Similarly, if we consider a delta function of the form

$$\delta^{(s)}(\mathbf{r} - \mathbf{r}') = \delta(x - x'),$$ (6.66)

a sifting integral over all space results in the expression

$$\int f(x', y', z')\delta^{(s)}(\mathbf{r} - \mathbf{r}')d\tau' = \int_{-\infty}^{\infty}\int_{-\infty}^{\infty} f(x, y', z')dy'dz'.$$ (6.67)

The resulting integral is a *surface integral* over the yz-plane. We may therefore use delta functions to express surface integrals in terms of volume integrals. This is typically most helpful in curvilinear coordinates. For example, a delta function which constrains a volume integral to a surface of constant r in spherical coordinates may be written as

$$\delta^{(r)}(\mathbf{r} - \mathbf{r}') = \delta(r - r').$$ (6.68)

Similarly, a delta function which constrains a volume integral to a surface of constant ρ in cylindrical coordinates may be written as

$$\delta^{(\rho)}(\mathbf{r} - \mathbf{r}') = \delta(\rho - \rho').$$ (6.69)

As a general rule, a delta function for a particular *coordinate* in a curved coordinate system may be written as

$$\delta^{(q_i)}(\mathbf{r} - \mathbf{r}') = \frac{\delta(q_i - q_i')}{h_i}.$$ (6.70)

Finally, we note that we may also write the three-dimensional delta function in spherical coordinates using the element of solid angle, i.e.

$$\delta^{(3)}(\mathbf{r} - \mathbf{r}') = \delta(r - r')\delta(\Omega - \Omega'),$$ (6.71)

such that

$$\int_{4\pi} f(\Omega')\delta(\Omega - \Omega')d\Omega' = f(\Omega). \tag{6.72}$$

This is a useful shorthand for problems involving spherical symmetry.

6.7 Additional reading

- J.-P. Marchand, *Distributions: an Outline* (Amsterdam, North-Holland, 1962). This short text covers the theory of distributions from a mathematician's perspective. Also in Dover.
- H. J. Wilcox and D. L. Myers, *An Introduction to Lebesgue Integration and Fourier Series* (Huntington, New York, Robert E. Krieger Publishing Company, 1978). This book gives an introduction to the theory of Lebesgue integration, discussed briefly in this chapter in the context of the definition of a function. Also in Dover.

6.8 Exercises

1. A sawtooth wave function is a periodic function $f(x)$ such that $f(x + 2\pi) = f(x)$, the first cycle of which is illustrated below.

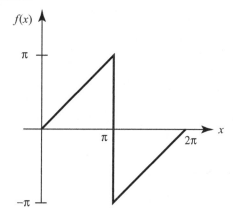

   Determine as a distribution the derivative of this function for all values of x.

2. Consider the function

$$f(x) = \begin{cases} x^2 + 1, & x < 0, \\ 4x, & x \geq 0. \end{cases}$$

   Evaluate, as a distribution, the derivative of this function.

3. Evaluate, as distributions, the first and second derivatives of the functions $f(x) = |x|\sin x$ and $f(x) = |x|\cos x$.

4. Evaluate, as distributions, the first and second derivatives of the functions $f(x) = |\cos x|$ and $f(x) = |\sin x|$.

5. Calculate the values of the following integrals:

   (a) $I_1 = \int_{-\infty}^{\infty} \delta(x - \pi/4)\tan(x)dx$,

   (b) $I_2 = \int_0^4 \delta(x - 3)\left[x^2 + 4x + 2\right]dx$,

(c) $I_3 = \int_0^\infty \delta(x+1)\left[x^3+4\right]dx$,

(d) $I_4 = \int_{-\infty}^\infty \delta'(x-\pi)\sin(x)dx$.

6. Calculate the values of the following integrals:

   (a) $I_1 = \int_{-\infty}^\infty \delta(x-\pi/2)\cos(2x)dx$,

   (b) $I_2 = \int_{-4}^0 \delta(x-4)[2x+1]dx$,

   (c) $I_3 = \int_0^\infty \delta''(x-\pi/2)\sin(x+\pi)dx$,

   (d) $I_4 = \int_{-4}^4 \delta'(x)[x^3+4x^2+x+1]dx$.

7. Calculate the values of the following integrals:

   (a) $I_1 = \int_{-\infty}^\infty (x^3+2)\delta(x^2+2x+2)dx$,

   (b) $I_2 = \int_{-4}^0 \cos(\pi x)\delta(x^3+1)dx$,

   (c) $I_3 = \int_0^\infty \delta(4x^2-1)\sin^2(\pi x)dx$,

   (d) $I_4 = \int_{-\pi/2}^{\pi/2} \delta(x^4+5x^2+4)[x^2+4]dx$.

8. Using the sifting property, and delta sequences when appropriate, show that the distribution $g(x)\delta'(x)$ may be written as

$$g(x)\delta'(x) = g(0)\delta'(x) - g'(0)\delta(x). \tag{6.73}$$

9. Using the sifting property, and delta sequences when appropriate, show that the distribution $g(x)\delta''(x)$ may be written as

$$g(x)\delta''(x) = g''(0)\delta(x) - 2g'(0)\delta'(x) + g(0)\delta''(x). \tag{6.74}$$

10. Simplify the expression for the distribution $\delta[x^2-a^2]$. Describe what happens as $a \to 0$.
11. Simplify the expression for the distribution $\delta[x^3-2x^2+2x]$.
12. Determine the behavior of the distribution $\delta[\sin(x)]$ for all values of x.
13. Determine the behavior of the distribution $\delta'[x^2-4]$.
14. Parabolic cylindrical coordinates are described by (σ,τ,z), where

$$x = \sigma\tau,$$
$$y = \frac{1}{2}(\tau^2-\sigma^2),$$
$$z = z.$$

Write an expression for the three-dimensional Dirac delta function in this coordinate system.

15. Prolate spheroidal coordinates are described by (μ,ν,ϕ), with $\mu \geq 0, 0 \leq \nu \leq \pi$, and $0 \leq \phi < 2\pi$, where

$$x = a\sinh\mu\sin\nu\cos\phi,$$
$$y = a\sinh\mu\sin\nu\sin\phi,$$
$$z = a\cosh\mu\cos\nu.$$

Write an expression for the three-dimensional Dirac delta function in this coordinate system.

16. Consider the sequence of functions

$$
f_n(x) = \begin{cases} g(n)[x+1/n], & -1/n \le x \le 0, \\ g(n)[1/n - x], & 0 < x \le 1/n, \\ 0, & \text{otherwise.} \end{cases}
$$

What should the function $g(n)$ be in order that $f_n(x)$ is a delta sequence?

17. Consider the sequence of functions

$$
f_n(x) = \begin{cases} g(n)[1 - (nx)^2], & |x| \le 1/n, \\ 0, & |x| > 1/n. \end{cases}
$$

What should the function $g(n)$ be in order that $f_n(x)$ is a delta sequence?

18. We consider the function $F(x) = S(x)\cos x$, where $S(x)$ is the Heaviside function. Show that the derivative $F'(x)$ may be expressed as a distribution of the form

$$
F'(x) = \delta(x) - S(x)\sin x.
$$

19. We can derive a step function from a delta sequence, as well as the other way around. Demonstrate that

$$
\lim_{n \to \infty} \int_{-\infty}^{x} \delta_n(x')dx' = S(x),
$$

where

$$
\delta_n(x) = \frac{n}{\pi} \frac{1}{1 + n^2 x^2}.
$$

7

Infinite series

7.1 Introduction: the Fabry–Perot interferometer

High precision interferometers typically employ reflecting surfaces to interfere an optical beam with itself multiple times. One of the earliest of these, developed by Charles Fabry and Alfred Perot, is also one of the most persistently useful and versatile interferometeric devices. First introduced in 1897 as a technique for measuring the optical thickness of a slab or air or glass [FP97], the device found its most successful application only two years later as a spectroscopic device [FP99]. In its simplest incarnation, the interferometer is a pair of parallel, partially reflecting and negligibly thin mirrors separated by a distance d; it is illustrated in Fig. 7.1. A plane wave incident from the left will be partially transmitted through the device, and partially reflected; the amount of light transmitted depends in a nontrivial way upon the properties of the interferometer, namely the mirror separation d, the mirror reflectivity r and transmissivity t, and the wavenumber k of the incident light. In this section we will study the transmission properties of the interferometer and show that a solution to the problem requires the summation of an infinite series.

The most natural way to analyze the effects of the Fabry–Perot is to follow the possible paths of the plane wave through the system and track all of its possible behaviors. We begin with a monochromatic plane wave incident from the left of the form $U(z,t) = A \exp[ikz - i\omega t]$. The time factor $\exp[-i\omega t]$ will be the same for all components of the field, and so we suppress it and focus on the spatial dependence of the field, $U(z) = A \exp[ikz]$. At the first mirror, part of the incident plane wave will be reflected and lost, and part will be transmitted. The factor t represents the fraction of the field amplitude which is transmitted, so immediately beyond the first mirror the wave may be written as $U(-d) = At \exp[-ikd]$. On propagating to the plane $z = 0$, the field acquires a phase $\exp[ikd]$, and part of this field gets transmitted by the second mirror, incurring an additional factor of t. Beyond the second mirror, the field may be written as $U_0(z) = At^2 \exp[ikz]$. This chain of events is shown in Fig. 7.2(a).

This is not the only contribution to the transmitted wave, however. Part of the wave gets *reflected* at the second mirror, propagates to the first mirror, gets reflected again, and finally is transmitted through the second mirror. This twice-reflected wave [illustrated in Fig. 7.2(b)] will differ in phase from the directly transmitted wave as follows. At the second

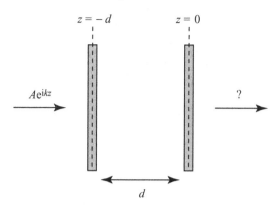

Figure 7.1 The Fabry–Perot interferometer, which consists of two partially reflecting mirrors separated by a distance d. A plane wave incident from the left is partially transmitted through the device, and partially reflected.

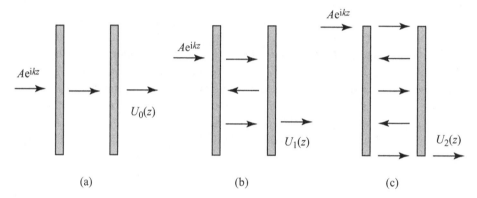

Figure 7.2 Contributions to the transmitted field in the interferometer. (a) The directly transmitted wave. (b) The twice-reflected wave. (c) The four-times-reflected wave.

mirror, part of the directly-transmitted wave is reflected, so that immediately upon reflection the wave has the form $U(0) = Atr$. On propagating back to the first mirror, it acquires a phase $\exp[ikd]$, and on reflection at the first mirror it has the form $U(-d) = Atr^2 \exp[ikd]$. On propagating back to the second mirror, this wave picks up an additional phase factor $\exp[ikd]$, and on transmission through the mirror may be written as $U_1(z) = At^2r^2 \exp[2ikd]$. This wave differs from the directly transmitted wave by a factor of $r^2 \exp[2ikd]$.

This is still not the end of the story, however; part of the twice-reflected wave gets reflected at the second mirror *again*, and makes another round trip through the interferometer, becoming a four-times-reflected wave $U_2(z) = At^2r^4 \exp[4ikd]$ [Fig. 7.2(c)]. Part of this wave will also be reflected back through the system, becoming a six-times-reflected wave U_3, and so on. Each round trip through the cavity modifies the field by an additional factor $r^2 e^{2ikd}$. The total output of the interferometer will be the combination of this in principle *infinite* sum of terms, and so the field at the output may be written as

$$U_{\text{out}}(z) = U_0(z) + U_1(z) + U_2(z) + U_3(z) + \cdots$$

$$= At^2 \left[1 + r^2 e^{2ikd} + r^4 e^{4ikd} + r^6 e^{6ikd} + \cdots \right] e^{ikz}. \tag{7.1}$$

The term in brackets is an *infinite series*. It can be written in a more compact form as

$$\sum_{n=0}^{\infty} \left(r^2 e^{2ikd} \right)^n = 1 + r^2 e^{2ikd} + r^4 e^{4ikd} + r^6 e^{6ikd} + \cdots, \tag{7.2}$$

where the $\sum$ denotes summation, n represents an integer index of summation, and the upper and lower limits on the summation sign represent the first and last terms of the series. For this example, each term of the series represents the contribution to the transmitted field which results from n round trips through the cavity. Happily, this series is well-understood and referred to as the *geometric series*, and we will show in this chapter that it can be summed to the closed functional form

$$T(k) = \sum_{n=0}^{\infty} \left(r^2 e^{2ikd} \right)^n = \frac{1}{1 - r^2 e^{2ikd}}. \tag{7.3}$$

The intensity of the field transmitted by the interferometer will be proportional to the absolute square of the field, $I \propto |U_{\text{out}}|^2$. The transmitted intensity is therefore proportional to $|T(k)|^2$, which has the form

$$|T(k)|^2 = \frac{1}{1 + r^4 - 2r^2 \cos(2kd)}. \tag{7.4}$$

An example of the transmitted intensity as a function of wavelength $\lambda = 2\pi/k$ is shown in Fig. 7.3. It can be seen that the device only transmits over narrow, well-spaced wavelength ranges which are approximately periodic. The effectiveness of the interferometer depends on two parameters: the width of the individual peaks, which determines the wavelength resolution of the device, and the separation of adjacent peaks, which determines the effective spectral range of the device. The latter quantity is known as the *free spectral range*, and is approximately given by

$$\Delta\lambda \approx \frac{\lambda^2}{2d}. \tag{7.5}$$

The width $\delta\lambda$ of a single peak is characterized by the *finesse* $\mathcal{F}$, given by

$$\mathcal{F} \equiv \frac{\Delta\lambda}{\delta\lambda} \approx \frac{\pi r^{1/2}}{1 - r}. \tag{7.6}$$

It can be seen that a higher reflectivity r corresponds to narrower peaks (higher finesse).

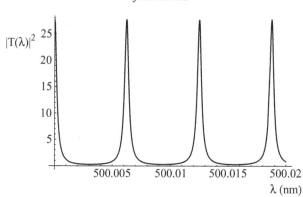

Figure 7.3 Transmission properties of a Fabry–Perot interferometer. Here $r = 0.9$ (the mirrors reflect 90% of the field amplitude) and $d = 2$ cm.

Infinite series occur quite often in optical and physical applications. An analysis similar to the Fabry–Perot solution can be used in many problems involving interference through multiple reflection, such as reflection from thin films and Newton's rings. However, very few series can be summed explicitly like the geometric series. Furthermore, there is no guarantee that the sum of the series will converge to a finite number, even for seemingly well-posed physical problems. For these reasons, among others, a study of the properties of infinite series is justified.

7.2 Sequences and series

An infinite *sequence* is a semi-infinite set of numbers which are assigned a particular order; this order is labeled by an index n which typically ranges from zero to infinity, and the nth term in the sequence is denoted by a_n. The members of the sequence may in general be complex numbers. An example of such a sequence is

$$a_n = \frac{1}{n}. \tag{7.7}$$

The *limit* of a sequence is the value to which the sequence approaches as $n \to \infty$. For example, the limit of the sequence defined in Eq. (7.7) is $A = \lim_{n \to \infty} a_n = 0$. A formal definition looks a little more intimidating, but is conceptually simple.

Definition 7.1 *The number A is the limit of the sequence a_n if, for every $\epsilon > 0$, there exists a fixed $N = N(\epsilon)$ such that*

$$|A - a_n| < \epsilon, \quad for \quad n > N. \tag{7.8}$$

What this means is that as we move further along the sequence (as n gets larger and larger), the points of the sequence tend to move closer and closer to the limit. Mathematically, we

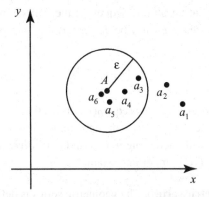

Figure 7.4 Defining the limit of a sequence of complex numbers. In this picture, "for $n > 2$, $|A - a_n| < \epsilon$."

quantify this by drawing a "circle" around the limit, of radius ϵ; for any radius we choose, we can find a number $N(\epsilon)$ such that all members of the sequence for which $n > N$ lie within the circle. This concept is illustrated for a sequence of complex numbers in a plane in Fig. 7.4.

It is to be noted that a sequence does not have to possess a limit; for instance, the sequence

$$a_n = (-1)^n \tag{7.9}$$

does not approach a single number, but flips between $+1$ and -1. A sequence such as

$$a_n = n \tag{7.10}$$

also does not have a limit, as the members of the sequence move further and further apart.

Finding the limit of a sequence is often a nontrivial task. For instance, consider the sequence defined as follows:

$$a_n = \left(1 + \frac{x}{n}\right)^n, \tag{7.11}$$

where x is a real-valued number. The limit of this sequence can be shown to be the exponential function

$$\lim_{n \to \infty} \left(1 + \frac{x}{n}\right)^n = e^x. \tag{7.12}$$

This can be demonstrated by equating the left-hand side of the formula with the Taylor series of the exponential function (to be discussed in Section 7.5).

With the preceding description of a sequence in mind, we now consider a *sequence of partial sums* s_n, defined as

$$s_n = \sum_{i=1}^{n} a_i, \tag{7.13}$$

where the terms a_i represent an infinite sequence; the collection s_n is therefore itself an infinite sequence. An *infinite series* may be defined as the limit of the sequence of partial sums,

$$s_\infty = \lim_{n \to \infty} s_n = \sum_{i=1}^{\infty} a_i. \tag{7.14}$$

This series is said to be *convergent* if the sequence s_n has a limit. If the sequence approaches no limit but instead oscillates between two or more values, it is said to be *oscillatory*. If the sequence tends to infinity with increasing n, it is said to be *divergent*.

We now consider a few (very important) examples.

Example 7.1 (The geometric series) The geometric series is defined as

$$S = a + ar + ar^2 + ar^3 + \cdots + ar^n + \cdots = a \sum_{n=0}^{\infty} r^n. \tag{7.15}$$

We can explicitly sum this series by noting that

$$s_n = a + ar + \cdots + ar^n = a + rs_{n-1} = s_{n-1} + ar^n. \tag{7.16}$$

Rearranging the latter two parts of this equation, we find that

$$s_{n-1} = a\frac{1 - r^n}{1 - r}. \tag{7.17}$$

For $|r| < 1$, the term $r^n \to 0$ as $n \to \infty$. We thus have

$$S = \lim_{n \to \infty} s_n = a\frac{1}{1 - r}. \tag{7.18}$$

It is relatively rare to find a series whose sum can be determined analytically. Returning to the definition of the geometric series, it is clear that it diverges for $|r| > 1$. This highlights an important observation about infinite series: a *necessary* condition for the series to converge is that $\lim_{n \to \infty} a_n = 0$. This does not guarantee convergence, however, as the next example shows.

◇

Example 7.2 (The harmonic series) The harmonic series is defined as

$$S = \sum_{n=1}^{\infty} \frac{1}{n} = 1 + \frac{1}{2} + \frac{1}{3} + \cdots. \tag{7.19}$$

The sequence which defines the series has the limit $\lim_{n \to \infty} 1/n = 0$, but this series in fact diverges.

To prove this, let us group the terms of this series in the logical order,

$$S = 1 + \frac{1}{2} + \left(\frac{1}{3} + \frac{1}{4}\right) + \left(\frac{1}{5} + \frac{1}{6} + \frac{1}{7} + \frac{1}{8}\right) + \cdots . \tag{7.20}$$

For $n > 0$, each grouping of terms may be written as

$$b_n = \sum_{p=1}^{2^{n-1}} \frac{1}{2^{n-1} + p}. \tag{7.21}$$

Defining $b_0 = 1$, we may rewrite the harmonic series in the form

$$S = \sum_{n=0}^{\infty} b_n. \tag{7.22}$$

Each term b_n may be written explicitly as

$$b_n = \frac{1}{2^{n-1} + 1} + \frac{1}{2^{n-1} + 2} + \cdots + \frac{1}{2^{n-1} + 2^{n-1}}. \tag{7.23}$$

For $p < 2^{n-1}$, the following inequality holds:

$$\frac{1}{2^{n-1} + p} > \frac{1}{2^{n-1} + 2^{n-1}} = \frac{1}{2^n}. \tag{7.24}$$

This in turn leads to

$$b_n > \sum_{p=1}^{2^{n-1}} \frac{1}{2^n} = \frac{2^{n+1}}{2^n} = \frac{1}{2}. \tag{7.25}$$

Every term b_n is therefore greater than $1/2$. Since there are an infinite number of these terms, we find that the series diverges.

This example shows that we need quantitative tests to determine the convergence of a given infinite series. We describe some such tests in the next section.

◇

7.3 Series convergence

There are a number of tests which may be used to determine whether a given series converges. But before we discuss them, it is useful to limit our attention to a specific class of infinite series, for reasons which will become clear momentarily.

We first consider a specific example of an infinite series, the so-called *alternating harmonic series*,

$$S = \sum_{n=1}^{\infty} (-1)^{n-1} n^{-1} = 1 - \frac{1}{2} + \frac{1}{3} - \frac{1}{4} + \cdots . \tag{7.26}$$

Infinite series

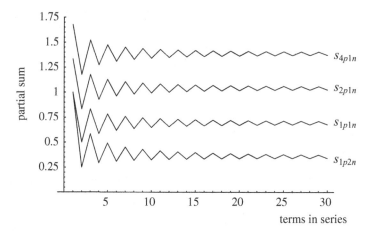

Figure 7.5 Convergence of the alternating harmonic series. Different groupings of terms lead to different limits of the series.

If the terms of this series are summed in order as shown above, the limit can be shown to be given by $S = \log 2$, where log is the natural logarithm (a demonstration of this can be found in the section on Taylor series, Section 7.5). However, let us consider what seems on the surface to be a reasonable idea, the regrouping of the terms of the series. For a finite summation of terms, addition is order-independent, i.e. $a + b + c = c + b + a$ and so forth; is this true for an infinite summation? As a test, we consider several infinite rearrangements of the terms of the series: grouping two positive terms and one negative term together, two negative terms and one positive term together, and four positive terms and one negative term, i.e.

$$S_{2p1n} = \left(1 + \frac{1}{3}\right) - \frac{1}{2} + \left(\frac{1}{5} + \frac{1}{7}\right) - \frac{1}{4} + \cdots, \tag{7.27}$$

$$S_{1p2n} = 1 - \left(\frac{1}{2} + \frac{1}{4}\right) + \frac{1}{3} - \left(\frac{1}{6} + \frac{1}{8}\right) + \cdots, \tag{7.28}$$

$$S_{4p1n} = \left(1 + \frac{1}{3} + \frac{1}{5} + \frac{1}{7}\right) - \frac{1}{2} + \left(\frac{1}{9} + \frac{1}{11} + \frac{1}{13} + \frac{1}{15}\right) - \frac{1}{4} \cdots. \tag{7.29}$$

We look at the convergence of these series numerically in Fig. 7.5.

Remarkably, each of these different ways of summing a series produces a different limit for the series! It can in fact be proven[1] that this series may be made to converge to any value, or even to diverge. Evidently different groupings of the terms of the alternating harmonic series are not equivalent.

---

[1] This was originally shown by Riemann [Rie68], and is not quite so difficult to prove as one might expect. An English description of the proof can be found in the excellent book by Knopp [Kno51, §44].

To understand this, we recall that in the previous section we saw that the harmonic series, defined as

$$S = \sum_{n=1}^{\infty} \frac{1}{n} = 1 + \frac{1}{2} + \frac{1}{3} + \cdots, \tag{7.30}$$

diverges. The elements of the series apparently decrease in magnitude too slowly to allow their sum to converge. It follows that the alternating harmonic series only converges because its terms alternate in sign, allowing a partial cancellation of the positive and negative terms. By rearranging the order of the terms, we change the nature of this partial cancellation and hence the limit of the series.

It would be nice to say that an infinite series which represents a physical quantity should converge regardless of how we sum the series, but in Section 7.4 we will see that this is not necessarily true. Tests of convergence are therefore extremely important. In essence, there are two effects which can cause a series to converge: its coefficients a_n may approach zero sufficiently rapidly, or partial cancellation of terms may be created by terms of alternating sign. To rule out situations when the latter effect holds sway, we introduce the concept of absolute convergence.

Definition 7.2 *A series* $\sum a_n$ *is said to be* absolutely convergent *if the series* $\sum |a_n|$ *converges.*

The sum of an absolutely convergent series behaves in a manner consistent with the addition of finite sums of numbers. For instance, it can be shown that an absolutely convergent series has a sum which is independent of the order of summation. Furthermore, two absolutely convergent series may be multiplied together and the limit of the product will be the product of the individual limits; i.e. if A and B represent the limits of the absolutely convergent series with terms a_n and b_n, then

$$\left(\sum a_n \right) \left(\sum b_n \right) = AB. \tag{7.31}$$

In other words, we may perform ordinary algebraic operations on absolutely convergent series without trouble. A series such as the alternating harmonic series which is not absolutely convergent is said to be *conditionally convergent*.

We will restrict ourselves to absolutely convergent series. There are many tools available to determine if a series is absolutely convergent, and we describe a number of them.

1. *Comparison test.* The easiest convergence test for an infinite series with terms a_n is as follows.

   If $|a_n| \leq |b_n|$ for all $n \geq N$, where N is an integer, and b_n form an absolutely convergent series, then a_n also form an absolutely convergent series. If $|a_n| > |b_n|$ for all $n \geq N$, and b_n form a divergent series, then a_n also form a divergent series.

   The demonstration of convergence follows because $\sum |a_n| \leq \sum |b_n| < \infty$. Similarly, the demonstration of divergence follows because $\sum |a_n| > \sum |b_n| = \infty$. It is to be noted that the inequalities may be violated for any finite number of terms, as a finite number of terms cannot affect the convergence property of the series. As long as one of the inequalities strictly holds for $n \geq N$, we can demonstrate convergence/divergence.

Example 7.3 (Dirichlet series) The Dirichlet series is defined as

$$D = \sum_{n=1}^{\infty} n^{-p}. \tag{7.32}$$

For $p < 1$, $n^{-p} > n^{-1}$ and since we know that $a_n = n^{-1}$ is divergent, it follows that the Dirichlet series diverges for $p \leq 1$. It is important to note that this test does not specify what happens when $p > 1$.

◇

This test is perhaps the most powerful and versatile of all the convergence tests. Many seemingly complicated series can be reduced with one or more simplifications to a series which can be directly compared. Furthermore, as we will see, many of the other convergence tests are in fact disguised special cases of the comparison test.

Example 7.4 As an example of a nontrivial comparison, we consider the series

$$\sum_{n=0}^{\infty} \frac{n+1}{n(n^3+1)^{1/3}}. \tag{7.33}$$

It is to be noted that, for large n, the terms of this series behave roughly as $a_n \sim 1/n$, as can be seen by neglecting the constant terms. We expect it to diverge; however, as we have seen, appearances can be deceiving. Let us make a series of comparisons to prove the series diverges. First, we note that

$$\frac{n+1}{n(n^3+1)^{1/3}} > \frac{n}{n(n^3+1)^{1/3}} = \frac{1}{(n^3+1)^{1/3}}. \tag{7.34}$$

We now note that $n^3 + 1 < n^3 + 3n^2 + 3n + 1 = (n+1)^3$, which implies that

$$\frac{1}{(n^3+1)^{1/3}} > \frac{1}{n+1}. \tag{7.35}$$

But this latter term is just one of the terms of the harmonic series! We therefore have

$$\frac{n+1}{n(n^3+1)^{1/3}} > \frac{1}{n+1}, \tag{7.36}$$

i.e. every term of the series is greater than a matched term of the harmonic series, which we know diverges. Therefore we know that our test series diverges, by comparison.

◇

2. *Cauchy root test.* The root test is defined as follows.

If $|a_n|^{1/n} \leq r < 1$ as $n \to \infty$, with r independent of n, then $\sum a_n$ converges absolutely.

This test can be confirmed by taking the nth power of each side of the test equation,

$$|a_n| \le r^n < 1. \tag{7.37}$$

The term r^n is just the nth term in a convergent geometric series, so by the comparison test, $\sum a_n$ converges absolutely. The root test is essentially the comparison test applied to the geometric series. If $\lim_{n\to\infty} |a_n|^{1/n} > 1$, the series must diverge by the comparison test. If $\lim_{n\to\infty} |a_n|^{1/n} = 1$, the test is inconclusive.

Example 7.5 We use the root test to investigate the convergence of the series with terms $a_n = [2n/(2+n)]^n$. Then $|a_n|^{1/n} = 2n/(2+n)$, and the limit $\lim_{n\to\infty} |a_n|^{1/n} = 2$. This series is divergent. A series which seems only superficially different, $a_n = [2n/(2+3n)]^n$, is found by the root test to satisfy $\lim_{n\to\infty} |a_n|^{1/n} = 2/3$, and this series converges.

◇

This test is primarily useful when dealing with power series such as Taylor series. It is usually difficult to manipulate quantities within a root, and the root test usually results in just such a situation.

3. *Cauchy ratio test.* The ratio test is defined as follows.

We consider the limit of the ratio $|a_{n+1}/a_n|$ as $n \to \infty$. If

$$\lim_{n\to\infty} \frac{|a_{n+1}|}{|a_n|} \begin{cases} < 1, & \text{convergence,} \\ > 1, & \text{divergence,} \\ = 1, & \text{inconclusive.} \end{cases} \tag{7.38}$$

The ratio test is perhaps the most consistently useful test of convergence, but also has a tendency to fail (i.e. be inconclusive) when it is needed most. The proof again follows by a comparison with the geometric series; for the geometric series, $|a_{n+1}/a_n| = r$ for all n, and the series converges for $r < 1$. If, for sufficiently large n, another series has a ratio $|a_{n+1}/a_n| < r$, then it must converge as well.

Example 7.6 We use the ratio test to investigate the convergence of the series with terms $a_n = n^2/n!$; the ratio is given by

$$\frac{a_{n+1}}{a_n} = \frac{1}{n+1} \left(\frac{n+1}{n} \right)^2 = \frac{1}{n} + \frac{1}{n^2}. \tag{7.39}$$

This series converges. As an example for which the test is inconclusive, we consider the harmonic series, $a_n = 1/n$. In this case, $a_{n+1}/a_n = n/(n+1)$; the limit as $n \to \infty$ is unity. We already know by other means that this series diverges, but the ratio test is inconclusive.

◇

4. *Integral test.* It is, in fact, rare to find an infinite series whose sum can be determined exactly. However, it is frequently possible to find an integral which can be evaluated exactly and can be compared to the series. The limit of the integral can then be used to determine convergence.

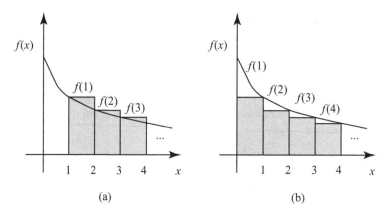

Figure 7.6 Illustration of the (a) lower bound, and (b) upper bound of the integral test.

Let $f(x)$ be a continuous, monotonically decreasing function in which $f(n) = |a_n|$. Then $\sum |a_n|$ converges if $\int_1^\infty f(x)\mathrm{d}x$ is finite and diverges if the integral is infinite.

This test can be understood by considering Fig. 7.6. The area under the function is always smaller than the area under the blocks on the left-hand side of the figure; we therefore have that the ith partial sum of the series satisfies the inequality

$$s_i > \int_1^{i+1} f(x)\mathrm{d}x. \tag{7.40}$$

From the right-hand side of the figure, it is clear that

$$s_i - |a_1| < \int_1^i f(x)\mathrm{d}x. \tag{7.41}$$

In the limit as $i \to \infty$, we have the following bounds on the series:

$$\int_1^\infty f(x)\mathrm{d}x \le \sum_{n=1}^\infty a_n \le \int_1^\infty f(x)\mathrm{d}x + |a_1|. \tag{7.42}$$

From this expression, we can see that the series will converge or diverge as the corresponding integral converges or diverges. The integral test is another application of the comparison test, comparing the series in question to a series based on an integrable function. Note that the requirement that the function $f(x)$ is monotonically decreasing is crucial here; if the function is oscillating up and down, the figures which lead to Eqs. (7.40) and (7.41) are no longer valid.

This test can be particularly useful in the numerical summation of an infinite series, as it can be used to set bounds on the remainder of a series which has been summed up to the Nth term. Thus, if

$$\sum_{n=1}^\infty a_n = \sum_{n=1}^N a_n + \sum_{m=N+1}^\infty a_m, \tag{7.43}$$

we have

$$\int_{N+1}^{\infty} f(x)dx \leq \sum_{m=N+1}^{\infty} a_m \leq \int_{N+1}^{\infty} f(x)dx + a_{N+1}. \tag{7.44}$$

Example 7.7 (Harmonic series) The integral test can be used to demonstrate the divergence of the harmonic series $a_n = 1/n$. We can use the function $f(x) = 1/x$, whose integral is of the form

$$\int_1^{\infty} f(x)dx = \int_1^{\infty} \frac{dx}{x} = \log(x)|_1^{\infty} \to \infty. \tag{7.45}$$

The integral diverges, as does the corresponding series.

◇

5. *Kummer's test.* This test is a general one in which a series is compared to a modified version of itself.

Suppose $b_n > 0$, $a_n, b_n \to 0$ as $n \to \infty$, and

$$\lim_{n\to\infty} \left[b_n \frac{|a_n|}{|a_{n+1}|} - b_{n+1} \right] = C, \tag{7.46}$$

where C is a constant. The series $\sum |a_n|$ converges if $C > 0$, and diverges if $\sum 1/b_n$ diverges and $C < 0$.

To prove the convergent case, we note that Eq. (7.46) implies that there exists an integer N such that, for $n \geq N$,

$$\left[b_n \frac{|a_n|}{|a_{n+1}|} - b_{n+1} \right] > \epsilon_N, \tag{7.47}$$

where $\epsilon_N < C$ is a positive number which depends upon N. In other words, for large enough N, the quantity in brackets is always larger than a nonzero positive number. This equation may be rewritten as

$$\epsilon_N |a_n| < b_{n-1}|a_{n-1}| - |a_n|b_n. \tag{7.48}$$

Let us sum this inequality for indices ranging from N to n; we then have

$$\epsilon_N \sum_{i=N}^{n} |a_i| < b_{N-1}|a_{N-1}| - b_N|a_N| + b_N|a_N| - b_{N+1}|a_{N+1}| + \cdots$$

$$+ b_{n-2}|a_{n-2}| - b_{n-1}|a_{n-1}| + b_{n-1}|a_{n-1}| - b_n|a_n|. \tag{7.49}$$

On inspection, all of the intermediate terms cancel and we may write

$$\epsilon_N \sum_{i=N}^{n} |a_i| \leq b_{N-1}|a_{N-1}| - b_n|a_n|. \tag{7.50}$$

In the limit $n \to \infty$, $b_n|a_n| \to 0$ and we may write

$$\sum_{i=N}^{n} |a_i| \leq b_{N-1}|a_{N-1}|/\epsilon_N. \tag{7.51}$$

The right-hand side of this inequality is finite, and the left-hand side only excludes a finite number of terms for $i < N$ which have a presumably finite sum. Therefore the series converges.

To prove the divergent case, we return to Eq. (7.46) and note that, for $C < 0$, it implies that there exists an integer N such that, for $n \geq N$,

$$\left[b_n \frac{|a_n|}{|a_{n+1}|} - b_{n+1} \right] < 0. \tag{7.52}$$

In other words, for large enough N, the bracketed quantity is always negative. We consider the case $n = N$ and $n = N + 1$ and rewrite Eq. (7.52) in the form

$$|a_{N+1}| > \frac{|a_N| b_N}{b_{N+1}}, \tag{7.53}$$

and also as

$$|a_{N+2}| > \frac{|a_{N+1}| b_{N+1}}{b_{N+2}}. \tag{7.54}$$

We may substitute from Eq. (7.53) into Eq. (7.54); performing this process multiple times, we may write

$$|a_{N+M}| > \frac{|a_N| b_N}{b_{N+M}}. \tag{7.55}$$

Thus we may write, for sufficiently large n, that $|a_n| > \alpha/b_n$, where α is a positive constant. We know that the series $\alpha \sum b_n^{-1}$ diverges, however, so by the comparison test $|a_n|$ diverges.

If we make the specific choice $b_n = 1$, Kummer's test reduces to the ratio test. If we make the choice $b_n = n$, Kummer's test reduces to Raabe's test, given next.

6. *Raabe's test.* In situations when the ratio test gives an indeterminate result, Raabe's test often will work. It is given by:

if

$$\lim_{n \to \infty} n \left(\left| \frac{a_n}{a_{n+1}} \right| - 1 \right) = K, \tag{7.56}$$

the series converges for $K > 1$ and diverges for $K < 1$. The test is indeterminate if $K = 1$.

As noted, Raabe's test is a special case of Kummer's test, with the choice $b_n = n$. We can see this by direct substitution into Eq. (7.46),

$$\lim_{n \to \infty} \left[b_n \frac{|a_n|}{|a_{n+1}|} - b_{n+1} \right] = \lim_{n \to \infty} \left[n \frac{|a_n|}{|a_{n+1}|} - n - 1 \right] = C. \tag{7.57}$$

This equation may be rearranged as

$$\lim_{n \to \infty} n \left[\frac{|a_n|}{|a_{n+1}|} - 1 \right] = C + 1 = K. \tag{7.58}$$

On comparison with Kummer's test, we have convergence for $C > 0$, or $K > 1$, and divergence for $C < 0$, or $K < 1$.

Example 7.8 We consider the series $a_n = 1/[n(n+1)]$; the ratio test gives

$$\frac{a_{n+1}}{a_n} = \frac{n(n+1)}{(n+1)(n+2)} = \frac{n}{n+2} \to 1 \quad \text{as} \quad n \to \infty, \tag{7.59}$$

and is inconclusive. Raabe's test gives

$$n\left(\frac{a_n}{a_{n+1}} - 1\right) = n\left(\frac{n+2}{n} - 1\right) = n+2-n = 2 \to 2 \quad \text{as} \quad n \to \infty. \tag{7.60}$$

Thus Raabe's test indicates that this series is convergent.

◇

7. *Gauss' test.* A test even more sensitive than Raabe's test is Gauss' test, which may be stated as follows.

Suppose $a_n > 0$ and that

$$\frac{a_n}{a_{n+1}} = 1 + \frac{h}{n} + \frac{B(n)}{n^2}, \tag{7.61}$$

where $B(n)$ is a bounded function as $n \to \infty$. The series $\sum a_n$ converges for $h > 1$ and diverges for $h \le 1$.

It is to be noted that Gauss' test has no indeterminate case. The results for $h > 1$ and $h < 1$ follow directly from Raabe's test. For $h = 1$, Raabe's test is indeterminate but we can use Kummer's test with $b_n = n \log n$ to determine that the series diverges.

Even more tests exist for determining series convergence. More sophisticated tests, however, are difficult to apply, whereas the easier tests can give an indeterminate result. In the end, there is no sure-fire systematic method for determining if a given series converges or not; one must simply try various tests until one of them gives a definite result. Some guidelines, however, are suggested below.

1. Does the series look similar to one whose convergence properties are known (such as the harmonic series or geometric series)? If so, try the comparison test.
2. Does the series involve only terms to the nth power, such as $a_n = x^n/b^{2n}$? If so, the root test is a good option.
3. Does the series involve factorials or rational products of n? If so, the ratio test should be your first choice.
4. Does the ratio test fail? Try using Raabe's test instead.
5. Is the term a_n monotonically decreasing and does it look like an integrable function? Then the integral test might work.
6. Can one estimate the sum of the series directly?

The last suggestion is worth exploring as an example.

Example 7.9 Consider the following variation on the idea of the harmonic series:

$$a_n = \frac{1}{n} - \frac{1}{n+1}. \tag{7.62}$$

It is to be noted that all terms of the series are non-negative, so we do not need to worry about the problem of conditional convergence. By grouping terms under a common fraction, one can readily find via the comparison test that this series is convergent; however, this can also be shown by the following argument. Let us sum three terms of the series as follows,

$$a_{n+1} + a_n + a_{n-1} = \frac{1}{n+1} - \frac{1}{n+2} + \frac{1}{n} - \frac{1}{n+1} + \frac{1}{n-1} - \frac{1}{n}. \tag{7.63}$$

It is to be noted that the terms with $n+1$ and n in the denominator completely cancel! This argument can be extended to a range of terms from $n = p$ to $n = q$,

$$\sum_{n=p}^{n=q} a_n = \frac{1}{p} - \frac{1}{q+1}. \tag{7.64}$$

If we take the limit $p = 1$ and $q \to \infty$, we readily find that the sum of the series is 1. As already noted, it is somewhat rare to find series that can be summed exactly in this manner, but in some cases (such as the harmonic series itself, for instance), one must think beyond the tests provided here and find other ways of assessing convergence.

◇

7.4 Series of functions

Up to this point, we have been focusing on series whose terms depend only upon the index n. For most physical problems which are amenable to series solution, the elements of the series also depend on one or more independent variables; an example is the Fabry–Perot transmission function $T(k)$ considered in Section 7.1. Of course, one of the most useful series expansions, the Taylor series, is an expansion of a function $f(x)$ into a series of polynomials x^n. In this section we consider some general properties of infinite series whose terms may each be a function of a variable x, i.e. $u_n = u_n(x)$.

Returning to the concept of the sequence of partial sums, the partial sum of such a sequence of functions will also be a function of x,

$$s_n(x) = u_1(x) + u_2(x) + \cdots + u_n(x), \tag{7.65}$$

as will be the series sum, which is as before the limit of the sequence of partial sums,

$$S(x) = \sum_{n=1}^{\infty} u_n(x). \tag{7.66}$$

The process is straightforward and usually does not lead to any problems. However, there is one issue that arises in the theory of series of functions and should be addressed: the idea of *uniform convergence*.

We have seen in the previous section that some series converge to their limit more rapidly than others; that is, for some series $a_n \to 0$ with respect to increasing n faster than others.

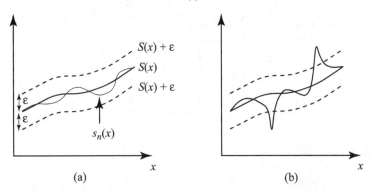

Figure 7.7 An illustration of the ideas of (a) uniform convergence and (b) nonuniform convergence.

In approximating an infinite series by a sum of a finite number of terms N, this rate of convergence dictates how large N must be to get a good approximation: a slower rate of convergence requires a larger N. When dealing with a series of functions which depend on x, this rate of convergence may vary appreciably with x. This is a potential problem when approximating a function by an infinite series, since we will always use a finite number of terms to approximate $S(x)$. For some values of x, this approximation may be much worse than others.

Ideally, we want the series to converge to its limit $S(x)$ *at essentially the same rate* for all x. If it does so, it is called a *uniformly convergent series*.

Definition 7.3 (Uniform convergence) *If for any $\epsilon > 0$ there exists a number N, independent of x in the interval $a \leq x \leq b$, such that*

$$|S(x) - s_n(x)| < \epsilon \quad \text{for all} \quad n \geq N, \tag{7.67}$$

the series is said to be uniformly convergent in $[a,b]$.

The idea of uniform convergence is illustrated in Fig. 7.7.

In less technical terms, a uniformly convergent series has some $n \geq N$ for which the partial sums all lie within the upper and lower ϵ-boundaries. It is to be noted that this definition implies that there must exist some regions of x within which the series *never* converges; otherwise, we could always define an exceedingly conservative value of ϵ to encompass the lagging parts of the series.

This concept is rather abstract and is usually not a concern in practical problems. However, it occasionally shows up in surprising places and can lead to quantitative errors if not treated properly. An example of this is in the application of Fourier series (which will be discussed in Chapter 8). We consider the series representation of a periodic square wave, illustrated in Fig. 7.8.

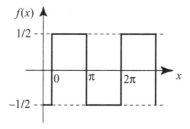

Figure 7.8 A periodic square wave.

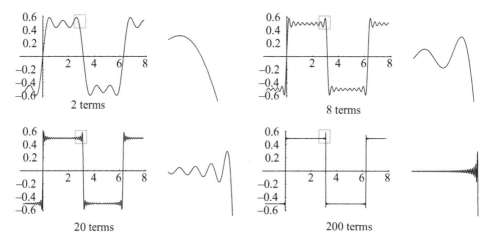

Figure 7.9 The Fourier series of a periodic square wave, with the Gibbs phenomenon of each set of terms illustrated.

This function can be represented by a Fourier expansion in the form

$$f(x) = \frac{2}{\pi} \sum_{n=1}^{\infty} \frac{\sin[(2n+1)x]}{(2n+1)}. \tag{7.68}$$

We present two sets of plots of the partial sums of this series in Fig. 7.9, one showing one full cycle, the second zoomed in on the edge of the square wave.

We can see that the series is approaching its limit everywhere except at the edges of the square wave, where it "overshoots" the correct answer. Even as the number of summed terms approaches infinity, one can show that an overshoot of roughly 18% exists near these edges. This is called the *Gibbs phenomenon* and is a potential problem when using such series for precise numerical work.

An exact calculation of the Gibbs "overshoot" will be discussed later, in Section 8.7. We may make one interesting connection before moving on, however; consider the series representation of the square wave at $x = \pi/2$, i.e.

$$f\left(\frac{\pi}{2}\right) = \frac{2}{\pi} \sum_{n=1}^{\infty} \frac{(-1)^n}{(2n+1)}. \tag{7.69}$$

The second half of this series has terms whose magnitudes roughly drop off as $1/n$, meaning that this series is conditionally convergent! We can see this by noting that

$$\frac{1}{(2n+1)} > \frac{1}{(2n+2)} = \frac{1}{2}\frac{1}{n+1}. \tag{7.70}$$

It turns out that whenever one works with the Fourier series of a function with a discontinuity, the series representation is conditionally convergent. The singular behavior of the function (the discontinuity) is in some sense expressed in the singular behavior of the series representation (conditional convergence).

7.5 Taylor series

One of the most beautiful and useful mathematical tools in all of mathematical physics is the Taylor expansion, in which a continuous function $f(x)$ is represented by a series of powers of $(x - x_0)$; the series is said to be an expansion about the point x_0. In the neighborhood of this point, the function is well-approximated by only a few terms of the Taylor series.

There are a number of ways to derive the Taylor expansion; one of the simplest (and least rigorous) is to *assume* that the function $f(x)$ may be expanded about the point x_0 in the form

$$f(x) = \sum_{n=0}^{\infty} a_n (x - x_0)^n. \tag{7.71}$$

If we let $x = x_0$, all powers of n except $n = 0$ vanish, and we have

$$a_0 = f(x_0). \tag{7.72}$$

If we instead take the first derivative of both sides of equation (7.71), we find that

$$\frac{d}{dx} f(x) \bigg|_{x=x_0} = 1 \times a_1. \tag{7.73}$$

Likewise, we can take the mth derivative of both sides of the equation, and find that

$$\frac{d^m}{dx^m} f(x) \bigg|_{x=x_0} = m! a_m, \tag{7.74}$$

from which we may write

$$f(x) = \sum_{n=0}^{\infty} \frac{1}{n!} \frac{d^n}{dx^n} f(x) \bigg|_{x=x_0} (x - x_0)^n. \tag{7.75}$$

This is the definition of the Taylor series. It is to be noted that the method we used to derive it *assumed* that such an expansion exists, so it does not *prove* that this expansion exists. We may derive the series in a more rigorous manner, which has the additional advantages of being valid for finite expansions and giving an estimate of the series remainder. We proceed as follows: assume that the kth derivative of the function $f(x)$ exists in the interval $a \leq x \leq b$ (all lower derivatives must therefore also exist and be continuous). Let us integrate this kth derivative k times, from the point x_0 to the point x each time; the first integration gives

$$\int_{x_0}^{x} f^{(k)}(x')dx' = f^{(k-1)}(x) - f^{(k-1)}(x_0).$$ (7.76)

The integration of Eq. (7.76) gives,

$$\int_{x_0}^{x} dx'' \int_{x_0}^{x''} f^{(k)}(x')dx' = f^{(k-2)}(x) - f^{(k-2)}(x_0) - (x - x_0)f^{(k-1)}(x_0).$$ (7.77)

Continuing this process through k integrations, it can be shown that

$$f(x) = \sum_{n=0}^{k-1} \frac{(x-a)^n}{n!} f^n(x_0) + R_k,$$ (7.78)

where the remainder R_k is given by

$$R_k = \int_{x_0}^{x} dx_k \cdots \int_{x_0}^{x_2} f^{(k)}(x_1)dx_1.$$ (7.79)

If the kth derivative of f is also continuous, this remainder can be made clearer by using the mean value theorem of integral calculus,

$$\int_{x_0}^{x} g(x)dx = (x - x_0)g(\xi),$$ (7.80)

where $x_0 \leq \xi \leq x$. Applying this to the remainder, and integrating the remaining $k-1$ times, we find that

$$R_k = \frac{(x-x_0)^k}{k!} f^{(k)}(\xi).$$ (7.81)

Even if the kth derivative is not continuous, we may still provide an estimate of the remainder provided the derivative exists. If it exists, it must be bounded over some finite range of x values, i.e.

$$-\alpha \leq f^{(k)}(x) \leq \alpha, \text{ for } a \leq x \leq b.$$ (7.82)

Applying this to the remainder equation, and performing the integrations, we may write

$$|R_k| \leq \alpha \frac{|x - x_0|^k}{k!}.$$ (7.83)

It is to be noted that the function R_k has a functional dependence of $(x - x_0)^k$. We classify this mathematically by saying that the reminder is of order $O[(x - x_0)^k]$ as $x \to x_0$. As $(x - x_0)^k$ goes to zero much more rapidly than $(x - x_0)^m$, where $m < k$, as $x \to x_0$, we may neglect this remainder for points x sufficiently close to x_0. The order parameter O is defined rigorously in Section 21.2.

If the function $f(x)$ is such that we are able to let $k \to \infty$, i.e. it is infinitely differentiable, we recover Taylor's series, Eq. (7.75). The remainder term can often be used to determine the region of convergence of a Taylor series; however, the convergence properties can typically be determined in a much easier manner either by using the convergence tests described earlier or by using methods of complex analysis, to be discussed in Chapter 9.

It is common to use Taylor series expansions about the origin $x_0 = 0$; in this case, we may write

$$f(x) = \sum_{n=0}^{\infty} f^{(n)}(0) \frac{x^n}{n!}. \tag{7.84}$$

This is often referred to as the *Maclaurin series*.

We now consider a few examples.

Example 7.10 (Exponential function) If $f(x) = e^x$, it is easy to show that $f^{(n)}(0) = 1$. From this we have

$$e^x = \sum_{n=0}^{\infty} \frac{x^n}{n!}. \tag{7.85}$$

If we sum only k terms, the remainder of this finite sum is given by

$$R_k = \frac{x^k}{k!} e^{\xi}, \quad 0 \le |\xi| \le x. \tag{7.86}$$

It follows that

$$|R_k| \le \frac{x^k e^x}{k!}. \tag{7.87}$$

For any finite value of x, $\lim_{k \to \infty} x^k / k! \to 0$. Therefore $R_k \to 0$, and the Taylor series of e^x converges for all finite x.

$\diamond$

Example 7.11 (Sine and cosine functions) For $f(x) = \sin(x)$, $f^{(1)}(x) = \cos(x)$, $f^{(2)}(x) = -\sin(x)$, $f^{(3)}(x) = -\cos(x)$, and $f^{(4)}(x) = \sin(x)$. The process therefore repeats, with $f^{(n+4)}(x) = f^{(n)}(x)$. For $x = 0$, we have $f(0) = 0$, $f^{(1)}(0) = 1$, $f^{(2)}(0) = 0$, and $f^{(3)}(0) = -1$. Putting all these pieces together, we find the Taylor series for the sine function is

$$\sin(x) = \sum_{n=0}^{\infty} (-1)^n \frac{x^{2n+1}}{(2n+1)!}. \tag{7.88}$$

Infinite series

Similarly,

$$\cos(x) = \sum_{n=0}^{\infty} (-1)^n \frac{x^{2n}}{(2n)!}. \tag{7.89}$$

These can be related directly to the Taylor series for the exponential function as follows. Let us consider the Taylor series of a *complex* exponential function, i.e.

$$e^{\mathrm{i}x} = \sum_{n=0}^{\infty} \frac{(\mathrm{i}x)^n}{n!}, \tag{7.90}$$

where of course $\mathrm{i}^2 = -1$. This series can be separated into terms which are purely real (even n) and terms which are purely imaginary (odd n). The result is

$$e^{\mathrm{i}x} = \sum_{m=0}^{\infty} \frac{(\mathrm{i})^{2m} x^{2m}}{(2m)!} + \sum_{m=0}^{\infty} \frac{(\mathrm{i})^{2m+1} x^{2m+1}}{(2m+1)!}. \tag{7.91}$$

Using $\mathrm{i}^{2m} = (-1)^m$ and $\mathrm{i}^{2m+1} = \mathrm{i}(-1)^m$, and the Taylor series expansions for $\sin(x)$ and $\cos(x)$, we readily find that

$$e^{\mathrm{i}x} = \cos x + \mathrm{i}\sin x. \tag{7.92}$$

This expression is Euler's formula.

◇

Example 7.12 (Binomial series) We now consider $f(x) = (1+x)^m$, where m may be negative and/or non-integer,

$$(1+x)^m = 1 + mx + \frac{m(m-1)}{2!}x^2 + \frac{m(m-1)(m-2)}{3!}x^3 + \cdots. \tag{7.93}$$

This series is known as the *binomial series*. For integer values of m, this expression may be written compactly using factorials,

$$(1+x)^m = \sum_{n=0}^{\infty} \frac{m!}{n!(m-n)!}x^n, \tag{7.94}$$

provided one defines $r! = \infty$ for $r < 0$. The remainder of this series can be shown to be

$$R_n = \frac{x^n}{n!}(1+\xi)^{m-n} m(m-1)\cdots(m-n+1). \tag{7.95}$$

For large enough $n > m$, $(1+\xi)^{m-n}$ has its maximum when $x = 0$. Using this in the above equation, we have

$$R_n \leq \frac{x^n}{n!} m(m-1)\cdots(m-n+1). \tag{7.96}$$

If m is non-integer and/or negative, this term is approximately x^n for large enough n; for R_n to vanish and the series to converge, we therefore must have $|x| < 1$.

This series is common enough that it is often written in the condensed form

$$(1+x)^m = \sum_{n=0}^{\infty} \binom{m}{n} x^n, \tag{7.97}$$

where $\binom{m}{n} = m!/[n!(m-n)!]$ is called the *binomial coefficent*. This coefficient is still valid for noninteger m, provided we replace the factorials by their equivalent gamma function representations, i.e. $m! = \Gamma(m+1)$. The gamma function is discussed in Appendix A.

◇

Example 7.13 (Natural logarithm $\log(1+x)$) Finally, let us consider the Taylor expansion of the function $f(x) = \log(1+x)$ about $x = 0$. The nth derivative of the function may be written as

$$\frac{d^n}{dx^n} \log(1+x) = \frac{(-1)^{n+1}}{(1+x)^n} \frac{n!}{n}. \tag{7.98}$$

Using this in the expression for the Taylor series, we find that

$$\log(1+x) = \sum_{n=1}^{\infty} (-1)^{n+1} \frac{x^n}{n}. \tag{7.99}$$

Applying the ratio test, one can establish that the series converges for $-1 < x < 1$, with clear divergence at $x = -1$ and an unclear convergence at $x = 1$. The series at $x = 1$, however, takes on the familiar form,

$$\log(2) = \sum_{n=1}^{\infty} \frac{(-1)^{n+1}}{n}, \tag{7.100}$$

which is simply the alternating harmonic series, which is known to be conditionally convergent! The series expansion of the function $\log(1+x)$ therefore gives us the limit of the alternating harmonic series as $\log(2)$.

It is perhaps interesting to note that we now have a connection between three types of singular behavior: a conditionally convergent series (1) represents the behavior of a Fourier series with a discontinuity (2) as well as the behavior of a Taylor series on the boundary of its region of convergence (3).

◇

It is helpful to note a "shortcut" which may be applied in some circumstances for calculating more complicated Taylor series expansions. Suppose we wish to determine the Taylor series expansion of a function such as $g(x) = \exp[-x^2]$ about the point $x = 0$, given that we know already the Taylor series expansion of $f(x) = \exp[-x]$. Since $g(x) = f(x^2)$, we can find the expansion of $g(x)$ by simply substituting x^2 into the argument of the Taylor series for $f(x)$. This shortcut works because the resulting series is already in powers of x and is therefore a valid Taylor series expansion. The shortcut would not work, for instance, if we tried to determine the series expansion of $g(x) = \exp[\sin(x)]$, given the expansion of $f(x) = \exp[x]$. The resulting substitution would result in a series of powers of $\sin(x)$, not powers of x, and would not be a valid Taylor series expansion. In such a case one would need to calculate the series expansion of $g(x)$ using the standard formula (7.75).

7.6 Taylor series in more than one variable

The idea of a Taylor series may be expanded to functions of more than one variable. We have already made use of such expansions in formulating our integral definitions of the divergence and curl in Chapter 2. The process for deriving the expression for the finite Taylor series for a many-variable function follows the process of the previous section.

We consider first the path integral of the form

$$I(\mathbf{r}) = \int_0^{\mathbf{r}} F^{(n)}(\mathbf{r}')dr',\tag{7.101}$$

where the path is a straight line from the origin of the coordinate system to the point $\mathbf{r}$ and the function $F^{(n)}$ is yet to be defined. We define $\mathbf{n}$ as the unit vector pointing along this path. We now take the function $F^{(n)}(\mathbf{r})$ to be of the form

$$F^{(n)}(\mathbf{r}) = [\mathbf{n} \cdot \nabla]^n f(\mathbf{r}),\tag{7.102}$$

where the nth partial derivatives of $f(\mathbf{r})$ are assumed to exist. We may rewrite $F^{(n)}(\mathbf{r})$ in the form

$$F^{(n)}(\mathbf{r}) = [\mathbf{n} \cdot \nabla]F^{(n-1)}(\mathbf{r}).\tag{7.103}$$

Equation (7.101) then takes on the form

$$I(\mathbf{r}) = \int_0^{\mathbf{r}} [\mathbf{n} \cdot \nabla']F^{(n-1)}(\mathbf{r}')dr' = \int_0^{\mathbf{r}} \nabla'F^{(n-1)}(\mathbf{r}') \cdot d\mathbf{r}'.\tag{7.104}$$

However, returning to our definition of the gradient operator from Section 2.3, we may write

$$\nabla'F^{(n-1)}(\mathbf{r}') \cdot d\mathbf{r}' = dF^{(n-1)},\tag{7.105}$$

where $dF^{(n-1)}$ is the differential of $F^{(n-1)}$. Our integral therefore takes on the form

$$\int_0^{\mathbf{r}} F^{(n)}(\mathbf{r}')dr' = \int_{F^{(n-1)}(0)}^{F^{(n-1)}(\mathbf{r})} dF^{(n-1)} = F^{(n-1)}(\mathbf{r}) - F^{(n-1)}(0).\tag{7.106}$$

Let us suppose we perform another path integral of this function from the origin to position $\mathbf{r}$; we have

$$\int_0^{\mathbf{r}} \int_0^{\mathbf{r}'} F^{(n)}(\mathbf{r}'') dr'' dr' = \int_0^{\mathbf{r}} F^{(n-1)}(\mathbf{r}') dr' - \int_0^{\mathbf{r}} F^{(n-1)}(0) dr'$$

$$= F^{(n-2)}(\mathbf{r}) - F^{(n-2)}(0) - r F^{(n-1)}(0). \qquad (7.107)$$

Continuing this process for a total of n path integrals, we have

$$\int_0^{\mathbf{r}} \cdots \int_0^{\mathbf{r}_2} F^{(n)}(\mathbf{r}_1) dr_n \cdots dr_1 = f(\mathbf{r}) - \sum_{j=0}^{n-1} \frac{F^{(j)}(0)}{j!} r^j. \qquad (7.108)$$

Using the definition of $F^{(j)}$ and noting that $\mathbf{r} = r\mathbf{n}$, we may rewrite this expression in the form

$$f(\mathbf{r}) = \sum_{j=0}^{n-1} \frac{1}{j!} \left[\mathbf{r} \cdot \nabla' \right]^j f(\mathbf{r}') \big|_{\mathbf{r}'=0} + R_n(\mathbf{r}), \qquad (7.109)$$

where R_n is a remainder term, to be discussed momentarily. This expression, which was derived without explicit reference to the number of variables, holds for a function of any dimensionality.

The remainder term follows immediately from Eq. (7.108) above, in the form

$$R_n(\mathbf{r}) = \int_0^{\mathbf{r}} \cdots \int_0^{\mathbf{r}_2} F^{(n)}(\mathbf{r}_1) dr_n \cdots dr_1. \qquad (7.110)$$

Recalling our assumption that the nth partial derivatives of $f(\mathbf{r})$ exist, this implies that $F^{(n)}$ is bounded, i.e. $|F^{(n)}| \leq \alpha$. Applying this inequality to the remainder, and integrating the function n times, we have

$$|R_n(\mathbf{r})| \leq \alpha \frac{r^n}{n!}. \qquad (7.111)$$

The remainder is bounded by r^n; we write this shorthand by stating that the remainder is of order $\mathbf{r}^n$, or $R_n = O(\mathbf{r}^n)$. In the limit $\mathbf{r} \to 0$, the remainder term becomes negligible compared to the regular terms of the series.

A function may be expanded about an arbitrary vector point $\mathbf{r}_0$ as

$$f(\mathbf{r}) = \sum_{j=0}^{n-1} \frac{1}{j!} \left[(\mathbf{r} - \mathbf{r}_0) \cdot \nabla' \right]^j f(\mathbf{r}') \big|_{\mathbf{r}'=\mathbf{r}_0} + R_n(\mathbf{r} - \mathbf{r}_0). \qquad (7.112)$$

The first three terms of the Taylor series are typically the most important in calculations, and it is to be noted that they can be written in a matrix form,

$$f(\mathbf{r}) = f(\mathbf{r}_0) + \mathbf{J}(\mathbf{r}_0) |\mathbf{r} - \mathbf{r}_0\rangle + \langle \mathbf{r} - \mathbf{r}_0 | \mathbf{H}(\mathbf{r}_0) | \mathbf{r} - \mathbf{r}_0 \rangle, \qquad (7.113)$$

where $\mathbf{J}$ is the *Jacobian vector*, defined in N variables as

$$\mathbf{J} = \begin{bmatrix} \dfrac{\partial f}{\partial x_1} \\ \vdots \\ \dfrac{\partial f}{\partial x_N} \end{bmatrix}, \tag{7.114}$$

and $\mathbf{H}$ is the *Hessian matrix*, defined as

$$\mathbf{H} = \begin{bmatrix} \dfrac{\partial^2 f}{\partial x_1^2} & \dfrac{\partial^2 f}{\partial x_1 x_2} & \cdots & \dfrac{\partial^2 f}{\partial x_1 x_N} \\ \dfrac{\partial^2 f}{\partial x_1 x_2} & \dfrac{\partial^2 f}{\partial x_2^2} & \cdots & \dfrac{\partial^2 f}{\partial x_2 x_N} \\ \vdots & \vdots & \ddots & \\ \dfrac{\partial^2 f}{\partial x_1 x_N} & \dfrac{\partial^2 f}{\partial x_2 x_N} & \cdots & \dfrac{\partial^2 f}{\partial x_N^2} \end{bmatrix}. \tag{7.115}$$

7.7 Power series

The Taylor series may be considered to be a special case of what is referred to as a power series, defined by

$$f(x) = a_0 + a_1 x + a_2 x^2 + \cdots = \sum_{n=0}^{\infty} a_n x^n. \tag{7.116}$$

It is important to note the relationship between the idea of a Taylor series and a power series. A Taylor series is a series expansion of a *closed form* differentiable function. Many functions, such as solutions of numerous differential equations, cannot be expressed in a closed form, but they can almost always be expressed as a power series.

Power series are extremely common, and have a number of useful properties. Among them are the following.

1. Radius of convergence. If, using the ratio test, the series a_n (without the power of x) satisfies the equation

$$\lim_{n \to \infty} \frac{a_{n+1}}{a_n} = R^{-1}, \tag{7.117}$$

   it follows that the power series $\sum a_n x^n$ converges for $-R < x < R$. The series may or may not converge for $|x| = R$; convergence at that point must be determined by other methods.
2. Uniform and absolute convergence. Within the radius of convergence, $-R < x < R$, it can be shown that the power series converges uniformly and absolutely.
3. Continuity. Within the radius of convergence, $f(x)$ is a continuous function.
4. Differentiation and integration. Within the radius of convergence, the series may be differentiated and integrated term by term, and the radius of convergence of the new series is the same as the original series.

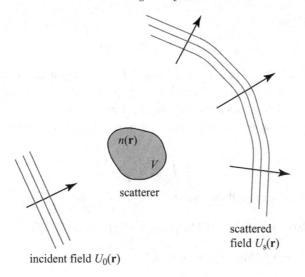

Figure 7.10 The scatttering of a scalar wave $U_0(\mathbf{r})$ from an object with inhomogeneous refractive index $n(\mathbf{r})$.

5. Uniqueness of power series. One can show that the power series representation of a function $f(x)$ is unique; that is, there is only one set of numbers a_n which represent a given function $f(x)$.

We will see more of power series when we deal with complex analysis and differential equations. In particular, many differential equations can only be solved analytically by use of a power series.

7.8 Focus: convergence of the Born series

Problems involving wave scattering from inhomogeneous objects are important in many branches of physics, notably quantum mechanics, acoustics, and optics. We consider an example of such a problem described as follows: a scalar wave $U_0(\mathbf{r})$ traveling in free space is incident upon an object, confined to a volume V, of refractive index $n(\mathbf{r})$. We are interested in the form of the scattered field, denoted by $U_s(\mathbf{r})$; the *total* field in the region of the scatterer is denoted by $U(\mathbf{r}) = U_0(\mathbf{r}) + U_s(\mathbf{r})$. The geometry is illustrated in Fig. 7.10.

The total field is assumed to satisfy the Helmholtz equation with non-uniform refractive index,

$$\left[\nabla^2 + k^2 n^2(\mathbf{r})\right] U(\mathbf{r}) = 0. \tag{7.118}$$

The solution to this equation will be shown in Section 19.10 to take the form of an integral equation,

$$U(\mathbf{r}) = U_0(\mathbf{r}) + \int_V F(\mathbf{r}')U(\mathbf{r}')\frac{e^{ikR}}{R}\mathrm{d}^3 r', \tag{7.119}$$

where $R \equiv |\mathbf{r} - \mathbf{r}'|$ and

$$F(\mathbf{r}) \equiv \frac{k^2}{4\pi}[n^2(\mathbf{r}) - 1]. \tag{7.120}$$

The quantity $F(\mathbf{r})$ is referred to as the *scattering potential*. Integral equations are in general difficult to solve and this one in particular has no readily found solution, due to the presence of the total field $U(\mathbf{r})$ both inside and outside of the integrand.

A formal solution can be found in series form, however. Let us define the *functional operator* $K[g](\mathbf{r})$ as

$$K[g] \equiv \int_V g(\mathbf{r}') \frac{e^{ikR}}{R} d^3 r'. \tag{7.121}$$

The solution to Eq. (7.119) is then given by

$$U(\mathbf{r}) = U_0(\mathbf{r}) + K[FU_0] + K[FK[FU_0]] + K[FK[FK[FU_0]]] + \cdots. \tag{7.122}$$

Abbreviating $K[f] = Kf$, this series may be written as

$$U(\mathbf{r}) = \sum_{n=0}^{\infty} (KF)^n U_0 = \sum_{n=0}^{\infty} U_n(\mathbf{r}). \tag{7.123}$$

This series solution to the scattering problem is known as the *Born series*. We defer derivation of the integral equation (7.119) until Chapter 19, and derivation of Eq. (7.123) until Chapter 21. However, we can demonstrate that it is a formal solution to the integral equation by borrowing a result from the theory of Green's functions,

$$[\nabla^2 + k^2] \frac{e^{ikR}}{R} = -4\pi \delta^{(3)}(\mathbf{r} - \mathbf{r}'), \tag{7.124}$$

where $\delta^{(3)}$ is the three-dimensional delta function of Section 6.6. In terms of the functional K, we may write

$$[\nabla^2 + k^2]K[g] = 4\pi g. \tag{7.125}$$

We may then write

$$[\nabla^2 + n^2 k^2]U(\mathbf{r}) = [\nabla^2 + k^2 + 4\pi F]U(\mathbf{r})$$

$$= [\nabla^2 + k^2] \sum_{n=0}^{\infty} KF(KF)^{n-1} U_0 + 4\pi F \sum_{n=0}^{\infty} (KF)^n U_0. \tag{7.126}$$

Using Eq. (7.125) above, this may be written as

$$[\nabla^2 + n^2 k^2]U(\mathbf{r}) = (-4\pi F + 4\pi F) \sum_{n=0}^{\infty} (KF)^n U_0 = 0. \tag{7.127}$$

Equation (7.123) represents a solution to the scattering integral equation (7.119), *provided the series converges*; we return to this point momentarily.

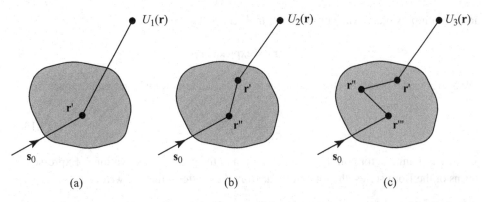

Figure 7.11 The physical significance of the terms of the Born series. (a) U_1 contains all single-scattering events, (b) U_2 contains all double-scattering events, (c) U_3 contains all triple-scattering events, and so forth.

The Born series can be interpreted in a very physical and intuitive way. Noting that each Green's function $G(R)$ represents radiation of a scattered field from a point scatterer, $U_1(\mathbf{r})$ represents all contributions in which the incident field is scattered *once* within the scatterer. Likewise, $U_2(\mathbf{r})$ represents all contributions in which the incident field is scattered *twice* within the scatterer, and so on. This process is illustrated in Fig. 7.11.

The U_0 term of this series is simply the incident field; the U_1 term may be written as

$$U_1(\mathbf{r}) \equiv \int_V F(\mathbf{r}')U_0(\mathbf{r}')\frac{e^{ikR}}{R}\mathrm{d}^3r'. \tag{7.128}$$

If the object is very weakly scattering (i.e. $F(\mathbf{r})$ is very close to zero), an approximate solution to the integral equation may be written in the form $U \approx U_0 + U_1$. This is known as the *Born approximation*.[2]

But does the complete series always converge? It is not immediately obvious why it would not. However, let us consider the case when the scattering potential $F(\mathbf{r})$ is very large; then is it possible that U_1 is *larger* than U_0, or more generally that U_{n+1} is larger than U_n. In such a case the series would clearly diverge, but what is not clear is what the quantitative conditions for convergence or divergence are.

This question was partly answered by Davies [Dav60] for a limited class of scattering potentials by an elegant application of the concept of absolute convergence and use of the comparison test. Let us assume that the scattering potential is spherically symmetric, i.e. $F(\mathbf{r}) = F(r)$, real-valued and positive $F(r) > 0$. Furthermore, we consider the incident field

---

[2] Born himself was rather irritated with the term "Born approximation", which is used far more often than the term "Born series". He once remarked, "I developed in that paper the whole perturbation expansion for the scattered field, valid to all orders, yet I am only given credit for the first term in the series!" [Wol83]

to be a complex plane wave propagating in direction $\mathbf{k}_0$, i.e.

$$U_0(\mathbf{r}) = \exp[i\mathbf{k}_0 \cdot \mathbf{r}]. \tag{7.129}$$

We also consider points of observation far from the scatterer, for which we may approximate

$$\frac{e^{ikR}}{R} \approx \frac{e^{ikr}}{r} e^{-i k \mathbf{s} \cdot \mathbf{r}}, \tag{7.130}$$

where $\mathbf{s}$ is a unit vector pointing in the direction of the point of observation $\mathbf{r}$. Expressed in terms of the Born series, the total field far from the scatterer may be written as

$$U(\mathbf{r}) \approx U_0(\mathbf{r}) + \frac{e^{ikr}}{r} f(k\mathbf{s}), \tag{7.131}$$

where

$$f(k\mathbf{s}) = \sum_{n=1}^{\infty} HF(KF)^{n-1} U_0, \tag{7.132}$$

and H is a new functional of the form

$$H[g] = \int_V g(\mathbf{r}') e^{-i k \mathbf{s} \cdot \mathbf{r}'} d^3 r'. \tag{7.133}$$

The function $f(k\mathbf{s})$ is called the *scattering amplitude* of the scattered field. Let us consider the absolute convergence of the radiation pattern, i.e. the convergence of the series $\tilde{f}$ defined as

$$\tilde{f} \equiv \sum_{n=1}^{\infty} \tilde{H} F(\tilde{K} F)^{n-1}, \tag{7.134}$$

where

$$\tilde{H}[g] = \int_V g(r') d^3 r', \tag{7.135}$$

$$\tilde{K}[g] = \int_V \frac{g(r')}{|\mathbf{r} - \mathbf{r}'|} d^3 r'. \tag{7.136}$$

Because the function $g(r')$ is spherically symmetric, we may evaluate the angular integrations in these two functionals,

$$\tilde{H}[g] = 4\pi \int_0^{\infty} g(r') r'^2 dr', \tag{7.137}$$

$$\tilde{K}[g] = 4\pi \int_0^{\infty} g(r') \min[1/r, 1/r'] r'^2 dr', \tag{7.138}$$

where $\min[x, y]$ represents the minimum of x and y.

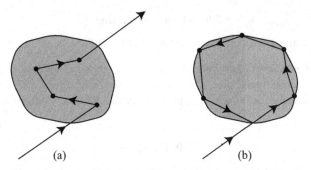

Figure 7.12 Scattering states versus bound states. (a) A multiple scattering state, in which the field scatters a finite number of times before leaving the object, is covered by the Born approximation. (b) A bound state, in which the field is in principle permanently trapped within the object (with an infinite number of scattering events) is not covered by the Born approximation.

At this stage, the derivation requires an appreciable understanding of the theory of integral equations, to be discussed much later in Section 19.7; we sketch out the important points and conclusions, and refer to the original article for the details.

The salient observation is that Eq. (7.134) is exactly equal to the scattered field for S-wave scattering in the limit of long wavelengths ($k \to 0$). S-wave scattering is that component of the scattered field which radiates equally in all directions. But the S-wave scattering series can in turn be compared (via what is essentially the comparison test) to the expression for *bound state* solutions to the scattering equation.

Bound states are those solutions to the wave equation which are confined to, or "trapped in", the scattering potential. By comparison of the bound state solutions to the scattering equation and the Born series, it can be shown that the *Born series is convergent if the scattering potential cannot support a bound state*.

This result is intuitively elegant and demonstrates how seemingly innocuous assumptions about the solution to a physical problem can lead to incorrect, even divergent, results. Though the Born series is in principle an exact solution, it is derived from the assumption that the scattered field exits the scattering object after a finite number of events and cannot get trapped within the object. A bound state, which may be roughly considered a solution to scattering problem in which the field bounces around infinitely within the object, is not encompassed by the assumption that leads to the Born series. The idea of a bound state versus a finite scattering state is shown in Fig. 7.12.

It should be emphasized that this result only applies to a relatively limited class of scattering potentials that are real-valued, negative, and spherically symmetric; for more complicated potentials the correlation between convergence and bound states is not exact. Nevertheless, this example shows that an understanding of series convergence is necessary to properly interpret certain physical problems and can provide a surprising amount of insight.

7.9 Additional reading

There are several nice books on infinite series, at different levels.

- J. M. Hyslop, *Infinite Series* (New York, Dover, 2006). A very nice little book on infinite series.
- K. Knopp, *Theory and Application of Infinite Series* (New York, Dover, 1990). This book is the most thorough reference on infinite series that one can find. It is a very difficult read, however.

7.10 Exercises

1. The Fabry–Perot interferometer also reflects light in a nontrivial way. In a manner similar to that of the introduction, sum the contributions to the reflected field and determine the behavior of the reflected intensity. Show that $|T(k)|^2 + |R(k)|^2 = 1$.

2. Determine whether the following series converge absolutely:

   (a) $\sum_{n=0}^{\infty} \frac{n^\alpha}{n!}$, with $\alpha > 0$, (b) $\sum_{n=1}^{\infty} \log\left(\frac{n+1}{n}\right)$, (c) $\sum_{n=2}^{\infty} \frac{1}{(\log n)^n}$,

   (d) $\sum_{n=0}^{\infty} \sqrt{\frac{n}{n^4+1}}$, (e) $\sum_{n=0}^{\infty} \frac{n}{\sqrt{2n^3+1}}$, (f) $\sum_{n=2}^{\infty} \frac{1}{n\log n}$,

   (g) $\sum_{n=0}^{\infty} \frac{2^n n!}{n^n}$.

3. Determine whether the following series converge absolutely:

   (a) $\sum_{n=0}^{\infty} \frac{3^n n!}{n^n}$, (b) $\sum_{n=0}^{\infty} \frac{n}{(n^2+4)^{3/2}}$, (c) $\sum_{n=1}^{\infty} \frac{1}{\sqrt{n(n+1)}}$,

   (d) $\sum_{n=0}^{\infty} \frac{1}{\sqrt{n}+\sqrt{n+1}}$, (e) $\sum_{n=0}^{\infty} \frac{(n!)^2}{4^n n^{2n}}$, (f) $\sum_{n=0}^{\infty} \sqrt{n+1}-\sqrt{n}$,

   (g) $\sum_{n=0}^{\infty} \frac{(n!)^2}{(2n)!} 3^{-2n}$.

4. The power series expansion of the function $e^x \sin x$ may be written in the form

$$e^x \sin x = \sum_{n=0}^{\infty} \frac{a_n}{n!} x^n.$$

   Find a *closed form* expression for a_n.

5. Determine the radius of convergence for the following power series:

   (a) $\sum_{n=0}^{\infty} \frac{n!}{\alpha^{n^2}} x^n, \alpha > 1$, (b) $\sum_{n=2}^{\infty} \frac{(n!)^3}{(3n)!} x^n$.

6. Determine the radius of convergence for the following power series:

   (a) $\sum_{n=0}^{\infty} \alpha^{n^2} x^n, 0 < \alpha < 1$, (b) $\sum_{n=0}^{\infty} \left[2n^2/(1+3n^2)\right]^n x^n$.

7. Find the Maclaurin series of the following functions by using "shortcuts": $\exp[-x^2]$, $\cos[5x^2]$, $\log[1+x^3]$.

8. Find the Maclaurin series of the following functions by using "shortcuts": $x\cos[x^3]$, $(1 + 2x)\exp[-x^2]$, $(1+8x^3)^{3/2}$.

9. Taylor series can often be found by taking derivatives of known series expansions. With this observation, find the Taylor series of

$$f(x) = \frac{1}{(1-x^2)^2}.$$

10. Taylor series can often be found by taking derivatives of known series expansions. With this observation, find the Taylor series of

$$f(x) = \frac{x}{(1-x^2)^2}.$$

11. Many Taylor series converge very slowly; for instance, $\log(1+x)$ has terms which decay as $1/n$. Show that, with an appropriate choice of constant c, the Taylor series of

$$(1+cx)\log(1+x)$$

has terms which decay as $1/n^2$. Similarly, find constants a and b such that the Taylor series of

$$(1+ax+bx^2)\log(1+x)$$

has terms which decay as $1/n^3$.

12. Find the first three nonzero terms of the Maclaurin series of the following functions:
    (a) $f(x) = (x^2+2)^{-1/2}$,
    (b) $f(x) = \cos x/(x^2+1)$.

13. Find the first three nonzero terms of the Maclaurin series of the following functions:
    (a) $f(x) = \log(1+x^2)/(1+x^2)$,
    (b) $f(x) = \exp(\sin x)$.

14. Suppose we consider an optical system which consists of two Fabry–Perot interferometers of thicknesses d_1 and d_2 and identical reflections r and transmissions t, separated by a distance l, as shown below.

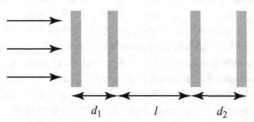

Determine the reflection and transmission properties of the system for a monochromatic plane wave of wavenumber k. (Hint: consider the overall reflection and transmission of each interferometer alone, then determine the behavior of the waves which pass between them.)

15. In Planck's theory of blackbody radiation, it is assumed that the energy of individual oscillators in matter is quantized, i.e. it can only take on values $nh\nu_0$, with n an integer, h Planck's constant,

and ν_0 a fundamental frequency. The average energy of an oscillator is given by the expression

$$\langle E \rangle = \frac{\sum_{n=1}^{\infty} nh\nu_0 \exp[-nh\nu_0/kT]}{\sum_{n=0}^{\infty} \exp[-nh\nu_0/kT]},$$

where k is Boltzmann's constant and T is the temperature. By rewriting the numerator and denominator in terms of the geometric series and its derivatives, show that

$$\langle E \rangle = \frac{h\nu_0}{\exp[h\nu_0/kT] - 1},$$

and that $\langle E \rangle \to kT$ in the limit $kT \gg h\nu_0$.

16. When light illuminates a thin film of refractive index n and thickness d, it undergoes multiple reflections within the surface, as illustrated below.

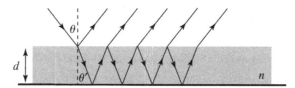

   Assuming the wave is incident upon the film at an angle θ, and that the reflection and transmission at the surface are r and t, respectively, find the amplitude of a monochromatic field of wavenumber k reflected from the system. (Hint: each time a wave is internally reflected, it suffers a phase delay because it must travel an additional distance before leaving the film. Don't forget that Snell's law applies inside the film, and that the wavenumber is kn inside the film.)

17. The paper by V. A. Markel and J. C. Schotland, On the convergence of the Born series in optical tomography with diffuse light, *Inverse Problems* **23** (2007), 1445–1465, studies the convergence of the Born series for a scattering problem similar to that discussed in Section 7.8. Describe the differences between the two wave models used in the paper and the earlier section, and summarize the condition for the convergence of the Born series in the paper.

18. It is known that, in highly disordered and strongly scattering media, electrons and even light can be trapped, in a process known as *Anderson localization*. Read the paper by P. W. Anderson, Absence of diffusion in certain random lattices, *Phys. Rev.* **109** (1958), 1492–1505. What is the explanation of Anderson localization, and how are infinite series used to describe the phenomenon?

19. One version of an *infinity mirror* is a pair of mirrors on opposite walls; an observer standing between the mirrors sees multiple images of himself, stretching off to "infinity". Assuming that the separation of the mirrors is d, and that the apparent width of an image decays as $1/R^2$, where R is the apparent distance to the image, argue that the total area of all images observed is finite.

20. In a simple one-dimensional model of a crystal, ions of charges $\pm q$ are arranged alternately along a lattice with separation R, as shown below.

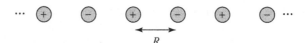

Given that the potential energy of ion i due to ion j is given by

$$V_{ij} = \frac{q_i q_j}{|r_i - r_j|},$$

determine the contribution of the ith ion to the total potential energy of the crystal. The expression for potential energy may be written in the form $V = \alpha q^2 / R$, where α is referred to as the *Madelung constant*.

What is the contribution of two adjacent ions to the total energy of the crystal?

21. Consider a series of the form

$$\sum_{n=1}^{\infty} u_n = \frac{1}{1^\alpha} + \frac{1}{2^\beta} + \frac{1}{3^\alpha} + \frac{1}{4^\beta} + \cdots,$$

where $1 < \alpha < \beta$. Show that this series converges, but that $\lim_{n\to\infty} u_{2n+1}/u_{2n} \to \infty$. Does this series violate the ratio test? Justify your answer.

8

Fourier series

8.1 Introduction: diffraction gratings

When a polychromatic beam of light is shined upon a translucent planar material etched with a periodic pattern known as a *diffraction grating*, the different colors of the beam of light propagate in different directions. Such a diffraction grating can be used as a monochromator of light or as a spectrometer, as schematically illustrated in Fig. 8.1.

The physics of a diffraction grating can be understood by a simple interference argument: we suppose that light of wavelength λ is normally incident upon a thin opaque screen containing a large number of periodically arranged narrow apertures (slits) separated by a distance Δ. Because the light is normally incident upon the grating, the phase of the field is the same within each aperture. Each aperture produces a spherical wave, and we consider the overlap of the fields from two such elements in a direction θ from the normal to the surface; this is illustrated in Fig. 8.2.

As can be seen from the picture, the wave emanating from aperture 2 has to travel a distance $\Delta \sin \theta$ farther than the wave emanating from aperture 1. This translates to a phase difference between the two waves of $k \Delta \sin \theta$, where $k = 2\pi/\lambda$ is the wavenumber of the light. These two waves, and consequently those from all other apertures, will constructively interfere if

$$k \Delta \sin \theta = 2\pi m, \tag{8.1}$$

where m is an integer. For angles θ not satisfying the above condition, field contributions from successive apertures will be progressively further out of phase; if the number of apertures is large enough (i.e. a large number of periods of the grating are illuminated), the combined waves from all apertures will then tend to cancel. The result is that the diffracted light will concentrate in bright narrow peaks centered at angles satisfying Eq. (8.1).

It is to be noted that in general this equation may be satisfied for different values of the integer m. A monochromatic beam of light incident on a diffraction grating is therefore split into multiple narrow beams, of *order m*, each propagating in a different direction. A polychromatic beam will in general produce multiple images of the spectrum of the incident light field.

The simple calculation given above, however, tells us nothing about the relative *amplitudes* of the various diffraction orders. Typically the central ($m = 0$) order receives the

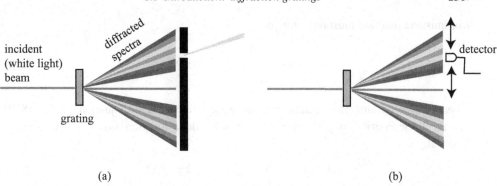

Figure 8.1 Use of a diffraction grating (a) as a monochromator and (b) as a spectrometer.

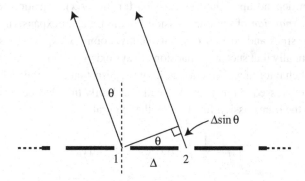

Figure 8.2 Simple calculation of the behavior of a diffraction grating.

greatest concentration of light; this central order, however, does not have any frequency sensitivity and cannot be used for making a monochromator or spectrometer. One would ideally like to design a grating which concentrates light instead into one of the higher orders; we use diffraction theory to look more closely at this possibility.

We assume that a monochromatic plane wave is normally incident upon a one-dimensional grating structure located in the plane $z = 0$. We assume that the grating possesses a transmission function $t(x)$, such that the field immediately beyond the grating is $U_0(x) = t(x)$. The field far from the grating, at distance r and direction $\mathbf{u}$, is described by Fraunhofer diffraction (to be discussed in Section 11.1), and is proportional to

$$U(r\mathbf{u}) \sim \int_{-\infty}^{\infty} t(x)e^{-iku_x x'}\,dx', \tag{8.2}$$

where $u_x = \sin\theta$ and $u_z = \cos\theta$. What sort of function $t(x)$ will result in a field which is present only in directions which satisfy Eq. (8.1)? A little thought suggests that the transmission function must be of the form

$$t(x) = \sum_{m=-\infty}^{\infty} c_m e^{i2\pi mx/\Delta}, \tag{8.3}$$

where c_m are constants to be determined. If we substitute from this equation into Eq. (8.2), and recall the Fourier integral representation of the delta function, Eq. (6.41), we readily find that

$$U(r\mathbf{u}) \sim \sum_{m=-\infty}^{\infty} c_m \delta(ku_x - 2\pi m/\Delta). \tag{8.4}$$

This (idealized) equation exhibits the proper behavior of the radiation pattern. From it, we see that the relative amplitudes of the different diffraction orders depend upon the coefficients c_m, which in turn depend upon the transmission function $t(x)$. Equation (8.3) represents a *Fourier series expansion* of the transmission function, i.e. an expansion of the periodic function $t(x)$ into sines and cosines or, equivalently, complex exponentials. We have not proven mathematically that such an expansion always exists; rather, we have demonstrated that the physical behavior of a diffraction grating suggests that every periodic transmission function may be expressed in this form. We will see shortly that the coefficients c_n may be determined from the transmission function by the integral

$$c_m = \frac{1}{\Delta} \int_0^{\Delta} t(x') e^{-i2\pi mx'/\Delta} dx'. \tag{8.5}$$

We now calculate the behavior of these coefficients for a grating of a particular shape. Let us assume that the grating consists of a sawtoothed collection of ridges, of refractive index n, length Δ, and angle θ_0. The geometry is illustrated in Fig. 8.3.

We assume that the grating is sufficiently thin, so that the influence of the grating on the incident wave may be described as a geometric phase delay. Over the range $0 \le x < \Delta$, the wave travels through a distance $h' = x \tan \theta_0$ of the grating material and a distance $h - h' = (\Delta - x) \tan \theta_0$ of air. The phase of the field upon clearing the grating can be written as

$$\phi(x) = knx \tan \theta_0 + k(\Delta - x) \tan \theta_0. \tag{8.6}$$

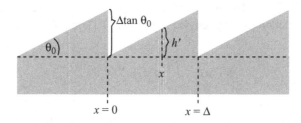

Figure 8.3 A blazed grating.

The transmission function for the grating therefore has the form

$$t(x) = e^{i\phi(x)} = e^{ik\Delta\tan\theta_0}e^{ik\tan\theta_0(n-1)x}. \tag{8.7}$$

Letting $\alpha \equiv k\tan\theta_0(n-1)$, and neglecting the constant exponential in the above expression, we may readily calculate the Fourier coefficients to be

$$c_m = e^{i(\alpha\Delta - 2\pi m)}\text{sinc}\,(\alpha\Delta/2 - \pi m), \tag{8.8}$$

where

$$\text{sinc}(y) \equiv \frac{\sin(y)}{y}. \tag{8.9}$$

The sinc function has a peak at $y = 0$, which suggests the largest Fourier coefficient c_m will be the one with m closest to the value

$$m = \frac{\alpha\Delta}{2\pi} = \frac{k\Delta\tan\theta_0(n-1)}{2\pi}. \tag{8.10}$$

For a given wavenumber of light k and grating index n, we can choose the period Δ and the angle θ_0 of the grating to direct the most light into any existing diffraction order; the only other constraint is that the order must satisfy the inequality

$$m \leq \frac{k\Delta}{2\pi}, \tag{8.11}$$

which is the requirement that the angle of the diffraction order be less than $90°$. A grating of the form shown in Fig. 8.3, which directs most of the diffracted energy into a higher order, is known as a *blazed grating*. An example of the Fourier coefficients of a blazed grating is shown in Fig. 8.4.

This simple calculation shows how the Fourier series can be used to understand the properties of periodic systems such as diffraction gratings. In this chapter we develop in detail the theory of the Fourier series.

8.2 Real-valued Fourier series

Definition 8.1 (Fourier series) *A Fourier series is an expansion of a function $f(t)$ over a finite interval $0 \leq t \leq T$ in terms of sines and cosines, in the form:*

$$f(t) = \frac{a_0}{2} + \sum_{n=1}^{\infty}[a_n\cos(n\omega_0 t) + b_n\sin(n\omega_0 t)], \tag{8.12}$$

where

$$\omega_0 = \frac{2\pi}{T} \tag{8.13}$$

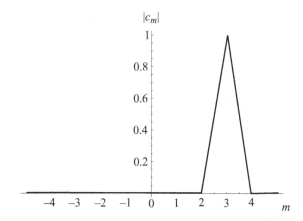

Figure 8.4 The coefficients c_m for a blazed grating. Here we have taken $k\Delta = 10\pi$, which implies that diffraction orders of up to $|m| = 5$ will be observable. Assuming $n = 1.50$ (crown glass), we take $\theta_0 = 50.2°$ to optimize the $m = 3$ order.

is called the fundamental frequency *of the series. The coefficients a_n and b_n are given by*

$$a_m = \frac{2}{T} \int_0^T f(t)\cos(m\omega_0 t)dt, \tag{8.14}$$

$$b_m = \frac{2}{T} \int_0^T f(t)\sin(m\omega_0 t)dt. \tag{8.15}$$

The first term of the series is a constant; the terms of higher n in the series oscillate at frequencies which are integer multiples of the fundamental frequency. The period T_n of the nth term of the Fourier series is given by

$$T_n = \frac{2\pi}{\omega_0 n}. \tag{8.16}$$

Terms of higher n therefore oscillate faster than the fundamental frequency.

There are many potential applications of the Fourier series; we will see in Section 15.3 that they appear quite naturally in the solution of partial differential equations using separation of variables. Also, it is to be noted from Eq. (8.12) that the Fourier series as a whole is periodic, with period T; a Fourier series may therefore also be used to represent an arbitrary periodic function of period T. One could say that there are two broad applications of Fourier series:

1. it may be used to represent an arbitrary function over a finite domain $0 \le t \le T$;
2. it may be used to represent a periodic function of period T over an infinite domain $-\infty \le t \le \infty$.

The coefficients a_n and b_n may be determined by using the following elementary properties of sine and cosine functions

$$\int_0^1 \sin(2\pi mx)\sin(2\pi nx)dx = \frac{1}{2}\delta_{mn}, \tag{8.17}$$

$$\int_0^1 \sin(2\pi mx)\cos(2\pi nx)dx = 0, \tag{8.18}$$

$$\int_0^1 \cos(2\pi mx)\cos(2\pi nx)dx = \frac{1}{2}\delta_{mn}. \tag{8.19}$$

These relations are easy to show using the exponential form of the trigonometric functions; for instance, with $m \neq n$, we may write

$$\int_0^1 \sin(2\pi mx)\sin(2\pi nx)dx = \int_0^1 \frac{e^{2\pi imx} - e^{-2\pi imx}}{2i}\frac{e^{2\pi inx} - e^{-2\pi inx}}{2i}dx$$

$$= -\frac{1}{4}\int_0^1 \{2\cos[2\pi(m+n)x] - 2\cos[2\pi(m-n)x]\}dx$$

$$= 0, \tag{8.20}$$

using $\sin[2\pi(m+n)] = 0$ and $\sin[2\pi(m-n)] = 0$. With $m = n$, we have

$$\int_0^1 \sin(2\pi mx)\sin(2\pi mx)dx = \int_0^1 \frac{e^{2\pi imx} - e^{-2\pi imx}}{2i}\frac{e^{2\pi imx} - e^{-2\pi imx}}{2i}dx$$

$$= -\frac{1}{4}\int_0^1 [2\cos(4\pi mx) - 2\cos(0)]dx = \frac{1}{2}. \tag{8.21}$$

If we now change variables to $x = (\omega_0 t)/2\pi$, our sine and cosine integrals become:

$$\int_0^T \sin(m\omega_0 t)\sin(n\omega_0 t)dt = \frac{\pi}{\omega_0}\delta_{mn}, \tag{8.22}$$

$$\int_0^T \sin(m\omega_0 t)\cos(n\omega_0 t)dt = 0, \tag{8.23}$$

$$\int_0^T \cos(m\omega_0 t)\cos(n\omega_0 t)dt = \frac{\pi}{\omega_0}\delta_{mn}. \tag{8.24}$$

To use these relations to determine the Fourier components a_n, b_n, we first multiply both sides of Eq. (8.12) by $\sin(m\omega_0 t)$ and integrate from 0 to T,

$$\int_0^T f(t)\sin(m\omega_0 t)dt = \int_0^T \left\{ \frac{a_0}{2} + \sum_{n=1}^{\infty}[a_n\cos(n\omega_0 t) + b_n\sin(n\omega_0 t)] \right\}\sin(m\omega_0 t)dt. \tag{8.25}$$

Using the orthogonality relations above, all terms of the integral on the right-hand side of this equation vanish except the term which contains b_m,

$$b_m = \frac{2}{T} \int_0^T f(t) \sin(m\omega_0 t) \mathrm{d}t. \tag{8.26}$$

Similarly, by multiplying by $\cos(m\omega_0 t)$, we determine a_m,

$$a_m = \frac{2}{T} \int_0^T f(t) \cos(m\omega_0 t) \mathrm{d}t. \tag{8.27}$$

It is to be noted that this equation also works for a_0; this is the motivation for defining the constant term of the Fourier series as $a_0/2$.

8.3 Examples

It is worth exploring a few examples of Fourier series, to see how they are, in practice, determined.

Example 8.1 (Sawtooth wave) As a preliminary example, we consider the Fourier series representation of a sawtooth wave, given by

$$f(t) = \begin{cases} t & \text{for} \quad 0 \le t \le \pi, \\ t - 2\pi & \text{for} \quad \pi \le t \le 2\pi, \end{cases} \tag{8.28}$$

which is a series of period $T = 2\pi$, so that $\omega_0 = 1$. One period of this function is depicted in Fig. 8.5.

The Fourier series for the sawtooth wave is written as

$$f(t) = \frac{a_0}{2} + \sum_{n=1}^{\infty} [a_n \cos(nt) + b_n \sin(nt)], \tag{8.29}$$

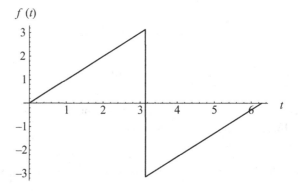

Figure 8.5 Depiction of one period of a sawtooth wave.

with

$$a_m = \frac{1}{\pi} \int_0^{2\pi} f(t)\cos(mt)dt, \tag{8.30}$$

$$b_m = \frac{1}{\pi} \int_0^{2\pi} f(t)\sin(mt)dt. \tag{8.31}$$

These integrals can be evaluated in a straightforward way,

$$a_m = \frac{1}{\pi} \int_0^{\pi} t\cos(mt)dt + \frac{1}{\pi} \int_\pi^{2\pi} (t - 2\pi)\cos(mt)dt$$

$$= \frac{1}{\pi} \int_0^{2\pi} t\cos(mt)dt - 2\int_\pi^{2\pi} \cos(mt)dt = 0, \tag{8.32}$$

$$b_m = \frac{1}{\pi} \int_0^{\pi} t\sin(mt)dt + \frac{1}{\pi} \int_\pi^{2\pi} (t - 2\pi)\sin(mt)dt$$

$$= \frac{1}{\pi} \int_0^{2\pi} t\sin(mt)dt - 2\int_\pi^{2\pi} \sin(mt)dt$$

$$= -\frac{2}{m} + \frac{2}{m}\cos(mt)\Big|_\pi^{2\pi} = (-1)^{m-1}\frac{2}{m}. \tag{8.33}$$

We may approximate the actual function by a finite number of terms in the series. We plot the results for several illustrative cases in Fig. 8.6.

It can be seen that near the jump in the value of the sawtooth, there always remains an "overshoot" of the function from its proper value. This is the Gibbs phenomenon, discussed in Section 7.4; it is an example of nonuniform convergence of a series.

◇

Example 8.2 (Square wave) Another common example to which Fourier techniques are applied is the square wave, defined by the relation

$$f(t) = \begin{cases} 1/2, & 0 \le t \le \pi, \\ -1/2, & \pi \le t \le 2\pi. \end{cases} \tag{8.34}$$

This function is illustrated in Fig. 8.7.

The coefficients of the series are given by

$$a_m = \frac{1}{\pi} \int_0^{2\pi} f(t)\cos(mt)dt = \frac{1}{2\pi} \int_0^{\pi} \cos(mt)dt - \frac{1}{2\pi} \int_\pi^{2\pi} \cos(mt)dt, \tag{8.35}$$

$$b_m = \frac{1}{\pi} \int_0^{2\pi} f(t)\sin(mt)dt = \frac{1}{2\pi} \int_0^{\pi} \sin(mt)dt - \frac{1}{2\pi} \int_\pi^{2\pi} \sin(mt)dt. \tag{8.36}$$

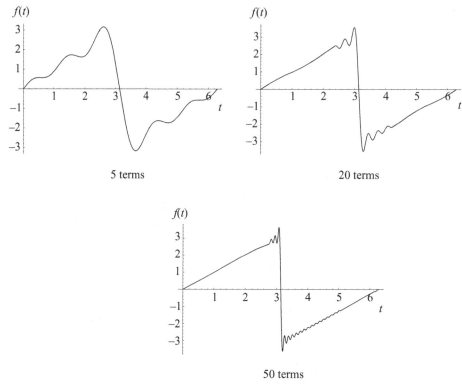

Figure 8.6 Illustration of the convergence of the Fourier series for a sawtooth wave.

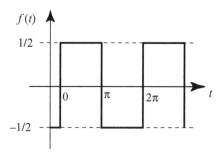

Figure 8.7 Depiction of several periods of a square wave.

It is not difficult to show that all the coefficients a_m vanish, and the coefficients b_m take on the form

$$b_m = 1 - \frac{(-1)^m}{\pi m}. \tag{8.37}$$

Noting that only odd powers of m contribute to the series, we may write the entire series in compact form as

$$f(t) = \frac{2}{\pi} \sum_{n=0}^{\infty} \frac{\sin\left[(2n+1)x\right]}{(2n+1)}. \tag{8.38}$$

$\diamond$

8.4 Integration range of the Fourier series

In discussing the Fourier series, it is to be noted that there exist numerous ways to define the Fourier expansion, using different domains or even different functions. As an example of the former, if we are interested in expanding a periodic function in the range $-T/2 \leq t \leq T/2$, we can write

$$f(t) = \frac{a_0}{2} + \sum_{n=1}^{\infty} [a_n \cos(n\omega_0 t) + b_n \sin(n\omega_0 t)], \tag{8.39}$$

but where the coefficients a_n and b_n are now given by

$$a_m = \frac{2}{T} \int_{-T/2}^{T/2} f(t) \cos(m\omega_0 t) dt, \tag{8.40}$$

$$b_m = \frac{2}{T} \int_{-T/2}^{T/2} f(t) \sin(m\omega_0 t) dt. \tag{8.41}$$

The advantage in expanding a function over a different domain is that one can often take advantage of symmetries in the problem to immediately determine some coefficients. For instance, the periodic sawtooth function described in the example above is an odd function about $t = 0$; if we use the Fourier expansion given above (center point $t = 0$), we can immediately see that, because $\cos(mt)$ is an even function, $a_n = 0$ for all n. Furthermore, the integral for b_n becomes easier,

$$b_n = \frac{1}{\pi} \int_{-\pi}^{\pi} t \sin(nt) dt = (-1)^{n-1} \frac{2}{n}. \tag{8.42}$$

On comparison with Eq. (8.33), we see that the coefficients of the series are the same. The Fourier series is independent of the specific choice of the integration range used to determine it.

8.5 Complex-valued Fourier series

We may develop an alternative form of the Fourier series using complex exponentials instead of sines and cosines. Recalling that

$$\cos(n\omega_0 t) = \frac{e^{i\omega_0 t} + e^{-i\omega_0 t}}{2}, \tag{8.43}$$

$$\sin(n\omega_0 t) = \frac{e^{i\omega_0 t} - e^{-i\omega_0 t}}{2i}, \tag{8.44}$$

we may rewrite the real-valued Fourier series in the form

$$f(t) = \frac{a_0}{2} + \sum_{n=1}^{\infty} \left[\frac{1}{2}(a_n - ib_n)e^{in\omega_0 t} + \frac{1}{2}(a_n + ib_n)e^{-in\omega_0 t} \right]. \tag{8.45}$$

Let us define new Fourier coeffiecnts c_n as

$$c_0 \equiv a_0/2, \quad c_n \equiv \frac{1}{2}(a_n - ib_n), \quad c_{-n} \equiv \frac{1}{2}(a_n + ib_n). \tag{8.46}$$

We may then write a new definition for a complex-valued Fourier series.

Definition 8.2 *A complex-valued Fourier series is an expansion of a function $f(t)$ over an interval $-T/2 \le t < T/2$ of the form*

$$f(t) = \sum_{n=-\infty}^{\infty} c_n e^{in\omega_0 t}, \tag{8.47}$$

where c_n is given by

$$c_n = \frac{1}{T} \int_{-T/2}^{T/2} f(t) e^{-in\omega_0 t} \mathrm{d}t. \tag{8.48}$$

This expression can in fact be directly connected to the Laurent series of complex analysis, to be described in Section 9.8. When the Fourier series is written in this form, the c_n are often referred to as the *spectrum* of the function, in obvious reference to the relevance of the coefficients to an exponential of a given frequency.

8.6 Properties of Fourier series

8.6.1 Symmetric series

In the previous section, we noted that if a function is odd about the point $t = 0$, its Fourier representation will only contain sine terms. Conversely, if the function is even about $t = 0$, its Fourier series will only contain cosine terms.

For cases with such symmetry, we can define Fourier sine and cosine series as follows.

Definition 8.3 (Fourier cosine series) *If a function is even about the point $t = 0$ within the domain $-T/2 \le t < T/2$, it may be expanded in a* Fourier cosine series *of the form*

$$f(t) = \frac{a_0}{2} + \sum_{n=1}^{\infty} a_n \cos(n\omega_0 t), \tag{8.49}$$

where

$$a_n = \frac{4}{T} \int_0^{T/2} f(t) \cos(n\omega_0 t) \mathrm{d}t. \tag{8.50}$$

Definition 8.4 (Fourier sine series) *If a function is odd about the point $t = 0$ within the domain $-T/2 \leq t < T/2$, it may be expanded in a* Fourier sine series *of the form*

$$f(t) = \sum_{n=1}^{\infty} b_n \sin(n\omega_0 t),$$ (8.51)

where

$$b_n = \frac{4}{T} \int_0^{T/2} f(t) \sin(n\omega_0 t) dt.$$ (8.52)

It is to be noted that the range of integration now goes from 0 to $T/2$, not from $-T/2$ to $T/2$. This simplification comes from the assumed symmetry of the function $f(t)$; for instance, for an even function, we have

$$a_n = \frac{2}{T} \int_{-T/2}^{T/2} f(t) \cos(n\omega_0 t) dt$$

$$= \frac{2}{T} \left[\int_{-T/2}^{0} f(t) \cos(n\omega_0 t) dt + \int_0^{T/2} f(t) \cos(n\omega_0 t) dt \right]$$

$$= \frac{2}{T} \left[\int_0^{T/2} f(-t) \cos(-n\omega_0 t) dt + \int_0^{T/2} f(t) \cos(n\omega_0 t) dt \right]$$

$$= \frac{2}{T} \left[\int_0^{T/2} f(t) \cos(n\omega_0 t) dt + \int_0^{T/2} f(t) \cos(n\omega_0 t) dt \right]$$

$$= \frac{4}{T} \int_0^{T/2} f(t) \cos(n\omega_0 t) dt,$$ (8.53)

where we have used the assumption that $f(-t) = f(t)$. For an odd function, Eq. (8.52) may be determined by using $f(-t) = -f(t)$.

With the introduction of Fourier sine and cosine series, we may make a curious observation. Let us suppose we have a piecewise continuous function $f(t)$ defined on the interval $0 \leq t < T$, which may be expanded in a Fourier series of the form of Eq. (8.12). We may also, however, define a new function $g(t)$ such that

$$g(t) = \begin{cases} f(t), & t \geq 0, \\ f(-t), & t < 0. \end{cases}$$ (8.54)

The function $g(t)$ is an even function defined on the domain $-T \leq t < T$, and may therefore be expanded in a Fourier cosine series. If we define a function $h(t)$ such that

$$h(t) = \begin{cases} f(t), & t \geq 0, \\ -f(-t), & t < 0, \end{cases}$$ (8.55)

it is an odd function which may be expanded in a Fourier sine series.

We therefore have three different ways in which a function $f(t)$, defined on $0 \leq t < T$, may be expanded in terms of periodic functions: a Fourier series of period T, a Fourier sine series of period $2T$, or a Fourier cosine series of period $2T$. Each series will have a different mathematical form, but all should sum to the function $f(t)$ on the domain $0 \leq t < T$.

When we use separation of variables to solve partial differential equations in Chapter 15, we will often encounter situations in which a general function is expanded in terms of a Fourier sine or cosine series alone.

8.6.2 Existence, completeness, and closure relations

In deriving the Fourier series, we made one major assumption which we have yet to justify – we *assumed* that a periodic function could be perfectly represented by a sum of sines and cosines. In other words, we assumed that the $\cos(n\omega_0 t)$, $\sin(n\omega_0 t)$, form a *basis* for all functions on the line $0 \leq t \leq T$.

In our discussion of distributions in Section 6.4, we suggested that an arbitrary function $f(t)$ might be considered a vector in an *infinite-dimensional* vector space. Such a vector space includes all piecewise-continuous functions defined over the domain $0 \leq t < T$. Whereas in three-dimensional space the components of a vector are labeled by the axes x, y, and z, the components of the function are labeled by the variable t.

However, just as in three-dimensional space we can write our vector $\mathbf{A}$ in a different basis, e.g. we can change coordinates from $\hat{\mathbf{x}}, \hat{\mathbf{y}}, \hat{\mathbf{z}}$ to $\hat{\mathbf{x}}', \hat{\mathbf{y}}', \hat{\mathbf{z}}'$, we expect that there are many different bases in which we can express our function $f(t)$. In deriving our Fourier series, we have postulated that one such basis is the collection of sines and cosines, $\cos(n\omega_0 t)$, $\sin(n\omega_0 t)$.

Can we prove that these functions represent a basis? As discussed in Section 6.4, a natural way to do this is to show that we can transform from our hypothetical basis, the sines and cosines, to a known basis, the delta functions. If we can construct a delta function from sines and cosines, evidently we can construct any function we want. It can in fact be shown that

$$\delta(t-a) = \frac{1}{T} + \sum_{n=1}^{\infty} \left[\frac{2}{T} \cos(n\omega_0 a) \cos(n\omega_0 t) + \frac{2}{T} \sin(n\omega_0 a) \sin(n\omega_0 t) \right]. \tag{8.56}$$

Such a relation is called a *closure relation* for the functions $\cos(n\omega_0 t)$, $\sin(n\omega_0 t)$. We can verify it nonrigorously by simple substitution of the delta function into Eqs. (8.15) and (8.14).

It should be said that the discussion of this section is somewhat oversimplified. First, the analogy between three-dimensional vectors and infinite-dimensional vectors is not perfect – there are some significant differences in their behavior. Second, we have already seen, in the Gibbs phenomenon, that the Fourier series representation of a discontinuous function *never* completely converges to the original function – there is always a small region around the discontinuities where the series differs by a finite amount. In talking about Fourier series convergence, then, it should be said that a Fourier series can represent a function *almost* everywhere, with the exception of a few unusual, discontinuous points which end up being insignificant.

8.7 Gibbs phenomenon and convergence in the mean

In Section 7.4, we introduced the Gibbs phenomenon in the context of uniform convergence. In 1899, J. W. Gibbs empirically discovered [Gib99] that the Fourier series representation of discontinuous functions always "overshoots" the exact value of the function in the neighborhood of the discontinuity. This overshoot converges to a finite value, but is increasingly confined to the region of the discontinuity. It is interesting to note that we can derive an analytic expression for the magnitude of the discrepancy, and the derivation highlights some of the techniques previously learned relating to infinite series in general and Fourier series in particular.

We return first to Eq. (8.12) which defines Fourier series, but consider the summation $S_N(x)$ of only a finite number of terms up to order N, i.e.

$$S_N(x) = \frac{a_0}{2} + \sum_{n=1}^{N} [a_n \cos(nx) + b_n \sin(nx)]. \tag{8.57}$$

For simplicity, we choose $-\pi < t < \pi$ as our domain of interest. Applying the definitions of a_n, b_n, given by Eqs. (8.14) and (8.15), and using the identity

$$\cos(A - B) = \cos A \cos B + \sin A \sin B, \tag{8.58}$$

we readily find that

$$S_N(x) = \frac{1}{\pi} \int_{-\pi}^{\pi} f(t) \left[\frac{1}{2} + \sum_{n=1}^{N} \cos[n(x - t)] \right] dt. \tag{8.59}$$

By expanding the cosine term into its complex exponential form, i.e.

$$\cos[n(x - t)] = \frac{e^{in(x-t)} + e^{-in(x-t)}}{2}, \tag{8.60}$$

we may express the partial sum as the partial sum of two finite geometric series. We then may use Eq. (7.17) to write these series in closed form, i.e.

$$\sum_{n=1}^{N} e^{in(x-t)} = e^{i(x-t)} \frac{1 - e^{iN(x-t)}}{1 - e^{i(x-t)}} = e^{i(x-t)/2} e^{iN(x-t)/2} \frac{\sin[N(x - t)/2]}{\sin[(x - t)/2]}. \tag{8.61}$$

On substitution into Eq. (8.59), we may write

$$S_N(x) = \frac{1}{\pi} \int_{-\pi}^{\pi} f(t) \left[\frac{1}{2} + \cos[(N + 1)(x - t)/2] \frac{\sin[N(x - t)/2]}{\sin[(x - t)/2]} \right] dt. \tag{8.62}$$

Finally, using the trigonometric relationship

$$\sin(A+B) + \sin(A-B) = 2\sin A \cos B, \tag{8.63}$$

this expression may be simplified to the form

$$S_N(x) = \frac{1}{2\pi} \int_{-\pi}^{\pi} f(t) \frac{\sin[(N+1/2)(x-t)]}{\sin[(x-t)/2]} dt. \tag{8.64}$$

We now consider the behavior of this partial sum for the special case of the square wave of Section 8.3. It may be written in the form

$$S_N(x) = -\frac{1}{4\pi} \int_{-\pi}^{0} \frac{\sin[(N+1/2)(x-t)]}{\sin[(x-t)/2]} dt + \frac{1}{4\pi} \int_{0}^{\pi} \frac{\sin[(N+1/2)(x-t)]}{\sin[(x-t)/2]} dt. \tag{8.65}$$

We now make a change of variable; we introduce $s = x - t$ as the variable of integration in the first integral, and $s = t - x$ in the second, which results in the expression

$$S_N(x) = -\frac{1}{4\pi} \int_{x}^{x+\pi} \frac{\sin[(N+1/2)s]}{\sin[s/2]} ds + \frac{1}{4\pi} \int_{-x}^{-x+\pi} \frac{\sin[(N+1/2)s]}{\sin[s/2]} ds. \tag{8.66}$$

The integrands of the two integrals differ only in sign, and there is an overlap between their domains; the contribution between the intermediate range from x to $\pi - x$ is negated, leaving

$$S_N(x) = \frac{1}{4\pi} \int_{-x}^{x} \frac{\sin[(N+1/2)s]}{\sin[s/2]} ds - \frac{1}{4\pi} \int_{\pi-x}^{\pi+x} \frac{\sin[(N+1/2)s]}{\sin[s/2]} ds. \tag{8.67}$$

The integrals are centered on the two discontinuities of the square wave, at $x = 0$ and $x = \pi$. If we focus our attention on the neighborhood around $x = 0$, particularly in the limit as $N \to \infty$, the second integral is well-behaved and has a vanishing contribution. We turn to the first integral, and note that

$$S_N(x) \approx \frac{1}{4\pi} \int_{-x}^{x} \frac{\sin[(N+1/2)s]}{\sin[s/2]} ds \tag{8.68}$$

as $N \to \infty$.

Let us restrict ourselves to $x \geq 0$ for the moment. The difference $\Delta_N(x)$ between the ideal step function and the Nth order approximation to the series is

$$\Delta_N(x) = \frac{1}{2} - \frac{1}{4\pi} \int_{-x}^{x} \frac{\sin[(N+1/2)s]}{\sin[s/2]} ds. \tag{8.69}$$

The largest value of $\Delta_N(x)$ occurs when its derivative vanishes, or when

$$\frac{\sin[(N+1/2)x]}{\sin[x/2]} = 0. \tag{8.70}$$

In particular, there are local maxima/minima of Δ_N at points x_k such that $(N+1/2)x_k = k\pi$, with k an integer, or

$$x_k = \frac{2k\pi}{2N+1}. \tag{8.71}$$

It is to be noted that the position of these extrema depend on the order of summation N; referring back to Fig. 7.9, we can see that the peaks and troughs of oscillation do in fact move as we add more terms to the series.

Returning to our value of $\Delta_N(x)$, we use one of the x_k as our value for x, and the integral becomes

$$\Delta_N(x_k) = \frac{1}{2} - \frac{1}{2\pi} \int_0^{x_k} \frac{\sin[(N+1/2)s]}{\sin[s/2]} ds. \tag{8.72}$$

We now make one final change of variable to $\rho \equiv (N+1/2)s$; then our integral takes on the form

$$\Delta_N(x_k) = \frac{1}{2} - \frac{1}{\pi} \int_0^{k\pi} \frac{\sin[\rho]}{(2N+1)\sin[\rho/(2N+1)]} d\rho. \tag{8.73}$$

We may rewrite the denominator in the form

$$(2N+1)\sin[\rho/(2N+1)] = \rho \frac{\sin[\rho/(2N+1)]}{\rho/(2N+1)}. \tag{8.74}$$

The latter part behaves as $\sin(x)/x$; the limit $N \to \infty$ is equivalent to taking $x \to 0$, or

$$\lim_{N \to \infty} (2N+1)\sin[\rho/(2N+1)] = \rho. \tag{8.75}$$

In terms of our integral, we have

$$\lim_{N \to \infty} \Delta_N(x_k) = \frac{1}{2} - \frac{1}{\pi} \int_0^{k\pi} \frac{\sin[\rho]}{\rho} d\rho. \tag{8.76}$$

Looking at the first extremum, $k = 1$, we may evaluate this integral numerically. We find that

$$\lim_{N \to \infty} \Delta_N(x_k) = -0.08949, \tag{8.77}$$

or the function undershoots the negative corner of the step by some 18%.

It is worth noting that the Gibbs phenomenon appears in any expansion of discontinuous functions into an orthogonal basis, and is not restricted to the Fourier series. Similar problems will arise in using Legendre polynomials or any other series of orthogonal functions.

It can be shown that the Fourier series of a continuous function $f(t)$ converges uniformly to the value of the function, $f(t)$. However, many functions of interest such as the sawtooth wave are discontinuous, and for such functions the series does not converge pointwise to the value of the represented function. It is helpful to have a more relaxed definition of convergence in such cases, and one generally refers to *convergence in the mean* for Fourier series.

Definition 8.5 (Convergence in the mean) *A series $\sum_{n=0}^{\infty} \alpha_n f_n(t)$ is said to converge in the mean to the function $f(t)$ over the domain $a \leq t \leq b$ if*

$$\lim_{N \to \infty} \int_a^b \left| f(t) - \sum_{n=0}^{N} \alpha_n f_n(t) \right|^2 dt \to 0. \tag{8.78}$$

It can be shown that the Fourier series of a piecewise continuous function converges in the mean to the function $f(t)$.

8.8 Focus: X-ray diffraction from crystals

In Section 1.6, we discussed the relationship between the structure of a crystal and the scattering of X-rays from that crystal. In this section, we use some basic scattering theory and concepts of Fourier series[1] to further quantify the relationship between the crystal structure and the scattered X-ray distribution.

We consider a crystal which has a periodic scattering potential $F(\mathbf{r})$ such that $F(\mathbf{r}+\mathbf{R}) = F(\mathbf{r})$, where $\mathbf{R}$ is a lattice vector. The crystal, though periodic, is assumed to be of finite volume V.

X-rays are highly energetic particles which typically undergo weak scattering on passing through a crystalline material; we are therefore justified in using the first Born approximation to describe the scattered field far from the crystal. Following the discussion of Section 7.8, the scattered field far from the scatterer in the first Born approximation may be written as

$$U_s(\mathbf{r}) = \frac{e^{ikr}}{r} \int_V F(\mathbf{r}')e^{-ik(\mathbf{s}-\mathbf{s}_0)\cdot\mathbf{r}'} d^3 r'. \tag{8.79}$$

The integral is a three-dimensional Fourier transform, to be discussed in detail in Chapter 11. To simplify this expression, we note that because $F(\mathbf{r})$ is a periodic function, it should be expressible as a Fourier series in three variables. This will be done numerous times in solving problems of separation of variables in rectangular geometries in Chapter 14. However, most crystal structures do not have a rectangular geometry. The natural solution is to expand the function in a Fourier series based on the reciprocal lattice vectors discussed in Section 1.6, i.e.

$$F(\mathbf{r}) = \sum_{\mathbf{G}} h_{\mathbf{G}} e^{i\mathbf{G}\cdot\mathbf{r}}, \tag{8.80}$$

where the summation is over the three indices of the reciprocal lattice vectors,

$$\mathbf{G} = [k_1\mathbf{g}_1 + k_2\mathbf{g}_2 + k_3\mathbf{g}_3]. \tag{8.81}$$

The expression (8.80) has the proper symmetry under translations with respect to the lattice vectors $\mathbf{R}$. We defer, for the moment, the explicit determination of the coefficients $h_{\mathbf{G}}$.

---

[1] We must also use some concepts from the theory of Fourier transforms, to be discussed in Chapter 11. Those unfamiliar with the basics of Fourier transforms can safely put off this section until that chapter has been covered.

We may substitute from Eq. (8.80) into Eq. (8.79); the result is

$$U_s(\mathbf{r}) = \frac{e^{ikr}}{r} \sum_G h_G \int_V e^{-i(k\mathbf{s}-k\mathbf{s}_0+\mathbf{G})\cdot\mathbf{r}'} d^3 r'. \tag{8.82}$$

This expression reproduces much of what we already know about X-ray crystal scattering from Section 1.6. For instance, let us suppose that the crystal is, in effect, of infinite extent. Then an integration over the Cartesian coordinates x', y', and z' results in a three-dimensional delta function, and we may write

$$U_s(\mathbf{r}) = (2\pi)^3 \frac{e^{ikr}}{r} \sum_G h_G \delta^{(3)}(k\mathbf{s} - k\mathbf{s}_0 + \mathbf{G}). \tag{8.83}$$

In this limit, one will only see a scattered field in well-defined directions specified by the expression

$$k\mathbf{s} - k\mathbf{s}_0 + \mathbf{G} = 0. \tag{8.84}$$

On comparison with Eq. (1.69), we see that this is the von Laue condition for X-ray diffraction. This condition depends upon the periodicity of the crystal and we may state that *the location of X-ray diffraction peaks is determined by the periodic structure of the crystal.*

Let us now assume that we have a large rectangular crystal confined to the domain $-\Delta x/2 \le x' \le \Delta x/2$, $-\Delta y/2 \le y' \le \Delta y/2$, and $-\Delta z/2 \le z' \le \Delta z/2$. The three-dimensional volume integral in Eq. (8.82) is separable into three single-variable Fourier integrals, each of the form

$$\int_{-\Delta x/2}^{\Delta x/2} e^{iK_x x'} dx' = \Delta x \left[\frac{\sin(K_x \Delta x/2)}{K_x \Delta x/2} \right] = \Delta x j_0(K_x \Delta x/2). \tag{8.85}$$

The j_0 function is the zeroth-order spherical Bessel function (discussed in Chapter 16) and is proportional to one of the delta sequences discussed in Section 6.2. It is a peaked, oscillating function whose width is inversely proportional to the width of the crystal. The product of all three Bessel functions is centered on the directions specified by the von Laue condition. We may now state that *the width of X-ray diffraction peaks is determined by the overall size of the crystal.*

It can be seen from Eq. (8.82) that the X-ray diffraction pattern also depends upon the coefficients h_G. In particular, the h_G represent the amplitudes of the individual X-ray diffraction peaks. To evaluate them, we try integration over one unit volume of the crystal,

$$\int_0^1 \int_0^1 \int_0^1 F(\mathbf{a}_1 p + \mathbf{a}_2 q + \mathbf{a}_3 r) e^{-i(k_1 \mathbf{g}_1 + k_2 \mathbf{g}_2 + k_3 \mathbf{g}_3)\cdot(\mathbf{a}_1 p + \mathbf{a}_2 q + \mathbf{a}_3 r)} dp \, dq \, dr$$

$$= \sum_{k_1', k_2', k_3'} \int_0^1 \int_0^1 \int_0^1 h_G e^{-i[(k_1-k_1')\mathbf{g}_1 + (k_2-k_2')\mathbf{g}_2 + (k_3-k_3')\mathbf{g}_3]\cdot(\mathbf{a}_1 p + \mathbf{a}_2 q + \mathbf{a}_3 r)} dp \, dq \, dr,$$

$$\tag{8.86}$$

where p, q and r are dimensionless integration variables. Noting that $\mathbf{a}_i \cdot \mathbf{g}_j = 2\pi\delta_{ij}$, we may write

$$\int_0^1 \int_0^1 \int_0^1 F(\mathbf{a}_1 p + \mathbf{a}_2 q + \mathbf{a}_3 r) e^{-i(k_1\mathbf{g}_1 + k_2\mathbf{g}_2 + k_3\mathbf{g}_3)\cdot(\mathbf{a}_1 p + \mathbf{a}_2 q + \mathbf{a}_3 r)} \mathrm{d}p\mathrm{d}q\mathrm{d}r$$

$$= \sum_{k_1', k_2', k_3'} h_{\mathbf{G}} \int_0^1 \int_0^1 \int_0^1 e^{-2\pi i[(k_1 - k_1')p + (k_2 - k_2')q + (k_3 - k_3')]} \mathrm{d}p\mathrm{d}q\mathrm{d}r. \qquad (8.87)$$

This latter integral may be written as

$$\int_0^1 \int_0^1 \int_0^1 e^{-2\pi i[(k_1 - k_1')p + (k_2 - k_2')q + (k_3 - k_3')]} \mathrm{d}p\mathrm{d}q\mathrm{d}r = \delta_{k_1 k_1'} \delta_{k_2 k_2'} \delta_{k_3 k_3'}. \qquad (8.88)$$

We may therefore write

$$h_{\mathbf{G}} = \int_0^1 \int_0^1 \int_0^1 F(\mathbf{r}) e^{-i\mathbf{G}\cdot\mathbf{r}} \mathrm{d}p\mathrm{d}q\mathrm{d}r. \qquad (8.89)$$

The coefficients $h_{\mathbf{G}}$ depend upon the specific form of the potential within each unit cell of the crystal, which in turn depends upon the distribution of atoms, and their electrons, within the cell. We have shown that *the amplitudes of the X-ray diffraction peaks are determined by the behavior of atoms and electrons within a unit cell of the crystal*. The coefficient $h_{\mathbf{G}}$ is typically referred to as a *structure factor*.

The structure factors $h_{\mathbf{G}}$ are usually simplified further by assuming that the unit cell consists of n individual atoms, labeled by index j, each of which has an electronic charge distribution $\rho_j(\mathbf{r})$, centered on the position $\mathbf{r}_j$. Each atom therefore has associated with it a quantity known as the *atomic form factor*, of the form

$$f_j(\mathbf{G}) \equiv -\frac{1}{e} \int_0^1 \int_0^1 \int_0^1 \rho_j(\mathbf{r} - \mathbf{r}_j) e^{-i\mathbf{G}\cdot\mathbf{r}} \mathrm{d}p\mathrm{d}q\mathrm{d}r, \qquad (8.90)$$

with e the fundamental unit of charge. With the form factors, the structure factor of the crystal may be written as

$$h_{\mathbf{G}} = \sum_{j=1}^n f_j(\mathbf{G}) e^{-i\mathbf{G}\cdot\mathbf{r}_j}. \qquad (8.91)$$

An application of Fourier series analysis to the problem of X-ray crystal diffraction has given us the following picture of the nature of the peaks of X-ray diffraction.

1. The location of X-ray diffraction peaks is determined by the periodic structure of the crystal.
2. The width of X-ray diffraction peaks is determined by the overall size of the crystal.
3. The amplitudes of the X-ray diffraction peaks are determined by the behavior of atoms and electrons within a unit cell of the crystal.

In X-ray crystallography, this diffraction data is used to deduce both the periodic structure of the crystal and the detailed structure of the unit cell. We have used Fourier analysis to determine the nature of the relationship between structure and scattering.

8.9 Additional reading

Two books that discuss Fourier series and other orthogonal functions are:

- H. F. Davis, *Fourier Series and Orthogonal Functions* (Boston, Allyn and Bacon, 1963).
- D. Jackson, *Fourier Series and Orthogonal Polynomials* (Oberlin, OH, Mathematical Association of America, 1941).

8.10 Exercises

1. Determine the real-valued Fourier series of the triangular function

$$f(t) = \begin{cases} t+\pi, & -\pi \le t \le 0, \\ \pi - t, & 0 \le t < \pi. \end{cases}$$

2. Determine the real-valued Fourier series of the function

$$f(t) = \begin{cases} t+\pi, & -\pi \le t \le 0, \\ \pi, & 0 \le t < \pi. \end{cases}$$

3. Determine the complex-valued Fourier series of the function

$$f(t) = e^{-t},$$

over the domain $0 \le t < 2\pi$.

4. Determine the complex-valued Fourier series of the function

$$f(t) = \int_0^t e^{-u} u \, du,$$

over the domain $0 \le t < 2\pi$.

5. Convergence in the mean can be used as an alternative method to derive the Fourier coefficients. Consider the integral

$$\Gamma \equiv \int_{-L/2}^{L/2} \left| f(t) - \frac{a_0}{2} - \sum_{n=1}^{N} [a_n \cos(n\omega_0 t) + b_n \sin(n\omega_0 t)] \right|^2 dt.$$

Show that the conditions

$$\frac{\partial \Gamma}{\partial a_m} = 0, \frac{\partial \Gamma}{\partial b_m} = 0,$$

lead to the expected equations for a_m, b_m.

6. It is occasionally possible to simplify a problem by exploiting hidden symmetries. Determine the real-valued Fourier series of the function

$$f(t) = \begin{cases} -t, & -\pi \le t < 0, \\ 0, & 0 \le t < \pi, \end{cases}$$

by first decomposing it into the sum of an even function and an odd function.

7. It is occasionally possible to simplify a problem by exploiting hidden symmetries. Determine the real-valued Fourier series of the function

$$f(t) = \begin{cases} 2, & -\pi \le t < 0, \\ 0, & 0 \le t < \pi, \end{cases}$$

by first decomposing it into the sum of an even function and an odd function.

8. Subtle problems can arise when taking derivatives of Fourier series. For instance, the Fourier series of $f(t) = t$ on the domain $-\pi \le t < \pi$ is given by

$$t = 2 \sum_{n=1}^{\infty} (-1)^{n+1} \frac{\sin(nt)}{n}.$$

Taking the derivative with respect to t on both sides, however, gives the result

$$1 = 2 \sum_{n=1}^{\infty} (-1)^{n+1} \cos(nt),$$

which is a nonsensical result, as the series is not convergent. What has gone wrong?

9. Sometimes Fourier series can be used to explicitly determine the otherwise indeterminable sum of an infinite series. By calculating the Fourier series of $f(t) = t^2$ over the domain $0 \le t < 2\pi$, find a value for the series

$$\zeta(2) = \sum_{n=1}^{\infty} \frac{1}{n^2}.$$

This is the value of the so-called Riemann zeta function $\zeta(u)$ evaluated at $u = 2$.

10. Sometimes Fourier series can be used to explicitly determine the otherwise indeterminable sum of an infinite series. By calculating the Fourier series of

$$f(t) = \begin{cases} t(\pi + t), & -\pi \le t < 0, \\ t(\pi - t), & 0 \le t < \pi, \end{cases}$$

over the domain $0 \le t < 2\pi$, find a value for the series,

$$\beta(3) = \sum_{n=0}^{\infty} \frac{(-1)^n}{(2n+1)^3}.$$

The function $\beta(3)$ is related to the Riemann zeta function.

11. A bit of a challenge: construct a function of period 2π whose Fourier series allows the determination of the value of π^4. Furthermore, the series must have coefficients a_n, b_n that decay *faster* than $1/n$.

12. We consider the term by term integration of the Fourier series of Eq. (8.12). Is the result necessarily a Fourier series itself?

13. We consider the term by term integration of the square wave given by Eq. (8.34), over the integration range from $t' = 0$ to $t' = t$. Show, by direct integration of both the square wave and its Fourier representation, that the result is proportional to a sawtooth wave.

14. Read the paper by D. W. Diehl and N. George, Analysis of multitone holographic interference filters by use of a sparse Hill matrix method, *Appl. Opt.* **43** (2004), 88–96. Describe what type of system is being studied in the paper, and how a Fourier series is used to characterize this system.

15. Read the paper by R. Albanese, J. Penn, and R. Medina, Short-rise-time microwave pulse propagation through dispersive biological media, *J. Opt. Soc. Am. A* **6** (1989), 1441–1446. Why is a Fourier series an appropriate tool for the system being considered, and what conclusions are drawn from the results?

16. Explain the technique described in the paper by L. Hu, L. Xuan, D. Li, Z. Cao, Q. Mu, Y. Liu, Z. Peng, and X. Lu, Real-time liquid-crystal atmosphere turbulence simulator with graphic processing unit, *Opt. Exp.* **17** (2009), 7259–7268. How is a Fourier series employed in the simulations?

17. The paper by A. A. Krokhin, P. Halevi, and J. Arriaga, Long-wavelength limit (homogenization) for two-dimensional photonic crystals, *Phys. Rev. B* **65** (2002), 115208, looks at a specific property of so-called photonic crystals. Explain what a photonic crystal is, why Fourier series are an appropriate tool for their characterization, and what property of such crystals is investigated by this paper.

18. In a fascinating paper by G. A. Schott, The electromagnetic field of a moving uniformly and rigidly electrified sphere and its radiationless orbits, *Phil. Mag.* **15** (1933), 752–761, he demonstrates a very counter-intuitive result in electromagnetic radiation theory. Explain the nature of this result. What property of the system is the Fourier series used to characterize?

19. We may also introduce a two-variable Fourier series, where a function $f(x,y)$ is expressed on the domain $-\pi \leq x < \pi, -\pi \leq y < \pi$ as

$$f(x,y) = \sum_{n=-\infty}^{\infty} \sum_{m=-\infty}^{\infty} c_{nm} e^{inx} e^{iny},$$

where

$$c_{nm} = \frac{1}{(\pi)^2} \int_{-\pi}^{\pi} \int_{-\pi}^{\pi} f(x,y) e^{-inx} e^{-iny} dxdy.$$

Evaluate the Fourier series coefficients c_{nm} for the function $f(x,y) = x + y$.

20. Evaluate the Fourier coefficients c_{nm} using the definitions of the previous problem for the function $f(x,y) = \sin(x+y)$.

9

Complex analysis

9.1 Introduction: electric potential in an infinite cylinder

We begin by considering the solution of a familiar problem in electrostatics: to determine the potential on the inside of a single infinite cylinder; the potential on the boundary at $\rho = \rho_0$ is taken to be $V_0(\phi)$. Inside the cylinder, the potential satisfies the differential equation (in Cartesian coordinates),

$$\frac{\partial^2 V(x,y)}{\partial x^2} + \frac{\partial^2 V(x,y)}{\partial y^2} = 0. \tag{9.1}$$

The standard technique for solving this equation would be to use separation of variables in polar coordinates, which we will undertake in Chapter 15. For now, we take a different approach: let us assume that $V(x,y)$ is a function of a single, combined variable w, i.e. $V(x,y) \equiv V(w)$, where

$$w(x,y) = ax + by, \tag{9.2}$$

and a and b are constants, and see if we can deduce some features about the solution. Using the chain rule, we may rewrite Eq. (9.1) in the form

$$\left(\frac{\partial w}{\partial x}\right)^2 \frac{\partial^2 V}{\partial w^2} + \frac{\partial^2 w}{\partial x^2}\frac{\partial V}{\partial w} + \left(\frac{\partial w}{\partial y}\right)^2 \frac{\partial^2 V}{\partial w^2} + \frac{\partial^2 w}{\partial y^2}\frac{\partial V}{\partial w} = 0. \tag{9.3}$$

By use of Eq. (9.2), we may simplify this equation to the form

$$a^2 \frac{\partial^2 V}{\partial w^2} + b^2 \frac{\partial^2 V}{\partial w^2} = 0. \tag{9.4}$$

This equation can only be satisfied if $a^2 = -b^2$. Clearly this cannot be done with real-valued numbers, but let us suppose $a = 1$, $b = \pm i$, where $i = \sqrt{-1}$ is a so-called "imaginary" number; a *complex number* is defined as a number that may be written as $z = x + iy$, with x and y real-valued. If we allow ourselves to make this transformation, we see that the solution of Laplace's equation must be of the general form

$$V(x,y) = f(x + iy) + g(x - iy). \tag{9.5}$$

In polar coordinates, $x = \rho \cos \phi$ and $y = \rho \sin \phi$, and we may use Euler's formula to write

$$x + \mathrm{i}y = \rho \mathrm{e}^{\mathrm{i}\phi}, \quad x - \mathrm{i}y = \rho \mathrm{e}^{-\mathrm{i}\phi}, \tag{9.6}$$

so that the solution may also be written in the form

$$V(x,y) = f(\rho \mathrm{e}^{\mathrm{i}\phi}) + g(\rho \mathrm{e}^{-\mathrm{i}\phi}). \tag{9.7}$$

Our solution must match the potential on the outer rim of the cylinder, $V_0(\phi)$. We may use the methods of the previous chapter and expand this potential in terms of a complex Fourier series,

$$V_0(\phi) = \sum_{n=-\infty}^{\infty} c_n \left(\mathrm{e}^{\mathrm{i}\phi}\right)^n = \sum_{n=0}^{\infty} c_n \left(\mathrm{e}^{\mathrm{i}\phi}\right)^n + \sum_{n=1}^{\infty} c_{-n} \left(\mathrm{e}^{\mathrm{i}\phi}\right)^{-n}, \tag{9.8}$$

where the c_n can be derived in the usual way from $V_0(\phi)$. We can evidently satisfy Laplace's equation and match our boundary conditions if we choose

$$f(\rho \mathrm{e}^{\mathrm{i}\phi}) = c_0 + \sum_{n=1}^{\infty} \frac{c_n}{\rho_0^n} \left(\rho \mathrm{e}^{\mathrm{i}\phi}\right)^n, \tag{9.9}$$

$$g(\rho \mathrm{e}^{-\mathrm{i}\phi}) = \sum_{n=1}^{\infty} \rho^n c_{-n} \left(\rho_0 \mathrm{e}^{\mathrm{i}\phi}\right)^{-n}. \tag{9.10}$$

If the potential on the boundary is taken to be real-valued, it follows that $c_{-n} = c_n^*$. We may then further simplify the solution to

$$V(x,y) = c_0 + \mathrm{Re} \left\{ \sum_{n=1}^{\infty} \frac{c_n}{\rho_0^n} (x + \mathrm{i}y)^n \right\} = c_0 + \mathrm{Re} \left\{ \sum_{n=1}^{\infty} \frac{c_n}{\rho_0^n} z^n \right\}, \tag{9.11}$$

where $\mathrm{Re}\{\cdots\}$ represents the real part of the quantity in brackets, and $z = x + \mathrm{i}y$. This is the general solution for the potential on the interior of a cylinder, expressed entirely in terms of the complex number z. Furthermore, Eq. (9.11) has the form of a power series in z, and we are able to solve Laplace's equation quite generally and easily.

Complex analysis is the study of the algebra of complex numbers and the calculus of functions of complex numbers. As we will see, it is a beautiful and powerful branch of mathematics which will solve many problems and answer many questions for us. We have said that the quantity i is referred to as an "imaginary" number, but if the usefulness of a concept is any indication of its reality, the term "imaginary" is grossly inappropriate.[1] Some of the topics which the theory of this chapter can be applied to are as follows.

---

[1] The term was evidently first coined by René Descartes in his 1637 book *La Géométrie*, and referred to the lack of geometric meaning for such numbers as the roots of quadratic equations. He wrote, "Neither the true nor the false roots are always real; sometimes they are imaginary; that is, while we can always conceive of as many roots for each equation as I have already assigned, yet there is not always a definite quantity corresponding to

1. The solution of differential equations. As we have just seen, solutions of $\nabla^2\phi = 0$ are intimately connected with the theory of complex variables, and complex analysis can be used to readily find and interpret such solutions. Furthermore, series solutions of ordinary differential equations are best understood from the context of complex analysis.

2. The evaluation of definite integrals. Complex analysis can be used to evaluate many difficult integrals, often reducing the integration to a matter of simple algebra. Many otherwise unevaluable Fourier transforms (discussed in Chapter 11) can be determined in this way.

3. The inverse Laplace transform. This transform (described in Section 12.3) is only defined using integration in the complex plane.

4. Properties of special functions. Many useful representations of special functions (Bessel functions, Hermite functions, Legendre functions) result from expressing the special function as an integral in the complex plane. Some of these will be discussed in Chapters 16 and 17.

5. Analytic properties of light fields. The observation that optical fields can be expressed as *analytic* functions leads to many significant physical and practical consequences. This property is of crucial concern in the theory of *inverse problems*. This will be touched upon in Section 10.6.

6. Discrete transforms. Numerical simulations of wave propagation can often be dealt with effectively with the help of the z-transform, a complex analysis technique to be discussed in Section 13.7.

In this chapter we begin by discussing the *algebra* of complex numbers: their addition, multiplication, etc., and then move on to the calculus of functions of a complex variable z. At the end of the chapter we discuss the *residue theorem*, which allows the evaluation of many complicated real-valued integrals.

9.2 Complex algebra

A *complex number* is a quantity of the form

$$z = x + iy, \tag{9.12}$$

where x and y are real numbers and i is the imaginary number such that $i^2 = -1$. We refer to x as the *real* part of z, $\text{Re}(z)$, and y as the *imaginary* part of z, $\text{Im}(z)$. If $y = 0$, we say that z is real, and if $x = 0$, we say that z is pure imaginary.

Graphically, we can represent a complex number in a two-dimensional coordinate system called the *complex plane*, illustrated in Fig. 9.1.

A complex number $z = x + iy$ may therefore be interpreted as a two-dimensional vector (x, y); the set of all complex numbers satisfies the requirements of a vector space (Section 1.5). It is to be noted that i has a role comparable to that of a unit vector, in that it distinguishes between the real and imaginary components.

---

each root so conceived of. Thus, while we may conceive of the equation $x^3 - 6x^2 + 13x - 10 = 0$ as having three roots, yet there is only one real root, 2, while the other two, however we may increase, diminish, or multiply them in accordance with the rules just laid down, remain always imaginary." [Des54] For a history of the use of complex numbers, see [Nah98].

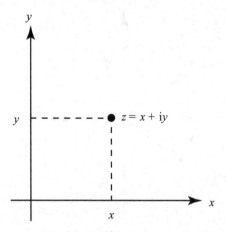

Figure 9.1 Illustrating the concept of the complex plane.

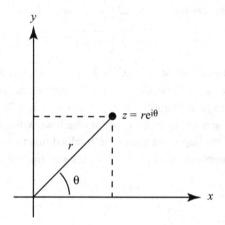

Figure 9.2 Polar coordinate representation of complex numbers.

We may also represent complex numbers in polar coordinates by writing

$$x = r\cos\theta, \quad y = r\sin\theta, \tag{9.13}$$

which is shown in Fig. 9.2.
    The complex number z may then be written as

$$z = x + iy = r(\cos\theta + i\sin\theta) = re^{i\theta}, \tag{9.14}$$

where we have used Euler's formula,

$$e^{i\theta} = \cos\theta + i\sin\theta, \tag{9.15}$$

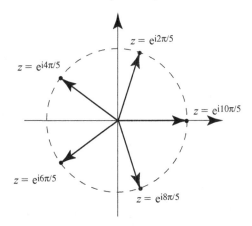

Figure 9.3 Illustration of the 5th roots of $z = 1$ in the complex plane.

in the last step. In this expression, r is called the *modulus* of z, and is given by

$$r = |z| = \sqrt{x^2 + y^2}, \tag{9.16}$$

and θ is referred to as the *argument* of z, often denoted arg(z). It is to be noted that there are many choices for the angle θ, because $\cos\theta$ and $\sin\theta$ are 2π-periodic; thus if θ_1 is an appropriate argument for a given complex number, so is $\theta_2 = \theta_1 + 2\pi n$, where n is an integer. The argument is said to be *multivalued*, a concept which we will see has more significance than one might expect. One important use of multivalued functions is the determination of the nth root, $\sqrt[n]{z}$, of a complex number $z = re^{i\theta}$ (with n an integer); writing

$$z = re^{i(\theta + 2\pi m)}, \tag{9.17}$$

with m an integer, we have

$$z^{1/n} = r^{1/n} e^{i\theta/n} e^{i2\pi m/n}. \tag{9.18}$$

For $m = 0, 1, \ldots, n-1$, we get different values for the quantity $z^{1/n}$, and for higher values of m, these values repeat. The n distinct roots of $z^{1/n}$ are thus given by Eq. (9.18). For example, the 5th roots of $z = 1$ in the complex plane are given by $1^{1/5} = e^{i2\pi m/5}$; this is illustrated in Fig. 9.3.

Addition, subtraction, multiplication, and division of complex numbers can all be defined in a straightforward way, keeping in mind that $i^2 = -1$; thus

$$z_1 z_2 = (x_1 + iy_1)(x_2 + iy_2) = (x_1 x_2 - y_1 y_2) + i(x_1 y_2 + x_2 y_1), \tag{9.19}$$

$$z_1 + z_2 = (x_1 + x_2) + i(y_1 + y_2), \tag{9.20}$$

$$\frac{z_1}{z_2} = \frac{x_1 + iy_1}{x_2 + iy_2} = \frac{x_1 + iy_1}{x_2 + iy_2} \frac{(x_2 - iy_2)}{(x_2 - iy_2)} = \frac{x_1 x_2 + y_1 y_2}{x_2^2 + y_2^2} + i\frac{x_2 y_1 - x_1 y_2}{x_2^2 + y_2^2}. \tag{9.21}$$

In the latter relation we employ what is known as the *complex conjugate* of z,

$$z^* = x - iy, \tag{9.22}$$

i.e. the complex conjugate of a complex number is found by replacing i by $-i$ wherever it appears in a formula. For example, $(1 + ie^{ix})^* = (1 - ie^{-ix})$. If we continue the analogy between complex numbers and two-dimensional vectors, it follows that the "dot product" of two complex numbers is given by

$$z_1 \cdot z_2 \equiv \mathrm{Re}(z_1^* z_2) = x_1 x_2 + y_1 y_2. \tag{9.23}$$

The modulus of a number z is then simply given by

$$|z| = \sqrt{z \cdot z} = \sqrt{z^* z} = \sqrt{x^2 + y^2}. \tag{9.24}$$

The real and imaginary parts of a complex number may be written using the complex conjugate,

$$\mathrm{Re}(z) = \frac{z + z^*}{2} = x, \quad \mathrm{Im}(z) = \frac{z - z^*}{2i} = y. \tag{9.25}$$

One elementary and important property of complex numbers is given by the so-called *triangle inequality*,

$$|z_1 + z_2| \le |z_1| + |z_2|. \tag{9.26}$$

If we consider the complex numbers as vectors in the complex plane, this inequality represents the geometric property that no side of a triangle is greater in length than the sum of the other two sides, as illustrated in Fig. 9.4. Referring back to the definition of a metric in a metric space (Section 1.5), we see that the modulus of z satisfies the conditions of a metric, making the set of all complex numbers a metric space as well as a vector space.

The inequality (9.26) can be proven by looking at the quantity

$$|z_1 + z_2|^2 = (z_1 + z_2)(z_1^* + z_2^*) = z_1 z_1^* + z_2 z_2^* + z_1 z_2^* + z_2 z_1^*$$
$$= |z_1|^2 + |z_2|^2 + 2\mathrm{Re}(z_1 z_2^*). \tag{9.27}$$

If we add $2|z_1||z_2|$ to each side of this equation, we may rewrite it as

$$|z_1 + z_2|^2 + 2|z_1||z_2| = |z_1|^2 + |z_2|^2 + 2|z_1||z_2| + 2\mathrm{Re}(z_1 z_2^*)$$
$$= (|z_1| + |z_2|)^2 + 2\mathrm{Re}(z_1 z_2^*). \tag{9.28}$$

Rearranging terms, we have

$$|z_1 + z_2|^2 - (|z_1| + |z_2|)^2 = 2\mathrm{Re}(z_1 z_2^*) - 2|z_1||z_2| = 2\mathrm{Re}(z_1 z_2^*) - 2|z_1 z_2^*|. \tag{9.29}$$

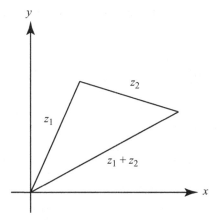

Figure 9.4 The triangle inequality for complex numbers.

The term of the right is always less than or equal to zero, because $\mathrm{Re}(z) = x \leq |z| = \sqrt{x^2 + y^2}$; we thus have

$$|z_1 + z_2|^2 - (|z_1| + |z_2|)^2 \leq 0. \tag{9.30}$$

Moving the second term to the right of the inequality, and taking the positive square root of both sides, we arrive at the triangle inequality.

9.3 Functions of a complex variable

Up to now we have only considered the properties of complex variables by themselves. We now look at defining functions of a complex variable, $f = f(z)$, and generalize various elementary functions such as the sine and cosine functions to situations in which the argument is complex.

The simplest nontrivial function of a complex variable which we can look at is the exponential function, e^z, which is formally defined by replacing the x in e^x by the complex number z. This can be given meaning as follows:

$$e^z = e^{x+iy} = e^x e^{iy} = e^x(\cos y + i \sin y). \tag{9.31}$$

In a similar way, we may define complex versions of $\sin x$ and $\cos x$, noting that

$$\cos x = \frac{e^{ix} + e^{-ix}}{2}, \quad \sin x = \frac{e^{ix} - e^{-ix}}{2i}. \tag{9.32}$$

Since we have already described the generalization of the exponential to complex arguments z, we can generalize the trigonometric functions in a straightforward way,

$$\cos z = \frac{e^{iz} + e^{-iz}}{2}, \quad \sin z = \frac{e^{iz} - e^{-iz}}{2i}. \tag{9.33}$$

Other trigonometric functions such as the tangent, cotangent, and secant may be defined using the above relations.

What about non-trigonometric functions, such as $\log(z)$? This may be evaluated using the polar representation of a complex number and the identity $\log(AB) = \log(A) + \log(B)$,

$$\log z = \log r e^{i(\theta + 2\pi n)} = \log r + \log\left[e^{i(\theta + 2\pi n)}\right] = \log r + i\theta + i2\pi n. \tag{9.34}$$

It is to be noted that we include the $2\pi n$ in the argument of z; this does not change the value of z, but *does* change the value of the logarithm of z. Evidently, since n can be any integer, $\log z$ is a *multivalued function*, having an infinite number of values for any given z. To avoid ambiguity, we will for now restrict the argument to have a value between 0 and 2π, and choose $n = 0$. We will return to multivalued functions in Section 9.10; this ambiguity in the logarithm of a complex number is actually of great significance in optics, in particular in the field of so-called *singular optics*, the study of wavefields near points where the field intensity is zero [SV01].

It is to be noted that in all the examples given, $f(z)$ has a real and imaginary part; the function may always be written in the form

$$f(z) = u(x, y) + iv(x, y), \tag{9.35}$$

where $u(x, y)$ and $v(x, y)$ are real-valued functions. We will find this representation useful when we consider complex derivatives of $f(z)$.

The original proof of Euler's formula [Eul48] consisted of showing that the complex Taylor series representation of the complex exponential function is related to those of the sine and cosine functions. In a similar manner, we will see that we can extend most complicated functions into the complex plane by using their power series representations (even a Taylor series, but we need to define complex derivatives before we can define such an expansion),

$$f(z) = \sum_{n=0}^{\infty} a_n (z - z_0)^n. \tag{9.36}$$

In describing the behavior of various special functions with complex argument, such as the Bessel functions (to be described in Chapter 16), we will employ their power series representations.

The radius of convergence of a series, discussed in the previous chapter, becomes crucial for such definitions. Using the ratio test, the radius of convergence of a power series is

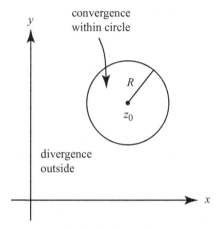

Figure 9.5 Illustrating the radius of convergence of a power series in z.

given by

$$R = \lim_{n\to\infty} \left| \frac{a_n}{a_{n+1}} \right|. \tag{9.37}$$

In the context of complex variables, R is a true radius in complex space: it describes the distance around the center point z_0 within which the series converges. This is illustrated in Fig. 9.5.

Some of the elementary functions already described, e^z, $\sin z$, $\cos z$, have an infinite radius of convergence, and may be represented by their power series expansions for all finite z, e.g.

$$e^z = \sum_{n=0}^{\infty} \frac{z^n}{n!}, \quad \sin z = \sum_{n=0}^{\infty} \frac{(-1)^n z^{2n+1}}{(2n+1)!}, \quad \cos z = \sum_{n=0}^{\infty} \frac{(-1)^n z^{2n}}{(2n)!}. \tag{9.38}$$

Even though a power series may have a finite radius of convergence, it is often possible to determine the behavior of a function beyond this radius. For instance, the power series

$$\sum_{n=0}^{\infty} \binom{-1}{n} z^n = \sum_{n=0}^{\infty} (-z)^n \tag{9.39}$$

has a radius of convergence of $|z| < 1$; however, we know that this series is a representation of the function $(1+z)^{-1}$, which is only undefined for the point $z = -1$. The function $(1+z)^{-1}$ may be considered a "continuation" of the power series given by Eq. (9.39) to more values of z; we will make such statements more rigorous in Section 10.2.

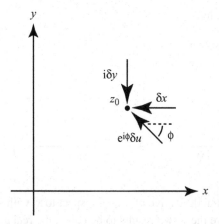

Figure 9.6 Paths used to determine the derivative of a function.

9.4 Complex derivatives and analyticity

Once we can define a function $f(z)$ of a complex number, the next natural question is whether we can perform calculus on such a function, i.e. whether we can perform integration and differentiation.

To answer this question, we first recall the definition of a derivative of a real function,

$$f'(x) \equiv f_x(x) \equiv \lim_{\delta x \to 0} \frac{f(x + \delta x) - f(x)}{\delta x}. \tag{9.40}$$

It is to be noted that $f(x)$ must be continuous for this definition to make sense; if $\lim_{\delta x \to 0} f(x + \delta x) \neq f(x)$, then the limit of Eq. (9.40) does not exist, at least not in the sense of ordinary functions.

We may try to define a derivative with respect to a complex variable z in a similar manner, i.e.

$$f'(z) = \lim_{\delta z \to 0} \frac{f(z + \delta z) - f(z)}{\delta z}, \tag{9.41}$$

but we immediately run into a problem: there are many ways to take the limit $\delta z \to 0$, because $\delta z = \delta x + i \delta y$ is a function of δx and δy. We may let $\delta z = \delta x$, and consider the limit $\delta x \to 0$, or we may let $\delta z = i \delta y$, and consider $\delta y \to 0$; we could also let $\delta z = e^{i\phi} \delta u$, where δu is real-valued and ϕ is the angle with respect to the x-axis, and take the limit $\delta u \to 0$. These three paths of approach are illustrated in Fig. 9.6.

The complex derivative of a function at a point in the complex plane only has meaning, then, *when it is independent of the way in which the limit is taken*. Let us quantify this by considering this limit in two directions, so that $\delta z = \delta x$, and $\delta z = i \delta y$. Writing $f'(z)$ in terms of $u(x,y)$ and $v(x,y)$, we have

$$f'(z) = \lim_{\delta x \to 0} \left(\frac{u(x+\delta x, y) - u(x,y)}{\delta x} + i\frac{v(x+\delta x, y) - v(x,y)}{\delta x} \right) = u_x(x,y) + iv_x(x,y)$$

(9.42)

and

$$f'(z) = \lim_{\delta y \to 0} \left(\frac{u(x,y+\delta y) - u(x,y)}{i\delta y} + i\frac{v(x,y+\delta y) - v(x,y)}{i\delta y} \right) = -iu_y(x,y) + v_y(x,y),$$

(9.43)

where $u_x(x,y)$ is the partial derivative of u with respect to x, with similar meanings for the other terms. If we require these two results to be equal, the real and imaginary parts must match, i.e.

$$u_x = v_y, \quad v_x = -u_y.$$

(9.44)

It turns out that these equations, known as the *Cauchy–Riemann conditions*, are *necessary* conditions for the derivative of $f(z)$ to exist. It is not difficult to show that the C-R conditions, together with the condition that the partial derivatives of u and v are continuous, are *sufficient* for the derivative to exist. We state these results as two distinct theorems.

Theorem 9.1 (The necessary Cauchy–Riemann conditions) *The function $f(z) = u(x,y) + iv(x,y)$ is differentiable at a point $z = x + iy$ of a region in the complex plane only if the partial derivatives u_x, u_y, v_x, v_y exist and satisfy the Cauchy–Riemann conditions, Eqs. (9.44), at that point.*

Theorem 9.2 (The sufficient Cauchy–Riemann conditions) *The function $f(z) = u(x,y) + iv(x,y)$ is differentiable at a point $z = x + iy$ of a region in the complex plane if the partial derivatives u_x, u_y, v_x, v_y are continuous and satisfy the Cauchy–Riemann conditions, Eqs. (9.44), at that point.*

If $f(z)$ is differentiable, the derivative may be taken just as if z were a real variable: $d(e^z)/dz = e^z$, $d(\sin z)/dz = \cos z$, $d(\cos z)/dz = -\sin z$, for instance. This follows immediately from the structural similarities between Eqs. (9.40) and (9.41).

The only difference between the necessary and sufficient Cauchy–Riemann conditions is the additional requirement, in the sufficient conditions, that the partial derivatives of u and v be continuous. This leaves open the possibility that functions exist which are differentiable but whose first partial derivatives are discontinuous. We will see that this is, in fact, not possible, but for now we may only say that a function is differentiable only if it satisfies the Cauchy–Riemann conditions and its partial derivatives exist.

It is important to note that it is possible to find pathological functions which satisfy the Cauchy–Riemann conditions at a point but which are not, in fact, differentiable at that point. The continuity of the derivatives is an important ingredient in the differentiability of a function. One can also find functions which are differentiable at a single point in the complex plane; for instance,

$$f(z) = |z|^2 = x^2 + y^2. \tag{9.45}$$

For this function we have $u(x,y) = x^2 + y^2$ and $v(x,y) = 0$. The derivatives of interest are

$$u_x = 2x, \quad v_x = 0, \quad u_y = 2y, \quad v_y = 0. \tag{9.46}$$

Only at $z = 0$ is the function differentiable. We are generally interested, however, in functions which are differentiable in a finite region of space, for instance in the solution of partial differential equations such as Laplace's equation. We therefore introduce the concept of *analyticity*.

Definition 9.1 (Analyticity) *When a complex function is differentiable at a point z_0 and in a neighborhood of that point, it is said that the function is* analytic *at that point. When a complex function is differentiable throughout a region $\mathcal{R}$ of the complex plane, it is said to be analytic in $\mathcal{R}$. If such a complex function is analytic throughout the entire finite complex plane, it is said to be* entire, *or* entire analytic.

A *neighborhood* of a point z_0 is a circular domain centered on that point, i.e. all points z such that $|z - z_0| < \epsilon$, where ϵ is a small positive number.

We will primarily be interested in complex functions which are analytic everywhere except perhaps at isolated points in the z-plane, which are called *singular points*. As should be clear from the above, analyticity is, in essence, a statement that a function is differentiable not just at a point but throughout some finite area of the complex plane.

Example 9.1 As an example of the Cauchy–Riemann conditions in action, we consider $f(z) = e^z = e^x(\cos y + i \sin y)$; the partial derivatives of this function are given by

$$\begin{aligned} u_x &= e^x \cos y = v_y, \\ u_y &= -e^x \sin y = -v_x. \end{aligned} \tag{9.47}$$

The C-R conditions are satisfied for all finite x, y; the function e^z is differentiable for all finite z, and it is entire analytic. However, if we consider $f(z) = e^{z^*} = e^x(\cos y - i \sin y)$, we have

$$\begin{aligned} u_x &= e^x \cos y = -v_y, \\ u_y &= -e^x \sin y = v_x. \end{aligned} \tag{9.48}$$

The C-R conditions are not satisfied for e^{z^*}; this function is not analytic anywhere in the complex plane.

◇

If the second-order partial derivatives of an analytic function exist and are continuous, we may derive a simple and rather surprising result. If we take the partial derivative of the first C-R equation with respect to x, and the second with respect to y, we have

$$u_{xx} = v_{yx}, \quad v_{xy} = -u_{yy}. \tag{9.49}$$

Because the order of differentiation is unimportant, $v_{yx} = v_{xy}$, and we may substitute from the second equation into the first,

$$u_{xx} + u_{yy} = \frac{\partial^2 u}{\partial x^2} + \frac{\partial^2 u}{\partial y^2} = 0. \tag{9.50}$$

The function $u(x,y)$ is a solution of Laplace's equation! Similarly, we can find that $v(x,y)$ also satisfies Laplace's equation,

$$\frac{\partial^2 v}{\partial x^2} + \frac{\partial^2 v}{\partial y^2} = 0. \tag{9.51}$$

A function which satisfies Laplace's equation in a domain D is often called a *harmonic function* in D. The Cauchy–Riemann conditions indicate that there is a strong relationship between the real and imaginary parts of an analytic function, which are referred to as *harmonic conjugates* of each other. This relationship is so strong, in fact, that u can be almost completely determined from v, and vice versa, as the next example shows.

Example 9.2 Suppose that an analytic function has a real part $u(x,y) = x^2 - y^2 + 4x + 2$. Find an expression for its harmonic conjugate.

From the C-R conditions, we know that $u_x = v_y$ and $u_y = -v_x$; on substitution from our expression for u into these differential equations, we get a pair of first-order partial differential equations for v,

$$v_y = 2x + 4, \tag{9.52}$$

$$v_x = 2y. \tag{9.53}$$

Each of these equations can be directly integrated, which results in two equations for v,

$$v = 2xy + 4y + c(x), \tag{9.54}$$

$$v = 2xy + d(y), \tag{9.55}$$

where $c(x)$ and $d(y)$ are functions only of x and y, respectively. Comparing the two equations, it is clear that they can be made consistent with the choice $c(x) = c_0$, a constant independent of x and y, and $d(y) = 4y + c_0$. Our expression for $v(x,y)$ is therefore

$$v(x,y) = 2xy + 4y + c_0. \tag{9.56}$$

If we put together $u + iv$, we can readily find an expression for the function $f(z)$,

$$u(x,y) + iv(x,y) = x^2 + 2ixy - y^2 + 4(x + iy) + 2 + c_0 = z^2 + 4z + 2 + c_0. \qquad (9.57)$$

It is to be noted that the C-R equations can only determine a harmonic conjugate to within a constant value.

◇

Put together, Eqs. (9.50) and (9.51) show that $f(z)$ itself is a solution of Laplace's equation. We have already seen that we can use complex analysis methods to solve physical problems involving Laplace's equation in two dimensions quite elegantly. Furthermore, since $f(z)$ satisfies Laplace's equation, it follows that $f(z)$ represents a "potential" of sorts for a conservative force in two dimensions. Many of the results that we derived for conservative forces will also apply to analytic functions, including the path independence of complex integrals and, more significantly, the "averaging" property of a conservative potential. In order to demonstrate these results, however, we first need to define integrals in the complex plane.

9.5 Complex integration and Cauchy's integral theorem

We wish to ascribe meaning to an integral of the form

$$S \equiv \int_{z_a}^{z_b} f(z)\mathrm{d}z. \qquad (9.58)$$

Points z_a and z_b represent points in the complex plane, which by analogy to vector calculus represent points in two-dimensional space. The quantity $\mathrm{d}z$, by analogy, represents an infinitesmal path element, and the integral S represents an integral along a generally curved path in the complex plane. Such an integral is referred to (as in ordinary vector calculus) as a *path integral* or *contour integral*, and it can be evaluated by dividing the contour into a large number of small line segments, each of which is approximately straight, illustrated in Fig. 9.7.

This may be represented by a partial sum,

$$S_n = \sum_{j=0}^{n} f(\zeta_j)(z_{j+1} - z_j), \qquad (9.59)$$

where ζ_j is a point on the segment between z_j and z_{j+1}. If we consider the limit of this sequence of partial sums, $n \to \infty$, which implies that $|z_{j+1} - z_j| \to 0$, we have

$$S \equiv \lim_{n \to \infty} S_n = \int_{z_a}^{z_b} f(z)\mathrm{d}z. \qquad (9.60)$$

Complex analysis

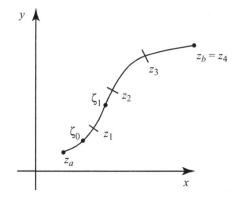

Figure 9.7 Technique for defining a path integral.

This definition is a bit mysterious; we can, however, reduce it to something more familiar. In terms of u, v, and dx and dy, Eq. (9.58) may be written as

$$\int_{z_a}^{z_b} f(z)dz = \int_{x_a,y_a}^{x_b,y_b} [u(x,y)+iv(x,y)][dx+idy]$$

$$= \int_{x_a,y_a}^{x_b,y_b} [u(x,y)dx - v(x,y)dy] + i\int_{x_a,y_a}^{x_b,y_b} [v(x,y)dx + u(x,y)dy]. \quad (9.61)$$

What general statements can we make about such a complex integration? An observant reader will note that both integrals in Eq. (9.61) may be written as *real-valued* contour integrals. If we define a pair of vectors

$$\mathbf{v}_1 = u(x,y)\hat{\mathbf{x}} - v(x,y)\hat{\mathbf{y}}, \quad (9.62)$$

$$\mathbf{v}_2 = v(x,y)\hat{\mathbf{x}} + u(x,y)\hat{\mathbf{y}}, \quad (9.63)$$

and a path element $\mathbf{dr} = dx\hat{\mathbf{x}} + dy\hat{\mathbf{y}}$, the complex integral may be written as

$$\int_{z_a}^{z_b} f(z)dz = \int_{x_a,y_a}^{x_b,y_b} \mathbf{v}_1 \cdot \mathbf{dr} + i\int_{x_a,y_a}^{x_b,y_b} \mathbf{v}_2 \cdot \mathbf{dr}, \quad (9.64)$$

where the two vector integrals may be evaluated as in Section 2.2. Let us consider a contour integral about a closed path in the complex plane, i.e. $z_a = z_b$. Then the complex integral becomes,

$$\oint f(z)dz = \oint \mathbf{v}_1 \cdot \mathbf{dr} + i\oint \mathbf{v}_2 \cdot \mathbf{dr}. \quad (9.65)$$

Let us for the moment assume that the first partial derivatives of the functions u and v exist and are continuous. The integrals may then be evaluated by the use of Stokes' theorem, discussed in Section 2.8,

$$\oint \mathbf{v} \cdot d\mathbf{r} = \int_S \nabla \times \mathbf{v} \cdot d\mathbf{a}, \tag{9.66}$$

where S is the area of the surface enclosed by the curve. Applying Stokes' theorem to each of the integrals of Eq. (9.65), we have

$$\oint f(z)dz = -\int_S \left(\frac{\partial v}{\partial x} + \frac{\partial u}{\partial y} \right) dxdy + i \int_S \left(\frac{\partial u}{\partial x} - \frac{\partial v}{\partial y} \right) dxdy. \tag{9.67}$$

If the function is analytic within and on the contour, the C-R equations are satisfied: $u_x = v_y$, $v_x = -u_y$; substituting from these into Eq. (9.67), we immediately find

$$\oint f(z)dz = 0. \tag{9.68}$$

Therefore the integral of an analytic function over a closed path vanishes! This result is known as *Cauchy's theorem*. From it, we can readily deduce (as we did for conservative forces) that contour integrals of analytic functions are independent of the path.

There is an important mathematical catch to this, however. We have *assumed* that the partial derivatives of u and v are continuous, but this is typically *proven* with Cauchy's theorem. This leaves open the possibility that there may exist analytic functions whose first partial derivatives are not continuous, and for which Cauchy's theorem does not hold. The loophole can in fact be closed using a more general proof known as the Cauchy–Goursat theorem, which does not require the assumption of continuous partial derivatives. Details can be found in [Det65, Chapter 3] and in [Cop35, Chapter 4].

Formally, Cauchy's theorem may be written as follows.

Theorem 9.3 (Cauchy's theorem) *If $f(z)$ is analytic in a simply connected domain D, then along a simple closed contour C in D,*

$$\oint f(z)dz = 0. \tag{9.69}$$

Two terms in this formal description need to be explained. The first is what we mean by a "simple" closed contour: by this we mean a contour which does not cross itself. This condition is illustrated in Fig. 9.8. The requirement of simple contours, though not essential for Cauchy's theorem, is significant for later, related, theorems.

The second term we need to explain is what we mean by a "simply connected domain". A simply connected domain is a domain without any nonanalytic "holes" in it, as illustrated in Fig. 9.9.

We will need to learn to cope with multiply connected domains because we want to consider integrals of functions which are *not* analytic at isolated points: the domain of analyticity does not include the isolated points and is therefore multiply connected.

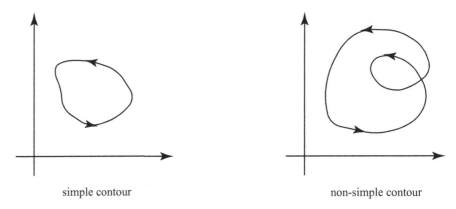

simple contour   non-simple contour

Figure 9.8 The meaning of a simple closed contour.

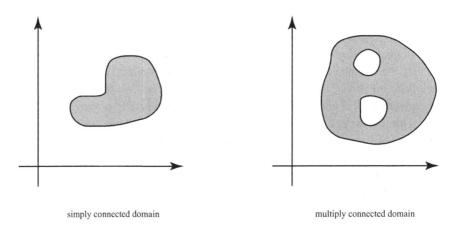

simply connected domain   multiply connected domain

Figure 9.9 Simply and multiply connected domains.

Can we apply Cauchy's theorem to a multiply connected domain? Yes, if we choose a contour as shown in Fig. 9.10.

The function is analytic in the area enclosed by this contour, so Cauchy's theorem applies,

$$\int_{C_+} f(z)\mathrm{d}z + \int_{C_-} f(z)\mathrm{d}z + \int_{L_+} f(z)\mathrm{d}z + \int_{L_-} f(z)\mathrm{d}z = 0. \qquad (9.70)$$

In the limit that L_+ and L_- move arbitrarily close to each other, their contributions cancel out due to the continuity of $f(z)$ and we have

$$\int_{C_-} f(z)\mathrm{d}z = -\int_{C_+} f(z)\mathrm{d}z. \qquad (9.71)$$

Thus the integrals about the two contours which enclose the "hole" differ only in sign. If we define a new contour C'_- which is simply in the opposite direction of C_-, Cauchy's

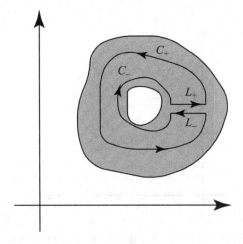

Figure 9.10 The application of Cauchy's theorem to a multiply connected domain.

theorem becomes

$$\int_{C'_-} f(z)\mathrm{d}z = \int_{C_+} f(z)\mathrm{d}z. \tag{9.72}$$

In other words, two path integrals of an analytic function in the same direction about a "hole" in a multiply connected domain have the same value; this value is *not* in general equal to zero. We will, unless otherwise specified, take all closed path integrals in the counter-clockwise direction from now on.

It is to be noted that the method we have used to prove Cauchy's theorem for a multiply connected domain is similar to that used to prove Gauss' law, in Section 6.1.

9.6 Cauchy's integral formula

In the previous section, we used Cauchy's theorem to show that two path integrals about a "hole" in an analytic function have the same value, if taken in the same direction. Now we determine the value of such an integral.

Let us consider a path integral which encloses the point $z = z_0$, and the function to be integrated is

$$g(z) = \frac{f(z)}{z - z_0}, \tag{9.73}$$

where $f(z)$ is analytic on and within the contour of integration C. Because of the singularity of the function $g(z)$ at $z = z_0$, the function is non-analytic at that point. The domain therefore has a "hole" at $z = z_0$. We have seen in the previous section that the integral over any closed path about the "hole" gives the same result; we therefore consider a path C_0 which is a circle of radius r centered on the point z_0, as illustrated in Fig. 9.11.

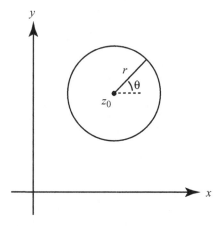

Figure 9.11 Integration in the immediate vicinity of a "hole" in an analytic function.

We may use polar coordinates to evaluate this integral, letting $z - z_0 = re^{i\theta}$ and $dz = ire^{i\theta} d\theta$; the integral is thus

$$\int_{C_0} \frac{f(z)}{z - z_0} dz = \int_{C_0} \frac{f(z_0 + re^{i\theta})}{re^{i\theta}} ire^{i\theta} d\theta = i \int_{C_0} f(z_0 + re^{i\theta}) d\theta. \qquad (9.74)$$

We now let the radius of this circle go to zero, $r \to 0$, so that

$$\int_{C_0} \frac{f(z)}{z - z_0} dz = if(z_0) \int_{C_0} d\theta = 2\pi i f(z_0). \qquad (9.75)$$

We have found a remarkable result, which is usually referred to as Cauchy's integral formula.

Theorem 9.4 (Cauchy's integral formula) *Let $f(z)$ be analytic inside and on a simple closed contour C. Then at any point z inside the contour,*

$$f(z) = \frac{1}{2\pi i} \oint_C \frac{f(z')}{z' - z} dz'. \qquad (9.76)$$

We may extend this formula to arbitrary powers of $1/(z' - z)$ in a non-rigorous way by taking the nth derivative of each side of Eq. (9.76). By doing so, we find that

$$f^{(n)}(z) = \frac{n!}{2\pi i} \oint_C \frac{f(z')}{(z' - z)^{n+1}} dz'. \qquad (9.77)$$

We may combine this formula and Cauchy's theorem to write an expression of some generality,

$$\frac{1}{2\pi i}\oint_C \frac{f(z')}{(z'-z)^{n+1}}dz' = \begin{cases} \frac{f^{(n)}(z)}{n!} & \text{for } z \text{ inside } C, \\ 0 & \text{for } z \text{ not inside } C. \end{cases} \tag{9.78}$$

This formula gives us the ability to evaluate a large number of complex integrals, in particular integrals of rational functions of polynomials, where

$$h(z) = \frac{P(z)}{Q(z)}, \tag{9.79}$$

where $P(z)$ and $Q(z)$ are polynomials of order p and q, respectively. By factorizing the denominator and reducing the fraction, it can be reduced into a number of terms which can be evaluated by the Cauchy theorem and Cauchy formula. We will see a more systematic method of evaluating such integrals in Section 9.11.

Example 9.3 Determine the value of the integral,

$$I = \oint \frac{dz}{(z-1)(z-3)}, \tag{9.80}$$

over the counterclockwise circular path defined by $|z| = 2$.

The interior of the path is $|z| < 2$; the integrand contains two singularities at $z = 1$ and $z = 3$, but only $z = 1$ lies within the contour of integration. We may write our integrand as $f(z)/(z-1)$, where $f(z) = 1/(z-3)$ is analytic on and within the contour. By Cauchy's integral formula, this integral is equal to $2\pi i f(1)$, or $I = -i\pi$.

◇

An important consequence of Cauchy's formula is that it can be used to demonstrate that an analytic function is infinitely differentiable! We can roughly see this from Eq. (9.77): the numerator of the integrand, $f(z')$, is continuous and bounded at all points on the boundary, and the denominator $1/(z'-z)^{n+1}$ is also continuous and bounded, as z is assumed to not lie on the boundary. The expressions for the derivatives are all finite, and this suggests continuity of all the derivatives. We therefore have the very unusual situation: by assuming the analyticity of $f(z)$, we have been led to the observation that all derivatives of $f(z)$ exist!

9.7 Taylor series

With the Cauchy formula of the previous section, we can now prove that an analytic function has a Taylor series expansion. We start with Eq. (9.76),

$$f(z) = \frac{1}{2\pi i}\oint_C \frac{f(z')}{z'-z}dz', \tag{9.81}$$

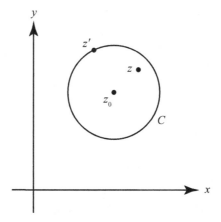

Figure 9.12 The scheme used to derive Taylor's theorem in the complex plane.

where the point z lies within the contour C, which is taken to be a circle. Let us suppose we are interested in a Taylor expansion about the point z_0, which is at the center of C; this is illustrated in Fig. 9.12.

It is to be noted that $|z' - z_0| > |z - z_0|$, since z' is on the contour and z lies within it. We rewrite the denominator of Cauchy's formula as

$$\frac{1}{z' - z} = \frac{1}{(z' - z_0) - (z - z_0)} = \frac{1}{z' - z_0} \frac{1}{1 - a}. \tag{9.82}$$

where $a = (z - z_0)/(z' - z_0)$ and $|a| < 1$. The latter term is the limit of the infinite geometric series, defined in Example 7.1. We may write

$$\frac{1}{1 - a} = \sum_{n=0}^{\infty} a^n = \sum_{n=0}^{\infty} \left(\frac{z - z_0}{z' - z_0}\right)^n. \tag{9.83}$$

On substituting from this formula into Eq. (9.81), we find that

$$f(z) = \frac{1}{2\pi i} \sum_{n=0}^{\infty} (z - z_0)^n \oint \frac{f(z')}{(z' - z_0)^{n+1}} dz. \tag{9.84}$$

The integral is essentially Cauchy's integral formula, Eq. (9.77). On substituting from that equation into Eq. (9.84), we have

$$f(z) = \sum_{n=0}^{\infty} \frac{f^{(n)}(z_0)}{n!} (z - z_0)^n, \tag{9.85}$$

which is the Taylor series of $f(z)$! Just as for Taylor series of real variables, we can use the ratio test to establish a radius of convergence (though now that we are working in two

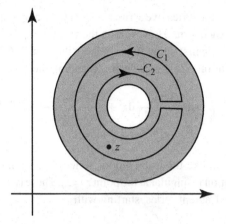

Figure 9.13 The contour used to derive the Laurent expansion of a function.

variables, x and y, we have a genuine radius). Within the radius of convergence, the series is uniformly and absolutely convergent, and can be integrated and differentiated term by term.

The formulation of the Taylor series in the complex plane actually answers a question raised in Chapter 7: what determines the radius of convergence of a Taylor series? It can be shown that the Taylor series expansion of a function has a radius of convergence that is no larger than the radial distance to the nearest singularity. The radius of convergence of a real-valued Taylor series expansion is restricted by the "hidden" singularities in the complex plane. This will be elaborated on in Section 10.2 on analytic continuation.

The derivation of the Taylor series follows from Cauchy's integral formula, which follows directly from the assumption of analyticity. The Taylor series, however, is infinitely differentiable. Again we see that, if a function is analytic, it is infinitely differentiable.

9.8 Laurent series

It turns out that there is a more general expansion than a Taylor series, which can accommodate expansions about singular points. Such an expansion is valid within an annular region about the singularity or singularities, and is called a *Laurent expansion*.

To derive this expansion, we assume that we have a function $f(z)$ which is analytic in an annular region, and consider Cauchy's formula for the contour shown in Fig. 9.13.

It is to be noted that C_2 is defined as the *contourclockwise* path around the circle. We take the limit that the segments which join the two circles come arbitrarily close together, so that their contributions cancel. Cauchy's formula then becomes:

$$f(z) = \frac{1}{2\pi i} \int_{C_1} \frac{f(z')}{z' - z} dz' - \frac{1}{2\pi i} \int_{C_2} \frac{f(z')}{z' - z} dz'. \qquad (9.86)$$

Just as we did for the Taylor series, we write $z' - z = (z' - z_0) - (z - z_0)$ and consider the geometric expansion of each term. However, on the curve C_1, $|z' - z_0| > |z - z_0|$, while for C_2, $|z' - z_0| < |z - z_0|$. In the first integral, we pull out the term $1/(z' - z_0)$ and in the second integral we pull out the term $1/(z - z_0)$. Then, expanding the geometric term, we have

$$f(z) = \frac{1}{2\pi i} \sum_{n=0}^{\infty} (z - z_0)^n \oint_{C_1} \frac{f(z')}{(z' - z_0)^{n+1}} dz' + \frac{1}{2\pi i} \sum_{n=0}^{\infty} (z - z_0)^{-n-1} \oint_{C_2} (z' - z_0)^n f(z') dz'.$$

(9.87)

The first summation looks like the usual Taylor series expansion of an analytic function, while the second summation is singular at the point $z = z_0$. These two series may be combined into one, referred to as a Laurent series, starting with

$$f(z) = \sum_{n=-\infty}^{\infty} a_n (z - z_0)^n,$$

(9.88)

where

$$a_n = \frac{1}{2\pi i} \oint_C \frac{f(z')}{(z' - z_0)^{n+1}} dz',$$

(9.89)

and C is now any counterclockwise contour within the annulus. This expansion is valid within an annular region whose inner radius is specified by the location of the interior singularities, and whose outer radius is specified by the location of the closest exterior singularity. It is important to note that a_n is *not* given by $f^{(n)}(z_0)/n!$, unless the function is analytic within the contour of integration.

Example 9.4 Determine the series expansion (Taylor or Laurent) of the function

$$f(z) = \frac{z^2 - 2z + 5}{(z - 2)(z^2 + 1)}$$

about the point $z = 2$ and in the annulus $1 < |z| < 2$.

This problem can be simplified immensely if we note that

$$z^2 - 2z + 5 = z^2 + 1 - 2(z - 2).$$

We may then write that

$$f(z) = \frac{1}{z - 2} - \frac{2}{z^2 + 1}.$$

Now the annular region, which is centered on $|z| = 0$, is easier to handle, so we do this expansion first. We may write each term in a geometric series, starting with

$$\frac{1}{z - 2} = -\frac{1}{2} \frac{1}{1 - z/2} = -\frac{1}{2} \sum_{n=0}^{\infty} \left(\frac{z}{2}\right)^n.$$

The other denominator is trickier, because we are interested in the *annular* region $1 < |z| < 2$. We therefore have $z > 1$, and need to take this into account,

$$\frac{1}{1+z^2} = \frac{1}{z^2}\frac{1}{1-(-1/z^2)} = \frac{1}{z^2}\sum_{n=0}^{\infty}\left(\frac{-1}{z^2}\right)^n = \frac{1}{z^2}\sum_{n=0}^{\infty}(-1)^n z^{-2n}.$$

Putting everything together, we have

$$f(z) = -\frac{1}{2}\sum_{n=0}^{\infty}\left(\frac{z}{2}\right)^n - \frac{2}{z^2}\sum_{n=0}^{\infty}(-1)^n z^{-2n}. \qquad (9.90)$$

We may now attempt the same thing for expansion about the point $z = 2$; this means we need to expand everything in powers of $z - 2$. We use a partial fraction expansion on the second term, so that

$$\frac{1}{1+z^2} = \frac{-1}{2i}\frac{1}{z+i} + \frac{1}{2i}\frac{1}{z-i}.$$

The two partial fractions may be written in the form

$$\frac{1}{z+i} = \frac{1}{(z-2)+2+i} = \frac{1}{2+i}\frac{1}{1+\frac{z-2}{2+i}} = \frac{1}{2+i}\sum_{n=0}^{\infty}(-1)^n\left(\frac{z-2}{2+i}\right)^n,$$

$$\frac{1}{z-i} = \frac{1}{(z-2)+2-i} = \frac{1}{2-i}\frac{1}{1+\frac{z-2}{2-i}} = \frac{1}{2-i}\sum_{n=0}^{\infty}(-1)^n\left(\frac{z-2}{2-i}\right)^n.$$

With everything in powers of $z - 2$, we only have to put it back together! We get

$$f(z) = \frac{1}{z-2} - 2\left[-\frac{1}{2i}\sum_{n=0}^{\infty}(-1)^n\frac{(z-2)^n}{(2+i)^{n+1}} + \frac{1}{2i}\sum_{n=0}^{\infty}(-1)^n\frac{(z-2)^n}{(2-i)^{n+1}}\right]. \qquad (9.91)$$

Pretty messy! It is to be noted that Eqs. (9.90) and (9.91) are completely different – the coefficients of the Laurent series depend on the central point of the expansion.

◇

The Laurent series is of crucial importance in understanding what follows, when we classify the singularities of analytic functions.

9.9 Classification of isolated singularities

We have seen that a function which is analytic in an annular region about a point $z = z_0$ may be expanded in a Laurent series, given by

$$f(z) = \sum_{n=-\infty}^{\infty} a_n(z - z_0)^n. \tag{9.92}$$

The point $z = z_0$ is referred to as an *isolated singular point* if $f(z)$ is not analytic at that point but is analytic at neighboring points. We classify the singular points of an analytic function by considering the Laurent expansion about that point.

1. **Removable singularity** Some points of an analytic function appear to be singular at first glance but are in fact perfectly regular. We call such a singularity *removable* if the coefficients $a_n = 0$ for all $n < 0$.

   Example 9.5 An example of this is the function

   $$f(z) = \frac{\sin z}{z}, \tag{9.93}$$

   which superficially appears to have a singularity at the point $z = 0$. However, if we consider the Taylor expansion of $\sin z$ about $z = 0$, we may write $f(z)$ in the form

   $$f(z) = \sum_{n=0}^{\infty} (-1)^n \frac{z^{2n}}{(2n+1)!}. \tag{9.94}$$

   This series has no terms of the form $1/z^n$ with $n > 0$, so the function is analytic at that point and the singularity is removable.

   ◇

   One can think of removable singularities as an artifact of the notation we use to write functions.

2. **Simple pole** If $a_{-1} \neq 0$ and $a_n = 0$ for all $n < -1$, the singularity is said to be a *simple pole*. The Laurent series has the form

   $$\frac{a_{-1}}{z - z_0} + \sum_{n=0}^{\infty} a_n(z - z_0)^n. \tag{9.95}$$

   We will most often encounter simple poles in our dealings with analytic functions.

   Example 9.6 An example of a function which has simple poles at $z = 0$ and $z = 1$ is $f(z) = 1/[z(z-1)]$; the expansion about $z = 0$ is given by

   $$f(z) = \frac{1}{z(z-1)} = \frac{-1}{z} + \frac{1}{z-1} = \frac{-1}{z} - \sum_{n=0}^{\infty} z^n. \tag{9.96}$$

The expansion about the point $z = 1$ is found similarly,

$$f(z) = \frac{-1}{z} + \frac{1}{z-1} = \frac{1}{z-1} + \frac{-1}{1+(z-1)} = \frac{1}{z-1} - \sum_{n=0}^{\infty}(-1)^n(z-1)^n. \qquad (9.97)$$

It is to be emphasized again that the coefficients of the Laurent series depend on which point one expands about.

◇

3. **Higher-order poles** If $a_n = 0$ for $n < -m$, with $m > 0$, the point z_0 is a *pole* of order m. A simple pole is a pole of order 1.
4. **Essential singularity** If the $a_n \neq 0$ even as $n \to -\infty$, then the singularity is said to be an *essential singularity*.

Example 9.7 An example of a function with an essential singularity is

$$f(z) = e^{1/z}. \qquad (9.98)$$

We know that the exponential function $\exp[w]$ can be expanded in a Taylor series representation valid for all finite w, so we may write $w = 1/z$ and

$$e^{1/z} = \sum_{n=0}^{\infty} \frac{1}{n!} \frac{1}{z^n}, \qquad (9.99)$$

which has an essential singularity at $z = 0$. The function is illustrated in Fig. 9.14. We will say more about this figure momentarily.

◇

In the neighborhood of an essential singularity, the function $f(z)$ is extremely poorly behaved. Two theorems, one due to Weierstrass [Wei76] and one due to Picard [Pic79], characterize this misbehavior.

Theorem 9.5 (Weierstrass' theorem) *In every neighborhood of an isolated essential singularity, there exists a point at which the function differs by as little as we wish from any previously assigned number.*

Theorem 9.6 (Picard's theorem) *In every neighborhood of an isolated essential singularity, there exists a point at which the function attains any given value with at most one exception.*

Both of these theorems suggest that, right around the point of an isolated essential singularity, the function $f(z)$ takes on *almost every complex value*. Referring to Fig. 9.14 can give us at least a rough intuition about how this occurs. It can be seen in the plot that the real and imaginary parts of the function have "tails" of extremely large and small values which spiral in towards the singular point. These tails are not perfectly overlapping, so one can

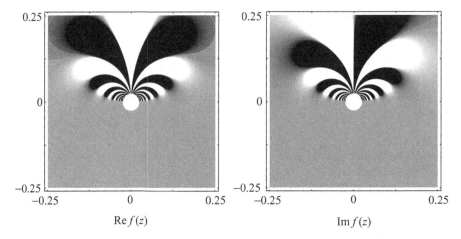

Figure 9.14 The behavior of the function $f(z) = e^{1/z}$ in the neighborhood of its essential singularity. Lighter colors indicate larger values. A region about the central singularity (white circle) has been excluded.

find regions where the real part is arbitrarily large, and the imaginary part small, and vice versa, as well as any other combination of real and imaginary parts.

A word on the figure: because the function behaves so poorly in the neighborhood of an essential singularity, one cannot in general plot its behavior in a meaningful way. One can get a feel for the behavior, however, by excluding the immediate neighborhood of the singularity, where the behavior is the worst.

One might expect that such pathological functions are not particularly common in physics, but this isn't exactly the case: many well-behaved functions have an essential singularity at the point at infinity. For instance, the behavior of the function $f(z) = \exp[-z]$ may be studied in the limit as $z \to \infty$ by using the coordinate transformation $w = 1/z$; we may then study the behavior of the function $f(w) = \exp[-1/w]$ as $w \to 0$, with the point $w = 0$ clearly an essential singularity. The field of asymptotics, to be discussed in Chapter 21, is often equivalent to the study of essential singularities. Further discussion of this point can be found in [BH75].

9.10 Branch points and Riemann surfaces

One class of singularity cannot be classified using an ordinary Laurent series. Consider the function

$$f(z) = z^{1/2}. \tag{9.100}$$

We already know that when dealing with the real function $f(x) = \sqrt{x}$, we must choose between two possible roots, $+\sqrt{x}$ and $-\sqrt{x}$. Likewise, we may find the roots of a complex

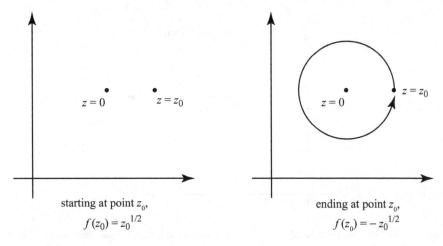

starting at point z_0,

$f(z_0) = z_0^{1/2}$

ending at point z_0,

$f(z_0) = -z_0^{1/2}$

Figure 9.15 The multivalued nature of the function $f(z) = z^{1/2}$ about the branch point at $z = 0$.

number z by writing $z = ze^{2\pi ni}$,

$$z^{1/2} = z^{1/2}e^{\pi ni} = \begin{cases} +z^{1/2} \\ \text{or} \\ -z^{1/2}. \end{cases} \tag{9.101}$$

There are two possible choices for the square root of z, *except* at the point $z = 0$, where the only possible value is 0. If we consider following a seemingly closed circular path around the point $z = 0$, starting at the point $z = z_0$, we find that we end at a different value than we started,

$$f(z) \to z_0^{1/2}e^{\pi i} = -z_0^{1/2}. \tag{9.102}$$

This process is illustrated in Fig. 9.15. Because of this ambiguity, we cannot develop a Taylor or Laurent series about the point $z = 0$. Such a point is called a *branch point*, presumably because the multiple values of the function "branch out" from the singularity. The function $z^{1/2}$ has two "branches", the positive and negative branches. We have already seen another multivalued function with an infinite number of branches, the logarithm,

$$\log z = \log r + i\theta + 2\pi in, \tag{9.103}$$

which has an infinite number of possible values, characterized by integer n.

How do we cope with this ambiguous behavior? We may simply *define* our multivalued function by one of its branches, e.g. we may always choose the positive branch of $f(z) = z^{1/2}$, but we will always have one line in the complex plane along which the function $f(z)$ is discontinuous; one possibility is illustrated in Fig. 9.16.

Such a line is usually referred to as a *cut line*. Its location depends on how the phase of a complex number is defined: if $0 \le \arg(z) < 2\pi$, the cut line runs from the origin to infinity

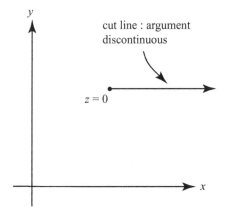

Figure 9.16 The cut line introduced by defining a multivalued function by a branch defined by $0 < \arg(z) < 2\pi$.

along the positive x-axis; if $-\pi < \arg(z) \le \pi$, the cut line runs from the origin to infinity along the negative x-axis. Other choices for the range of arguments will produce cut lines that stretch towards infinity along other directions. The function $f(z)$ is non-analytic along this cut line, which means we cannot use Cauchy's theorem across the line. This could be a problem in a number of situations.

Another way to deal with multivaluedness, due to the mathematician Riemann, is to consider both branches together. As we have noted, our choice of the location of the cut line is arbitrary for the function $f(z) = z^{1/2}$, which suggests that it is an artifact which arises from looking at the problem of multivaluedness too narrowly. Riemann suggested that the proper domain of the function $f(z)$ is a pair of complex planes which are joined together along the cut line, as opposed to a single complex plane. This geometry is illustrated in Fig. 9.17.

One can imagine constructing such a surface by cutting each of the two complex planes along their cut lines, and then connecting the cut edges of one plane to the opposing edges of the other. Conceptually, this results in a function $f(z)$ which is analytic everywhere except the point $z = 0$, although the word "everywhere" now applies to a pair of complex planes. Collectively, this linked set of planes is referred to as a *Riemann surface*, and each individual complex plane is called a *Riemann sheet*; the theory of Riemann surfaces is elegant and beautiful, and numerous books have been written on the topic.[2] It is to be noted that the Riemann surfaces cannot be constructed in three-dimensional space without the surfaces crossing.

If we consider the function $f(z) = z^{1/3}$, we find that the Riemann surface consists of three complex planes joined together along a cut line. The function $\log(z)$, which we have seen

---

[2] See, for instance, [Spr81].

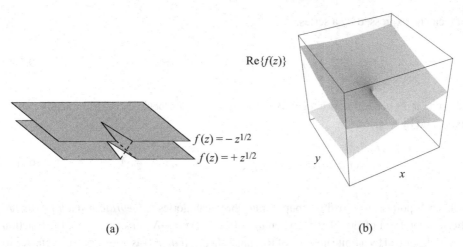

Figure 9.17 (a) A schematic of the Riemann surfaces of the multivalued function $f(z) = z^{1/2}$. (b) The real part of $\sqrt{z}$, showing how the two surfaces are tied together.

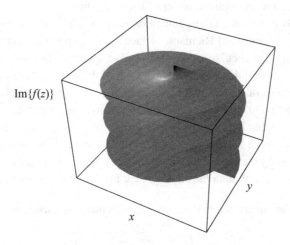

Figure 9.18 (a) The imaginary part of $f(z) = \log(z)$, showing the multiple sheets of the Riemann surface.

may be written as

$$\log(z) = \log(r) + i\theta + 2\pi in, \tag{9.104}$$

where n is an integer, consists of an infinite number of complex planes joined together. This is illustrated in Fig. 9.18.

An ordinary Laurent series cannot be used to describe a branch point; however, one can construct a more general Laurent series which can be used to characterize the behavior of the singularity. For instance, a branch point which has behavior related to a fractional power of z can be expressed by a series,

$$f(z) = (z - z_0)^\alpha \sum_{n=-\infty}^{\infty} a_n (z - z_0)^n, \qquad (9.105)$$

where α is in general non-integer. A branch point which behaves logarithmically can be expressed by a series,

$$f(z) = (z - z_0)^\alpha \log(z - z_0) \sum_{n=-\infty}^{\infty} a_n (z - z_0)^n. \qquad (9.106)$$

Branch points are broadly grouped into three categories. *Algebraic branch points* are those of the form of Eq. (9.105), when $a_n = 0$ for all $n < -N$. In such a case, the function can be described by a finite number of Riemann sheets. If $a_n \neq 0$ as $n \to -\infty$, it is referred to as a *transcendental branch point*, and it is the multivalued analogue of an isolated essential singularity. *Logarithmic branch points* are those which have behavior such as in Eq. (9.106), or more generally have an infinite number of Riemann sheets.

One may readily construct analytic functions with a large number of branch points, and a correspondingly complicated Riemann surface.[3] It is important to note, however, that a branch point with a particular order must always appear with what one might call its "topological counterpart"; that is, if a branch point of the form $z^{1/n}$ appears in a function, there must also be a second branch point of the form $z^{-1/n}$ somewhere else in the complex plane. This second point, however, may be at infinity. For instance, $f(z) = z^{1/2}$ has a branch point at $z = 0$, but by making the transformation $w = 1/z$, we see that the function $f(w) = w^{-1/2}$ also has a branch point at $w = 0$, or $z = \infty$. If we attempt to describe a Riemann surface using the idea of cut lines, the cut lines are required to connect between topological counterparts, though the choice of cut lines may not be unique.

Example 9.8 We are interested in describing the Riemann surface of the function

$$f(z) = \sqrt{z(z - 1)}. \qquad (9.107)$$

The first step in this effort is to determine the location of the branch points. It seems clear from inspection that there are branch points at $z = 0$ and $z = 1$; this can be confirmed by the following construction, illustrated in Fig. 9.19. Let us consider the behavior of the function $f(z)$ as we follow a small counterclockwise circular path around the point $z = 0$ of radius r_0 and polar angle θ_0, such that $z = r_0 \exp[i\theta_0]$. At the same time, the quantity $z - 1$ may also be written in polar form $z - 1 = r_1 \exp[i\theta_1]$, though it is to be noted that r_1 is not a constant. As we increase θ_0 continuously from 0 to 2π, the argument of z increases by 2π, $z \to z e^{2\pi i}$. The argument θ_1, however, decreases from $\theta_1 = \pi$ to a smaller value, and then

---

[3] I like to describe multilevel parking garages as an example of a Riemann-like surface: by driving in circles around one or more central points of the structure, one can change levels.

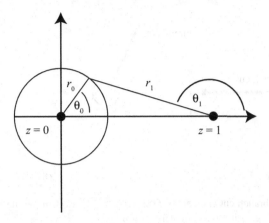

Figure 9.19 Construction used to evaluate the behavior of a branch point.

returns to $\theta_1 = \pi$ at the end of the path, so that $z - 1 \to z - 1$. The overall change in $f(z)$ as we make a complete circle around $z = 0$ is $f(z) \to f(z)e^{\pi i}$; that is, the argument of the function increases by π. If we make a counterclockwise circle around $z = 1$, we find that the argument also increases by π by similar reasoning. We may therefore state that $z = 0$ and $z = 1$ are branch points of the form $z^{1/2}$.

There are two overall possible values of the function, and two Riemann sheets, since two orbits around either branch point returns us to the original value of the function. The manner in which the sheets are connected is determined by the location of the branch cuts; a path which crosses a branch cut results in a transition to the next Riemann sheet. The choice of possible branch cuts is illustrated in Fig. 9.20. We may interpret the branch cut as passing directly between the two branch points, or we may imagine two branch cuts which join each of the branch points with its complement at infinity. These two cases correspond to the Riemann sheets being joined between the two branch points, or joined everywhere except between the two branch points.

There is some flexibility in choosing the branch cuts in this case because $z^{1/2}$ can act as its own topological counterpart: a counterclockwise rotation around $z^{1/2}$, increasing the phase of the function by π, is the same effect as a counterclockwise rotation around $z^{-1/2}$, which decreases the phase of the function by π. This happy circumstance is only true for functions with square root type branch points, as the next example shows.

◇

Example 9.9 We now consider the behavior of the function

$$f(z) = [z(z-1)]^{1/3}. \tag{9.108}$$

By using reasoning similar to the previous example, we find that the phase of the function increases by $2\pi/3$ when a transit is made around either the point $z = 0$ or $z = 1$,

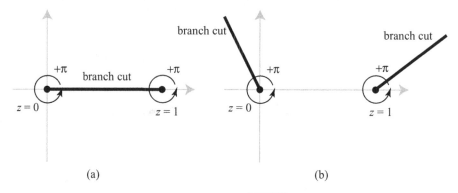

Figure 9.20 Possible branch cut choices for $f(z) = \sqrt{z(z-1)}$. (a) Cut joining the two singularities. (b) Cut joining each singularity to its counterpart at infinity.

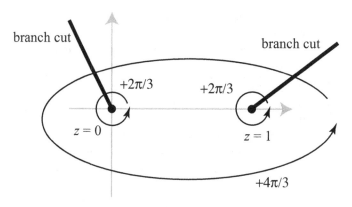

Figure 9.21 Branch cuts for $f(z) = [z(z-1)]^{1/3}$.

which suggests that there are three Riemann sheets. However, a circle which encloses both $z = 0$ and $z = 1$ results in the phase of the function increasing by $4\pi/3$. This implies that the branch cut does not lie between the two points, but must rather connect each of these points with its topological counterpart at infinity. The geometry is illustrated in Fig. 9.21.

◇

Some understanding of branch points and cut lines is necessary when we evaluate real integrals which involve noninteger powers of x. Furthermore, as described in Section 9.15, there is a close connection between branch points in complex analysis and phase singularities in optical wavefields.

9.11 Residue theorem

Let us consider a function which has a Laurent expansion about the point $z = z_0$, and calculate the integral about a closed path (a circle of radius r) centered on this point, i.e.

$$\oint \sum_{n=-\infty}^{\infty} a_n (z - z_0)^n dz = \sum_{n=-\infty}^{\infty} a_n \oint (z - z_0)^n dz. \tag{9.109}$$

We may evaluate this integral term by term because the series is uniformly convergent. If we do so, we find that the integrals of all terms for which $n \geq 0$ vanish, by Cauchy's theorem. For $n = -m < 0$, we find

$$a_{-m} \oint (z - z_0)^{-m} dz = a_{-m} \int_0^{2\pi} r^{-m} e^{-im\theta} i r e^{i\theta} d\theta = i a_{-m} r^{-m+1} \int_0^{2\pi} e^{-i(m-1)\theta} d\theta. \tag{9.110}$$

This integral is identically zero, except for $m = 1$, or $n = -1$. The contour integral about the point $z = z_0$ thus reduces to the form

$$\oint f(z) dz = 2\pi i a_{-1}. \tag{9.111}$$

This is a surprising and remarkable result! The contour integral only depends on the value a_{-1}, which is called the *residue* of the function $f(z)$ at $z = z_0$, which we will write as $\text{Res}(z_0)$.

What happens if our contour encloses several singular points of $f(z)$? In such a case, we may perform our usual contour integration tricks to redraw the contour in the form shown in Fig. 9.22.

From this contour, and Cauchy's integral theorem, we have

$$\oint_C f(z) dz + \oint_{C_1} f(z) dz + \oint_{C_2} f(z) dz + \cdots = 0. \tag{9.112}$$

As in the derivation of Eq. (9.72), we have taken the limit in which the paths connecting C_i to C move together and have canceling contributions. The contours C_i are all clockwise; rewriting them as counterclockwise contours C_i', and rearranging terms in the above equation, we have

$$\oint_C f(z) dz = \oint_{C_1'} f(z) dz + \oint_{C_2'} f(z) dz + \cdots. \tag{9.113}$$

Each of the circular contours may be evaluated by Eq. (9.111), and combining these results, we have

$$\oint_C f(z) dz = 2\pi i \left[\sum_i \text{Res}(z_i) \right]. \tag{9.114}$$

The above equation may be stated as a theorem.

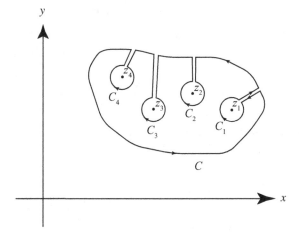

Figure 9.22 A contour which can be used to evaluate a path integral enclosing numerous isolated singularities.

Theorem 9.7 (Residue theorem) *The integral of a function $f(z)$ which is analytic on and within a closed contour except at a finite number of isolated singular points is given by $2\pi i$ times the sum of the residues enclosed by the contour.*

The residue theorem is an extremely powerful tool for integration. It reduces the evaluation of a contour integral of the function $f(z)$ to the mostly algebraic determination of the residues of the function. Many otherwise intractable integrals can be solved in a straightforward, almost trivial way.

To find the residue of a function at a point $z = z_0$, one generally has to find the Laurent expansion of the function about that point. This procedure simplifies greatly, however, if the singularity is a simple pole or low-order pole. If the singularity is a simple pole, it has a Laurent expansion of the form

$$f(z) = \frac{a_{-1}}{z - z_0} + \sum_{n=0}^{\infty} a_n(z - z_0)^n. \tag{9.115}$$

Let us multiply both sides of this equation by $(z - z_0)$ and take the limit as $z \to z_0$,

$$\lim_{z \to z_0} (z - z_0)f(z) = a_{-1} + \lim_{z \to z_0} \sum_{n=0}^{\infty} a_n(z - z_0)^{n+1} = a_{-1} = \mathrm{Res}(z_0). \tag{9.116}$$

Therefore the residue of a simple pole can be found by

$$\mathrm{Res}(z_0) = \lim_{z \to z_0} (z - z_0)f(z). \tag{9.117}$$

How can one tell that a singularity is a simple pole? If, about a point $z = z_0$, the function can be separated into the form,

$$f(z) = \frac{g(z)}{z - z_0}, \tag{9.118}$$

where $g(z)$ is analytic about $z = z_0$, then the function has a simple pole.

Higher-order poles are more challenging to work with, but less common. Let us determine the residue of a function which has a second-order pole, so that

$$f(z) = \frac{a_{-2}}{(z - z_0)^2} + \frac{a_{-1}}{z - z_0} + \sum_{n=0}^{\infty} a_n(z - z_0)^n. \tag{9.119}$$

We could try multiplying by $(z - z_0)^2$ and taking the limit $z \to z_0$, but this would give us a_{-2}, and only a_{-1} contributes to the value of the integral. Let us multiply by $(z - z_0)^2$, *take the derivative of the product with respect to z, and then take the limit as $z \to z_0$*:

$$\lim_{z \to z_0} \frac{d}{dz} \left[(z - z_0)^2 f(z) \right] = a_{-1}. \tag{9.120}$$

We can derive in a similar manner formulas for all higher-order residues. We have

$$a_{-1} = \lim_{z \to z_0} \frac{1}{(n-1)!} \left(\frac{d}{dz} \right)^n \left[(z - z_0)^n f(z) \right]. \tag{9.121}$$

Example 9.10 Let us determine the singularities and residues of the function

$$f(z) = \frac{3z + 1}{z(z - 1)^3}. \tag{9.122}$$

The function clearly has singularities at $z = 0$ and $z = 1$. Because $zf(z)$ is regular at $z = 0$, $z = 0$ is a simple pole. Because $(z - 1)^3 f(z)$ is regular at $z = 1$, $z = 1$ is a third-order pole. We may find the residues in a straightforward manner,

$$\text{Res}(z = 0) = \lim_{z \to 0} zf(z) = \lim_{z \to 0} \frac{3z + 1}{(z - 1)^3} = -1, \tag{9.123}$$

$$\text{Res}(z = 1) = \lim_{z \to 0} \frac{1}{2!} \frac{d^2}{dz^2} (z - 1)^3 f(z) = \lim_{z \to 0} \frac{1}{2!} \frac{d^2}{dz^2} \frac{3z + 1}{z} = +1. \tag{9.124}$$

Once the residues have been calculated, we can readily find the values of a variety of path integrals. A number of paths for the integral

$$\oint_{C_m} f(z) dz = 2\pi i \left[\sum_i \text{Res}(z_i) \right] \tag{9.125}$$

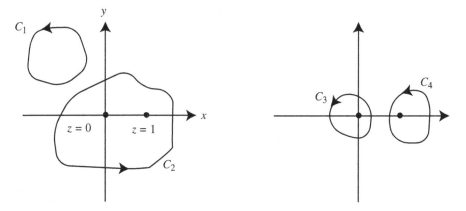

Figure 9.23 Contours for the residue theory example.

are shown in Fig. 9.23. The integral along path C_1, which encloses no singularities, has value 0. The integral along path C_2, which encloses both singularities, has value 0 as well. The integral along path C_3, which encloses only the singularity at $z = 0$, has value $-2\pi i$. The integral along path C_4, which encloses only the singularity at $z = 1$, has value $2\pi i$.

◇

9.12 Evaluation of definite integrals

We finally come to a direct application of complex analysis tricks to a practical problem: the evaluation of certain classes of real-valued integrals. There are in fact a large number of integrals that can be evaluated in a variety of different ways using complex analysis; we restrict ourselves to several classes which are relatively common.

9.12.1 Definite integrals $\int_0^{2\pi} f(\sin\theta, \cos\theta)\mathrm{d}\theta$

If we have an integral of the form

$$I = \int_0^{2\pi} f(\sin\theta, \cos\theta)\mathrm{d}\theta, \tag{9.126}$$

where f is finite for all values of θ and is a rational function of $\sin\theta$ and $\cos\theta$, we may evaluate this integral by conversion to a complex integral. Let

$$z = e^{\mathrm{i}\theta}, \quad \mathrm{d}z = \mathrm{i}e^{\mathrm{i}\theta}\mathrm{d}\theta; \tag{9.127}$$

we may then write

$$\mathrm{d}\theta = -\mathrm{i}\frac{\mathrm{d}z}{z}, \quad \sin\theta = \frac{z - z^{-1}}{2\mathrm{i}}, \quad \cos\theta = \frac{z + z^{-1}}{2}. \tag{9.128}$$

If we substitute these results into our original integral, we have that

$$I = -i \oint f \left(\frac{z - z^{-1}}{2i}, \frac{z + z^{-1}}{2} \right) \frac{dz}{z},$$ (9.129)

where the path of integration is now the unit circle. Because the function f is assumed to be a rational function of $\sin\theta$ and $\cos\theta$, it is also a rational function of z and therefore analytic everywhere except at a finite collection of poles.

Near the beginning of this chapter, we evaluated certain complex integrals on a circle by converting them to polar form; now, we reverse the process and evaluate certain complicated polar form integrals by converting them into a path on the unit circle.

With the above integral, we may apply the residue theorem to find that

$$I = (-i)2\pi i \sum \text{residues of } f \left[(z - z^{-1})/2i, (z + z^{-1})/2 \right] /z \text{ within the unit circle.}$$
(9.130)

Example 9.11 As an example, consider

$$I' = \int_0^{2\pi} \frac{d\theta}{A + B\sin\theta},$$ (9.131)

where $A^2 > B^2$ and $A > 0$. On making the above substitutions, we have that

$$I' = \oint \frac{2dz}{2iAz + B(z^2 - 1)} = \frac{2}{B} \oint \frac{dz}{z^2 + 2iAz/B - 1} = \frac{2}{B} \oint \frac{dz}{(z - z_1)(z - z_2)},$$ (9.132)

where the roots are given by

$$z_1 = \frac{-iA + i\sqrt{A^2 - B^2}}{B},$$ (9.133)

$$z_2 = \frac{-iA - i\sqrt{A^2 - B^2}}{B}.$$ (9.134)

Only one of these roots lies within the unit circle; since $A > 0$, it follows that $|z_1| < |z_2|$ and z_1 lies within the circle. Using the residue theorem, then, we have

$$I = 2\pi i \frac{2}{B} \frac{1}{z_1 - z_2} = \frac{2\pi}{\sqrt{A^2 - B^2}}.$$ (9.135)

◇

9.12.2 Definite integrals $\int_{-\infty}^{\infty} e^{iax} f(x) dx$

We now consider integrals of the form

$$\int_{-\infty}^{\infty} e^{iax} f(x) dx,$$ (9.136)

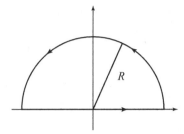

Figure 9.24 The semicircular path used for the evaluation of certain definite integrals.

where a is a positive number. This integral represents a Fourier transform, to be discussed in Chapter 11. We first assume that the function $|f(z)| < M\,e^{ay}$, where M is a finite constant. This implies that $|e^{iaz}f(z)| \to 0$ as $y \to \infty$. We then consider the integral,

$$I = \oint_C f(z)e^{iaz}\mathrm{d}z, \tag{9.137}$$

where the path is given in Fig. 9.24.

We may then write

$$I = \int_{-R}^{R} f(x)e^{iax}\mathrm{d}x + \oint_{C_{\mathrm{semi}}} f(z)e^{iaz}\mathrm{d}z, \tag{9.138}$$

where C_{semi} is the semicircle of radius R. If we consider the limit as $R \to \infty$, the contribution of the semicircular piece will evidently vanish, because of the exponential decay of the integrand as $y \to \infty$, i.e. $\exp[iaz] = \exp[iax]\exp[-ay]$. We will be left with

$$I = \oint_C f(z)e^{iaz}\mathrm{d}z = \int_{-\infty}^{\infty} f(x)e^{iax}\mathrm{d}x. \tag{9.139}$$

However, because of the residue theorem, we have that

$$\int_{-\infty}^{\infty} f(x)e^{iax}\mathrm{d}x = 2\pi i \sum \text{residues of } f(z)e^{iaz} \text{ within the infinite positive half space.}$$

$$\tag{9.140}$$

By making similar arguments, and assuming that $f(z)$ decays rapidly in the $-y$ direction, we can write

$$\int_{-\infty}^{\infty} f(x)e^{-iax}\mathrm{d}x = -2\pi i \sum \text{residues of } f(z)e^{-iaz} \text{ within the infinite negative half space.}$$

$$\tag{9.141}$$

The minus sign in the above equation comes from the reversal of direction of the contour in the negative half space, shown in Fig. 9.25.

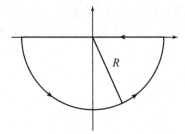

Figure 9.25 The semicircular path used for the evaluation of certain definite integrals.

The one weakness in the calculation we have just described is the argument that the integral over the semicircular piece vanishes as $R \to \infty$. To truly justify that the residue calculus can be used to evaluate integrals of the form of Eq. (9.136), we must rigorously show that the contribution of the semicircle becomes negligible, using what is known as Jordan's lemma.

Lemma 9.1 (Jordan's lemma) *Let us suppose that the function $f(z) \to 0$ uniformly on the upper semicircular path C_{semi} as $R \to \infty$. Then*

$$\lim_{R \to \infty} \int_{C_{\text{semi}}} e^{iaz} f(z) \mathrm{d}z = 0, \qquad (9.142)$$

with $a > 0$. A similar result holds for the lower semicircular path with $a < 0$.

"Uniform convergence" of the function $f(z) \to 0$ in this sense means that, for any R, there exists a constant A_R such that $|f(z)| \leq A_R$ for all z on the contour and $A_R \to 0$ as $R \to \infty$. With this result, we may write that

$$I_{\text{semi}} = \left| \int_{C_{\text{semi}}} e^{iaz} f(z) \mathrm{d}z \right| \leq \int_0^\pi e^{-aR\sin\theta} A_R R \mathrm{d}\theta, \qquad (9.143)$$

where we have used the triangle inequality and $y = R\sin\theta$. The latter integral may be rewritten as

$$I_{\text{semi}} = 2A_R R \int_0^{\pi/2} e^{-aR\sin\theta} \mathrm{d}\theta. \qquad (9.144)$$

But $\sin\theta \geq 2\theta/\pi$, which means that $\exp[-aR\sin\theta] \leq \exp[-2aR\theta/\pi]$, and we have

$$I_{\text{semi}} \leq 2A_R R \int_0^{\pi/2} e^{-2aR\theta/\pi} \mathrm{d}\theta = \frac{2A_R R\pi}{2aR} \left[1 - e^{-aR} \right]. \qquad (9.145)$$

In the limit $R \to \infty$, $I_{\text{semi}} \to 0$ because of the assumption $A_R \to 0$, and Jordan's lemma is proven.

Jordan's lemma implies that the semicircular piece only vanishes if $f(z) \to 0$ as $R \to \infty$. Our initial assumption that $|f(z)| < Me^{ay}$ turns out to be too weak a condition to ensure convergence.

Example 9.12 As an example, consider

$$I' = \int_{-\infty}^{\infty} \frac{e^{ikx}}{x^2+4}dx, \tag{9.146}$$

with $k > 0$. We may write this integral as

$$I' = \int_C f(z)e^{ikz}dz = \int_C \frac{e^{ikz}}{z^2+4}dz, \tag{9.147}$$

because it is clear that as $y \to \infty, f(z) \to 0$. The denominator may be factorized

$$I' = \int_C \frac{e^{ikz}}{(z-2i)(z+2i)}dz. \tag{9.148}$$

The only singularity in the upper half space is at $z = 2i$; the residue of the function at this singularity is

$$\text{Res}(z = 2i) = \frac{e^{ik2i}}{(2i+2i)} = \frac{e^{-2k}}{4i}. \tag{9.149}$$

We therefore have

$$I' = \frac{\pi}{2}e^{-2k}. \tag{9.150}$$

◇

9.12.3 Definite integrals $\int_{-\infty}^{\infty} f(x)dx$

It is also possible to use residue theory to evaluate integrals of the form

$$\int_{-\infty}^{\infty} f(x)dx \tag{9.151}$$

by using a semicircular contour as in Fig. 9.24; however, the function must satisfy a relatively strict constraint in order that the semicircular contribution vanishes. We state this constraint as a theorem.

Theorem 9.8 *Let us suppose that $|f(z)| \leq A/R^2$ on the semicircle C_{semi} as $R \to \infty$. Then the integral*

$$\lim_{R\to\infty} \int_{C_{\text{semi}}} f(z)dz = 0. \tag{9.152}$$

The proof of this is relatively straightforward; we have

$$\left| \lim_{R \to \infty} \int_{C_{\text{semi}}} f(z) dz \right| \leq \lim_{R \to \infty} \int_{C_{\text{semi}}} |f(z)| R d\theta \leq \lim_{R \to \infty} \frac{\pi A}{R} = 0. \tag{9.153}$$

With Theorem 9.8 satisfied, we may make the association

$$\int_{-\infty}^{\infty} f(x) dx = \oint_C f(z) dz, \tag{9.154}$$

where C is the closed upper semicircular path. It is to be noted that one can also usually use the lower semicircular path to evaluate the integral.

It would seem that Theorem 9.8 is somewhat too strict, as it seems that if $|f(z)| \leq A/R^{\alpha}$, with $\alpha > 1$, the semicircular contribution would also vanish. With $1 < \alpha < 2$, however, the function will necessarily possess branch points, and a simple contour such as that of Fig. 9.24 cannot be used.

Example 9.13 We consider the evaluation of the integral

$$I'' = \int_{-\infty}^{\infty} \frac{dx}{x^3 - ia^3}. \tag{9.155}$$

This function clearly satisfies the conditions of Theorem 9.8, so we may evaluate the integral as

$$I'' = \oint_C f(z) dz, \tag{9.156}$$

with C being the closed upper semicircular contour. There are three simple poles in the complex plane; writing $z^3 = a^3 e^{i\pi/2} e^{i2\pi m}$, the three poles are at $z_1 = a \exp[i\pi/6]$, $z_2 = \exp[i3\pi/2]$, and $z_3 = \exp[i5\pi/6]$. Two of these, z_1 and z_3, are in the upper half space; their residues are

$$\text{Res}(z_1) = \frac{1}{(z_1 - z_2)(z_1 - z_3)}, \tag{9.157}$$

$$\text{Res}(z_3) = \frac{1}{(z_3 - z_1)(z_3 - z_2)}. \tag{9.158}$$

The total value of the integral is therefore

$$I'' = 2\pi i \left[\frac{1}{(z_1 - z_2)(z_1 - z_3)} + \frac{1}{(z_3 - z_1)(z_3 - z_2)} \right]$$

$$= \frac{2\pi i}{z_1 - z_3} \left[\frac{1}{(z_1 - z_2)} - \frac{1}{(z_3 - z_2)} \right]. \tag{9.159}$$

If we simplify this expression by putting everything over a common denominator, we have

$$I'' = \frac{2\pi i}{z_1 - z_3} \frac{z_3 - z_1}{(z_1 - z_2)(z_3 - z_2)} = -2\pi i \frac{1}{(z_2 - z_1)(z_2 - z_3)}. \tag{9.160}$$

It is to be noted that we may also use the lower semicircular contour to evaluate the integral; we immediately find

$$I'' = -2\pi i \operatorname{Res}(z_2) = -2\pi i \frac{1}{(z_2 - z_1)(z_2 - z_3)}, \tag{9.161}$$

which is in agreement with the result derived using the upper semicircular contour.

◇

9.12.4 Other integrals

Many other types of integral may be evaluated, including integrals of functions with branch cuts, using an appropriate choice of contour.

Example 9.14 (Fresnel integrals) In the diffraction of light by a rectangular aperture in the Fresnel approximation (see Section 12.1), one encounters integrals of the form

$$\int_0^\beta \cos(tx^2)dx, \quad \int_0^\beta \sin(tx^2)dx, \tag{9.162}$$

where β is related to the radius of the aperture and t is a propagation constant. These integrals cannot be evaluated analytically, except in the limit $\beta \to \infty$, in which they take the form

$$C \equiv \int_0^\infty \cos(tx^2)dx, \quad S \equiv \int_0^\infty \sin(tx^2)dx. \tag{9.163}$$

We consider the evaluation of each of these by combining them into the single complex exponential integral

$$\mathcal{E} \equiv \int_0^\infty e^{itx^2}dx. \tag{9.164}$$

The integrand is analytic throughout the finite complex plane, and we note that we can write

$$\oint_C e^{itz^2}dz = 0, \tag{9.165}$$

where C is a contour to be given. The contour which allows calculation of the Fresnel type integrals is a "pie slice" of radius R, shown in Fig. 9.26.

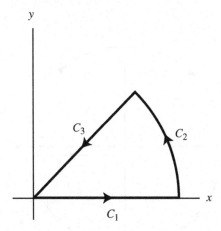

Figure 9.26 The pie slice contour used to evaluate the complex Fresnel integral.

We may then write

$$\oint_C e^{itz^2} dz = \int_{C_1} e^{itz^2} dz + \int_{C_2} e^{itz^2} dz + \int_{C_3} e^{itz^2} dz = 0. \qquad (9.166)$$

The integral along C_1 is simply the desired integral $\mathcal{E}$. The integral along C_2 can be eliminated using a Jordan lemma-type argument, since the integral will include an exponential term decaying with increasing R. For C_3, we note that the path is given by $z = re^{i\pi/4}$, which implies that

$$\int_{C_3} e^{itz^2} dz = \int_R^0 e^{-tr^2} e^{i\pi/4} dr = -e^{i\pi/4} \frac{1}{2} \sqrt{\frac{\pi}{t}}. \qquad (9.167)$$

On substitution, one readily finds that

$$\mathcal{C} = \mathcal{S} = \frac{1}{2} \sqrt{\frac{\pi}{2t}}. \qquad (9.168)$$

For future reference, it is also useful to have the value of the following complex integral

$$\int_{-\infty}^{\infty} e^{itx^2} dx = e^{i\pi/4} \sqrt{\frac{\pi}{t}}, \qquad (9.169)$$

which shows up in the evaluation of certain integral transforms.

◇

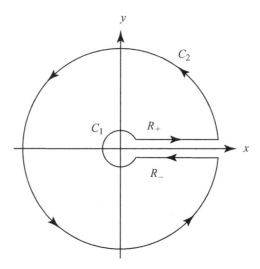

Figure 9.27 The "keyhole" contour used to evaluate Eq. (9.171).

Example 9.15 (Branch points) When the function to be integrated includes a branch point, we must be exceedingly clever in our choice of contour. We consider an integral of the form

$$I = \int_0^\infty \frac{x^{a-1}}{1+x} dx, \tag{9.170}$$

where $0 < a < 1$. Integrals of this form appear in problems related to the gamma function, discussed in Appendix A. Extending this integral to the complex plane, we have a branch point at $z = 0$ due to the presence of the non-integer exponent a,

$$I = \int_C \frac{z^{a-1}}{1+z} dz, \tag{9.171}$$

We may choose a contour as shown in Fig. 9.27, consisting of two circular pieces C_1 and C_2 and two semi-infinite lines R_+ and R_-. Since $a > 0$, the contribution of C_1 becomes negligible in the limit of vanishing radius. Also, since $a < 1$, the contribution of C_2 becomes negligible in the limit that its radius becomes infinite.

We take the argument of z on the curve R_+ to be 0; the argument of z on the curve R_- is therefore $\arg(z) = 2\pi(a-1)$, using the construction of Eq. (9.18). The residue theorem therefore results in the equation

$$\int_0^\infty \frac{x^{a-1}}{1+x} dx - e^{i2\pi(a-1)} \int_0^\infty \frac{x^{a-1}}{1+x} dx = 2\pi i \operatorname{Res}\{z = -1\}. \tag{9.172}$$

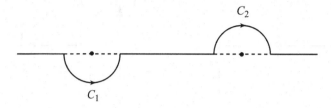

Figure 9.28 The semicircular paths used to avoid singularities directly on the path of integration.

The residue is readily determined to be $(-1)^{a-1}$; a rearrangement of terms gives

$$\int_0^\infty \frac{x^{a-1}}{1+x}\,dx = 2\pi i \frac{(-1)^{a-1}}{1 - e^{i2\pi(a-1)}}. \tag{9.173}$$

With further rearrangement, we find that

$$\int_0^\infty \frac{x^{a-1}}{1+x}\,dx = \frac{\pi}{\sin(\pi a)}. \tag{9.174}$$

◇

9.13 Cauchy principal value

Occasionally one runs into contour integrals for which a simple pole lies on the path of integration. Such a case is not covered by the methods described previously but may be given meaning by considering a construction in which the contour is deformed into a semicircle which is centered on the singular path, illustrated in Fig. 9.28.

The integrals for these two cases may be evaluated explicitly, using polar coordinates, with $f(z) = g(z)/(z - z_0)$,

$$\int_{C_1} \frac{g(z)dz}{z - z_0} = ig(z_0) \int_\pi^{2\pi} d\phi = i\pi \operatorname{Res}(z_0), \tag{9.175}$$

$$\int_{C_2} \frac{g(z)dz}{z - z_0} = ig(z_0) \int_\pi^0 d\phi = -i\pi \operatorname{Res}(z_0). \tag{9.176}$$

We therefore approach the limit of having a simple pole on the contour by having the contour wrap itself around the singular point, from one side or the other. This would seem to be a problem, since we have two distinct values for what should be the same integral. However, let us consider what happens when the integral is taken over a closed contour, as shown in Fig. 9.29.

This integral may be written as

$$\oint f(z)dz = \int_C f(z)dz + \int_{C_j} f(z)dz, \tag{9.177}$$

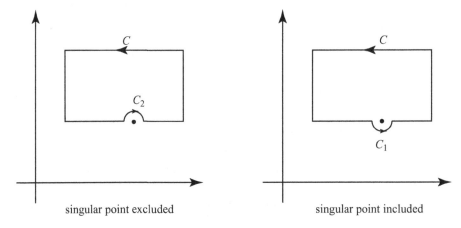

singular point excluded   singular point included

Figure 9.29 Behavior of contour integrals with a singularity on a closed path.

where $j = 2$ if the contour *excludes* the singular point, and $j = 1$ if the contour *includes* the singular point. If the contour includes the singular point, we have

$$\oint f(z)\mathrm{d}z = \int_C f(z)\mathrm{d}z + \mathrm{Res}(z_0)\pi\mathrm{i} = 2\pi\mathrm{i}\mathrm{Res}(z_0), \qquad (9.178)$$

and if the contour excludes the singular point, we have

$$\oint f(z)\mathrm{d}z = \int_C f(z)\mathrm{d}z - \mathrm{Res}(z_0)\pi\mathrm{i} = 0, \qquad (9.179)$$

by Cauchy's integral theorem. If we take the limit as the semicircle becomes arbitrarily small, we find that

$$\int_C f(z)\mathrm{d}z = \mathrm{Res}(z_0)\pi\mathrm{i}, \qquad (9.180)$$

for both cases! Evidently if the simple pole is on the boundary, it is irrelevant which limit one takes. This method of taking the limit around a singular point is known as taking the *Cauchy principal value* of the integral, usually denoted by writing a P in front of the integral sign. In essence, it involves "approaching" the singular point at the same rate from all directions on the path.

Using this result, we may generalize the residue theorem to make the following statement.

Theorem 9.9 (Extended residue theorem) *The integral of a function $f(z)$ which is analytic on and within a closed contour except at a finite number of isolated points within the contour and possibly at a finite number of simple poles on the contour is given by*

$$P \oint f(z)\mathrm{d}z = 2\pi\mathrm{i}\sum \text{enclosed residues} + \pi\mathrm{i}\sum \text{residues on boundary}. \qquad (9.181)$$

This extended version of the residue theorem may be used to evaluate real integrals which have singularities on the path of integration. Such a case appears in physical problems, most notably in the derivation of dispersion relations in optics, to be discussed in Section 9.14.

9.14 Focus: Kramers–Kronig relations

An important part of any optics education, beyond simply being able to do the math, is acquiring a "big picture" view of the properties of light. The behavior of an electromagnetic wave is constrained to satisfy Maxwell's equations, and much of the optical research done since the formulation of these equations has focused on understanding what electromagnetic waves can and cannot do.

Complex analysis can help immensely in this regard, and investigations of the analytic properties of wavefields lead to broad statements about their behavior. In particular, a wavefield has analytic properties associated with its *temporal* evolution, and this observation leads to a very powerful connection between analyticity and causality. In this section we discuss this connection.[4]

We consider an input wavefield $f(t)$ which propagates through a *linear, time-shift invariant, causal* system. The wave $f_0(t)$ output from such a system may be written in the form

$$f_0(t) = \int_\tau^\infty f(\tau) g(t - \tau) d\tau, \tag{9.182}$$

where $g(t)$ is the *response function* which characterizes the system response to an input. We also add the assumption that $g(t)$ is square integrable, i.e.

$$\int_{-\infty}^\infty |g(t)|^2 dt < \infty, \tag{9.183}$$

which implies that it may be expressed as a Fourier integral.

"Linearity" implies that the response of the system is linear with respect to the input, i.e. if we superpose two inputs $f^{(1)}(t)$ and $f^{(2)}(t)$ in the system, the total output is simply the superposition of the individual outputs, $f_0(t) = f_0^{(1)}(t) + f_0^{(2)}(t)$. "Time-shift invariant" implies that the response of the system does not change with time, i.e. the only effect in changing the origin of time of the input pulse $f(t) \to f(t + t_0)$ is a corresponding change of origin of the output, $f_0(t) \to f_0(t + t_0)$. "Causality" implies that there is no output from the system before there is an input, i.e. if $f(t) = 0$ for $t < 0$, then $f_0(t) = 0$ for $t < 0$. It is characterized by the lower limit of τ in the integral or, equivalently, by letting $g(t) = 0$ for $t < 0$.

---

[4] This section is inspired by a wonderful lecture entitled, "Analyticity, causality and dispersion relations," which Professor Emil Wolf gave for many years in his own mathematical methods class. A printed version of this lecture, which gives more detail and historical background, may be found in [Wol01].

This equation can be applied to a large variety of wave systems, and is of broad physical interest. In optics, $f(t)$ might represent a wave pulse incident upon a Fabry-Perot interferometer (Section 7.1), and $g(t)$ would then represent the properties of the interferometer. Equation (9.182) can also be used to describe pulse propagation in a linear optical medium; in such a case, $g(t)$ would be connected to the dispersive properties of the medium.

Equation (9.182) represents a convolution operation, to be discussed in Section 11.5. If we look at the same equation in the frequency domain via a Fourier transform, we readily find that

$$\tilde{f}_0(\omega) = 2\pi \tilde{f}(\omega)\tilde{g}(\omega), \tag{9.184}$$

where, for instance,

$$\tilde{g}(\omega) = \frac{1}{2\pi}\int_0^\infty g(t)e^{i\omega t}dt. \tag{9.185}$$

The imposition of causality sets the lower limit of the integral in Eq. (9.185). This, in turn, implies that the function $\tilde{g}(\tilde{\omega})$, where $\tilde{\omega} = \omega_R + i\omega_I$ is a complex number, is an analytic function in the upper half of the complex plane ($\omega_I \geq 0$). A rigorous proof of this statement is rather involved (it can be extracted from a careful reading of [Det65], Section 4.9), and we only demonstrate its plausibility here.

With complex variable $\tilde{\omega}$, it is clear that the functions $\exp[i\tilde{\omega}t]$ are analytic for all finite values of the variable t. We then note that the expression (9.185) takes on the form

$$\tilde{g}(\tilde{\omega}) = \frac{1}{2\pi}\int_0^\infty g(t)e^{-\omega_I t}e^{i\omega_r t}dt. \tag{9.186}$$

Because $g(t)$ is assumed square integrable, it is also bounded, i.e. $|g(t)| < M$ for all t, where M is a positive constant. For positive values of ω_I, then, the integral converges faster than the integral of the exponential function $\exp[-\omega_I t]$. The integral of the latter function is analytic for $\omega_I > 0$ which suggests (though, again, this is not a rigorous proof) that $\tilde{g}(\tilde{\omega})$ is analytic over at least that domain.

The analyticity of $\tilde{g}(\tilde{\omega})$ suggests that the real and imaginary parts of the function cannot be independent, even when we consider its behavior on the real axis. This is encapsulated in a pair of relations known as the *Kramers–Kronig relations*, given by

$$\tilde{g}_R(\omega) = \frac{1}{\pi}P\int_{-\infty}^\infty \frac{\tilde{g}_I(\omega')}{\omega' - \omega}d\omega', \tag{9.187}$$

$$\tilde{g}_I(\omega) = -\frac{1}{\pi}P\int_{-\infty}^\infty \frac{\tilde{g}_R(\omega')}{\omega' - \omega}d\omega'. \tag{9.188}$$

To demonstrate these relations, we apply the contour of Fig. 9.24 to the following integral

$$I = \oint_C \frac{\tilde{g}(z')}{z' - z}dz'. \tag{9.189}$$

Because of the exponential decay in the positive-ω_I direction, the contribution of the semi-circular part of the path becomes negligible in the limit $R \to \infty$. For the part lying along the real axis, we apply the extended residue theorem of Section 9.13, taking the Cauchy principal value of the integral, i.e.

$$P \oint_C \frac{\tilde{g}(z')}{z'-z}dz' = \pi i \sum \text{residues of } \frac{\tilde{g}(z')}{z'-z} \text{ on real axis.} \qquad (9.190)$$

Because $\tilde{g}(z)$ is analytic in the upper half-space, the only singularity is on the boundary, at $z' = z$, and the residue at this point is $\tilde{g}(z)$. From this, we may write

$$I = P \int_{-\infty}^{\infty} \frac{\tilde{g}(\omega')}{\omega'-\omega}d\omega' = \pi i \tilde{g}(\omega). \qquad (9.191)$$

Taking the real and imaginary parts of this expression, we get the Kramers–Kronig relations.

With a slight modification, the Kramers–Kronig relations apply to the refractive index of a linear causal medium. Let us suppose that we have a pulse $U(z,t)$ propagating in the z-direction into a medium with frequency-dependent refractive index $n(\omega)$. By straightforward Fourier analysis, we can show that the field in the medium is given by the formula

$$U(z,t) = \int_{-\infty}^{\infty} A(\omega)e^{-i\omega t}e^{in(\omega)\omega z/c}d\omega, \qquad (9.192)$$

where $A(\omega)$ is the temporal Fourier transform of the field at the entrance $z=0$ to the medium, i.e.

$$A(\omega) = \frac{1}{2\pi} \int_{-\infty}^{\infty} U(0,t')e^{i\omega t'}dt'. \qquad (9.193)$$

On substitution from this equation into Eq. (9.192), we can rearrange terms such that

$$U(z,t) = \int_{-\infty}^{\infty} G(z,t-t')U(0,t')dt', \qquad (9.194)$$

where

$$G(z,\tau) = \frac{1}{2\pi} \int_{-\infty}^{\infty} e^{-i\omega\tau}e^{in(\omega)\omega z/c}d\omega. \qquad (9.195)$$

Equation (9.194) is in the form of a linear, time-invariant optical system. It is clear that, if this system is causal as well, it takes a finite amount of time for the signal to propagate from $z=0$ to position z, or that $G(z,\tau) = 0$ for $\tau < T$, with T a finite value. With an inverse Fourier transform with respect to τ and a coordinate transformation, we may write

$$e^{in(\omega)\omega z/c} = e^{i\omega T} \int_0^{\infty} G(z,t'+T)e^{i\omega t'}dt'. \qquad (9.196)$$

This expression is comparable to Eq. (9.185), and suggests in turn that $\exp[in(\omega)\omega z/c]$ must be an analytic function in the positive half-space. This then suggests that $\omega n(\omega)$ must be analytic in the same half-space, as must $n(\omega)$. To apply the Kramers–Kronig relations, however, we need a function which is also square integrable. Physical considerations suggest that $n(\omega) \to 1$ as $\omega \to \infty$, i.e. the medium looks transparent at arbitrarily large frequencies. The function $n(\omega) - 1$ will properly vanish for large values of ω, and this function must satisfy the Kramers–Kronig relations. Substituting this in Eq. (9.191) for $\tilde{g}(\omega)$, we find that

$$n_R(\omega) = 1 + \frac{1}{\pi} P \int_{-\infty}^{\infty} \frac{n_I(\omega')}{\omega' - \omega} d\omega', \tag{9.197}$$

$$n_I(\omega) = -\frac{1}{\pi} P \int_{-\infty}^{\infty} \frac{n_R(\omega') - 1}{\omega' - \omega} d\omega'. \tag{9.198}$$

The Kramers–Kronig formulas can be used to make a number of broad statements about the behavior of a medium. For instance, it follows from them that there is no material which is non-absorbing $n_I(\omega) = 0$ for all frequencies ω.

For our purposes, we have now seen that there is a close connection between the concept of causality and the concept of analyticity.

9.15 Focus: optical vortices

In Section 9.10, we discussed multivalued functions, the branch points associated with them, and the Riemann surfaces used to characterize and describe the behavior of such functions. It is easy to imagine that such abstract concepts as Riemann surfaces play a small role in physical applications, but in fact they play a crucial role in optics, in the subfield of *singular optics*.

When coherent monochromatic wavefields are brought together to form an interference pattern, they typically produce regions of complete destructive interference where the field amplitude vanishes. The most familiar example of this effect is the interference pattern produced in Young's two-pinhole experiment, illustrated in Fig. 9.30.

In Young's experiment, the regions of zero field amplitude are lines on the measurement plane; if we consider a three-dimensional region of space, the zero regions are nearly planar surfaces. Typically, when one thinks of interference, the pattern produced by Young's interferometer is the first picture that comes to mind.

If one considers a Young-type interferometer with three or more pinholes, however, the interference pattern is of a very different nature. The geometry and interference pattern of a three-pinhole interferometer is illustrated in Fig. 9.31. The regions of zero field amplitude are now points on the measurement plane (arranged in a hexagonal pattern), and in three-dimensional space manifest as lines of zero field amplitude. If one considers an experiment with the interference of light from four or more pinholes, one finds that the situation is qualitatively similar to the three-pinhole case: the zeros manifest as lines in three-dimensional space. Evidently, a typical interference pattern consists of these zero lines, and the zero

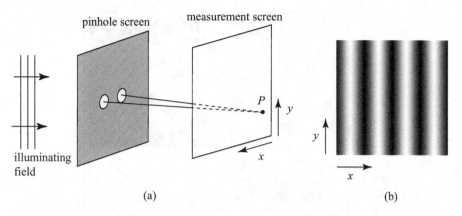

Figure 9.30 (a) An illustration of Young's famous two-pinhole experiment. (b) A simulation of the interference pattern produced on the measurement screen.

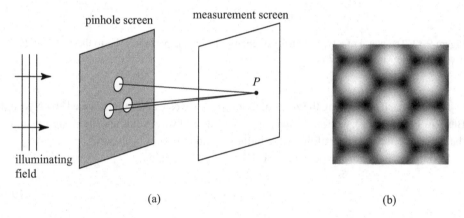

Figure 9.31 (a) An illustration of a three-pinhole interference experiment. (b) A simulation of the intensity of the interference pattern produced on the measurement screen.

surfaces of Young's two-pinhole experiment are in some sense unusual. This is borne out by experimental evidence: an example of a naturally-produced interference pattern is the speckle pattern produced when a laser scatters from a rough surface; the zeros manifest as points in any planar projection of the speckle pattern, and as lines in a three-dimensional region containing the reflected light field.

These observations represent a unique philosophical departure from the typical manner in which optics problems are investigated. Instead of asking "What are the *possible* behaviors of optical wavefields?", we may now ask "What are the *most common* behaviors of optical wavefields?" The "most common" behaviors of wavefields are usually referred to as *generic* features.

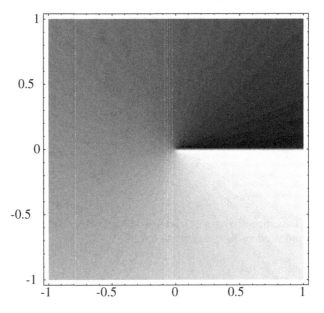

Figure 9.32 The phase of a typical phase singularity in a plane perpendicular to the zero line. Black is $\phi = 0$, while white is $\phi = 2\pi$.

This approach to studying the zeros of interfering waves was first introduced[5] by Nye and Berry in 1974 [NB74], who described the typical zeros of wavefields and, more significantly, the typical phase structure of the wave in the neighborhood of such zeros.

A complex monochromatic wavefield $U(\mathbf{r})$ may be written as

$$U(\mathbf{r}) = A(\mathbf{r})e^{i\phi(\mathbf{r})}, \tag{9.199}$$

where $A(\mathbf{r})$ is the real-valued amplitude of the wavefield and $\phi(\mathbf{r})$ is the real-valued phase. Equation (9.199) is a functional analogy to the polar representation of complex numbers, $z = re^{i\theta}$. The representation is well-defined at all points $\mathbf{r}$ except those for which the amplitude vanishes, $A(\mathbf{r}) = 0$; at such points, there is no unique definition of the phase $\phi(\mathbf{r})$ and it is said to be *singular*. These points are generally referred to as *phase singularities*.

The singular behavior of the phase might, at first glance, seem to be an mathematical artifact of our use of Eq. (9.199), and of no physical significance. A close look at a phase structure of a wave in the neighborhood of a phase singularity, however, shows nontrivial and unusual behavior, illustrated in Fig. 9.32. The phase of the wavefield increases continuously by 2π as one traverses a closed circular path which encloses the phase singularity. Loosely

---

[5] Arnold Sommerfeld briefly looked at wavefields with a similar philosophy in his *Optics* [Som64], but he seems to have considered the behavior of the zeros as relatively insignificant, stating, "However, just because the amplitude vanishes there, they do not produce any stronger effect than other points of varying intensity."

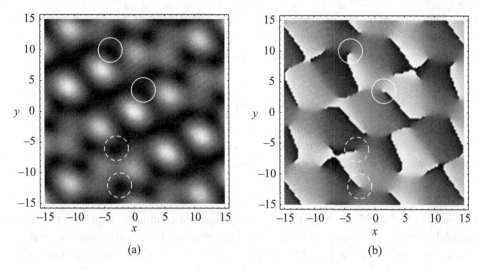

(a)   (b)

Figure 9.33 The (a) intensity and (b) phase of a speckle pattern produced by the interference of four randomly-oriented monochromatic plane waves. The circles each enclose a single optical vortex in the pattern.

speaking, the phase swirls around the central zero of amplitude, like water circling a drain; for this reason, such phase singularities are typically referred to as *optical vortices*.

An inspection of the intensity and phase of any speckle pattern shows numerous phase singularities; a simulation of such a pattern is shown in Fig. 9.33. Numerous optical vortices can be seen in the phase plot; the circles enclose some typical examples. It can be seen that the vortices are each connected to a zero of the intensity of the wavefield.

Looking at the two dashed circles at the bottom of the figure, it is to be noted that the phase *increases* by 2π as one traverses a counterclockwise closed path around the lower circle, but it *decreases* by 2π as one traverses a counterclockwise closed path around the upper circle. There are evidently two generic types of vortices that appear in interference patterns; the former are referred to as *left-handed vortices*, while the latter are *right-handed vortices*.

It is possible to determine the functional form $U(\mathbf{r})$ of a wavefield in the immediate neighborhood of a phase singularity; in doing so, we begin to make a connection between such objects and the analytic functions described in this chapter. We consider a solution of the Helmholtz equation in the immediate neighborhood of a phase singularity, and consider a coordinate system in which the zero line intersects the origin and is parallel to the z-direction. The Helmholtz equation may be written as

$$\frac{\partial^2 U}{\partial x^2} + \frac{\partial^2 U}{\partial y^2} + \frac{\partial^2 U}{\partial z^2} + k^2 U = 0. \qquad (9.200)$$

For points sufficiently close to the axis of the zero line, $U(\mathbf{r}) \approx 0$. Also, because the zero line runs along the z-axis, the field changes very slowly along this axis, which implies that $\frac{\partial^2 U}{\partial z^2} \approx 0$. This means that, in the immediate neighborhood of the phase singularity,

$$\frac{\partial^2 U}{\partial x^2} + \frac{\partial^2 U}{\partial y^2} = 0, \tag{9.201}$$

or $U(\mathbf{r})$ satisfies the Helmholtz equation in x and y! Following the argument of Section 9.1, this suggests that the wavefield in the immediate vicinity of the singularity, and a plane of fixed z, is well-approximated by the expression

$$U(x,y) = f(x+iy) + g(x-iy).$$

Considering that solutions of the Helmholtz equation are continuous to at least their first partial derivatives, this implies that $U(x,y)$ may be written as the sum of an analytic function $f(z)$ and the complex conjugate of an analytic function, $g(z^*)$. We may consider the Taylor expansion of each of these functions; assuming that the first nonzero term of the series is the sth power, we may write

$$U(x,y) \approx a(x+iy)^s + b(x-iy)^s, \tag{9.202}$$

where a and b are constants which are in general complex. We restrict ourselves, for simplicity, to the cases $a=1$, $b=0$, and $a=0$, $b=1$. In polar coordinates, the singularity may then be written in the form

$$U(r,\theta) = r^{|s|} e^{is\theta}, \tag{9.203}$$

where s is now a positive integer for $a=1$, $b=0$ and a negative integer for $a=0$, $b=1$. The zero line is said to be a *vortex of order* s; if we traverse a closed path containing the origin, we find that the phase changes continuously by the amount $2\pi s$.

To put this in the context of complex analysis, we consider a new complex phase defined by the relation $U(\mathbf{r}) = \exp[\psi(\mathbf{r})]$; writing this function for the vortex of Eq. (9.203), we find that

$$\psi(r,\theta) = |s| \log r + is\theta. \tag{9.204}$$

On comparison with Eq. (9.34), we see that *the complex phase behaves locally as a logarithmic branch point*. This implies that the complex phase of a wavefield can only be described self-consistently in the neighborhood of an optical vortex by treating it as a multivalued Riemann surface. This has important consequences for applications which require phase retrieval, as we discuss in a moment.

The quantity s is referred to as the *topological charge* or *winding number* of the optical vortex; it can be represented in integral form as

$$s \equiv \frac{1}{2\pi} \oint_S \nabla\phi(\mathbf{r}) \cdot d\mathbf{r}, \tag{9.205}$$

where $\phi(\mathbf{r})$ is the phase of the wavefield and S is a path that encloses the phase singularity. As noted, the topological charge is a measure of the number of times that the phase increases by 2π as one traverses a path around the singularity.

It can be shown that only vortices with topological charge $s = \pm 1$ are generic. Vortices of topological charge $|s| > 1$ may be thought of as s vortices of unit topological charge superimposed upon one another. Such higher-order vortices are unstable under amplitude and phase variations of the field, and typically break into s first-order vortices when perturbed; possible perturbations include reflection from a rough surface and passing through a random phase plate.

General observations about logarithmic branch points may be applied almost directly to optical vortices. Just as we noted that each branch point must be accompanied by its topological counterpart in a sort of "conservation of branch points", the total topological charge of an optical wavefield must be conserved under perturbations of the amplitude and phase of the wavefield. This in turn implies that vortices can only be "created" or "destroyed" in pairs of opposite topological charge, in rough analogy with the creation of particle/anti-particle pairs in high-energy particle physics. It is to be noted, however, that the conservation of topological charge potentially includes vortices at infinity, just as we had branch points at infinity.

The connection between vortices and branch points is not quite exact, as the wavefield is only a solution of Laplace's equation in the immediate neighborhood of a vortex. However, the wavefield is still analytic in a more general sense in all of three-dimensional space, as noted in Section 10.6.

The conservation of topological charge under perturbations of the wavefield has led a number of authors to suggest that vortices might be used as information carriers for free-space optical communications systems [GCP+04, GT08].

The circulation of phase in optical vortices can in many cases be associated with circulation of momentum in the light field and hence orbital angular momentum. Laser beams possessing a high-order vortex, such as the Laguerre–Gauss beams to be discussed in Section 18.3.6, have been used to trap and rotate microscopic particles, making an "optical spanner" (see, for instance, [ABP03]). This ability to trap particles has been used to make microscopic light-driven devices such as a microoptomechanical pump [LG04].

The existence of optical vortices causes serious difficulty in phase reconstruction problems, in which the phase of a wavefield is reconstructed from measurements of the intensity of the wavefield. First, it is to be noted that the presence of a vortex automatically makes the reconstruction process non-unique; considering Eq. (9.203), it can be seen that a vortex with topological charge $+s$ has exactly the same intensity profile $|U(\mathbf{r})|^2$ as a vortex with topological charge $-s$. Furthermore, the existence of these vortices implies that the phase $\phi(\mathbf{r})$ to be reconstructed is not a single-valued function but rather an infinite collection of Riemann sheets tied together in a non–trivial way.

The generic properties of optical vortices can be used to alleviate these problems somewhat. Because vortices in a wavefield are typically of charge $+1$ or -1, one can make educated guesses as to the nature of any vortices in the reconstruction of a phase function;

this has been done with some success [AFN$^+$01]. In highly randomized wavefields, e.g. speckle fields, there are additional relationships between neighboring vortices that might be used to improve the guesses [Fre98].

The principles of singular optics have been extended to other types of "phase singularities" of wavefields, including singularities in the Poynting vector, singularities in the state of polarization of the wavefield, and singularities in the correlation function. Reviews of these topics can be found in [DOP09, SV01].

9.16 Additional reading

- M. J. Ablowitz and A. S. Fokas, *Complex Variables* (Cambridge, Cambridge University Press, 1997). A very readable and thorough book on complex analysis and its applications.
- J. W. Dettman, *Applied Complex Variables* (New York, Macmillan, 1965). This book is a very rigorous treatment of complex analysis, but is consequently a little harder to read.
- E. T. Copson, *An Introduction to the Theory of Functions of a Complex Variable* (Oxford, Clarendon Press, 1935). A classic older text on the subject.
- K. Knopp, *Theory of Functions* (New York, Dover, 1996). A very short and sweet discussion of the theory of complex variables, with two volumes of material totaling 145 pages!

9.17 Exercises

1. Can a function $f(z)$ which is purely real-valued (i.e. $\text{Im}[f(z)] = 0$) be analytic? Describe the behavior of such functions.
2. In explaining Cauchy's theorem, we limited ourselves to simple contours. How would an extension to complex contours affect later theorems such as the residue theorem?
3. Read the paper by E. Wolf, Is a complete determination of the energy spectrum possible from measurements of the degree of coherence?" *Proc. Phys. Soc.* **80** (1962), 1269–1272, in which the properties of an analytic function are used to solve a phase problem. What function is analytic, and why? Under what circumstances is it not possible to solve the phase problem?
4. Read the paper by D. L. Fried, Branch point problem in adaptive optics, *J. Opt. Soc. Am. A* **15** (1998), 2759–2767. Describe the purpose of adaptive optics, and how the presence of branch points (optical vortices) complicates the problem. What solutions are proposed?
5. Assuming an analytic function of the form $f = u + iv$, find the harmonic conjugates of the functions
   (a) $u = x^2 - y^2 + 2x + y$,
   (b) $v = xe^x \cos y - ye^x \sin y$.
6. Show that the choice $\delta z = \delta u e^{i\phi}$ in the definition of a complex derivative, with δu real-valued and ϕ an arbitrary angle results in the Cauchy–Riemann conditions.
7. Assuming an analytic function of the form $f = u + iv$, find the harmonic conjugates of the functions
   1. $v = 2xy + 2x$,
   2. $u = (x^2 - y^2)\cos x e^{-y} - 2xy \sin x e^{-y}$.
8. Evaluate the integral $\oint_C f(z)dz$ for the following functions using Cauchy's integral formula:

1. $f(z) = \frac{1}{z^2+1}$, with C the path $|z| = 2$,

2. $f(z) = \frac{\sin z}{z^2(z-1)}$, with C the path $|z-1| = 1/2$.

9. Evaluate the integral $\oint_C f(z)dz$ for the following functions using Cauchy's integral formula:

   1. $f(z) = \frac{e^z}{z(z+2)}$, with C the path $|z| = 1$,

   2. $f(z) = \frac{\cos z}{(z+\pi)^2}$, with C the path $|z| = 2\pi$.

10. Suppose that the function $f(z)$ is analytic on and within a closed contour C and that z_0 lies within C. Show that

$$\oint_C \frac{f'(z)}{z-z_0}dz = \oint_C \frac{f(z)}{(z-z_0)^2}dz.$$

11. Prove that

$$\delta_{mn} = \frac{1}{2\pi i}\oint_C z^{m-n-1}dz,$$

where δ_{mn} is the Kronecker delta, m and n are integers and C is a closed contour containing the origin.

12. Find the radii of convergence of the following series:

   1. $\sum_{n=1}^{\infty} \frac{z^n}{n}$,
   2. $\sum_{n=1}^{\infty} \frac{n}{2^n} z^n$.

13. Find the radii of convergence of the following series:

   1. $\sum_{n=1}^{\infty} n^n z^n$,
   2. $\sum_{n=1}^{\infty} [3+(-1)^n]^n z^n$.

14. Determine the series expansion (Taylor or Laurent) of the function

$$f(z) = \frac{z^2}{(z+1)(z-2)}$$

about the point $z = 1$ and in the annulus $1 < |z| < 2$.

15. Determine the series expansion (Taylor or Laurent) of the function

$$f(z) = \frac{z+1}{z^2-2}$$

on the circle $|z-\sqrt{2}| = 1$ and on the circle $|z| = 2$.

16. The most general series expansion we considered in this chapter is of the form

$$f(z) = z^\alpha \log z \sum_{n=-\infty}^{\infty} a_n z^n.$$

Give an example of a function that cannot be described by such an expansion.

17. Find and classify the singular points of the following functions:

   1. $\frac{1}{z(z^2+4)^2}$,
   2. $\frac{1-e^z}{1+e^z}$,
   3. $\frac{\sin[\sqrt{z}]}{\sqrt{z}}$.

18. Find and classify the singular points of the following functions:

    1. $\frac{z^4}{1+z^4}$,

    2. $\frac{1}{\sin z - \sin a}$, with a an arbitrary real constant,

    3. $\frac{z-1}{\sqrt{z^2-1}}$.

19. Find the location of the singularities of the following functions and their residues:

    1. $\frac{1}{z^3-z^5}$,

    2. $\frac{\sin 2z}{(z+1)^3}$.

20. Find the location of the singularities of the following functions and their residues:

    1. $\frac{z^2+z-1}{z^2(z-1)}$,

    2. $z^n \sin(1/z)$, with n a positive integer.

21. Evaluate the integral

$$\int_0^{2\pi} \frac{d\theta}{1-2a\cos\theta+a^2}.$$

22. Evaluate the integral

$$\int_0^{2\pi} \frac{d\theta}{(a+b\cos\theta)^2}.$$

23. Evaluate the integral

$$\int_{-\infty}^{\infty} \frac{\cos\alpha x}{x^2+\beta^2}\,dx,$$

    with α and β real.

24. Evaluate the integral

$$\int_{-\infty}^{\infty} \frac{x\sin x}{x^2+2x+2}\,dx.$$

25. Find the principal value of the integral

$$\int_{-\infty}^{\infty} \frac{\sin x}{(x^2+9)(x-2)}\,dx.$$

26. Find the principal value of the integral

$$\int_{-\infty}^{\infty} \frac{\cos x}{(x^2-4)(x+1)}\,dx.$$

27. Describe the Riemann surfaces of the following functions, giving the location of the branch points, the number of surfaces and their connections:

    1. $f(z) = \log(z)\sqrt{z+2}$,

    2. $f(z) = [(z-1)^2(z+1)]^{1/4}$.

28. Describe the Riemann surfaces of the following functions, giving the location of the branch points, the number of surfaces and their connections:

1. $f(z) = [(z-1)(z+1)^2]^{1/3}$,
2. $f(z) = (z-1)^{1/3} + (z-2)^{1/4}$.

29. Determine the topological charge of an optical vortex which has the mathematical form

$$U(x,y) = a(x+iy) + b(x-iy),$$

in its immediate neighborhood, with a and b complex constants.

10

Advanced complex analysis

10.1 Introduction

As we have noted, ideas and techniques of complex analysis will be useful in nearly all subjects to be discussed in the remainder of this book. Here we cover a few advanced topics of complex analysis that are not encompassed by the later chapters.

10.2 Analytic continuation

We have seen how, given an analytic function $f(z)$, we can determine a Taylor series representation $f_{\text{taylor}}(z)$ of that function. This representation will typically be analytic (convergent) over a region smaller than that of the function $f(z)$ itself. For instance, the function $1/(1-z)$ is analytic everywhere except the point $z = 1$; the Taylor series of this function about the point $z = 0$ is simply the geometric series $\sum_{n=0}^{\infty} z^n$, which is analytic only for $|z| < 1$.

A natural problem to investigate, then, is the inverse of the above process: given a Taylor series representation $f_{\text{taylor}}(z)$ of a function, can we find the values of the "original" function $f(z)$ in regions beyond the domain of convergence of the Taylor series? The surprising thing about analytic functions is that it is in fact possible to do so, using a process called *analytic continuation*.

As an example, we will consider the geometric series $\sum_{n=0}^{\infty} z^n$, which is the Taylor expansion of the function $1/(1-z)$ about the point $z = 0$. We already mentioned that this series only converges for $|z| < 1$. However, the function $1/(1-z)$ is analytic for all finite points, save $z = 1$. Therefore the series converges all the way up to the singularity, as illustrated in Fig. 10.1.

This is in agreement with the statement of Section 9.7 that the radius of convergence is dictated by the location of the nearest singularity. We may further demonstrate this by performing a Taylor expansion about a different point, say $z = i$,

$$\frac{1}{1-z} = \sum_{n=0}^{\infty} \frac{1}{(1-i)^{n+1}} (z-i)^n. \tag{10.1}$$

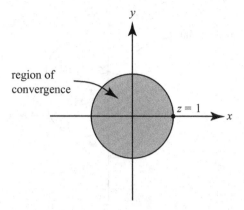

Figure 10.1 Taylor series convergence of $f(z) = \sum_{n=0}^{\infty} z^n$ to the nearest singularity.

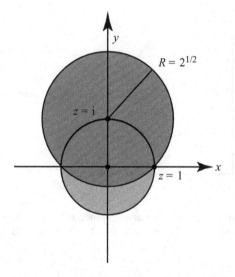

Figure 10.2 Comparison of convergence radii for Taylor series about different points.

If we use the ratio test to determine the radius of convergence,

$$R = \lim_{n \to \infty} \left| \frac{a_n}{a_{n+1}} \right| = |1 - i| = \sqrt{2}. \tag{10.2}$$

In Fig. 10.2, we find that our suspicion is confirmed.

This example suggests a procedure for finding the original function that our Taylor series represents. Starting with our original series, Taylor expand about different points within the radius of convergence; if chosen well, the radius of convergence of the new power series has points that lie outside of the original radius of convergence – we have "continued" the

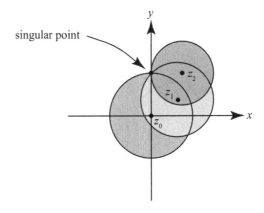

Figure 10.3 Extending or continuing our function beyond its radius of convergence.

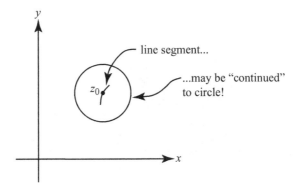

Figure 10.4 The least amount of information about a function you need to analytically continue. The behavior of an analytic function on a line segment may be "continued" to the interior of a circle.

function outside of its original radius of convergence. This procedure may then be repeated, Taylor expanding about different points, and the value of the function can in principle be determined everywhere the original function is analytic, as illustrated in Fig. 10.3.

How much information is required to analytically continue an analytic function? It turns out that one may perform analytic continuation if one knows the value of $f(z)$ on any finite length line segment in the complex plane, as illustrated in Fig. 10.4.

This is intuitively reasonable, because a Taylor series expansion requires the value of a function and all of its derivatives at a point. To determine those derivatives, one needs the value of the function along a line segment, and then the limit definition, Eq. (9.41), may be used.

Interestingly enough, from this reasoning it immediately follows that if an analytic function is zero on any finite length line segment, it is zero everywhere within the domain of

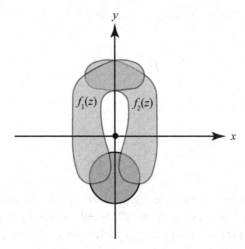

Figure 10.5 The continued analytic functions $f_1(z)$ and $f_2(z)$ will be equal in their overlap region – unless a branch point is located in the enclosed region between their domains.

analyticity. Furthermore, if an analytic function is constant on any finite length line segment, it is constant everywhere within the domain of analyticity.

There are a few caveats to the application of analytic continuation. An analytic continuation is in general unique, and different paths of continuation typically result in the same function. The notable exception to this is the case when two distinct paths of continuation encircle a branch point in opposite senses, as illustrated in Fig. 10.5. The continued functions $f_1(z)$ and $f_2(z)$ are not equal in their overlap region, because the original function has been continued to two different Riemann surfaces.

It is also possible to find functions that cannot be continued beyond a finite region of the complex plane. An illustrative example is the analytic function

$$f(z) = \sum_{n=0}^{\infty} z^{2^n}. \tag{10.3}$$

Applying the ratio test to this function, we find that the radius of convergence is unity, so the function is defined for $|z| < 1$. To analytically continue onto the unit circle, we need to find a continuous range of points on the circle for which $f(z)$ is analytic. Clearly, however, the function is non-analytic at $z = 1$, at which the sum is infinite. We now note that the function $f(z)$ satisfies the equation

$$f(z^2) = \sum_{n=0}^{\infty} (z^2)^{2^n} = \sum_{n=1}^{\infty} z^{2^n} = f(z) - z. \tag{10.4}$$

Continuing this process for $f(z^4)$, and so on, we find that

$$f\left(z^{2^m}\right) = f(z) - \sum_{i=0}^{m-1} z^{2^i}, \tag{10.5}$$

or, rearranging the equation, that

$$f(z) = f\left(z^{2^m}\right) - \sum_{i=0}^{m-1} z^{2^i}. \tag{10.6}$$

Let us consider the value of the function at one of the 2^m roots of unity, i.e. $z_m = e^{2\pi ni/2^m}$, where $n = 0, \ldots, m-1$. For instance, for $m = 2$, $z_2 = 1, -1, i, -1$. According to Eq. (10.6), however, the function $f(z)$ is non-analytic at these points, because $f\left(z^{2^m}\right)$ is the sum of $f(1)$, which is infinite, and a finite sum of terms. This process can be extended to $m \to \infty$, in which case we find that the boundary $|z| = 1$ is densely packed with singular points, and there is no path through the boundary by which one can perform analytic continuation. This is an example of a *natural boundary*; such boundaries occasionally appear in mathematical problems.

Though it may seem rather abstract, analytic continuation has great significance in optical problems, in particular for inverse problems, in which an optical image of an object is reconstructed using limited information. If the optical image is represented by an analytic function, it is in principle possible to reconstruct additional features of the image from limited starting information. However, in practice the process of analytic continuation is unstable and very susceptible to small errors in the original function. The result of this is that the function cannot be continued very far beyond its original domain.

The implications of analytic continuation for optics are discussed further in Section 10.6.

10.3 Stereographic projection

In our previous discussions of the complex plane, "infinity" plays a special, somewhat ambiguous role. However, it is possible to create an alternative representation in which "infinity" is represented as a single point which may be treated as any other point in the plane. This representation is quite natural and, as we will see, quite useful in many situations.

We consider a sphere of unit radius whose south pole is situated at the origin of the complex plane; a point on the sphere is defined by coordinates (ξ, η, ζ). For any point z in the plane, we may draw a straight line that intersects both the point z and the north pole of the sphere. This line will intersect the sphere at one additional point (ξ, η, ζ). This construction defines a *mapping* of the complex plane onto the unit sphere; each point z in the finite complex plane is mapped uniquely to a point (ξ, η, ζ) on the surface of the sphere. For instance, the point $z = 0$ corresponds to the south pole of the sphere. This mapping of

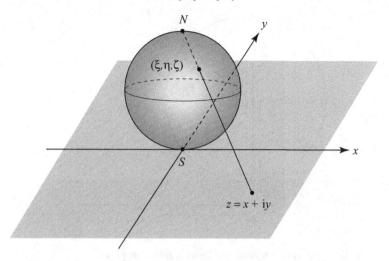

Figure 10.6 The definition of a stereographic projection. A point z in the complex plane is mapped onto the sphere by the straight line connecting z to the north pole.

the plane onto the sphere is known as *stereographic projection*; it is illustrated in Fig. 10.6. The sphere itself is referred to as the *Riemann sphere*.

With points in the finite complex plane mapped, we may consider the nature of infinity. As $z \to \infty$, *in any direction*, the intersection point between the line and the sphere approaches the north pole. We formally define infinity as the point at the top of the sphere, known as the *point at infinity*; curiously, this projection suggests that all paths to infinity in the complex plane arrive at the same place. With the point at infinity included, all points on the sphere are defined and this sphere is referred to as the *compactified complex plane*.

On the Riemann sphere, the point at infinity has a role no different from any other point. This will be a useful philosophical view in the next section, when we discuss conformal mapping, in which the point at infinity can be mapped to any other point in the complex plane, and vice versa.

We may derive explicit formulas for the mapping by using the similar triangles illustrated in Fig. 10.7. From the figure, we may determine that

$$\tan\theta = \frac{2-\zeta}{\rho} = \frac{2}{r},\qquad(10.7)$$

where $\rho = \sqrt{\xi^2 + \eta^2}$ and $r = |z| = \sqrt{x^2 + y^2}$. Furthermore, the coordinates on the sphere must satisfy the equation

$$\xi^2 + \eta^2 + (\zeta - 1)^2 = 1.\qquad(10.8)$$

This may be rewritten as

$$\rho^2 = 1 - (\zeta - 1)^2.\qquad(10.9)$$

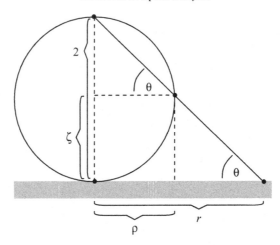

Figure 10.7 Similar triangles to determine the stereographic mapping.

We may combine Eqs. (10.7) and (10.9) into the equation

$$(2-\zeta)^2 = \frac{4}{r^2}[1-(\zeta-1)^2]. \tag{10.10}$$

This quadratic equation has two roots, $\zeta = 2$ (the intersection with the north pole) and, with some effort,

$$\zeta = \frac{2|z|^2}{|z|^2+4}. \tag{10.11}$$

With this solution for ζ, we may substitute back into Eq. (10.9) to write

$$\rho = \frac{4/r}{1+4/r^2}. \tag{10.12}$$

We may also come up with expressions for x and y using the same arguments about similar triangles, i.e.

$$\frac{x}{2} = \frac{x-\xi}{\zeta}, \tag{10.13}$$

$$\frac{y}{2} = \frac{y-\eta}{\zeta}. \tag{10.14}$$

Solving for ξ and η, and using the expression for ζ, we arrive at

$$\xi = \frac{4x}{|z|^2+4}, \tag{10.15}$$

$$\eta = \frac{4y}{|z|^2+4}. \tag{10.16}$$

We may test these expressions and Eq. (10.11) by confirming the behavior at $z = 0$ and as $z \to \infty$. As $z \to \infty$, we find that $(\xi, \eta, \zeta) \to (0, 0, 2)$, and as $z \to 0$, we find that $(\xi, \eta, \zeta) \to (0, 0, 0)$.

The inverse mapping may be determined from the equations for the forward mapping. The result is

$$x = \frac{2\xi}{2 - \zeta}, \tag{10.17}$$

$$y = \frac{2\eta}{2 - \zeta}. \tag{10.18}$$

The most significant result from the derivation of the explicit formulas for the stereographic projection is the demonstration that circles and lines in the complex plane are mapped onto circles on the Riemann sphere. A circle in the complex plane is described by an equation of the form

$$x^2 + y^2 + \alpha x + \beta y + \delta = 0, \tag{10.19}$$

where α and β depend on the position of the center of the circle and δ depends on this position and the circle's radius. Let us convert this equation to (ξ, η, ζ) coordinates,

$$\frac{4\xi^2}{(2 - \zeta)^2} + \frac{4\eta^2}{(2 - \zeta)^2} + \frac{2\alpha\xi}{2 - \zeta} + \frac{2\beta\eta}{2 - \zeta} + \delta = 0. \tag{10.20}$$

From Eq. (10.9), it follows that $\xi^2 + \eta^2 = \zeta(2 - \zeta)$; we may then simplify the equation for the circle to

$$\frac{4\zeta}{2 - \zeta} + \frac{2\alpha\xi}{2 - \zeta} + \frac{2\beta\eta}{2 - \zeta} + \delta = 0. \tag{10.21}$$

Multiplying by $2 - \zeta$, we have

$$2\alpha\xi + 2\beta\eta + (4 - \delta)\zeta + 2\delta = 0. \tag{10.22}$$

This equation is linear in ξ, η, and ζ, and has the form of a dot product; that is, we may write the expression as

$$\boldsymbol{\lambda} \cdot \boldsymbol{\alpha} = 2\delta, \tag{10.23}$$

where $\boldsymbol{\lambda} = (\xi, \eta, \zeta)$ and $\boldsymbol{\alpha} = [2\alpha, 2\beta, (4 - \delta)]$. This dot product defines a plane in (ξ, η, ζ), and the intersection of a plane with a sphere results in a circle.

Similarly, the equation for a line in the complex plane,

$$y = ax + b, \tag{10.24}$$

can also be rewritten as the equation of a plane in (ξ, η, ζ). Lines in the complex plane extend to infinity, and are therefore mapped into circles on the sphere that include the north pole.

In Section 1.5, we introduced the idea of a metric space, a set of elements that have a unique non-negative distance (metric) $\rho(z_1, z_2)$ defined between any pair of elements z_1

and z_2. In the complex plane, this metric is typically taken to be $\rho(z_1, z_2) = |z_1 - z_2|$; on the Riemann sphere, we may introduce the *chordal metric*, which represents the straight line distance between two points on the sphere. Denoting $\lambda = (\xi, \eta, \zeta)$, we define the metric on the sphere as

$$\rho(\lambda_1, \lambda_2) = \sqrt{(\xi_1 - \xi_2)^2 + (\eta_1 - \eta_2)^2 + (\zeta_1 - \zeta_2)^2}. \tag{10.25}$$

By the use of our mapping formulas (10.11), (10.15), and (10.16), we may write this new metric in terms of the complex variables z_1 and z_2 as

$$\rho(z_1, z_2) = \frac{4|z_1 - z_2|}{\sqrt{|z_1|^2 + 4}\sqrt{|z_2|^2 + 4}}. \tag{10.26}$$

With a little effort, it can be shown that this function satisfies the definition of a metric as described in Section 1.5. In the limit that $z_2 \to \infty$, the metric has the form

$$\rho(z_1, \infty) = \frac{4}{\sqrt{|z_1|^2 + 4}}. \tag{10.27}$$

In the chordal metric, we may assign a finite value for the distance between z_1 and infinity.

One of the advantages of stereographic projections is the ability to depict the Riemann surfaces of multivalued functions in a compact and elegant way. By the use of such projections, and more general topological transformations,[1] we may construct very simple models of various Riemann surfaces; the discussion of this section follows [Spr81].

We noted in Section 9.10 that it is not possible to construct a Riemann surface of a function such as $f(z) = z^{1/2}$ which does not cross itself in three-dimensional space. We now attempt to do this by mapping each Riemann sheet of the function to a sphere. The result is a pair of spheres, each of which has a "rip" from the north pole to the south pole. The rip must be at the same longitude for both spheres, but otherwise is arbitrary; these are illustrated in Fig. 10.8(a).

With the rip, however, it is possible to map each of these spheres onto a hemisphere; physically, one can imagine the process as slicing the rind of an orange from top to bottom and then pulling it around the sides of the orange until one is left with a hemispherical rind. The result is that each Riemann sheet may be drawn as a single hemisphere, as shown in Fig. 10.8(b).

To construct the Riemann surface, we need only join the two surfaces along their appropriate cut lines; the result is a single sphere, missing only the north and south poles! This is illustrated in Fig. 10.8(c). Evidently the Riemann surface of the function $f(z) = z^{1/2}$ is topologically equivalent to a sphere missing two points.

Furthermore, any function with two Riemann sheets and two branch points can also be mapped to a single sphere. A function such as $f(z) = \sqrt{(z-a)(z-b)}$, with $b \neq a$, has two

---

[1] *Topology* is the study of those spatial properties of an object that are preserved under "stretching" but no "tearing" or "gluing". A solid ball, for instance, is topologically equivalent to a solid cube, but not equivalent to a hollow ball.

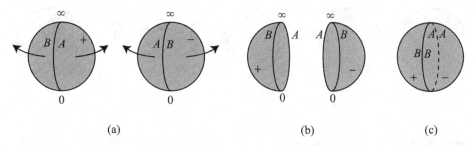

Figure 10.8 Construction of the Riemann surface for $f(z) = z^{1/2}$. (a) The two sheets (labeled '+' and '−') projected onto spheres, each with a rip from the north to south pole. The cut edges A must be joined together, likewise for B. (b) Distorting of the ripped spheres into hemispheres. (c) Joining of the hemispheres into a single sphere with two holes, at the north and south poles.

Riemann sheets and branch points as $z = a$ and $z = b$. Treating the branch cut as lying between the two points, we may "stretch" the rip on each sphere to the points $z = 0$ and $z = \infty$, and then the preceding construction may be applied directly.

A function with any finite number of Riemann sheets and only two branch points may also be mapped to a single sphere. For instance, $f(z) = z^{1/3}$ has three sheets, each of which may be projected to a sphere with a rip from the north to south pole. Each of these ripped spheres may be peeled to form one third of a sphere, and the thirds may be joined together to form a whole sphere.

Functions with more than two branch points result in more complicated topological structures. For instance, the function

$$f(z) = \sqrt{(z-a)(z-b)(z-c)(z-d)}, \qquad (10.28)$$

with $a \neq b \neq c \neq d$, has four branch points at $z = a, b, c, d$, but only two Riemann surfaces (each branch point can only change the entire function by $e^{\pi i}$). Each surface becomes a sphere with two rips in it, which we take to run from $a \to b$ and $c \to d$. We note that a sphere with two rips in it is topologically comparable to an open-ended cylindrical surface, or a "tube". By converting one of the Riemann spheres into a tube, we may construct the Riemann surface as a sphere with a "handle" sticking out of it; the process is illustrated in Fig. 10.9. Alternatively, we may make both Riemann spheres into tubes, in which case they connect together to form a torus.

It is to be noted that the stereographic projection described in this section is not unique. Another common projection results from allowing the origins of the unit sphere and the complex plane to coincide. The mapping is still defined by the intersection of the sphere and the line which connects the north pole to the point z. In fact, one can define a stereographic projection by allowing the sphere to lie with its north pole at any finite distance above the plane. All the results of this section remain valid, notably the mapping of circles and lines into circles on the sphere, though the specific functional form of the mapping is different for different sphere positions.

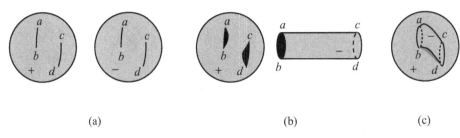

(a)   (b)   (c)

Figure 10.9 Construction of the Riemann surface for $f(z) = \sqrt{(z-a)(z-b)(z-c)(z-d)}$. (a) The two sheets (labeled '+' and '−') projected onto spheres, each with two rips (branch cuts). (b) One of the ripped spheres is distorted into a tube. (c) Joining of the tube and hemisphere along the appropriate cut lines, resulting in a "ball with a handle".

Example 10.1 (Poincaré sphere) A number of physical quantities which would otherwise be very complicated and hard to visualize can be represented quite naturally on a spherical surface. An example of this is the state of polarization of a light wave, which can be represented by Jones vectors as described in Section 4.1. By use of a stereographic projection, this state of polarization can be projected onto a sphere known as the *Poincaré sphere* [Poi92, Chapter 12]. This representation has proven very useful in analyzing the changes in polarization arising on propagation through optical systems; here we follow the discussion introduced in [Jer54].

We begin with Eq. (4.2) for the complex electric field vector of an electromagnetic plane wave, with components

$$E_x(t) = \epsilon_x e^{i\phi_x} e^{-i\omega t}, \tag{10.29}$$

$$E_y(t) = \epsilon_y e^{i\phi_y} e^{-i\omega t}. \tag{10.30}$$

We determine the behavior of this complex field vector by considering the ratio of these components,

$$\frac{E_y}{E_x} = \frac{\epsilon_y}{\epsilon_x} \exp[i\Delta] = \frac{\epsilon_y}{\epsilon_x} \cos\Delta + i\frac{\epsilon_y}{\epsilon_x} \sin\Delta \equiv u + iv, \tag{10.31}$$

where $\Delta \equiv \phi_y - \phi_x$ and u, v characterize the location of this ratio in the complex plane.

The physically observable electric field $[\mathcal{E}_x(t), \mathcal{E}_y(t)]$ is simply determined by the real parts of Eqs. (10.29) and (10.30), i.e. $\mathcal{E}_x(t) = \mathrm{Re}\{E_x(t)\}$, $\mathcal{E}_y(t) = \mathrm{Re}\{E_y(t)\}$. Eliminating time from these field components results in an expression for the path of the electric field vibrations,

$$\frac{\mathcal{E}_x^2}{\epsilon_x^2} + \frac{\mathcal{E}_y^2}{\epsilon_y^2} - 2\frac{\mathcal{E}_x \mathcal{E}_y}{\epsilon_x \epsilon_y} \cos\Delta = \sin^2\Delta. \tag{10.32}$$

This is the equation for an ellipse that is inscribed in a box of side length $2\epsilon_x$ and $2\epsilon_y$, as illustrated in Fig. 10.10. It is clear that the box depends on the specific choice of coordinate system, and we would prefer to characterize the ellipse by its orientation ϕ and ellipticity θ;

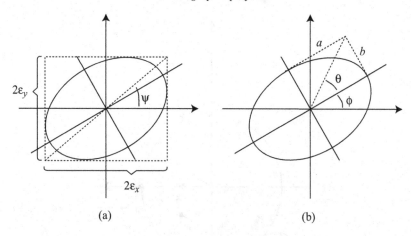

Figure 10.10 (a) The polarization ellipse, as characterized by its inscribing rectangle. (b) The polarization ellipse, defined by its orientation, ellipticity, and major and minor axes.

unfortunately, it is in general a non trivial task to transform from ϵ_x, ϵ_y, and Δ into the orientation and ellipticity.

If we return to Eq. (10.31), we may, however, use a stereographic projection to map the complex state variable $u + iv$ onto a unit sphere, known as the *Poincaré sphere*. By convention, the sphere is oriented such that the origins of the (u, v)-plane and the sphere coincide, the v-axis is parallel to the z-axis, and the u-axis is antiparallel to the y-axis; this is shown in Fig. 10.11.

With a significant amount of mathematical effort, one can show that the coordinates on the Poincaré sphere are directly related to the ellipticity and orientation of the polarization ellipse by the equations

$$x = \cos(2\theta)\cos(2\phi), \tag{10.33}$$

$$y = \cos(2\theta)\sin(2\phi), \tag{10.34}$$

$$z = \sin(2\theta). \tag{10.35}$$

In geometric terms, the tangent of half the latitude on the sphere represents the ellipticity of the polarization ellipse, while the tangent of half the longitude on the sphere represents the orientation.

Different regions of the Poincaré sphere may be directly related to different states of polarization. The equator of the sphere corresponds to $v = 0$, and represents linearly polarized light ($\Delta = 0$). The north pole of the sphere corresponds to $v = 1$, $u = 0$, and represents right circularly polarized light ($\Delta = \pi/2$). The south pole of the sphere corresponds to $v = -1$, $u = 0$, and represents left circularly polarized light ($\Delta = -\pi/2$).

The Poincaré sphere can also be interpreted in terms of another set of parameters commonly used to characterize the state of polarization, the *Stokes parameters*. These four

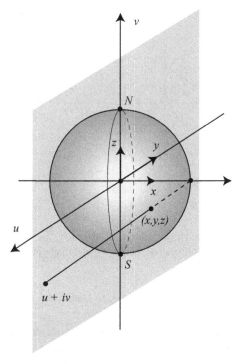

Figure 10.11 The stereographic projection to the Poincaré sphere. It is to be noted that the projection is taken from the "back" side of the complex (u, v) plane.

parameters are defined as

$$S_0 = E_x^* E_x + E_y^* E_y, \tag{10.36}$$

$$S_1 = E_x^* E_x - E_y^* E_y, \tag{10.37}$$

$$S_2 = E_x^* E_y + E_y^* E_x, \tag{10.38}$$

$$S_3 = -i(E_x^* E_y - E_y^* E_x). \tag{10.39}$$

These parameters represent observable properties of the field that can be measured with simple polarization-sensitive optical elements; for instance, S_0 represents the total intensity of the light wave. It can be shown, with lengthy calculations, that the latter three Stokes parameters can be written as

$$S_1 = S_0 \cos(2\theta) \cos(2\phi), \tag{10.40}$$

$$S_2 = S_0 \cos(2\theta) \sin(2\phi), \tag{10.41}$$

$$S_3 = S_0 \sin(2\theta), \tag{10.42}$$

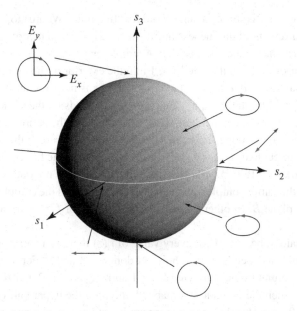

Figure 10.12 The relationship between the state of polarization, the Poincaré sphere, and the Stokes parameters. Figure courtesy of Professor Taco Visser of Delft University.

which shows that the Stokes parameters are directly related to points on a Poincaré sphere of radius S_0. This relation is illustrated in Fig. 10.12.

Further discussion of the Stokes parameters and their relation to the Poincaré sphere can be found in [Bro98, Wol07].

The Poincaré sphere is an elegant way to visualize the state of polarization of an optical field; furthermore, the effects of optical devices on the state of polarization can be interpreted geometrically through the use of the sphere. The Poincaré sphere has also been used to derive [Ara92] a simple proof of Pancharatnam's theorem [Pan56], in which a polarization vector acquires a geometric phase in traversing a closed path in its state of polarization.

◇

10.4 Conformal mapping

The mapping of the extended complex plane onto the Riemann sphere is but one of an infinite number of transformations of the complex plane that can be used in applications. In this section, we discuss a broad class of such transformations that are known as *conformal mappings* that are especially useful in finding solutions to Laplace's equation.

To make things more precise, we introduce a proper definition of a mapping for the complex plane.

Definition 10.1 (Mapping) *A mapping $f(z)$ is a rule that assigns to each element z of a domain A in the complex plane a unique number $f(z)$ in a domain B in the complex plane.*

In essence, "mapping" is simply a fancy word for "function". We are more restrictive here, however; the requirement of uniqueness means that $f(z) = z^2$ is a mapping but $f(z) = z^{1/2}$ is not unless a particular branch is specified. A simple mapping is the function $f(z) = 1/z$, that maps the interior of the unit circle $|z| < 1$ into the exterior of the unit circle $|f(z)| > 1$.

A mapping is said to be *one-to-one* if every value of z is mapped into a unique value of $f(z)$, i.e. $f(z_1) = f(z_2)$ implies $z_1 = z_2$. This property depends on the choice of domains A and B. For instance, if A is the upper half-space $z > 0$, and B is the entire complex plane, the function $f(z) = z^2$ is one-to-one. This can be seen by letting $z = re^{i\theta}$; then $z^2 = r^2 e^{2i\theta}$ and every point in the upper half-space ($0 < \theta < \pi$) is mapped into the full complex plane, with the exception of the line $\text{Re}(z^2) \geq 0$. The same function $f(z)$ is not one-to-one if A and B are both taken as the entire complex plane; in this case, z^2 maps the complex plane A *twice* into the complex plane B. In other words, for every value of $f(z)$ there are two possible values of z.

A mapping is said to be *onto* if for every value of $f(z)$ there is a corresponding value of z. Again this definition depends on the choice of domains A and B. For instance, if A is the first quadrant of the unit circle satisfying $|z| < 1$ and $0 \leq \theta < \pi/2$ and B is the complete unit circle $|z| < 1$, then the domain A is mapped only into the upper half of the unit circle by $f(z) = z^2$, i.e. $0 \leq \arg[f(z)] < \pi$; the mapping is not onto. If A is upper half of the unit circle satisfying $|z| < 1$ and $0 \leq \theta < \pi$ and B is the complete unit circle $|z| < 1$, then the domain A is mapped entirely into the unit circle, and the mapping is onto.

Theorem 10.1 (Inverse mapping) *The inverse $f^{-1}(z)$ of a mapping $f(z)$ exists if and only if $f(z)$ is one-to-one and onto. The inverse satisfies $f^{-1}[f(z)] = z$.*

If a function is not one-to-one, then there exist regions of B in which there are multiple values of z for a given value of $f(z)$, and it is not possible to come up with a unique inverse. If a function is not onto, then there exist regions of B that have no counterpart in A, and it is not possible to define an inverse mapping. These cases are illustrated in Fig. 10.13. This is not a rigorous proof, but illustrates the necessity of one-to-one and onto for the existence of an inverse. We will generally require our mappings to be invertible, for reasons which will become clear.

Let us now consider the problem of solving Laplace's equation,

$$\frac{\partial^2 V}{\partial x^2} + \frac{\partial^2 V}{\partial y^2} = 0, \tag{10.43}$$

in a general domain D subject to a boundary condition $V_0(x,y)$. For a general boundary, this problem is in general difficult to solve, but let us suppose there exists an invertible mapping $w(z)$ that maps the domain D onto the unit circle, and *the mapping preserves the form of Laplace's equation*. This mapping will also transform the boundary condition into the form $V_0(\phi)$. In Section 9.1, we saw that we can always find an analytic series solution to Laplace's equation on the interior of the unit circle. Once this is done, we may use the inverse mapping $w^{-1}(z)$ to map the solution back into the original domain D.

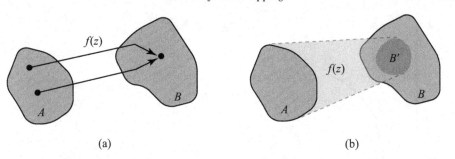

Figure 10.13 (a) A function which is not one-to-one; two points in A map to the same point in B. (b) A function which is not onto; the domain A maps to a domain B' that is smaller than B.

What conditions must the mapping $w(z) = u + iv$ satisfy, apart from being invertible? We use the chain rule to write the derivatives of V with respect to x and y in terms of derivatives with respect to u and v,

$$\frac{\partial V}{\partial x} = \frac{\partial V}{\partial u}\frac{\partial u}{\partial x} + \frac{\partial V}{\partial v}\frac{\partial v}{\partial x}, \tag{10.44}$$

$$\frac{\partial V}{\partial y} = \frac{\partial V}{\partial u}\frac{\partial u}{\partial y} + \frac{\partial V}{\partial v}\frac{\partial v}{\partial y}. \tag{10.45}$$

Extending this to second derivatives, we have (writing $u_x \equiv \frac{\partial u}{\partial x}$ for brevity)

$$\frac{\partial^2 V}{\partial x^2} = V_{uu}u_x^2 + V_{uv}u_x v_x + V_u u_{xx} + V_{vv}v_x^2 + V_{vu}u_x v_x + V_v v_{xx}, \tag{10.46}$$

$$\frac{\partial^2 V}{\partial y^2} = V_{uu}u_y^2 + V_{uv}u_y v_y + V_u u_{yy} + V_{vv}v_y^2 + V_{vu}u_y v_y + V_v v_{yy}. \tag{10.47}$$

If we assume that our function $w(z)$ is analytic, then we immediately note that $w_{xx} + w_{yy} = 0$ and, by the Cauchy–Riemann conditions, $u_x v_x = -u_y v_y$, $(u_x)^2 = (v_y)^2$ and $(u_y)^2 = (v_x)^2$. Laplace's equation then takes on the form

$$\frac{\partial^2 V}{\partial x^2} + \frac{\partial^2 V}{\partial y^2} = |f'(z)|^2 \left(\frac{\partial^2 V}{\partial u^2} + \frac{\partial^2 V}{\partial v^2} \right) = 0. \tag{10.48}$$

We find that the form of Laplace's equation is preserved under the mapping $w(z)$, *provided that the function $w(z)$ is analytic and $|f'(z)| \neq 0$.*

Mappings that are analytic and satisfy $|f'(z)| \neq 0$ can readily be shown to be *conformal*.

Definition 10.2 (Conformal mapping) *A* conformal map *is a map that preserves the magnitude of the angle between two intersecting curves, and their sense.*

The idea of a conformal map is illustrated in Fig. 10.14. If two curves A and B intersect in the z-plane such that their tangent lines form an angle θ, their intersection will form the

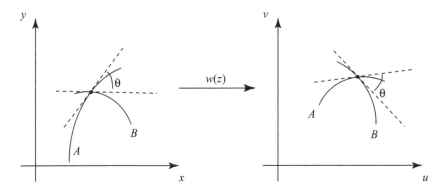

Figure 10.14 Illustrating the preservation of magnitude and sense of angle in a conformal transformation.

same angle θ in the w-plane. Also, if A is more counterclockwise than B in the z-plane, it remains more counterclockwise in the w-plane; the "sense" of the angle is preserved.

We may show that an analytic map satisfying $|f'(z)| \neq 0$ is conformal by the following argument. Let us consider the point z_0 and consider two parametric curves $z_1(t)$ and $z_2(t)$ passing through that point, such that $z_1(0) = z_2(0) = z_0$. The tangent curves at the point z_0 are

$$z_1(t) = z_0 + z_1'(0)t, \tag{10.49}$$

$$z_2(t) = z_0 + z_2'(0)t. \tag{10.50}$$

As the slope of the line z_1 is simply given by $\arg(z_1')$, we note that the angle between the two curves is $\arg[z_2'(0)] - \arg[z_1'(0)]$. We now consider a mapping $w(z)$; in the neighborhood of z_0, this mapping has the approximate form (via a Taylor series representation)

$$w(z) \approx w(z_0) + w'(z_0)(z - z_0). \tag{10.51}$$

This implies that the tangent curve for z_1 now has the form

$$w_1(z) \approx w(z_0) + w'(z_0)z_1'(0)t. \tag{10.52}$$

The angle that this curve makes with the u-axis is simply $\arg(w_1) = \arg[w'(z_0)] + \arg[z_1'(0)]$; a similar result holds for the tangent curve z_2. The relative angle between the tangent curves is simply $\arg(w_2) - \arg(w_1) = \arg[z_2'(0)] - \arg[z_1'(0)]$, which is the same angle the tangent curves had in the z-plane. An analytic map satisfying $|f'(z)| \neq 0$ is therefore conformal.

From the above argument, one may also show that this mapping magnifies distances by a factor $|w'(z_0)|$. From Eq. (10.51), we can readily see that

$$\frac{|w(z) - w(z_0)|}{|z - z_0|} \approx |w'(z_0)|. \tag{10.53}$$

The usefulness and power of conformal maps was demonstrated by Riemann, who proved that one can map nearly any domain in the complex plane onto the unit circle.

Theorem 10.2 (Riemann mapping theorem) *Let D be a simply connected domain of the complex plane that is not the finite complex plane or the completed complex plane (with point at infinity). Then there exists a conformal analytic function $w = f(z)$ that provides a one-to-one map onto the unit circle $|w| < 1$. This mapping is unique if a point z_0 is specified that maps into the origin, i.e. $w(z_0) = 0$, and the direction of a curve passing through the point z_0 is specified in the w-plane.*

Somewhat hidden in this statement is that the mapping is "onto" the unit circle, which implies that it is an invertible mapping, as well. Proof of the Riemann mapping theorem is somewhat involved and not useful for our purposes, so we omit it here and refer to [Det65, Chapter 6]. It is to be noted, however, that Riemann's theorem is *non-constructive*; that is, it does not tell us *how* to find a mapping onto the unit circle, only that such a mapping exists. As noted in the introduction to this section, this theorem suggests that Laplace's equation in two dimensions may always be mapped into the unit circle, at which point the solution is readily found.

Because $f(z)$, and hence $f'(z)$, are analytic functions whose zeros must be isolated, this suggests that an analytic function (excluding the trivial constant function) is non-conformal only at isolated points, which may be referred to as singular points of the mapping. What is the behavior in the neighborhood of such a singular point z_0? The function must have a Taylor series representation at the point of the form

$$f(z) = a_0 + a_2(z - z_0)^2 + a_3(z - z_0)^3 + \cdots. \tag{10.54}$$

Let us consider the function $f(z) = z^2$ at the non-conformal point $z = 0$; a straight line passing through the origin has the form $z(t) = te^{i\phi}$, with ϕ a constant. This line is mapped into a curve whose local tangent has an argument 2ϕ; the argument is doubled. A pair of lines intersecting at an angle θ passing through the origin will not have the angle θ preserved under the transformation.

There are a large number of mappings one can consider. Two broad classes are worth special attention: bilinear transformations and Schwarz–Christoffel transformations.

Definition 10.3 (Bilinear transformation) *A* bilinear transformation *(also known as a* linear fractional *or* Möbius transformation*) is a transformation of the form*

$$w = f(z) = \frac{az + b}{cz + d}, \tag{10.55}$$

with a, b, c, d complex, subject to $ad - bc \neq 0$.

These transformations have a large number of significant properties. This map is conformal at all points in the complex plane, including the point at infinity. The inverse transformation can be readily found to be

$$w^{-1}(z) = \frac{dz - b}{-cz + a}. \tag{10.56}$$

On substituting from Eq. (10.55), one can readily find that $w^{-1}[f(z)] = z$.

All bilinear transformations have one or two fixed points z_0 that do not change under transformation, i.e. $f(z_0) = z_0$. This expression results in a quadratic equation,

$$cz_0^2 + (d - a)z_0 + b = 0. \tag{10.57}$$

This equation has two roots, except for the case $c = 0$ (in which case the transformation is linear) and the case $(d - a)^2 + 4bc = 0$, in which the roots are redundant.

Perhaps the most important property of a bilinear transformation is that it maps circles and lines into other circles or lines. To see this, we write the equation of a line as

$$Ax + By + C = 0, \tag{10.58}$$

where A, B, and C are real constants. We may use Eqs. (9.25) to write this equation in the form

$$Dz + D^*z^* + C = 0, \tag{10.59}$$

where $D \equiv (A + iB)/2$. The equation of a circle may be written as

$$|z - z_0|^2 = R^2, \tag{10.60}$$

where z_0 is the center of the circle and R its radius. This may be expanded into the general form

$$zz^* + E^*z + Ez^* + F = 0. \tag{10.61}$$

Let us apply a bilinear transformation of the form of Eq. (10.56) to Eq. (10.59). We end up with the cumbersome expression

$$(|c|^2 - 2\text{Re}[Dc^*d])ww^* + 2\text{Re}[(Da^*d + Dc^*b - Ca^*c)w] - 2\text{Re}[Da^*b] + |a|^2C = 0. \tag{10.62}$$

Provided $|c|^2 - 2\text{Re}[Dc^*d] \neq 0$, this expression is the form of a circle in w. If $|c|^2 - 2\text{Re}[Dc^*d] = 0$, it has the form of a line. Similarly, one can show that circles map into lines and circles.

A consequence of this observation is the conclusion that transformations exist that can map a half-space of the complex plane into the unit circle, and vice versa. We may therefore solve half-space potential problems by mapping them into the unit circle, as well.

As noted, there is no general procedure for determining the transformation from a general domain to the unit circle or half-space. If the domain D is the interior of an N-sided polygon, however, we may apply the so-called Schwarz–Christoffel transformation to map D into the upper half of the z-plane.

$z = a_1$

O

(a)   (b)

Figure 10.15 Illustrating a mapping from (a) the upper half of the z-plane to (b) a wedge domain in the w-plane with opening angle $\alpha_1 \pi$.

Theorem 10.3 (Schwarz–Christoffel transformation) *Let D be the interior of an N-sided polygon with boundary C in the w-plane, and let the interior angles at the ith vertex of the polygon be $\alpha_i \pi$. Then the transformation*

$$\frac{dw}{dz} = \gamma (z - a_1)^{\alpha_1 - 1}(z - a_2)^{\alpha_2 - 1} \cdots (z - a_N)^{\alpha_N - 1}, \qquad (10.63)$$

where a_i are real numbers and γ is a complex constant, maps C onto the real axis of the complex z-plane and D to the upper half of the z-plane. The vertices of the polygon are mapped to the points a_i. The mapping is one-to-one and conformal.

We will again not rigorously prove this theorem, and instead illustrate its reasonableness with specific examples. Let us consider the mapping illustrated in Fig. 10.15, in which the upper half space $|z| \geq 0$ is mapped into a semi-infinite wedge of opening angle $\alpha_1 \pi$, centered at the origin. The appropriate mapping function can be readily found by considering a polar representation,

$$z - a_1 = r e^{i\theta}. \qquad (10.64)$$

The line $\theta = \pi$ will be mapped to $\theta = \alpha_1 \pi$ by the mapping

$$w(z) = (z - a_1)^{\alpha_1} = r^{\alpha_1} e^{i\theta \alpha_1}, \qquad (10.65)$$

provided we take the principal branch of the function. It is to be noted that, provided the wedge angle is less than $\pi°$, $\alpha_1 < 1$. Furthermore, the mapping is conformal everywhere except at $z = a_1$.

If we multiply the mapping function by a complex constant $\beta = d e^{i\phi}$, the net effect is to rotate the wedge by an angle ϕ and stretch it by a length d. If we add a constant A to the mapping function, it translates the origin of the wedge to coordinate A.

The derivative of this mapping function is then given by

$$\frac{dw}{dz} = \gamma (z - a_1)^{\alpha_1 - 1}, \qquad (10.66)$$

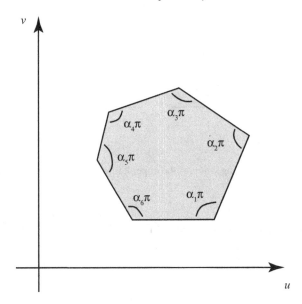

Figure 10.16 A six-sided polygon in the w-plane.

where γ is another complex constant. Evidently, any mapping function that maps the point a_1 to the tip of a wedge in the domain of w must locally have a derivative of the form of Eq. (10.66). This in turn suggests that a mapping to a geometric figure with N-sides such as shown in Fig. 10.16 must be of the form

$$\frac{\mathrm{d}w}{\mathrm{d}z} = \gamma(z - a_1)^{\alpha_1 - 1}(z - a_2)^{\alpha_2 - 1} \cdots (z - a_N)^{\alpha_N - 1}. \tag{10.67}$$

It is not immediately clear that Eq. (10.67) produces a polygon with straight sides. If we consider points on the real z-axis, however, we can see that the argument of $\mathrm{d}w/\mathrm{d}z$, and consequently the argument of $\mathrm{d}w$, only changes when z crosses one of the points a_i. When $\mathrm{d}w$ has a constant argument, w follows a straight line path. It follows that the boundary of the half-space $z \geq 0$ maps onto an N-sided polygon with straight sides.

As noted, this is not a rigorous proof, only a plausibility argument. It is also to be noted that it requires some effort to construct an actual mapping to a specific polynomial from Eq. (10.63), which is only a derivative relation.

10.5 Significant theorems in complex analysis

There are a number of theorems in complex analysis that are very important and possess broad consequences but are not typically used in applications, at least not directly. In this section we prove a number of these theorems and their consequences.

10.5.1 Liouville's theorem

Liouville's theorem is derived by the use of Cauchy's integral formula, Eq. (9.77),

$$f^{(n)}(z) = \frac{n!}{2\pi i} \oint_C \frac{f(z')}{(z'-z)^{n+1}} dz'. \tag{10.68}$$

Let us consider the situation where the path of integration C is a circle centered on the point z, such that $|z - z'| = R$, and furthermore suppose that the function $f(z)$ is bounded in that region, i.e. $|f(z)| \leq M$ on and within the path C. If we look at the modulus of $f^{(n)}(z)$, we may readily find, using the triangle inequality, that

$$|f^{(n)}(z)| \leq \frac{n!}{2\pi} \oint_C \frac{|f(z')|}{|z'-z|^{n+1}} |dz'|. \tag{10.69}$$

Using the definition of the path and the boundedness of $|f(z)|$, we may write

$$|f^{(n)}(z)| \leq \frac{n!M}{2\pi i R^{n+1}} \oint_C |dz'|. \tag{10.70}$$

The integral of $|dz'|$ is simply $2\pi R$, which gives us the inequality

$$|f^{(n)}(z)| \leq \frac{n!M}{R^n}. \tag{10.71}$$

Let us suppose that the function $f(z)$ is entire analytic (analytic throughout the finite z-plane) and bounded in this plane, $|f(z)| \leq M$ for all finite z. By Eq. (10.71), we may write

$$|f'(z)| \leq \frac{M}{R}, \tag{10.72}$$

for every finite z. However, because we may make R arbitrarily large, we immediately find that $|f'(z)| = 0$ throughout the finite z-plane. This in turn implies that the function $f(z)$ must be a constant! We have just proven Liouville's theorem.

Theorem 10.4 (Liouville's theorem) *If $f(z)$ is entire analytic and is bounded throughout the z-plane, then $f(z)$ is a constant.*

Liouville's theorem, in essence, tells us that a non-trivial analytic function must "blow-up" somewhere: if there are no singular points in the finite z-plane, there must be a singularity at infinity.

10.5.2 Morera's theorem

When we discussed potential theory in Section 2.9, we were able to establish the equivalence between conservative potentials and the path independence of work integrals; that is, (a) the integral of a conservative potential is path independent and, conversely, (b) any function whose integral is path independent for all paths represents a conservative potential.

We have already seen that there is a strong connection between analytic functions and conservative potentials. We have already shown that (a) the integral of an analytic function is path independent; our analogy suggests that we should also be able to prove that (b) any function $f(z)$ whose integral is path independent for all paths in a domain D is an analytic function. This is the basis of Morera's theorem, which is the converse of Theorem 9.3, Cauchy's theorem.

Theorem 10.5 (Morera's theorem) *Let us suppose that $f(z)$ is continuous in a domain D and that*

$$\oint_C f(z)\mathrm{d}z = 0 \tag{10.73}$$

for every simple closed contour C within D. Then $f(z)$ is analytic in D.

The strategy in the proof is to demonstrate that the integral of $f(z)$ over an open path defines an analytic function $F(z)$, and that the derivative of that analytic function is $f(z)$. It then follows that $f(z)$ must also be analytic.

Because the integral of $f(z)$ vanishes over any closed path, it immediately follows that the integral of $f(z)$ of a non-closed path is path independent, and may be written as

$$\int_z^{z_0} f(z')\mathrm{d}z' = F(z) - F(z_0). \tag{10.74}$$

We attempt to construct a derivative of $F(z)$ in the usual manner,

$$\frac{\mathrm{d}F(z)}{\mathrm{d}z} \equiv \frac{F(z+\Delta z) - F(z)}{\Delta z} = \frac{1}{\Delta z}\int_{z_0}^{z+\Delta z} f(z')\mathrm{d}z' - \frac{1}{\Delta z}\int_{z_0}^{z} f(z')\mathrm{d}z'. \tag{10.75}$$

This may immediately be simplified to

$$\frac{F(z+\Delta z) - F(z)}{\Delta z} = \frac{1}{\Delta z}\int_z^{z+\Delta z} f(z')\mathrm{d}z'. \tag{10.76}$$

We add and subtract the expression $f(z)[\int_z^{z+\Delta z}\mathrm{d}z']/\Delta z$ to the right-hand side of this equation to find

$$\frac{F(z+\Delta z) - F(z)}{\Delta z} = \frac{f(z)}{\Delta z}\int_z^{z+\Delta z}\mathrm{d}z' + \frac{1}{\Delta z}\int_z^{z+\Delta z}[f(z')-f(z)]\mathrm{d}z'. \tag{10.77}$$

The first integral is simply Δz. We may furthermore observe that

$$|f(z')-f(z)| \leq |\max[f(z')-f(z)]|, \tag{10.78}$$

where $\max[g(z)]$ is the maximum value of the function $g(z)$ on the domain $z \leq z' \leq z + \Delta z$. This inequality can be applied, with the help of the triangle inequality, to the second integral of Eq. (10.77), resulting in

$$\left| \frac{1}{\Delta z} \int_z^{z+\Delta z} [f(z') - f(z)] dz' \right| \leq |\max[f(z') - f(z)]|. \tag{10.79}$$

In the limit that $\Delta z \to 0$, $|\max[f(z') - f(z)]| \to 0$ because $f(z)$ is taken to be a continuous function. We therefore find that the derivative of $F(z)$ exists in the domain D, and $F(z)$ is consequently analytic, and furthermore that

$$\frac{dF(z)}{dz} = f(z). \tag{10.80}$$

Because the derivative of an analytic function is itself analytic, we have proven Morera's theorem.

Morera's theorem therefore demonstrates that complex integral path independence and analyticity are equivalent.

10.5.3 Maximum modulus principle

The maximum modulus principle is another statement, like Liouville's theorem, that describes the general behavior of analytic functions.

Theorem 10.6 (Maximum modulus principle) *We consider a function $f(z)$ that is analytic in a domain D and continuous on the boundary C. The modulus $|f(z)|$ assumes its maximum value only on the boundary of the region. Furthermore, $|f(z)|$ cannot have a maximum in D unless $f(z)$ is a constant function.*

We may prove this result by considering Cauchy's integral formula again, in the form

$$f(z) = \frac{1}{2\pi i} \oint_C \frac{f(z')}{(z' - z)} dz'. \tag{10.81}$$

Let us assume that the point z_0 is a maximum that lies in the domain D, with $|f(z')| \leq |f(z_0)| = M$. We consider first a circular domain of radius R centered on z_0 that lies entirely within D; we therefore have $z' - z_0 = re^{i\theta}$. After some manipulation, the integral formula for this domain may be written as

$$f(z_0) = \frac{1}{2\pi} \int_0^{2\pi} f(z_0 + Re^{i\theta}) d\theta. \tag{10.82}$$

The integral formula holds, however, for any circle of radius $r < R$, as well,

$$f(z_0) = \frac{1}{2\pi} \int_0^{2\pi} f(z_0 + re^{i\theta}) d\theta. \tag{10.83}$$

We multiply this latter formula by $r\,dr$ and integrate from $r = 0$ to $r = R$; we then have

$$f(z_0)\frac{R^2}{2} = \frac{1}{2\pi}\int_A f(z_0 + re^{i\theta})\,da,\qquad(10.84)$$

where A is the area of the circle of radius R and $da = r\,dr\,d\theta$.

If we consider the absolute value of $|f(z)|$ we may write this expression as

$$|f(z_0)| \le \frac{1}{\pi R^2}\int_A |f(z_0 + re^{i\theta})|\,da,\qquad(10.85)$$

with help from the triangle inequality. However, the right-hand integral is simply the average value of $|f(z')|$ throughout the circular domain. We therefore find that the quantity $|f(z_0)|$ must be less than or equal to the average value of $|f(z')|$, which leads to only two possibilities. Either $|f(z')|$ must have values larger than M, or $|f(z)|$ must be constant throughout the circle, and equal to $|f(z_0)|$.

In the former case, we have proven by contradiction that the point z_0 is not a maximum of $|f(z)|$. This argument is valid for any point in the interior of D, which means that the only possible place for a maximum is on the boundary C.

In the latter case, we have a function whose modulus is constant throughout a finite area. The only analytic function which satisfies that condition is a constant function. By analytic continuation, we can demonstrate that $f(z)$ must be constant throughout the domain D.

The maximum modulus principle has its analogue in potential theory, as one might expect. A potential function $V(x, y)$ can have no local maxima or minima in a domain in which Laplace's equation is satisfied; this is readily shown by considering said equation,

$$\frac{\partial^2 V}{\partial x^2} + \frac{\partial^2 V}{\partial y^2} = 0.\qquad(10.86)$$

A local maximum along the x-direction would require $\frac{\partial^2 V}{\partial x^2} < 0$; however, this implies that the same function would have a local minimum along the y-direction, $\frac{\partial^2 V}{\partial y^2} > 0$. The critical point is neither a maximum nor a minimum, but rather a saddle point.

10.5.4 Arguments, zeros, and Rouché's theorem

As we have seen in the application of the Cauchy residue theorem, the behavior of singularities, namely poles, in an analytic function dictates the behavior of the function under integration. Zeros of an analytic function may be considered complementary to poles, as the zeros of $f(z)$ are the poles of $1/f(z)$. With this in mind, there are a number of theorems that describe the behaviors of zeros and poles of analytic functions; in this section we consider some of these.

It is to be noted in the discussion that follows that a *zero of order* n_s is a zero z_s for which $f(z)$ has the Taylor series representation

$$f(z) = \sum_{n=n_s}^{\infty} a_n(z - z_s)^n, \qquad (10.87)$$

just as a pole of order n_s is a pole for which $f(z)$ has the Laurent series representation

$$f(z) = \sum_{n=-n_s}^{\infty} a_n(z - z_s)^n. \qquad (10.88)$$

Furthermore, the *multiplicity* of a zero or pole is the value n_s. The summed multiplicity M of a collection of N zeros in a contour C is the sum

$$Z = \sum_{i=1}^{N} n_{is}, \qquad (10.89)$$

where n_{is} is the multiplicity of the ith zero within the contour; a similar definition exists for the summed multiplicity P of the poles within the contour.

Theorem 10.7 (Argument principle) *Let $f(z)$ be an analytic function possessing only poles within the closed contour C, with no zeros or poles on the contour. Then the following integral relation holds,*

$$\frac{1}{2\pi i} \oint_C \frac{f'(z)}{f(z)} dz = Z - P, \qquad (10.90)$$

where Z and P are the summed multiplicities *of zeros and poles, respectively, of $f(z)$ inside C.*

Let us first suppose that $z = z_s$ is a zero or pole of order n_s. It is straightforward to show that the function $f'(z)/f(z)$ has the following Laurent series expansion about $z = z_s$

$$\frac{f'(z)}{f(z)} = \frac{\pm n_s}{z - z_s} + g(z), \qquad (10.91)$$

where $g(z)$ is a regular analytic function and the plus/minus sign corresponds to a zero/pole, respectively.

Each zero or pole of $f(z)$ is therefore a simple pole of $f'(z)/f(z)$, and by the residue theorem, the residue of each pole is simply $\pm n_s$. The total contribution to the residue due to the zeros within the contour is Z, while the total contribution to the residue due to the poles within the contour is $-P$. From this, we derive the argument principle.

The argument principle can be interpreted as characterizing the amount by which the argument of $f(z)$ increases upon following the closed path. This can be shown by explicitly evaluating the following integral

$$\frac{1}{2\pi i} \int_{z_a}^{z_b} \frac{f'(z)}{f(z)} dz = \frac{1}{2\pi i} \{\log[f(z_b)] - \log[f(z_a)]\}. \tag{10.92}$$

We have seen in Section 9.10 that the logarithm function $\log[f(z)]$ possesses a branch point at any point z_0 for which $f(z_0)$ possesses a pole or a zero. A path integral whose path encircles a branch point will end up on a different Riemann sheet than where it began.

Using the definition of the logarithm,

$$\frac{1}{2\pi i} \int_{z_a}^{z_b} \frac{f'(z)}{f(z)} dz = \frac{1}{2\pi i} \{\log[|f(z_b)|] - \log[|f(z_a)|]\} + \frac{1}{2\pi} \{\arg[f(z_b)] - \arg[f(z_a)]\}. \tag{10.93}$$

Let us suppose that the path consists of a loop which starts and ends at point z_a such that $z_b = z_a$. The value of the above integral will reflect the amount by which the argument of $f(z)$ has changed in traversing that closed path. For every zero of order n_s enclosed, the argument will increase by $2\pi n_s$; for every pole of order n_s enclosed, the argument will decrease by $2\pi n_s$. For future reference, we refer to the change in the argument of a function $f(z)$ as one traverses a closed path C as $\Delta \arg[f(z)]$.

Related to the argument principle is another theorem known as Rouché's theorem, which expresses what might loosely be referred to as a "conservation law of zeros".

Theorem 10.8 (Rouché's theorem) *Let functions $f(z)$ and $g(z)$ be analytic on and within a closed contour C. If $|f(z)| > |g(z)|$ for all points on C, then the functions $f(z)$ and $f(z) + g(z)$ have the same number of zeros inside the contour C.*

To prove this theorem, we note that because $|f(z)| > |g(z)|$ on C, the function $|f(z)| \neq 0$ on C. We may define a new function $w(z)$ such that

$$w(z) = \frac{f(z) + g(z)}{f(z)}, \tag{10.94}$$

and this function is analytic on C. Furthermore, the integral

$$\frac{1}{2\pi i} \oint_C \frac{w'(z)}{w(z)} dz \tag{10.95}$$

exists and characterizes the multiplicity of zeros and poles of $w(z)$.

We now note that $w(z) = 1 + g(z)/f(z)$, and illustrate the following identity concerning the argument of $f + g$,

$$\arg[f + g] = \arg[f(1 + g/f)] = \arg[fw] = \arg[f] + \arg[w]. \tag{10.96}$$

The argument of $f + g$ differs from the argument of f by the argument of w. We may then ask how the argument of w changes as we traverse our path C. Because $w(z) - 1 = g(z)/f(z)$, and $|g(z)|/|f(z)| < 1$ on the boundary C, it follows that

$$|w(z) - 1| < 1 \text{ for all } z \in C. \tag{10.97}$$

This implies that, as z traces out the path C, $w(z)$ follows a path that lies *within* a circle of unit radius centered on the point $w = 1$. The argument of w never goes beyond the fourth and first quadrants of the complex plane, which is equivalent to stating that $-\pi/2 \leq \arg(w) \leq \pi/2$. Because the argument of w never completes a 2π-circuit, it returns to its original value at the end of C. The change in the argument of w upon traversing the path is zero, which leads to the conclusion

$$\Delta \arg[f + g] = \Delta \arg[f]. \tag{10.98}$$

Because f and g are analytic within the contour C, there are no poles for either function and no P contribution to the arguments. Therefore the number of zeros of $f + g$ must equal the number of zeros of f.

The argument principle and Rouché's theorem have both been used to prove a number of other theorems of fundamental importance. Rouché's theorem is perhaps best known as a method to prove the fundamental theorem of algebra.

Theorem 10.9 (Fundamental theorem of algebra) *A polynomial of degree N has exactly N complex roots, counting degenerate roots (multiplicities).*

We consider a polynomial of the form

$$P(z) = z^N + \sum_{n=0}^{N-1} a_n z^n. \tag{10.99}$$

We define $f(z) = z^N$ and $g(z) = \sum_{n=0}^{N-1} a_n z^n$, so that $f + g = P$. By the use of the triangle inequality, we readily find that

$$|g(z)| \leq \sum_{n=0}^{N-1} |a_n||z|^n. \tag{10.100}$$

If we consider the case $|z| > 1$, we may further note that $|z|^n \leq |z|^{N-1}$ for $n = 0, \ldots, N-1$. We then have

$$|g(z)| \leq |z|^{N-1} \sum_{n=0}^{N-1} |a_n|. \tag{10.101}$$

If we choose as our contour C a circle of radius $R > 1$, then $|f(z)| > |g(z)|$ when R is larger than both unity (to satisfy $|z| > 1$) and $\sum_{n=0}^{N-1} |a_n|$ (to satisfy $|z| > \sum_{n=0}^{N-1} |a_n|$). In this case, $P(z)$ has the same number of zeros in the circle as $f(z) = z^N$; the latter has an N-fold degenerate root at $z = 0$. Because R can be made arbitrarily large without changing the argument, it follows that $P(z)$ has exactly N complex roots, counting degeneracies.

This proof also serves a practical purpose: we conclude that all the zeros of a polynomial must lie within a circle of radius R given by

$$R = \max \left(1, \sum_{n=0}^{N-1} |a_n| \right). \tag{10.102}$$

This estimate may be too broad – that is to say, the zeros may lie completely within a circle of much smaller size – but it is an excellent bound on the location of the zeros.

This ability to characterize and even manipulate the zeros of a complex function can be applied in many physical problems. In an optical context, we have already noted (in Section 9.15) that the phase singularities of an optical wavefield bear a strong analogy to the logarithmic branch points of analytic functions. In that context, the argument principle is analogous to the observation that the integral of Eq. (9.205) gives back the sum of the topological charges of the individual optical vortices.

Rouché's theorem also has an optical analogy related to vortices. Just as one can move the zeros of an analytic function $f(z)$ by the addition a new function $g(z)$, one can move the vortices of an optical wavefield by superimposing that wavefield with an appropriate plane wave.

10.6 Focus: analytic properties of wavefields

An important part of any optics education, beyond simply being able to do the math, is acquiring a big picture view of the properties of light. The behavior of an electromagnetic wave is constrained to satisfy Maxwell's equations, and much of the optical research done since their formulation has focused on understanding what electromagnetic waves can and cannot do.

In Section 9.14, we noted that there is a strong connection between the temporal evolution of a wavefield and its corresponding analytic properties. One can also show, if we restrict our attention to a scalar monochromatic wave $U(\mathbf{r})$, that $U(\mathbf{r})$ *represents the boundary value of a multivariable analytic function* – in several different senses. From this observation we can make some broad statements about the behavior of monochromatic wavefields, and can furthermore gain a broad understanding of several *inverse problems* in optics. In this section we first discuss several mathematical results relating to analytic functions, and then discuss their significance to monochromatic wavefields and inverse problems.

Let us begin by considering the Fourier transform (to be discussed in more detail in Chapter 11) of a function $f(x)$ which is of *finite support*, i.e. $f(x) = 0$ outside the domain $a \leq x \leq b$. The transform $F(k)$ may be written as

$$F(k) = \frac{1}{2\pi} \int_a^b f(x) e^{-ikx} dx. \tag{10.103}$$

A theorem by Plancherel and Polya [PP37] describes the analytic properties of such a transform.

Table 10.1 *Properties of analytic functions of more than one variable.*

FT is...	Object is of finite...	FT is analytic in...	Continued from...	Zeros are...
1-dimensional	length	1 complex variable	a line segment	points
2-dimensional	area	2 complex variables	an area	lines
3-dimensional	volume	3 complex variables	a volume	surfaces

Theorem 10.10 (Plancherel–Polya theorem) *If $f(x)$ is a well-behaved function of finite support, its Fourier transform $F(k)$ represents the boundary value of an entire analytic function in complex k.*

In other words, if we replace k in $F(k)$ by complex $\kappa \equiv k_1 + ik_2$, then $F(\kappa)$ is an analytic function in κ. The proof of this theorem is rather subtle, and will not be considered in detail here; we refer to [Red53] for a discussion. It follows from the fact that $\exp[ikx]$ is an analytic function of the complex variable k for all finite x. The function $F(k)$ is therefore in essence a sum of analytic functions, an identification which can be made formal by approximating the integral by a Riemann-type sum. Any finite sum of analytic functions is itself analytic, and all that remains is to show that the Riemann sum uniformly converges to the integral $F(k)$ as the number of pieces is made infinite.

Because $F(k)$ is an analytic function of k, it is uniquely determined (can be analytically continued from) values of the function on any finite line segment. This in turn suggests that $F(k)$ either vanishes only at isolated points in the complex k-plane, or vanishes everywhere.

We can extend this theorem to N-dimensional Fourier transforms of functions of N variables. The two-dimensional Fourier transform $F(k_x, k_y)$ of a function $f(x,y)$ of finite support is an analytic function in the complex variables k_x, and k_y, and can be analytically continued from values of the function on any finite *surface* in the 4-variable complex space. The function $F(k_x, k_y)$ vanishes at most on finite line segments in the complex (k_x, k_y)-space, or vanishes everywhere. The three-dimensional Fourier transform $F(k_x, k_y, k_z)$ of a function $f(x,y,z)$ of finite support is an analytic function in the complex variables k_x, k_y and k_z, and can be analytically continued from values of the function on any finite *volume* in the 6-variable complex space. The function $F(k_x, k_y, k_z)$ vanishes at most on finite surfaces in the complex (k_x, k_y, k_z)-space, or else vanishes everywhere. The results are summarized in Table 10.1.

What is the significance of these results for optics? First, we consider the case of aperture diffraction, as illustrated in Fig. 10.17. A field $U^{(i)}(x', y')$ is incident upon an opaque screen in the plane $z' = 0$ with an aperture $\mathcal{A}$. As noted in Section 11.1 in the next chapter, the field far from the aperture plane (the Fraunhofer diffraction regime), is proportional to the

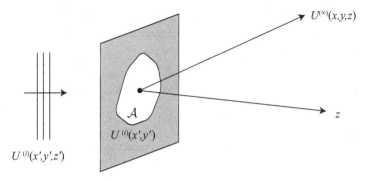

Figure 10.17 The notation relating to aperture diffraction.

two-dimensional Fourier transform of the field in the source plane,

$$U^{(\infty)}(x,y,z) \sim \tilde{U}^{(i)}\left(\frac{k}{z}x, \frac{k}{z}y\right), \tag{10.104}$$

where we define $k_x \equiv kx/z$ and $k_y \equiv ky/z$, and

$$\tilde{U}(k_x,k_y) = \frac{1}{(2\pi)^2}\int_A U(x',y')e^{-i(k_x x'+k_y y')}dx'dy'. \tag{10.105}$$

The quantity $\tilde{U}^{(i)}(k_x,k_y)$ characterizes the non trivial structure of the field far from the source plane. It is an analytic function of k_x and k_y or, equivalently, of x and y. This suggests that the field in the far zone can only vanish on lines in (k_x,k_y)-space. This is borne out by typical examples of radiation patterns, illustrated in Fig. 10.18. Figure 10.18(a) illustrates the radiation pattern for a Young's double-slit interferometer, while Fig. 10.18(b) illustrates the radiation pattern from a single circular aperture. The zeros of the radiation pattern in the Young's interferometer are vertical lines; the zeros of the radiation pattern for a circular aperture are circles.

A more sophisticated analysis of solutions to the Helmholtz equation[2] demonstrates that its solutions are analytic in all three spatial variables x, y, and z. This suggests that a monochromatic field can be zero on at most a surface in three-dimensional space, or it must vanish identically. This is in agreement with the behavior of the radiation pattern described previously.

An understanding of the analytic properties of wavefields is also useful in understanding *inverse problems* in optics, in which one attempts to deduce the properties of a source or scattering object from the light emitted/scattered by the object.

---

[2] This is discussed in [CK98], Theorem 2.2.

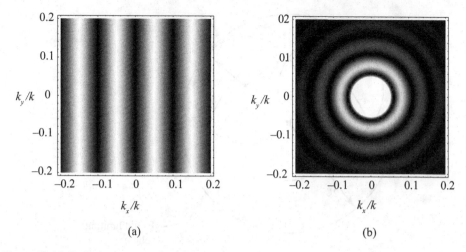

Figure 10.18 Radiation patterns for (a) Young's double-slit interferometer, with slit separation $d = 10\lambda$, (b) a single circular aperture, aperture radius $a = 10\lambda$.

We consider the case of a plane wave $U_0(\mathbf{r})$ scattering from a weakly-scattering object of refractive index $n(\mathbf{r})$, with scattered field given by Eq. (7.128),

$$U_1(\mathbf{r}) \equiv \int_V F(\mathbf{r}')U_0(\mathbf{r}')\frac{e^{ikR}}{R}d^3r', \tag{10.106}$$

where $F(\mathbf{r})$ is the scattering potential given by Eq. (7.120) and $R = |\mathbf{r} - \mathbf{r}'|$. Looking in the far-zone of the scatterer, we may approximate

$$\frac{e^{ikR}}{R} \sim \frac{e^{ikr}}{r}e^{-ik\mathbf{s}\cdot\mathbf{r}'}, \tag{10.107}$$

where $\mathbf{s}$ is the direction of observation. We also consider the incident field to be of the form

$$U_0(\mathbf{r}) = U_0e^{ik\mathbf{s}_0\cdot\mathbf{r}}, \tag{10.108}$$

where $\mathbf{s}_0$ is a unit vector specifying the direction of the incident plane wave. On substitution, one readily finds that

$$U_1(\mathbf{r}) \approx (2\pi)^3\frac{e^{ikr}}{r}\tilde{F}[k(\mathbf{s} - \mathbf{s}_0)], \tag{10.109}$$

where $\tilde{F}$ is the three-dimensional spatial Fourier transform of the scattering potential,

$$\tilde{F}(\mathbf{K}) \equiv \frac{1}{(2\pi)^3}\int_D F(\mathbf{r}')e^{-i\mathbf{K}\cdot\mathbf{r}'}d^3r'.$$

For a single direction of illumination $\mathbf{s}_0$, the far-zone scattered field is described by the functional behavior of $\tilde{F}$ with respect to $\mathbf{s}$. This represents a sphere in three-dimensional

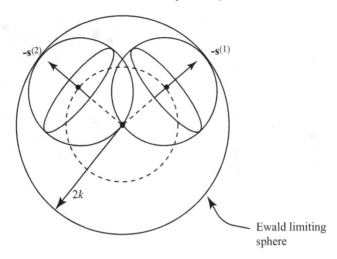

Figure 10.19 Illustration of a pair of Ewald spheres for fixed directions of incidence $\mathbf{s}^{(1)}$ and $\mathbf{s}^{(2)}$, and the Ewald limiting sphere.

$\mathbf{K}$-space, referred to as an *Ewald sphere*. If one collects scattered field information for all directions of illuminations $\mathbf{s}_0$ and scattering $\mathbf{s}$, one has information within a spherical volume known as the *Ewald limiting sphere*; this is illustrated schematically in Fig. 10.19.

In principle, this is the most information which can be acquired from scattering data. With it, one can perform an inverse Fourier transform to reconstruct a bandlimited version of the scattering potential $F(\mathbf{r})$. But what, at least in principle, is the minimum amount of information required to reconstruct the object? Because the far-zone radiation pattern is the three-dimensional Fourier transform of an object of finite volume, it follows that $\tilde{F}(\mathbf{K})$ is an analytic function of three variables K_x, K_y, and K_z. We can perform analytic continuation on this function if we know its value over a finite volume of $\mathbf{K}$-space. Such a finite volume can only be acquired if we perform scattering measurements for a continuous range of both incident and scattered field directions, $\mathbf{s}_0$ and $\mathbf{s}$. In other words, we cannot uniquely determine the structure of the scattering object from measurements from a single direction of illumination, or even from measurements for a finite number of directions of illumination.

This observation has important and even unusual consequences. It has been shown that, in theory, one can construct scattering objects which are invisible for a finite number of directions of illumination [Dev78]. Conversely, it has been shown that one can uniquely reconstruct the scattering object if measurements are made for a continuous range of illuminating and scattering directions [WH93, Nac88]. The uniqueness of inverse scattering problems, such as computed tomography, can therefore be connected to the analytic properties of the wavefield.

The analytic behavior of the radiation pattern also suggests the possibility of performing analytic continuation on the function to determine those Fourier components which lie outside the Ewald sphere. Methods have been proposed for doing this, some based directly upon analytic continuation [Ram85] and some which are based on an understanding of wave propagation [SD92]. Such techniques are extremely sensitive to noise in the measured data, which puts a practical limit on how much improvement can be made to the resolution.

A detailed discussion of the analytic properties of scattered wavefields can be found in [WNV85].

10.7 Focus: optical cloaking and transformation optics

In recent years, advances in materials science have allowed unprecendented control of the structure of materials down to the nanometer scale. This in turn potentially gives researchers the ability to fabricate materials with almost any desired optical properties, including properties not typically found in nature, such as a negative refractive index.

The possibility of negative refractive index materials was first suggested by Vesalago in 1968 [Ves68], though this work received little attention until 2000, when Pendry created a furor by speculating that a slab of negative index material could be used as a perfect lens [Pen00]. This result spawned a massive and ongoing effort to create materials with unusual optical effects through engineering of their structure on the nanometer scale; such materials are now broadly referred to as *metamaterials*.

In 2006, two papers appeared side by side in *Science Magazine* [Leo06, PSS06], each suggesting the possibility that metamaterials could be used to create a "cloaking device". The optical properties of the device would guide light rays around an interior region and return them to their original trajectories on the opposite side. The cloak, and any object within it, would be effectively invisible. The cloaking papers sparked great excitement in both the scientific community and the general public, and this was further fueled by a crude experimental demonstration of cloaking performed soon after [SMJ⁺06]. Since then numerous strategies for creating an optical cloak have been proposed, each with its strengths and weaknesses.

The observations and techniques introduced in the first cloaking papers have formed the foundation of a new branch of physical optics referred to as *transformation optics* [LP09], in which the refractive index profile of an object is treated as a transformation of space, or equivalently a mapping of the coordinate system. The first of the two original cloaking papers [Leo06] considered cloaking in a two-dimensional geometry, and used techniques of conformal mapping to construct the refractive index distribution. We show in this section how this construction is done.

We consider the solution of the Helmholtz equation in two dimensions, of the form

$$\left[\partial_x^2 + \partial_y^2 + n^2(\mathbf{r})k^2\right] U(\mathbf{r}) = 0, \tag{10.110}$$

with $\mathbf{r} = (x, y)$. It is assumed for simplicity that geometrical optics is valid, which implies that the refractive index $n(\mathbf{r})$ varies slowly over the distance of a wavelength $\lambda = 2\pi/k$.

Introducing the complex variable $z = x + iy$, one can show that

$$\partial_x = \partial_z + \partial_z^*, \tag{10.111}$$

$$\partial_y = i\partial_z - i\partial_z^*. \tag{10.112}$$

We may rewrite the Helmholtz equation in terms of z as

$$[4\partial_z^* \partial_z + n^2 k^2] U(z, z^*) = 0. \tag{10.113}$$

Let us now introduce a new complex coordinate w such that it is an analytic function of z, i.e. $w = w(z)$. This map is conformal and we may write the Helmholtz equation in terms of w as

$$[4\partial_w^* \partial_w + n'^2 k^2] U(w, w^*) = 0, \tag{10.114}$$

where

$$n'^2 = n^2 / |dw/dz|^2. \tag{10.115}$$

Let us suppose that the refractive index is chosen such that $n = |dw/dz|$ for some conformal map $w(z)$, and that the map $w(z) \to z$ for large values of w. This choice of refractive index results in the conditions that (a) $U(w, w^*)$ satisfies the *free-space* Helmholtz equation, with $n' = 1$, and (b) the transformation $w(z) = z$ far from the origin. In w-space, then, light follows straight-line paths, and in z-space, the light follows the *same* straight-line paths when sufficiently far from the origin. The net result is that in z-space, light rays approach the "cloak" along straight-line paths, are distorted in some centralized region, and then return to their original trajectories; the refractive index distribution is effectively "invisible" to an outside observer. A rough idea of the process is illustrated in Fig. 10.20.

The choice of a mapping $w = w(z)$ is nontrivial, however, and necessarily requires the use of Riemann surfaces. Consider, for instance, the simple map of the form

$$w(z) = z + \frac{a^2}{z}, \tag{10.116}$$

where a is a positive real number. This function is analytic and $w(z) = z$ as $w \to \infty$, and corresponds to a refractive index of the form

$$n(\mathbf{r}) = \left| 1 - a^2 / z^2 \right|. \tag{10.117}$$

It is to be noted that the refractive index is not rotationally symmetric about the central axis $\mathbf{r} = 0$.

Though this distribution satisfies our rough conditions for an optical cloak, it in fact does not provide a good cloaking effect. To see this, we consider the inverse transformation,

$$z = \frac{1}{2} \left[w \pm \sqrt{w^2 - 4a^2} \right]. \tag{10.118}$$

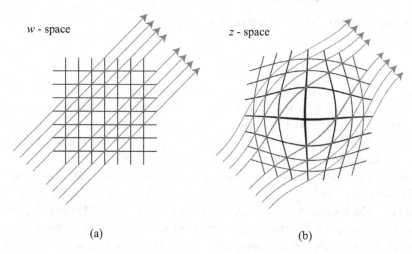

w - space   z - space

(a)   (b)

Figure 10.20 An illustration of the idea of cloaking. Light rays in (a) w-space travel in straight lines, while the rays in (b) z-space are bent around the central region by the inverse transformation from w-space.

which is a multi valued function with two Riemann surfaces, with branches distinguished by the $\pm$. It can be seen from Eq. (10.118) that the region $|z| < a$ is mapped to one Riemann surface (with "$-$") of w, to be called the "interior" surface, and the region $|z| > a$ is mapped to the other Riemann surface (with "$+$") of w, to be called the "exterior" surface. There are two branch points to this function, at $w = \pm 2a$, and a branch cut connecting the two points. In z-space, this branch cut is the unit circle itself.

We may look at the consequences of this branch cut both in w-space and in z-space. First, we consider straight-line rays in w-space which begin on the exterior surface and satisfy the parametric equation $w = f(t)$, with $-\infty < t < \infty$. Provided the ray does not cross the branch cut, we remain on the exterior surface and find that $z \to f(t)$ in the limit as $t \to \infty$. However, if the ray passes the branch cut, it passes onto the other Riemann sheet, and we instead find that the ray asymptotically approaches the point

$$\lim_{t \to \infty} z(t) = \frac{1}{2}\left[f(t) - \sqrt{f(t)^2 - 4a^2} \right] = 0.$$

In z-space, we find that the circle $|z| = a$ acts as a sort of "event horizon", crudely analogous to the event horizon of a black hole. Rays which do not cross this event horizon curve around the central region and eventually return to their original trajectories; rays which pass the event horizon spiral to the origin where they are inevitably absorbed. This absorption results in the cloak casting a shadow, and it is far from invisible.

The simplest solution to this conundrum is to replace the refractive index of the interior Riemann sheet by one which guides the rays back around the branch cut a second time, and returns them to real space at the same position and in the same direction that they entered.

A choice that does this is the Kepler potential, of the form

$$n'^2 = \frac{r_0}{|w - 2a|} - 1. \tag{10.119}$$

This potential is equivalent to a gravitational potential centered on the branch point at $w = 2a$, and it is well-known that such potentials produce Keplerian periodic elliptical orbits. The quantity r_0 defines the radius at which n'^2 becomes negative; outside of this circle, light cannot propagate at all and this is the "cloaked" region. For consistency, this circle r_0 must be taken large enough to encompass the branch cut.

We may therefore construct our medium $n(\mathbf{r})$ as follows. Outside the circle of radius a, the refractive index satisfies Eq. (10.117). Inside the circle, we have the potential given by Eq. (10.119), which we convert to real space by the use of Eqs. (10.115) and (10.116).

Even this construction does not make a perfect cloak; refraction at the boundary $r = a$ can distort the invisibility effect, though this distortion can be minimized. The analysis described here is also limited to geometrical optics, meaning that wave effects may also break the cloaking illusion.

The first cloaking papers were especially surprising because of earlier research that suggested that perfect invisibility is in fact impossible [Nac88]. The authors of the cloaking papers pointed out a number of loopholes in that argument: Pendry *et al.* [PSS06] observed that the earlier work did not apply to anisotropic optical materials, while Leonhardt [Leo06] noted that the absence of "perfect" invisibility does not exclude the possibility of creating invisibility arbitrarily close to perfection. Both of these observations have been pursued, and researchers have demonstrated both theoretically ideal cloaking for electromagnetic waves in anisotropic materials [CWZK07] and schemes for producing good but not ideal cloaking[AE09].

10.8 Exercises

1. Demonstrate explicitly that lines in the complex plane are mapped into circles on the Riemann sphere, and show that this circle includes the north pole of the sphere.
2. Suppose we define a stereographic projection with a sphere of radius a whose south pole lies a distance b above the complex plane. Find the circle in the complex plane that is mapped onto the equator of the Riemann sphere.
3. Find a mapping $w(z)$ that maps the semicircle $|z| < 1$, $\text{Im}(z) > 0$ onto the upper half-plane such that $w(-1) = 0$, $w(0) = 1$, $w(1) = \infty$.
4. Find a mapping $w(z)$ that maps the semicircle $|z| < 1$, $\text{Im}(z) > 0$ onto the disk $|w| < 1$ such that $w(\pm 1) = \pm 1$, $w(0) = -i$.
5. A laser beam has a state of polarization characterized by $E_x = E_0 e^{i\pi/2}$, $E_y = 2E_0$. Show, by stereographic projection to the Poincaré sphere, that the location on the sphere properly characterizes the ellipticity and orientation of the elliptically polarized field.
6. A laser beam has a state of polarization characterized by $E_x = E_0 e^{i\pi/4}$, $E_y = E_0 e^{i\pi/2}$. Show, by stereographic projection to the Poincaré sphere, that the location on the sphere properly characterizes the ellipticity and orientation of the elliptically polarized field.

7. Find a mapping $w(z)$ that maps the upper half-space into the right triangle with corners $w = 0$, $w = i$, $w = 1$, with $w(0) = 0$, $w(1) = 1$, and $w(-1) = i$.
8. Find a mapping $w(z)$ that maps the upper half-space into the right triangle with corners $w = 1+i$, $w = i$, $w = 1$, with $w(0) = 1+i$, $w(1) = i$, and $w(-1) = 1$.
9. Find a mapping $w(z)$ that maps the upper half-space into the rectangle with corners $w = 0$, $w = i$, $w = 2+i$, $w = 2$, with $w(1) = 2$, $w(0) = 0$, $w(-1) = i$, $w(-2) = 2+i$.
10. Find the domain onto which the function $w(z) = \frac{1}{2}(z+1/z)$ maps the domains
    (a) $|z| < 1$,
    (b) $|z| > 1$,
    (c) $\text{Im}(z) > 1$.
11. Find the domain onto which the function $w(z) = (z-1)/(z+2)$ maps the domains
    (a) $|z| < 1$,
    (b) $|z| > 1$,
    (c) $\text{Im}(z) > 1$.
12. Demonstrate that the argument principle holds inside the domain $|z| < 3$ for the function

$$f(z) = \frac{(z^2 - 2z + 1)}{z^2 + 1},$$

both by explicitly determining the multiplicities of the zeros and poles and by evaluating the integral

$$\frac{1}{2\pi i} \oint_C \frac{f'(z)}{f(z)} dz.$$

13. Demonstrate that the argument principle holds inside the domain $|z| < 2$ for the function

$$f(z) = \frac{(z^2 + 2z - 3)}{z^2 + 2iz - 1},$$

both by explicitly determining the multiplicities of the zeros and poles and by evaluating the integral

$$\frac{1}{2\pi i} \oint_C \frac{f'(z)}{f(z)} dz.$$

14. Read the paper by I. J. Cox and C. J. R. Sheppard, Information capacity and resolution in an optical system, *J. Opt. Soc. Am. A* **3** (1986), 1152–1158. What is meant by the "information capacity of an optical system", and what are the two types of "superresolution" described? Explain how analytic continuation is used to achieve superresolution, and describe the type of systems the authors apply it to.
15. Consider the analytic function

$$f(z) = \frac{2z + 1}{z^2 + z - 2}.$$

Determine the Taylor series expansion of the function at points $z = -1$, $z = 2i$. Would it be possible to directly determine the second series from the first, via analytic continuation? How about from the first series to the second?

11

Fourier transforms

11.1 Introduction: Fraunhofer diffraction

One of the most fundamental problems in the theory of physical optics is the diffraction of light from an aperture in an opaque screen. The Rayleigh–Sommerfeld (R-S) solution to the problem (which we will derive much later, in Chapter 19) may be described as follows. Let us assume that a scalar monochromatic wavefield $U^{(i)}(\mathbf{r})$ of frequency ω and wavenumber k is incident from the negative half-space $z < 0$ onto an opaque screen with an aperture $\mathcal{A}$; this is illustrated in Fig. 11.1. The field $U(\mathbf{r})$ diffracted into the positive half-space $z > 0$ is given by the formula

$$U(\mathbf{r}) = \frac{1}{2\pi} \int_{\mathcal{A}} U^{(i)}(\boldsymbol{\rho}',0) \frac{\partial}{\partial z'} [G(R)]_{z'=0} \, \mathrm{d}^2 \rho', \tag{11.1}$$

where the integral is over the aperture area in the plane $z' = 0$. In this equation we have taken $\mathbf{r} = (\boldsymbol{\rho}, z)$, with $\boldsymbol{\rho} = (x, y)$, $G(R)$ is the Green's function of the Helmholtz equation,

$$G(R) = \frac{\mathrm{e}^{\mathrm{i}kR}}{R}, \tag{11.2}$$

and

$$R \equiv |\mathbf{r} - \mathbf{r}'| = \sqrt{(x - x')^2 + (y - y')^2 + (z - z')^2}. \tag{11.3}$$

Equation (11.1), known as the *Rayleigh–Sommerfeld diffraction integral of the first kind*, is a combination of exact mathematical reasoning and crude physical approximation. (The approximation primarily lies in assuming that the field within the aperture is *exactly* the unperturbed incident field.) The Rayleigh–Sommerfeld integral works remarkably well at modeling the field away from the neighborhood of the source plane $z' = 0$.

We may evaluate the derivative in the integrand, which then takes on the form

$$\frac{\partial}{\partial z'} G(R)|_{z'=0} = \left(\mathrm{i}k + \frac{1}{R_0} \right) \frac{\mathrm{e}^{\mathrm{i}kR_0}}{R_0} \cos\theta, \tag{11.4}$$

where $R_0 = R|_{z'=0}$ and $\theta(\hat{\mathbf{z}}, \mathbf{r})$ represents the angle between the normal to the aperture plane $\hat{\mathbf{z}}$ and the field point $\mathbf{r}$.

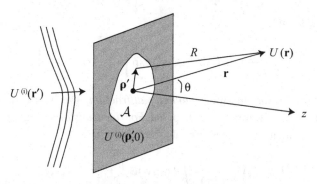

Figure 11.1 Illustration of the notation related to aperture diffraction.

The R-S integral cannot in general be evaluated analytically. In many circumstances, however, we are interested in the behavior of the wavefield extremely far from the aperture, in which case we may make an asymptotic approximation to the integrand. Let us suppose that z is much larger than all values of $x - x'$ and $y - y'$; this implies that the field is observed only within a narrow cone around the z-axis. Recalling Example 7.12, we may then use a binomial expansion to write

$$R_0 = z\sqrt{1 + \frac{(x-x')^2 + (y-y')^2}{z^2}} \approx z\left(1 + \frac{(x-x')^2 + (y-y')^2}{2z^2}\right). \tag{11.5}$$

We keep the only the first term of this expression for the denominator of Eq. (11.4), and both terms for the exponent. The R-S integral then takes on the form

$$U(\mathbf{r}) \approx \frac{i}{\lambda} \frac{e^{ikz}}{z} \int_A U^{(i)}(\boldsymbol{\rho}') e^{ik[(x-x')^2 + (y-y')^2]/2z}\, dx'dy'. \tag{11.6}$$

This is the *Fresnel diffraction integral*, to be discussed further in Section 12.1. Let us expand out the quadratic terms of the integrand,

$$U(\mathbf{r}) = \frac{e^{ikz}e^{i\frac{k}{2z}(x^2+y^2)}}{z} \frac{1}{i\lambda} \int_A U^{(i)}(x',y') e^{i\frac{k}{2z}(x'^2+y'^2)} \exp\left\{-i\frac{k}{z}[xx'+yy']\right\} dx'dy'. \tag{11.7}$$

A further simplification can be made if we assume that the field is being measured at a distance z sufficiently far from the aperture such that

$$\max\left[\frac{k}{2}(x'^2+y'^2)\right] \ll z. \tag{11.8}$$

When this condition is satisfied, then the quadratic terms in the exponent of Eq. (11.7) will vary over less than a radian and not contribute significantly to the integral. This equation then takes on the form

$$U(\mathbf{r}) = \frac{e^{ikz}e^{i\frac{k}{2z}(x^2+y^2)}}{z} \frac{1}{i\lambda} \int_A U^{(i)}(x',y')\exp\left\{-i\frac{k}{z}[xx'+yy']\right\} dx'dy', \tag{11.9}$$

which is known as the formula for *Fraunhofer diffraction*. The field $U(\mathbf{r})$ in this case is directly related to the two-dimensional spatial *Fourier transform* of the field within the aperture; the two-dimensional transform of $f(x,y)$ is defined as

$$\tilde{f}(k_x,k_y) = \frac{1}{(2\pi)^2}\int f(x',y')e^{-i(k_xx'+k_yy')}dx'dy'. \tag{11.10}$$

With this definition, we may write

$$U(\mathbf{r}) = \frac{e^{ikz}e^{i\frac{k}{2z}(x^2+y^2)}}{z} \frac{1}{i\lambda}(2\pi)^2\tilde{U}^{(i)}\left(\frac{k}{z}x, \frac{k}{z}y\right). \tag{11.11}$$

This Fourier relation gives us a simple picture of the behavior of the diffracted wavefield. From the Fourier uncertainty relation, to be derived in Section 11.5.3, it follows that length inequalities tend to be *reversed* by Fourier transformation; that is, if the x-width of the aperture is greater than the y-width of the aperture, the x-width of the diffraction pattern is *less* than the y-width of the diffraction pattern. This is illustrated in Fig. 11.2.

Fraunhofer diffraction is but the first example of the relationship between monochromatic wavefields and Fourier transforms, which is now considered its own subfield of *Fourier optics*. We will consider several examples of applying this relationship to practical problems in wave propagation and image processing at the end of the chapter.

11.2 The Fourier transform and its inverse

In Chapter 8, we developed the theory of Fourier series. The Fourier series may be used to expand a function $f(t)$ defined on a finite line segment $a \le t \le b$ into a sum of complex exponentials which are periodic on that interval. In essence, it is a decomposition of a function into a fundamental set of oscillations. If we are given the spectrum (the coefficients c_n) of the function, we can reconstruct the function by assembling the Fourier series. We will see that it is extremely useful to do so, to solve both ordinary differential equations and partial differential equations.

When the function $f(t)$ is defined on the infinite line, $-\infty < t < \infty$, we may perform a similar decomposition, the *Fourier transform*.

Definition 11.1 (Fourier transform) *The Fourier transform $\tilde{f}(\omega)$ of a function $f(t)$ defined on the domain $-\infty < t < \infty$ is given by*

$$\tilde{f}(\omega) = \frac{1}{2\pi}\int_{-\infty}^{\infty} f(t)e^{i\omega t}dt. \tag{11.12}$$

aperture shape diffraction pattern

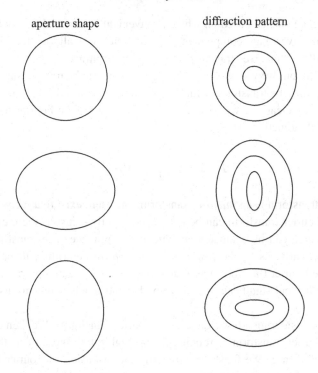

Figure 11.2 Schematic of the Fraunhofer diffraction pattern for elliptical apertures of various shapes. "Tall and thin" apertures result in "short and fat" diffraction patterns, and vice versa.

We will also refer to the Fourier transform $\tilde{f}(\omega)$ as the *Fourier spectrum* of the function. The importance of Fourier transforms cannot be overestimated, especially in the field of optics. Not only are they useful for the practical solution of problems (differential equations, again), but, as we have seen, they arise quite naturally in the study of diffraction phenomena. In fact, the study of all of classical optics can in a very real sense be described as the study of Fourier transforms.

The usefulness of Fourier transforms can be directly correlated with the existence of an *inverse Fourier transform*.

Definition 11.2 (Inverse Fourier transform) *Given the Fourier transform $\tilde{f}(\omega)$ of a function $f(t)$, the function $f(t)$ may be determined from its transform by the integral relation*

$$f(t) = \int_{-\infty}^{\infty} \tilde{f}(\omega) e^{-i\omega t} d\omega. \tag{11.13}$$

It is to be noted that this equation is structurally similar to Eq. (11.12), with three differences: (1) the variable of integration, (2) the factor of 2π, and (3) the sign of the exponential. We will see that the placement of the exponential sign and the factor of 2π is quite flexible, and can be often be chosen based on what is convenient for the task at hand. The Fourier transform and its inverse are complementary, and which we call the "forward" transform and what we call the "inverse" transform is somewhat arbitrary.

A word on notation: we will typically denote a function to be transformed by lowercase letters, e.g. $f(t)$, and the transformed function by the use of a tilde, e.g. $\tilde{f}(\omega)$. For reasons which will become clear, we will also find it useful to write the Fourier transform as an operation $\mathcal{F}$ on the function $f(t)$, i.e.

$$\mathcal{F}[f(t)] = \frac{1}{2\pi} \int_{-\infty}^{\infty} f(t) e^{i\omega t} dt. \tag{11.14}$$

The Fourier transform is an *integral* transform, and there exist relatively few functions for which the Fourier transform can be calculated exactly – just as there exist relatively few integrals which can be evaluated analytically. In practice such transforms are often evaluated numerically, using the *fast Fourier transform* algorithm, to be discussed in Chapter 13. The Fourier transform has a large number of intriguing properties, with practical and even physical significance, and we will spend much of our discussion on these properties.

We begin by considering a few examples of Fourier transforms. We then discuss useful properties of Fourier transforms, grouping them roughly into those of mathematical and of physical significance. We finish by providing the extension of Fourier transforms to higher-dimensional functions, and discuss some applications.

11.3 Examples of Fourier transforms

Example 11.1 (Top hat function) We consider first the Fourier transform of a so-called "top hat" function of the form

$$f(t) = \begin{cases} 1, & |t| \le a/2, \\ 0, & |t| > a/2. \end{cases} \tag{11.15}$$

This function is plotted in Fig. 11.3 (a). Its transform may be calculated easily by the use of Eq. (11.12),

$$\tilde{f}(\omega) = \frac{1}{2\pi} \int_{-a/2}^{a/2} e^{i\omega t} dt. \tag{11.16}$$

Because the function is identically zero outside the range $|t| \le a/2$, we have restricted the integration range appropriately. The value of this integral is readily found to be

$$\tilde{f}(\omega) = \frac{\sin(\omega a/2)}{\pi \omega} = \frac{a}{2\pi} j_0(\omega a/2), \tag{11.17}$$

where we have used the definition of the first spherical Bessel function, $j_0(x) = \sin(x)/x$, in the last step; these functions will be discussed in Chapter 16.

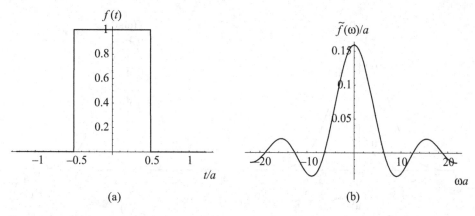

Figure 11.3 Illustration of (a) the step function $f(t)$ and (b) its Fourier transform $\tilde{f}(\omega)$.

The transform is plotted in Fig. 11.3 (b). This particular function and its transform appears in Fraunhofer diffraction theory when the diffracting aperture is rectangular.

One should note the presence of the scaling factor a in the plots of the function and its inverse. If a is doubled (the width of the hat is doubled), the width of its corresponding transform is halved. The opposite is also true: if the width of the hat is halved, the width of the corresponding transform is doubled. This foreshadows what we will later see to be a general property of Fourier transforms: the reciprocal nature of the width of the function and its transform.

Example 11.2 (Gaussian) We now consider the Fourier transform of the Gaussian function

$$f(t) = e^{-t^2/2\sigma^2}, \tag{11.18}$$

where σ is the width of the Gaussian. This function is illustrated in Figure 11.4 (a). The transform may be written as

$$\tilde{f}(\omega) = \frac{1}{2\pi} \int_{-\infty}^{\infty} e^{-t^2/2\sigma^2} e^{i\omega t} dt = \frac{1}{2\pi} \int_{-\infty}^{\infty} e^{-t^2/2\sigma^2 + i\omega t} dt. \tag{11.19}$$

To evaluate this integral, we begin by completing the square in the exponential, i.e.

$$-t^2/2\sigma^2 + i\omega t = -t^2/2\sigma^2 + i\omega t + \omega^2\sigma^2/2 - \omega^2\sigma^2/2$$

$$= -\left(\frac{t}{\sqrt{2}\sigma} - i\frac{\omega\sigma}{\sqrt{2}}\right)^2 - \omega^2\sigma^2/2. \tag{11.20}$$

On substituting from this into the formula for the Fourier transform, Eq. (11.19), we have

$$\tilde{f}(\omega) = \frac{1}{2\pi} e^{-\omega^2\sigma^2/2} \int_{-\infty}^{\infty} e^{-\left(\frac{t}{\sqrt{2}\sigma} - i\frac{\omega\sigma}{\sqrt{2}}\right)^2} dt. \tag{11.21}$$

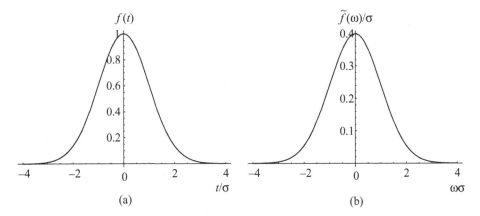

Figure 11.4 Illustration of (a) the Gaussian function and (b) its Fourier transform.

Now, let us suppose we can make the following coordinate transformation,

$$t' = t - i\omega\sigma^2.$$ (11.22)

Our integral then takes on the form

$$\tilde{f}(\omega) = \frac{1}{2\pi} e^{-\omega^2\sigma^2/2} \int_{-\infty}^{\infty} e^{-t'^2/2\sigma^2} dt',$$ (11.23)

which now looks like the standard Gaussian integral,

$$\int_{-\infty}^{\infty} e^{-t^2/(2\sigma^2)} dt = \sqrt{2\pi}\,\sigma,$$ (11.24)

and we therefore find that

$$\tilde{f}(\omega) = \frac{\sigma}{\sqrt{2\pi}} e^{-\omega^2\sigma^2/2}.$$ (11.25)

The Fourier transform of a Gaussian function is another Gaussian function, with a width $1/\sigma$. This is illustrated in Fig. 11.4 (b). This seems simple enough; the mathematically rigorous reader, however, will be suspicious of the coordinate transformation defined by Eq. (11.22). It seems highly dubious, given that it shifts the integral into the complex domain and off the real axis. Does such a transformation leave the value of the integral unchanged?

We can justify this transformation by using the now familiar tricks of complex analysis. We consider the following contour integral,

$$I = \oint_C e^{-z^2/2\sigma^2} dz = 0,$$ (11.26)

where the contour C is taken as shown in Fig. 11.5. The integral is readily shown to be zero using the residue theorem, as there are no singularities within the rectangle. The integral on

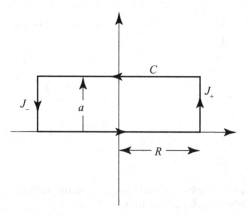

Figure 11.5 Illustration of the contour used to justify coordinate transformation, Eq. (11.22).

the upward vertical path can be written out explicitly as

$$J_+ = i \int_0^a e^{-(R+iy)^2/2\sigma^2} dy = i e^{-R^2/2\sigma^2} \int_0^a e^{-iyR/\sigma^2 + y^2/2\sigma^2} dy, \qquad (11.27)$$

with a similar expression for J_-. In the limit $R \to \infty$, the contribution of this integral to the overall integral disappears, due to the Gaussian factor in R^2. The contribution of the two "endcaps" therefore vanish in the limit as $R \to \infty$ and our contour integral (11.26) reduces to the equation

$$\int_{-\infty}^{\infty} e^{-x^2/2\sigma^2} dx = \int_{-\infty}^{\infty} e^{-(x+ia)^2/2\sigma^2} dx. \qquad (11.28)$$

Therefore the shifted Gaussian integral is simply equal to the original Gaussian integral, and our coordinate transformation is justified.

Again, it is to be noted that the widths of the Gaussian function and its Fourier transform are inversely related. The Gaussian function is used quite often in both optics and quantum mechanics. It appears quite naturally on its own accord, for instance as the radial behavior of the ground state wavefunction of the hydrogen atom and as the transverse profile of the lowest-order laser mode in laser resonator theory. Furthermore, because it is a well-behaved function (smooth, highly localized with a well-defined width) with a simple and well-behaved Fourier transform, it is often used as a simple model of other localized signals, such as electrical or optical pulses and quantum mechanical wave packets.

◇

Example 11.3 (Exponential) We finally consider the Fourier transform of the exponential function

$$f(t) = e^{-|t|/\sigma}. \qquad (11.29)$$

This function is illustrated in Fig. 11.6 (a). The quantity σ is the effective width of the exponential. Because of the absolute value in the exponential, we must separate the Fourier

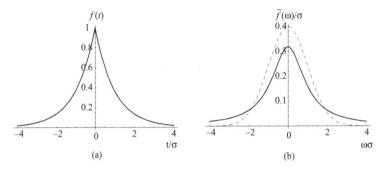

Figure 11.6 Illustration of (a) the exponential function and (b) its Fourier transform. A Gaussian function is shown as a dashed line for comparison.

integral into two parts, over $t \geq 0$ and $t < 0$,

$$\tilde{f}(\omega) = \frac{1}{2\pi} \left[\int_0^\infty e^{-t/\sigma} e^{i\omega t} dt + \int_{-\infty}^0 e^{t/\sigma} e^{i\omega t} dt \right]. \tag{11.30}$$

Grouping the exponentials together, we may rewrite this integral in the form

$$\tilde{f}(\omega) = \frac{1}{2\pi} \left[\int_0^\infty e^{(-1/\sigma + i\omega)t} dt + \int_{-\infty}^0 e^{(1/\sigma + i\omega)t} dt \right]. \tag{11.31}$$

These integrals can be readily calculated, and we find that

$$\tilde{f}(\omega) = \frac{1}{2\pi} \left[-\frac{1}{-1/\sigma + i\omega} + \frac{1}{1/\sigma + i\omega} \right]. \tag{11.32}$$

This function can be simplified by putting both terms over a common denominator, and the final result is

$$\tilde{f}(\omega) = \frac{1}{\pi} \frac{\sigma}{1 + \omega^2 \sigma^2}. \tag{11.33}$$

This function is referred to as a *Lorentian*. Its behavior is illustrated in Fig. 11.6 (b), where it is compared with a Gaussian function of comparable width.

The inverse transform can be calculated using residue theory, as described in Section 9.12. For complex ω, there are two simple poles at $\omega = \pm i/\sigma$. By using a semicircular contour as in Fig. 9.24, only the residue of the positive pole contributes.

◇

11.4 Mathematical properties of the Fourier transform

The discussion of Fourier transforms thus far has been limited to specific examples. Much of their usefulness, however, comes from properties which apply to broad classes of functions.

In this section we consider a number of general mathematical properties of the Fourier transform.

11.4.1 Existence conditions for Fourier transforms

It can be readily shown that the Fourier transform integral does not converge for all functions. In Eq. (11.12), we defined the Fourier transform of a function by the equation

$$\tilde{f}(\omega) = \frac{1}{2\pi} \int_{-\infty}^{\infty} f(t)e^{i\omega t} dt. \tag{11.34}$$

It is almost immediately clear that not all functions $f(t)$ possess a well-defined Fourier transform. For instance, if we try to take the transform of the Heaviside step function, introduced in Section 6.5,

$$S(t) = \begin{cases} 1 & \text{for } t \geq 0, \\ 0 & \text{for } t < 0, \end{cases} \tag{11.35}$$

we immediately run into difficulties. Let us rewrite the Fourier transform definition in a limit form,

$$\tilde{f}(\omega) = \frac{1}{2\pi} \lim_{A \to \infty} \int_{-A}^{A} f(t)e^{i\omega t} dt. \tag{11.36}$$

Then the Fourier transform of the Heaviside function is given by

$$\tilde{f}(\omega) = \frac{1}{2\pi} \lim_{A \to \infty} \int_{0}^{A} e^{i\omega t} dt = \frac{1}{2\pi} \lim_{A \to \infty} \frac{1}{i\omega} \left(e^{i\omega A} - 1 \right). \tag{11.37}$$

The limit does not exist (at least in the sense of ordinary functions), because the function $\exp(i\omega A)$ oscillates as $A \to \infty$. For what sort of functions, then, does the Fourier transform exist? We introduce a few new definitions to help answer this question.

Definition 11.3 (Absolutely integrable) *A function* $f(t)$ *is called* absolutely integrable *if* $\int_{-\infty}^{\infty} |f(t)| dt < \infty.$

A function being absolutely integrable is comparable to a series being absolutely convergent (recall Section 7.3). If a function is absolutely integrable, then its Fourier transform is finite-valued, because

$$|\tilde{f}(\omega)| = \left| \int_{-\infty}^{\infty} f(t)e^{i\omega t} dt \right| \leq \int_{-\infty}^{\infty} \left| f(t)e^{i\omega t} \right| dt = \int_{-\infty}^{\infty} |f(t)| dt < \infty. \tag{11.38}$$

The first inequality is derived from the triangle inequality. Intuitively, absolute integrability suggests that the absolute value of the function $f(t)$ must decay at a rate faster than $1/t$ as $t \to \infty$. Looking back again at the Heaviside step function, we see that it is not an absolutely integrable function.

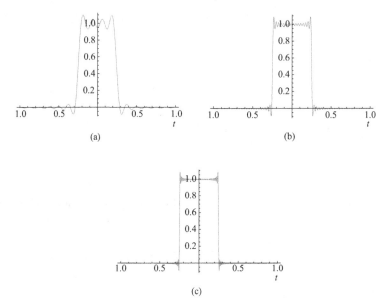

Figure 11.7 Numerical evaluation of the inverse Fourier transform for a top hat function with $a = 0.5$, with an integration step size $\Delta\omega = 0.05$. (a) $A = 1000$, (b) $A = 4000$, (c) $A = 8000$.

It is tempting to take absolute integrability as our standard requirement for the existence of the Fourier transform. However, this requirement excludes important functions that we have already considered. For instance, the top hat function has a Fourier transform that behaves as $1/\omega$ as $\omega \to \infty$, and this Fourier transform is not absolutely convergent.

The resolution to this problem is analogous to our discussion of infinite series and Fourier series in Chapters 7 and 8. Let us suppose that a function $f(t)$ that is absolutely convergent has a Fourier transform $\tilde{f}(\omega)$. We introduce the inverse Fourier transform as a limiting process,

$$I(t) = \lim_{A\to\infty} \int_{-A}^{A} \tilde{f}(\omega)e^{-i\omega t}d\omega. \tag{11.39}$$

It can be shown that the function $I(t)$ converges uniformly to the function $f(t)$ if $\tilde{f}(\omega)$ is absolutely integrable. If $\tilde{f}(\omega)$ is not absolutely integrable, then the function $I(t)$ only converges to $f(t)$ in the sense of convergence in the mean, discussed in Section 8.7.

As one might expect, the function $I(t)$ exhibits the Gibbs phenomenon in the limit $A \to \infty$. This is illustrated in Fig. 11.7.

It can be shown that the Fourier transform converges in the mean if the function $f(t)$ is square integrable, i.e.

$$\int_{-\infty}^{\infty} |f(t)|^2 dt < \infty. \tag{11.40}$$

It is possible to relax existence conditions even further if one treats the transforms as distributions as in Chapter 6. In particular, we will see that we may consider the Fourier transform of a *constant* function to be the delta function,

$$\frac{1}{2\pi} \int_{-\infty}^{\infty} e^{i\omega t} dt = \delta(\omega). \tag{11.41}$$

The Fourier transform of the Heaviside function may also be described as a distribution.

11.4.2 The Fourier inverse and delta functions

We have defined a Fourier transform as

$$\tilde{f}(\omega) = \frac{1}{2\pi} \int_{-\infty}^{\infty} f(t) e^{i\omega t} dt, \tag{11.42}$$

and the inverse Fourier transform as

$$f(t) = \int_{-\infty}^{\infty} \tilde{f}(\omega) e^{-i\omega t} d\omega. \tag{11.43}$$

If these two transforms are true inverses of each other, we should be able to substitute from Eq. (11.42) into Eq. (11.43), and the result of the two integral operations should be the original function $f(t)$. Denoting the inverse transform by $I(t)$, we again write it in the limit form

$$I(t) = \lim_{A \to \infty} \int_{-A}^{A} \tilde{f}(\omega) e^{-i\omega t} d\omega. \tag{11.44}$$

We wish to see if $I(t) = f(t)$. On substituting from Eq. (11.42) into this equation, we have

$$I(t) = \lim_{A \to \infty} \int_{-A}^{A} \left[\frac{1}{2\pi} \int_{-\infty}^{\infty} f(t') e^{i\omega t'} dt' \right] e^{-i\omega t} d\omega. \tag{11.45}$$

On interchanging the order of integration (and we assume that we are allowed to do so), we have

$$I(t) = \int_{-\infty}^{\infty} f(t') \left[\frac{1}{2\pi} \int_{-A}^{A} e^{i\omega(t'-t)} d\omega \right] dt'. \tag{11.46}$$

The quantity in brackets can be integrated, to find that

$$\frac{1}{2\pi} \int_{-A}^{A} e^{i\omega(t'-t)} d\omega = \frac{A}{\pi} \frac{\sin[A(t'-t)]}{A(t'-t)}. \tag{11.47}$$

But this function is the delta sequence defined by Eq. (6.20)! This can be proven explicitly by looking at the integral,

$$\lim_{n \to \infty} \int_{-\infty}^{\infty} f(x) \frac{\sin nx}{\pi x} dx = \lim_{n \to \infty} \int_{-\infty}^{\infty} f(y/n) \frac{\sin y}{\pi y} dy = f(0) \int_{-\infty}^{\infty} \frac{\sin y}{\pi y} dy = f(0). \tag{11.48}$$

We may therefore write

$$\lim_{A \to \infty} \frac{1}{2\pi} \int_{-A}^{A} e^{i\omega(t'-t)} d\omega = \frac{1}{2\pi} \int_{-\infty}^{\infty} e^{i\omega(t'-t)} d\omega = \delta(t - t'), \qquad (11.49)$$

and on substituting from this equation into Eq. (11.46), we find, as expected, that

$$I(t) = f(t). \qquad (11.50)$$

We have therefore demonstrated that the inverse Fourier transform recovers the original function $f(t)$, and have in the process derived the Fourier representation of the delta function, Eq. (11.49).

11.4.3 Other forms of the Fourier transform

It is to be noted that many different authors write Fourier transforms in different forms. Each of these forms is equivalent but, depending on the particular application at hand, one may be more convenient than another. We briefly describe some of these transformations.

First, let us consider the basic Fourier transform formula, Eq. (11.12),

$$\tilde{f}(\omega) = \frac{1}{2\pi} \int_{-\infty}^{\infty} f(t) e^{i\omega t} dt. \qquad (11.51)$$

We make a change of coordinates to the scaled variable $t' = t/2\pi$. Then the Fourier transform may be written as

$$\tilde{f}(\omega) = \int_{-\infty}^{\infty} f(t') e^{2\pi i \omega t'} dt'. \qquad (11.52)$$

In this expression $f(t') \equiv f(t)$. We may write the inverse transform in terms of the scaled variable as well, as

$$f(t') = \int_{-\infty}^{\infty} \tilde{f}(\omega) e^{-2\pi i \omega t'} d\omega. \qquad (11.53)$$

Equations (11.52) and (11.53) are a more symmetric form of the Fourier relations. The two transforms have the same structural form, only differing in the sign of the exponent. This form of the Fourier transform is very convenient for numerical implementation; we will see it in the discussion of discrete Fourier transforms. The extra factors of 2π in the exponents can be cumbersome, however, for purely theoretical calculations.

Another form which is commonly used can be derived by letting $t' = t/\sqrt{2\pi}$ and $\omega' = \sqrt{2\pi}\omega$ in Eqs. (11.12) and (11.13). We then find that

$$\tilde{f}(\omega') = \frac{1}{\sqrt{2\pi}} \int_{-\infty}^{\infty} f(t') e^{i\omega' t'} dt', \qquad (11.54)$$

$$f(t') = \frac{1}{\sqrt{2\pi}} \int_{-\infty}^{\infty} \tilde{f}(\omega') e^{-i\omega' t'} d\omega'. \qquad (11.55)$$

These equations are also in a more symmetric form, in that they both have the same prefactor to the integral.

One can also generate other forms by different scalings of the variables. In the end, though, all forms are equivalent – it usually does not matter which convention of the Fourier transform is used, as long as the corresponding inverse transform is used.

11.4.4 Linearity of the Fourier transform

The Fourier transform is *linear*; that is, the Fourier transform of a sum of functions is equal to the sum of the Fourier transforms of the individual functions. Defining

$$\mathcal{F}[f](\omega) = \frac{1}{2\pi} \int_{-\infty}^{\infty} f(t)e^{i\omega t}dt, \tag{11.56}$$

linearity implies that, for functions $f(t)$ and $g(t)$ and constants a and b,

$$\mathcal{F}[af(t)+bg(t)] = a\mathcal{F}[f]+b\mathcal{F}[g]. \tag{11.57}$$

In other words, the Fourier transform of $af(t)+bg(t)$ is $a\tilde{f}(\omega)+b\tilde{g}(\omega)$.

11.4.5 Conjugation

If the function $f(t)$ is complex, we may consider how the Fourier transform of $f^*(t)$ is related to the Fourier transform of $f(t)$. We have

$$\frac{1}{2\pi} \int_{-\infty}^{\infty} f^*(t)e^{i\omega t}dt = \frac{1}{2\pi} \int_{-\infty}^{\infty} \left[f(t)e^{-i\omega t}\right]^* dt = \left[\frac{1}{2\pi} \int_{-\infty}^{\infty} f(t)e^{-i\omega t}dt\right]^* = [\tilde{f}(-\omega)]^*. \tag{11.58}$$

If the function $f(t)$ is real-valued [$f^*(t) = f(t)$], this equation reduces to the simple form

$$\tilde{f}(\omega) = [\tilde{f}(-\omega)]^*. \tag{11.59}$$

This latter result implies that the negative frequency components of a real-valued function are directly related to the positive frequency components, i.e. that knowing the positive components is equivalent to knowing the negative components.

11.4.6 Time shift

If we consider the Fourier transform of a function shifted in the time domain by a time a, i.e. $f(t-a)$, we can easily find how the Fourier transform of this function relates to the transform of $f(t)$,

$$\mathcal{F}[f(t-a)] = \frac{1}{2\pi} \int_{-\infty}^{\infty} f(t-a)e^{i\omega t}dt. \tag{11.60}$$

If we change variables of integration to $t' = t - a$, we have

$$\mathcal{F}[f(t-a)] = \frac{1}{2\pi} \int_{-\infty}^{\infty} f(t')e^{i\omega(t'+a)}dt' = e^{i\omega a}\tilde{f}(\omega). \tag{11.61}$$

We therefore find the result that *if a function f (t) is shifted by a time a in the time domain, its Fourier spectrum is multiplied by a factor* $e^{i\omega a}$.

11.4.7 Frequency shift

We may determine the converse of the preceding result by considering the Fourier transform of the function $f(t)e^{-ibt}$,

$$\mathcal{F}\left[e^{-ibt}f(t)\right] = \frac{1}{2\pi} \int_{-\infty}^{\infty} f(t)e^{-ibt}e^{i\omega t}dt = \frac{1}{2\pi} \int_{-\infty}^{\infty} f(t)e^{i(\omega-b)t}dt = \tilde{f}(\omega - b). \tag{11.62}$$

We therefore find that *if a function f (t) is multiplied by a factor* e^{-ibt}, *its Fourier spectrum is frequency shifted, i.e. the Fourier transform is* $\tilde{f}(\omega - b)$.

11.4.8 Scaling property

We now consider the effect of scaling on the Fourier transform of a function, i.e. we consider the Fourier transform of the function $f(ct)$, where $c > 0$,

$$\mathcal{F}[f(ct)] = \frac{1}{2\pi} \int_{-\infty}^{\infty} f(ct)e^{i\omega t}dt. \tag{11.63}$$

We make a change of variables to the variable $\tau = ct$, so that

$$\mathcal{F}[f(ct)] = \frac{1}{2\pi c} \int_{-\infty}^{\infty} f(\tau)e^{i\omega\tau/c}d\tau = \frac{1}{c}\tilde{f}\left(\frac{\omega}{c}\right). \tag{11.64}$$

If $c < 0$, we make make the same coordinate transform, though the transformation introduces an additional minus sign because of the reversal of the range of integration. Putting the two results together gives

$$\mathcal{F}[f(ct)] = \frac{1}{|c|}\tilde{f}\left(\frac{\omega}{c}\right). \tag{11.65}$$

A special case of scaling is *time reversal*, for which $c = -1$. In this case, we find that

$$\mathcal{F}[f(-t)] = \tilde{f}(-\omega). \tag{11.66}$$

Such scaling properties become especially significant in the study of *wavelet transforms*, which we will return to in Section 12.6.

11.4.9 Differentiation in the time domain

The property of Fourier transforms most useful for the solution of differential equations is their effect on derivatives of functions. Let us consider the Fourier transform of the function $f'(t)$,

$$\mathcal{F}[f'(t)] = \frac{1}{2\pi} \int_{-\infty}^{\infty} f'(t)e^{i\omega t}dt. \tag{11.67}$$

We may perform integration by parts on this equation, to find

$$\mathcal{F}[f'(t)] = \frac{1}{2\pi}\left[f(t)e^{i\omega t}\right]_{-\infty}^{\infty} - i\omega \int_{-\infty}^{\infty} f(t)e^{i\omega t}dt = -i\omega \tilde{f}(\omega), \tag{11.68}$$

where we have used the fact that $\lim_{t\to\pm\infty} f(t) = 0$ (recall Section 11.4.1: we only consider functions which are absolutely integrable and which must therefore vanish at infinity).
We therefore have the result

$$\mathcal{F}[f'(t)] = -i\omega \tilde{f}(\omega). \tag{11.69}$$

In other words, the Fourier transform of a function eliminates the derivative of the function, replacing it by a factor $-i\omega$. We will see how we can use this to solve both ordinary differential equations and partial differential equations. We may also extend this argument to higher-order derivatives, i.e.

$$\mathcal{F}\left[f^{(n)}(t)\right] = (-i\omega)^n \tilde{f}(\omega). \tag{11.70}$$

11.4.10 Differentiation in the frequency domain

We may also consider the complementary case of looking at how the derivative of the Fourier transform of a function affects the inverse transform. In this case, we have

$$\tilde{f}'(\omega) = \frac{1}{2\pi} \int_{-\infty}^{\infty} f(t)\frac{d}{d\omega}e^{i\omega t}dt = \frac{1}{2\pi} \int_{-\infty}^{\infty} it f(t)e^{i\omega t}dt. \tag{11.71}$$

We therefore have the relation

$$\mathcal{F}[it f(t)] = \tilde{f}'(\omega). \tag{11.72}$$

We may again extend this to arbitrary orders of derivatives, to find that

$$\mathcal{F}\left[(it)^n f(t)\right] = \tilde{f}^{(n)}(\omega). \tag{11.73}$$

11.5 Physical properties of the Fourier transform

We now turn to several results concerning Fourier transforms which are of broad physical interest and have significant consequences for any research problem in which the Fourier transform plays a role. Each of these results is important not only mathematically, but

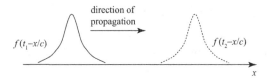

Figure 11.8 Illustration of a pulse propagating along the positive x-axis. At time $t_2 > t_1$, the peak of the pulse has moved a distance $c(t_2 - t_1)$.

physically as well, and we will couch the discussion in terms of a specific, simple physical system.

We consider a wave train propagating to the right (positive-x direction) at speed c. This wave train may represent a pulse of light in an optical fiber, or a wave on a string. The pulse will satisfy the wave equation

$$\frac{\partial^2 f}{\partial x^2} - \frac{1}{c^2}\frac{\partial^2 f}{\partial t^2} = 0. \tag{11.74}$$

The solutions to this equation represent right-going or left-going waves traveling at constant speed without a change in shape; we restrict ourselves to solutions which are entirely right-going; this is illustrated in Fig. 11.8. This wave may be described by a general function $f(t-x/c)$. If at $t = t_1$, the peak of the wave is located at $x = 0$, then when $t = t_2 > t_1$, the peak of the wave is at $x = c(t_2 - t_1)$. Detailed discussion of solutions of the wave equation will be delayed until Chapter 15; for now, we consider the properties of this pulse as measured at the point $x = 0$. At this point, the behavior of the wave is simply described by the function $f(t)$. The Fourier transform of this pulse, $\tilde{f}(\omega)$, gives its frequency content, and the squared modulus of the Fourier transform, $|\tilde{f}(\omega)|^2$, gives its energy spectrum (how the energy of the pulse is distributed in frequency). The Fourier transform operation can be used to determine the frequency content of a pulse from its temporal shape, or vice versa. With this picture in mind, we now note that each of the following results from Fourier theory gives us some insight into the general physics of pulse propagation.

11.5.1 The convolution theorem

The *convolution* of two functions $f(t)$ and $g(t)$, denoted by $f*g$ or $\mathcal{C}_{fg}$, is defined by the formula

$$\mathcal{C}_{fg}(t) = (f*g)(t) = \int_{-\infty}^{\infty} f(\tau)g(t-\tau)\mathrm{d}\tau, \tag{11.75}$$

provided that the integral exists. Convolution can be used to represent the effect of a *linear optical system* on the behavior of a pulse; that is, suppose a pulse $f(t)$ is input into a device such as a Fabry–Perot interferometer, as illustrated in Fig. 11.9. Some discussion of such systems was previously given in the context of causality and analyticity in Section 9.14. We assume for the moment that the temporal response of the Fabry–Perot can be modeled by a function $g(t)$, and the output of the interferometer is simply given by the convolution of

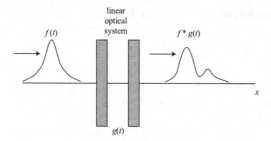

Figure 11.9 The effect of a linear optical system on a pulse. If the linear optical system is characterized by a function $g(t)$, the output of the system will be the convolution of $f(t)$ and $g(t)$.

$f(t)$ and $g(t)$. Clearly the Fourier spectrum of the field $f(t)$ input into the interferometer is $\tilde{f}(\omega)$, but what is the spectrum of light which exits the interferometer? To determine this, we begin by substituting into Eq. (11.75) the inverse Fourier transforms of $f(t)$ and $g(t-\tau)$,

$$f(\tau) = \int_{-\infty}^{\infty} \tilde{f}(\omega')e^{-i\omega'\tau}d\omega', \tag{11.76}$$

$$g(t-\tau) = \int_{-\infty}^{\infty} \tilde{g}(\omega'')e^{-i\omega''(t-\tau)}d\omega''. \tag{11.77}$$

One should note that we have written the integration variables as ω' and ω'' to distinguish them from each other. On substitution, we find that

$$C_{fg}(t) = \int_{-\infty}^{\infty}\int_{-\infty}^{\infty}\int_{-\infty}^{\infty} \tilde{f}(\omega')\tilde{g}(\omega'')e^{-i\omega'\tau}e^{-i\omega''(t-\tau)}d\omega'\,d\omega''\,d\tau. \tag{11.78}$$

We may group the exponentials in this formula, and rearrange the order of integration. Because τ now only appears in the exponentials, we may evaluate this integral first,

$$C_{fg}(t) = \int_{-\infty}^{\infty}\int_{-\infty}^{\infty} \tilde{f}(\omega')\tilde{g}(\omega'') \left[\int_{-\infty}^{\infty} e^{-i\tau(\omega'-\omega'')}d\tau\right] e^{-i\omega''t}d\omega'\,d\omega''. \tag{11.79}$$

As we have seen in Section 11.4.2, the integral over τ results in a delta function,

$$\int_{-\infty}^{\infty} e^{-i\tau(\omega'-\omega'')}d\tau = 2\pi\delta(\omega'-\omega''). \tag{11.80}$$

On substitution from this expression into Eq. (11.79), and applying the sifting properties of the delta function, we find that

$$C_{fg}(t) = 2\pi \int_{-\infty}^{\infty} \tilde{f}(\omega')\tilde{g}(\omega')e^{-i\omega't}d\omega'. \tag{11.81}$$

This expression is essentially the inverse Fourier transform of the product of $\tilde{f}(\omega')$ and $\tilde{f}(\omega'')$. If we consider the frequency content of the output pulse, by taking the Fourier transform of the convolution function, we find that

$$\tilde{C}_{fg}(\omega) = \int_{-\infty}^{\infty} \int_{-\infty}^{\infty} \tilde{f}(\omega') \tilde{g}(\omega') e^{-i(\omega'-\omega)t} d\omega' dt. \tag{11.82}$$

The integral over t is proportional to a delta function again! On substituting in for this delta function, and applying the sifting property, we have

$$\tilde{C}_{fg}(\omega) = 2\pi \tilde{f}(\omega) \tilde{g}(\omega). \tag{11.83}$$

Here we have an extremely important result, referred to generally as the *convolution theorem*.

Theorem 11.1 (Convolution theorem) *Let $f(t)$ and $g(t)$ be functions with Fourier representations $\tilde{f}(\omega)$ and $\tilde{g}(\omega)$. If $C_{fg}(t)$ represents the convolution of f and g, then*

$$\tilde{C}_{fg}(\omega) = 2\pi \tilde{f}(\omega) \tilde{g}(\omega). \tag{11.84}$$

In other words, the Fourier transform of the convolution of two functions is proportional to the product of the Fourier transforms of the two functions.

Physically, the convolution theorem implies that a linear optical system acts as a filter on an input field – the frequency content of the incident field is multiplied by a filter function $\tilde{g}(\omega)$, which is simply the frequency dependence of the system response function.

Of course, the frequency content of the system response function must be found by other means. In the case of a Fabry–Perot interferometer, we derived its transmission function $T(k)$ in Section 7.1. As the wavenumber $k = \omega/c$ in free space, the function $T(k) = \tilde{g}(\omega)$. By putting together the convolution theorem and the results of Section 7.1, we can develop a good description of the effect of an interferometer on a pulse of arbitrary shape.

11.5.2 Parseval's theorem and Plancherel's identity

Let us invert the argument of the previous section: what is the Fourier transform of a pair of functions in the time domain, $h(t) = f(t)g^*(t)$? By use of formulas for Fourier inversion, and the delta function, we have

$$\tilde{h}(\omega) = \frac{1}{2\pi} \int_{-\infty}^{\infty} f(t) g^*(t) e^{i\omega t} dt = \int_{-\infty}^{\infty} \tilde{f}(\omega') \tilde{g}^*(\omega - \omega') d\omega'. \tag{11.85}$$

This is, in fact, the convolution theorem again, showing that the convolution of two functions in the frequency domain is equal to to the product of the Fourier transform of the two functions in the time domain.

Let us consider the case $\omega = 0$. Then we find that

$$\frac{1}{2\pi} \int_{-\infty}^{\infty} f(t) g^*(t) dt = \int_{-\infty}^{\infty} \tilde{f}(\omega') \tilde{g}^*(-\omega') d\omega'. \tag{11.86}$$

Using the result (described in Section 11.4.5 above) that $\tilde{f}^*(-\omega) = \left[\tilde{f}(\omega)\right]^*$, Eq. (11.86) reduces to the form

$$\frac{1}{2\pi} \int_{-\infty}^{\infty} f(t) g^*(t) \mathrm{d}t = \int_{-\infty}^{\infty} \tilde{f}(\omega') [\tilde{g}(\omega')]^* \mathrm{d}\omega'. \tag{11.87}$$

This result is generally known as *Parseval's theorem*. It will be important when we introduce new integral transforms related to the Fourier transform in Chapter 12.

If we specialize Parseval's theorem to the case $g(t) = f(t)$, we readily find that

$$\int_{-\infty}^{\infty} |f(t)|^2 \mathrm{d}t = 2\pi \int_{-\infty}^{\infty} |\tilde{f}(\omega')|^2 \mathrm{d}\omega'. \tag{11.88}$$

This equation is referred to as *Plancherel's identity*. Mathematically, it tells us that the integral of the squared modulus of $f(t)$ is proportional to the integral of the squared modulus of its transform, $\tilde{f}(\omega)$. This formula is a mathematical representation of a statement which is physically obvious: if $|f(t)|^2$ represents the energy in the pulse at time t, and $|\tilde{f}(\omega)|^2$ represents the energy in the pulse at frequency ω, the sum of the pulse energy over all times is equal to the sum of the pulse energy over all frequencies. Another way to state this is that, within a multiplicative constant, the squared-modulus area of the pulse is the same in the time and frequency domains.

Theorem 11.2 (Parseval's theorem) *Given a function $f(t)$ that has a Fourier representation, the following relation holds*

$$\frac{1}{2\pi} \int_{-\infty}^{\infty} f(t) g^*(t) \mathrm{d}t = \int_{-\infty}^{\infty} \tilde{f}(\omega') [\tilde{g}(\omega')]^* \mathrm{d}\omega'. \tag{11.89}$$

Theorem 11.3 (Plancherel's identity) *Given functions $f(t)$ and $g(t)$ which have Fourier representations, the following relation holds*

$$\int_{-\infty}^{\infty} |f(t)|^2 \mathrm{d}t = 2\pi \int_{-\infty}^{\infty} |\tilde{f}(\omega')|^2 \mathrm{d}\omega'. \tag{11.90}$$

11.5.3 Uncertainty relations

We have seen, for specific examples, that there is an inverse relationship between the width of a function in the time domain and its width in the frequency domain. We now show that this observation holds quite generally for Fourier transform pairs. The expression which results is mathematically identical to the famous Heisenberg uncertainty relation.

We consider the following integral

$$\int_{-\infty}^{\infty} \left| \alpha t f(t) + \frac{\mathrm{d}}{\mathrm{d}t} f(t) \right|^2 \mathrm{d}t \geq 0, \tag{11.91}$$

where α is a real-valued parameter. The fact that this integral is always nonnegative can be seen by inspection, because the integrand is always nonnegative. If we expand this integrand, we have the result

$$\alpha^2 \int_{-\infty}^{\infty} t^2 |f(t)|^2 dt + \alpha \int_{-\infty}^{\infty} \left[t f^*(t) \frac{df}{dt} + t f(t) \frac{df^*}{dt} \right] dt + \int_{-\infty}^{\infty} \left| \frac{df}{dt} \right|^2 dt. \quad (11.92)$$

The central term can be simplified by noting that it may be rewritten as a total derivative,

$$\alpha \int_{-\infty}^{\infty} \left[t f^*(t) \frac{df}{dt} + t f(t) \frac{df^*}{dt} \right] dt = \alpha \int_{-\infty}^{\infty} t \frac{d}{dt} |f(t)|^2 dt = -\alpha \int_{-\infty}^{\infty} |f(t)|^2 dt, \quad (11.93)$$

where the last step was derived using integration by parts. The last term of Eq. (11.92) may be rewritten using Plancherel's identity, Eq. (11.90), in the form

$$\int_{-\infty}^{\infty} \left| \frac{df}{dt} \right|^2 dt = 2\pi \int_{-\infty}^{\infty} \left| \mathcal{F} \left(\frac{df}{dt} \right) \right|^2 d\omega. \quad (11.94)$$

This may be further simplified by using the result of Section (11.4.9) for the Fourier transform of a derivative,

$$\mathcal{F}\left[f'(t) \right] = -i\omega \tilde{f}(\omega). \quad (11.95)$$

The last term then becomes

$$\int_{-\infty}^{\infty} \left| \frac{df}{dt} \right|^2 dt = 2\pi \int_{-\infty}^{\infty} \omega^2 |\tilde{f}(\omega)|^2 d\omega, \quad (11.96)$$

and our inequality becomes

$$\alpha^2 \int_{-\infty}^{\infty} t^2 |f(t)|^2 dt - \alpha \int_{-\infty}^{\infty} |f(t)|^2 dt + 2\pi \int_{-\infty}^{\infty} \omega^2 |\tilde{f}(\omega)|^2 d\omega \geq 0. \quad (11.97)$$

This equation may be written in a more physical form. First, we divide the inequality by the integral contained in the center term, leaving

$$\alpha^2 \frac{\int_{-\infty}^{\infty} t^2 |f(t)|^2 dt}{\int_{-\infty}^{\infty} |f(t)|^2 dt} - \alpha + 2\pi \frac{\int_{-\infty}^{\infty} \omega^2 |\tilde{f}(\omega)|^2 d\omega}{\int_{-\infty}^{\infty} |f(t)|^2 dt} \geq 0. \quad (11.98)$$

Using Plancherel's identity, we may also write

$$\int_{-\infty}^{\infty} |f(t)|^2 dt = 2\pi \int_{-\infty}^{\infty} |\tilde{f}(\omega')|^2 d\omega', \quad (11.99)$$

so that the above equation may be written as

$$\alpha^2 \frac{\int_{-\infty}^{\infty} t^2 |f(t)|^2 dt}{\int_{-\infty}^{\infty} |f(t)|^2 dt} - \alpha + \frac{\int_{-\infty}^{\infty} \omega^2 |\tilde{f}(\omega)|^2 d\omega}{\int_{-\infty}^{\infty} |\tilde{f}(\omega)|^2 d\omega} \geq 0. \quad (11.100)$$

The first and last terms of this inequality may be given a physical meaning. Given that the function $f(t)$ is centered on the point $t = 0$, the first term represents the *squared width* of the function $f(t)$ in the time domain. Likewise, given that the function $\tilde{f}(\omega)$ is centered on the point $\omega = 0$, the last term represents the squared width of the function $\tilde{f}(\omega)$ in the frequency domain. We may see this by defining

$$(\Delta t)^2 \equiv \frac{\int_{-\infty}^{\infty} t^2 |f(t)|^2 \, dt}{\int_{-\infty}^{\infty} |f(t)|^2 \, dt}, \tag{11.101}$$

and

$$(\Delta \omega)^2 \equiv \frac{\int_{-\infty}^{\infty} \omega^2 |\tilde{f}(\omega)|^2 \, d\omega}{\int_{-\infty}^{\infty} |\tilde{f}(\omega)|^2 \, d\omega}. \tag{11.102}$$

The first equation represents the average value of t^2 for the function $f(t)$, and the second equation represents the average value of ω^2. Our original inequality then becomes

$$\alpha^2 (\Delta t)^2 - \alpha + (\Delta \omega)^2 \geq 0. \tag{11.103}$$

This equation is necessarily positive for all values of α, from the way we defined it; this puts a constraint on the behavior of $\Delta \omega$ and Δt. The easiest, nonrigorous, way to find this constraint is to observe that if this inequality is true for all α, it must be true for the value of α which minimizes the left of the inequality: this value can be found by the condition

$$\frac{\mathrm{d}}{\mathrm{d}\alpha} \left(\alpha^2 (\Delta t)^2 - \alpha + (\Delta \omega)^2 \right) = 2\alpha (\Delta t)^2 - 1 = 0. \tag{11.104}$$

Thus the value of α which minimizes the left side is

$$\alpha_{\min} = \frac{1}{2(\Delta t)^2}. \tag{11.105}$$

On substituting this back into our inequality, we have

$$\frac{1}{4(\Delta t)^2} - \frac{1}{2(\Delta t)^2} + (\Delta \omega)^2 \geq 0. \tag{11.106}$$

Rearranging terms, we have the result

$$(\Delta \omega)^2 (\Delta t)^2 \geq \frac{1}{4}. \tag{11.107}$$

This formula is usually referred to as an *uncertainty relation*; in quantum mechanics, it is usually described as *Heisenberg's uncertainty relation*. We state it as a theorem below.

Theorem 11.4 (Fourier uncertainty relation) *Given a function $f(t)$ which is continuous, the width Δt of the function in the time domain is related to the width $\Delta \omega$ of its Fourier transform $\tilde{f}(\omega)$ in the frequency domain by the relation*

$$(\Delta \omega)^2 (\Delta t)^2 \geq \frac{1}{4}. \tag{11.108}$$

The requirement of continuity in the above theorem deserves a special mention. If the function $f(t)$ is discontinuous, the derivative df/dt will have Dirac delta function singularities at each point of discontinuity, and the square integral of such a delta function is infinite. It will be shown in Section 21.3 that this is equivalent to stating that the function $\tilde{f}(\omega)$ decays only as $1/\omega$ as $\omega \to \infty$. The uncertainty relation does not hold for discontinuous functions.

What does this uncertainty relation mean in the context of our optical problem? The product of the width of the function in time and the width of the function in space has a minimum value; the function cannot be made arbitrarily narrow in both time and frequency. Furthermore, the wider a function is in time, the narrower it is in frequency, and vice versa. This is exemplified by the example of a Gaussian function, given by

$$f(t) = e^{-t^2/2\sigma^2}, \tag{11.109}$$

whose Fourier transform we found to be

$$\tilde{f}(\omega) = \frac{\sigma}{\sqrt{2\pi}} e^{-\omega^2\sigma^2/2}. \tag{11.110}$$

If the width of the Gaussian in the time domain is σ, the width in the frequency domain is proportional to $1/\sigma$.

It is worth emphasizing that the uncertainty relation is not exclusively a quantum-mechanical effect: it is a property of all continuous functions which are Fourier transform pairs. The unique nature of Heisenberg's work involves the *interpretation* of the uncertainty relation in the context of quantum mechanics.

Referring back to Section 11.1, the Fourier uncertainty relation explains the radiation patterns of Fig. 11.2. The Fourier relation between the aperture and the radiation pattern, in both the horizontal and vertical directions, suggests that a tall aperture will produce a short radiation pattern, and a wide aperture will produce a narrow radiation pattern. Taking the relations for the horizontal and vertical directions together, a tall thin aperture must produce a short wide radiation pattern, and so forth.

11.6 Eigenfunctions of the Fourier operator

We have often described the Fourier transform as an operation $\mathcal{F}$ upon a function $f(t)$. In this sense, we may consider $f(t)$ to be an infinite-dimensional vector which is operated upon by an infinite-dimensional matrix $\mathcal{F}$. It is natural to ask, then, if one can define eigenfunctions and eigenvalues of the Fourier operator, of the form

$$\mathcal{F}[f](\omega) = \frac{1}{\sqrt{2\pi}} \int_{-\infty}^{\infty} f(t)e^{i\omega t}dt = \lambda f(\omega), \tag{11.111}$$

where λ is an eigenvalue. We assume, for simplicity, that t and ω are dimensionless variables.

It can be shown that the eigenfunctions $\psi_n(t)$ of this Fourier operation are related to the Hermite polynomials $H_n(t)$, in the form

$$\psi_n(t) = e^{-t^2/2} H_n(t), \tag{11.112}$$

defined by

$$H_n(t) = (-1)^n e^{t^2} \frac{d^n}{dt^n} [e^{-t^2}], \tag{11.113}$$

where n is an integer value ranging from 0 to ∞. The first eigenfunction, with $n = 0$, is simply the Gaussian function, which we have already seen to be its own Fourier transform. It can be shown, by the use of the Hermite polynomial generating function (to be discussed in more detail in Chapter 16) that the eigenfunctions are orthogonal and that the eigenfunctions satisfy the following relationship,

$$\frac{1}{\sqrt{2\pi}} \int_{-\infty}^{\infty} \psi_n(t) e^{i\omega t} dt = i^n \psi_n(\omega). \tag{11.114}$$

Therefore the eigenfunctions of the Fourier operator are $\psi_n(t)$, and the eigenvalues are $i^n = e^{in\pi/2}$.

The eigenfunctions can also be shown to form a complete basis of square-integrable functions on $-\infty < t < \infty$. This description of Fourier transforms is useful in defining the fractional Fourier transform, discussed in Section 12.4.

11.7 Higher-dimensional transforms

In physical problems we will often have need of not only Fourier transforms in one variable but also Fourier transforms in two or three variables, i.e. two or three dimensional Fourier transforms. Because of the linearity of the Fourier transform, multidimensional Fourier transforms may be defined straightforwardly as multiple applications of the one-dimensional transform. For instance, a two-dimensional Fourier transform is given by

$$\tilde{f}(k_1, k_2) = \frac{1}{(2\pi)^2} \iint_{-\infty}^{\infty} f(x_1, x_2) e^{-i(k_1 x_1 + k_2 x_2)} dx_1 dx_2. \tag{11.115}$$

It is to be noted that we now define the forward transform with a minus sign in the exponential. This is a convention which is used in optics, where x_i represents a spatial variable and k_i represents the corresponding spatial frequency.

The inverse transform is also developed as simply the combination of two inverse transforms,

$$f(x_1, x_2) = \iint_{-\infty}^{\infty} \tilde{f}(k_1, k_2) e^{i(k_1 x_1 + k_2 x_2)} dk_1 dk_2. \tag{11.116}$$

We may express a similar result for a three-dimensional Fourier transform, or in fact for an n-dimensional Fourier transform,

$$\tilde{f}(k_1,\ldots,k_n) = \frac{1}{(2\pi)^n} \underbrace{\int_{-\infty}^{\infty} \cdots \int_{-\infty}^{\infty}}_{n\ \text{integrals}} f(x_1,\ldots,x_n)e^{-i(k_1x_1+\cdots+k_nx_n)}dx_1\cdots dx_n, \quad (11.117)$$

with a corresponding inverse,

$$f(x_1,\ldots,x_n) = \underbrace{\int_{-\infty}^{\infty} \cdots \int_{-\infty}^{\infty}}_{n\ \text{integrals}} \tilde{f}(k_1,\ldots,k_n)e^{i(k_1x_1+\cdots+k_nx_n)}dk_1\cdots dk_n, \quad (11.118)$$

The results of the previous sections, i.e. all the properties of Fourier transforms, may be extended to Fourier transforms in more than one variable. For instance, Plancherel's theorem in two dimensions becomes

$$\iint_{-\infty}^{\infty} |f(x_1,x_2)|^2 dx_1 dx_2 = (2\pi)^2 \iint_{-\infty}^{\infty} |\tilde{f}(k_1,k_2)|^2 dk_1 dk_2. \quad (11.119)$$

It is worth mentioning an additional class of integral transforms which are a special case of the two-dimensional Fourier transform. Let us assume that the function $f(x_1,x_2)$ is rotationally symmetric about the origin, i.e. $f(x_1,x_2) \equiv f(\rho)$, where $\rho \equiv \sqrt{x_1^2 + x_2^2}$,

$$\tilde{f}(k_1,k_2) = \frac{1}{(2\pi)^2} \iint_{-\infty}^{\infty} f(\rho)e^{-i(k_1x_1+k_2x_2)}dx_1 dx_2. \quad (11.120)$$

The quantity $k_1x_1 + k_2x_2$ is simply equal to the dot product of the vectors $\mathbf{x}$ and $\mathbf{k}$, which may be rewritten as

$$k_1x_1 + k_2x_2 = k\rho\cos\theta, \quad (11.121)$$

where $k \equiv \sqrt{k_1^2 + k_2^2}$ and θ is the angle between $\mathbf{x}$ and $\mathbf{k}$. Writing the integral above in polar coordinates, we have

$$\tilde{f}(k_1,k_2) = \tilde{f}(k) = \frac{1}{(2\pi)^2} \int_0^{\infty} \int_0^{2\pi} f(\rho)e^{-ik\rho\cos\theta}\rho\,d\rho\,d\theta. \quad (11.122)$$

It turns out that the integral over θ results in the zeroth order Bessel function $J_0(x)$,

$$J_0(x) = \frac{1}{2\pi} \int_0^{2\pi} e^{ix\cos\theta'}d\theta', \quad (11.123)$$

to be discussed in Chapter 16.

Using the formula for the Bessel function, we may write

$$\tilde{f}(k) = \frac{1}{2\pi} \int_0^{\infty} f(\rho)J_0(k\rho)\rho\,d\rho. \quad (11.124)$$

Similarly, we may find the inverse transform is given by

$$f(\rho) = 2\pi \int_0^\infty \tilde{f}(k) J_0(k\rho) k \, dk. \tag{11.125}$$

We have developed a new pair of one-dimensional integral transforms which are a special case of what are referred to as *Hankel transforms* or *Fourier–Bessel transforms*. Such transforms are significant in solving wave and vibration problems in circularly symmetric systems.

It is worth mentioning that such Fourier–Bessel transforms give us another representation for a delta function. Just as we derived a Fourier transform expression for a delta function by the substituting the Fourier transform into the inverse transform, we may do the same for Fourier–Bessel transforms. One can show that

$$\frac{1}{\rho}\delta(\rho - \rho') = \int_0^\infty J_0(k\rho) J_0(k\rho') k \, dk. \tag{11.126}$$

11.8 Focus: spatial filtering

At the beginning of the chapter, we introduced the idea of the Fourier transform in the context of the theory of diffraction. We noted that the field observed in the aperture plane $z = 0$ is related via a Fourier transform to the field observed at distances extremely far from the aperture. In principle, Fraunhofer diffraction therefore gives a straightforward all-optical technique for calculating the Fourier transform of a wavefield, but one which is limited by the long propagation distances required. However, we will show in this section that the field in the focal plane of a thin lens is also directly related to the Fourier transform of the field incident upon the lens. This observation allows one to manipulate the properties of an optical image by physically manipulating the Fourier transform of a light field in the focal plane, a technique known as *spatial filtering*.

The Fresnel diffraction formula is given by Eq. (11.6) as

$$U(x,y,z) = \frac{e^{ikz}}{z}\frac{1}{i\lambda}\int_{\mathcal{A}} U_0(x',y')\exp\left\{i\frac{k}{2z}\left[(x-x')^2 + (y-y')^2\right]\right\} dx'dy', \tag{11.127}$$

where $\mathcal{A}$ is the region of the aperture and U_0 is the field illuminating the aperture. Let us suppose that there is a thin lens within the plane of the aperture, so that the field immediately beyond the aperture is equal to the field incident upon it, $U^{(i)}(x',y')$, modified by a transmission function $T(x',y')$, i.e.

$$U_0(x',y') = U^{(i)}(x',y')T(x',y'). \tag{11.128}$$

In the Fresnel approximation, a thin lens constructed of spherical surfaces induces a quadratic curvature in the wavefront, i.e.

$$T(x,y) = \exp\left[-i\frac{k}{2f}(x^2 + y^2)\right], \tag{11.129}$$

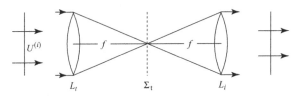

Figure 11.10 The spatial filtering system.

where f is the focal length of the lens.

Let us substitute this expression into the Fresnel diffraction formula, and consider what happens in the geometric focal plane $(z = f)$. In this case, the quadratic phase factors cancel out, and we are left with the formula

$$U(x, y; f) = e^{ikf} \frac{e^{ik(x^2 + y^2)/2f}}{i\lambda f} \int_{\mathcal{A}} U^{(i)}(x', y') e^{-ik(xx' + yy')/f} dx' dy'. \tag{11.130}$$

This formula, with the exclusion of the quadratic phase term prefactor, is simply the spatial Fourier transform of the incident field within the aperture, i.e.

$$U(x, y; f) = e^{ikf} \frac{e^{ik(x^2 + y^2)/2f}}{i\lambda f} (2\pi)^2 \tilde{U}^{(i)} \left(\frac{kx}{f}, \frac{ky}{f} \right), \tag{11.131}$$

where

$$\tilde{f}(k_x, k_y) = \frac{1}{(2\pi)^2} \int f(x', y') e^{-i(k_x x' + k_y y')} dx' dy'. \tag{11.132}$$

The field in the focal plane of the lens is essentially the *Fraunhofer diffraction pattern*.

With this in mind, we consider the optical system shown in Fig. 11.10. An incident field is focused by lens L_t into the focal plane Σ_t, which we have seen from the previous calculation to be the Fourier transform of the incident field. Another lens L_i placed at focal length f from the plane Σ_t is used to convert the field back into its incident form. If the system is ideal, the field output from lens L_i (the image lens) will be a perfect reproduction of the field input into lens L_t (the transform lens).

In the plane Σ_t (the transform plane), however, the field is the Fourier transform of the incident field. By attenuating or blocking various spatial locations in this plane, we may alter the spatial frequency content of the image on the output lens L_i; this technique is known as *spatial filtering*. For example, by placing an aperture in the transform plane, we block all the high frequency components of the image and the output of the system is a low-pass filtered image of the original object.

A simulation of spatial filtering is shown in Fig. 11.11. An image is given for which the high spatial frequencies were filtered out and an image is given for which the low spatial

Figure 11.11 A simulation of spatial filtering. Figure (a) is the original image of sleeping Komodo dragons from the Rotterdam zoo. Figure (b) is a low-pass filtered version of the image. Figure (c) is a high-pass filtered version of the image.

frequencies were suppressed. The low-pass filtering can be used to smooth out graininess in an image, while high-pass filtering can enhance details in a blurred image.

11.9 Focus: angular spectrum representation

The theory of diffraction was originally developed using the Huygens–Fresnel principle, i.e. the idea that a wavefield can be mathematically decomposed into a collection of secondary spherical waves. When light is propagating in a homogeneous medium, it is also possible to decompose the field into a collection of plane waves, in what is known as an *angular spectrum representation* of wavefields.

11.9.1 Angular spectrum representation in a half-space

We consider again a monochromatic scalar wavefield $V(\mathbf{r},t) = U(\mathbf{r})e^{-i\omega t}$ within the half-space $z > 0$, where $\mathbf{r} = (x,y,z)$. There exist no sources within the half-space, and the medium is assumed to be vacuum. The space-dependent part of the field satisfies the Helmholtz equation,

$$(\nabla^2 + k^2)U(\mathbf{r}) = 0, \tag{11.133}$$

where $k = \omega/c$.

We make the reasonable assumption that within any plane of constant z, the field may be represented as a two-dimensional Fourier integral, i.e.

$$U(x,y,z) = \iint_{-\infty}^{\infty} \tilde{U}(u,v;z)e^{i(ux+vy)}\,du\,dv. \tag{11.134}$$

The corresponding inverse representation is

$$\tilde{U}(u,v;z) = \frac{1}{(2\pi)^2}\iint_{-\infty}^{\infty} U(x,y,z)e^{-i(ux+vy)}\,dx\,dy. \tag{11.135}$$

What does this assumption tell us about the form of the field? If we substitute the above formula into the Helmholtz equation, we find that

$$\iint_{-\infty}^{\infty} (\nabla^2 + k^2)\tilde{U}(u,v;z)e^{i(ux+vy)}\,du\,dv = 0, \tag{11.136}$$

where we have interchanged the order of integration and differentiation. The derivatives with respect to x and y may now be taken directly, and we are left with the equation

$$\iint_{-\infty}^{\infty}\left[(-u^2 - v^2 + k^2)\tilde{U}(u,v;z) + \frac{\partial^2 \tilde{U}}{\partial z^2}\right]e^{i(ux+vy)}\,du\,dv = 0. \tag{11.137}$$

The Helmholtz equation must hold for all values of x and y, and Eq. (11.137) must therefore hold for each (u,v) pair. This implies that the function $\tilde{U}(u,v;z)$ must satisfies the differential equation

$$\frac{\partial^2 \tilde{U}(u,v;z)}{\partial z^2} + w^2\tilde{U}(u,v;z) = 0, \tag{11.138}$$

where

$$w^2 = k^2 - u^2 - v^2. \tag{11.139}$$

It is to be noted that there are values of w which are imaginary; we therefore pick a particular branch of w,

$$w = \begin{cases} (k^2 - u^2 - v^2)^{1/2} & \text{when } u^2 + v^2 \leq k^2, \\ i(u^2 + v^2 - k^2)^{1/2} & \text{when } u^2 + v^2 > k^2. \end{cases} \tag{11.140}$$

Equation (11.138) is simply the harmonic oscillator equation, and has the solution

$$\tilde{U}(u,v;z) = A(u,v)e^{iwz} + B(u,v)e^{-iwz}, \tag{11.141}$$

where A and B are functions that characterize the behavior of a given wavefield. The general solution of the Helmholtz equation may be written in the form

$$U(x,y,z) = \iint_{-\infty}^{\infty} A(u,v)e^{i(ux+vy+wz)}\,du\,dv + \iint_{-\infty}^{\infty} B(u,v)e^{i(ux+vy-wz)}\,du\,dv. \tag{11.142}$$

This formula represents the solution to the Helmholtz equation as a superposition of four types of plane waves. These types are:

1. $e^{i(ux+vy+wz)}$, with $u^2 + v^2 \leq k^2$. These are (homogeneous) plane waves that propagate in the positive z-direction.
2. $e^{i(ux+vy+wz)}$, with $u^2 + v^2 > k^2$. Because $w = i(u^2 + v^2 - k^2)^{1/2}$, the z-component of this wave is exponentially decaying in the positive z-direction, i.e. $e^{iwz} = e^{-|w|z}$. This is referred to an an inhomogeneous plane wave or *evanescent wave*.
3. $e^{i(ux+vy-wz)}$, with $u^2 + v^2 \leq k^2$. These are (homogeneous) plane waves that propagate in the negative z-direction.
4. $e^{i(ux+vy-wz)}$, with $u^2 + v^2 > k^2$. This is also an evanescent wave, but one whose z-component is exponentially decaying in the negative z-direction, i.e. $e^{iwz} = e^{|w|z}$.

Equation (11.142) represents the decomposition of an arbitrary field into a collection of (homogeneous and inhomogeneous) plane waves, and is referred to as an *angular spectrum representation* of the wavefield.

Why is it called an "angular spectrum"? The direction of a particular plane wave is completely specified by the values of u and v. These coordinates are equivalent to specifying the angle at which the plane wave is propagating; hence, angular spectrum.

Physically, the first two classes consist of waves propagating in the positive z-direction, while the last two classes are waves propagating in the negative z-direction. If we are interested only in waves which are diffracted from the plane $z = 0$ into the positive half-space, we may set $B(u,v) = 0$. We are then left with

$$U(x,y,z) = \iint_{-\infty}^{\infty} A(u,v)e^{i(ux+vy+wz)}\,du\,dv. \tag{11.143}$$

The wavefield propagating into a half-space can be represented as a sum of homogeneous and inhomogeneous plane waves propagating in the positive z-direction. Because $u^2 + v^2 = k^2$ is the boundary between homogeneous and inhomogeneous waves, it is useful to write the angular spectrum representation using slightly different variables,

$$U(x,y,z) = \iint_{-\infty}^{\infty} a(p,q)e^{ik(px+qy+mz)}\,dp\,dq, \tag{11.144}$$

where $u = kp$, $v = kq$, $w = km$, $a(p,q) = k^2 A(u,v)$, and

$$m = \begin{cases} (1 - p^2 - q^2)^{1/2} & \text{when } p^2 + q^2 \leq 1, \\ \mathrm{i}(p^2 + q^2 - 1)^{1/2} & \text{when } p^2 + q^2 > 1. \end{cases} \tag{11.145}$$

The field $U(x,y,0)$ in the plane $z = 0$ and the *spectral amplitude* $a(p,q)$ are related simply by a two-dimensional Fourier transform. First we define

$$\tilde{U}_0(u,v) = \frac{1}{(2\pi)^2} \iint_{-\infty}^{\infty} U(x,y,0) \mathrm{e}^{-\mathrm{i}(ux+vy)} \mathrm{d}x \mathrm{d}y. \tag{11.146}$$

From the angular spectrum representation (11.144), we immediately get an equation for the field at $z = 0$,

$$U(x,y,0) = \iint_{-\infty}^{\infty} a(p,q) \mathrm{e}^{\mathrm{i}k(px+qy)} \mathrm{d}p \mathrm{d}q. \tag{11.147}$$

If we plug this formula into the Fourier transform formula above, and use the relation

$$\frac{1}{2\pi} \int_{-\infty}^{\infty} \mathrm{e}^{\mathrm{i}(u-u')x} \mathrm{d}x = \delta(u - u'), \tag{11.148}$$

we immediately find that

$$\tilde{U}_0(u,v) = \frac{1}{k^2} a\left(\frac{u}{k}, \frac{v}{k}\right), \tag{11.149}$$

or

$$a(p,q) = k^2 \tilde{U}_0(kp, kq). \tag{11.150}$$

This tells us that the spectral amplitude of each plane wave mode of the angular spectrum representation is specified by a single Fourier component of the boundary value of the field in the plane $z = 0$.

It is to be noted that the coordinates u, v are typically referred to as *spatial frequencies*, as they represent the rate of spatial variation of their Fourier component.

This plane wave representation of an arbitrary field is incredibly useful in optical physics problems because the evolution of a plane wave through a system can often be calculated in a straightforward manner. For instance, exact formulas exist for the reflection and refraction of plane waves through stratified media.

11.9.2 The Weyl representation of a spherical wave

We now have two formulas which may be used to study the diffraction of wavefields: (a) the Rayleigh–Sommerfeld diffraction integral of Section 11.1, based on the Huygens–Fresnel principle, which expresses wavefields in terms of spherical waves, and (b) the angular spectrum representation, which expresses wavefields in terms of plane waves. We can make a connection between these two representations using a result due to H. Weyl (circa 1919), in which a spherical wave is represented in terms of an angular spectrum of plane waves. We seek an angular spectrum representation of the function

$$G(\mathbf{r}) = \frac{e^{ikr}}{r}, \tag{11.151}$$

where $r = \sqrt{x^2 + y^2 + z^2}$. It is to be noted that this function represents a spherical wave radiating from a point source at the origin; to the right of the origin ($z > 0$), the field is moving entirely to the right, while to the left of the origin, the field is moving entirely to the left. It is expected that we will need to evaluate each side of the origin separately.

Starting with the region $z > 0$, we make the identification

$$\frac{e^{ikr}}{r} = \iint_{-\infty}^{\infty} a(p,q) e^{ik(px+qy+mz)} dp\,dq. \tag{11.152}$$

In the limit $z \to 0$, this reduces to

$$\frac{e^{ik\sqrt{x^2+y^2}}}{\sqrt{x^2+y^2}} = \iint_{-\infty}^{\infty} a(p,q) e^{ik(px+qy)} dp\,dq. \tag{11.153}$$

If we take the Fourier transform of both sides of this equation, we get the expression

$$a(p,q) = \frac{k^2}{(2\pi)^2} \iint_{-\infty}^{\infty} \frac{e^{ik\sqrt{x^2+y^2}}}{\sqrt{x^2+y^2}} e^{-ik(px+qy)} dx\,dy. \tag{11.154}$$

This is an evaluable integral, if we change to polar coordinates and define $\rho = \sqrt{x^2 + y^2}$, $\beta = \sqrt{p^2 + q^2}$, and ϕ as the angle between (x,y) and (p,q). We then have

$$a(p,q) = \frac{k^2}{(2\pi)^2} \int_0^\infty \int_0^{2\pi} \frac{e^{ik\rho}}{\rho} e^{-ik\beta\rho\cos\phi} \rho\,d\rho\,d\phi. \tag{11.155}$$

The integral over ϕ results in a Bessel function, as noted in Section 11.7,

$$\int_0^{2\pi} e^{-ik\beta\rho\cos\phi} d\phi = 2\pi J_0(k\beta\rho). \tag{11.156}$$

The integral for $a(p,q)$ is then given by

$$a(p,q) = \frac{k^2}{2\pi} \int_0^\infty e^{ik\rho} J_0(k\beta\rho) d\rho. \tag{11.157}$$

To evaluate this remaining integral, we require two well-known Hankel transform formulas,

$$\int_0^\infty \cos ax J_0(xy) dx = \begin{cases} 0, & 0 < y < a, \\ \left(\frac{1}{y^2 - a^2}\right)^{1/2}, & a < y < \infty, \end{cases} \tag{11.158}$$

$$\int_0^\infty \sin ax J_0(xy)\mathrm{d}x = \begin{cases} \left(\frac{1}{a^2-y^2}\right)^{1/2}, & 0 < y < a, \\ 0, & a < y < \infty \end{cases} \tag{11.159}$$

From these we may write

$$\int_0^\infty e^{iax} J_0(xy)\mathrm{d}x = \begin{cases} \dfrac{i}{(a^2-y^2)^{1/2}}, & 0 < y < a, \\ \dfrac{1}{(y^2-a^2)^{1/2}}, & a < y < \infty. \end{cases} \tag{11.160}$$

In terms of our original integral, we have

$$a(p,q) = \frac{k^2}{2\pi} \begin{cases} \dfrac{i}{(k^2-(k\beta)^2)^{1/2}}, & \beta^2 < 1, \\ \dfrac{1}{((k\beta)^2-k^2)^{1/2}}, & \beta^2 > 1. \end{cases} \tag{11.161}$$

Using the definitions of β and m, the latter given by Eq. (11.145), we may simply write

$$a(p,q) = \frac{ik}{2\pi} \frac{1}{m}. \tag{11.162}$$

This gives us our result for the angular spectrum representation of a spherical wave,

$$\frac{e^{ikr}}{r} = \frac{ik}{2\pi} \int\!\!\int_{-\infty}^\infty \frac{1}{m} e^{ik(px+qy+mz)}\mathrm{d}p\mathrm{d}q. \tag{11.163}$$

It is to be noted the only thing that distinguishes this calculation from the calculation for the left-hand space is the sign of m, which is negative for $z < 0$. We may combine the left and right half-space results to get the Weyl representation,

$$\frac{e^{ikr}}{r} = \frac{ik}{2\pi} \int\!\!\int_{-\infty}^\infty \frac{1}{m} e^{ik(px+qy+m|z|)}\mathrm{d}p\mathrm{d}q. \tag{11.164}$$

It is to be noted that the integrand of this equation contains a singularity; this is to be expected, since the spherical wave itself is singular at the origin.

11.10 Additional reading

- R. J. Beerends, H. G. ter Morsche, J. C. van den Berg and E. M. van de Vrie, *Fourier and Laplace Transforms* (Cambridge, Cambridge University Press, 2003). A very up-to-date introduction to the theory of Fourier transforms.
- J. F. James, *A Student's Guide to Fourier Transforms* (Cambridge, Cambridge University Press, 2003, 2nd edition). A very popular and easy to read introductory text.
- I. N. Sneddon, *Fourier Transforms* (New York, McGraw-Hill Book Company, 1951). A book that focuses heavily on the physical applications of Fourier transforms.

11.11 Exercises

1. Find the Fourier transform of the triangular function

$$f(t) = \begin{cases} A(1 - c|t|), & |x| \leq 1/c, \\ 0, & |x| > 1/c. \end{cases}$$

2. Determine the Fourier transform of the function

$$f(t) = \frac{1}{t^2 - 2t + 2}.$$

3. Find the Fourier transform of the function

$$f(t) = \frac{\cos(\alpha t)}{t^2 + \beta^2}.$$

4. Calculate the Fourier transform of the function

$$f(t) = \begin{cases} -1 & \text{for} \quad -T < t < 0, \\ 1 & \text{for} \quad 0 \leq t \leq T, \\ 0 & \text{otherwise.} \end{cases}$$

5. Find the Fourier transform of the function

$$f(t) = \begin{cases} 0, & |t| > \pi, \\ 1, & -\pi \leq t \leq 0, \\ \cos t, & 0 < t \leq \pi. \end{cases}$$

6. Take the Fourier transform of the equation

$$\frac{d^2 V}{dt^2} - \omega_0^2 V = f(t),$$

and show that the solution can be written as

$$V(t) = -\int_{-\infty}^{\infty} \frac{e^{-i\omega t} \tilde{f}(\omega)}{\omega^2 + \omega_0^2} d\omega.$$

7. A wave $u(x,t)$ propagating in a one-dimensional system over $-\infty < x < \infty$ satisfies the wave equation

$$\frac{\partial^2 u}{\partial x^2} - \frac{1}{c^2} \frac{\partial^2 u}{\partial t^2} = 0, \tag{11.165}$$

where c is the wave speed. Show that the general solution to this equation may be written as $u(x,t) = f(x - ct) + g(x + ct)$, with $f(z)$ and $g(z)$ arbitrary functions, by (a) taking the Fourier transform with respect to x, (b) solving the simple differential equation in t, and (c) taking the inverse Fourier transform.

Now suppose the wave satisfies the initial conditions $u(x,0) = F(x)$, $u_t(x,0) = 0$. Write the solution of the wave equation.

8. A Gaussian function $g(t) = e^{-t^2/2}$ has a Fourier transform $F(\omega) = e^{-\omega^2/2}/\sqrt{2\pi}$. Using elementary properties of Fourier transforms (scaling, time shift, etc.), find the Fourier transforms of:

1. $f_1(t) = t^3 g(t)$,
2. $f_2(t) = e^{-2t^2}$,
3. $f_3(t) = (t^2 - 4t + 4)e^{-2t^2 + 8t - 8}$.

9. An exponential function $g(t) = e^{-|t|}$ has a Fourier transform $F(\omega) = 1/\pi(1 + \omega^2)$. Using elementary properties of Fourier transforms (scaling, time shift, etc.), find the Fourier transforms of:

1. $f_1(t) = t^2 g(t)$,
2. $f_2(t) = e^{i\beta t}e^{-4|t|}$,
3. $f_3(t) = (t^2 - 4t + 4)e^{-|2t - 4|}$.

10. Sometimes Fourier transforms can simplify an integral. Find the Fourier transform of the function

$$F(t) = \int_{-\infty}^{t} e^{-(t-t')}g(t')dt'$$

in terms of the Fourier transform of $g(t)$. (Hint write $g(t')$ as the inverse Fourier transform of $\tilde{g}(\omega)$, and then evaluate the integral over t' first.)

11. A problem involving Fourier uncertainty relations. Consider the function

$$f(t) = \sqrt{\frac{2a^3}{\pi}} \frac{1}{t^2 + a^2}.$$

Calculate Δt, $\Delta\omega$, and $\Delta t \cdot \Delta\omega$. (Hint use the residue theorem to calculate all integrals involving $|f(t)|^2$, and the Fourier transform of $f(t)$. It is important to note that, for $\omega < 0$, the path of integration is the lower half-space.)

12. The triangular wave function,

$$f(x) = \begin{cases} x, & 0 \leq x \leq \pi, \\ -x, & -\pi \leq x \leq 0, \\ 0 & \text{elsewhere}, \end{cases}$$

can be expanded in either a Fourier series or by a Fourier transform. Calculate the Fourier series representation over $-\pi \leq x \leq \pi$ and calculate the Fourier transform of this function over $-\infty \leq x \leq \infty$. Obviously the two representations are different, even though both are representations of a function in terms of its harmonic components. Why are they different? (Hint: what does the Fourier series look like outside the domain it was calculated on?)

13. Consider the paper by G. W. Forbes and M. A. Alonso, Measures of spread for periodic distributions and the associated uncertainty relations, *Am. J. Phys.* **69** (2001), 340–347. Describe how the authors adapt the idea of a Fourier uncertainty relation to a periodic function. Can you think of any applications of this sort of uncertainty relation?

14. Read the paper by I. McNulty *et al.*, High-resolution imaging by Fourier transform X-ray holography, *Science* **256** (1992), 1009–1012. Explain the idea of Fourier transform holography, and describe what role the Fourier transform plays in data acquisition. What sort of specimens are to be measured with this technique, and what resolution can be achieved?

15. Read the paper by J. R. Fienup, Reconstruction of an object from the modulus of its Fourier transform, *Opt. Lett.* **3** (1978), 27–29. Describe the technique for determining the image of an object that is presented. For what sort of system would only the modulus of the Fourier transform of the signal be available?
16. Read the paper by N. Bleistein and J. K. Cohen, Nonuniqueness in the inverse source problem in acoustics and electromagnetics, *J. Math. Phys.* **18** (1977), 194–201. Give the expressions for the radiation field of a scalar and an electromagnetic wave. What does it mean for a source to be "nonradiating"? Give the conditions for nonradiation. Explain what an "inverse source problem" is, and explain what the existence of nonradiating sources implies about such a problem.
17. In the waist plane of a laser beam of wavenumber k, its field has a Gaussian shape,

$$U(x,y) = U_0 \exp[-(x^2+y^2)/\sigma^2].$$

   1. Determine the angular spectrum of the wavefield.
   2. Assuming $k\sigma \gg 1$, show that the field retains a Gaussian shape as it propagates in the z-direction.
18. Use the Fourier transform of the top hat function, and Plancherel's identity, to determine the value of the integral

$$\int_{-\infty}^{\infty} \left(\frac{\sin x}{x}\right)^2 dx.$$

19. Determine the Fourier transform of the triangular function

$$f(t) = \begin{cases} A(1-c|t|), & |x| \le 1/c, \\ 0, & |x| > 1/c. \end{cases}$$

   and with the help of Plancherel's identity determine the value of the integral

$$\int_{-\infty}^{\infty} \left(\frac{\sin x}{x}\right)^4 dx.$$

20. Prove that the function $\psi_2(t)$, given by

$$\psi_2(t) = e^{t^2/2} \frac{d^2}{dt^2}[e^{-t^2}],$$

   is an eigenfunction of the Fourier transform operator, i.e.

$$\frac{1}{\sqrt{2\pi}} \int_{-\infty}^{\infty} \psi_2(t) e^{i\omega t} dt = -\psi_2(\omega). \tag{11.166}$$

12

Other integral transforms

12.1 Introduction: the Fresnel transform

In Section 11.1, we noted that the Rayleigh–Sommerfeld diffraction integral, given by

$$U(\mathbf{r}) = \frac{1}{2\pi} \int_{\mathcal{A}} U^{(i)}(\boldsymbol{\rho}',0) \frac{\partial}{\partial z'} [G(R)]_{z'=0} \, \mathrm{d}^2 \rho', \tag{12.1}$$

cannot in general be evaluated analytically. At distances far from the aperture plane, the separation R can be approximated as a quadratic function,

$$R = z\sqrt{1 + \frac{(x-x')^2 + (y-y')^2}{z^2}} \approx z\left(1 + \frac{(x-x')^2 + (y-y')^2}{2z^2}\right), \tag{12.2}$$

in which case the field may be written as

$$U(\mathbf{r}) \approx \frac{\mathrm{i}}{\lambda} \frac{\mathrm{e}^{\mathrm{i}kz}}{z} \int_{\mathcal{A}} U^{(i)}(\boldsymbol{\rho}',0) \mathrm{e}^{\mathrm{i}k[(x-x')^2 + (y-y')^2]/2z} \mathrm{d}x' \mathrm{d}y'. \tag{12.3}$$

This expression is known as the *Fresnel diffraction integral*, and is an approximation to the mathematically rigorous Rayleigh–Sommerfeld integral.

Under even stricter conditions the Fresnel integral can be simplified further to derive the *Fraunhofer diffraction integral*. As we have seen in Chapter 11, in this situation the effect of diffraction is reduced essentially to the application of a Fourier transform. Recognizing this, the general properties of Fourier transforms – the existence of an inverse, scaling, differentiation, etc. – may be directly applied to the diffraction problem.

A similar, though less well-known, situation applies to the Fresnel diffraction integral. Fresnel diffraction may be interpreted as the action of a different integral transform, known as the *Fresnel transform*, which possesses general properties that are analogous to the Fourier transform.

386

Definition 12.1 (Fresnel transform) *The Fresnel transform $F_\alpha(x)$ of a function $f(x)$ is defined as*

$$F_\alpha(x) = \mathcal{E}_\alpha\{f\}(x) = \sqrt{-i\alpha/\pi} \int_{-\infty}^{\infty} f(x')\exp\left[i\alpha(x-x')^2\right]dx', \qquad (12.4)$$

where α is a fixed parameter of the transform.

The Fresnel transform is a linear transform, and the integral clearly exists for a broad range of functions. As seen in Section 9.12, the transform of a constant function can be defined without the use of distributions, unlike the Fourier transform. A number of other fundamental properties of the Fresnel transform are described below.[1]

12.1.1 Shift

We consider the effect of multiplying the function $f(x)$ by a linear complex exponential on the Fresnel transform of the function, i.e.

$$\mathcal{E}_\alpha\left\{f(x')e^{ikx'}\right\}(x) = \sqrt{-i\alpha/\pi} \int_{-\infty}^{\infty} f(x')e^{ikx'}\exp\left[i\alpha(x-x')^2\right]dx'. \qquad (12.5)$$

We can reduce this to the basic form of the Fresnel transform by completing the square of the exponent, i.e.

$$kx' + \alpha(x-x')^2 = \alpha\left[x^2 + x'^2 - 2(x - k/2\alpha)x'\right]$$

$$= \alpha\left[x' - (x - k/2\alpha)\right]^2 - \frac{(k-2\alpha x)^2}{4\alpha} + \alpha x^2. \qquad (12.6)$$

Our transform then reduces to the form

$$\mathcal{E}_\alpha\left\{f(x')e^{ikx'}\right\}(x) = F_\alpha(x - k/2\alpha)\exp\left[i(kx - k^2/4\alpha)\right]. \qquad (12.7)$$

This expression may be compared to Eq. (11.62), the analogous result for Fourier transforms. It can be seen that, in both cases, multiplication by a complex exponential results in a shift of the origin of the transform.

12.1.2 Scaling property

We next consider the effect of scaling the argument of the function by a constant c, i.e. $f(cx)$. From the definition of the Fresnel transform, we have

$$\mathcal{E}_\alpha\{f(cx')\}(x) = \sqrt{-i\alpha/\pi} \int_{-\infty}^{\infty} f(cx')\exp\left[i\alpha(x-x')^2\right]dx'. \qquad (12.8)$$

---

[1] Even more properties can be found in the excellent article, "Why is the Fresnel transform so little known?" [Gor94].

We may make a simple change of coordinates to evaluate the integral, namely $\xi = cx$ and $\xi' = cx'$, from which we may immediately write

$$\mathcal{E}_\alpha \{f(cx')\}(x) = \sqrt{-i\alpha/c^2\pi} \int_{-\infty}^{\infty} f(\xi') \exp\left[i(\alpha/c^2)(\xi - \xi')^2\right] d\xi'. \tag{12.9}$$

This may be written, using the definition of the Fresnel transform, as

$$\mathcal{E}_\alpha \{f(cx')\}(x) = F_{\alpha/c^2}(cx). \tag{12.10}$$

12.1.3 Differentiation

We now consider the effect of differentiation on the form of the Fresnel transform. We consider the transform of the function df/dx, i.e.

$$\mathcal{E}_\alpha \{df(x')/dx'\}(x) = \sqrt{-i\alpha/\pi} \int_{-\infty}^{\infty} \frac{df}{dx'}(x') \exp\left[i\alpha(x - x')^2\right] dx'. \tag{12.11}$$

A straightforward integration by parts, assuming the function $f(x)$ and its derivative vanish as $|x| \to \infty$, leads one to the formula

$$\mathcal{E}_\alpha \{df(x')/dx'\}(x) = \sqrt{-i\alpha/\pi} \int_{-\infty}^{\infty} [2i\alpha(x - x')]f(x') \exp\left[i\alpha(x - x')^2\right] dx'. \tag{12.12}$$

However, this is simply equal to the derivative of the Fresnel transform with respect to x,

$$\mathcal{E}_\alpha \{df(x')/dx'\}(x) = \frac{d}{dx}\left\{\sqrt{-i\alpha/\pi} \int_{-\infty}^{\infty} \frac{df}{dx'}(x') \exp\left[i\alpha(x - x')^2\right] dx'\right\} = \frac{dF_\alpha}{dx}. \tag{12.13}$$

12.1.4 Inverse Fresnel transform

The Fresnel transform may be inverted, in a manner comparable to that of the Fourier transform.

Definition 12.2 (Inverse Fresnel transform) *The inverse Fresnel transform is defined as*

$$f(x) = \sqrt{i\alpha/\pi} \int_{-\infty}^{\infty} F_\alpha(x'') \exp\left[-i\alpha(x - x'')^2\right] dx''. \tag{12.14}$$

The inverse transform differs from the forward transform only in the reversal of the sign of α. To test that this is the proper inverse, we substitute from Eq. (12.4) into Eq. (12.14),

$$f(x) = \sqrt{i\alpha/\pi} f(x)$$

$$= \sqrt{-i\alpha/\pi} \int_{-\infty}^{\infty} \int_{-\infty}^{\infty} f(x') \exp\left[i\alpha(x'' - x')^2\right] \exp\left[-i\alpha(x - x'')^2\right] dx' dx''. \tag{12.15}$$

Combining the terms in the exponents, we find that

$$f(x) = \frac{\alpha}{\pi} \int_{-\infty}^{\infty} \int_{-\infty}^{\infty} f(x') \exp[-2i\alpha x'x'' + i\alpha x'^2 - i\alpha x^2 + 2i\alpha xx''] dx' dx''. \qquad (12.16)$$

The integral over x'' is the Fourier transform of a constant function, which results in a delta function,

$$\int_{-\infty}^{\infty} \exp[-2i\alpha x'x'' + 2i\alpha xx''] dx'' = 2\pi\delta\left[2\alpha(x-x')\right] = \frac{\pi}{\alpha}\delta(x-x'), \qquad (12.17)$$

where the last form of the equation was derived from the scaling property of delta functions, Eq. (6.25). On substitution into Eq. (12.16), we immediately find that it is satisfied, and that our inversion formula is correct.

12.1.5 Convolution

We may also consider the effect of the Fresnel transform on a convolution operation. Recalling Eq. (11.75), we write the convolution of two functions as

$$f * g(x') = \int_{-\infty}^{\infty} f(x'')g(x'-x'')dx'', \qquad (12.18)$$

and we consider the Fresnel transform of this integral,

$$\mathcal{E}_\alpha\left\{f * g(x')\right\}(x) = \sqrt{-i\alpha/\pi} \int_{-\infty}^{\infty} \int_{-\infty}^{\infty} f(x'')g(x'-x'') \exp\left[i\alpha(x-x')^2\right] dx' dx''. \qquad (12.19)$$

This may be expressed in two useful and equivalent forms. First, if we introduce a change of variables $x' = x''' + x''$, we may write our integral in the form

$$\mathcal{E}_\alpha\left\{f * g(x')\right\}(x) = \sqrt{-i\alpha/\pi} \int_{-\infty}^{\infty} \int_{-\infty}^{\infty} f(x'')g(x''') \exp\left[i\alpha(x-x''-x''')^2\right] dx'' dx'''. \qquad (12.20)$$

The integral over x''' is simply a Fresnel transform, which results in the expression

$$\mathcal{E}_\alpha\left\{f * g(x')\right\}(x) = \int_{-\infty}^{\infty} f(x'')G_\alpha(x-x'')dx'' = f * G_\alpha(x). \qquad (12.21)$$

Alternatively, we may introduce a change of variables $x' = x'' + x - x'''$, and then we may write Eq. (12.19) in the form

$$\mathcal{E}_\alpha\left\{f * g(x')\right\}(x)$$
$$= -\sqrt{-i\alpha/\pi} \int_{-\infty}^{\infty} \int_{-\infty}^{\infty} f(x'')g(x-x''') \exp\left[i\alpha(x'''-x'')^2\right] dx'' dx'''. \qquad (12.22)$$

The integral over x'' is now a Fresnel transform, and we may write

$$\mathcal{E}_\alpha \left\{ f * g(x') \right\}(x) = -\int_{-\infty}^{\infty} F_\alpha(x''')G_\alpha(x - x''')dx'' = -F_\alpha * g(x). \qquad (12.23)$$

Therefore the Fresnel transform of a convolution results in another convolution, which may be written as

$$\mathcal{E}_\alpha \left\{ f * g(x') \right\}(x) = f * G_\alpha(x) = -F_\alpha * g(x). \qquad (12.24)$$

12.1.6 Plancherel's identity

In Section 11.5.2, we proved the Plancherel identity for Fourier transforms,

$$\int_{-\infty}^{\infty} |f(x)|^2 dx = 2\pi \int_{-\infty}^{\infty} |\tilde{f}(k)|^2 dk. \qquad (12.25)$$

In diffraction theory, the wavefield in the aperture and the wavefield in the far-zone are related by a Fourier transformation, which allows us to interpret the Plancherel identity as an energy conservation formula. As the Fresnel transform can be used to represent the diffracted wavefield in an intermediate zone, it is natural to expect that energy conservation, and hence Plancherel's identity, should hold for these transforms as well. We begin by writing the integral of $|f(x)|^2$ in terms of its inverse Fresnel transform,

$$\int_{-\infty}^{\infty} |f(x)|^2 dx$$
$$= \frac{\alpha}{\pi} \int_{-\infty}^{\infty} \int_{-\infty}^{\infty} \int_{-\infty}^{\infty} F_\alpha^*(x') \exp\left[i\alpha(x - x')^2\right] F_\alpha(x'') \exp\left[-i\alpha(x - x'')^2\right] dx' dx'' dx.$$
$$(12.26)$$

The integral over x, however, results in a delta function, which contracts our integral to the form

$$\int_{-\infty}^{\infty} |f(x)|^2 dx = \int_{-\infty}^{\infty} |F_\alpha(x')|^2 dx'. \qquad (12.27)$$

12.1.7 Observations

It is to be noted that the Fresnel transform shares many similarities with the Fourier transform, and many of its properties could be derived with an appropriate use of Fourier transform tricks. This is an observation which will hold for most of the transforms to be discussed in this chapter. In fact, one could derive the Fresnel transform results simply by treating it as a Fourier transform with a complex filter $\exp[i\alpha x'^2]$ applied to the input function. However, such an analysis does not provide much intuition as to the general behavior of the transform.

In addition to its application in diffraction theory, the Fresnel transform can be used in the study of pulse propagation in dispersive media, as well as in the study of the evolution of a quantum wavepacket. Other transforms which are discussed in this chapter have also proven their value because of physical considerations. However, we will see that the different transforms described often provide different ways to interpret and understand the nature of a wave signal and its behavior.

12.2 Linear canonical transforms

The Fourier transform and the Fresnel transform may both be considered special cases of a general set of transforms known as *linear canonical transforms* (LCTs).

Definition 12.3 (Linear canonical transform) *A linear canonical transform $F_M(x)$ of a function $f(x)$ is defined as*

$$F_M(x) \equiv \int_{-\infty}^{\infty} f(x') K_M(x, x') dx', \qquad (12.28)$$

where $K_M(x, x')$ is the kernel *of the transform, given by*

$$K_M(x, x') \equiv (2\pi b)^{-1/2} \exp[-i\pi/4] \exp\left[i(ax'^2 - 2xx' + dx^2)/2b\right], \qquad (12.29)$$

subject to the constraint $ad - bc = 1$.

The subscript M refers to the matrix associated with numbers a, b, c, d, which we will consider again momentarily. With the choice $a = d = 0$, $b = -1$, the kernel reduces to the kernel of the Fourier transform, while with the choice $a = d = 1$, $b = 1$, the kernel reduces to the kernel of the Fresnel transform.

The motivation for this kernel comes from the derivative-reducing behavior of the Fourier transform. We have seen (in Section 11.4.10) that the Fourier transform of the function $xf(x)$ is related to the derivative of the Fourier transform of the function $f(x)$, i.e. if

$$\tilde{f}(p) = \frac{1}{2\pi} \int_{-\infty}^{\infty} f(x) e^{ipx} dx, \qquad (12.30)$$

the Fourier transform of $xf(x)$ is

$$\mathcal{F}\{xf(x)\}(p) = -i\frac{d}{dp}\tilde{f}(p). \qquad (12.31)$$

Let us define an operator $\mathcal{X}$ such that

$$\mathcal{X}f(x) = xf(x), \quad \mathcal{X}\tilde{f}(p) = p\tilde{f}(p), \qquad (12.32)$$

and an operator $\mathcal{P}$ such that

$$\mathcal{P}f(x) = -i\frac{d}{dx}f(x), \quad \mathcal{P}\tilde{f}(p) = -i\frac{d}{dp}\tilde{f}(p). \tag{12.33}$$

The effect of a Fourier transform on these operators may then be represented as

$$\mathcal{F}\mathcal{X}f = \mathcal{P}\mathcal{F}f, \tag{12.34}$$

$$\mathcal{F}\mathcal{P}f = \mathcal{X}\mathcal{F}f. \tag{12.35}$$

If we apply the identity $\mathcal{F}^{-1}\mathcal{F} = \mathbf{I}$ within these expressions, they take the form

$$\mathcal{F}\mathcal{X}\mathcal{F}^{-1} = \mathcal{P}, \tag{12.36}$$

$$\mathcal{F}\mathcal{P}\mathcal{F}^{-1} = \mathcal{X}. \tag{12.37}$$

The similarity transformation [recall Eq. (4.121)] of the operator $\mathcal{X}$ results in $\mathcal{P}$, and vice versa.

We now seek a more general operator $\mathcal{K}$ which performs a similarity transformation of the form

$$\mathcal{K}\mathcal{X}\mathcal{K}^{-1} = d\mathcal{X} - b\mathcal{P}, \tag{12.38}$$

$$\mathcal{K}\mathcal{P}\mathcal{K}^{-1} = -c\mathcal{X} + a\mathcal{P}. \tag{12.39}$$

where we introduce the constraint that $ad - bc = 1$. It is to be noted that with $a = d = 1$ and $b = c = 0$ we have the identity operation. We can show that Eqs. (12.38) and (12.39) are satisfied by the linear canonical transformation. For instance, let us consider,

$$\mathcal{K}\mathcal{P}f = -i\int_{-\infty}^{\infty} \frac{d}{dx'}f(x')K_M(x,x')dx' = i\int_{-\infty}^{\infty} f(x')\frac{d}{dx'}K_M(x,x')dx', \tag{12.40}$$

where we have assumed $f(x') \to 0$ as $x' \to \infty$, and used integration by parts. The derivative evaluates to

$$i\frac{d}{dx'}K_M(x,x') = -\left(\frac{ax'}{b} - \frac{x}{b}\right)K_M(x,x'). \tag{12.41}$$

In the second term in parenthesis, we apply $ad - bc = 1$, and may rearrange terms as

$$i\frac{d}{dx'}K_M(x,x') = -\left[cx + a\left(\frac{x'}{b} - \frac{dx}{b}\right)\right]K_M(x,x'). \tag{12.42}$$

The parenthetical term may be compared with the derivative of the kernel with respect to x,

$$-i\frac{d}{dx}K_M(x,x') = \left(\frac{dx}{b} - \frac{x'}{b}\right)K_M(x,x'). \tag{12.43}$$

We immediately find that $\mathcal{K}\mathcal{P}f = (-c\mathcal{X} + a\mathcal{P})\mathcal{K}f$.

The idea of the linear canonical transform provides a great unifying structure for many of the transforms we have and will consider. The Fourier transform, Fresnel transform, fractional Fourier transform, and even the Laplace transform (to an extent) may be treated as special cases of the LCT. With this in mind, we consider a few general properties.

12.2.1 Inversion

Definition 12.4 (Inverse LCT) *The inverse linear canonical transform is given by the expression*

$$f(x) \equiv \int_{-\infty}^{\infty} F_M(x'') K_M^*(x'', x) dx''. \tag{12.44}$$

where $K_M(x'', x)$ is the kernel of the forward LCT.

The kernel for the inverse transform is simply the complex conjugate of the forward kernel, though it is to be noted that the integration is over the first variable, not the second. The proof of this can be found by substitution from Eq. (12.28),

$$\int_{-\infty}^{\infty} F_M(x'') K_M^*(x'', x) dx'' = \int_{-\infty}^{\infty} \int_{-\infty}^{\infty} f(x') K_M(x'', x') K_M^*(x'', x) dx' dx''. \tag{12.45}$$

Expansion of the kernels provides the formula

$$K_M(x'', x') K_M^*(x'', x) = \frac{-i}{2\pi b} \exp\left\{i\left[a(x'^2 - x^2) + 2x''(x - x')\right]/2b\right\}. \tag{12.46}$$

The integral over x'' provides $2\pi i\delta[(x - x')/b]$, and the integral becomes

$$\int_{-\infty}^{\infty} F_M(x'') K_M^*(x'', x) dx'' = \int_{-\infty}^{\infty} f(x') \exp\left[ia(x'^2 - x^2)/2b\right]\delta[x - x'] dx' = f(x). \tag{12.47}$$

The LCT is therefore invertible.

12.2.2 Plancherel's identity

In a strictly similar manner to that used to demonstrate the existence of an inverse, one can show that Plancherel's identity is satisfied for LCTs,

$$\int_{-\infty}^{\infty} |f(x')|^2 dx' = \int_{-\infty}^{\infty} |F_M(x)|^2 dx. \tag{12.48}$$

12.2.3 Composition

We now consider the effect of two consecutive LCTs on a function. If we were considering a problem of Fresnel diffraction, for instance, the two LCTs might represent the propagation of a wave from the plane $z = 0$ to $z = z_0$, and then the propagation from $z = z_0$ to $z = z_1$. We therefore consider

$$\mathcal{C}_1\mathcal{C}_0 f(x) = \int_{-\infty}^{\infty} F_{M_0}(x'') K_{M_1}(x,x'') \mathrm{d}x''$$

$$= \int_{-\infty}^{\infty} \int_{-\infty}^{\infty} f(x') K_{M_0}(x'',x') K_{M_1}(x,x'') \mathrm{d}x' \mathrm{d}x'', \tag{12.49}$$

where M_0 and M_1 refer to the set of constants a_i, b_i, c_i, d_i, with $i = 0, 1$.

The calculation is rather lengthy, but one can show that

$$\mathcal{C}_1\mathcal{C}_0 = \mathcal{C}_2, \tag{12.50}$$

where the coefficients of the transformation $\mathcal{C}_2$ can be found by matrix multiplication,

$$\mathbf{M}_2 = \mathbf{M}_1 \mathbf{M}_0 = \begin{bmatrix} a_1 & b_1 \\ c_1 & d_1 \end{bmatrix} \begin{bmatrix} a_0 & b_0 \\ c_0 & d_0 \end{bmatrix} = \begin{bmatrix} a_1 a_0 + b_1 c_0 & a_1 b_0 + b_1 d_0 \\ c_1 a_0 + d_1 c_0 & c_1 b_0 + d_1 d_0 \end{bmatrix}. \tag{12.51}$$

We therefore have the remarkable result that the effect of multiple LCTs can be reduced to the calculation of a single LCT whose coefficients are determined by matrix multiplication.

12.2.4 Identity limit

Looking back at Eqs. (12.38) and (12.39), it seems that with the choice $a = d = 1$ and $b = c = 0$, the transformation has no effect on the operations $\mathcal{X}$ and $\mathcal{P}$. This in turn suggests that the transformation simply returns the input function unchanged, and that the kernel itself must become a delta function. This is not immediately obvious, but let us look at the kernel with $a = d = 1$ and $b = \epsilon$,

$$K_\epsilon(x,x') \equiv (2\pi\epsilon)^{-1/2} \exp[-i\pi/4] \exp[i(x - x')^2/2\epsilon]. \tag{12.52}$$

This function looks suspiciously like a delta sequence, and in fact it can be shown to be one. To do so, we apply the method of stationary phase, to be discussed in Section 21.4. This method applies to integrals of the form

$$F(k) = \int_a^b f(x') e^{ikg(x')} \mathrm{d}x, \tag{12.53}$$

where $g(x)$ is a real-valued function and k is a parameter. In the limit when k gets very large, the integral may be approximated by the form

$$F(k) \approx f(x_0) \exp[ikg(x_0)] \exp[i\pi/4] \sqrt{\frac{2\pi}{g''(x_0)k}}, \tag{12.54}$$

where x_0 is a point at which $g'(x) = 0$, i.e. a stationary point. For our integral, we have $k = 1/\epsilon$, $g(x') = (x - x')^2/2$, $x_0 = x$, and we readily find that

$$\lim_{\epsilon \to 0} F_\epsilon(x) = f(x). \tag{12.55}$$

12.2.5 Observations

We have already noted that the Fourier transform and Fresnel transform are special cases of the LCT. The two transforms which follow, the Laplace transform and the fractional Fourier transform, are also special cases, although the Laplace transform extends the definition of the LCT to complex parameters a, b, c, d.

All of these transforms, because of their ability to "remove" derivatives, can be applied to the solution of differential equations. We will discuss the use of integral transforms in ordinary and partial differential equations in Chapters 14 and 15.

A thorough discussion of linear canonical transforms and many other useful integral transforms can be found in [Wol79].

12.3 The Laplace transform

The Laplace transform may be considered the special case of a LCT in which $a = d = 0$, $b = c = i$. The kernel which results, however, diverges for negative values of the argument, and it is generally only applied to causal functions for which $f(t) = 0$ for $t < 0$. In a more limited sense, the Laplace transform may be considered a complex form of the Fourier transform, and it shares many properties with its kin. The use of complex coefficients for the LCT, however, introduces new challenges, particularly in the inverse transformation. The Laplace transform is regularly used in the solution of ordinary differential equations, and it is worth describing its properties in detail.

Definition 12.5 (Laplace transform) *The Laplace transform $F(s)$ of a function $f(t)$ which is specified on the interval $0 \leq t < \infty$ is defined as*

$$F(s) = \int_0^\infty f(t) e^{-st} dt. \tag{12.56}$$

In general, we will allow s to be a complex number, i.e. $s = \sigma + i\omega$; in fact, we will *require* this in order to define the inverse transform. As in the case of the Fourier transform and the LCT, it will also be helpful to write the Laplace transform in an operator notation, i.e.

$$\mathcal{L}[f(t)] = \int_0^\infty f(t) e^{-st} dt. \tag{12.57}$$

We consider a few examples, assuming $\mathrm{Re}\{s\} > 0$.

Example 12.1 The Laplace transform of a constant function $f(t) = C_0$ is readily found to be

$$F(s) = \frac{C_0}{s}. \tag{12.58}$$

The Laplace transform of a function $f(t) = t^n$, where n is a positive integer, can be determined via integration by parts,

$$F(s) = \int_0^\infty t^n e^{-st} dt = \frac{n!}{s^{n+1}}. \tag{12.59}$$

We may also evaluate the Laplace transform of trigonometric functions such as $f(t) = \sin(kt)$,

$$F(s) = \int_0^\infty \sin(kt) e^{-st} dt = \int_0^\infty \frac{e^{ikt} - e^{-ikt}}{2i} e^{-st} dt = \frac{k}{s^2 + k^2}. \tag{12.60}$$

◇

We have noted that the Laplace transform is structurally similar to the Fourier transform, in that the integration kernel is an exponential function with a linear exponent. In fact, if s is complex, i.e. $s = \sigma + i\omega$, the Laplace transform is analogous to the Fourier transform of the function $\hat{f}(t) e^{-\sigma t}$, where

$$\hat{f}(t) = \begin{cases} f(t), & t \geq 0, \\ 0, & t < 0. \end{cases} \tag{12.61}$$

In other words, we may write

$$\mathcal{L}[f(t)] = \int_0^\infty \left[f(t) e^{-\sigma t} \right] e^{-i\omega t} dt = \mathcal{F}\left[\hat{f}(t) e^{-\sigma t} \right]. \tag{12.62}$$

Consequently, many of the results relating to Fourier transforms have a directly analogous result for Laplace transforms. We discuss several of these, saving the inverse Laplace transform for last.

12.3.1 Existence of the Laplace transform

From the examples given above, it can be seen that the Laplace transform exists even for some functions which diverge as $t \to \infty$. In an attempt to quantify the existence criteria for the Laplace transform, we consider the function $f(t) = e^{at}$, where a is a real constant, i.e.

$$\mathcal{L}[e^{at}] = \int_0^\infty e^{(a-s)t} dt. \tag{12.63}$$

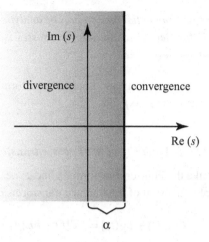

Figure 12.1 The region of convergence for the Laplace transform.

Let us at first restrict ourselves to real-valued s. For $s < a$, the exponent will be positive and the integral will clearly diverge, while for $s > a$

$$\mathcal{L}[e^{at}] = \frac{1}{s-a}. \tag{12.64}$$

We may extend this result to complex values of s. The addition of an imaginary part to s makes the integrand oscillatory; however, for $\mathrm{Re}(s) < a$, this oscillation is exponentially increasing and the integral is undefined. We may therefore say that the Laplace transform exists and converges for values of s such that $\mathrm{Re}(s) > a$. This argument may be extended to a more general class of functions.

Definition 12.6 (Exponential order) *A function $f(t)$ is said to be of* exponential order α *if* $|\exp[-\alpha t]f(t)| \leq M$ *for all* $\alpha < t < \infty$.

In other words, $f(t)$ increases no faster than an exponential of the form $\exp[\alpha t]$. In such a case, we may state that the Laplace transform of $f(t)$ converges for $\mathrm{Re}(s) > \alpha$. The domain of convergence is illustrated in Fig. 12.1. It should be noted that this is a *sufficiency* condition; functions exist which are not of exponential order but have Laplace transforms, but such functions are not typically encountered in practice.

Considering Eq. (12.64) again, we note a curious discrepancy. The Laplace transform of the function $\exp[-at]$ exists only for $\mathrm{Re}(s) > a$, but the function $1/(s-a)$ is in fact analytic everywhere in the complex s-plane except at the point $s = a$. This is an example of *analytic continuation*, which was discussed in Section 10.2: the function $\mathcal{L}[e^{at}]$ is a *representation* of the function $1/(s-a)$, the latter of which only possesses a single simple pole. The following theorem makes a more general argument.

Theorem 12.1 *A function of exponential order α may be analytically continued to a function which is regular analytic for $Re(s) > \alpha$ and only possesses isolated singularities (poles, branch points) in the domain $Re(s) \leq \alpha$.*

This theorem is important, as the inverse Laplace transform will be developed from a contour which encloses the space $Re(s) < \alpha$.

12.3.2 Linearity of the Laplace transform

The Laplace transform, like the Fourier transform, is linear; i.e. the Laplace transform of a sum of functions is equal to the sum of the Laplace transforms of the individual functions,

$$\mathcal{L}[af(t) + bg(t)] = a\mathcal{L}[f] + b\mathcal{L}[g]. \tag{12.65}$$

12.3.3 Time shift

The effect of a time shift must be handled with a little care, because the integration range of the Laplace transform is from 0 to ∞. We assume that the function $f(t) = 0$ for $t < 0$; then the time-shifted function $f(t - b)$ vanishes for $t < b$. We may then write

$$\mathcal{L}[f(t-b)] = \int_b^\infty f(t-b)e^{-st}\mathrm{d}t = e^{-sb}\int_0^\infty f(t)e^{-st}\mathrm{d}t = e^{-sb}\mathcal{L}[f(t)]. \tag{12.66}$$

This result can be compared with the analogous Fourier result, Eq. (11.61).

12.3.4 Scaling property

It is also straightforward to consider the effect of a Laplace transform on a function $f(ct)$. Assuming $c > 0$, we have

$$\mathcal{L}[f(ct)] = \int_0^\infty f(ct)e^{-st}\mathrm{d}t = \frac{1}{c}\int_0^\infty f(t')e^{-st'/c}\mathrm{d}t' = \frac{1}{c}F(s/c). \tag{12.67}$$

12.3.5 Differentiation in the time domain

We may evaluate the Laplace transform of $f'(t)$ by the use of integration by parts,

$$\mathcal{L}\left[f'(t)\right] = \int_0^\infty f'(t)e^{-st}\mathrm{d}t = \left[f(t)e^{-st}\right]_0^\infty + s\int_0^\infty f(t)e^{-st}\mathrm{d}t. \tag{12.68}$$

The result can be written as

$$\mathcal{L}\left[f'(t)\right] = sF(s) - f(0). \tag{12.69}$$

This argument can be extended to higher-order derivatives; for second derivatives, we have

$$\mathcal{L}[f''(t)] = s^2 F(s) - sf(0) - f'(0). \tag{12.70}$$

In Section 14.10, we will see how the "derivative-reducing" ability of Laplace transforms can be used to solve ordinary differential equations and automatically incorporate the initial conditions.

12.3.6 Inverse transform

To be a worthwhile and practical integral transform, we must be able to determine the function $f(t)$ from the function $F(s)$. It is not immediately obvious, however, how one would go about performing the inversion. In the case of the Fourier transform, the kernel for the inverse transform, $\exp[-i\omega t]$, has an exponent with the opposite sign of the kernel of the forward transform, $\exp[i\omega t]$. This is problematic for the Laplace transform, because the obvious guess for the inverse transform is a kernel of the form $\exp[st]$, which is divergent for positive real s. Let us try such a kernel anyway, but write the inverse transform as an indefinite integral over *complex* values of $s = \sigma + i\omega$, i.e.

$$f(t) = C \int e^{(\sigma+i\omega)t} \int_0^\infty e^{-(\sigma+i\omega)t'} f(t') dt' ds$$

$$= C \int \int_0^\infty e^{\sigma(t-t')} e^{i\omega(t-t')} f(t') dt' ds. \tag{12.71}$$

In this equation C is a constant whose appropriate value is to be determined, and ds is in general an element over a path in the complex s-plane. Noting the complex exponential $\exp[i\omega(t - t')]$, if we choose to integrate along a path of constant σ in the complex plane, i.e. $ds = id\omega$, we may use the Fourier relation,

$$\delta(t - t') = \frac{1}{2\pi} \int_{-\infty}^\infty e^{i\omega(t-t')} d\omega. \tag{12.72}$$

With a choice $C = 1/2\pi i$, Eq. (12.71) properly reconstructs the function $f(t)$.

We may evidently write the inverse Laplace transform as a path integral in the complex plane.

Definition 12.7 (Inverse Laplace transform) *The inverse Laplace transform is an integral in the complex domain of the form*

$$f(t) = \frac{1}{2\pi i} \int_{\omega=-i\infty}^{\omega=+i\infty} F(s) e^{st} (id\omega), \tag{12.73}$$

where $s = \sigma + i\omega$ and $ds = id\omega$.

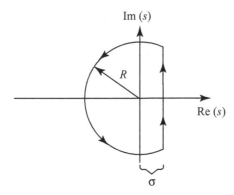

Figure 12.2 The contour for the inverse Laplace transform.

The inverse transform therefore involves extending the function $F(s)$ into the complex plane, and integrating it along a contour parallel to the imaginary axis. It can be shown that the function $F(s)$ is analytic, and we may therefore evaluate the inverse transform by using the residue theorem with a contour as shown in Fig. 12.2, as $R \to \infty$. It should be noted that the semicircular part of the contour must necessarily extend into the $\mathrm{Re}(s) < 0$ half-space to avoid divergence of $\exp[st]$. Occasionally one will find a function $F(s)$ which is multivalued or possesses an infinite number of singularities in the half-space, in which case one must choose the contour with care. An example of this will arise in Section 15.7.

Example 12.2 To demonstrate the application of the inverse transform, we consider the function

$$F(s) = \frac{k}{s^2 + k^2}, \tag{12.74}$$

which we have seen above is the Laplace transform of $\sin(kt)$. The inverse transform is of the form

$$f(t) = \frac{1}{2\pi \mathrm{i}} \oint e^{st} \frac{k}{s^2 + k^2} \mathrm{d}s = \sum \mathrm{Res} \left\{ e^{st} \frac{k}{s^2 + k^2} \right\}. \tag{12.75}$$

Two simple poles exist at the points $s = +\mathrm{i}k$ and $s = -\mathrm{i}k$, with corresponding residues $\exp[\mathrm{i}kt]/2\mathrm{i}$ and $-\exp[-\mathrm{i}kt]/2\mathrm{i}$. Adding these together immediately produces $f(t) = \sin(kt)$.
◇

12.4 Fractional Fourier transform

We have seen that the linear canonical transform encompasses operations including the regular Fourier transform as well as the identity operation (i.e. no effect). In 1980, Victor Namias demonstrated [Nam80] that one could define a transform, called the *fractional*

Fourier transform (fracFT), characterized by a single continuous parameter which encompasses both the regular FT and the identity as its extremes. In the mid-1990s, this transform was shown to be relevant to a number of optical systems [MO93, OM93, OM95, PF94]. In more recent years, the initial excitement about the fracFT seems to have died down, but it is still useful in characterizing optical systems and in signal processing, and generally illustrative of some of the broader points we have made in this chapter.

The original derivation of the fracFT consisted of modifying the known eigenfunction relationship for the regular Fourier transform. As discussed in Section 11.6, the functions $\psi_n(t) = \exp[-t^2/2]H_n(t)$ are eigenfunctions of the Fourier operator which satisfy the following relationship

$$\mathcal{F}\{\psi_n\} = \frac{1}{\sqrt{2\pi}} \int_{-\infty}^{\infty} \psi_n(t)e^{i\omega t}dt = e^{in\pi/2}\psi_n(\omega). \tag{12.76}$$

The effect of Fourier transformation is to multiply the eigenfunction by a complex exponential of the form $\exp[in\pi/2]$. Namias then introduced a fractional transformation $\mathcal{F}_\alpha$ such that

$$\mathcal{F}_\alpha\{\psi_n\} \equiv e^{in\alpha}\psi_n(\omega), \tag{12.77}$$

where α is a parameter ranging from 0 to $\pi/2$. The limit $\alpha = 0$ is the identity operation, while $\alpha = \pi/2$ is the Fourier transform. Successive fractional transformations on a general function $f(t)$ will have an additive effect, i.e.

$$\mathcal{F}_{\alpha_2}\mathcal{F}_{\alpha_1}f = \mathcal{F}_{\alpha_1+\alpha_2}f. \tag{12.78}$$

With some effort, it can be shown that the fracFT can be written in integral form.

Definition 12.8 (Fractional Fourier transform) *The integral form of the fractional Fourier transform of order α is given by*

$$\mathcal{F}_\alpha f = \int_{-\infty}^{\infty} K_\alpha(t,t')f(t')dt', \tag{12.79}$$

where t and t' are taken to be dimensionless variables, and

$$K_\alpha(t,t') = \frac{e^{i\pi/4}e^{-i\alpha/2}}{\sqrt{2\pi \sin(\alpha)}} \exp\left[-\frac{i}{2}\cot(\alpha)t^2\right]\exp\left[\frac{i}{\sin(\alpha)}tt'\right]\exp\left[-\frac{i}{2}\cot(\alpha)t'^2\right]. \tag{12.80}$$

It can be seen that this represents a linear canonical transform, with $b = \sin(\alpha)$, $a = d = \cos(\alpha)$.

The fracFT has an interesting interpretation, which will be elaborated on in later sections. Let us suppose our initial function $f(t)$ represents the behavior of a signal in the time domain; its Fourier transform, $\tilde{f}(\omega)$, therefore represents the behavior of the signal in the frequency

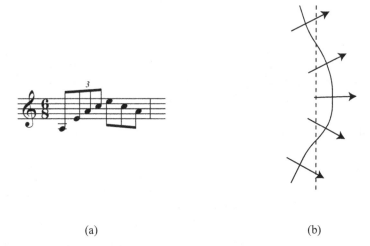

(a)   (b)

Figure 12.3 Examples where a simple Fourier transform gives an unsatisfying depiction of the behavior of a wavefield: (a) a musical score, (b) quasi-geometrical wave propagation.

domain. The operation $\mathcal{F}_\alpha$, which lies between these two extremes, evidently represents a mixed representation of the function in both time and frequency.

Applications of the fracFT to optics have taken advantage of this mixed nature. Ozaktas and Mendlovic [OM95] showed that the fracFT can be used to describe the behavior of a wavefield at planes intermediate between the initial image plane and the focal plane. Gbur and Wolf [GW01] suggested that the fracFT can be used to characterize the strength of diffraction effects in a tomographic imaging system. Alonso and Forbes [AF97] have applied the fracFT to wave theory in the geometric limit, and have used it to avoid caustics in the direct and Fourier domains.

It is worth noting that the fracFT is but one parametric path from the identity operator to the Fourier transform. Any continuous variation of the LCT parameters a,b,c,d, subject to $ad - bc = 1$ and with start and end points $a = d = 1$, $b = 0$ and $a = d = 0$, $b = -1$, respectively, will meet the same goals. The fracFT has a special geometric significance, however, to be discussed in Section 12.7.

12.5 Mixed domain transforms

Considering again the Fourier transform, we noted in Section 11.5 that the Fourier transform may be considered a measure of the frequency content, or spectrum, of the complete wavefield. If we consider a few real-world examples of wavefields, however, it becomes clear that the Fourier transform cannot be the *only* description of a wave's spectral properties, and by no means the best. Let us consider a musical score, as illustrated in Fig. 12.3(a). Sheet music describes the pitch (frequency) that must be played as a function of time, i.e. it is a description of the behavior of a wavefield simultaneously in both time and frequency.

In the spatial frequency domain, we consider another example: the propagation of a wavefield and its description using the angular spectrum representation, as discussed in Section 11.9. This example is illustrated in Fig. 12.3(b). We can either describe a optical wave by its functional behavior in space by Huygens' principle or in direction by its plane wave decomposition. In the limit that geometrical optics is valid, however, we should be able to, at least approximately, describe *locally* both the amplitude of the wavefield and its directionality.

The Fourier transform is a poor way to characterize a musical score because the function $\tilde{f}(\omega)$ is a measure of the contribution of frequency ω to the *entire* musical composition. Information about the temporal distribution of frequencies in the musical piece is hidden in the complex phase of the Fourier transform, as indicated in Section 11.4.6. Ideally, we would like an integral transform which allows us to simultaneously describe the temporal and frequency behavior of a wavefield.

We attempt to do this by introducing a general linear integral transform $\mathcal{G}_{t_0,\omega_0}$, which may be written as follows

$$\mathcal{G}_{t_0,\omega_0}\{f\} = F_{t_0,\omega_0} = \int_{-\infty}^{\infty} f(t')G_{t_0,\omega_0}^*(t')\mathrm{d}t', \tag{12.81}$$

where G_{t_0,ω_0}^* represents the kernel of the integral transform, and t_0 and ω_0 are central points in time and frequency, to be discussed momentarily.

The transform given by Eq. (12.81) is a broad definition which encompasses all the transforms of this chapter, as well as the Fourier transform. The problem, as it were, with the Fourier kernel, $G_{t_0,\omega_0}^*(t) = \exp[i\omega_0 t]$, is that it is poorly localized in time. We may attempt to introduce time localization by modifying the Fourier kernel with a "window", i.e.

$$G_{t_0,\omega_0}^*(t) = g(t - t_0)\exp[i\omega_0 t], \tag{12.82}$$

where the function $g(t)$ is taken to be a function localized in time, for example a Gaussian function,

$$g(t - t_0) = \exp[-(t - t_0)^2/2\sigma^2]. \tag{12.83}$$

A transform with a kernel given by Eq. (12.82) is known as a *windowed Fourier transform*, and was first introduced by Dennis Gabor for the spectral analysis of sound waves [Gab46]. An example of the kernel is shown in Fig. 12.4. The transform given by Eq. (12.81) therefore provides information about the Fourier spectra of the function $f(t)$ within a region of width σ centered on the time t_0.

This localization comes with a cost, however. Equation (12.81) can also be written in the frequency domain with the aid of the Parseval theorem, Eq. (11.89), in the form

$$\int_{-\infty}^{\infty} f(t')G_{t_0,\omega_0}^*(t')\mathrm{d}t' = 2\pi \int_{-\infty}^{\infty} \tilde{f}(\omega')\tilde{G}_{t_0,\omega_0}^*(\omega')\mathrm{d}\omega'. \tag{12.84}$$

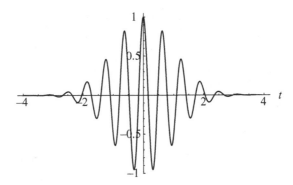

Figure 12.4 The real part of the kernel for the windowed Fourier transform, with $\sigma = 1$ and $\omega_0 = 10$.

The value of $\tilde{G}^*_{t_0,\omega_0}(\omega')$ can be readily found to be

$$\tilde{G}^*_{t_0,\omega_0}(\omega') = \frac{\sigma}{\sqrt{2\pi}}e^{-(\omega-\omega_0)^2\sigma^2/2}. \tag{12.85}$$

The same integral transform provides information about the spectrum within a region roughly of width $1/\sigma$ centered on the frequency ω_0. It can be seen that, for a given value of σ, the shape of the function is fixed in both time and frequency and changes in t_0, ω_0 change only the origin of the function.

Because $G^*_{t_0,\omega_0}(t)$ and its Fourier transform are subject to the Fourier uncertainty relation, Eq. (11.107),

$$(\Delta\omega)(\Delta t) \geq \frac{1}{2}, \tag{12.86}$$

our simultaneous measurement of a signal's temporal and frequency behavior is restricted to a region in (t,ω)-space of area $1/2$ or greater. In the specific case of a Gaussian window, our measurement is only certain to within a box of temporal width $\sigma/\sqrt{2}$ and frequency height $1/\sqrt{2}\sigma$. Several of these boxes are illustrated in Fig. 12.5, along with the limiting case of a pure (unwindowed) Fourier transform and the identity operation. The pure Fourier transform is a box of infinite width in the time domain and vanishing width in the frequency domain, while the identity operation is a box of vanishing width in time and infinite width in frequency. Following the notation of [Mal98], we refer to the collection of functions $G^*_{t_0,\omega_0}$ of fixed form and for all t_0, ω_0 as *time–frequency atoms*.

Like most integral transforms, the usefulness of a windowed transform is directly related to the ability to find an inverse transform. We postulate an inverse transform $I(t)$ of the form

$$I(t) = \int_{-\infty}^{\infty} \int_{-\infty}^{\infty} F_{t_0,\omega_0}g^*(t-t_0)e^{-i\omega_0 t}dt_0d\omega_0. \tag{12.87}$$

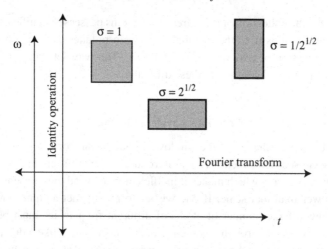

Figure 12.5 Time–frequency atoms of various sizes.

It is to be noted that the inverse transformation requires an integral over both time t_0 and frequency ω_0, as the function F_{t_0,ω_0} is localized in both time and frequency. To demonstrate the inverse, we substitute from Eq. (12.81) into the above,

$$I(t) = \int_{-\infty}^{\infty}\int_{-\infty}^{\infty}\int_{-\infty}^{\infty} g(t'-t_0)e^{i\omega_0 t'}g^*(t-t_0)e^{-i\omega_0 t}f(t')dt'dt_0 d\omega_0. \qquad (12.88)$$

The integral over ω_0 may be directly evaluated to a delta function, which contracts the integration over t' to t,

$$I(t) = 2\pi f(t)\int_{-\infty}^{\infty}|g(t-t_0)|^2 dt_0 = 2\pi f(t)\int_{-\infty}^{\infty}|g(t'')|^2 dt''. \qquad (12.89)$$

Provided the function $g(t)$ is square integrable, the resulting function $I(t)$ is directly proportional to $f(t)$. With an appropriate choice of normalization for $g(t)$, we can make $I(t)$ the exact inverse transform of the windowed Fourier transform.

Several comments are in order here. First, it is to be noted that the choice of window size greatly effects the form of the transform: there is no uniquely-defined window for measuring the time–frequency behaviors of a system, and often the method of analysis is determined by the experimental system. For instance, the Fabry–Perot interferometer discussed in Section 7.1 filters an input signal with respect to frequency, and the output is measured in time: it represents, in essence, a time–frequency analyzer. Second, it is be noted that the time-frequency atoms are nonorthogonal, i.e. the integral of two distinct overlapping atoms is nonzero,

$$\int_{-\infty}^{\infty} g^*(t-t_0)e^{-i\omega_0 t}g(t-t_1)e^{i\omega_1 t}dt \neq 0. \qquad (12.90)$$

This suggests that the collection of time–frequency atoms possesses significant redundancy, which is perhaps obvious from the fact that the transform takes us from a function of one variable t to a function of two variables, t_0 and ω_0. The wavelet transforms in the next section provide an elegant solution to these difficulties.

12.6 The wavelet transform

We have noted that the behavior of the windowed Fourier transform depends upon the size of the window chosen. However, this also introduces a length scale into the transform – namely, the window size – which makes it inefficient when used to represent objects much wider or narrower than that scale. If one wishes to reconstruct a signal much narrower than the window, a large number of atoms of different frequencies must be used which produce the narrow shape through destructive interference, much like a delta function can be constructed by an infinite superposition of harmonic waves. Conversely, if one wishes to reconstruct a signal much wider than the window, a large number of atoms of different times must be superimposed to produce the wide shape.

The windowed Fourier transform is most efficient, then, when it is used to reconstruct signals with features of size comparable to the window. Ideally, however, one would like a time–frequency representation of a field which does not introduce a characteristic length into the problem.

We therefore introduce a family of time–frequency atoms of the form,

$$\psi_{u,s}(t) = \frac{1}{\sqrt{s}} \psi \left(\frac{t-u}{s} \right),$$ (12.91)

where $\psi_{u,s}(t)$ is a square-integrable function with zero average,

$$\int_{-\infty}^{\infty} \psi(t) \mathrm{d}t = 0,$$ (12.92)

and unit normalization,

$$\int_{-\infty}^{\infty} |\psi(t)|^2 \mathrm{d}t = 1.$$ (12.93)

The family of atoms is time-shifted by a delay u and *scaled* by a length s. The shapes of such *wavelets* are unchanged under this process, as illustrated in Fig. 12.6 for the so-called Mexican hat wavelet, defined as

$$\psi(t) = \frac{2}{\pi^{1/4}\sqrt{3\sigma}} \left(\frac{t^2}{\sigma^2} - 1 \right) \exp[-t^2/2\sigma^2].$$ (12.94)

For small values of s, the wavelet represents a rapidly-varying, localized function, while for large values of s, the wavelet represents a slowly-varying, non-local function. Wavelets

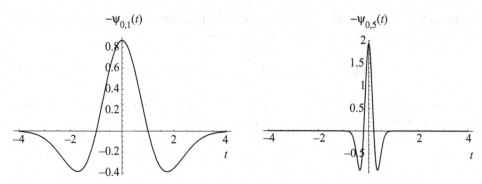

Figure 12.6 Two scaled versions of the Mexican hat wavelet, with $\sigma = 1$.

can be real-valued, or can be complex analytic, such that their Fourier representation $\hat{\psi}(\omega)$ vanishes for $\omega < 0$.

Definition 12.9 (Wavelet transform) *The wavelet transform of a real-valued function $f(t)$ is an expansion of that function into time and scale, of the form*

$$\mathcal{S}\{f\}(u,s) = \int_{-\infty}^{\infty} f(t)\psi_{u,s}(t)dt. \tag{12.95}$$

We introduce the inverse wavelet transform by means of a theorem.

Theorem 12.2 *Let $\psi(t)$ be a real-valued, square-integrable function such that*

$$\beta \equiv \int_0^{\infty} \frac{|\tilde{\psi}(\omega)|^2}{\omega} d\omega < \infty. \tag{12.96}$$

Then a square-integrable function $f(t)$ may be represented in a wavelet expansion of the form

$$f(t) = \frac{1}{\beta} \int_0^{\infty} \int_{-\infty}^{\infty} \mathcal{S}\{f\}(u,s)\psi_{u,s}(t)du \frac{ds}{s^2}. \tag{12.97}$$

We can prove this theorem by transforming Eq. (12.97) into the Fourier domain, i.e.

$$\tilde{f}(\omega) = \frac{1}{\beta} \int_0^{\infty} \int_{-\infty}^{\infty} \mathcal{S}\{f\}(u,s)\tilde{\psi}_{u,s}(\omega)du \frac{ds}{s^2}. \tag{12.98}$$

Furthermore, by using the shift and scaling properties of the Fourier transform, Eqs. (11.61) and (11.65) respectively, we may write

$$\tilde{\psi}_{u,s}(\omega) = \sqrt{s}\exp[i\omega u]\tilde{\psi}(\omega s). \tag{12.99}$$

The wavelet transform itself is the convolution of the function $f(t)$ with the function $\psi_{u,s}(t)$. We may expand it in the Fourier domain in the form

$$\mathcal{S}\{f\}(u,s) = 2\pi\sqrt{s}\int_{-\infty}^{\infty}\tilde{f}(\omega')\tilde{\psi}^*(s\omega')e^{-i\omega'u}d\omega'. \tag{12.100}$$

On substitution from Eqs. (12.99) and (12.100) into Eq. (12.98), the integration over u becomes proportional to a delta function. We are left with

$$\tilde{f}(\omega) = \frac{(2\pi)^2}{\beta}\tilde{f}(\omega)\int_0^{\infty}|\tilde{\psi}(s\omega)|^2\frac{ds}{s}. \tag{12.101}$$

By a change of variable to $\xi = s\omega$, the latter integral can be transformed into

$$\int_0^{\infty}|\tilde{\psi}(\xi)|^2\frac{d\xi}{\xi} = \beta < \infty. \tag{12.102}$$

Provided our wavelet $\psi(t)$ satisfies Eq. (12.96), known as the *admissibility condition*, the integral is well-behaved and the inverse exists.

We may also introduce a version of Plancherel's identity for wavelets

Theorem 12.3 *For a real-valued square integrable function $f(t)$,*

$$\int_{-\infty}^{\infty}|f(t)|^2dt = \frac{1}{\beta}\int_0^{\infty}\int_{-\infty}^{\infty}|\mathcal{S}\{f\}(u,s)|^2du\frac{ds}{s^2}. \tag{12.103}$$

There are obviously a large number of functions which could be chosen as wavelets. Specific choices depend upon the desired signal processing application.

We have noted that the wavelets are non-orthogonal, i.e. two wavelets separated by a time interval du and a scale interval ds will have a nonzero overlap integral. This redundancy can be useful in the reduction of calculational noise. In other cases, a set of orthogonal wavelets is desired. One may in fact find a discrete basis for wavelets, made up of discrete scalings and time shifts, and even an orthogonal basis for wavelets. The simplest example of such a basis is the so-called *Haar basis*, where

$$\psi(t) = \begin{cases} 1, & 0 \leq t < 1/2, \\ -1, & 1/2 \leq t < 1, \\ 0, & \text{otherwise.} \end{cases} \tag{12.104}$$

We may define a set of scaled and shifted functions as

$$\psi_{m,n}(t) = 2^{-m/2}\psi(2^{-m}t - n), \tag{12.105}$$

and it can be shown that these functions form an orthonormal basis for all square-integrable functions; that is, an arbitrary square-integrable function $f(t)$ may be written as

$$f(t) = \sum_{m=-\infty}^{\infty}\sum_{n=-\infty}^{\infty}c_{mn}\psi_{m,n}(t), \tag{12.106}$$

with

$$c_{mn} = \int_{-\infty}^{\infty} \psi_{m,n}(t) f(t) \mathrm{d}t. \qquad (12.107)$$

This set is not optimal because it consists of discontinuous functions, but it has been shown that one can construct orthonormal bases of wavelets with finite support and as much regularity as desired; we refer the reader to [Dau88] for the details. It has been shown, however, that the Haar basis is the only orthonormal wavelet basis with a fundamental wavelet of simple symmetry.

12.7 The Wigner transform

The windowed Fourier transform and the wavelet transform can be used to decompose a *field* $U(t)$ into a simultaneous time–frequency representation. However, observable quantities in optics,[2] such as the intensity, are *second-order* in the field, i.e. $I \equiv |U|^2$. The *Wigner transform* [Wig32], first introduced to aid in the quantum-mechanical analysis of thermodynamic systems, is a representation of such a second-order quantity in a time–frequency form. The Wigner transform of a function $f(t)$ may be defined as follows.

Definition 12.10 (Wigner transform) *The Wigner transform of a square-integrable function $f(t)$ is defined as*

$$W(t,\omega) = \int_{-\infty}^{\infty} f(t+t'/2) f^*(t-t'/2) \exp[-i\omega t'] \mathrm{d}t'. \qquad (12.108)$$

This distribution represents the second-order properties of the field in terms of both time and frequency. Returning to the examples of Fig. 12.3, it would seem that this distribution would be ideal for characterizing the instantaneous power spectrum of a wavefield and the simultaneous spatial and directional properties of a wavefield. The integral of the Wigner function over time and frequency gives results consistent with this interpretation,

$$\frac{1}{(2\pi)^2} \int_{-\infty}^{\infty} W(t,\omega) \mathrm{d}\omega = |f(t)|^2, \qquad (12.109)$$

$$\int_{-\infty}^{\infty} W(t,\omega) \mathrm{d}t = |\tilde{f}(\omega)|^2. \qquad (12.110)$$

In other words, the integral of the Wigner function over all frequencies gives the total power of the wavefield at time t, and the integral over all times gives the Fourier power spectrum.

However, the Wigner function fails to work as a proper instantaneous spectrum for one very important reason. Any definition of a power spectrum should be a non-negative

---

[2] One of the major themes of optical coherence theory is the formulation of optics entirely in terms of observable, i.e. measurable, quantities. See, for instance, [Wol54].

quantity, but the Wigner function can be shown to typically take on both positive and negative values.

The failure of the Wigner function in this respect comes from the assumption that such a thing as an "instantaneous power spectrum" exists for a wave signal. As a consequence of the Fourier uncertainty relations, a reasonable measurement of the frequency of a signal requires it to be measured over many cycles; any attempt to reduce this measurement time results in an increase of the uncertainty in frequency. No instantaneous power spectrum may be defined which satisfies all expected properties of a power spectrum.

A suitable averaging of the Wigner function, however, can produce a quantity which has all the properties of a power spectrum. Let us consider the quantity

$$I_{t,\omega} \equiv \left| \int_{-\infty}^{\infty} f(t')\phi_{t,\omega}^*(t')\mathrm{d}t' \right|^2, \qquad (12.111)$$

where $\phi_{t,\omega}$ is a time–frequency atom such as the windowed Fourier transform or the wavelet transform. $I_{t,\omega}$ represents the power in the signal $f(t')$ associated with the atom centered on t and ω. With some application of Fourier relations, one readily finds that

$$I(t,\omega) = \int_{-\infty}^{\infty} \int_{-\infty}^{\infty} W(u',\xi') W_{t,\omega}(u',\xi')\mathrm{d}u'\mathrm{d}\xi', \qquad (12.112)$$

where W is the Wigner transform of f and $W_{t,\omega}$ is the Wigner transform of the time–frequency atom $\phi_{t,\omega}$. Because $I(t,\omega)$ is initially a non-negative quantity by definition, we find that a time-frequency averaging of the Wigner distribution results in a non-negative quantity. It has been well-argued that one can only talk sensibly about a instantaneous power spectrum of a wavefield in the context of a specific experimental apparatus [EW77].

A large number of properties can be associated with the Wigner transform; [Bas78, Loh93] describe many of these properties in the context of optics. It is interesting to note that the Wigner transform can be used to give a geometrical interpretation of the fractional Fourier transform; a fractional transform through angle ϕ transforms the Wigner function to the form

$$W_0(t,\omega) \to W_0(t\cos\phi - \omega\sin\phi, \omega\cos\phi + t\sin\phi). \qquad (12.113)$$

In other words, a fracFT performs a rotation of the Wigner function in the time–frequency plane.

In spite of its limitations, the Wigner function has found use in optics in elucidating the foundations of geometrical optics and radiometry [Wol78].

12.8 Focus: the Radon transform and computed axial tomography (CAT)

We now consider an application where integral transforms and their inverses play an important role: X-ray computed tomography, also known as computed axial tomography, or CAT

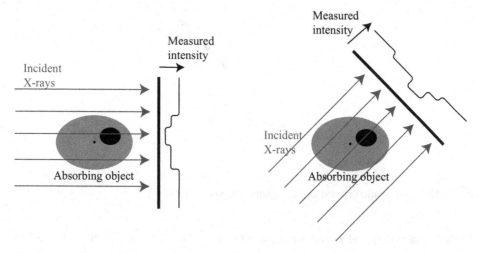

Figure 12.7 An illustration of a basic measurement scheme for computed tomography. Parallel X-rays pass through the human body, undergoing absorption but not diffraction, and the remaining intensity is measured at a detector plane.

scan. CAT scans are now a standard tool in medical imaging for determining the internal structure of the human body. In computed tomography, X-rays are used to probe the absorption properties of an object (assumed to be two-dimensional for simplicity). We refer to the measured intensity profile behind the object as a "shadowgram"; the relevant geometry is shown in Fig. 12.7.

A shadowgram is a projection of the two-dimensional object absorption properties onto a one-dimensional image. The problem before us is to reconstruct the two-dimensional absorption properties of the object using measurements of the object's one-dimensional shadowgrams for all directions of incident X-rays.

This is a type of problem known as mathematically as an *inverse problem*. Broadly speaking, inverse problems represent the determination of the "cause" of a physical phenomenon from measurements of the "effect" of the phenomenon. In our case, the "cause" is the absorption properties of the object of interest, and the "effect" is the absorption of X-rays by this object. Every inverse problem is built upon a *direct* problem which represents the usual physical cause–effect relationship.

In order to solve this inverse problem, we first need a model for how the X-rays are affected by the object being illuminated. X-rays are high-energy photons which are rarely deflected by the low-density material of the human body; their path of propagation is well-approximated by straight lines. The only influence of human tissue on the X-rays is a slight attenuation of their intensity through absorption. This attenuation is well-approximated by a general form of what is known as *Beer's law*,

$$I_2 = I_1 \exp\left\{-\int_{P_1}^{P_2} f(\rho) dl\right\}. \tag{12.114}$$

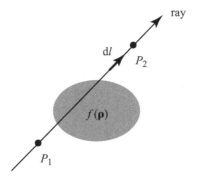

Figure 12.8 The notation related to the implementation of Beer's law.

In this equation, I_1 and I_2 represent the intensity of the X-ray at positions P_1 and P_2, respectively, $f(\rho)$ represents the position-dependent absorption coefficient of the object, and dl represents an infinitesmal path element along a straight line. This geometry is illustrated in Fig. 12.8.

It is to be noted that if the X-ray does not intersect the object, its intensity is unchanged, for $f(\rho) = 0$ outside the object. Beer's law is an empirical property of X-ray propagation; we will assume that it is true, though it is to be noted that one can derive it through the use of Maxwell's equations in certain asymptotic limits (in particular, the wavelength has to be extremely small).

If Beer's law is satisfied, the line integral of the absorption coefficient through the object can be found from measurements of the attenuated intensity through a logarithm,

$$\int_{P_0}^{P} f(\rho)\mathrm{d}l = -\log\left[\frac{I(P)}{I(P_0)}\right], \tag{12.115}$$

where $I(P)$ is the intensity at the detector at point P and $I(P_0)$ is the intensity at the X-ray source at point P_0.

To attempt to solve the inverse problem, we need to set up an appropriate coordinate system. We consider the arrangement depicted in Fig. 12.9.

A particular shadowgram is comprised of the image of a large number of rays, all propagating parallel to each other and perpendicular to a unit vector $\hat{\mathbf{n}}$. The position of that ray along the $\hat{\mathbf{n}}$ direction is given by p. The attenuation of a particular ray is defined as $F(\hat{\mathbf{n}},p)$, and this can be shown to be related to the absorption coefficient by

$$F(\hat{\mathbf{n}},p) = \iint_{-\infty}^{\infty} f(\rho)\delta(p - \hat{\mathbf{n}}\cdot\rho)\mathrm{d}^2\rho. \tag{12.116}$$

We have written the line integral of Eq. (12.115) in terms of a two-dimensional area integral of a one-dimensional delta function, as in Eq. (6.65). The delta function selects only that ray for which $\hat{\mathbf{n}}\cdot\rho = p$. An individual shadowgram is defined by the function $F(\hat{\mathbf{n}},p)$, with

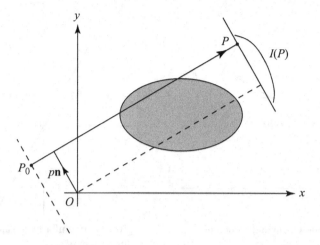

Figure 12.9 Coordinate system for performing the computed tomography inverse problem.

$\hat{\mathbf{n}}$ fixed. Equation (12.116) represents another integral transform, like the Fourier transform, and is generally referred to as the *Radon transform*. We now wish to attempt to reconstruct $f(\rho)$, given the projections $F(\hat{\mathbf{n}}, p)$, i.e. we wish to find the inverse Radon transform.

Fourier analysis can be used to find the appropriate relationships. First, we note that the delta function of Eq. (12.116) may be written in terms of a Fourier representation,

$$\delta(p - \hat{\mathbf{n}} \cdot \boldsymbol{\rho}) = \frac{1}{2\pi} \int_{-\infty}^{\infty} e^{iu(p - \hat{\mathbf{n}} \cdot \boldsymbol{\rho})} du. \tag{12.117}$$

On substituting from this equation into our expression for $F(\hat{\mathbf{n}}, p)$, we find that

$$F(\hat{\mathbf{n}}, p) = \frac{1}{2\pi} \iint_{-\infty}^{\infty} \int_{-\infty}^{\infty} f(\boldsymbol{\rho}) e^{iu(p - \hat{\mathbf{n}} \cdot \boldsymbol{\rho})} d^2 \rho \, du. \tag{12.118}$$

Now it is to be noted that the integral over ρ may be evaluated; it is simply related to a two-dimensional Fourier transform of $f(\rho)$,

$$\tilde{f}(\mathbf{k}) \equiv \frac{1}{(2\pi)^2} \iint_{-\infty}^{\infty} f(\boldsymbol{\rho}) e^{-i\mathbf{k} \cdot \boldsymbol{\rho}} d^2 \rho. \tag{12.119}$$

Our equation for the shadowgram may be written as

$$F(\hat{\mathbf{n}}, p) = 2\pi \int_{-\infty}^{\infty} \tilde{f}(u\hat{\mathbf{n}}) e^{iup} du. \tag{12.120}$$

But now the right-hand side of this equation looks like a one-dimensional Fourier transform with respect to u! If we take a one-dimensional transform of $F(\hat{\mathbf{n}}, p)$ with respect to p, i.e.

$$\tilde{F}(\hat{\mathbf{n}}, u) \equiv \frac{1}{2\pi} \int_{-\infty}^{\infty} F(\hat{\mathbf{n}}, p) e^{-iup} du, \tag{12.121}$$

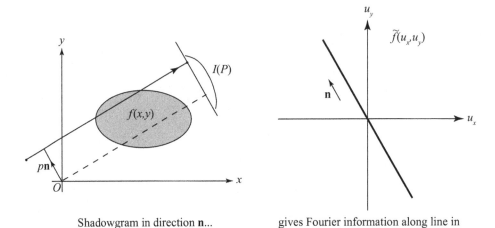

Shadowgram in direction **n**...

gives Fourier information along line in direction of **n**!

Figure 12.10 The relationship between one shadowgram and the two-dimensional Fourier transform of the absorption coefficient.

we immediately find that

$$\tilde{F}(\hat{\mathbf{n}}, u) = \tilde{f}(u\hat{\mathbf{n}}). \tag{12.122}$$

This is a remarkable result. It tells us that *the one-dimensional Fourier transform of a shadowgram is related to the two-dimensional Fourier transform of the absorption coefficient of the function!* In other words, the shadowgram related to direction $\hat{\mathbf{n}}$ contains information about the Fourier components of $\tilde{f}$ along the line in Fourier space in direction $\hat{\mathbf{n}}$, as illustrated in Fig. 12.10.

This statement is usually referred to as the *projection-slice theorem* of computed tomography. The projection of the absorption coefficient along a direction perpendicular to $\hat{\mathbf{n}}$ gives us a "slice" of the object in Fourier space, along the line parallel to $\hat{\mathbf{n}}$.

We may use further Fourier transform tricks to determine a formula for $f(\boldsymbol{\rho})$ based on its projections. First, we write

$$\hat{\mathbf{n}} = \hat{\mathbf{x}}\cos\phi + \hat{\mathbf{y}}\sin\phi. \tag{12.123}$$

Then we may write

$$\tilde{f}(u\hat{\mathbf{n}}) = \tilde{f}(u\cos\phi, u\sin\phi). \tag{12.124}$$

The inverse Fourier transform of $\tilde{f}$ in polar coordinates may be written as

$$f(\boldsymbol{\rho}) = \iint_{-\infty}^{\infty} \tilde{f}(u_x, u_y)\mathrm{e}^{\mathrm{i}\mathbf{u}\cdot\boldsymbol{\rho}}\,\mathrm{d}u_x\mathrm{d}u_y = \int_0^{\infty}\int_0^{2\pi} \tilde{f}(u\cos\phi, u\sin\phi)\mathrm{e}^{\mathrm{i}\mathbf{u}\cdot\boldsymbol{\rho}}u\mathrm{d}u\mathrm{d}\phi. \tag{12.125}$$

If we substitute from Eq. (12.122) into this equation, we find that

$$f(\boldsymbol{\rho}) = \int_0^\infty \int_0^{2\pi} \tilde{F}(\hat{\mathbf{n}}, u) \mathrm{e}^{-iu\hat{\mathbf{n}}\cdot\boldsymbol{\rho}} u \, du \, d\phi. \tag{12.126}$$

This formula is, in essence, the inverse Radon transform. It combines the one-dimensional Fourier transforms of measured shadowgrams in a two-dimensional inverse Fourier transform to reconstruct the original absorption coefficient $f(\boldsymbol{\rho})$.

This equation can be written in a more compact form by separating the angular integral into two pieces,

$$f(\boldsymbol{\rho}) = \int_0^\infty \int_0^{\pi} \tilde{F}(\hat{\mathbf{n}}, u) \mathrm{e}^{-iu\hat{\mathbf{n}}\cdot\boldsymbol{\rho}} u \, du \, d\phi + \int_0^\infty \int_\pi^{2\pi} \tilde{F}(\hat{\mathbf{n}}, u) \mathrm{e}^{-iu\hat{\mathbf{n}}\cdot\boldsymbol{\rho}} u \, du \, d\phi \tag{12.127}$$

Because $\cos(\phi + \pi) = -\cos\phi$ and $\sin(\phi + \pi) = -\sin\phi$, we may change the range of integration in the second integral,

$$f(\boldsymbol{\rho}) = \int_0^\infty \int_0^{\pi} \tilde{F}(\hat{\mathbf{n}}, u) \mathrm{e}^{-iu\hat{\mathbf{n}}\cdot\boldsymbol{\rho}} u \, du \, d\phi + \int_0^\infty \int_0^{\pi} \tilde{F}(-\hat{\mathbf{n}}, u) \mathrm{e}^{iu\hat{\mathbf{n}}\cdot\boldsymbol{\rho}} u \, du \, d\phi. \tag{12.128}$$

It follows from the definition of the Radon transform that $F(-\hat{\mathbf{n}}, p) = F(\hat{\mathbf{n}}, -p)$. On substitution, we have

$$f(\boldsymbol{\rho}) = \int_0^\infty \int_0^{\pi} \tilde{F}(\hat{\mathbf{n}}, u) \mathrm{e}^{-iu\hat{\mathbf{n}}\cdot\boldsymbol{\rho}} u \, du \, d\phi + \int_0^\infty \int_0^{\pi} \tilde{F}(\hat{\mathbf{n}}, -u) \mathrm{e}^{iu\hat{\mathbf{n}}\cdot\boldsymbol{\rho}} u \, du \, d\phi. \tag{12.129}$$

If we change variables of integration in the second integral from u to $-u$, we are led to the result,

$$f(\boldsymbol{\rho}) = \int_{-\infty}^\infty \int_0^{\pi} \tilde{F}(\hat{\mathbf{n}}, u) \mathrm{e}^{-iu\hat{\mathbf{n}}\cdot\boldsymbol{\rho}} |u| \, du \, d\phi. \tag{12.130}$$

This formula is the basis of what is known as the *filtered back-projection algorithm* for parallel rays. The reconstruction problem in computed tomography is thus:

1. measure the projections (shadowgrams) for a large number of directions of incident X-rays;
2. calculate the Fourier transform of each of these projections;
3. put these individual transform shadowgrams together into a single function $\tilde{F}(\hat{\mathbf{n}}, u)$;
4. multiply this function by a *ramp filter*, $|u|$ (hence the "filtered" in "filtered back-projection");
5. take the inverse Fourier transform of the filtered function to determine the object profile.

We have dealt with the simplest modality of computed tomography in this section, the case for which the X-rays for a given shadowgram are all parallel to each other. It is to be noted that one can also solve this problem when the incident X-rays emanate from a point-like source. This method is easier to implement because each shadowgram can be constructed using radiation from a single source. In performing parallel-ray tomography, one can only create a parallel-ray system by scanning a well-collimated source across a range of positions. The so-called *fan-beam problem* is more involved theoretically than the parallel-ray case and we will not solve it here.

It is worth noting that knowledge of the Radon transform is not essential to computed tomography. The original clinical CAT scans [Hou73] were performed by formulating the problem as matrix multiplication and using a computer to perform the necessary matrix inversion (hence the "computed" in computed tomography). It was only later that Radon's related theoretical formalism [Rad17] was rediscovered by practitioners. The formalism has been extremely important, however, in reducing redundancy in the measured dataset and in understanding and eliminating artifacts in the reconstruction process.

12.9 Additional reading

- S. Mallat, *A Wavelet Tour of Signal Processing* (San Diego, Academic Press, 1998). A detailed introduction to wavelets with an emphasis on their use in signal processing.
- G. Kaiser, *A Friendly Guide to Wavelets* (Boston, Birkhäuser, 1994). As the title says, a very user-friendly introduction to wavelets, with interesting chapters on the application of wavelets to physical problems such as electromagnetics. Be sure to read Chapter 1, with its specialized notation, to avoid getting lost later in the book.
- F. Natterer, *The Mathematics of Computerized Tomography* (New York, Wiley, 1986). A great mathematical discussion of computed tomography, including issues involved in practical reconstruction.
- A. C. Kak and M. Slaney, *Principles of Computerized Tomographic Imaging* (New York, IEEE Press, 1988). Detailed discussion of many different types of computed imaging, including with diffracting wavefields.
- K. B. Wolf, *Integral Transforms in Science and Engineering* (Plenum Press, New York, 1979). This is a thorough book that covers pretty much every integral transform one might wish to consider.

12.10 Exercises

1. Read the paper by C. L. Epstein and J. C. Schotland, The bad truth about Laplace's transform, *SIAM Review* **50** (2008), 504–520. Explain what properties of the Laplace transform are being studied, and the techniques used to do so. What conclusions are drawn about the inverse Laplace transform?
2. Read the paper by A. T. Friberg, Energy transport in optical systems with partially coherent light, *Appl. Opt.* **25** (1986), 4547–4556. What quantity is being represented by the Wigner function, and what problem is it being used to solve? Write the Wigner expression used in the paper, making sure to define all terms in the expression.
3. One of the few Fresnel transforms that can be calculated analytically is for the Gaussian function, $f(x) = \exp[-x^2/2\sigma^2]$. Show that the Fresnel transform is given by

$$F_\alpha(x) = \sqrt{\frac{2i\sigma^2\alpha}{1 - 2i\sigma^2\alpha}} \exp\left[\frac{x^2}{1 - 2i\sigma^2\alpha}\right].$$

4. The Fresnel transform of the Gaussian function $f(x) = \exp[-x^2/2]$ is given by

$$F_\alpha = \sqrt{\frac{2i\alpha}{1-2i\alpha}} \exp\left[\frac{x^2}{1-2i\alpha}\right].$$

Using elementary properties of the Fresnel transform (scaling, shifting, etc.), determine the Fresnel transforms of

1. $f(x) = \exp[-x^2/2]\exp[2ix]$,
2. $f(x) = (1-4x^2)\exp[-2x^2]$.

5. The Fresnel transform of the Gaussian function $f(x) = \exp[-x^2/2]$ is given by

$$F_\alpha = \sqrt{\frac{2i\alpha}{1-2i\alpha}} \exp\left[\frac{x^2}{1-2i\alpha}\right].$$

Using elementary properties of the Fresnel transform (scaling, shifting, etc.), determine the Fresnel transforms of

1. $f(x) = x\exp[-x^2/2]$,
2. $f(x) = \exp[-2x^2]\exp[ix]$.

6. Prove the composition property of LCTs, specifically Eq. (12.51), by evaluating the integral of Eq. (12.49) with respect to x''.

7. Prove Plancherel's identity for LCTs.

8. The Laplace transform of the derivative of a function is of the form $\mathcal{L}[f'] = s\mathcal{L}[f] - f(0)$. Derive the Laplace transform of $f(x) = \cos(kx)$, and use the derivative relation to derive the Laplace transform of

$$g(x) = -\frac{1}{k^2}\frac{d^2}{dx^2}\cos(kx),$$

and confirm that they give the same result.

9. Determine the Laplace transform of the following functions, and specify the domain in which they converge:

1. $f(t) = 5t^3 - 2t^2 + 1$,
2. $f(t) = e^{-2t}\cosh(2t)$,
3. $f(t) = 5^t$.

10. Determine the Laplace transform of the following functions, and specify the domain in which they converge:

1. $f(t) = \sinh^2(2t)$,
2. $f(t) = 2t^2 - 3t$,
3. $f(t) = t^{1/2}$.

11. Determine the inverse Laplace transform of the following functions, with $a > 0$:

1. $F(s) = n!/(s-a)^{n+1}$,
2. $F(s) = s/(s^2 - a^2)$.

12. Determine the inverse Laplace transform of the following functions, with $a, b > 0$:

1. $F(s) = b/[(s-a)^2 + b^2]$,
2. $F(s) = 1/(s-a) + 4/(s-b)^2$.

13. Determine the inverse Laplace transform of the following functions:

1. $F(s) = e^{-s}/[s(s^2+1)]$,
2. $F(s) = (s^2 + 2s + 2)/(s^3 + 1)$.

14. Read the paper by G. Gbur and E. Wolf, Relation between computed tomography and diffraction tomography, *J. Opt. Soc. Am. A* **18** (2001), 2132–2137. What is the physical system being considered in the problem, and what property of the system is being characterized by a fracFT?

15. Read the paper by M. Brunel, S. Coetmellec, M. Lelek, and F. Louradour, Fractional-order Fourier analysis for ultrashort pulse characterization, *J. Opt. Soc. Am. A* **24** (2007), 1641–1646. What is the system being studied, what is the fracFT being used to characterize, and why is it an ideal signal-processing tool for the problem?

16. Suppose the function $f(t)$ has a Wigner distribution $W(t, \omega)$. Show that
    (a) $e^{i \alpha t^2} f(t)$ has the Wigner distribution $W(t, \omega - 2 \alpha t)$,
    (b) $\frac{1}{\sqrt{s}} f(t/s)$ has the Wigner distribution $W(t/s, s\omega)$.

17. Prove Eq. (12.113), that if a function $f(t)$ has a Wigner distribution $W_0(t, \omega)$, the fracFT of $f(t)$ through angle ϕ has the distribution $W_0(t \cos \phi - \omega \sin \phi, \omega \cos \phi + t \sin \phi)$.

18. Another second-order property used in radar theory is the *ambiguity function*, defined as

$$A(t', \omega') \equiv \int_{-\infty}^{\infty} f(t + t'/2) f^*(t - t'/2) e^{-i\omega' t'} dt'.$$

Show that the ambiguity function is related to the Wigner function through the Fourier relation

$$A(t', \omega') = \int_{-\infty}^{\infty} \int_{-\infty}^{\infty} W(t, \omega) e^{-i(\omega' t + \omega t')} dt d\omega.$$

19. Read the paper by E. Wolf, Three-dimensional structure determination of semi-transparent objects from holographic data, *Opt. Commun.* **1**, 153–156 (1969), which describes a different type of inverse scattering problem. How does the physical problem being investigated differ from that of computed tomography? Describe the types of integral transform used in the inverse problem, and write the equation comparable to Eq. (12.122). Be sure to explain the meaning of all terms in the equation.

20. Show that the Haar wavelet defined by Eq. (12.104) satisfies the admissibility condition, Eq. (12.96).

21. A diffracting structure has a circularly symmetric *amplitude* transmission function given by

$$t(r) = \begin{cases} \frac{1}{2}[1 + \cos(\gamma r^2)], & r \leq R, \\ 0, & r > R. \end{cases}$$

In the limit of Fresnel diffraction, describe the manner in which the diffracting structure acts like a lens when a plane wave is incident upon it, and determine the effective focal length.

22. Use the Mexican hat wavelet, defined by Eq. (12.94) with $\sigma = 1$, to calculate the continuous wavelet transform of a Gaussian pulse of the form

$$f(t) = \exp[-(t - t_0)^2 / \delta^2] \exp[-i\omega_0 t]. \tag{12.131}$$

Show that the wavelet transform has a peak in the expected location.

13

Discrete transforms

13.1 Introduction: the sampling theorem

Because the number of functions whose integral transforms can be calculated analytically is limited, one must inevitably turn to numerical techniques. The general method for doing so is to discretize the function, replacing it by a finite number of steps. The integral, in turn, is approximated by a sum of a discrete and finite number of terms.

It is natural to ask what effect discretization has on the value of the integral. Loosely speaking, how much "information" about the original function is lost in the discretization process? A fundamental theorem in this regard is known as the *sampling theorem*, which we discuss here.

We consider a function $f(t)$ in the time domain which is *bandlimited*, i.e. a function whose Fourier transform vanishes for frequencies greater than a critical frequency $\Delta/2$,

$$\tilde{f}(\omega) = 0 \quad \text{for} \quad |\omega| > \Delta/2. \tag{13.1}$$

What does this imply about the behavior of the function $f(t)$ itself? Higher frequencies in the Fourier transform of a function correspond to more rapid changes in the function itself. If the function is bandlimited, this suggests that the function is slowly varying over timescales $T = 2\pi/\Delta$; this is illustrated schematically in Fig. 13.1.

From this observation, an intriguing possibility presents itself. If the function is bandlimited, changes in the time domain must occur slowly over a time $2\pi/\Delta$. This suggests that if we sample (i.e. measure) such a function at discrete times separated by $2\pi/\Delta$, we are getting a good approximation to the behavior of the continuous function. Surprisingly, though, we can show that a function which is sampled at or above this rate can in principle be reconstructed *exactly* from the collection of samples.

Theorem 13.1 (Sampling theorem) *A function f (t) which is bandlimited to frequencies* $|\omega| < \Delta/2$ *may be reconstructed exactly from samples taken at times* $t = nt_0$, *where n is an integer and* $t_0 = 2\pi/\Delta$, *by the formula*

$$f(t) = \sum_{n=-\infty}^{\infty} f(nt_0) \frac{\sin[\Delta(t - nt_0)/2]}{\Delta(t - nt_0)/2}. \tag{13.2}$$

Equation (13.2) is known as the Whittaker–Shannon sampling formula.

419

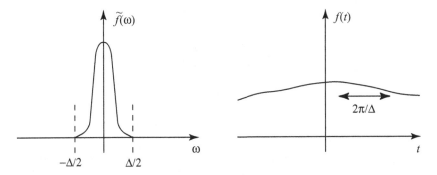

Figure 13.1 Illustration of the idea of a bandlimited function. A function bandlimited to a frequency range $|\omega| < \Delta/2$ will be slowly varying on the scale of times $T = 2\pi/\Delta$.

To prove this, we first write the function $f(t)$ in terms of its inverse Fourier transform,

$$f(t) = \int_{-\infty}^{\infty} \tilde{f}(\omega) e^{-i\omega t} d\omega. \tag{13.3}$$

Because $\tilde{f}(\omega)$ is bandlimited, we may change the limits of integration to include only the nonzero regions of the function,

$$f(t) = \int_{-\Delta/2}^{\Delta/2} \tilde{f}(\omega) e^{-i\omega t} d\omega. \tag{13.4}$$

Now our knowledge of Fourier *series* becomes significant. Because we are only interested in the behavior of the function $\tilde{f}(\omega)$ over a finite frequency domain, we may expand this function in a Fourier series in frequency, i.e.

$$\tilde{f}(\omega) = \sum_{n=-\infty}^{\infty} c_n e^{in\omega t_0}, \tag{13.5}$$

where $t_0 = 2\pi/\Delta$ and

$$c_n = \frac{1}{\Delta} \int_{-\Delta/2}^{\Delta/2} \tilde{f}(\omega) e^{-in\omega t_0} d\omega. \tag{13.6}$$

These formulas can be derived from the usual formulas of Fourier series given in Chapter 8 by exchanging t and ω and replacing T by Δ. On substituting from Eq. (13.5) into Eq. (13.4), we find that

$$f(t) = \int_{-\Delta/2}^{\Delta/2} \sum_{n=-\infty}^{\infty} c_n e^{-i\omega(t-nt_0)} d\omega. \tag{13.7}$$

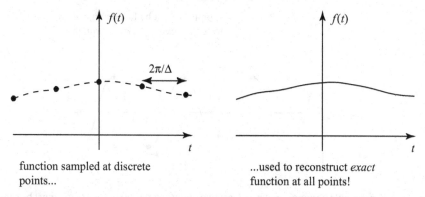

function sampled at discrete points... ...used to reconstruct *exact* function at all points!

Figure 13.2 Illustrating the idea of the sampling theorem. For a bandlimited function of bandwidth Δ, sampling at points separated by a distance $t_0 = 2\pi/\Delta$ is in principle sufficient for reconstruction.

We may interchange the order of integration and summation, and immediately evaluate the integral. We find that

$$f(t) = \Delta \sum_{n=-\infty}^{\infty} c_n \frac{\sin[\Delta(t - nt_0)/2]}{\Delta(t - nt_0)/2}. \tag{13.8}$$

Looking back at the definition of c_n, it is essentially the inverse Fourier transform of $\tilde{f}(\omega)$, evaluated at $t = nt_0$. This reduces our formula for $f(t)$ to the form

$$f(t) = \sum_{n=-\infty}^{\infty} f(nt_0) \frac{\sin[\Delta(t - nt_0)/2]}{\Delta(t - nt_0)/2}, \tag{13.9}$$

which is exactly the sampling formula (13.2).

The sampling theorem demonstrates that a bandlimited function may be *exactly reconstructed* using discrete time samples equally spaced at a distance $t_0 = 2\pi/\Delta$, referred to as the *Nyquist distance*. The frequency Δ is usually referred to as the *Nyquist frequency*. The Nyquist frequency represents the *minimum* sampling frequency needed to exactly reconstruct the signal; this is illustrated in Fig. 13.2.

Example 13.1 As a simple example, let us assume that the field to be sampled is a cosine wave, i.e.

$$f(t) = \cos(\omega_0 t). \tag{13.10}$$

This function will have a Fourier representation

$$\tilde{f}(\omega) = \frac{1}{2}[\delta(\omega - \omega_0) + \delta(\omega + \omega_0)]. \tag{13.11}$$

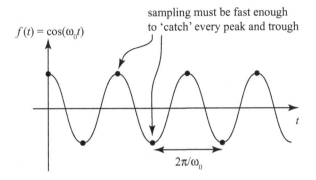

Figure 13.3 For a simple harmonic wave, at least two points per period of oscillation must be sampled to make a valid reconstruction.

The Fourier transform has delta function peaks at $\pm\omega_0$; the Nyquist frequency is therefore $\Delta = 2\omega_0$. This means that the Nyquist distance is $t_0 = \pi/\omega_0$, which is one half the period of the original cosine wave. Intuitively, this is reasonable – the spacing between sampling points (the Nyquist distance) must be small enough to at least detect the individual peak and trough of the harmonic wave, as illustrated in Fig. 13.3.

$\diamond$

The sampling theorem forms a fundamental component of the field of information theory: it gives a criterion for constructing continuous functions from a discrete set of sampled points in signal processing applications.

Compact disc players take advantage of the Nyquist frequency in their operation. Because human hearing can only register signals with frequencies less than about 20 kHz, audible sound is effectively bandlimited, and the Nyquist frequency is approximately $\Delta = 80\,000\pi$ radians/s. Compact disc players, which read a sampled version of a digital audio signal, must perform this sampling at or above the Nyquist frequency.[1]

The sampling theorem forms a nice bridge between the continuous transforms of the previous chapters and discrete versions of those transforms. In formulating a discrete version of the Fourier transform, the result will only be meaningful if the signal is sampled at a rate above the Nyquist frequency.

In this chapter we consider the properties of discrete transforms, starting with the discrete Fourier transform (DFT). The mathematics of discrete Fourier transforms (DFTs) is nearly as involved as that of continuous Fourier transforms, and introduces its own complications. Speed is an important consideration in computational work, and we will also discuss the so-called *fast Fourier transform*, an algorithm which allows an astonishing improvement in the rate of calculation of DFTs. The chapter is concluded with a discussion

[1] See [Poh92], particularly Chapter 2, for a non-technical discussion of how sampling applies to CD audio.

of the z-transform, which can be applied to determine the convolution of discrete signals of unspecified duration.

13.2 Sampling and the Poisson sum formula

It is illustrative to break down the sampling formula (13.2) into a series of easy to understand, experimentally realizable steps. This process requires us to introduce another significant formula in the relationship between discrete and continuous transforms, the *Poisson sum formula*, which we discuss first.

Let us consider a general periodic function of the form

$$G(t) = \sum_{n=-\infty}^{\infty} g(t - nt_0). \tag{13.12}$$

It is assumed that $g(t)$ is an integrable function which decays rapidly for large t (specifically, $g(t)$ must decay faster than $1/t$). The function $G(t)$ is periodic with period t_0, so we may consider the Fourier series representation of $G(t)$, which has the form

$$G(t) = \sum_{k=-\infty}^{\infty} c_k e^{2\pi i k t/t_0}, \tag{13.13}$$

with

$$c_k = \frac{1}{t_0} \int_0^{t_0} G(t) e^{-2\pi i k t/t_0} dt. \tag{13.14}$$

With a substitution from Eq. (13.12) into Eq. (13.14), we may write

$$c_k = \frac{1}{t_0} \int_0^{t_0} \sum_{n=-\infty}^{\infty} g(t - nt_0) e^{-2\pi i k t/t_0} dt. \tag{13.15}$$

With the assumption that the function $g(t)$ is well-behaved, we may interchange the order of integration and summation. Furthermore, we may change variables for each term of the sum to a new variable $t' = t - nt_0$,

$$c_k = \frac{1}{t_0} \sum_{n=-\infty}^{\infty} \int_{-nt_0}^{-(n-1)t_0} g(t') e^{-2\pi i k t'/t_0} e^{-2\pi i k n} dt'. \tag{13.16}$$

The quantities $\exp[-2\pi i k n] = 1$ for all k and n. The integrals are all with respect to the function $g(t') \exp[-2\pi i k t' t_0]$ over domains of width t_0 which together cover the entire real line. We may therefore write

$$c_k = \frac{1}{t_0} \int_{-\infty}^{\infty} g(t') e^{-2\pi i k t'/t_0} dt' = \frac{1}{t_0} \tilde{g}(k/t_0), \tag{13.17}$$

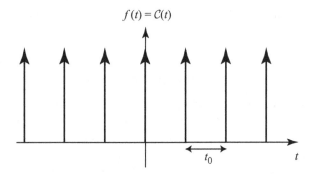

Figure 13.4 The "comb function": a distribution which consists of an infinite array of equally-spaced delta functions.

where

$$\tilde{g}(\omega) = \frac{1}{2\pi} \int_{-\infty}^{\infty} g(t')e^{-i\omega t'}\,dt'. \tag{13.18}$$

On substitution into Eq. (13.13), we finally may write

$$G(t) = \sum_{n=-\infty}^{\infty} g(t - nt_0) = \frac{2\pi}{t_0} \sum_{k=-\infty}^{\infty} \tilde{g}(2\pi k/t_0)e^{2\pi ikt/t_0}. \tag{13.19}$$

This equation is the *Poisson sum formula*. In essence, it relates the behavior of a periodic function in the time domain, constructed from a fundamental function $g(t)$, to the properties of $g(t)$ in the frequency domain.

The Fourier transform of the periodic function $G(t)$ will be of use to us. We may write

$$\tilde{G}(\omega) = \frac{1}{2\pi} \int_{-\infty}^{\infty} G(t')e^{-i\omega t'}\,dt' = \frac{2\pi}{t_0} \sum_{k=-\infty}^{\infty} \tilde{g}(2\pi k/t_0)\delta(\omega - 2\pi k/t_0). \tag{13.20}$$

We now return to an investigation of the sampling theorem. The sampling process itself may be treated as the multiplication of our trial function $f(t)$ by a so-called "comb function" $\mathcal{C}(t)$, given by

$$\mathcal{C}(t) = \sum_{n=-\infty}^{\infty} \delta(t - nt_0). \tag{13.21}$$

This function is an infinite array of evenly-spaced delta functions which looks roughly like a comb; this is illustrated in Fig. 13.4. Our sampled function, $f_s(t)$, is then the product of the original function and the comb,

$$f_s(t) \equiv \mathcal{C}(t)f(t) = f(t) \sum_{n=-\infty}^{\infty} \delta(t - nt_0). \tag{13.22}$$

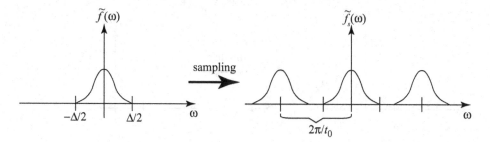

Figure 13.5 The effect of sampling on the spectrum of the function.

What is the Fourier spectrum of this sampled function? By the use of the convolution theorem, we may immediately write

$$\tilde{f}_s(\omega) = \int_{-\infty}^{\infty} \tilde{f}(\omega')\tilde{C}(\omega - \omega')d\omega'. \tag{13.23}$$

The function $C(t)$ is of the form $G(t)$ used in the Poisson sum formula, and its Fourier transform may be written via Eq. (13.20), with $g(t) = \delta(t)$, as

$$\tilde{C}(\omega - \omega') = \frac{1}{t_0} \sum_{k=-\infty}^{\infty} \delta(\omega - \omega' - 2\pi k/t_0). \tag{13.24}$$

On substitution into Eq. (13.23), and integrating, we readily find that

$$\tilde{f}_s(\omega) = \frac{1}{t_0} \sum_{k=-\infty}^{\infty} \tilde{f}(\omega - 2\pi k/t_0). \tag{13.25}$$

This formula indicates that the effect of sampling on the Fourier transform of a function is to add an infinite number of copies of the original spectrum, spaced by distances $2\pi/t_0$; this is illustrated schematically in Fig. 13.5. If these copies overlap (i.e. if t_0 is large), we cannot extract the Fourier spectrum of the original function from its sampled version. However, if $2\pi/t_0 > \Delta$, the individual copies of the spectrum do not overlap and we may "filter" out all copies save the original, as illustrated in Fig. 13.6.

This filtering operation may be simulated by a top hat function in the frequency domain,

$$S(\omega) = \begin{cases} t_0, & |\omega| \leq \Delta/2, \\ 0, & |\omega| > \Delta/2, \end{cases} \tag{13.26}$$

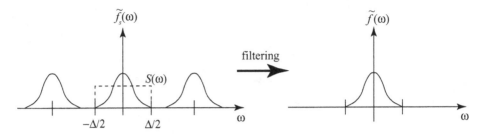

Figure 13.6 The final stage of sampling reconstruction: filtering of the spectrum of the sampled field.

with a scaling factor of t_0 to recover exactly the original Fourier spectrum. We reconstruct our original function by taking the inverse transform of our filtered sampled function spectrum,

$$f(t) = \int_{-\infty}^{\infty} \tilde{f}_s(\omega) S(\omega) e^{-i\omega t} d\omega$$

$$= \frac{1}{(2\pi)^2} \int_{-\infty}^{\infty} \int_{-\infty}^{\infty} \int_{-\infty}^{\infty} C(t') f(t') e^{i\omega t'} dt' \Delta t_0 \frac{\sin \Delta t''/2}{\Delta t''/2} e^{i\omega t''} dt'' e^{-i\omega t} d\omega. \quad (13.27)$$

We may perform the integral over ω, to find

$$f(t) = \frac{1}{2\pi} \int_{-\infty}^{\infty} C(t') f(t') \Delta t_0 \frac{\sin \Delta(t-t')/2}{\Delta(t-t')/2} dt'. \quad (13.28)$$

Using the definition of the comb function, and that $\Delta = 2\pi/t_0$ ($\Delta t_0 = 2\pi$), this reduces to

$$f(t) = \sum_{-\infty}^{\infty} f(nt_0) \frac{\sin \Delta(t-nt_0)/2}{\Delta(t-nt_0)/2}. \quad (13.29)$$

This is exactly the sampling theorem again! We have thus seen that the sampling theorem may be summarized as follows (1) Sampling the function changes the function's Fourier spectrum, creating an infinite number of shifted copies of the original spectrum. (2) By filtering the sampled function spectrum appropriately, we can recover the original spectrum of the function. Looking at Figs. 13.5 and 13.6 above, it may be said that sampling and filtering are complementary operations on the Fourier spectrum of a function: sampling widens the spectrum of a wavefield, while filtering narrows the spectrum. In this sense, the process of sampling is another type of integral transform, just like the Fourier and Laplace transform: sampling is the forward transform, and the filtering process is the inverse transform. The important distinction is that the sampling transform is only invertible for bandlimited functions.

This description of the sampling theorem also gives some insight into the problem of *aliasing*, in which artifacts appear in the reconstruction of an insufficiently sampled signal. When a signal is sampled at too low a rate (large t_0), the process produces copies of the

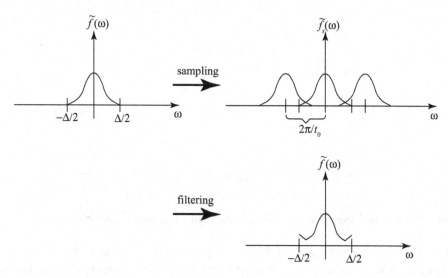

Figure 13.7 The problem of aliasing. When a signal is sampled at too low a frequency, the copies produced by sampling overlap and the final filtering process does not recover the original signal.

original spectrum that overlap, as illustrated in Fig. 13.7. The final filtering process does not recover the original signal, and the aliased signal contains distortions associated with the overlap.

The problem can be alleviated by pre-filtering the signal *before* sampling with an *anti-aliasing filter*. Such a filter removes those frequencies above the desired Nyquist frequency and prevents severe aliasing effects.

13.3 The discrete Fourier transform

The sampling theorem suggests that it is possible to perform Fourier analysis and get accurate results even when a function is approximated by its discrete samples. We are now interested in making an approximation to the usual Fourier transform formula that can be implemented on a computer. There are two fundamental changes which must be made: (1) the function $f(t)$ can only be sampled at discrete points, and (2) the domain of "integration" must be finite. Let us assume for simplicity that the function $f(t)$ is effectively nonzero only within the range $0 \leq t \leq T$. We attempt to evaluate the Fourier transform by using the Riemann integral approximation, breaking up the domain T into N subdomains, over each of which the function and the Fourier exponent approximated by a constant; the process is illustrated in Fig. 13.8.

We may therefore write

$$\tilde{f}(\omega) = \frac{1}{2\pi} \int_{-\infty}^{\infty} f(t)e^{i\omega t}dt \approx \frac{1}{2\pi} \sum_{n=0}^{N-1} f[n]e^{i\omega nT/N} \frac{T}{N}, \tag{13.30}$$

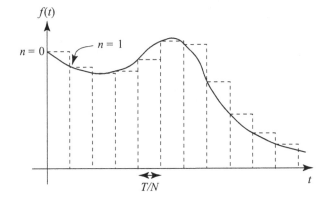

Figure 13.8 Riemann integral approximation of the Fourier transform of a function.

where T/N is the width of each subdomain, $f[n] \equiv f(n\Delta t)$, and $\Delta t \equiv T/N$.

If we do this, we immediately encounter a problem. From our experience with Fourier series, it is clear that the approximated $\tilde{f}(\omega)$ is a periodic function, which repeats over a distance $\Delta \omega = 2\pi N/T$,

$$\tilde{f}(\omega + 2\pi N/T) = \frac{1}{2\pi} \sum_{n=0}^{N-1} f[n] e^{i(\omega + 2\pi N/T)nT/N} \frac{T}{N}$$

$$= \frac{1}{2\pi} \sum_{n=0}^{N-1} f[n] e^{i2\pi n} e^{i\omega nT/N} \frac{T}{N} = \tilde{f}(\omega). \tag{13.31}$$

Of course, the actual Fourier transform of the function $f(t)$ is not periodic; the periodicity is an artifact of the two changes mentioned above. More specifically, the equation (13.30) has only N independent variables; from matrix theory we expect that we can only derive N independent equations from these variables. It is therefore reasonable to expect that we should only calculate the Fourier transform at N distinct frequencies. Choosing these frequencies to be equally spaced within the region $\Delta \omega = 2\pi N/T$, we therefore specify $\omega = 2\pi k/T$, with k an integer. We define the discrete Fourier transform by the formula

$$\tilde{f}[k] = \sum_{n=0}^{N-1} f[n] e^{i2\pi kn/N}, \tag{13.32}$$

where we have eliminated a number of constants in going from the continuous transform to the discrete one. The discrete Fourier transform converts N sampled points of a function $f(t)$ into N sampled points of its Fourier transform, as illustrated in Fig. 13.9.

Turning to the inverse transform, we begin by constructing a "delta function" for the DFT. We first define

$$w \equiv e^{2\pi i/N}. \tag{13.33}$$

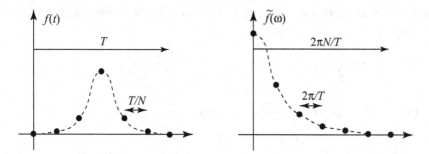

Figure 13.9 The sampling of a function and its Fourier transform.

Then we consider the following summation,

$$\frac{1}{N}\sum_{k=0}^{N-1}e^{2\pi ink/N} = \frac{1}{N}\sum_{k=0}^{N-1}w^{nk}.$$ (13.34)

This is a finite geometric series, whose sum can be readily found from Eq. (7.17) to be

$$\frac{1}{N}\sum_{k=0}^{N-1}w^{nk} = \begin{cases} 1, & n \text{ an integer multiple of } N, \\ \frac{1}{N}\frac{1-w^{nN}}{1-w^n}=0, & \text{otherwise.} \end{cases}$$ (13.35)

The latter result follows from the observation that $w^{nN} = \exp[2\pi in] = 1$. Since we are only interested in values of n such that $n < N$, this function behaves as a Kronecker delta function. We define this function as

$$\delta_N[n] = \frac{1}{N}\sum_{k=0}^{N-1}e^{2\pi ink/N}.$$ (13.36)

It is to be noted that this expression is comparable to the continuous Fourier transform formula for a delta function,

$$\frac{1}{2\pi}\int_{-\infty}^{\infty}e^{i\omega(t'-t)}d\omega = \delta(t-t').$$ (13.37)

Now let us define the inverse DFT by the formula

$$f[n] = \frac{1}{N}\sum_{k=0}^{N-1}\tilde{f}[k]e^{-i2\pi kn/N}.$$ (13.38)

Does this formula really work? We may find out by substituting from Eq. (13.32) into the above, to find

$$f[n] = \frac{1}{N} \sum_{k=0}^{N-1} \sum_{n'=0}^{N-1} f[n'] e^{i2\pi kn'/N} e^{-i2\pi kn/N} = \frac{1}{N} \sum_{k=0}^{N-1} \sum_{n'=0}^{N-1} f[n'] e^{2\pi ik(n'-n)}. \tag{13.39}$$

Interchanging the order of summations in the above, we find that

$$f[n] = \sum_{n'=0}^{N-1} f[n'] \left[\frac{1}{N} \sum_{k=0}^{N-1} e^{2\pi ik(n'-n)} \right] = \sum_{n'=0}^{N-1} f[n'] \delta_N[n'-n] = f[n]. \tag{13.40}$$

We therefore define the DFT and its inverse as follows.

Definition 13.1 (Discrete Fourier transform) *The discrete Fourier transform $f[k]$ of a discrete function $f[n]$ with N elements is given by the expression,*

$$\tilde{f}[k] = \sum_{n=0}^{N-1} f[n] e^{i2\pi kn/N}, \tag{13.41}$$

where $\tilde{f}[k]$ takes on N distinct values. The inverse transform is

$$f[n] = \frac{1}{N} \sum_{k=0}^{N-1} \tilde{f}[k] e^{-i2\pi kn/N}. \tag{13.42}$$

If we consider the function $f[n]$ as a vector of length N, it is clear that we can treat the DFT and its inverse as matrix operations.

13.4 Properties of the DFT

In this section we make a few observations about the DFT and its inverse.

13.4.1 Nyquist frequency

One immediate connection can be made between the DFT and the sampling theorem. In our derivation of the DFT, we assumed that the data was sampled at points separated by a distance $\Delta t = T/N$. The maximum frequency that we calculate using our DFT is $\omega_{max} = 2\pi N/T = 2\pi/\Delta t$, which is the Nyquist frequency. Our DFT has built into it the sampling theorem observation that if we sample a signal at a particular rate, the most Fourier information we can recover is within the Nyquist frequency. Outside that frequency, the DFT repeats itself.

This does not imply that the DFT gives an accurate representation of the continuous Fourier transform, however. N must be taken large enough to match the Nyquist frequency of the continuous function $f(t)$.

13.4.2 Shift in the n-domain

Let us first assume that our function $f[n]$ is periodic with respect to the variable n, and repeats itself after N integers. This provides some complementarity between the function, $f[n]$, and its transform, $\tilde{f}[k]$, which we know to be periodic.

We note the following property of the sum of a periodic function

$$\sum_{n=j}^{j+N-1} g[n] = \sum_{n=0}^{N-1} g[n]. \tag{13.43}$$

This follows because of the periodicity of the function $g[n]$: shifting the start point of the summation only shifts the *order* in which the terms are summed, which does not affect the sum. For example, the series $1, 2, 3, 1, 2, 3, \ldots$ is periodic with period 3. If we sum three consecutive terms, starting anywhere in the series, we always get the same result. For instance, if we start at the beginning, we get $1 + 2 + 3 = 6$, while if we start with the third term we get $3 + 1 + 2 = 6$.

Using this property of periodic functions, we consider the DFT of the shifted function $f[n-l]$,

$$\sum_{n=0}^{N-1} f[n-l]e^{2\pi i n k/N} = \sum_{n=-l}^{N-1-l} f[n]e^{2\pi i (n+l)k/N}$$

$$= e^{2\pi i l k/N} \sum_{n=0}^{N-1} f[n]e^{2\pi i n k/N} = e^{2\pi i l k/N}\tilde{f}[k], \tag{13.44}$$

where in the last step we used Eq. (13.43). This result is comparable to that used in Section 11.4.6 in the chapter on Fourier transforms, involving time shifting. That result was

$$\mathcal{F}[f(t-a)] = e^{i\omega a}\tilde{f}(\omega). \tag{13.45}$$

The results of this section indirectly imply that one may choose any origin for the DFT; that is, one can begin the summation at a different point, e.g.

$$\tilde{f}[k] = \sum_{-M}^{M} f[n]e^{2\pi i n k/N}, \tag{13.46}$$

for $N = 2M + 1$. Where one begins the summation often depends on the symmetry of the problem of interest.

13.4.3 Shift in the k-domain

Similarly, one can show that the multiplication of the function in the n-domain by a complex exponential leads to a shift in the k-domain,

$$\sum_{n=0}^{N-1} e^{2\pi i nk/N} \left\{ f[n] e^{-2\pi i nl/N} \right\} = \tilde{f}[k-l]. \tag{13.47}$$

This is analogous to the discussion of frequency shifts in Section 11.4.7.

13.5 Convolution

In the chapter on Fourier transforms, we developed the convolution theorem, Eq. (11.84),

$$\tilde{C}_{fg}(\omega) = 2\pi \tilde{f}(\omega)\tilde{g}(\omega), \tag{13.48}$$

where

$$C_{fg}(t) = \int_{-\infty}^{\infty} f(\tau)g(t-\tau)d\tau. \tag{13.49}$$

We may use the convolution theorem to derive an expression for convolution with discrete transforms; we assume that the discrete Fourier transform of a convolution operator results in the function

$$\tilde{C}_{fg}[k] = \tilde{f}[k]\tilde{g}[k]. \tag{13.50}$$

We again assume that the signals of interest $f[n]$, $g[n]$ are periodic in n, i.e. $f[n+N]=f[n]$, because of the time shift naturally incorporated in convolutions. We may calculate the inverse DFT of this expression by straightforward Fourier expansion of $\tilde{f}[k]$, $\tilde{g}[k]$; with the application of Eq. (13.36), we find that

$$C_{fg}[n] = \sum_{n'=0}^{N-1} f[n']g[n-n']. \tag{13.51}$$

With the convolution theorem demonstrated for DFTs, we may readily derive Parseval's theorem for them as well,

$$\sum_{n=0}^{N-1} f[n]g^*[n] = \frac{1}{N}\sum_{k=0}^{N-1} \tilde{f}[k]\tilde{g}^*[k]. \tag{13.52}$$

13.6 Fast Fourier transform

In calculating DFTs for large values of N, it is very important to attempt to minimize the number of mathematical operations taken. Looking at the formula for the DFT,

$$\tilde{f}[k] = \sum_{n=0}^{N-1} f[n]e^{i2\pi kn/N} = f[0] + f[1]e^{i2\pi k/N} + \cdots + f[N-1]e^{i2\pi k(N-1)/N}, \quad (13.53)$$

it is clear that to calculate the DFT for a single value of k will require $N - 2$ complex multiplications (we do not count the term $\exp[0] = 1$ and $\exp[i\pi] = -1$, which can be regarded as complex additions) and $N - 1$ complex additions. Overall, calculating the DFT for a single k requires $2N - 3$ elementary mathematical operations. This does not seem that bad, but if we want the DFT for all k values, this requires a total of $2N^2 - 3N$ mathematical operations, which for large N is approximately N^2 mathematical operations. This number grows quite rapidly: for $N = 1000$, a not unusual value, the number of mathematical operations required to calculate the DFT is approximately a million.

It turns out, however, that this method of taking a DFT is inefficient and fails to take into account symmetries of the process. A method which does take this into account is called the *fast Fourier transform* and reduces the number of elementary operations from N^2 to $N \log_2 N$. The savings for large N is nothing short of spectacular; for $N = 1000$, $N^2 = 1000000$, while $N \log_2 N \approx 10000$, two orders of magnitude difference!

Numerous software packages exist for implementing the FFT [PTVF92], and using them typically requires little understanding of their inner workings. The FFT, however, highlights an important point: symmetries in a computational problem may often be used to shorten the computation time significantly. This is still worth understanding in detail, as researchers are developing other "fast" algorithms for transforms related to the DFT (such as fractional Fourier transforms, and wavelet transforms).

Let us assume that N is an integer power of 2, i.e. $N = 2^m$, where m is an integer. Using the notation

$$w \equiv e^{2\pi i/N}, \quad (13.54)$$

we may write the DFT in the form

$$\tilde{f}[k] = \sum_{n=0}^{N-1} f[n]w^{kn}. \quad (13.55)$$

Let us consider the specific example $N = 8$. Then this formula may be written as

$$\tilde{f}[k] = f[0] + f[1]w^k + f[2]w^{2k} + f[3]w^{3k} + f[4]w^{4k} + f[5]w^{5k} + f[6]w^{6k} + f[7]w^{7k}. \quad (13.56)$$

We may rearrange the terms of this expression to the form

$$\tilde{f}[k] = \left\{ f[0] + f[4]w^{4k} \right\} + w^{2k} \left\{ f[2] + f[6]w^{4k} \right\} + w^k \left\{ f[1] + f[5]w^{4k} \right\}$$
$$+ w^{3k} \left\{ f[3] + f[7]w^{4k} \right\}. \tag{13.57}$$

The significant thing about this equation is that each term in brackets only contains the term $w^{4k} = (-1)^k$, which follows from the definition of w. This equation may then be written as

$$\tilde{f}[k] = \left\{ f[0] + (-1)^k f[4] \right\} + w^{2k} \left\{ f[2] + (-1)^k f[6] \right\}$$
$$+ w^k \left\{ f[1] + (-1)^k f[5] \right\} + w^{3k} \left\{ f[3] + (-1)^k f[7] \right\}. \tag{13.58}$$

Each bracketed term has only two possible values, and each of these values can be determined by a single complex addition. If we define

$$f_{04}^+ = f[0] + f[4], \tag{13.59}$$

$$f_{04}^- = f[0] - f[4], \tag{13.60}$$

with similar definitions for f_{26}^+, f_{26}^-, and so on, we may write our Fourier transform as

$$\tilde{f}[k] = \begin{cases} f_{04}^+ + w^{2k} f_{26}^+ + w^k f_{15}^+ + w^{3k} f_{37}^+, & k \text{ even}, \\ f_{04}^- + w^{2k} f_{26}^- + w^k f_{15}^- + w^{3k} f_{37}^-, & k \text{ odd}. \end{cases} \tag{13.61}$$

We can already see at this point that we have made improvements, because we have only used two operations (addition and subtraction) to deal with all combinations of $f[0]$ and $f[4]$ for all k. At this point, it will be useful to keep a running tally of the number of elementary operations (additions and multiplications) we use to calculate our DFT. Labeling this running tally as $\mathcal{N}_{\text{operations}}$, we have

$$\mathcal{N}_{\text{operations}} = 8, \quad 1 \text{ addition to construct each of } f_{04}^+, f_{04}^-, f_{26}^+, f_{26}^-, f_{15}^+, f_{15}^-,$$
$$f_{37}^+, f_{37}^-. \tag{13.62}$$

Returning to Eq. (13.61), let us break up the results into k being even, $k = 2m$, and the case k being odd, $k = 2m + 1$. We may therefore write

$$\tilde{f}[2m] = f_{04}^+ + w^{4m} f_{26}^+ + w^{2m} \left\{ f_{15}^+ + w^{4m} f_{37}^+ \right\}, \tag{13.63}$$

$$\tilde{f}[2m+1] = f_{04}^- + w^{4m+2} f_{26}^- + w^{2m+1} \left\{ f_{15}^- + w^{4m+2} f_{37}^- \right\}, \tag{13.64}$$

where now $m = 0, 1, 2, 3$. As before, in each of these equations, we have $w^{4m} = (-1)^m$. We may rewrite the two equations as

$$\tilde{f}[2m] = f_{04}^+ + (-1)^m f_{26}^+ + w^{2m} \left\{ f_{15}^+ + (-1)^m f_{37}^+ \right\}, \tag{13.65}$$

$$\tilde{f}[2m+1] = f_{04}^- + (-1)^m w^2 f_{26}^- + w^{2m+1} \left\{ f_{15}^- + (-1)^m w^2 f_{37}^- \right\}. \tag{13.66}$$

Again, many of the complex exponentials have effectively vanished from the equations, simplifying our calculations considerably. Let us now define

$$f_{0426}^{++} = f_{04}^+ + f_{26}^+, \quad f_{0426}^{+-} = f_{04}^+ - f_{26}^+, \quad f_{0426}^{-+} = f_{04}^- + w^2 f_{26}^-,$$

$$f_{0426}^{--} = f_{04}^- - w^2 f_{26}^-, \tag{13.67}$$

with similar definitions for $f_{1537}^{++}, f_{1537}^{+-}, f_{1537}^{-+}, f_{1537}^{--}$. The calculation of $f_{0426}^{++}, f_{0426}^{+-}, f_{1537}^{++}, f_{1537}^{+-}$ each require a single addition, while the calculation of terms $f_{0426}^{-+}, f_{0426}^{--}, f_{1537}^{-+}, f_{1537}^{--}$ requires two complex multiplications and four complex additions. Calculation of all these combinations thus requires 10 operations, bringing our running total to 18:

$$\mathcal{N}_{\text{operations}} = 8 + 10 = 18, \quad \text{8 additions, 2 multiplications to construct}$$

4-point transforms. $\tag{13.68}$

Our Fourier transform may be written in the extended form, writing $m = 2l$ if m is even, $m = 2l + 1$ if m odd, with $l = 0, 1$,

$$\tilde{f}[k] = \begin{cases} f_{0426}^{++} + w^{4l} f_{1537}^{++}, & k \text{ even}, k = 2m, m \text{ even}, m = 2l, \\ f_{0426}^{+-} + w^{4l+2} f_{1537}^{+-}, & k \text{ even}, k = 2m, m \text{ odd}, m = 2l+1, \\ f_{0426}^{-+} + w^{4l+1} f_{1537}^{-+}, & k \text{ odd}, k = 2m+1, m \text{ even}, m = 2l, \\ f_{0426}^{--} + w^{4l+3} f_{1537}^{--}, & k \text{ odd}, k = 2m+1, m \text{ odd}, m = 2l+1. \end{cases} \tag{13.69}$$

Each of these terms may be further broken down into the choice l even, or l odd. Defining a third string of summations,

$$f_{04261537}^{+++} = f_{0426}^{++} + f_{1537}^{++} = \tilde{f}[0], \tag{13.70}$$

$$f_{04261537}^{-++} = f_{0426}^{-+} + w^1 f_{1537}^{-+} = \tilde{f}[1], \tag{13.71}$$

$$f_{04261537}^{+-+} = f_{0426}^{+-} + w^2 f_{1537}^{+-} = \tilde{f}[2], \tag{13.72}$$

$$f_{04261537}^{--+} = f_{0426}^{--} + w^3 f_{1537}^{--} = \tilde{f}[3], \tag{13.73}$$

$$f_{04261537}^{++-} = f_{0426}^{++} - f_{1537}^{++} = \tilde{f}[4], \tag{13.74}$$

$$f_{04261537}^{-+-} = f_{0426}^{-+} - w^1 f_{1537}^{-+} = \tilde{f}[5], \tag{13.75}$$

$$f_{04261537}^{+--} = f_{0426}^{+-} - w^2 f_{1537}^{+-} = \tilde{f}[6], \tag{13.76}$$

$$f_{04261537}^{---} = f_{0426}^{--} - w^3 f_{1537}^{--} = \tilde{f}[7]. \tag{13.77}$$

Calculating these terms requires eight additions and three multiplications. Putting these terms together to get our Fourier transform, our final running total becomes,

$$\mathcal{N}_{\text{operations}} = 18 + 11 = 29, \quad 8 \text{ additions, 3 multiplications to construct}$$

$$8\text{-point transform.} \tag{13.78}$$

What is the calculation total for an ordinary DFT? With $N = 8$, and the total number of operations being $\mathcal{N}_{\text{DFT}} = (2N - 3)N$, we have

$$\mathcal{N}_{\text{DFT}} = 13 \times 8 = 104. \tag{13.79}$$

By performing an FFT, we have reduced the number of required operations from 104 to 29! It is to be noted that the approximate number of operations required for an FFT is predicted to be $N \log_2 N = 24$, in good agreement with our final result.

The technique of the FFT is therefore to rearrange the terms of the DFT, then combine adjacent pairs in two-point DFTs, then combine the remaining terms in pairs in two-point DFTs, and so on. The method is relatively straightforward, the only real question remaining is how to order the initial terms. Looking back at the eight-point transform, let us write the original series and the reordered series in binary format,

$$\tilde{f}[k] = f[000] + f[001]w^k + f[010]w^{2k} + f[011]w^{3k} + f[100]w^{4k} + f[101]w^{5k}$$

$$+ f[110]w^{6k} + f[111]w^{7k}, \tag{13.80}$$

and

$$\tilde{f}[k] = \left\{ f[000] + f[100]w^{4k} \right\} + w^{2k} \left\{ f[010] + f[110]w^{4k} \right\}$$

$$+ w^k \left\{ f[001] + f[101]w^{4k} \right\} + w^{3k} \left\{ f[011] + f[111]w^{4k} \right\}. \tag{13.81}$$

Comparing the order of terms in the two series, we see that

$$f[000] \to f[000], \quad f[001] \to f[100], \quad f[010] \to f[010], \quad f[011] \to f[110],$$
$$f[100] \to f[001], \quad f[101] \to f[101], \quad f[110] \to f[011], \quad f[111] \to f[111]. \tag{13.82}$$

Evidently, to find the ordering of the terms used to perform the FFT, we simply write the number of each term in binary, invert the order of the bits, and place the term in the slot given by the resulting binary number.

With this bit reordering technique, it is possible to derive a systematic algorithm for calculating fast Fourier transforms. In conclusion, the moral of this section is that there are sometimes surprising methods for simplifying numerical calculations.

A fast convolution can also be calculated in a straightforward manner via the convolution theorem, Eq. (13.50). The FFTs of the two functions to be convolved are taken, their product taken, and then the inverse FFT performed. Although this calculation involves the calculation of three FFTs, it is still faster, for large N, than calculating the convolution directly. For instance, a direct convolution of functions with $N = 1000$ would require a million operations, while a convolution via three FFTs would require some $30\,000$ operations.

13.7 The z-transform

The usefulness of Fourier and Laplace transforms lies in large part in their ability to reduce complicated convolution formulas in the time domain to product formulas in the frequency domain via the convolution theorem. When one is dealing with discrete, sampled signals, a similar strategy is desired. One can in principle use the discrete Fourier transform to evaluate the convolution, but the DFT is periodic in step n, while in many applications the signal of interest is "open-ended", i.e. has an indeterminate end point. In such cases a new transform, the z-transform, proves most helpful.

Let us assume we have a discrete signal $x[n] = x(nT)$, with sampling interval T, which may be specified for all values $-\infty < n < \infty$. The z-transform of this signal is defined as

$$\mathcal{Z}\{x\}(z) = X(z) = \sum_{n=-\infty}^{\infty} x[n]z^{-n}. \tag{13.83}$$

The z-transform maps each sampled time nT to an inverse power of z.

This expansion looks very much like a Laurent series expanded about the point $z = 0$, as described in Section 9.8. The region of convergence will therefore be an annulus centered on $z = 0$ in the complex plane. The inner and outer radii of this annular region can be calculated by the usual series methods, if the function $x[n]$ is specified.

To write expressions for these radii, we first break the transform into two parts, the *causal* and the *anti-causal* pieces, defined as

$$X(z) = \sum_{0}^{\infty} x[n]z^{-n}, \quad \text{causal,} \tag{13.84}$$

$$X(z) = \sum_{-\infty}^{-1} x[n]z^{-n}, \quad \text{anti-causal.} \tag{13.85}$$

The causal piece is a power series in $w = 1/z$, with coefficients $x[n]$. According to Section 7.7, it will converge in a radius

$$R_o > \frac{1}{\lim_{n \to \infty} \frac{x[n+1]}{x[n]}}. \tag{13.86}$$

Similarly, the anti-causal piece is a power series in z, with coefficients $x[-n]$, which will converge in a radius

$$R_i < \frac{1}{\lim_{n \to \infty} \frac{x[-(n+1)]}{x[-n]}}. \tag{13.87}$$

Provided the annular region of analyticity exists, an inverse transform may be simply constructed using the residue theorem,

$$x[n] = \frac{1}{2\pi i} \oint_C X(z) z^{n-1} dz, \tag{13.88}$$

where C is a path that lies in the annular region. We will see that in most cases of interest this inverse does not need to be calculated formally.

Much of the time we will be interested in causal signals, for which $x[n] = 0$, $n < 0$. In such a case the z-transform converges in an infinite domain outside the radius R_o. We will restrict ourselves to causal signals for the remainder of this section.

As always, it is worth looking at a few examples.

Example 13.2 (Unit pulse) We consider the discrete analogy to the delta function, a unit pulse defined as

$$\delta[n] = \begin{cases} 1, & n = 0, \\ 0, & n \neq 0. \end{cases} \tag{13.89}$$

The z-transform of this function is

$$\mathcal{Z}\{\delta\}(z) = 1. \tag{13.90}$$

This is analogous to the observation that the Fourier or Laplace transform of a Dirac delta function centered on the origin is a constant. Clearly the z-transform is convergent for all values of z.

◇

Example 13.3 (Step function) A discrete version of the step function is defined as

$$S[n] = \begin{cases} 1, & n \geq 0, \\ 0, & n < 0. \end{cases} \tag{13.91}$$

Immediately on substitution into the definition of the z-transform we find that

$$\mathcal{Z}\{S\}(z) = \sum_{n=0}^{\infty} z^{-n}. \tag{13.92}$$

By use of the geometric series, Example 7.1, we may write this expression in a closed form as

$$Z\{S\}(z) = \frac{1}{1 - z^{-1}}. \tag{13.93}$$

This expression only converges for $|z| > 1$.

◇

Example 13.4 (Exponential function) A discrete exponential function, with decay constant α and time step T, is defined as

$$x[n] = \exp[-\alpha nT], \tag{13.94}$$

with $\alpha > 0$. The z-transform of this function may also be written in a closed form by use of the geometric series

$$X(z) = \sum_{n=0}^{\infty} e^{-\alpha nT} z^{-n} = \sum_{n=0}^{\infty} \left(e^{-\alpha T} z^{-1} \right)^n = \frac{1}{1 - e^{-\alpha T} z^{-1}}. \tag{13.95}$$

The series converges for $|z| > e^{-\alpha T}$.

◇

As with our other transforms, we may introduce many general properties of the z-transform.

13.7.1 Relation to Laplace and Fourier transform

We consider the Laplace transform of a function $f(t)$, which was defined in Section 12.3 as

$$F(s) = \int_0^\infty f(t) e^{-st} dt. \tag{13.96}$$

We now consider a time-sampled function $f(t)$, which may be written as

$$f(t) = \sum_{n=0}^{\infty} f(nT) \delta(t - nT). \tag{13.97}$$

On substitution into the definition of the Laplace transform, we find that the Laplace transform of a time-sampled function may be written as

$$F(s) = \sum_{n=0}^{\infty} f(nT) e^{-snT}. \tag{13.98}$$

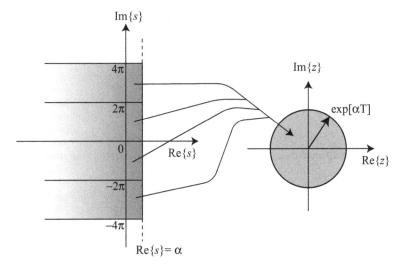

Figure 13.10 Multiple mappings of $z = \exp[sT]$ into the interior of a circle. Each strip for which Im$\{s\}$ covers a range of 2π will fill the unit circle.

If we allow s to be complex ($s = \sigma + i\omega$), we find that the function $\exp[sT]$ is a mapping of the complex plane into itself. That is, the function maps all values of s in the complex plane into different values in the complex plane. In particular, the half-space Re$\{s\} < \alpha$ ($\alpha > 0$) is mapped multiple times into the interior of a circle of radius $r < \exp[\alpha T]$; this is illustrated in Fig. 13.10. If we define $z = \exp[sT]$, we may rewrite our complex Laplace transform as

$$F(z) = \sum_{n=0}^{\infty} f[n]z^{-n}, \tag{13.99}$$

which is exactly the definition of the causal z-transform. The causal z-transform, then, may be considered as a discrete version of the Laplace transform.

The inverse Laplace transform is defined in Section 12.3.6 as

$$f(t) = \frac{1}{2\pi i} \int_{s=\alpha-i\infty}^{s=\alpha+i\infty} F(s)e^{st}(id\omega), \tag{13.100}$$

where $s = \sigma + i\omega$ and the integral is taken over a closed loop which includes the line Re$\{s\} = \sigma$ and encloses the space Re$\{s\} < \sigma$. When we consider the mapping $z = \exp[st]$, however, we note that this half-space represents an infinite collection of identical mappings into a circle in the complex z-plane centered on the origin and of radius $\exp[\alpha T]$. We include only one of these in our inverse transform integral when we express it in terms of z. Also restricting ourselves to $t = nT$, we have

$$f(nT) = \frac{1}{2\pi i T} \oint_{|z|<\exp[\alpha T]} F(z)z^{n-1}dz, \tag{13.101}$$

where we have used $z = \exp[sT]$ to derive $dz = Tizd\omega$. On comparison with Eq. (13.88), we see that we have equated the inverse Laplace transform for discrete functions with the inverse z-transform. The extra factor of T in Eq. (13.101) comes from our definition of $f(t)$ in terms of the Dirac delta function, with its dimensions of inverse time. The singular delta function behavior of $f(t)$ has been suppressed by only considering the singularities present in one strip of the s-plane, as illustrated in Fig. 13.10.

The z-transform is therefore an elegant technique for performing the Laplace transform of a discrete causal function.

In a similar manner, we may relate the z-transform to the Fourier transform of a discrete function. This is perhaps not surprising in light of the structural similarity between the Fourier and Laplace transforms. Assuming again a discrete function $f(t)$ of the form of Eq. (13.97), we may write the Fourier transform of a discrete function as

$$F(\omega) = \frac{1}{2\pi} \int_{-\infty}^{\infty} f(t)e^{-i\omega t} dt = \frac{1}{2\pi} \sum_{-\infty}^{\infty} f(nT)e^{-i\omega nT}. \tag{13.102}$$

If we define a mapping $z = \exp[i\omega T]$, where $\omega = \omega_R + i\omega_I$ is a complex frequency, we may write

$$z = e^{i\omega_R T} e^{-\omega_I T}. \tag{13.103}$$

We find that the upper half-space of ω ($\omega_I > 0$) is mapped multiple times into the unit circle in the z-plane, as illustrated in Fig. 13.11. Applying the mapping to z, we have

$$F(z) = \frac{1}{2\pi} \sum_{-\infty}^{\infty} f[n]z^{-n}, \tag{13.104}$$

which is again the z-transform, with an extra factor of 2π. The inverse Fourier transform is defined as

$$f(t) = \int_{-\infty}^{\infty} F(\omega)e^{i\omega t} d\omega. \tag{13.105}$$

For causal functions $f(t)$, the function $F(\omega)$ is analytic in the upper half-space (see Section 9.14), and we choose as our path of integration the semicircle which encloses the said space, as discussed in Section 9.12.2. The integral will enclose an infinite number of vertical strips on which the function $F(\omega)$ has the same value. On moving the inverse transform into the z-plane, we keep only one of these. The result is an integral upon the unit circle of the form

$$f(t) = \frac{1}{iT} \oint_{|z|=1} F(z)z^{n-1} dz, \tag{13.106}$$

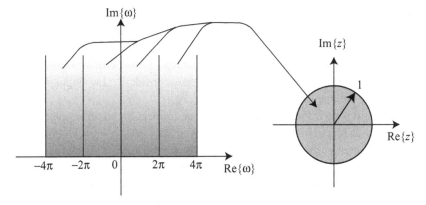

Figure 13.11 Multiple mappings of $z = \exp[i\omega T]$ into the interior of the circle. Each strip for which ω_R covers a range of 2π will fill the unit circle.

where we have used $z = \exp[i\omega T]$ to evaluate the derivatives. This is again comparable to the inverse transform given by Eq. (13.88), with the only significant difference being that the factor of 2π is buried in the definition of the forward z-transform.

As we will see, the connection between the Fourier and Laplace transforms and the z-transform is extremely useful in deriving discrete approximations to differential equations, particularly when dispersive effects are included.

It is perhaps worth comparing Figs. 13.10 and 13.11 to Fig. 13.5, which shows how the process of sampling produces an infinite number of frequency-shifted copies of the original spectrum. We are, in essence, seeing the same process in the complex Laplace and Fourier transforms of sampled signals: sampling divides the complex spectrum into a series of repeating "strips". When we perform the inverse z-transform, we take a path integral enclosing a single strip, which is equivalent to filtering out all duplicates of the spectrum in the sampling theorem.

13.7.2 Convolution theorem

Let us suppose we have a discrete (infinite domain) convolution formula of the form

$$y[n] = \sum_{i=0}^{\infty} h[n-i]x[i]. \tag{13.107}$$

Such a formula is a discrete representation of a linear system, such as discussed in Section 11.5. The z-transform may be readily calculated for both sides of this formula,

$$Y(z) = \sum_{n=0}^{\infty} y[n]z^{-n} = \sum_{n=0}^{\infty}\sum_{i=0}^{\infty} h[n-i]x[i]z^{-n}. \tag{13.108}$$

We may simplify the right-hand side by multiplying each term by $z^{-i}z^i$, which gives

$$Y(z) = \sum_{i=0}^{\infty} x[i]z^{-i} \sum_{n=0}^{\infty} h[n-i]z^{-n+i}. \tag{13.109}$$

We may now introduce a new parameter $k = n - i$, so that

$$Y(z) = \sum_{i=0}^{\infty} x[i]z^{-i} \sum_{k=-i}^{\infty} h[k]z^{-k}. \tag{13.110}$$

Assuming that $h[k]$ is causal, it vanishes for $k < 0$, and we may then write

$$Y(z) = \sum_{i=0}^{\infty} x[i]z^{-i} \sum_{k=0}^{\infty} h[k]z^{-k}. \tag{13.111}$$

The right-hand side of this expression, however, is now the product of z-transforms,

$$Y(z) = X(z)H(z). \tag{13.112}$$

This is the convolution theorem for z-transforms.

13.7.3 Time shift

Suppose we consider the z-transform of a signal which has been shifted in time by a number of steps k. The z-transform of such a function takes on the form

$$\mathcal{Z}\{x[n-k]\}(z) = \sum_{n=-\infty}^{\infty} x[n-k]z^{-n}. \tag{13.113}$$

By a change of summation parameter, this formula may be rewritten as

$$\mathcal{Z}\{x[n-k]\}(z) = \sum_{n=-\infty}^{\infty} x[n]z^{-k}z^{-n} = z^{-k}X(z). \tag{13.114}$$

A shift in time by step k results in a multiplication by z^{-k}. This result may be directly compared to the time shift result for Fourier transforms, Eq. (11.61).

This result also holds for causal signals, as then $x[n] = 0$ for $n < 0$.

13.7.4 Initial and final values

One interesting property of the causal z-transform is the ability to determine the initial and final (asymptotic) values of the original sampled functions by taking the appropriate limits of the transform. We state these as theorems.

Theorem 13.2 *The initial value $x[0]$ of a discrete causal function $x[n]$ may be found by taking the limit,*

$$x[0] = \lim_{z \to \infty} X(z). \qquad (13.115)$$

This is immediately evident from the definition of the causal z-transform, Eq. (13.84). As $z \to \infty$, all terms associated with negative powers of z become negligibly small, leaving only the z^0 term, $x[0]$.

Theorem 13.3 *The asymptotic value as $n \to \infty$ of a discrete causal function $x[n]$ may be found by taking the limit,*

$$x(\infty) = \lim_{z \to 1} (1 - z^{-1}) X(z). \qquad (13.116)$$

There are several proofs of this, all of which are somewhat subtle. We will use one that is due to Atre [Atr75], which illustrates nicely how one can use the relationship between the discrete and continuous transforms to solve problems.

As noted in Section 12.3.5, the Laplace transform can reduce the derivatives of functions to polynomial expressions. The Laplace transform of $f'(t)$, for instance, is

$$\mathcal{L}\left[f'(t)\right] = \int_0^\infty f'(t) e^{-st} dt = sF(s) - f(0). \qquad (13.117)$$

The value of the function $f(t)$ as $t \to \infty$ can be readily found by letting $s \to 0$,

$$\lim_{s \to 0} sF(s) = \lim_{s \to 0} \int_0^\infty f'(t) e^{-st} dt + f(0) = \int_0^\infty f'(t) dt + f(0) = \lim_{t \to \infty} f(t). \qquad (13.118)$$

Assumptions must be made about the behavior of the function $f'(t)$, namely that it is well-behaved as $t \to \infty$. In Section 12.3 we demonstrated that the Laplace transform of the step function $S(t)$ is simply $1/s$. From this, we may write

$$sF(s) = \frac{\mathcal{L}[f(t)]}{\mathcal{L}[S(t)]}. \qquad (13.119)$$

We may replace $f(t)$ by its sampled version, Eq. (13.97), which we will denote $\hat{f}(t)$. From this, we may write

$$\lim_{n \to \infty} \hat{f}(nT) = \lim_{s \to 0} \frac{\mathcal{L}\left[\hat{f}(t)\right]}{\mathcal{L}\left[\hat{S}(nT)\right]}. \qquad (13.120)$$

We now consider the mapping $z = \exp[sT]$ as in Section 13.7.1, and the conversion of the Laplace transform to the z-transform. The limit $s \to 0$ is equivalent to $z \to 1$, and the

z-transform of the step function is given by Eq. (13.93), which leads to the result

$$\lim_{n\to\infty} \hat{f}(nT) = \lim_{z\to 1} \frac{F(z)}{1/(1-z^{-1})}. \tag{13.121}$$

Simplifying the expression leads directly to the final value theorem.

It is interesting to note that the final value theorem suggests that the asymptotic behavior of $f[n]$ is given by the behavior of the function $F(z)$ at $z = 1$.

13.8 Focus: z-transforms in the numerical solution of Maxwell's equations

The z-transform is particularly useful in providing a shortcut in formulating numerical solutions to Maxwell's equations using the finite difference time domain (FDTD) method. In this section we briefly describe FDTD and explain how the z-transform can be used to simplify the analysis of a problem. The z-transform technique described here was first suggested by Sullivan [Sul92, Sul00].

We consider first the simple case of an electromagnetic plane wave propagating in the ζ-direction in a Cartesian coordinate system. (We use ζ instead of z to avoid confusion with the z-transform.) In this system, the two curl equations of Maxwell's are most significant,

$$\frac{\partial \mathbf{E}}{\partial t} = c\nabla \times \mathbf{H}, \quad \frac{\partial \mathbf{H}}{\partial t} = -c\nabla \times \mathbf{E}, \tag{13.122}$$

where $\mathbf{E}(\zeta,t)$ is the electric field and $\mathbf{H}(\zeta,t)$ is the magnetic field of the wave. Assuming without loss of generality that the electric field is polarized along the x-direction, these equations may be written as

$$\frac{\partial E_x}{\partial t} = -c\frac{\partial H_y}{\partial \zeta}, \tag{13.123}$$

$$\frac{\partial H_y}{\partial t} = -c\frac{\partial E_x}{\partial \zeta}. \tag{13.124}$$

Curiously, these equations are easily solved analytically but it is not immediately obvious how one would numerically solve them, as there are two field variables E_x and H_y which simultaneously evolve in both space and time. The solution was provided by Yee [Yee66], who suggested that the electric and magnetic fields be *interleaved* in space and time. Let us choose a fundamental time step Δt; we assume that the magnetic field is defined on integer multiples of this step, i.e. at times $t = n\Delta t$, but we assume that the electric field is defined on half-integer multiples of this step, i.e. at times $t = (n+1/2)\Delta t$. We furthermore assume a fundamental space step $\Delta\zeta$, and assume that the electric field is defined on integer multiples of this step, i.e. $\zeta = k\Delta\zeta$, while the magnetic field is defined on half-integer multiples of this step, i.e. $\zeta = (k+1/2)\Delta\zeta$. We approximate each of the derivatives in Eqs. (13.123) and (13.124) using a central difference approximation, i.e.

$$\frac{\partial E_x}{\partial t}(\zeta,t) \approx \frac{E_x(k\Delta\zeta,(n+1/2)\Delta t) - E_x(k\Delta\zeta,(n-1/2)\Delta t)}{\Delta t}, \tag{13.125}$$

and so on. We write $E_x(k\Delta\zeta, n\Delta t) = E_x[k,n]$ for brevity from now on. Maxwell's equations then take on the form

$$\frac{E_x[k,n+1/2] - E_x[k,n-1/2]}{\Delta t} = -c\frac{H_y[k+1/2,n] - H_y[k-1/2,n]}{\Delta\zeta}, \tag{13.126}$$

$$\frac{H_y[k+1/2,n+1] - H_y[k+1/2,n]}{\Delta t} = -c\frac{E_x[k+1,n+1/2] - E_x[k,n+1/2]}{\Delta\zeta}. \tag{13.127}$$

If we specify the values of E_x over all space at time $n - 1/2$, and specify the values of H_y over all space at time n, we may determine the value of E_x over all space at time $n + 1/2$ using Eq. (13.126). We may then use this value to determine the value of H_y over all space at time $n + 1$ using Eq. (13.127). The process may be continued iteratively to evaluate the field throughout space for all future times.

These formulas form the basis of FDTD for a one-dimensional free-space propagation problem. The interleaving process is illustrated in Fig. 13.12. A full account of FDTD is very involved and well outside the scope of this book; see [Sul00] and [TH05]. We note here, however, that the relative size $\Delta\zeta/\Delta t$ of the steps is important; for a one-dimensional problem, $\Delta t = \Delta\zeta/2c$ is a good working number.

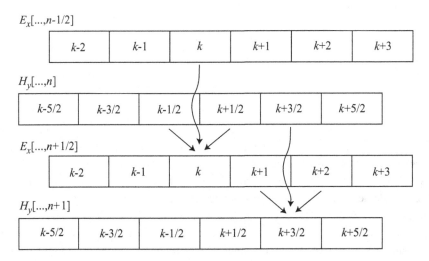

Figure 13.12 Illustration of the interleaving process. The value of $E_x[k,n+1/2]$ is determined by the values $E_x[k,n-1/2]$ and the values $H_y[k-1/2,n]$, $H_y[k+1/2,n]$.

If we now turn to a wave propagating in a uniform but dispersive media, things immediately become more complicated. We now must evaluate the series of three equations

$$\frac{\partial D_x}{\partial t}(\zeta,t) = -c\frac{\partial H_y}{\partial \zeta}(z,t), \tag{13.128}$$

$$\frac{\partial H_y}{\partial t}(\zeta,t) = -c\frac{\partial E_x}{\partial \zeta}(\zeta,t), \tag{13.129}$$

$$\tilde{D}_x(\zeta,\omega) = \epsilon(\omega)\tilde{E}_x(\zeta,\omega), \tag{13.130}$$

where $\epsilon(\omega)$ is the frequency-dependent permittivity of the medium and D_x is the displacement field. The difficulty here is that the formula (13.130) is expressed in the frequency domain, while the two curl formulas are written in the time domain. By use of the convolution theorem (11.81), we may write Eq. (13.130) in the time domain in an integral form,

$$D_x(\zeta,t) = \frac{1}{2\pi}\int_{-\infty}^{\infty} \tilde{\epsilon}(t-\tau)E_x(\zeta,\tau)\mathrm{d}\tau. \tag{13.131}$$

Assuming the medium is causal ($\tilde{\epsilon}(t) = 0$ for $t < 0$), we still find that the displacement field depends upon the value of the electric field *at all earlier times*. To numerically solve Maxwell's equations in a dispersive medium, we must be able to take the inverse transform of $\epsilon(\omega)$, and then be able to properly discretize the convolution integral which arises.

Here, however, is where the z-transform helps immensely. Because there is a direct relationship between the Fourier transform of a discrete causal function and the z-transform of that function, we may replace any function $\tilde{f}(\omega)$ by its z-transform equivalent, $F(z)$. Instead of performing an inverse Fourier transform via Eq. (13.105) to determine the function in the time domain, we perform an inverse z-transform via Eq. (13.88) to determine the sampled form of the function in the time domain.

A list relating common Fourier transforms to their corresponding z-transforms is shown in Table 13.1. It is to be noted that the Fourier transforms of $S(t)$ and $tS(t)$ are only well-defined for complex ω in the upper half-space. From the complex exponential transforms, one can readily determine the transforms of sine and cosine functions as well.

As an illustration, we consider propagation of a wave in a metal, for which the permittivity may be written as

$$\epsilon(\omega) = \epsilon_c + i\frac{4\pi\sigma}{\omega}. \tag{13.132}$$

Here ϵ_c is the real-valued dielectric constant of the metal and σ is the conductivity of the metal. Consulting Table 13.1, we find that we may write the z-transform of this function as

$$\epsilon(z) = 2\pi\epsilon_c - \frac{8\pi^2\sigma}{1-z^{-1}}. \tag{13.133}$$

Table 13.1 *A list of functions in time, frequency, sampled time and z-space, with sampled times t = nT.*

$f(t)$	$\tilde{f}(\omega)$	$f[n]$	$F(z)$
$\delta(t)$	$\frac{1}{2\pi}$	$\delta[n]$	1
$S(t)$	$\frac{1}{2\pi i\omega}$	$S[n]$	$\frac{1}{1-z^{-1}}$
$tS(t)$	$\frac{1}{2\pi(i\omega)^2}$	$nTS[n]$	$\frac{Tz^{-1}}{(1-z^{-1})^2}$
$\exp[-\alpha t]S(t)$	$\frac{1}{2\pi}\frac{1}{\alpha+i\omega}$	$\exp[-\alpha nT]S[n]$	$\frac{1}{1-\exp[-\alpha T]z^{-1}}$
$\exp[\pm i\beta t]\exp[-\alpha t]S(t)$	$\frac{1}{2\pi}\frac{1}{\alpha+i\omega\mp i\beta}$	$\exp[-\alpha nT]\exp[\pm i\beta nT]S[n]$	$\frac{1}{1-z^{-1}\exp[\alpha T]\exp[\mp i\beta T]}$

We may now write Eq. (13.130) for the displacement field as

$$\tilde{D}_x(\zeta,z) = \left[2\pi\epsilon_c - \frac{8\pi^2\sigma}{1-z^{-1}}\right]\tilde{E}_x(\zeta,z). \tag{13.134}$$

We now expand the right-hand term in a geometric series. Because each power of z^{-1} represents a shift in time T, we may immediately write our expression for D_x in sampled time space as

$$D_x(\zeta,nT) = 2\pi\epsilon_c E_x(\zeta,nT) - 8\pi^2\sigma\sum_{m=0}^{n} E_x(\zeta,mT). \tag{13.135}$$

This equation may be rewritten in a useful form by extracting the $m = n$ term from the series,

$$E_x(\zeta,nT) = \frac{D_x(\zeta,nT) + 8\pi^2\sigma\sum_{m=0}^{n-1}E_x(\zeta,mT)}{8\pi^2\sigma - 2\pi\epsilon_c}. \tag{13.136}$$

Here we have an expression which allows us to calculate the current value of E_x from the current value of D_x and all previous values of E_x. We may completely discretize Eq. (13.136), and Eqs. (13.128) and (13.129) as well, to arrive at a new set of discrete formulas to calculate the electromagnetic field,

$$D_x[k,n+1/2] = D_x[k,n-1/2] - c\frac{\Delta t}{\Delta\zeta}(H_y[k+1/2,n] - H_y[k-1/2,n]), \tag{13.137}$$

$$I[k,n-1/2] = 8\pi^2\sigma\sum_{m=0}^{n-1/2} E_x[k,m], \tag{13.138}$$

$$E_x[k,n+1/2] = \frac{D_x[k,n+1/2] + 8\pi^2\sigma I[k,n-1/2]}{8\pi^2\sigma - 2\pi\epsilon_c}, \tag{13.139}$$

$$H_y[k+1/2,n+1] = H_y[k+1/2,n] - c\frac{\Delta t}{\Delta\zeta}(E_x[k+1,n+1/2] - E_x[k,n+1/2]). \tag{13.140}$$

We may calculate the next value of D_x from previous values of H_y. We may then calculate the next value of E_x from previous values of E_x and the current value of D_x. We may then calculate the next value of H_y from the current values of E_x.

The z-transform provides a great shortcut for calculating the time-domain response of a dispersive medium. For a medium with extremely complicated frequency response, the savings in effort can be significant.

13.9 Focus: the Talbot effect

In Section 8.1, we analyzed the behavior of diffraction gratings through a combination of Fourier series and Fourier transform techniques. This analysis was limited to the behavior of the diffracted light field far from the grating, in the Fraunhofer diffraction regime. With the advanced integral transform techniques developed in the previous few chapters, however, we may now investigate the behavior of the diffracted light field in the region immediately behind the grating. The diffracted field shows surprising and complicated behavior, most obvious the self-imaging of the diffraction grating at regular distances from the plane of the grating, a phenomenon now referred to as the *Talbot effect*.

In 1836, H. F. Talbot [Tal36] was observing the field of a diffraction grating with a magnifying glass. Starting with the grating itself in focus, he noted that as he moved the glass away from the grating, the image of the grating would become defocused and periodically find itself back in focus. He described it as follows:

I then viewed the light which had passed through this grating with a lens of considerable magnifying power. The appearance was very curious, being a regular alternation of numerous lines or bands of red and green colour, having their direction parallel to the lines of the grating. On removing the lens a little further from the grating, the bands gradually changed their colours, and became alternately blue and yellow. When the lens was a little more removed, the bands again became red and green. And this change continued to take place for an indefinite number of times, as the distance between the lens and grating increased. ...It was very curious to observe that though the grating was greatly out of the focus of the lens, yet the appearance of the bands was perfectly distinct and well defined.

Talbot's observations went mostly unnoticed until 1881, when Lord Rayleigh rediscovered them in the context of copying diffraction gratings [Ray81]. By placing a photographic plate in one of the imaging planes of the diffraction grating, one can record a near-perfect reproduction of it.

We begin the analysis of the Talbot effect by considering a one-dimensional grating with period L that is invariant along the y-direction. The grating extends along the x-direction and lies in the plane $z = 0$, and we consider the propagation of the field into the half-space $z > 0$. We do not consider the detailed structure of the grating, but simply note that the wavefield $U_0(x)$ immediately beyond it is periodic and may be represented by a Fourier series,

$$U_0(x) = \sum_{n=-\infty}^{\infty} c_n e^{i2\pi nx/L}, \tag{13.141}$$

where we may write

$$c_n = \frac{1}{L} \int_0^L U_0(x') e^{-i2\pi nx'/L} dx'. \tag{13.142}$$

The field at an arbitrary z-plane may be determined by use of a two-dimensional form of the angular spectrum representation introduced in Section 11.9, such that

$$U(x,z) = \int_{-\infty}^{\infty} \tilde{U}_0(k_x) e^{i(k_x x + k_z z)} dk_x, \tag{13.143}$$

with

$$k_z(k_x) = \begin{cases} (k^2 - k_x^2)^{1/2} & \text{when } k_x^2 \leq k^2, \\ i(k_x^2 - k^2)^{1/2} & \text{when } k_x^2 > k^2. \end{cases} \tag{13.144}$$

and

$$\tilde{U}_0(k_x) = \frac{1}{2\pi} \int_{-\infty}^{\infty} U_0(x) e^{-ik_x x} dx. \tag{13.145}$$

For future convenience, we assume at this point that the diffraction is primarily in the forward direction, such that $\tilde{U}_0(k_x) \neq 0$ only when $k_x \ll k$. We may then approximate k_z by its lowest-order binomial series approximation,

$$k_z \approx k - \frac{k_x^2}{2k}. \tag{13.146}$$

With the choice of field given by Eq. (13.141), we find that the field at any $z > 0$ is given by

$$U(x,z) = e^{ikz} \sum_{n=-\infty}^{\infty} c_n e^{i2\pi nx/L} e^{-i(2\pi n/L)^2 z/2k}. \tag{13.147}$$

Clearly, when $z = 0$, Eq. (13.147) reduces to Eq. (13.141). However, it is also clear that when

$$\left(\frac{2\pi}{L}\right)^2 \frac{z}{2k} = 2\pi m, \qquad (13.148)$$

where m is an integer, the second exponent will take the form $\exp[-2\pi i n^2 m] = 1$, and the field will be nearly equal to Eq. (13.141), the only difference being an overall phase factor. We therefore find that the field produces a perfect image of the grating plane at distances such that

$$z = \frac{kL^2}{\pi} m, \qquad (13.149)$$

with m an integer. The distance $z_T \equiv kL^2/\pi$ is referred to as the *Talbot distance*.

Another set of simple images are formed at distances midway between these primary Talbot images. If we consider the nature of the field at distances

$$z_o = \frac{kL^2}{\pi} \frac{2m+1}{2}, \qquad (13.150)$$

the field takes on the form

$$U(x, z_o) = e^{ikz_o} \sum_{n=-\infty}^{\infty} c_n e^{i2\pi n x/L} e^{-i(2m+1)n^2\pi}. \qquad (13.151)$$

We can readily show that n^2 is even or odd coincident with n itself being even or odd. Similarly, $(2m+1)n^2$ is even or odd coincident with n. We may simplify the above equation to the form

$$U(x, z_o) = e^{ikz_o} \sum_{n=-\infty}^{\infty} c_n e^{i2\pi n x/L} e^{-in\pi}. \qquad (13.152)$$

We may compare this with the behavior of the field in the aperture plane when it is shifted by $L/2$, i.e.

$$U_0(x - L/2) = \sum_{n=-\infty}^{\infty} c_n e^{i2\pi n x/L} e^{-in\pi}. \qquad (13.153)$$

At distances midway between the primary Talbot images, we find a shifted set of Talbot images, displaced by half a period. These primary and secondary Talbot images are illustrated schematically in Fig. 13.13.

At other distances intermediate between the primary and secondary images, other images are formed. The intensity pattern formed over a region of (x,z)-space is beautiful and

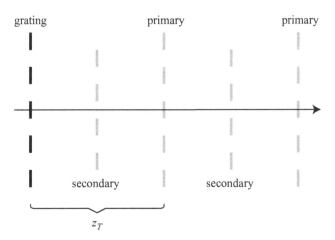

Figure 13.13 Illustrating the arrangement of primary and secondary Talbot images produced by a grating.

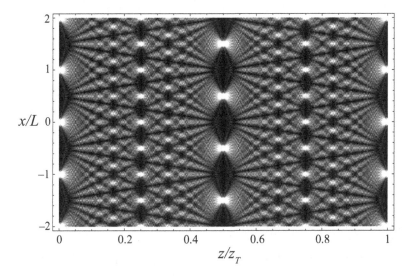

Figure 13.14 An example of a Talbot carpet. The grating was taken to be a series of apertures of width $L/16$, and $\lambda = L/50$.

intricate, and is referred to as a *Talbot carpet*. An example of such a carpet is shown in Fig. 13.14. It can be seen that intermediate planes seem to possess miniaturized images of the original grating, of various sizes.

To analyze the images at these intermediate distances, it is helpful to use Fresnel diffraction. We can readily show that the approximation of Eq. (13.146) is essentially equivalent to the Fresnel approximation through the use of the Poisson sum formula, Eq. (13.19),

$$\sum_{n=-\infty}^{\infty} g(x-nL) = \frac{2\pi}{L} \sum_{m=-\infty}^{\infty} \tilde{g}(2\pi m/L) e^{2\pi i m x/L}. \tag{13.154}$$

We compare the right-hand side of this expression with Eq. (13.147) for the field, which suggests the association

$$\tilde{g}(2\pi n/L) = \frac{L}{2\pi} c_n e^{ikz} e^{-i(2\pi n/L)^2 z/2k}. \tag{13.155}$$

Using the definition of c_n given by Eq. (13.142), this suggests that

$$\tilde{g}(\kappa) = \frac{1}{2\pi} e^{ikz} e^{-i\kappa^2 z/2k} \int_0^L U_0(x') e^{-i\kappa x'} dx', \tag{13.156}$$

or that

$$g(x) = \frac{1}{2\pi} e^{ikz} \int_0^L \int_{-\infty}^{\infty} e^{-i\kappa^2 z/2k} e^{-i\kappa x'} U_0(x') e^{i\kappa x} dx' d\kappa. \tag{13.157}$$

We may evaluate the integral over κ using the Fresnel integral of Section 9.12.4, which leads to the result

$$g(x) = e^{ikz} \sqrt{-i/\lambda z} \int_0^L U_0(x') \exp\left[\frac{ik}{2z}(x-x')^2\right] dx'. \tag{13.158}$$

Returning to the sum formula, this suggests that the field in an arbitrary z-plane may be written as

$$U(x,z) = \sum_{n=-\infty}^{\infty} e^{ikz} \sqrt{-i/\lambda z} \int_0^L U_0(x') \exp\left[\frac{ik}{2z}(x-nL-x')^2\right] dx', \tag{13.159}$$

or, treating $U_0(x)$ as a periodic function of period L,

$$U(x,z) = e^{ikz} \sqrt{-i/\lambda z} \int_{-\infty}^{\infty} U_0(x') \exp\left[\frac{ik}{2z}(x-x')^2\right] dx'. \tag{13.160}$$

Aside from the factor exp[ikz], this is simply a Fresnel transform, defined by Eq. (12.4), with $\alpha = \pi/\lambda z$.

We now consider the field arising from a delta comb of the form

$$C(x) = \sum_{n=-\infty}^{\infty} \delta(x-nL), \tag{13.161}$$

evaluated on a plane at distance

$$z_f = \frac{p}{q} z_T, \tag{13.162}$$

where p and q are integers and p/q is an irreducible fraction. We first note, using the Poisson sum formula, that the comb function may be written as

$$\mathcal{C}(x) = \frac{1}{L} \sum_{n=-\infty}^{\infty} \exp[2\pi i n x / L]. \tag{13.163}$$

Substituting this into the formula for the field on plane $z = z_f$, we have

$$U(x, z_f) = \exp[ikz] \sqrt{\frac{-iq}{p\lambda z_T}} \frac{1}{L} \int_{-\infty}^{\infty} \sum_{n=-\infty}^{\infty}$$

$$\times \exp[2\pi i n x'/L] \exp\left[\frac{i\pi}{2L^2} \frac{q}{p}(x-x')^2\right] dx'. \tag{13.164}$$

We may complete the square to make the integral a pure Fresnel integral. We have

$$\frac{\pi q}{2p} \frac{1}{L^2}\left(x'^2 - 2xx' + x^2\right) + 2\pi n x'/L$$

$$= \frac{\pi q}{2p} \frac{1}{L^2}\left[x' - \left(x - \frac{2pL}{q}n\right)\right]^2 + \frac{2\pi n x}{L} - \frac{2\pi p}{q}n^2. \tag{13.165}$$

Evaluating the Fresnel integral, we have

$$U(x, z_f) = e^{ikz} \frac{1}{L} \sum_{n=-\infty}^{\infty} \exp[2\pi i n x / L] \exp\left[-\frac{2\pi i p}{q} n^2\right]. \tag{13.166}$$

We may now take advantage of certain symmetries in the summation. Let

$$n = qN + m, \tag{13.167}$$

and replace the infinite summation over n with an infinite summation over N and a summation over m from $m = 0$ to $m = q - 1$. We then have

$$U(x, z_f) = e^{ikz} \frac{1}{L} \sum_{N=-\infty}^{\infty} \sum_{m=0}^{q-1} \exp[2\pi i (qN + m)x / L] \exp\left[-2\pi i \frac{p}{q}(qN + m)^2\right]. \tag{13.168}$$

We may expand the second exponent into three terms,

$$\exp\left[-i2\pi \frac{p}{q}(qN + m)^2\right] = \exp\left[-i\left(2\pi pqN^2 + 4\pi pmN + 2\pi \frac{p}{q}m^2\right)\right]. \tag{13.169}$$

The first and second of these terms are integer multiples of 2π, and do not change the value of the exponent. The expression for the field therefore simplifies to

$$U(x, z_f) = e^{ikz} \frac{1}{L} \sum_{N=-\infty}^{\infty} \sum_{m=0}^{q-1} \exp[2\pi i qNx/L] \exp[2\pi imx/L] \exp\left[-i2\pi \frac{p}{q} m^2\right]. \quad (13.170)$$

Only the first exponent is a function of N; using Eqs. (13.161) and (13.163), we may rewrite this summation as a comb function. Equation (13.170) may then be written as

$$U(x, z_f) = e^{ikz} \frac{1}{q} \sum_{N=-\infty}^{\infty} \delta\left(x - \frac{NL}{q}\right) \sum_{m=0}^{q-1} \exp[2\pi imx/L] \exp\left[-2\pi i \frac{p}{q} m^2\right]. \quad (13.171)$$

Due to the delta functions, we may replace x in the exponent by NL/q; the field at distance z_f may finally be written as

$$U(x, z_f) = e^{ikz} \sum_{N=-\infty}^{\infty} F[N]\delta\left(x - \frac{NL}{q}\right), \quad (13.172)$$

where

$$F[N] = \frac{1}{q} \sum_{m=0}^{q-1} \exp[2\pi imN/q] \exp\left[-2\pi i \frac{p}{q} m^2\right]. \quad (13.173)$$

Equation (13.172) illustrates a remarkable result: at the plane $z_f = pz_T/q$, the image of a comb function with period L consists of a contracted comb function of period L/q. The new comb, however, has a distinct phase associated with each image, given by $F[N]$. A general field in the grating plane may be represented by the convolution of the comb function with a generating function $\phi(x)$; the image in the plane z_f will consist of phased images contracted by a factor q. These contracted images are typically referred to as *fractional Talbot images*.

The structure of the fractional images is often a little more subtle than a superficial glance at Eq. (13.172) would indicate. For instance, for $p = 1$, $q = 2$, it is possible to show that half of the coefficients are $F[N]$ zero, with a similar result for $p = 1$, $q = 2M$, with M an integer. Referring back to Fig. 13.14, one can clearly see half-period Talbot images at $z/z_T = 1/4$ and $z/z_T = 3/4$, with $q = 4$.

One may also look at the behavior of the field at distances which are not rational fractions of the Talbot distance z_T. In such planes, it has been shown that the field has a fractal structure [BK96].

The Talbot effect represents a "revival" of the original grating wavefield after an evolution in space over a distance z_T. In quantum mechanics, similar revivals occur in the evolution of a quantum wavepacket in *time* in certain systems. This similarity is discussed in [BMS01].

The Talbot effect produces distributions of intensity which are periodic in three dimensional space; researchers have investigated the manipulation of the structure of Talbot carpets for possible application in generating photonic crystal structures lithographically [TSC06].

Related, but significantly different, images can be generated when the diffraction grating is a pure phase grating. In certain circumstances, a periodic square wave intensity profile can be generated from such a grating at fractions of the Talbot distance. These phase-generated images are referred to as Lohmann images, after the researcher who first proposed them [Loh88, Sul97].

13.10 Exercises

1. Can one use the sampling theorem on the function

$$f(t) = 2\frac{\sin[\pi(1 - t/\alpha)]}{\alpha - t}?$$

   If so, what is the minimum sampling rate needed to perform a good reconstruction?

2. We consider the effect of aliasing on the reconstruction of an undersampled signal. Let the spectrum of a signal be

$$\tilde{f}(\omega) = \begin{cases} \frac{1}{\omega_0}[1 - |\omega|/2\omega_0], & |\omega| \le \omega_0/2, \\ 0, & |\omega| > \omega_0/2. \end{cases}$$

   Determine the mathematical expression for (a) the original signal $f(t)$, and (b) a reconstructed signal which was sampled at a frequency $\omega_0/4$.

3. Read the paper by K. Khare and N. George, Direct sampling and demodulation of carrier-frequency signals, *Opt. Commun.* **211** (2002) 85–94. In this paper, it is shown that under the right circumstances a reconstruction of a sampled signal can be done even when sampling below the Nyquist frequency. What is the nature of the technique, and what sort of signals does it apply to?

4. Read the paper by A. Papoulis, Generalized sampling expansion, *IEEE Trans. Circuits Syst.* **CAS-24** (1977), 652–654. This article describes a technique for reconstructing a bandlimited signal from samples taken below the Nyquist frequency. Describe the technique, and explain what additional information is needed to perform the reconstruction.

5. Suppose that the function $f(t)$ is bandlimited; show that the convolution $f * g$ with any function $g(t)$ results also in a bandlimited signal. If $f(t)$ is limited to $|\omega| \le \Delta/2$, what is the convolution bandlimited to?

6. We may construct a periodic function $G(t)$ from a generating function $g(t)$ in the form

$$G(t) = \sum_{n=-\infty}^{\infty} g(t - nt_0).$$

   Not every function $g(t)$ will produce a well-behaved (i.e. finite-valued) function $G(t)$. Suggest a sufficiency condition for the function $g(t)$ to produce a well-behaved $G(t)$.

7. Explicitly calculate the coefficient $F[N]$ of Eq. (13.173) for the Talbot effect for the cases (a) $p = 1, q = 2$, (b) $p = 1, q = 4$. Show that the coefficient vanishes for certain N values, and hence the period of the fractional Talbot images is larger than L/q.

8. A function $f(t)$ may be written in the frequency domain as

$$\tilde{f}(\omega) = \frac{1}{2\pi} \frac{1}{\alpha^2 + (\omega - \beta)^2}. \tag{13.174}$$

Assuming the function is sampled in time with period T, determine the z-transform $F(z)$ of the sampled signal, and the behavior of the sampled function in time, $f[n]$.

9. A function $f(t)$ may be written in the frequency domain as

$$\tilde{f}(\omega) = \frac{1}{2\pi} \frac{1}{\omega(\omega^2 + \alpha^2)}. \tag{13.175}$$

Assuming the function is sampled in time with period T, determine the z-transform $F(z)$ of the sampled signal, and the behavior of the sampled function in time, $f[n]$.

10. The z-transform of a function is given by the expression

$$X(z) = \frac{1 + z}{1 - z}.$$

Determine the sampled function $x[n]$ associated with this transform.

11. The z-transform of a function is given by the expression

$$X(z) = \frac{z^2}{z^2 - 1}.$$

Determine the sampled function $x[n]$ associated with this transform.

14

Ordinary differential equations

14.1 Introduction: the classic ODEs

An ordinary differential equation (ODE) is an equation that relates a function $y(t)$ and one or more of its derivatives. Physicists are introduced to two ordinary differential equations early in their career, namely the equation of exponential decay and the harmonic oscillator equation. The equation for exponential decay is of the form

$$\frac{dy(t)}{dt} = -\alpha y(t), \tag{14.1}$$

where α is a constant independent of time t. This equation is readily solvable by dividing by $y(t)$, so that

$$\frac{y'}{y} = \frac{d\log(y)}{dt} = -\alpha. \tag{14.2}$$

Here we have written $y' = dy/dt$. Integrating, the solution is of the form $\log(y) = -\alpha t + c_0$, where c_0 is the constant of integration. This formula can be exponentiated to the expression

$$y(t) = C_0 \exp[-\alpha t], \tag{14.3}$$

where $C_0 = \exp(c_0)$ is a constant which is determined by the value of $y(t)$ at $t = 0$, i.e. $C_0 = y(0)$.

The equation of motion for the harmonic oscillator is

$$y'' + k^2 y = 0, \tag{14.4}$$

where k is a constant independent of t. The solution is typically derived in the form of an *ansatz* (guess), i.e. one tries a solution of the form

$$y(t) = e^{\beta t}, \tag{14.5}$$

where β is a constant to be determined. On substitution, one finds that

$$\beta^2 y + k^2 y = 0. \tag{14.6}$$

This suggests that $\beta^2 = -k^2$, or $\beta = \pm ik$. This leads us to two independent solutions,

$$y_1(t) = e^{ikt}, \quad y_2(t) = e^{-ikt}, \tag{14.7}$$

or, equivalently, the solutions

$$y_1(t) = \sin(kt), \quad y_2(t) = \cos(kt). \tag{14.8}$$

A general solution may then be written as

$$y(t) = A\cos(kt) + B\sin(kt), \tag{14.9}$$

where constants A and B are typically related to the initial behavior of the oscillator, i.e.

$$A = y(0), \quad B = y'(0)/k. \tag{14.10}$$

The solutions to these differential equations are straightforward in large part because the equations themselves are exceedingly simple. Even slight changes to an equation, however, can change the solutions significantly and make the simple methods described here invalid.

Throughout most of this chapter we will describe a variety of methods for solving ordinary differential equations. However, this emphasis can be somewhat misleading, because many of the ODEs of interest in modern physics cannot be solved in any exact form. One must either describe the behavior in a qualitative manner using phase space techniques, as discussed in Section 14.3, or turn to computational methods, as described in Section 14.12.

We begin by considering the classification of ODEs, as different strategies can be used to solve different classes of equations.

14.2 Classification of ODEs

The most general ordinary differential equation can be written in the form

$$F\left(y, \frac{dy}{dt}, \dots, \frac{d^N y}{dt^N}, t\right) = g(t), \tag{14.11}$$

where F is a function of the dependent variable $y(t)$, as well as all of its derivatives up to the Nth, and of the independent variable t. It is assumed that $F(0,0,\dots,0,t) = 0$. One usually finds such equations written in the slightly different form

$$\frac{d^N y}{dt^N} + G\left(y, \frac{dy}{dt}, \dots, \frac{d^{N-1}y}{dt^{N-1}}, t\right) = g(t), \tag{14.12}$$

where $G(0,0,\dots,0,t) = 0$. This form avoids the ambiguity inherent in equations such as

$$\left(\frac{dy}{dt}\right)^2 - y^2 = 0, \tag{14.13}$$

where two possible roots of the equation exist, $y' = y$, $y' = -y$, which have very different solutions.

If the function G contains only terms *linear* in $y(t)$ and all its derivatives, we say that it is a *linear equation*; otherwise, it is a *nonlinear equation*. Generally, it is not possible to find analytic solutions to nonlinear equations, and we will focus primarily on linear equations in this chapter.

The highest-order derivative present in the differential equation represents the *order* of the differential equation; the general equation above is an Nth order differential equation. Differential equations of different orders have very different behaviors. An equation of Nth order describes how the behavior of the Nth derivative of y depends on the lower-order derivatives of y. We will show that there are N independent solutions to an equation of Nth order, and a unique solution to an equation can only be found by specifying an additional N conditions on the solution, which we generally refer to as *boundary conditions*.

If the function $g(t) = 0$, we say that the equation is *homogeneous*, i.e. the equation is satisfied by the trivial solution $y(t) = 0$. If $g(t) \neq 0$, the equation is *inhomogeneous*. Inhomogeneous equations typically describe physical systems subject to some sort of external applied effect, e.g. an atom excited by an electric field or a string vibrated by an external force.

If the function $g(t) = 0$ and the function G does not explicitly depend upon t, it is said that the equation is *autonomous*. An autonomous equation where the independent variable represents time typically represents a system for which the configuration is unchanging in time. For instance, the harmonic oscillator equation,

$$y'' + k^2 y = 0, \tag{14.14}$$

is an autonomous equation. If we consider a pendulum whose length is slowly changing in time, i.e. $k = k(t)$, the equation is now non-autonomous.

Ordinary differential equations in physics are typically of first or second order. We will focus most of our attention on such equations, but point out how the results can be extended to equations of higher order.

14.3 Ordinary differential equations and phase space

As we have noted, many important ordinary differential equations cannot be solved with a closed form analytic solution. One can nevertheless often make some very general statements about the solution by investigating its behavior in so-called *phase space*. In addition, the description of phase space provides an excellent unifying framework for the specific ODE solutions described afterwards, and also serves as a starting point for the numerical analysis of ODEs.

We first note that a general Nth-order ordinary differential equation may always be rewritten as a collection of N coupled first-order ordinary differential equations. Let us consider again expression (14.12) for a general Nth-order equation,

$$\frac{d^N y}{dt^N} + G\left(y, \frac{dy}{dt}, \ldots, \frac{d^{N-1}y}{dt^{N-1}}, t\right) = g(t). \tag{14.15}$$

Let us define new variables as follows

$$y_1 = y,$$
$$y_2 = y_1',$$
$$\cdots$$
$$y_{N-1} = y_{N-2}'$$
$$y_N = y_{N-1}' = \frac{d^{N-1}y_1}{dt^{N-1}}. \tag{14.16}$$

Our ODE takes on the form

$$y_N' + G(y_1, y_2, \ldots, y_N) = g(x). \tag{14.17}$$

We now have a first-order ODE for y_N that, combined with Eqs. (14.16), form a set of N coupled first-order ODEs. The solution of this set is not necessarily easier to solve than the original Nth-order ODE – an Nth-order nonlinear ODE becomes a set of N first order ODEs with nonlinear terms – but the decomposition has certain advantages in interpretation.

As an example of such a decomposition, we consider the damped harmonic oscillator equation,

$$y'' + \alpha y' + k^2 y = 0. \tag{14.18}$$

We define two new variables,

$$y_1 \equiv y, \tag{14.19}$$
$$y_2 \equiv y_1' = y'. \tag{14.20}$$

Our second-order ODE then takes on the form

$$y_2' + \alpha y_2 + k^2 y_1 = 0. \tag{14.21}$$

This equation, together with the equation

$$y_1' = y_2, \tag{14.22}$$

form a set of two coupled first-order ODEs. The specification of the initial values of y_1 and y_2 allows us to determine the derivatives of these functions, y_1' and y_2'.

From Eqs. (14.16) and (14.17), we can immediately see that the solution to an Nth order ODE is completely determined by specifying the value of the function and the first $N-1$

derivatives of the function at some initial point $t = t_0$. If these are given, the derivatives of $y_1, \ldots, y_N$ immediately follow, and the behavior of the function at some infinitesimally later time may be determined.

Let us restrict ourselves to autonomous equations. We define the *phase space* of a set of N first-order ODEs as an N-dimensional vector space with vectors $\mathbf{y} \equiv (y_1, \ldots, y_N)$. A point in this phase space completely specifies the state of the physical system under consideration. If we consider a solution $\mathbf{y}(t)$ to the system of differential equations for $t > 0$, given an initial condition $\mathbf{y}(0)$, we readily see that the solution traces out a line in phase space, typically called a *trajectory*. If we consider all possible initial conditions $\mathbf{y}(0)$, we get an infinite collection of trajectories which we will refer to as a *phase portrait*.

Points of special interest in the phase plane are those where all first derivatives vanish, i.e. $y_1' = 0$, $y_2' = 0$, and so on. These points represent *equilibrium points* of the system of equations, and will be seen to have special significance.

For an autonomous system, it becomes clear with some reflection that the trajectories of the system of equations cannot cross, and cannot meet except perhaps at an equilibrium point. Because the system is completely specified by the vector $\mathbf{y}$, only one path can lead away from any point $\mathbf{y}$ in the phase space. For example, once the position and velocity of a harmonic oscillator is specified, its behavior for all future times is uniquely determined. A non-autonomous system, however, may have many trajectories through a single point. For instance, let us consider a harmonic oscillator driven by an external force. Even though the system may, at two different times, have the same position and velocity, its evolution also depends upon the behavior of the driving force at those times.

In principle, a phase space can be constructed for any system of N first-order equations. We will restrict ourselves primarily to second-order systems, for several reasons. First, the most common ODEs encountered are second-order ODEs, e.g. the harmonic oscillator. Also, the phase space of a second-order ODE is a plane, and trajectories in a plane can be readily visualized and drawn. Finally, higher-order equations introduce an additional complication: the possibility of *chaotic motion*, which we address briefly at the end of the section.

The trajectories of a given system may be mapped out by considering the *vector field* defined by the system of differential equations. We can reduce our system of differential equations to a vector equation of the form

$$\mathbf{y}' = \mathbf{v}(\mathbf{y}), \tag{14.23}$$

where $\mathbf{v}$ is a vector in phase space which indicates the direction of evolution of the vector $\mathbf{y}$. By drawing a picture of the vector field in phase space, we can, at least roughly, follow that field to map out the trajectories.

Example 14.1 (Damped harmonic oscillator) We consider the damped harmonic oscillator of the form

$$y'' + \alpha y' + k^2 y = 0, \tag{14.24}$$

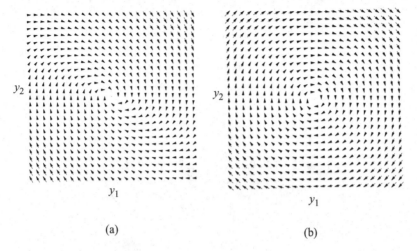

(a)   (b)

Figure 14.1 The vector field for the damped harmonic oscillator, with $k = 1$ and (a) $\alpha = 1$, (b) $\alpha = 0$. The latter case is an undamped oscillator.

with $\alpha > 0$. Though this equation can be solved exactly, we consider its behavior using phase space techniques. Defining $y_1 = y$ and $y_2 = y_1'$, we may write

$$y_2' = -\alpha y_2 - k^2 y_1, \tag{14.25}$$

$$y_1' = y_2 \tag{14.26}$$

as our system of equations. The vector $\mathbf{v}$ is of the form

$$\mathbf{v} = \begin{bmatrix} y_2 \\ -\alpha y_2 - k^2 y_1 \end{bmatrix}. \tag{14.27}$$

The vector field $\mathbf{v}(\mathbf{y})$ is sketched in Fig. 14.1, both with and without damping. From the pictures, we can clearly see the behavior of the oscillator. With damping, the solutions spiral towards the center equilibrium point at $\mathbf{y} = 0$. Without damping, the solutions follow circular paths around the origin.

Some typical trajectories have been determined from the exact solution of the equation and are shown in Fig. 14.2. It can be seen that the trajectories never cross.

It is to be noted that a simple sketch might not provide sufficient information to describe the detailed behavior of a system. For instance, a damped harmonic oscillator with $k = 1$, $\alpha \ll 1$ would result in a picture very similar to that drawn for the undamped oscillator, though their long-term behavior would be very different.

◇

Example 14.2 (Pendulum motion) As a less trivial example, we consider the behavior of a rigid pendulum of length l and angle θ with respect to the vertical. The equation of motion

Ordinary differential equations

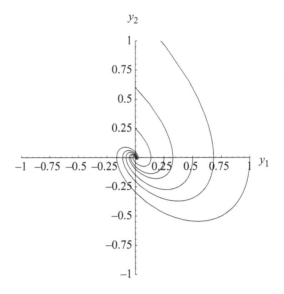

Figure 14.2 Some typical trajectories of the damped harmonic oscillator, with $k = 1$ and $\alpha = 1$.

may be written as

$$\theta'' + k^2 \sin\theta = 0, \tag{14.28}$$

where $k^2 = g/l$ and g is the gravitational constant. We may write this expression in terms of new variables $y_1 \equiv \theta$, $y_2 = \theta'$, so that we have two equations,

$$y_2' = -k^2 \sin y_1, \tag{14.29}$$

$$y_1' = y_2. \tag{14.30}$$

We again have a two-dimensional phase space, and can define a vector field for that space as $\mathbf{v} = (y_2, -k^2 \sin y_1)$. The vector field, shown in Fig. 14.3, is rather complicated. It is immediately obvious, however, that the equilibrium positions of the pendulum are at positions $(n\pi, 0)$, where n is an integer. Those equilibrium positions such that n is even gives those positions for which the pendulum lies at the bottom of its swing, and it can be seen that the pendulum oscillates around such a position when its initial amplitude and velocity are small. Those positions such that n is odd are the positions for which the pendulum is in an unstable equilibrium sitting at the top of its swing.

Around the unstable equilibrium positions, we can examine the behavior by either looking at a zoomed-in version of the vector field, as shown in Fig. 14.4, or by making a Taylor

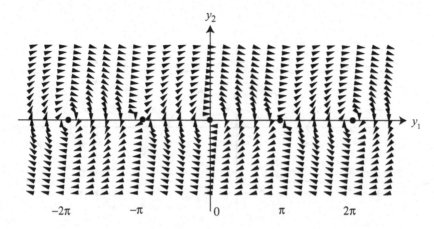

Figure 14.3 The vector field for the nonlinear pendulum.

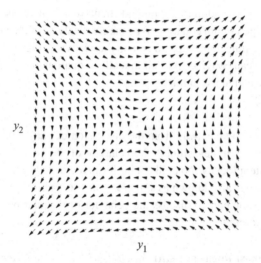

Figure 14.4 The vector field in a box of side 1 around the unstable equilibrium point $(y_1, y_2) = (\pi, 0)$.

series expansion of the vector field around the singular point,

$$y_2' \approx -k^2 z_1, \tag{14.31}$$

$$z_1' = y_2, \tag{14.32}$$

where $z_1 \equiv y_1 - \pi$. Mathematically, these equations represent a *saddle point* (discussed below), which is stable along one axis through its center and unstable along the other.

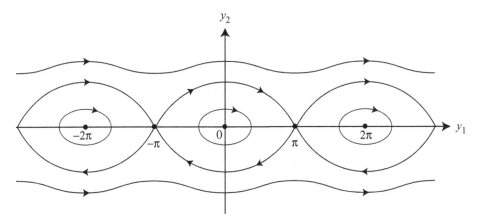

Figure 14.5 The phase portrait for a rigid pendulum.

The trajectories of the pendulum are more difficult to calculate, as there is not a closed form solution for the pendulum equation. By a careful analysis of the vector field, however, one can show that the trajectories appear roughly as shown in Fig. 14.5.

By inspection, one can convince oneself that all the qualitative behaviors expected from a rigid pendulum are contained in this figure.

◇

An important lesson to take away from the discussion of phase space is the importance of the equilibrium points in determining the behavior of the ODE. Almost everywhere in phase space, the system locally looks like a uniform vector field. The only exceptions are those equilibrium points, also known as *singular points* or *critical points*.

Let us take a few moments to discuss the behavior of singular points for second-order ODEs that are *almost linear*.

Definition 14.1 (Almost linear system) *An ODE with a phase space vector **y** is referred to as* almost linear *in the neighborhood of a singular point **y**$_0$ if the differential equation may be written in the form*

$$\mathbf{Y}' = \mathbf{A}\mathbf{Y} + \mathbf{v}(\mathbf{Y}), \tag{14.33}$$

where $\mathbf{Y} = \mathbf{y} - \mathbf{y}_0$, $\det(\mathbf{A}) \neq 0$, *and*

$$\lim_{\mathbf{Y} \to 0} |\mathbf{v}(\mathbf{Y})| / |\mathbf{Y}| = 0. \tag{14.34}$$

The idea of an almost linear system is straightforward: in the immediate neighborhood of a singular point, the system behaves essentially as a linear ordinary differential equation

Table 14.1 *Types of singular points for almost linear systems.*

r_1 and r_2	Type of singular point	Stability of singular point
$r_1 > r_2 > 0$	source	unstable
$r_1 > r_2 > 0$	sink	asymptotically stable
$r_1 > 0, r_2 < 0$ or $r_1 < 0, r_2 > 0$	saddle	unstable
$r_1 = r_2 > 0$	source or spiral point	unstable
$r_1 = r_2 < 0$	sink or spiral point	asymptotically stable
$r_1, r_2 = u + iv, u > 0$	spiral point	unstable
$r_1, r_2 = u + iv, u < 0$	spiral point	asymptotically stable
$r_1, r_2 = \pm iv$	center or spiral point	indeterminate

with constant coefficients of the form

$$\mathbf{Y}' = \mathbf{A}\mathbf{Y}, \tag{14.35}$$

and can be treated using the methods of Sections 14.5 or 14.11. Borrowing from those sections, we assume a solution of the form

$$\mathbf{Y} = \mathbf{Y}_0 e^{rt}. \tag{14.36}$$

On substitution into Eq. (14.35), we find that solutions must satisfy the homogeneous system of equations

$$(\mathbf{A} - r\mathbf{I})\mathbf{Y} = \mathbf{0}. \tag{14.37}$$

For a second-order ODE, this results in two values of r, which we label r_1 and r_2. Because the corresponding solutions for $\mathbf{Y}$ have the functional forms $e^{r_1 t}$ and $e^{r_2 t}$, we may classify the behavior of the solution in the neighborhood of the singular point by the values of r_1 and r_2; this is summarized in Table 14.1.

A *source* is a point for which all trajectories point away, while a *sink* is a point to which all trajectories flow. (These are also referred to in mathematics as *nodes*, with a distinction between *proper* and *improper* nodes that we will not concern ourselves with.) A *saddle point* is a point in the neighborhood of which the trajectories point inward along one axis and outward along another. A *spiral point* is a point in the neighborhood of which the trajectories spiral towards or away from. A *center* is a point around which the trajectories circulate. The various possibilities are illustrated in Fig. 14.6.

A singular point is an equilibrium point of the system, and the values of r_1 and r_2 also characterize the stability of the equilibrium. An equilibrium point is *stable* if a small displacement from equilibrium results in a trajectory that stays close to equilibrium; it is

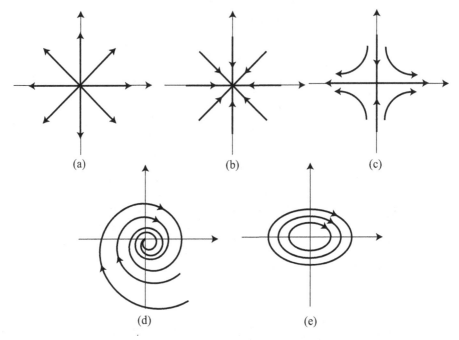

Figure 14.6 The different types of singular points: (a) source, (b) sink, (c) saddle, (d) spiral, (e) center.

asymptotically stable if a small displacement results in a trajectory that returns to equilibrium. Clearly, a source is unstable, while a sink is asymptotically stable; a saddle point is also unstable. A spiral point can be asymptotically stable or unstable, depending on the values of r_1 and r_2. A center is stable in a system that is exactly linear; in an almost linear system, the stability is then determined by the nonlinear properties of the system. The properties of these singular points can be used to characterize the overall behavior of the ODE.

Though the discussion given here can be extended in principle to higher-order systems of differential equations, new challenges arise in the analysis. In addition to the problem of visualizing phase spaces of three or higher dimensions, such higher-order systems can exhibit so-called *chaotic behavior*. Intuitively, most people have come to understand chaos as a high sensitivity to initial conditions: if we start two chaotic systems simultaneously with almost identical initial conditions, we will eventually find that the two systems are behaving drastically differently.

In strict mathematical terms, a system is only called chaotic if it satisfies three criteria (1) a sensitivity to initial conditions, already noted, (2) it exhibits topological mixing, and (3) it possesses dense periodic orbits. A detailed discussion of these properties is well outside the scope of this text, but we can qualitatively describe each of them. *Topological mixing* loosely describes a system where a phase trajectory essentially fills the phase space; that is, given enough time, the trajectory will move arbitrarily close to any specified point. *Dense periodic orbits* refers to a system which acts almost periodically – it returns regularly to a given neighborhood of phase space – and the set of those quasi-periodic orbits essentially fills the phase space. The orbit of the Earth around the Sun is a good example of a dense periodic orbit: because of the gravitational effects of the other planets, the Earth never repeats the same path around the Sun, though it still clearly follows a path that we conventionally think of as periodic.

Though chaos is a difficult subject to quantify, it is relatively straightforward to see why it can only manifest in systems of order greater than 2. In any autonomous system, we know that the trajectories cannot intersect. In a two-dimensional system, this means that each trajectory of the system provides a "barrier" of sorts for the others.[1] Once two trajectories are specified, any trajectory in between them must be remain between them for all time. In three or more dimensions, however, a single trajectory does not form any sort of "natural boundary" for the path of the others: depending on the specific characteristics of the system, two adjacent trajectories can evolve almost independently of one another.

In the next few sections of this chapter, we consider specific techniques for finding analytic solutions to ordinary differential equations. In the last section we consider the basics of numerical analysis of ODEs.

14.4 First-order ODEs

A first-order linear differential equation of the most general form can be written as

$$\frac{dy}{dx} = f(x, y),$$ (14.38)

where x is the independent variable and $y(x)$ is the dependent variable we wish to solve for. As already noted, the equation is called first-order because it only contains the first derivative of y. The function $f(x, y)$ is more or less arbitrary – if it depends only on the first power of y, the equation is said to be linear; otherwise, the equation is nonlinear. A variety of strategies exist for solution of the equation, depending on its exact form. To get a unique solution, one must specify the value of the function y at a specific point $x = x_0$.

14.4.1 Separable equation

If the ODE has the special form

$$\frac{dy}{dx} = -\frac{P(x)}{Q(y)},$$ (14.39)

---

[1] Those who are old enough to remember the 1982 movie TRON can envision the light-cycle races, in which each cycle creates a wall behind it that the other cycles cannot pass. A single light-cycle severely limits the motion of the others.

it is said to be *separable*. It may then be rewritten as

$$P(x)dx + Q(y)dy = 0. \tag{14.40}$$

This can be considered as the integrand of a path integral in two-dimensional space which may be integrated from the point (x_0, y_0) to the point (x, y). This gives us a solution to the equation,

$$\int_{x_0}^{x} P(\xi)d\xi + \int_{y_0}^{y} Q(\eta)d\eta = 0. \tag{14.41}$$

We may use this technique to solve certain nonlinear equations, provided the ODE is separable. It is to be noted, however, that the solution may be transcendental, i.e. it may not be solvable for $y(x)$.

Example 14.3 We consider the equation

$$y' = -\frac{\alpha y}{(x+1)^2}, \tag{14.42}$$

with the boundary condition $y(0) = y_0$. This equation is separable, and may be written in the form

$$\frac{dy}{y} = -\frac{\alpha dx}{(x+1)^2}. \tag{14.43}$$

Integrating both sides, we find that

$$\log y = \frac{\alpha}{x+1} + c_0, \tag{14.44}$$

or

$$y = \exp[c_0]\exp\left[\frac{\alpha}{x+1}\right]. \tag{14.45}$$

To satisfy the boundary condition, we note that

$$y(0) = \exp[c_0]\exp[\alpha] = y_0, \tag{14.46}$$

or $\exp[c_0] = \exp[-\alpha]y_0$. Our final solution is

$$y(x) = y_0 \exp[-\alpha]\exp\left[\frac{\alpha}{x+1}\right]. \tag{14.47}$$

◇

14.4.2 *Exact equation*

We now consider an ODE of the form

$$\frac{dy}{dx} = -\frac{P(x,y)}{Q(x,y)},$$

(14.48)

where now P and Q are both functions of both x and y, i.e. the equation is not separable. Let us suppose, however, that there exists a function $u(x,y)$ such that

$$du = P(x,y)dx + Q(x,y)dy$$

(14.49)

In this case, our differential equation represents an *exact* differential of the function $u(x,y)$, which may also be written as

$$du = \frac{\partial u}{\partial x}dx + \frac{\partial u}{\partial y}dy.$$

(14.50)

We may state this condition in an alternative form as

$$\frac{\partial u}{\partial x} = P(x,y),$$

(14.51)

$$\frac{\partial u}{\partial y} = Q(x,y).$$

(14.52)

In this case, our differential equation reduces to the simple equation

$$du = 0,$$

(14.53)

whose solution is readily found to be

$$u = \text{constant}.$$

(14.54)

Most ODEs are, unfortunately, not exact. We may readily develop a test for exactness, which we state as a theorem.

Theorem 14.1 *Suppose the functions P,Q, and their first partial derivativess with respect to x and y are continuous in some region of x, y space. The differential equation (14.48) is then exact in that region if and only if*

$$\frac{\partial P}{\partial y} = \frac{\partial Q}{\partial x}.$$

(14.55)

One can easily prove that an exact equation satisfies Eq. (14.55). This can be done by taking the first partial derivative of Eq. (14.51) with respect to y and Eq. (14.52) with respect to x, and noting their equality. To prove the converse, i.e. that Eq. (14.55) implies exactness, we must construct a function $u(x,y)$ which satisfies Eqs. (14.51) and (14.52) from only the assumption of Eq. (14.55). This proof is more involved and we omit it here.

Though most ODEs are not exact, one can always make them so by multiplying by an appropriate function $\alpha(x,y)$, called an *integrating factor*, such that

$$\alpha(x,y)P(x,y)dx + \alpha(x,y)Q(x,y)dy = 0 \tag{14.56}$$

is exact. The new equation will be exact if it satisfies Theorem 14.1, i.e.

$$\frac{\partial(\alpha P)}{\partial y} = \frac{\partial(\alpha Q)}{\partial x}. \tag{14.57}$$

This equation is a partial differential equation for $\alpha(x,y)$, which is in general difficult to solve. If the factor depends only upon x or y and not both, however, the problem simplifies considerably. Assuming $\alpha(x,y) = \alpha(x)$, we may write

$$\alpha\frac{\partial P}{\partial y} = \alpha\frac{\partial Q}{\partial x} + Q\frac{d\alpha}{dx}. \tag{14.58}$$

This equation can be solved as

$$\frac{d\alpha}{dx} = \frac{P_y - Q_x}{Q}\alpha, \tag{14.59}$$

which is a first-order ODE for $\alpha(x)$, which can be readily solved,

$$\log[\alpha] = \int \frac{P_y - Q_x}{Q}dx. \tag{14.60}$$

The quantity $(P_y - Q_x)/Q$ serves as a test of the behavior of α; if it depends only upon x, the integrating factor depends only upon x and it can be found by application of Eq. (14.60). A similar argument applies for an integrating factor which depends upon y alone.

Example 14.4 We consider the solution of the equation

$$y' = \exp[2x] + y - 1. \tag{14.61}$$

For this equation, we have $P(x,y) = \exp[2x] + y - 1$ and $Q(x,y) = -1$. It can readily be seen from application of Eq. (14.55) that the equation is not exact. An integrating factor which depends upon x alone can be found, however, as

$$\frac{P_y - Q_x}{Q} = -1. \tag{14.62}$$

By use of Eq. (14.60), we find that $\alpha(x) = \exp[-x]$. The function $u(x,y)$ which characterizes the solution must satisfy the equations

$$\frac{\partial u}{\partial x} = \exp[x] + y\exp[-x] - \exp[-x], \tag{14.63}$$

$$\frac{\partial u}{\partial y} = -\exp[-x]. \tag{14.64}$$

Integrating these equations directly results in the equations

$$u(x,y) = \exp[x] - y\exp[-x] + \exp[-x] + f(y), \tag{14.65}$$

$$u(x,y) = -y\exp[-x] + g(x), \tag{14.66}$$

where $f(y)$ and $g(x)$ are functions of integration. On comparing the equations, we find that $f(y) = 0$, $g(x) = \exp[x] + \exp[-x]$, and we may write the solution to our differential equation as

$$\exp[x] - y\exp[-x] + \exp[-x] = C_0, \tag{14.67}$$

where C_0 is the constant specified by the initial condition. This equation can be rewritten explicitly as

$$y(x) = \exp[2x] + 1 + C_0\exp[x]. \tag{14.68}$$

On substitution into the original Eq. (14.61), the solution is found to be valid.

◇

14.4.3 Linear equation

A linear equation is one of the form

$$\frac{dy}{dx} + p(x)y = q(x). \tag{14.69}$$

The most straightforward way to solve this equation is to look for an integrating factor, as in the previous section. If we multiply the above equation by $\alpha(x)$, we have

$$\alpha(x)\frac{dy}{dx} + \alpha(x)p(x)y = \alpha(x)q(x). \tag{14.70}$$

We want the function $\alpha(x)$ to satisfy the equation

$$\frac{d[\alpha(x)y]}{dx} = \alpha(x)q(x). \tag{14.71}$$

Expanding the derivative in this equation and equating it with (14.70), we find that we must have

$$\frac{d\alpha}{dx} = \alpha(x)p(x).$$ (14.72)

This is readily solvable,

$$\alpha(x) = \exp\left[\int_x p(\xi)d\xi\right].$$ (14.73)

We may therefore find a function $\alpha(x)$ for which the differential equation (14.69) reduces to the form (14.71). This latter equation may be directly integrated, and we have the final result,

$$y(x) = [\alpha(x)]^{-1}\left\{\int_x \alpha(\xi)q(\xi)d\xi + C_0\right\},$$ (14.74)

where C_0 is a constant of integration.

Example 14.5 We consider the solution of the equation

$$y' + \frac{2}{x}y = \exp[x]/x,$$ (14.75)

subject to the initial condition $y(1) = 2$. Here $p(x) = 2/x$, so the integrating factor may be found from Eq. (14.73) as

$$\alpha(x) = \exp\left[\int_x 2d\xi/\xi\right] = C_0 x^2.$$ (14.76)

From Eq. (14.74), we can find the solution for $y(x)$ by direct integration, with $q(x) = \exp[x]/x$,

$$y(x) = \exp[x]/x - \exp[x]/x^2 + C_0/x^2.$$ (14.77)

Applying the initial condition $y(1) = 2$, we find that $C_0 = 2$.

◇

14.5 Second-order ODEs with constant coefficients

Second-order ODEs are typically much more difficult to solve than first-order ODEs. It is in general not possible to directly integrate a second-order ODE for a solution, and such a strategy was the basis for all of our solutions of first-order equations. We restrict our attention to linear second-order equations of the form

$$A(x)\frac{d^2y}{dx^2} + B(x)\frac{dy}{dx} + C(x)y = D(x).$$ (14.78)

Often we may divide such an equation by the term $A(x)$ to arrive at a reduced equation (assuming, of course, that $A(x) \neq 0$):

$$y'' + P(x)y' + Q(x)y = R(x). \tag{14.79}$$

Such an equation is *homogeneous* if $R(x) = 0$; otherwise, it is *inhomogeneous*.

The solutions of a homogeneous linear second-order equation have many significant properties. As noted previously, one can always find *two* linearly independent solutions to such an equation; two functions are *linearly independent* if they are not multiples of each other, i.e. if

$$C_1 y_1(x) + C_2 y_2(x) = 0 \tag{14.80}$$

is satisfied for all x only if $C_1 = C_2 = 0$. This is analogous to the definition of linear independence given for vectors in Section 1.5. If any two linearly independent solutions to the differential equation are found, a linear combination of these is also a solution: i.e. if $y_1(x)$ and $y_2(x)$ satisfy Eq. (14.79), so does the function

$$y(x) = C_1 y_1(x) + C_2 y_2(x). \tag{14.81}$$

There are, then, an infinite number of solutions to the differential equation (14.79). We may find a unique solution if we specify a pair of *point conditions* (boundary conditions) on the function $y(x)$; typically, one is given the value of the function and its first derivative at a single point $x = x_0$,

$$y(x_0) = a, \quad y'(x_0) = b. \tag{14.82}$$

Let us restrict our attention first to homogeneous equations with constant real coefficients, of the form

$$Ay'' + By' + Cy = 0. \tag{14.83}$$

To find a general solution to this equation, we try, as in Section 14.1, the ansatz

$$y(x) = e^{\alpha x}, \tag{14.84}$$

where α is a constant to be determined. We substitute this form of the solution into Eq. (14.83), take the derivatives, and divide out the function $y(x)$ to arrive at the equation

$$A\alpha^2 + B\alpha + C = 0. \tag{14.85}$$

This is a quadratic equation, with solutions of the form

$$\alpha = \frac{-B \pm \sqrt{B^2 - 4AC}}{2A}. \tag{14.86}$$

There are therefore three distinct classes of solutions, depending on whether $B^2 - 4AC < 0$, $B^2 - 4AC = 0$ or $B^2 - 4AC > 0$. If $B^2 - 4AC > 0$, the solutions are exponentials, while if $B^2 - 4AC < 0$, the solutions are mixed oscillating/exponential functions.

What happens in the case $B^2 - 4AC = 0$? Our ansatz only gives us one solution, while we know that a second-order ODE has two linearly independent solutions. We try a solution of the form

$$y_2(x) = f(x)e^{\alpha x}, \tag{14.87}$$

where $f(x)$ is to be determined. Taking the derivatives, and substituting into our differential equation, we have

$$Af'' + (2A\alpha + B)f' + (A\alpha^2 + B\alpha + C)f = 0. \tag{14.88}$$

The latter term in parenthesis vanishes directly by Eq. (14.85). The middle term vanishes as well; Eq. (14.86), with $B^2 - 4AC = 0$, leads to $\alpha = -B/(2A)$. We are left with the condition $f'' = 0$, or $f \propto x$. For the case $B^2 - 4AC = 0$, then, the two independent solutions are

$$y_1(x) = e^{\alpha x}, \quad y_2(x) = xe^{\alpha x}. \tag{14.89}$$

Once a pair of general solutions has been found, we can find a specific solution by applying our boundary conditions (14.82).

14.6 The Wronskian and associated strategies

Turning to more general linear differential equations of second order, we find that the methods of solution become more complicated. We consider the solution of differential equations of the form

$$y'' + P(x)y' + Q(x)y = R(x). \tag{14.90}$$

How does one find the solutions for such an equation? There is no straightforward, systematic way to do so, but there are a number of tricks one can use to find a solution provided the problem has already been partly solved. The first of these involves the so-called *Wronskian*, which we now discuss.

Definition 14.2 *Let us suppose that $y_1(x)$ and $y_2(x)$ are two linearly independent solutions to the homogeneous form of the differential equation (14.90). We define the Wronskian as the function*

$$W[y_1(x), y_2(x)] \equiv y_1(x)y_2'(x) - y_2(x)y_1'(x). \tag{14.91}$$

Since both $y_1(x)$ and $y_2(x)$ satisfy the original differential equation, we have

$$y_1'' + P(x)y_1' + Q(x)y_1 = 0, \qquad (14.92)$$

$$y_2'' + P(x)y_2' + Q(x)y_2 = 0. \qquad (14.93)$$

We may multiply the first of these equations by y_2, the second by y_1, and subtract them. We then find that

$$y_1 y_2'' - y_2 y_1'' + P(x)[y_1 y_2' - y_2 y_1'] = 0. \qquad (14.94)$$

The first part of this differential equation is simply the first derivative of the Wronskian, and the second part involves the Wronskian itself. This equation may be rewritten as

$$\frac{dW}{dx} + P(x)W = 0. \qquad (14.95)$$

This first-order differential equation for the Wronskian may be solved using the methods of the previous section, and we find that

$$W(x) = W(x_0) \exp\left\{ -\int_{x_0}^{x} P(\xi)d\xi \right\}. \qquad (14.96)$$

The Wronskian therefore depends only on the function $P(x)$ of the differential equation! In essence, it is possible to find a functional of the solutions (the function W which depends on y_1 and y_2) without ever explicitly finding the solutions themselves.

The Wronskian can be used to find a second solution to the differential equation, provided the first is already known. To do so, we note that

$$\frac{d}{dx}\left(\frac{y_2}{y_1}\right) = \frac{y_1 y_2' - y_2 y_1'}{y_1^2} = \frac{W(x)}{y_1^2}, \qquad (14.97)$$

where we assume that $y_1(x)$ is already known. This is a linear first order equation whose right-hand side is already determined. We may therefore integrate this and find the solution to be

$$\frac{y_2}{y_1} = W(x_0) \int_{x_1}^{x} \exp\left\{ -\int_{x_0}^{\xi} P(\eta)d\eta \right\} \frac{d\xi}{[y_1(\xi)]^2}. \qquad (14.98)$$

In other words, we first calculate the Wronskian by integration, then we integrate again to determine the ratio of y_2 to y_1.

Example 14.6 The ordinary differential equation

$$y'' + \frac{2y'}{x} = 0 \qquad (14.99)$$

has an obvious solution $y(x) = 1$. The other solution may be found by applying Eq. (14.98). The solution for the Wronskian is $W(x) = x^{-2}$, and the second solution is found by integration to be $y(x) = x^{-1}$.

◇

14.7 Variation of parameters

If the equation is inhomogeneous ($R(x) \neq 0$), but we know two homogeneous solutions to the equation, y_1 and y_2, we may solve the problem using a method known as the *variation of parameters*. We assume the inhomogeneous equation has a solution of the form

$$u(x) = v_1(x)y_1(x) + v_2(x)y_2(x), \tag{14.100}$$

which looks very much like the homogeneous solution except that the multiplicative "constants" are now functions v_1 and v_2. The derivative of $u(x)$ is given by

$$u'(x) = v_1 y_1' + v_2 y_2' + v_1' y_1 + v_2' y_2. \tag{14.101}$$

Our functions v_1 and v_2 right now are in some sense arbitrary; it can be shown that there are many choices of v_1 and v_2 which satisfy equation (14.100). For instance, if $v_1(x)$ and $v_2(x)$ result in a solution to the inhomogeneous equation, so do the alternative functions,

$$\tilde{v}_1(x) = \frac{1}{2}v_1(x) + \frac{1}{2}\frac{v_2(x)y_2(x)}{y_1(x)}, \quad \tilde{v}_2(x) = \frac{1}{2}v_2(x) + \frac{1}{2}\frac{v_1(x)y_1(x)}{y_2(x)}. \tag{14.102}$$

This can be seen by direct substitution into Eq. (14.100). To restrict the possible values of v_1 and v_2, let us impose the additional constraint

$$v_1' y_1 + v_2' y_2 = 0, \tag{14.103}$$

in which case the derivative of $u(x)$ simplifies to the form

$$u'(x) = v_1 y_1' + v_2 y_2'. \tag{14.104}$$

We may calculate the second derivative of $u(x)$ as well,

$$u'' = v_1 y_1'' + v_2 y_2'' + v_1' y_1' + v_2' y_2'. \tag{14.105}$$

With these derivatives, we may substitute them back into the inhomogeneous equation,

$$u'' + P(x)u' + Q(x)u = R(x). \tag{14.106}$$

If we do so, and use the fact that y_1 and y_2 are solutions of the *homogeneous* version of the equation, we find that

$$v_1' y_1' + v_2' y_2' = R. \tag{14.107}$$

This equation, along with the additional constraint, Eq. (14.103),

$$v_1' y_1 + v_2' y_2 = 0, \tag{14.108}$$

is a system of two equations with two unknowns (v_1', v_2'). We may solve these, and one can show that

$$v_1' = -y_2 R/W, \quad v_2' = y_1 R/W. \tag{14.109}$$

Each of these equations can be integrated directly, and we find that

$$v_1(x) = -\int_{x_1}^{x} \frac{y_2(\xi) R(\xi)}{W(\xi)} d\xi, \quad v_2(x) = \int_{x_2}^{x} \frac{y_1(\xi) R(\xi)}{W(\xi)} d\xi, \tag{14.110}$$

where x_1 and x_2 are arbitrary constants which will be specified by any point conditions on the equation.

Example 14.7 We attempt to find the general solution to the inhomogeneous ODE

$$y'' + 4y = \sec^2(2x). \tag{14.111}$$

In this expression, $P(x) = 0$, $Q(x) = 4$, and $R(x) = \sec^2(2x)$. The solutions to the homogeneous equation are clearly

$$y_1(x) = \sin(2x), \quad y_2(x) = \cos(2x), \tag{14.112}$$

and the Wronskian can be readily found to be $W(x) = -2$. From these results we may readily find the functions $v_1(x)$ and $v_2(x)$,

$$v_1(x) = \int \frac{\cos(2\xi) \sec^2(2\xi)}{2} d\xi = \frac{1}{4} \log[\sec(2x) + \tan(2x)] + C_1,$$

$$\tag{14.113}$$

$$v_2(x) = -\int \frac{\sin(2\xi) \sec^2(2\xi)}{2} d\xi = -\frac{1}{4} \frac{1}{\cos(2x)} + C_2. \tag{14.114}$$

Applying these, we have a general solution of the form

$$y(x) = C_1 \sin(2x) + C_2 \cos(2x) + \frac{\sin(2x)}{4} \log[\sec(2x) + \tan(2x)] - \frac{1}{4}. \tag{14.115}$$

The constants C_1 and C_2 will depend on the choice of initial conditions.

◇

In summary, then, if we know one solution to a homogeneous linear second-order equation, we can find another homogeneous solution in integral form. If we know two homogeneous solutions, we can find an inhomogeneous solution by variation of constants. The real trick, then, is to find that first homogeneous solution, and the most general way to do that is by searching for a power series solution.

14.8 Series solutions

In the previous section we saw some very simple and elegant methods for solving second-order linear ODEs, provided a single solution of the homogeneous equation is known. However, finding that single solution can be quite a challenge, even for equations of seemingly simple form, such as

$$y'' + xy = 0. \tag{14.116}$$

Is there a way to find a general solution of such an equation, at least in a limited domain? Let us return first to the general form of a second-order linear equation,

$$y'' + P(x)y' + Q(x)y = R(x). \tag{14.117}$$

Suppose we replace x by a complex number z. If the functions $P(z)$, $Q(z)$ and $R(z)$ are all analytic functions in the neighborhood of some point $z = a$, it follows that the function $y(z)$ must be analytic in the same neighborhood. We must therefore be able to expand the function $y(z)$ in a Taylor series expansion around the point $z = a$, i.e.

$$y(z) = \sum_{n=0}^{\infty} c_n (z - a)^n. \tag{14.118}$$

Likewise, we may Taylor expand $P(z)$, $Q(z)$, and $R(z)$ about the point $z = a$ as well. The *power series method* of solving such a differential equation involves expanding all these functions in a power series, substituting them into the differential equation, and then equating like powers of z. Because different powers of z are linearly independent, it follows that the coefficients of each power must vanish identically. In this way, we can (in principle) determine all the unknown coefficients c_n of the Taylor series for $y(z)$.

Example 14.8 We consider the solution of the equation

$$y'' - k^2 y = 0 \tag{14.119}$$

about the point $z = 0$. This is similar to the harmonic oscillator equation except for the sign of the second term. We let

$$y(z) = \sum_{n=0}^{\infty} c_n z^n. \tag{14.120}$$

The derivatives can readily be found to be

$$y'(z) = \sum_{n=1}^{\infty} n c_n z^{n-1} = \sum_{n=0}^{\infty} (n+1) c_{n+1} z^n, \tag{14.121}$$

and

$$y''(z) = \sum_{n=2}^{\infty} n(n-1)c_n z^{n-2} = \sum_{n=0}^{\infty} (n+2)(n+1)c_{n+2} z^n. \qquad (14.122)$$

On substituting into our differential equation, we find that

$$\sum_{n=0}^{\infty} (n+2)(n+1)c_{n+2} z^n - k^2 \sum_{n=0}^{\infty} c_n z^n = 0. \qquad (14.123)$$

Grouping terms with equal powers of z, we find that

$$\sum_{n=0}^{\infty} \left[(n+2)(n+1)c_{n+2} - k^2 c_n \right] z^n = 0. \qquad (14.124)$$

The only way this will be satisfied for a continuous range of z values is if

$$c_{n+2} = \frac{k^2}{(n+2)(n+1)} c_n. \qquad (14.125)$$

This is a *recursion formula* for the coefficients c_n. It is to be noted that the even coefficients are specified by choosing c_0, while all the odd coefficients are specified by choosing c_1. The even and odd terms are therefore independent of each other, and the entire series may be written as

$$y(z) = c_0(1 + (kz)^2/2! + (kz)^4/4! + \cdots) + \frac{c_1}{k}(kz + (kz)^3/3! + (kz)^5/5! + \cdots). \qquad (14.126)$$

The first and second parts of this solution represents the two independent solutions of the differential equation, and an observant reader will note that

$$1 + (kz)^2/2! + (kz)^4/4! + \cdots = \cosh kz, \qquad (14.127)$$

$$kz + (kz)^3/3! + (kz)^5/5! + \cdots = \sinh kz. \qquad (14.128)$$

Of course, in general one cannot expect the solutions of ODEs to reduce to some simple elementary functions. This reality is, in effect, the whole motivation behind series solutions – most solutions have no closed-form expression.

◇

14.9 Singularities, complex analysis, and general Frobenius solutions

Let us return to the expression for a homogeneous second-order linear ODE,

$$y'' + P(z)y' + Q(z)y = 0. \qquad (14.129)$$

In the previous section, we argued that the solution $y(z)$ of the differential equation is analytic in the neighborhood of a point z_0, provided that the functions $P(z)$ and $Q(z)$ are analytic in that same neighborhood. However, it is quite common to encounter differential equations for which $P(z)$ and $Q(z)$ may be singular at the point $z = z_0$; for instance, Bessel's equation, to be discussed in Chapter 16, has the form

$$y'' + \frac{1}{z}y' + \left(1 - v^2/z^2\right)y = 0, \tag{14.130}$$

and both $P(z)$ and $Q(z)$ are singular at $z = 0$.

If the differential equation possesses singularities at $z = z_0$, it is natural to expect that the solution $y(z)$ will also possess a singularity at that same point. One would further expect that the solution can be generally represented by a Laurent series, and possibly by a branch point.

We distinguish between possible singular cases in the following definition.

Definition 14.3 *Given the second-order differential equation (14.129), the point $z = z_0$ is said to be an* ordinary point *if the functions $P(z)$ and $Q(z)$ are analytic in some neighborhood of z_0; otherwise it is referred to as a* singular point. *A singular point z_0 is said to be* regular singular *if $P(z)$ is at worst a simple pole and $Q(z)$ is at worst a second-order pole; otherwise, it is referred to as* irregular singular.

The solution of a differential equation around a regular singular point will typically be of the form

$$y(z) = z^s \sum_{n=0}^{\infty} c_n z^n = \sum_{n=0}^{\infty} c_n z^{n+s}, \tag{14.131}$$

where now the parameter s may be a positive integer, a negative integer, or a non-integer, corresponding to a zero, a pole, or a branch point of $y(z)$. Furthermore, as we will explain, under certain circumstances the solution may also take on a general logarithmic behavior.

We may apply this solution in a manner almost identical to that of the previous section, the only difference being the introduction of the parameter s. This manner of solution was originally introduced by Frobenius and therefore has his name.

Example 14.9 (Branch point at the origin) As an example, we consider the Bessel equation of order $v = 1/2$,

$$z^2 y'' + z y' + (z^2 - 1/4)y = 0. \tag{14.132}$$

We assume a series solution of the form

$$y(z) = \sum_{n=0}^{\infty} c_n z^{n+s}, \tag{14.133}$$

so that

$$y'(z) = \sum_{n=0}^{\infty} (s+n)c_n z^{n+s-1}, \tag{14.134}$$

$$y''(z) = \sum_{n=0}^{\infty} (s+n)(s+n-1)c_n z^{n+s-2}. \tag{14.135}$$

On substituting into the differential equation and rearranging terms, we find that

$$\sum_{n=0}^{\infty} c_n[(s+n)(s+n-1)+(s+n)-1/4]z^{s+n} + \sum_{n=0}^{\infty} c_n z^{s+n+2} = 0. \tag{14.136}$$

Note that the left sum of this equation has two powers of z which do not appear in the right series of the equation, z^s and z^{s+1}; the right-hand series does not contain these powers. The coefficients for $n=0$ and $n=1$ on the left-hand side must vanish by themselves; for $n=0$, we get the equation

$$s(s-1)+s-1/4 = 0, \tag{14.137}$$

which is a quadratic equation with roots $s_1 = -1/2$, $s_2 = 1/2$, referred to as the *indicial equation*. Let us consider the behavior of the solution for $s = s_1$. If we now look at the coefficients for $n=1$, we find the requirement that

$$c_1((s+1)s + (s+1) - 1/4) = 0, \tag{14.138}$$

which is in fact automatically satisfied by our root $s = s_1$. Therefore both c_0 and c_1 may be chosen to be nonzero numbers for $s = -1/2$, and these two numbers specify all higher-order coefficients of the series. Matching powers of z, we can find the recursion relation of the series to be

$$c_n = -\frac{1}{n(n-1)}c_{n-2}. \tag{14.139}$$

It follows that the general solution to this equation is given by

$$y_1(z) = c_0 z^{-1/2}(1 - z^2/2 + z^4/(2\cdot 3\cdot 4) - \cdots)$$
$$+ c_1 z^{-1/2}(z - z^3/(2\cdot 3) + z^5/(2\cdot 3\cdot 4\cdot 5) - \cdots). \tag{14.140}$$

Surprisingly, we already have here two linearly independent solutions to the differential equation – the first series, which has even powers of z, is clearly independent of the second series, which has odd powers of z. What happened to the root $s_2 = +1/2$? If we factorize out a power of z from the second series, it may be rewritten as

$$c_1 z^{+1/2}(1 - z^2/(2 \cdot 3) + z^4/(2 \cdot 3 \cdot 4 \cdot 5) - \cdots).$$ (14.141)

In other words, when we solved for the solution for the root $s = -1/2$, we automatically got the solution for $s = +1/2$! One can understand this by noting that the two roots of the equation differ by an integer power of z. A solution for the higher root follows automatically from a solution for the lower root, with an additional power of z to make up the difference. It turns out, however, that we got lucky in our solution, as the next example shows.

◇

Example 14.10 We now consider the differential equation

$$zy'' + y' = 0,$$ (14.142)

which we will try and solve about the point $z = 0$. The recursion relation for this equation can be shown to be

$$c_n(s+n)^2 = 0,$$ (14.143)

and the roots of the indicial equation are both $s = 0$. The only solution of the Frobenius type we can find, then, is $y_1(z) = c_0 = $ constant. Why can we not find another solution? It turns out that the second solution is the natural logarithm, $y_2(z) = \log z$, which can be verified by direct substitution. The solution has a logarithmic branch point at $x = 0$, which is why our simple Frobenius series failed.

◇

If the roots of the indicial equation differ by an integer, it can be shown that there is a chance of a redundancy in the solution which can only be resolved by introducing a logarithmic term. One solution $y_1(z)$ of an equation about a regular singular point is always of the Frobenius type, Eq. (14.131) with indicial root r_1; the second solution is typically of the form

$$y_2(z) = \beta y_1(z) \log(z) + z^{r_2} \sum_{n=0}^{\infty} c_n z^n,$$ (14.144)

where r_2 is the second root of the indicial equation and β is a constant.

We may classify the possible solutions of a general differential equation using the Frobenius method according to the roots of the corresponding indicial equation.

1. Two distinct roots, r_1 and r_2. The are two linearly independent solutions of Frobenius type.
2. A double root, $r_1 = r_2$. There is one power-type solution and one logarithmic-type solution, with $\beta = 1$.

3. Two distinct roots which differ by an integer, $r_1 - r_2 = N$. Either there are two Frobenius-type solutions, which may both be derived by use of the lower root, or the lower root does not provide a solution, in which case a logarithmic solution exists, with β to be determined.

Only practice (and much suffering) can help one decide how to best approach a given problem.

The solution of an ODE around an irregular singular point is much more difficult, and a series solution cannot in general be found.

14.10 Integral transform solutions

As we have seen in Chapters 11 and 12, Fourier transforms and Laplace transforms tend to have a derivative-removing action on functions; that is,

$$\mathcal{F}\left[f'(t)\right] = -i\omega F(\omega), \tag{14.145}$$

$$\mathcal{L}\left[f'(t)\right] = sF(s) - f(0). \tag{14.146}$$

This suggests another strategy for solving differential equations: by taking a Laplace transform of an ODE for $y(t)$, the equation can be reduced to an algebraic equation for the transformed function, $Y(s)$, which may then be inverted to find the solution. The advantage of this technique, as we will see, is that it naturally incorporates the initial conditions into the solution.

Example 14.11 We consider the Laplace transform of the harmonic oscillator equation,

$$m\frac{d^2x(t)}{dt^2} + kx(t) = 0, \tag{14.147}$$

with initial conditions

$$x(0) = x_0, \quad x'(0) = x_1. \tag{14.148}$$

The Laplace transform of this equation immediately takes on the form

$$m\left\{s\mathcal{L}[x'(t)] - x'(0)\right\} + kX(s) = 0, \tag{14.149}$$

where we have used Eq. (14.146) to reduce $x''(t)$ to $x'(t)$. Finishing the transformation by a second application of Eq. (14.146), the equation becomes

$$ms^2X(s) - msx(0) - mx'(0) + kX(s) = 0, \tag{14.150}$$

which has no derivatives at all anymore! We can solve this equation for $X(s)$, and also apply our initial conditions, which are automatically included in this formula,

$$X(s) = \frac{sx_0 + x_1}{s^2 + k/m}. \tag{14.151}$$

We then apply an inverse Laplace transform to this formula, which involves a contour integral where the contour encompasses the half-space $\text{Re}(s) < \sigma$, i.e.

$$x(t) = \frac{1}{2\pi} \int_{-\infty}^{\infty} X(s) e^{(\sigma+i\omega)t} d\omega. \tag{14.152}$$

The simple poles $s = \pm i\sqrt{k/m} \equiv i\alpha$ lie within this half-space, and the integral therefore has the value

$$x(t) = \left\{ \frac{-i\alpha x_0 + x_1}{-2i\alpha} e^{-i\alpha t} + \frac{i\alpha x_0 + x_1}{2i\alpha} e^{i\alpha t} \right\}. \tag{14.153}$$

Combining terms, we quickly find that

$$x(t) = x_0 \cos(\alpha t) + \frac{x_1}{\alpha} \sin(\alpha t). \tag{14.154}$$

Remarkably, we have found the solution to the harmonic oscillator equation without making any ansatz (guess) as to the form of the solution!

$\diamond$

This Laplace transform technique is typically only helpful for equations with constant coefficients. For equations with variable coefficients, the Laplace transform tends to replace one difficult equation with another. As a trivial example, consider the Laplace transform of the equation

$$f'(t) + t f(t) = 0. \tag{14.155}$$

The Laplace transform of this equation results in the following equation for $F(s)$:

$$sF(s) - \frac{d}{ds}F(s) = f(0), \tag{14.156}$$

which is essentially the same equation we started with, except it is now inhomogeneous!

Ordinary differential equations with constant coefficients are encountered quite often, however, in the theory of partial differential equations, and we will see that Laplace transforms can be very helpful in the solution of such problems.

14.11 Systems of differential equations

It is not uncommon to encounter systems of differential equations which are *coupled* together, i.e. each equation contains more than one dependent variable to solve for. The obvious example of this in the study of partial differential equations are Maxwell's equations, which couple the electric and magnetic fields together through their curl formulas,

$$\nabla \times \mathbf{E}(\mathbf{r},t) = -\frac{1}{c}\frac{d\mathbf{B}(\mathbf{r},t)}{dt}, \tag{14.157}$$

$$\nabla \times \mathbf{B}(\mathbf{r},t) = \frac{4\pi}{c}\mathbf{J}(\mathbf{r},t) + \frac{1}{c}\frac{d\mathbf{E}(\mathbf{r},t)}{dt}. \tag{14.158}$$

Both equations contain both $\mathbf{E}$ and $\mathbf{B}$, and neither can be solved individually. However, as noted in Section 2.10, in a source/current-free region one can combine these equations with the two divergence-based Maxwell's equations to produce individual wave equations for $\mathbf{E}$ and $\mathbf{B}$.

We noted in Section 14.3 that any higher-order ordinary differential equation can be converted into a system of coupled differential equations; in this context, systems of ordinary differential equations are ubiquitous. We will see that this conversion is the common strategy for numerically solving ODEs, in particular because numerical techniques for solving first-order ODEs are well-established.

Before looking at these techniques, we discuss the solution of systems of linear first-order ODEs with *constant* coefficients.

We consider a system of the form

$$\left|y'(t)\right\rangle = \mathbf{A}\left|y(t)\right\rangle, \tag{14.159}$$

where $|y(t)\rangle$ is a N-term vector of the form

$$|y(t)\rangle = \begin{bmatrix} y_1(t) \\ y_2(t) \\ \cdots \\ y_N(t) \end{bmatrix}, \tag{14.160}$$

and $\mathbf{A}$ is an $N \times N$ matrix whose coefficients a_{ij} are independent of t. Equation (14.159) represents a coupled system of ODEs which must be solved as a whole.

Let us suppose that the matrix $\mathbf{A}$ can be diagonalized. Let $\mathbf{B}$ be the matrix which converts $\mathbf{A}$ into its diagonal form $\tilde{\mathbf{A}}$. We substitute $\mathbf{B}^{-1}\mathbf{B} = \mathbf{I}$ into the equation three times, i.e.

$$\mathbf{B}^{-1}\mathbf{B}\left|y'(t)\right\rangle = \mathbf{B}^{-1}\mathbf{B}\mathbf{A}\mathbf{B}^{-1}\mathbf{B}\left|y(t)\right\rangle. \tag{14.161}$$

If we define $|z(t)\rangle = \mathbf{B}\,|y(t)\rangle$, we may multiply the above equation by $\mathbf{B}$ to find

$$\left|z'(t)\right\rangle = \tilde{\mathbf{A}}\left|z(t)\right\rangle. \tag{14.162}$$

It is important to note that the transformation of the left-hand side of the equation is only possible because $\mathbf{A}$, and hence $\mathbf{B}$, is a constant matrix. We then have $(\mathbf{B}\,|y\rangle)' = \mathbf{B}\,|y'\rangle$.

Because the matrix $\tilde{\mathbf{A}}$ is diagonal, i.e. $(\tilde{\mathbf{A}})_{ij} = a_i \delta_{ij}$, our series of equations reduce to a set of N separable equations,

$$z_i'(t) = a_i z_i(t). \tag{14.163}$$

The solution of each of these is readily found to be

$$z_i(t) = C_i \exp[a_i t], \tag{14.164}$$

where C_i is a constant determined by the initial conditions. Our solution for $|y(x)\rangle$ may be found by applying the transformation $\mathbf{B}^{-1}$ to the vector $|z(t)\rangle$,

$$|y(t)\rangle = \mathbf{B}^{-1} |y(t)\rangle. \tag{14.165}$$

It is to be noted that this elegant matrix technique no longer works if the matrix $\mathbf{A}$ is a function of t. In such a case, $(\mathbf{B}|y\rangle)' \neq \mathbf{B}|y'\rangle$ and the transformation begun with Eq. (14.161) will no longer produce a simple separated system of equations.

It is entirely possible to find a system of equations for which the matrix $\mathbf{A}$ cannot be diagonalized. The $N \times N$ matrix $\mathbf{A}$ therefore has fewer than N linearly independent eigenvectors. Such a case is analogous to that discussed at the end of Section 14.5, in which only a single solution of pure exponential form $\exp[at]$ can be found, and a second solution of the form $t \exp[at]$ must be considered. Similarly, more general solutions must be considered in the case of non-diagonal $\mathbf{A}$.

14.12 Numerical analysis of differential equations

We have seen that most solutions of systems of ordinary differential equations only apply when the system is *linear*; that is, the function $y(t)$ and its derivatives only appear in terms of the first power in the equation. Many interesting problems, however, require the analysis of nonlinear systems of equations, and for most of these equations the analytic methods described previously cannot be applied. When such problems are of interest, one must either turn to phase space to understand the qualitative behavior of the system or turn to numerical methods to investigate quantitative solutions.

Quite a large variety of techniques exist for numerically solving ordinary differential equations, most common among them the family of Runge–Kutta solutions. Detailed discussions of algorithmic implementation of these techniques can be found in numerous texts (see, for instance, [PTVF92]); we focus more on the explaining the motivation for some of the earliest and most common techniques. Further discussion along this line can be found in the excellent book by Iserles [Ise96].

We will consider numerical solutions of the familiar first-order ordinary differential equation

$$\frac{dy}{dt} = f(y, t), \tag{14.166}$$

subject to the initial condition $y(t_0) = y_0$. As we have noted previously, all Nth-order ODEs may be rewritten as a system of N coupled first-order ODEs, and the numerical solution of such higher-order ODEs is typically implemented as the solution of such a set of coupled first-order equations.

The goal of a numerical solution is to use the value of the function at a time $t = t_n$ to accurately determine the value after a finite time interval has passed, i.e. at a time $t_{n+1} = t_n + h$. This can be done formally by integrating Equation (14.166) over the interval $t_n < t < t_n + h$. The left-hand side of the equation becomes

$$\int_{t_n}^{t_n+h} \frac{dy}{dt}(t')dt' = y(t_n + h) - y(t_n), \tag{14.167}$$

and Eq. (14.166) may be written as

$$y(t_n + h) = y(t_n) + \int_{t_n}^{t_n+h} f[y(t'), t']dt'. \tag{14.168}$$

This expression, though exact, is not particularly helpful, since it depends upon a continuous range of values of $y(t)$ in the right-hand integrand. The crux of every computational method we will describe is applying approximations to the integral that allow us to calculate future values of $y(t)$ from previous, discrete, values.

There are a number of issues that arise in the construction of numerical approximations to Eq. (14.168). Perhaps the foremost is the issue of *convergence*: a numerical technique is valid only if, in the limit $h \to 0$, the numerical solution of the differential equation approaches the true solution. It is not difficult, for the methods to be considered here, to prove convergence, but the proofs are lengthy and somewhat irrelevant to the application of the techniques. We will only consider convergent methods, and will omit the proofs; if one is developing a new numerical method, however, demonstrating convergence should be the first priority.

Another clear concern is the error introduced in the computational solution of the differential equation. These are typically grouped into two classes: *round-off error* and *truncation error*. Round-off error is the error introduced by the finite precision of the computer used; though the specific nature of this error depends on the machine used, it is roughly inversely proportional to the step size h. Truncation error is the error introduced in using an approximate form of the integral in Eq. (14.168). This error is may be further subdivided into *local truncation error*, which is the error incurred in each finite step of the computation, and *global truncation error*, which is the accumulated truncation error in progressing through a finite number of steps. Local truncation error tends to accumulate as some power of step size, i.e. as h^{p+1}, and p is referred to as the *order* of the numerical method. The combination of round-off error and truncation error results in some optimum value for h which is small, but not too small.

Intuitively, it may seem that the concepts of convergence and order are somewhat redundant, in that a high-order numerical method should automatically be convergent; however, this intuition is wrong. Even a high-order method may fail to be convergent, and therefore useless. We emphasize that convergence for a numerical technique should always be confirmed, regardless of order.

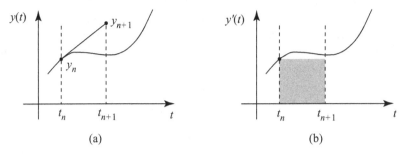

Figure 14.7 Illustrating the principle of Euler's method, which can be equivalently thought of as (a) extrapolating the tangent line, (b) approximating the integral by a rectangular area.

We now consider a few of the most illustrative and common techniques for numerically solving Eq. (14.166).

14.12.1 Euler's method

The simplest technique for numerically solving ODEs was first described by Leonhard Euler around 1768, some two hundred years before the first electronic computers were developed. In Euler's method, the integral in Eq. (14.168) is derived by replacing $f[y(t),t]$ by the lowest-order term in its Taylor expansion, i.e.

$$\int_{t_n}^{t_n+h} f[y(t'),t']dt' \approx \int_{t_n}^{t_n+h} f[y(t_n),t_n]dt' = hf[y(t_n),t_n]. \tag{14.169}$$

We may approximate the value of $y(t_{n+1})$ as

$$y(t_{n+1}) \approx y(t_n) + hf[y(t_n),t_n]. \tag{14.170}$$

Let us refer to the exact solution of a differential equation at time t_n by $y(t_n)$, and an approximate numerical solution at time t_n by y_n. Euler's method results in the following formula for the approximate value of y_{n+1}

$$y_{n+1} = y_n + hf[y_n,t_n]. \tag{14.171}$$

This formula may be interpreted in a number of ways. First, because $f[y(t_n),t_n] = y'(t_n)$, we may write Eq. (14.170) in the form

$$y(t_{n+1}) \approx y(t_n) + (t_{n+1} - t_n)y'(t_n). \tag{14.172}$$

Here $h = t_{n+1} - t_n$. This equation approximates the value of $y(t_{n+1})$ by extrapolating the tangent to the curve $y(t)$ at t_n to the point t_{n+1}. This is illustrated in Fig. 14.7(a); Euler's method is also referred to as the *tangent line method*.

Since Euler's method replaces the function $f[y(t), t]$ by a constant, it may be interpreted as approximating the integral of $y'(t)$ in Eq. (14.168) as a rectangular area, as illustrated in Fig. 14.7(b).

To determine the order of Euler's method, we note that the true solution at $y(t_{n+1})$ may be approximated by its Taylor series representation, i.e.

$$y(t_{n+1}) \approx y(t_n) + hf[y(t_n), t_n] + O(h^2), \tag{14.173}$$

where O represents an *order parameter* and $O(h^2)$ indicates that the remaining terms of the series have a quadratic or higher functional dependence on h. We introduced order parameters in our discussion of the divergence in Section 2.4. Assuming that $y(t_n) = y_n$, we can assess the truncation error $e(h)$ in a single step in Euler's method as

$$e(h) = y(t_{n+1}) - y_{n+1} = O(h^2). \tag{14.174}$$

For small values of h, the local truncation error of Euler's method will therefore be approximately proportional to h^2,

$$e(h) \approx Mh^2. \tag{14.175}$$

From the definition of the order of the method, we say that Euler's method is of order 1. We may estimate the global truncation error $E(h)$ by assuming that the local truncation error is the same for each step. Given N total steps, from $t = t_0$ to $t = t_N$, the global truncation error is given by

$$E(h) \approx Ne(h) = \frac{t_N - t_0}{h} e(h) = (t_N - t_0)Mh. \tag{14.176}$$

The global truncation error increases as h.

Example 14.12 As an example of the Euler method, we consider an ODE whose solution can be determined analytically,

$$y' = 1 - t^2 - y, \tag{14.177}$$

subject to the initial condition $y(0) = y_0$. The solution to this equation can be found by methods in Section 14.4.3 to be

$$y(t) = (y_0 + 1)e^{-t} - t^2 + 2t - 1. \tag{14.178}$$

We attempt to evaluate this expression using the Euler method for two different step sizes; the results are shown in Fig. 14.8. It can be seen that the numerical solution matches quite well for the smaller value of h.

◇

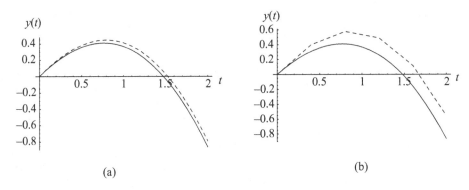

Figure 14.8 Applying Euler's method to Eq. (14.177), with $y(0) = 0$ and with (a) $h = 0.1$, (b) $h = 0.4$. Solid line is exact solution.

14.12.2 Trapezoidal method and implicit methods

Euler's method is known as an *explicit method*, in that the value y_{n+1} can be determined immediately from Eq. (14.171) without any additional mathematical steps. We next consider an *implicit method*, where the approximation to Eq. (14.168) results in general in a nonlinear equation for y_{n+1} which must be solved before computation.

Euler's method approximates the integrand in Eq. (14.171) by a constant function; a more sophisticated approximation is provided by the trapezoidal rule of calculus,

$$\int_a^b f(t)dt \approx (b-a)\frac{f(a)+f(b)}{2}. \tag{14.179}$$

The application of the trapezoidal rule approximates the integral as the area of the trapezoid formed using the endpoint values of the function, as illustrated in Fig. 14.9. This results in an approximate formula for $y(t_{n+1})$ of the form

$$y(t_{n+1}) \approx y(t_n) + \frac{h}{2}\{f[y(t_n),t_n]+f[y(t_{n+1}),t_{n+1}]\}, \tag{14.180}$$

and the comparable numerical relation,

$$y_{n+1} = y_n + \frac{h}{2}\{f[y_n,t_n]+f[y_{n+1},t_{n+1}]\}. \tag{14.181}$$

In this equation, however, y_{n+1} appears on *both* sides of the equation. In order to apply the trapezoidal method, one must solve this equation for y_{n+1}. It can be worth it, however, for the trapezoidal rule results in a higher-order numerical technique. We note that the Taylor expansion of $y(t_{n+1})$ may be written as

$$y(t_{n+1}) \approx y(t_n) + hy'(t_n) + \frac{h^2}{2}y''(t_n) + O(h^3). \tag{14.182}$$

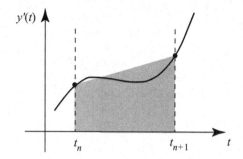

Figure 14.9 Illustration of the trapezoidal rule.

Furthermore, we note that $f[y(t_{n+1}), t_{n+1}] = y'(t_{n+1})$; we may then Taylor expand $y'(t_{n+1})$ to find that

$$f[y(t_{n+1}), t_{n+1}] \approx y'(t_n) + hy''(t_n) + O(h^2).\tag{14.183}$$

Putting this expression into the trapezoidal formula, Eq. (14.180), provides the approximate relation

$$y_{n+1} \approx y(t_n) + hy'(t_n) + \frac{h^2}{2}y''(t_n) + O(h^3).\tag{14.184}$$

We readily find that

$$y(t_{n+1}) - y_{n+1} = O(h^3).\tag{14.185}$$

Assuming the method is convergent, this suggests that the trapezoidal rule results in a numerical method of second order, an improvement over the Euler method.

Example 14.13 For comparison, we consider the ODE of the previous example,

$$y' = 1 - t^2 - y,\tag{14.186}$$

subject to the initial condition $y(0) = y_0$. This equation leads to a trapezoidal function of the form

$$y_{n+1} = y_n + \frac{h}{2}\left[1 - t_n^2 - y_n + 1 - t_{n+1}^2 - y_{n+1}\right].\tag{14.187}$$

Since this is a linear formula in y_{n+1}, however, we can readily solve for it,

$$y_{n+1} = \frac{1}{1 + h/2}\left[(1 - h/2)y_n + h(2 - t_n^2 - t_{n+1}^2)/2\right].\tag{14.188}$$

The trapezoid method is applied in Fig. 14.10. It can be seen that it matches the correct result much more closely than the Euler method.

◇

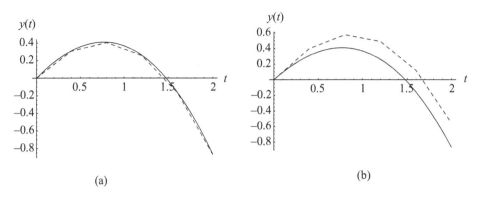

Figure 14.10 Illustration of the (a) the trapezoid method and (b) the Euler method, with $h = 0.4$. The dashed lines indicate the numerical result, the solid lines the exact analytic result.

The trapezoidal method is an implicit method, but it can also be made an explicit one by approximating y_{n+1} on the right-hand side of Eq. (14.181) with the result from the Euler method, i.e.

$$y_{n+1} = y_n + \frac{h}{2} \{f[y_n, t_n] + f[y_n + hf(y_n, t_n), t_{n+1}]\}. \tag{14.189}$$

One can show with a little work that this method is also of order 2; it is known as the *improved Euler method* or *Heun formula*. In essence, the Heun formula attempts to determine a better approximation for the function $f[y, t]$ by calculating the average of its value at the beginning of the interval and the approximate value at the end of the interval. This philosophy can be extended to derive approximations of even higher order, and this forms the basis for the Runge–Kutta methods described in the next section.

14.12.3 Runge–Kutta methods

As we have seen, the solution of ODEs is essentially an exercise in numerical integration. Euler's formula approximates the integrand $f[y, t]$ as a constant function in t, while the trapezoidal formula approximates the integral as a linear function in t. To continue the process, one typically applies the idea of *Gaussian quadratures* to evaluate the integral.[2] The basic idea is as follows: suppose we have an integral of the form

$$I \equiv \int_a^b w(t)f(t)\mathrm{d}t, \tag{14.190}$$

---

[2] The "Gaussian" here refers to the person, Carl Friedrich Gauss (1777–1855), who first developed the technique in 1814, not to the function which also bears his name.

where $f(t)$ is a general function and $w(t)$ is a *weight function* which we have some freedom in choosing. We approximate this integral by the weighted sum

$$I \approx \sum_{j=1}^{N} w_j f(t_j),$$

(14.191)

where the w_j are the *weights* and t_j the *abscissas* of the quadrature. By an appropriate choice of these constants it can be shown that Eq. (14.191) can be made *exact* for all functions $f(t)$ which are polynomials of some order less than p. It is to be noted that p may be greater than N, and in fact may be equal to $2N$. That is, it is possible to choose the weights and abscissas such that the quadrature formula (14.191) is exact for polynomials of the form

$$f(t) = \sum_{j=0}^{2N-1} \alpha_j t^j.$$

(14.192)

Example 14.14 This argument may seem rather hard to believe at first, so we construct a simple example. Suppose we wish to construct a quadrature which is exact for the integral

$$J = \int_0^1 f(t)dt,$$

(14.193)

when $f(t)$ is a polynomial of the form

$$f(t) = a + bx + cx^2 + dx^3.$$

(14.194)

The exact value of the integral is

$$J = a + \frac{b}{2} + \frac{c}{3} + \frac{d}{4}.$$

(14.195)

We have a polynomial of order 3, which suggests we should be able to construct an exact quadrature with $N = 2$. We assume a pair of weights (w_1, w_2) and a pair of abscissas (t_1, t_2), where we assume $t_1 < t_2$. Our quadrature formula is of the form

$$Q = w_1 f(t_1) + w_2 f(t_2).$$

(14.196)

We now try and determine the weights and abscissas to make the integral for a polynomial of the form of Eq. (14.194) exact. We find that

$$Q = (w_1 + w_2)a + (w_1 t_1 + w_2 t_2)b + (w_1 t_1^2 + w_2 t_2^2)c + (w_1 t_1^3 + w_2 t_2^3)d.$$

(14.197)

Equating this to Eq. (14.194), we find the system of equations

$$w_1 + w_2 = 1, \tag{14.198}$$

$$w_1 t_1 + w_2 t_2 = \frac{1}{2}, \tag{14.199}$$

$$w_1 t_1^2 + w_2 t_2^2 = \frac{1}{3}, \tag{14.200}$$

$$w_1 t_1^3 + w_2 t_2^3 = \frac{1}{4}. \tag{14.201}$$

This is a set of four equations with four unknowns. It can be solved, and one finds that

$$w_1 = \frac{1}{2}, \quad w_2 = \frac{1}{2}, \quad t_1 = \frac{1}{2} - \frac{\sqrt{3}}{6}, \quad t_2 = \frac{1}{2} + \frac{\sqrt{3}}{6}. \tag{14.202}$$

We have therefore represented the integral of a polynomial of order 3 by a quadrature formula of two terms. It is to be noted that the abscissas lie within the region of integration but are not equally spaced within it.

◇

It is typically convenient to choose orthogonal polynomials such as the Legendre polynomials or the Hermite polynomials as the basis for a quadrature formula. One motivation for doing this is that it can be shown that the abscissas for a quadrature of the form of Eq. (14.191) will be the zeros of the $N + 1$th-order polynomial. We will discuss such orthogonal polynomials in Chapter 18.

We now apply these ideas to the solution of the equation (14.168). Assuming an N-term quadrature formula, we may write

$$y(t_n + h) = y(t_n) + \sum_{j=1}^{N} w_j f[y(\tilde{t}_j), \tilde{t}_j], \tag{14.203}$$

where $t_n \leq \tilde{t}_j \leq t_n + h$ are the abscissas of our chosen quadrature. This formula is problematic, however, because it depends upon the values of the function $y(t)$ at multiple points throughout the integration interval. With $N = 1$, we have at best an implicit method (in fact, with the appropriate choice of weights and abscissas we have the trapezoidal method); for $N > 1$, we have a single equation and multiple unknowns.

The solution to this conundrum is to form a series of successive approximations to the function $y(t)$. At each step of approximation, we calculate a new estimate for $y'(t)$ based on values of the function $f[y, t]$ throughout the interval. In general, an N-step method may be written in a pseudo-matrix form as

$$|\mathbf{Y}_{n+1}\rangle = h\mathbf{A}\,|\mathbf{y}_n\rangle, \tag{14.204}$$

where

$$
|\mathbf{y}_n\rangle \equiv
\begin{bmatrix}
y_n \\
f(\xi_1, \tilde{t}_1) \\
\vdots \\
f(\xi_{N-1}, \tilde{t}_{N-1}) \\
f(\xi_N, \tilde{t}_N)
\end{bmatrix},
\quad
|\mathbf{Y}_{n+1}\rangle \equiv
\begin{bmatrix}
\xi_1 \\
\xi_2 \\
\vdots \\
\xi_N \\
y_{n+1}
\end{bmatrix},
\tag{14.205}
$$

and the matrix $\mathbf{A}$ has elements a_{ij}, for which $a_{1j} = \delta_{1j}$ and $a_{Nj} = w_j$ are the weights of the quadrature. This matrix form is used to organize the general terms of the approximation; the final element in the right-most vector is given by

$$
y_{n+1} = y_n + h \sum_{i=1}^{N} w_i f(\xi_i, \tilde{t}_i),
\tag{14.206}
$$

The other elements a_{ij} of the matrix $\mathbf{A}$ can be chosen somewhat independently, with the minimum constraint that they must be chosen to make the approximation agree with the exact result to as high an order as possible. Equation (14.204) form the basis of a general *Runge–Kutta method*.

The determination of the matrix $\mathbf{A}$, and the calculation of the order of a particular method, is too involved to derive here. We note, however, that the matrix $\mathbf{A}$ will be lower triangular ($a_{ij} = 0$ for $j \geq i$) for explicit Runge–Kutta methods. For an explicit method, each successive element of $|\mathbf{Y}_{n+1}\rangle$ is determined in turn, and the values of ξ_i which result are used in determining the next element of $|\mathbf{y}_n\rangle$. Such a method allows the derivation of ξ_i only from values ξ_j with $j < i$. Implicit methods may also be used, for which the matrix $\mathbf{A}$ is not restricted to lower triangular. In such a case, however, the series of equations given by Eq. (14.204) must be solved explicitly for each of the individual ξ_i.

The most common Runge–Kutta method described is typically written in the form

$$
k_1 = hf(y_n, t_n),
$$
$$
k_2 = hf(y_n + k_1/2, t_n + h/2),
$$
$$
k_3 = hf(y_n + k_2/2, t_n + h/2),
$$
$$
k_4 = hf(y_n + k_3, t_n + h),
$$
$$
y_{n+1} = y_n + \frac{k_1}{6} + \frac{k_2}{3} + \frac{k_3}{3} + \frac{k_4}{6} + O(h^5).
\tag{14.207}
$$

14.12.4 Higher-order equations and stiff equations

Each of the methods described previously can be applied to a higher-order differential equation by breaking it into a system of equations. If we consider a general equation of the form

$$y'' = f(y, y', t),$$ (14.208)

we define $z \equiv y'$ to get the system

$$y' = z,$$ (14.209)

$$z' = f(y, z, t).$$ (14.210)

It is straightforward to implement explicit numerical methods to such systems of equations by interleaving the steps used to calculate the different dependent variables. For instance, let us consider the Runge–Kutta method applied to the following system of two equations

$$y' = f(y, z, t),$$ (14.211)

$$z' = g(y, z, t).$$ (14.212)

The Runge–Kutta method could be applied as

$$k_1 = hf(y_n, z_n, t_n),$$

$$l_1 = hg(y_n, z_n, t_n),$$

$$k_2 = hf(y_n + k_1/2, z_n + l_1/2, t_n + h/2),$$

$$l_2 = hg(y_n + k_1/2, z_n + l_1/2, t_n + h/2),$$

$$k_3 = hf(y_n + k_2/2, z_n + l_2/2, t_n + h/2),$$

$$l_3 = hg(y_n + k_2/2, z_n + l_2/2, t_n + h/2),$$

$$k_4 = hf(y_n + k_3, z_n + l_3, t_n + h),$$

$$l_4 = hg(y_n + k_3, z_n + l_3, t_n + h),$$

$$y_{n+1} = y_n + \frac{k_1}{6} + \frac{k_2}{3} + \frac{k_3}{3} + \frac{k_4}{6},$$

$$z_{n+1} = z_n + \frac{l_1}{6} + \frac{l_2}{3} + \frac{l_3}{3} + \frac{l_4}{6}.$$ (14.213)

Example 14.15 We briefly demonstrate the application of the method to the harmonic oscillator,

$$y' = z,$$ (14.214)

$$z' = -k^2 y.$$ (14.215)

Figure 14.11 shows the result of a numerical simulation with $k = 1$. It can be seen that the method matches the exact solution very closely for step size $h = 0.1$. One needs to go to a very large step size, $h = 1$, in order to see significant deviations; this step size, however,

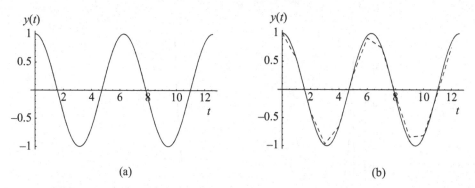

Figure 14.11 Illustration of the Runge–Kutta method for the harmonic oscillator with $k = 1$, and step size (a) $h = 0.1$, (b) $h = 1$.

is comparable to the period of oscillation, and it is not surprising that the method breaks down under such a circumstance.

◇

Another short example, however, illustrates a problem which can arise in the numerical solution of higher-order equations.

Example 14.16 The equations we consider are functionally very similar to those of the harmonic oscillator,

$$y' = z, \tag{14.216}$$

$$z' = +k^2 y, \tag{14.217}$$

the only difference being the change in sign for the z' equation. The general solution to this equation, with initial conditions $y(0) = y_0$ and $y'(0) = y_1$, is given by

$$y(t) = \frac{ky_0 - y_1}{2k} \exp[-kt] + \frac{ky_0 - y_1}{2k} \exp[kt]. \tag{14.218}$$

With $k = 5$, and $y_0 = 1$, $y_1 = -5$, the solution has the form

$$y(t) = \exp[-5t]. \tag{14.219}$$

Figure 14.12 illustrates the numerical solution. Though at first the solution matches very well the expected analytic solution, somewhere around $t = 7$ the solution becomes exponentially increasing, and is obviously wrong. What has happened?

◇

The second-order equation in the example has two independent solutions, one exponentially growing, one exponentially decaying. Even though we chose the initial conditions to

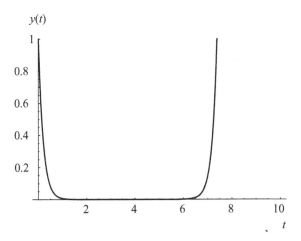

Figure 14.12 Catastrophic failure of the Runge–Kutta method for a stiff equation, with $h = 0.05$.

exclude the growing solution, it is built into the differential equations and always has the possibility to "leak" into our solution, especially once errors accumulate. Equations which have this built in instability are known as *stiff equations*. Such instability can occur even if both solutions of the equation are exponentially decaying; when one decay constant is much stronger than another, stiffness is likely.

One solution to the stiffness problem is to use *implicit* methods, i.e. the trapezoidal rule or an implicit Runge–Kutta solution. It can be shown that such implicit methods are stable and therefore resistant to stiffness. Another option is to use so-called *backwards differentiation formulas*.

14.12.5 Observations

This section dealt with some basic concepts involved in the numerical solution of differential equations. It is important to note, however, that both the theory and application of such methods is much more involved than can be covered in so brief a discussion.

Numerous other techniques for solving ODEs exist, each of which has its strengths and weaknesses. Where all of the techniques described here iterate a first-order ODE by using a single past value of the function, it is also possible to derive *multistep methods* which use multiple prior values of the function to determine a future value. Even the application of Runge–Kutta can be, and should be, improved by the use of *adaptive step-size techniques*, in which the value of h is varied depending on the local behavior of the function derivative.

See [PTVF92] for a detailed discussion of the algorithmic implementation of various methods. [But08] is a standard text on all aspects of numerical solution. Finally, [Ise96] gives more of the mathematician's perspective on the stability and accuracy of numerical techniques.

14.13 Additional reading

- W. E. Boyce and R. C. DiPrima, *Elementary Differential Equations and Boundary Value Problems* (New York, Wiley, 1992, 5th edition). A thorough and readable undergraduate textbook on solving ODEs.
- E. L. Ince, *Ordinary Differential Equations* (London, Longmans, Green and Co., 1926). A detailed and rigorous monograph on the subject of ODEs, and a classic in the field. Also available in Dover.
- A. Iserles, *A First Course in the Numerical Analysis of Differential Equations* (Cambridge, Cambridge University Press, 1996). A great and readable introduction to the mathematical theory behind numerical techniques.
- J. C. Butcher, *Numerical Methods for Ordinary Differential Equations* (Chichester, Wiley, 2008, 2nd edition). The standard textbook on numerical methods; very thorough.

14.14 Exercises

1. Classify the following ordinary differential equations, describing their order, if they are linear/nonlinear, and homogeneous/inhomogeneous.

$$\text{(a) } y'' + 4xy = \sin x, \text{ (b) } y'''' + y''' = 0, \text{ (c) } y'' + x^2 \sin y = 0.$$

2. Classify the following ordinary differential equations, describing their order, if they are linear/nonlinear, and homogeneous/inhomogeneous.

$$\text{(a) } y' + \sin x \sin y = \cos y - 1, \text{ (b) } y''' = e^x, \text{ (c) } y'' + 4y' + x^2 y = 0.$$

3. For the following system of equations, determine the location, type and stability of all critical points:

$$y_1' = y_1 + y_1 y_2, \quad y_2' = y_2 + y_1^2.$$

4. For the following system of equations, determine the location, type and stability of all critical points:

$$y_1' = 1 + y + 1, \quad y_2' = y_1^2 - y_2^2.$$

5. Find solutions for the following first order differential equations by any appropriate technique.
   1. $y' = \dfrac{2x}{y + x^2 y}, y(0) = -1,$
   2. $y' = \dfrac{2x - y}{x - 2y}, y(1) = 4,$
   3. $y' = y + xe^{2x}, y(0) = 2.$

6. Find solutions for the following first order differential equations by any appropriate technique.
   1. $(2x + 4y) + (2x - 2)y' = 0, y(0) = 2,$
   2. $xy' + 2y = e^{-x}, y(1) = 1,$
   3. $y' = \dfrac{2x}{1 + 2y}, y(2) = 1.$

7. Find a second linear independent solution to each of the following equations using the Wronskian and the given solution.
   1. $x^2 y'' + xy' + (x^2 - 1/4)y = 0$, $y_1 = \cos x/\sqrt{x}$,
   2. $x^2 y'' - 3xy' + 4y = 0$, $y_1 = x^2$.
8. Find a second linear independent solution to each of the following equations using the Wronskian and the given solution.
   1. $(1-x)y'' + xy' - y = 0$, $y_1 = x$,
   2. $x^2 y'' - 2y = 0$, $y_1 = x^2$.
9. Find a second solution to the homogeneous equation and solve the inhomogeneous equation using variation of parameters,

$$x^2 y'' - 2y = 3x^2 - 1,$$

with $y_1 = x^2$, $y(1) = 1$ and $y'(1) = 2$.
10. Find a second solution to the homogeneous equation and solve the inhomogeneous equation using variation of parameters,

$$xy'' - (1+x)y' + y = x^2 e^{2x},$$

with $y_1 = 1+x$, $y(1) = 0$ and $y'(1) = 4$.
11. Find a pair of linearly independent series solutions to the following equation:

$$y'' - xy' - y = 0,$$

around the point $x = 0$.
12. Find a pair of linearly independent series solutions to the following equation:

$$xy'' + y' + xy = 0,$$

around the point $x = 1$.
13. Find a pair of linearly independent series solutions to the following equation:

$$x^2 y'' + 2xy' + xy = 0,$$

around the point $x = 0$.
14. Find a pair of linearly independent series solutions to the following equation:

$$x^2 y'' + 4xy' + (2+x)y = 0,$$

around the point $x = 0$.
15. Determine the solution to the following equation using the Laplace transform:

$$y'' + 2y' + 2y = f(x),$$

with $y(0) = 0$, $y'(0) = 0$, and

$$f(x) = \begin{cases} 1, & \pi \le x \le 2\pi, \\ 0, & \text{otherwise.} \end{cases}$$

16. Determine the solution to the following equation using the Laplace transform:

$$y'' + y = f(x),$$

with $y(0) = 1$, $y'(0) = 0$, and

$$f(x) = \begin{cases} x, & 0 \leq x \leq 1, \\ 1, & t > 1. \end{cases}$$

17. Write the following second-order equations as a system of equations and find the solution using matrix methods.
    1. $y'' + 4y' + 4y = 0$, $y'(0) = 1$, $y(0) = 0$,
    2. $3y'' + y' - 2y = 0$, $y'(0) = 0$, $y(0) = 1$.
18. Write the following second-order equations as a system of equations and find the solution using matrix methods.
    1. $y'' - y = 0$, $y'(0) = 2$, $y(0) = 0$,
    2. $4y'' + y' - 2y = 0$, $y'(0) = 1$, $y(0) = 1$.
19. Read the paper by G. L. Lippie *et al.*, Phase space techniques for steering laser transients, *J. Opt. B* **2** (2000), 375–381. Describe the problem in laser physics that is being addressed and discuss how phase space techniques are used to solve the problem.
20. Solve the following second-order ODE:

$$y'' + 2y' - y = 0, \quad y(0) = 1, \quad y'(0) = 1.$$

21. Solve the following second-order ODE:

$$4y'' + y' + 4y = 0, \quad y(0) = -1, \quad y'(0) = 2.$$

22. We have seen that a second-order linear ODE has two linearly independent solutions; conversely, given two linearly independent functions it is possible to determine the second-order equation that they satisfy. Given the function

$$y(x) = c_1 x + c_2 x^2,$$

determine $y'(x)$ and $y''(x)$ and eliminate the arbitrary constants c_1 and c_2 to determine the second-order ODE that $y(x)$ satisfies.
23. It is often possible to directly solve a second-order ODE if it is completely independent of y, i.e.

$$y'' = f(x, y').$$

By defining $v \equiv y'$, we may convert the equation to a first-order equation of the form $v' = f(x, v)$ which may be solved by the methods described in this chapter. Find a general solution of the equation

$$y'' + x(y')^2 = 0.$$

24. It is often possible to solve a second-order ODE if it is completely independent of x, i.e.

$$y'' = f(y, y').$$

Let us define $v = y'$, so that we have the equation $v' = f(y,v)$. If we now treat y as the *independent* variable, we may use the chain rule to write

$$\frac{dv}{dx} = \frac{dv}{dy}\frac{dy}{dx} = v\frac{dv}{dy},$$

so that we need to solve the equation $v\,dv/dy = f(y,v)$. Use this technique to find a general solution to the equation

$$y'' + y(y')^3 = 0.$$

25. Certain second-order ODEs may also be considered *exact* in a similar manner to first-order ODEs. We say that the equation

$$Py'' + Qy' + Ry = 0$$

is exact if it may be rewritten in the form

$$(Py')' + [fy]' = 0,$$

where $f(x)$ is a function that depends on Q and R. In the case that an equation is exact, it may be integrated directly once to find a first-order ODE which may then be solved.

1. Show that a necessary condition for a second-order equation to be exact is that $P'' - Q' + R = 0$. (It can also be shown that this condition is sufficient.)
2. Find a general solution for the equation $x^2 y'' + xy' - y = 0$.

15

Partial differential equations

15.1 Introduction: propagation in a rectangular waveguide

An important class of physical problems involves of the guiding of waves, primarily electromagnetic waves, in a transversely confined space. Such waveguiding of light by optical fibers forms the basis of modern telecommunications, while metallic waveguides are used in free space microwave communications systems. A proper description of the guiding of electromagnetic waves in such systems would require the solution of Maxwell's equations; for simplicity we restrict ourselves to the solution of the Helmholtz equation,

$$\nabla^2 U(x,y,z) + k^2 U(x,y,z) = 0, \tag{15.1}$$

which describes the propagation of monochromatic scalar waves in three-dimensional space. The Helmholtz equation is a *partial differential equation* (PDE), which depends upon the partial derivatives of the function $U(x,y,z)$ in all three Cartesian coordinates via the Laplacian operator. We deal here with a partial solution of the Helmholtz equation applied to the problem of waveguiding, which introduces the basic concepts involved in PDEs.

Let us consider a wave propagating in a rectangular waveguide which is infinite in the z-direction, and for which waves are confined transversely to the region $0 \leq x \leq a$, $0 \leq y \leq b$. On the boundary of the waveguide, the wave amplitude is assumed to vanish, i.e. $U(0,y,z) = U(a,y,z) = 0$, and so on. The basic geometry is illustrated in Fig. 15.1.

To solve this equation, we take a hint from the rectangular geometry of the wave system, and look for solutions which are separable into pure functions of x, y, and z, i.e.

$$U(x,y,z) = X(x)Y(y)Z(z). \tag{15.2}$$

This assumption is the first part of the technique known as *separation of variables*, to be discussed in detail in this chapter. If we substitute back into Eq. (15.1), and divide by XYZ, the equation becomes

$$\frac{1}{X}\frac{d^2 X}{dx^2} + \frac{1}{Y}\frac{d^2 Y}{dy^2} + \frac{1}{Z}\frac{d^2 Z}{dz^2} + k^2 = 0. \tag{15.3}$$

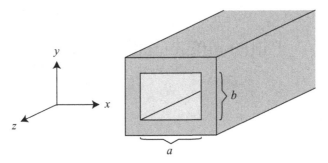

Figure 15.1 Illustrating the geometry related to waveguide propagation.

Excepting the constant, each term of this equation is only dependent on one variable. If we were to vary x, for instance, only the first part of the equation would change. Because the equation must hold for some continuous range of (x,y,z) values, each term of the equation must therefore be a constant. We may express this conclusion in the form of three quasi-independent equations,

$$\frac{1}{X}\frac{d^2X}{dx^2} = -L^2, \quad \frac{1}{Y}\frac{d^2Y}{dy^2} = -M^2, \quad \frac{1}{Z}\frac{d^2Z}{dz^2} = -N^2, \tag{15.4}$$

where L, M, and N are related by the condition

$$L^2 + M^2 + N^2 = k^2. \tag{15.5}$$

Equations (15.4) form a set of three *ordinary* differential equations, and each member of the set is simply the harmonic oscillator equation, i.e.

$$\frac{\partial^2 X}{\partial x^2} + L^2 X = 0. \tag{15.6}$$

A solution to the Helmholtz equation of the form of Eq. (15.2), characterized by the values of L, M, and N, may therefore be written as a product of sines and cosines or, equivalently, complex exponentials. For reasons which will become clear, we use complex exponentials to describe only the z-dependence,

$$U_{LMN}(x,y,z) = [A_L \sin(Lx) + B_L \cos(Lx)][[C_M \sin(My) + D_M \cos(My)]$$
$$\left[E_N e^{iNz} + F_N e^{-iNz} \right]. \tag{15.7}$$

The constants A_L, B_L, and so on are unspecified constants from the solution of the trio of second-order ODEs. Additional constraints on these constants may be determined using the conditions imposed on the boundary, known as *boundary conditions*. The condition that $U(0,y,z) = 0$ results in the requirement that $B_L = 0$, while the condition that $U(x,0,z) = 0$ results in the requirement that $D_M = 0$.

The conditions $U(a,y,z) = 0$ and $U(x,b,z) = 0$ are slightly trickier. The condition $U(a,y,z) = 0$ will only be satisfied if

$$\sin(La) = 0, \tag{15.8}$$

and this will hold only for $L = l\pi/a$, where l is a positive integer. Likewise, the condition $U(x,b,z) = 0$ leads to the constraint

$$\sin(Mb) = 0, \tag{15.9}$$

which will hold only for $M = m\pi/b$, where m is a positive integer. Consolidating the remaining constants together, we are left with a solution of the form

$$U_{lm}(x,y,z) = \sin(l\pi x/a)\sin(m\pi y/a)\left[E_{lm}e^{ik_z z} + F_{lm}e^{-ik_z z}\right], \tag{15.10}$$

where k_z is taken to be of the form

$$k_z = \pi\sqrt{k^2 - l^2/a^2 - m^2/b^2}, \tag{15.11}$$

and the remaining constants E_{lm} and F_{lm} depend upon the specific choice of l, m.

What about the z-component of the field? Assuming the wave is a harmonic wave with time dependence $\exp[-i\omega t]$, a plane wave traveling in the positive z-direction will be of the form

$$U(z,t) = e^{ik_z z - i\omega t}. \tag{15.12}$$

If we restrict our solution to only include positive-traveling waves, we must choose $F_N = 0$. Our (incomplete) solution to the partial differential equation, subject to the boundary conditions, is of the form

$$U_{lm}(x,y,z) = A_{lm}\sin(l\pi x/a)\sin(m\pi y/a)e^{ik_z z}. \tag{15.13}$$

Such a solution is usually referred to as a *mode* of the waveguide. The solution is not yet complete, however, because an arbitrary wave propagating in a waveguide may consist of a linear combination of modes of all values of l, m.

Even this partial solution has value, however. We note that if

$$l^2/a^2 + m^2/b^2 > k^2, \tag{15.14}$$

k_z can be written in the form

$$k_z = i\pi\sqrt{l^2/a^2 + m^2/b^2 - k^2}. \tag{15.15}$$

On substitution from this expression into Eq. (15.11), we see that propagation in the z-direction is then exponentially damped. Such modes are *nonpropagating*; that is, they dissipate away to nothing along a very short distance in the waveguide. A single mode waveguide is one for which only the $l = 1$, $m = 1$ mode is propagating.

To specify completely the solution to the Helmholtz equation, we would also need to specify an additional *boundary condition* – namely the value of the wavefield in some initial transverse plane of constant z. It is to be noted that often the general behavior as described above is more useful than the determination of a specific solution which meets the boundary conditions.

In this chapter we discuss the solution of a variety of partial differential equations. An important aspect of such solutions is an understanding of what types of auxiliary conditions are needed to determine a unique solution to a given differential equation.

15.2 Classification of second-order linear PDEs

In elementary physics courses, one typically encounters three partial differential equations of fundamental significance, each of which can be used to model a variety of physical systems. The first of these is Laplace's equation, which in two independent variables has the form

$$\nabla^2 \phi(x,y) = \frac{\partial^2 \phi}{\partial x^2} + \frac{\partial^2 \phi}{\partial y^2} = 0. \tag{15.16}$$

Laplace's equation can be used to describe the electric potential formed by a static collection of charges, as well as the steady-state distribution of heat in an object such as a heated sheet of metal.

The second archetypical equation is the diffusion equation, which in a single spatial variable has the form

$$\frac{\partial^2}{\partial x^2} \psi(x,t) - \frac{1}{a^2} \frac{\partial \psi}{\partial t} = 0. \tag{15.17}$$

The diffusion equation can be used to describe the diffusion of gas, heat, or even light throughout a system.

The third archetypical equation is the wave equation, which in a single spatial variable is written as

$$\frac{\partial^2}{\partial x^2} \psi(x,t) - \frac{1}{c^2} \frac{\partial^2 \psi}{\partial t^2} = 0. \tag{15.18}$$

The wave equation can be used to describe the propagation of acoustic waves, water waves, or electromagnetic waves.

It can be shown rigorously that these three equations encapsulate the qualitative behavior of *all* linear, second-order partial differential equations; that is, every linear second-order partial differential equation has "wave", "potential", or "diffusive" behavior in different domains. We first demonstrate this for a restricted class of PDEs, and then discuss the general case.

15.2.1 Classification of second-order PDEs in two variables with constant coefficients

We consider a linear second-order partial differential equation in two variables of the form

$$A\frac{\partial^2 u}{\partial t^2} + B\frac{\partial^2 u}{\partial x \partial t} + C\frac{\partial^2 u}{\partial x^2} = 0, \tag{15.19}$$

where A, B, and C are given constants. This equation is in general difficult to solve, but suppose there is a new set of coordinates ξ and η such that the equation can be rewritten in the form

$$\frac{\partial^2 u}{\partial \xi \partial \eta} = 0. \tag{15.20}$$

In this situation, the solution of the equation is trivial, because the equation can be integrated directly, with a solution given by

$$u(\xi, \eta) = F(\xi) + G(\eta), \tag{15.21}$$

with F and G arbitrary functions.

Let us try and find such a set of coordinates by making the substitution

$$\xi = \alpha x + \beta t, \quad \eta = \gamma x + \delta t. \tag{15.22}$$

We can employ the chain rule to express derivatives of the new coordinates in terms of the old; for instance,

$$\frac{\partial}{\partial x} = \frac{\partial \xi}{\partial x}\frac{\partial}{\partial \xi} + \frac{\partial \eta}{\partial x}\frac{\partial}{\partial \eta} = \alpha\frac{\partial}{\partial \xi} + \gamma\frac{\partial}{\partial \eta}. \tag{15.23}$$

Applying these relations to our original differential equation (15.19), we find that our equation takes on the following form in the new coordinate system:

$$(A\beta^2 + B\alpha\beta + C\alpha^2)\frac{\partial^2 u}{\partial \xi^2} + [2A\beta\delta + B(\alpha\delta + \beta\gamma) + 2C\alpha\gamma]\frac{\partial^2 u}{\partial \xi \partial \eta}$$

$$+ (A\delta^2 + B\gamma\delta + C\gamma^2)\frac{\partial^2 u}{\partial \eta^2} = 0. \tag{15.24}$$

For this equation to take on the form of Eq. (15.20), we require

$$A\beta^2 + B\alpha\beta + C\alpha^2 = 0, \quad A\delta^2 + B\gamma\delta + C\gamma^2 = 0. \tag{15.25}$$

There are many solutions to this problem, since we have only two equations but four unknowns. Two of the coefficients are simply scaling parameters, which do not effect the dynamics of the coordinates. Let us divide the first equation by α^2 and the second equation by γ^2; then we end up with two quadratic equations for the ratios of β/α and δ/γ,

$$A(\beta/\alpha)^2 + B(\beta/\alpha) + C = 0, \quad A(\delta/\gamma)^2 + B(\delta/\gamma) + C = 0. \tag{15.26}$$

The solutions to these quadratic equations are simply

$$\frac{\beta}{\alpha} = \frac{1}{2A}\left[-B \pm \sqrt{B^2 - 4AC}\right], \tag{15.27}$$

$$\frac{\delta}{\gamma} = \frac{1}{2A}\left[-B \pm \sqrt{B^2 - 4AC}\right]. \tag{15.28}$$

The choice of coordinate transformation depends, as one would expect, only upon the coefficients of the differential equation.

A few important observations may be made about the solutions. If we want our new coordinates ξ and η to be *different* from each other, the ratios β/α and δ/γ must be different. That means we must take the positive root in one of the formulas of Eq. (15.28) and the negative root in the other. Our ability to do so depends crucially on the behavior of the discriminant $B^2 - 4AC$, and there are three distinct cases:

1. $B^2 - 4AC > 0$. When this inequality is satisfied, we have two distinct *real* roots to the quadratic equations. The differential equation is said to be *hyperbolic*. If we make the coordinate transformation

$$\xi = 2Ax + [-B + \sqrt{B^2 - 4AC}]t, \tag{15.29}$$

$$\eta = 2Ax + [-B - \sqrt{B^2 - 4AC}]t, \tag{15.30}$$

we can reduce Eq. (15.19) to the form

$$\frac{\partial^2 u}{\partial \xi \partial \eta} = 0, \tag{15.31}$$

which is the form we desired. Equation (15.31) is referred to as the *canonical form* of hyperbolic equations.

There is another, more familiar, canonical form of a hyperbolic equation, which may be found by making a further transformation of Eq. (15.31) with the variables, $\mu \equiv \xi + \eta$, $\nu \equiv \xi - \eta$. The new canonical form is

$$\frac{\partial^2 u}{\partial \mu^2} - \frac{\partial^2 u}{\partial \nu^2} = 0. \tag{15.32}$$

An example of such an equation is the wave equation,

$$\frac{\partial^2 \psi}{\partial x^2} - \frac{1}{c^2}\frac{\partial^2 \psi}{\partial t^2} = 0, \tag{15.33}$$

which has $A = 1$, $B = 0$, $C = -1/c^2$. Thus $B^2 - 4AC = 4/c^2 > 0$. One can choose coordinates $\xi = x - ct$ and $\eta = x + ct$, which represent waves traveling in the positive and negative x directions with speed c.

2. $B^2 - 4AC = 0$. In this case, there is only one root to Eqs. (15.27) and (15.28). The differential equation is said to be *parabolic*. To see the effect of this, let us choose

$$\beta/\alpha = -\frac{B}{2A}.$$

(15.34)

If we substitute this into equation (15.24), we find that our differential equation reduces to the form

$$\frac{\partial^2 u}{\partial \eta^2} = 0,$$

(15.35)

where we have also used the fact that $B/2A = 2C/B$. We *cannot* reduce our differential equation to the form of Eq. (15.20), only to the form of Eq. (15.35), which is the canonical form of parabolic equations. Also, our choice of coordinate η is arbitrary; if we choose

$$\xi = 2Ax - Bt,$$

(15.36)

$$\eta = t,$$

(15.37)

we arrive at the form of Eq. (15.35). The solution to this equation is given by

$$u(\xi, \eta) = p(\xi) + \eta q(\xi).$$

(15.38)

An example of a parabolic equation is the diffusion equation,

$$\frac{\partial^2}{\partial x^2} \psi(x) - \frac{1}{a^2} \frac{\partial \psi}{\partial t} = 0,$$

(15.39)

which possesses only a single second-order derivative. The first-order derivative in t makes the solution of the equation nontrivial, and different from the solution of the simple canonical equation; we will talk about such solutions in an upcoming section.

3. $B^2 - 4AC < 0$. In this case, the roots to Eqs. (15.27) and (15.28) are *complex*. There is no real coordinate transformation that can reduce our differential equation to the form of Eq. (15.20). In this case, the differential equation is said to be *elliptic*. However, if we make the coordinate transformation

$$\xi = \frac{2Ax - Bt}{\sqrt{4AC - B^2}},$$

(15.40)

$$\eta = t,$$

(15.41)

we find that the differential equation takes on the form

$$\left[\frac{\partial^2 u}{\partial \xi^2} + \frac{\partial^2 u}{\partial \eta^2} \right] = 0,$$

(15.42)

which is Laplace's equation! The canonical form of an elliptic differential equation is Laplace's equation, and Laplace's equation in two spatial variables is itself an example of an elliptic

PDE,

$$\nabla^2 \phi(x, y) = 0. \tag{15.43}$$

Some solutions to this equation in special cases have already been seen in earlier chapters. We may use our coordinate transformations to make a connection to earlier discussions concerning Laplace's equation. For this equation, $A = 1$, $B = 0$, $C = 1$. If we try to find a coordinate transformation which reduces this equation to the form of Eq. (15.20), we see that

$$\xi = x + iy, \tag{15.44}$$

$$\eta = x - iy, \tag{15.45}$$

so that Laplace's equation has the solution

$$u(x, y) = F(x + iy) + G(x - iy). \tag{15.46}$$

The solution to Laplace's equation may be written in terms of functions of a complex number $z = x + iy$ and its complex conjugate, $z^* = x - iy$. The consequences of this observation were examined at the beginning of Chapter 9.1.

All second-order linear partial differential equations with constant coefficients A, B, and C may be divided into three classes, *hyperbolic*, *parabolic*, and *elliptic* equations, which each have their own unique behaviors. We may also consider equations which include the lower-order derivatives, such as

$$A\frac{\partial^2 u}{\partial t^2} + B\frac{\partial^2 u}{\partial x \partial t} + C\frac{\partial^2 u}{\partial x^2} + D\frac{\partial u}{\partial t} + E\frac{\partial u}{\partial x} + Fu = 0. \tag{15.47}$$

The overall behavior of the differential equation is still dominated by the second-order derivatives, which means that the discriminant $B^2 - 4AC$ still distinguishes between the three classes of equation.

It should be further noted that the discriminant can be used to classify equations for which the coefficients are functions of the independent variables, i.e. $A(x, t)$, etc. The classification scheme can be applied locally, and the differential equation can potentially have a different classification in different domains of the independent variables.

15.2.2 Classification of general second-order PDEs

It is possible to extend the analysis of the previous subsection to second-order partial differential equations with any number of independent variables. As we will see, such equations can again be reduced to hyperbolic, elliptic, and parabolic equations.

We consider a linear second-order equation of N independent variables $x_1, \ldots, x_N$,

$$\sum_{ij}^{N} A_{ij} \partial_i \partial_j u = 0, \tag{15.48}$$

where A_{ij} is a real, constant, symmetric matrix and $\partial_i = \partial / \partial x_i$.

To evaluate the behavior of this differential equation, we again consider its reduction to a canonical form. To simplify the calculations, however, we take a slightly different approach to that of the previous section. Let us assume that the Fourier transform of the function $u(\mathbf{x})$ exists, i.e.

$$\tilde{u}(\mathbf{k}) = \frac{1}{(2\pi)^N} \int u(\mathbf{x}) \exp[-i\mathbf{k} \cdot \mathbf{x}] d^N x. \tag{15.49}$$

By the derivative-removing action of Fourier transforms described in Section 11.4.9, we may transform Eq. (15.48) to the form

$$\left[\sum_{ij}^N A_{ij} k_i k_j \right] \tilde{u}(\mathbf{k}) = 0. \tag{15.50}$$

The quantity in brackets is a quadratic form, which may be expressed using matrices as

$$\sum_{ij}^N A_{ij} k_i k_j = \langle \mathbf{k} | \mathbf{A} | \mathbf{k} \rangle. \tag{15.51}$$

Because the matrix A_{ij} is real symmetric, it can be diagonalized, and the eigenvalues are real-valued. Let us suppose that a matrix $\mathbf{R}$ transforms the vector $|\mathbf{k}\rangle$ into the diagonal basis $|\boldsymbol{\xi}\rangle$,

$$\langle \mathbf{k} | \mathbf{A} | \mathbf{k} \rangle = \langle \mathbf{k} | \mathbf{R}^{-1} \mathbf{R} \mathbf{A} \mathbf{R}^{-1} \mathbf{R} | \mathbf{k} \rangle = \langle \boldsymbol{\xi} | \mathbf{A}' | \boldsymbol{\xi} \rangle. \tag{15.52}$$

Here $\mathbf{A}'$ is the diagonal form of $\mathbf{A}$, with diagonal elements equal to the eigenvalues A_i of the matrix $\mathbf{A}$; i.e. $(\mathbf{A}')_{ii} = A_i$. We may make a further coordinate transformation to vectors $|\boldsymbol{\eta}\rangle$ such that

$$(|\boldsymbol{\eta}\rangle)_i = \sqrt{A_i}(|\boldsymbol{\xi}\rangle)_i. \tag{15.53}$$

With these coordinates, our quadratic form may be written as

$$\langle \mathbf{k} | \mathbf{A} | \mathbf{k} \rangle = \langle \boldsymbol{\eta} | \tilde{\mathbf{I}} | \boldsymbol{\eta} \rangle, \tag{15.54}$$

where the matrix $\tilde{\mathbf{I}}$ is a diagonal matrix with diagonal elements either ± 1, or zero,

$$\tilde{\mathbf{I}} \equiv \begin{bmatrix} \pm 1 & 0 & 0 & \cdots & 0 \\ 0 & \pm 1 & 0 & \cdots & 0 \\ 0 & 0 & \pm 1 & \cdots & 0 \\ & & \vdots & \ddots & \\ 0 & 0 & 0 & \cdots & \pm 1 \end{bmatrix}. \tag{15.55}$$

Our quadratic form in Fourier space may be written as

$$\left[\sum_{ij}^{N} A_{ij} k_i k_j \right] \tilde{u}(\mathbf{k}) = \left[\sum_{i}^{N} \tilde{I}_i \eta_i^2 \right] \tilde{u}(\boldsymbol{\eta}) = 0, \qquad (15.56)$$

where $\tilde{I}_i$ may be ± 1 or zero. There evidently exists a set of coordinates $\boldsymbol{\zeta}$ in the direct space such that the exponent of our Fourier kernel may be written as

$$\mathbf{k} \cdot \mathbf{x} = \boldsymbol{\eta} \cdot \boldsymbol{\zeta}. \qquad (15.57)$$

An inverse Fourier transform of Eq. (15.56) will then leave us with the partial differential equation

$$\sum_{i}^{N} \tilde{I}_i \frac{\partial^2}{\partial \zeta_i^2} u(\boldsymbol{\zeta}) = 0. \qquad (15.58)$$

This is the canonical form of a general second-order PDE in N variables with constant coefficients. We may classify an equation as elliptic, hyperbolic, or parabolic depending on the values of the coefficients $\tilde{I}_i$.

1. All $\tilde{I}_i \neq 0$, and all $\tilde{I}_i$ have the same sign. In this case, the equation is said to be elliptic. Laplace's equation in any number of variables is an elliptic equation, as is the Helmholtz equation.
2. All $\tilde{I}_i \neq 0$, and some $\tilde{I}_i$ have different signs. If only a single $\tilde{I}_i$ has a sign different from the others, the equation is said to be hyperbolic. If more than one $\tilde{I}_i$ has a sign different from the majority, the equation is said to be ultrahyperbolic. The wave equation in any number of spatial variables is a hyperbolic equation; the time variable is the sole variable with different sign.
3. Some $\tilde{I}_i = 0$. Such an equation is parabolic. Typically, only a single $\tilde{I}_i = 0$, and the coordinate associated with that eigenvalue has a nonzero first partial derivative in the equation. The diffusion equation in any number of spatial variables is a parabolic equation.

As in the case of two variables, the classification still applies for a PDE which includes lower-order derivatives. Furthermore, when the coefficients of the equation depend upon the variables, the classification may still be applied *locally* to the differential equation. If we consider an equation of the form

$$\sum_{ij}^{N} A_{ij}(\mathbf{x}) \partial_i \partial_j u(\mathbf{x}) + \sum_{k}^{N} B_k(\mathbf{x}) \partial_k u(\mathbf{x}) + C(\mathbf{x}) u(\mathbf{x}) + D(\mathbf{x}) = 0, \qquad (15.59)$$

it can be shown that we may classify the behavior of this equation *in a given region of* **x**-*space* by reducing it to one of the standard forms (hyperbolic, parabolic, elliptic) by a *nonlinear* coordinate transformation. It is to be noted that it is possible to have PDEs which are of one class in one region but another class in another region.

15.2.3 Auxiliary conditions for second-order PDEs, existence, and uniqueness

The goal of classifying partial differential equations is in large part to emphasize that all second-order linear PDEs fall into one of three broad classes of behavior: hyperbolic (wave-like), parabolic (diffusion-like), and elliptic (potential-like). If we can understand the behaviors of these three classes, we have gone a long way to broadly understanding the behaviors of PDEs.

The nature of the PDE also specifies what sort of additional conditions are necessary to specify a unique solution to the problem. Typically, conditions which specify the value of the function or its derivatives on a spatial boundary are known as *boundary conditions*, while conditions which specify the value of the function or its derivatives at some initial time are known as *initial conditions*; they are collectively referred to as boundary conditions, or auxiliary conditions. Boundary conditions are classified into two types of boundary and three types of specific conditions. The boundary conditions may be specified on an *open* surface or a *closed* surface. The types of condition are:

1. Cauchy conditions. The value of the function and its normal derivative are specified on the boundary.
2. Dirichlet conditions. The value of the function is specified on the boundary.
3. Neumann conditions. The value of the normal derivative of the function is specified on the boundary.

Different classes of PDE require different types of boundary condition. We summarize these requirements in the list below, citing physics to justify them.

1. Elliptic equations. Laplace's equation can be used to specify the electric potential inside a volume with a fixed potential specified on the surface. Good boundary conditions are therefore specified on a *closed* surface, and one may specify either the value of the function (Dirichlet) or its derivative (Neumann) on the boundary; specifying both is too restrictive.
2. Parabolic equations. The diffusion equation is often used to specify the spread of heat through a metal rod. To solve a diffusion equation, we normally specify the initial distribution (the heat distribution at some initial time) and the fixed temperature of the rod at the endpoints. The surface is technically an *open* surface (we only specify the value of the function for one value of t – technically, the surface is a box with one end missing in (x, t) space). We may specify either the value of the function (Dirichlet) or its derivative (Neumann) on the boundary; specifying both is too restrictive.
3. Hyperbolic equations. The wave equation can be used to describe vibrations on a string. To solve it, we need to specify both the initial amplitude of the string and its initial speed (because waves may propagate in both directions on the string, the initial speed is required). We may also specify boundary conditions at the endpoints of the string. The surface is again an open surface, for the same reasons as given for parabolic equations, and we must specify both the amplitude and speed of the string (Cauchy boundary conditions).

An illustration of the different types of condition are given in Fig. 15.2.

It is important to note that, in our discussion so far, we have assumed that a solution to the differential equations under consideration exists (*existence*) and that, with the appropriate

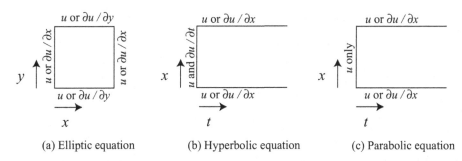

(a) Elliptic equation   (b) Hyperbolic equation   (c) Parabolic equation

Figure 15.2 An illustration of the required boundary conditions for elliptic, hyperbolic, and parabolic equations in two independent variables.

specification of boundary conditions, a single unique solution exists (*uniqueness*). The proof of this is involved, and we simply state the conclusion, known as the Cauchy–Kowalevski theorem.[1]

Theorem 15.1 (Cauchy–Kowalevski theorem) *Let us suppose we wish to find a function $u(\mathbf{x},t)$ which satisfies an Nth order partial differential equation in $M+1$ independent variables $(\mathbf{x},t)$ of the form*

$$\frac{\partial^N u}{\partial t^N} + F[u, \partial u, \mathbf{x}, t] = 0, \tag{15.60}$$

where ∂u represents all possible partial derivatives of u of order less than N in time. N initial conditions of the form

$$\frac{\partial^k u}{\partial t^k} = f_k(\mathbf{x}) \tag{15.61}$$

for all $k < N$ are specified. If the function F is analytic in some neighborhood of the point $F[f_0, f_k, \mathbf{x}_0, 0]$, and the functions representing the initial conditions are analytic in some neighborhood of the point $\mathbf{x}_0$, then the differential equation has a unique analytic solution in some neighborhood of $\mathbf{x}_0$ and $t = 0$.

It is to be noted that analyticity of the relevant functions is necessary for a solution to be found. When dealing with well-formulated linear PDEs based on a physical problem, it is usually safe to assume that the solution exists and is unique when subject to the appropriate boundary conditions described above.

---

[1] It is worth noting that this theorem was originally posed in a limited form by Cauchy in 1842, but was proven in general form by Sophie Kowalevski in 1875 [Kow75]. Kowalevski is considered to be the first major female Russian mathematician, and the first woman to be given a full professorship in northern Europe.

15.3 Separation of variables

The most important and useful (non-numerical) technique for solving partial differential equations is separation of variables. The strategy is to assume that the solutions of the PDE factorize into a product of functions of independent variables. In the introduction to the chapter, we considered solutions to the Helmholtz equation in Cartesian coordinates,

$$\frac{\partial^2 \psi}{\partial x^2} + \frac{\partial^2 \psi}{\partial y^2} + \frac{\partial^2 \psi}{\partial z^2} + k^2 \psi = 0. \tag{15.62}$$

By assuming a solution of the form

$$\psi(x, y, z) = X(x)Y(y)Z(z), \tag{15.63}$$

the Helmholtz equation can be reduced to the separated equation

$$\frac{1}{X}\frac{d^2 X}{dx^2} + \frac{1}{Y}\frac{d^2 Y}{dy^2} + \frac{1}{Z}\frac{d^2 Z}{dz^2} + k^2 = 0. \tag{15.64}$$

Each term of this equation is only dependent on one variable. In order for this equation to be satisfied, each term must be equal to a constant. We assume

$$\frac{1}{X}\frac{d^2 X}{dx^2} = -l^2, \frac{1}{Y}\frac{d^2 Y}{dy^2} = -m^2, \frac{1}{Z}\frac{d^2 Z}{dz^2} = -n^2. \tag{15.65}$$

This is a set of three ordinary differential equations, each of which is the harmonic oscillator equation:

$$\frac{\partial^2 X}{\partial x^2} + l^2 X = 0. \tag{15.66}$$

One "separable" solution to the PDE is therefore

$$\psi_{l,m,n}(x, y, z) = [a_l \sin(lx) + b_l \cos(lx)][c_m \sin(my) + d_m \cos(my)]$$
$$[e_n \sin(nz) + f_n \cos(nz)], \tag{15.67}$$

where we find, by substituting into the original differential equation, the additional constraint

$$k^2 = l^2 + m^2 + n^2. \tag{15.68}$$

The coefficients $a_l, b_l, c_m, d_m, e_n, f_n$ are determined by the boundary conditions. The general solution will in fact be determined by a combination of the separable solutions,

$$\psi(x, y, z) = \sum_{l,m,n} \psi_{l,m,n}(x, y, z). \tag{15.69}$$

We will see how this works in more detail in a moment. For now, we recall from Chapter 3 that there are 11 coordinate systems for which Laplace's equation can be separated, and a few additional systems for which the Helmholtz equation can be separated. If the geometry of a particular problem is compatible with one of these systems, we can perform separation of variables to solve it.

As another illustration, we consider the solution of the Helmholtz equation in a cylinder, for which cylindrical coordinates are suitable,

$$\nabla^2 \psi(\rho, \phi, z) + k^2 \psi(\rho, \phi, z) = 0. \tag{15.70}$$

Expressing the Laplacian in cylindrical coordinates, the Helmholtz equation may be written as

$$\frac{1}{\rho} \frac{\partial}{\partial \rho} \left(\rho \frac{\partial \psi}{\partial \rho} \right) + \frac{1}{\rho^2} \frac{\partial^2 \psi}{\partial \phi^2} + \frac{\partial^2 \psi}{\partial z^2} + k^2 \psi = 0. \tag{15.71}$$

We assume a solution of the form

$$\psi(\rho, \phi, z) = P(\rho) \Phi(\phi) Z(z), \tag{15.72}$$

and the differential equation becomes

$$\frac{1}{\rho P} \frac{d}{d\rho} \left(\rho \frac{dP}{d\rho} \right) + \frac{1}{\rho^2 \Phi} \frac{d^2 \Phi}{d\phi^2} + \frac{1}{Z} \frac{d^2 Z}{dz^2} + k^2 = 0. \tag{15.73}$$

The third term in this equation is the only one which depends on z – it must therefore be constant, leading to the equation

$$\frac{d^2 Z}{dz^2} = l^2 Z, \tag{15.74}$$

whose solutions are cosh and sinh functions,

$$Z_n(z) = A \cosh(lz) + B \sinh(lz). \tag{15.75}$$

With this solution, we let $k^2 + l^2 = n^2$, and our differential equation becomes

$$\frac{\rho}{P} \frac{d}{d\rho} \left(\rho \frac{dP}{d\rho} \right) + n^2 \rho^2 = -\frac{1}{\Phi} \frac{d^2 \Phi}{d\phi^2}. \tag{15.76}$$

The left-hand side of this equation depends only on ρ, and the right-hand side depends only on ϕ. Each side must be a constant; therefore, we let

$$\frac{d^2 \Phi}{d\phi^2} = -m^2 \Phi, \tag{15.77}$$

and for the ρ-dependent term we have

$$\rho \frac{d}{d\rho}\left(\rho \frac{dP}{d\rho}\right) + (n^2\rho^2 - m^2)P = 0. \tag{15.78}$$

The equation for Φ is again the harmonic oscillator equation. Because the function ϕ must repeat over 2π, i.e. $\Phi(\phi + 2\pi) = \Phi(\phi)$, m must be an integer, and the solution for Φ is of the form

$$\Phi_m(\phi) = Ce^{im\phi} + De^{-im\phi}, \tag{15.79}$$

where m is a positive integer. The general solution for the differential equation in cylindrical coordinates can be written as

$$\psi(\rho,\phi,z) = \sum_{mn} a_{mn}P_{mn}(\rho)\phi_m(\phi)Z_n(z), \tag{15.80}$$

where the exact nature of the solutions will again be dictated by the boundary conditions.

The equation (15.78) for the function $P_{mn}(\rho)$ is Bessel's differential equation, to be dealt with in Chapter 16. The solutions to Bessel's equation are the Bessel functions J_n, and the Neumann functions N_n. These functions cannot be expressed simply in terms of elementary functions, so in this chapter we will avoid problems where such "special functions" appear. We look instead at examples of elliptical, hyperbolic, and parabolic PDEs which we can solve completely using separation of variables with functions that are already known to us.

15.4 Hyperbolic equations

The archetypical hyperbolic equation is the wave equation,

$$u_{xx} - \frac{1}{c^2}u_{tt} = 0. \tag{15.81}$$

As noted in the previous section, a unique solution to the wave equation requires Cauchy boundary conditions specified on an open surface. The initial wave amplitude $u(x,0)$ and the initial wave speed $u_t(x,0)$ must be specified. In addition, one typically specifies conditions on the behavior of the wave at its spatial boundaries, as well. The types of conditions which can be specified are as follows.

1. Dirichlet condition. The function $u(x,t)$ is specified on the spatial boundary. For instance, a vibrating string which has its ends clamped at $x = \pm L$ will satisfy the pair of conditions $u(\pm L,t) = 0$.
2. Neumann condition. The normal derivative $u_x(x,t)$ is specified on the spatial boundary. For instance, a vibrating string which has an end free to slide at $x = L$ will satisfy the condition $u_x(L,t) = 0$.
3. Robin condition. A combination of the function and its normal derivative is specified on the boundary, of the form $u_x(L,t) + \alpha u(L,t) = g(t)$, where α is a constant.

Things become a little more complicated if the equation is to be solved on an infinite domain. One presumably must specify the behavior of the solution at the "endpoints" (infinity), but conditions of the Dirichlet, Neumann, or Robin type cannot be sensibly applied. Physical constraints, such as requiring the waves to be outgoing (going towards infinity) rather than ingoing (coming from infinity) are typically applied. On an infinite domain, integral equation techniques are often helpful.

Example 15.1 Wave propagation on a finite length string. Dirichlet boundary conditions

We consider the solution of the wave equation,

$$u_{xx} - \frac{1}{c^2} u_{tt} = 0 \tag{15.82}$$

on a string with ends at $x = \pm L$, subject to the boundary conditions

$$u(-L,t) = 0, \tag{15.83}$$

$$u(+L,t) = 0 \tag{15.84}$$

and the initial conditions

$$u(x,0) = \phi(x), \tag{15.85}$$

$$u_t(x,0) = \eta(x). \tag{15.86}$$

We assume a solution of the form

$$u(x,t) = X(x)T(t). \tag{15.87}$$

On substituting this into the wave equation, we have the result

$$X''(x)T(t) - \frac{1}{c^2}X(x)T''(t) = 0, \tag{15.88}$$

which we can simplify further by dividing each side of the equation by XT, and separating across the equal sign,

$$\frac{X''}{X} = \frac{T''}{c^2 T}. \tag{15.89}$$

Each side of this equation must be equal to a constant, say $-k^2$; we then have two differential equations,

$$T'' + (kc)^2 T = 0, \tag{15.90}$$

$$X'' + k^2 X = 0. \tag{15.91}$$

Each of these ODEs can be readily solved,

$$T(t) = A \sin(kct) + B \cos(kct), \tag{15.92}$$

$$X(x) = C \sin(kx) + D \cos(kx). \tag{15.93}$$

Any function of the form

$$u_k(x,t) = [A_k \sin(kct) + B_k \cos(kct)][C_k \sin(kx) + D_k \cos(kx)] \tag{15.94}$$

will satisfy the partial differential equation.

Now we attempt to satisfy our spatial boundary conditions. The requirement $u(0,t) = 0$ leads us to the conclusion that $D_k = 0$. Consolidating the constant C_k into A_k and B_k, we are led to the simplified solution

$$u_k(x,t) = \sin(kx)[A_k \sin(kct) + B_k \cos(kct)]. \tag{15.95}$$

Our other boundary condition leads to the requirement

$$\sin(kL) = 0, \tag{15.96}$$

which suggests that k must be an integer multiple of π/L. We define

$$k_n = \frac{\pi n}{L}, \quad n = 1, 2, 3, \ldots \tag{15.97}$$

and our separable solutions to the PDE are of the form

$$u_n(x,t) = \sin(n\pi x/L)[A_n \sin(n\pi ct/L) + B_n \cos(n\pi ct/L)], \tag{15.98}$$

where the individual separable solutions are now labeled by the index n. The most general solution of the PDE is of the form

$$u(x,t) = \sum_{n=1}^{\infty} \sin(n\pi x/L)[A_n \sin(n\pi ct/L) + B_n \cos(n\pi ct/L)]. \tag{15.99}$$

We now match the initial conditions. At $t = 0$, we require that $u(x,0) = \phi(x)$ and $u_t(x,0) = \beta(x)$; this leads us to the equations

$$\phi(x) = \sum_{n=1}^{\infty} B_n \sin(n\pi x/L), \tag{15.100}$$

$$\beta(x) = \sum_{n=1}^{\infty} A_n \frac{n\pi c}{L} \sin(n\pi x/L). \tag{15.101}$$

These are both Fourier sine series; the coefficients A_n and B_n can be isolated by an appropriate Fourier integral,

$$B_m = \frac{2}{L} \int_0^L \phi(x) \sin(n\pi x/L) dx, \tag{15.102}$$

$$A_m = \frac{2}{n\pi c} \int_0^L \beta(x) \sin(n\pi x/L) dx. \tag{15.103}$$

We therefore have our complete solution to the differential equation, which satisfies the initial conditions and the boundary conditions.

◇

Example 15.2 Wave propagation on a finite length string. Robin boundary conditions
We consider the solution of the wave equation,

$$u_{xx} - \frac{1}{c^2} u_{tt} = 0, \tag{15.104}$$

on a string with ends at $x = 0$ and $x = L$, subject to the boundary conditions

$$u(0,t) = 0, \tag{15.105}$$

$$u(L,t) + \alpha u_x(L,t) = 0 \tag{15.106}$$

and the initial conditions

$$u(x,0) = \phi(x), \tag{15.107}$$

$$u_t(x,0) = \eta(x). \tag{15.108}$$

This problem illustrates the nontrivial difficulties which arise when applying Robin boundary conditions to a PDE. Separation of variables is performed exactly as in the previous example; before applying boundary and initial conditions, we find that any function of the form

$$u_k(x,t) = [A_k \sin(kct) + B_k \cos(kct)][C_k \sin(kx) + D_k \cos(kx)] \tag{15.109}$$

will satisfy the partial differential equation.

Next we attempt to satisfy our spatial boundary conditions. The requirement $u(0,t) = 0$ leads us to the conclusion that $D_k = 0$. Consolidating the constant C_k into A_k and B_k, we are led to the simplified solution

$$u_k(x,t) = \sin(kx)[A_k \sin(kct) + B_k \cos(kct)]. \tag{15.110}$$

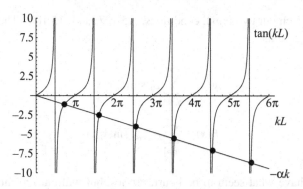

Figure 15.3 The solution of the transcendental equation for k, with $\alpha/L = 0.5$.

Here things become somewhat difficult. To satisfy our Robin boundary condition, the solution u_k must satisfy the equation

$$\sin(kL) + \alpha k \cos(kL) = 0, \tag{15.111}$$

which can be rewritten as

$$\tan(kL) = -\alpha k. \tag{15.112}$$

This is a transcendental equation which cannot be analytically solved for k. The solutions can be visualized graphically as the intersection between the curves $\tan(kL)$ and $-\alpha k$; these intersections are illustrated in Fig. 15.3. The number of intersections are clearly countably infinite; the value of the nth intersection is labeled k_n, where $n = 1, 2, 3, \ldots$. For large values of n, the values approximately take on the form $k_n \approx (2n+1)\pi/(2L)$.

The best analytic formula we may write, then, is of the form

$$u_n(x,t) = \sin(k_n x)\left[A_n \sin(k_n ct) + B_n \cos(k_n ct)\right], \tag{15.113}$$

with k_n determined numerically. A formal solution to the wave equation which satisfies our given boundary conditions is

$$u(x,t) = \sum_{n=1}^{\infty} \sin(k_n x)\left[A_n \sin(k_n ct) + B_n \cos(k_n ct)\right]. \tag{15.114}$$

We now need match our two initial conditions, (15.107) and (15.108). The two conditions lead us to the equations

$$\phi(x) = \sum_{n=1}^{\infty} B_n \sin(k_n x), \tag{15.115}$$

$$\beta(x) = \sum_{n=1}^{\infty} A_n k_n c \sin(k_n x). \tag{15.116}$$

Here we again have what seem to be Fourier series, but with unconventional frequency components k_n. It can be shown that the set of functions $\sin(k_n x)$ are complete, i.e. they can be used to construct all well-behaved functions on the line $0 \leq x \leq L$. We will not show this, but we can readily demonstrate the orthogonality of the functions:

$$\int_0^L \sin(k_n x) \sin(k_m x) dx = -\frac{1}{2} \int_0^L \{\cos[(k_n + k_m)x] - \cos[(k_n - k_m)x]\} dx. \tag{15.117}$$

With a little bit of algebraic and trigonometric manipulation, this formula takes on the form

$$\int_0^L \sin(k_n x) \sin(k_m x) dx = \begin{cases} \frac{\alpha}{2}\left[\cos^2(k_m L) + \frac{L}{\alpha}\right] \equiv \gamma_m, & m = n, \\ 0, & m \neq n. \end{cases} \tag{15.118}$$

We may write a normalization condition,

$$\frac{1}{\gamma_m} \int_0^L \sin(k_n x) \sin(k_m x) dx = \delta_{mn}. \tag{15.119}$$

We then have

$$B_m = \frac{1}{\gamma_m} \int_0^L \phi(x) \sin(k_m x) dx, \tag{15.120}$$

$$A_m = \frac{1}{\gamma_m k_m c} \int_0^L \beta(x) \sin(k_m x) dx. \tag{15.121}$$

We therefore have a solution, Eq. (15.114), to the differential equation which satisfies both the boundary conditions and the initial conditions.

This example illustrates an important observation: depending on the choice of boundary conditions, a variety of different sets of orthogonal functions such as $\sin(k_n x)$ may be needed to solve a problem. One typically can prove orthogonality using an equation such as Eq. (15.117); it is usually safe to assume that this set of functions is complete.

◇

15.5 Elliptic equations

The archetypical elliptic equation is Laplace's equation, which in two variables may be written in the form

$$\frac{\partial^2 V(x,y)}{\partial x^2} + \frac{\partial^2 V(x,y)}{\partial y^2} = 0. \tag{15.122}$$

A unique solution to an elliptic equation requires the specification of the value of the function (Dirichlet) or its derivative (Neumann) upon the boundary of the domain of interest.

Example 15.3 Laplace's equation in a circular domain

We consider the solution of Laplace's equation in a circular domain, with the potential $V(\rho,\phi)$ taking on the value $V_0(\phi)$ on the boundary $\rho = a$. Laplace's equation in polar coordinates can be found directly from the Laplacian in cylindrical coordinates, Eq. (3.44), neglecting the z-component,

$$\frac{1}{\rho}\frac{\partial}{\partial \rho}\left(\rho\frac{\partial V}{\partial \rho}\right) + \frac{1}{\rho^2}\frac{\partial^2 V}{\partial \phi} = 0. \tag{15.123}$$

We assume a solution of the form

$$V(\rho,\phi) = R(\rho)\Phi(\phi). \tag{15.124}$$

We substitute into the differential equation, divide by $R\Phi$ and multiply by ρ^2. We arrive at the equation

$$\frac{1}{R}\rho\frac{\partial}{\partial \rho}\left(\rho\frac{\partial R}{\partial \rho}\right) = \frac{1}{\Phi}\frac{\partial \Phi}{\partial \phi}. \tag{15.125}$$

Each side of this equation must be equal to a constant, which we label $-\lambda^2$. The right-hand side results in the expression

$$\Phi'' + \lambda^2\Phi = 0, \tag{15.126}$$

which is simply the harmonic oscillator equation again. Solutions are of the form

$$\Phi_\lambda(\phi) = C_\lambda e^{i\lambda\phi} + D_\lambda e^{-i\lambda\phi}. \tag{15.127}$$

This solution, however, suggests that λ cannot be completely arbitrary. A physical solution of Laplace's equation must be single-valued, which constrains λ to be an integer $m = 0, 1, 2, \ldots$. The requirement of single-valuedness is essentially a "hidden" boundary condition which arises from solving a differential equation in a curved coordinate system; we will see another such condition momentarily.

We now turn to the equation for R; let us consider the $m = 0$ case separately. For this case, the equation for R has the form

$$\rho R'' + R' = 0. \tag{15.128}$$

This equation can be solved in a straightforward way to find that

$$R_0(\rho) = a_0 + b_0 \log \rho. \tag{15.129}$$

Here we must impose another "hidden" boundary condition: our solution to Laplace's equation must be well-behaved, i.e. differentiable, throughout the unit circle. This immediately leads to the conclusion that $b_0 = 0$. It should be noted, however, that the logarithmic term cannot be neglected in solving Laplace's equation in an annular region.

With $\lambda = m > 0$, the equation for R takes on the form

$$\rho \frac{\partial}{\partial \rho}\left(\rho \frac{\partial R}{\partial \rho}\right) - m^2 R = 0. \tag{15.130}$$

This is a nontrivial ordinary differential equation, which one would typically solve using techniques of the previous chapter. In this case, however, we make a lucky guess and try a trial solution

$$R(\rho) = \rho^k, \tag{15.131}$$

with k to be determined. On substitution, one readily finds that $k^2 = m^2$; We therefore find that R may be written as

$$R_m(\rho) = A_m \rho^m + B_m \rho^{-m}. \tag{15.132}$$

Again we impose the "hidden" boundary condition that our solution to Laplace's equation must be differentiable throughout the unit circle. This leads to the conclusion that $B_m = 0$ for all m. Our particular solution is therefore of the form

$$V_m(\rho, \phi) = \rho^m \left(C_m e^{im\phi} + D_m e^{-im\phi}\right), \tag{15.133}$$

which includes the case $m = 0$. The general solution can be written as

$$V_m(\rho, \phi) = \sum_{m=0}^{\infty} C_m \rho^m e^{im\phi} + \sum_{m=1}^{\infty} D_m \rho^m e^{-im\phi}. \tag{15.134}$$

Our last requirement is to satisfy the boundary condition; at $\rho = a$, we must satisfy

$$V_0(\phi) = \sum_{m=0}^{\infty} C_m a^m e^{im\phi} + \sum_{m=1}^{\infty} D_m a^m e^{-im\phi}. \tag{15.135}$$

We again have a Fourier series, whose coefficients can be found in the usual way,

$$C_m = \frac{1}{a^m \pi} \int_0^{2\pi} e^{-im\phi} V_0(\phi) d\phi, \tag{15.136}$$

$$D_m = \frac{1}{a^m \pi} \int_0^{2\pi} e^{im\phi} V_0(\phi) d\phi. \tag{15.137}$$

With Eq. (15.134), our solution is complete.

◇

Example 15.4 Laplace's equation in three dimensions

We now consider the solution of Laplace's equation in three dimensions, in a rectangular region ($0 \leq x \leq a$, $0 \leq y \leq b$, $0 \leq z \leq c$). Laplace's equation may be written as

$$\frac{\partial^2 V(x,y,z)}{\partial x^2} + \frac{\partial^2 V(x,y,z)}{\partial y^2} + \frac{\partial^2 V(x,y,z)}{\partial z^2} = 0. \tag{15.138}$$

We first consider the solution of the problem for the case when the potential $V(x,y,z) = 0$ on all boundaries except the $x = 0$ boundary, on which it has the value $V_0(y,z)$; we will consider a more complicated condition at the end of this example.

We try a separated solution of the form

$$V(x,y,z) = X(x)Y(y)Z(z), \tag{15.139}$$

which leads us to the separated differential equation

$$\frac{X''}{X} + \frac{Y''}{Y} + \frac{Z''}{Z} = 0. \tag{15.140}$$

Each term of this equation is a function of an independent variable, and each of them must therefore be equal to a constant. Though, in the end, the solution of the differential equation must be independent of how we format the constants, it is easier to find the solution if we make an informed choice here. We expect that our final solution will involve a Fourier series relationship with the boundary value $V_0(x,y)$, so we would like our solutions in y and z to have the form of sines and cosines. We therefore make the choice

$$\frac{X''}{X} = L^2, \quad \frac{Y''}{Y} = -M^2, \quad \frac{Z''}{Z} = -N^2, \tag{15.141}$$

with $L^2 - M^2 - N^2 = 0$. The solutions for Y and Z are

$$Y_M(y) = A_M \cos(My) + B_M \sin(My), \tag{15.142}$$

$$Z_N(z) = C_N \cos(Nz) + D_N \sin(Nz), \tag{15.143}$$

while the family of solutions for X may be written as

$$X_L(x) = E_L e^{Lx} + F_L e^{-Lx}. \tag{15.144}$$

We now attempt to match the homogeneous boundary conditions. The requirement that $V(x,0,z) = 0$ immediately leads to $A_M = 0$; similarly, the requirement that $V(x,y,0) = 0$ leads to $C_N = 0$. The requirement that $V(x,b,z) = 0$ leads to

$$M = \frac{\pi m}{b}, \tag{15.145}$$

where m is an integer greater than zero, and the requirement that $V(x,y,c) = 0$ leads to

$$N = \frac{\pi n}{c}, \tag{15.146}$$

where n is a positive integer. With these conditions, our separated solution is of the form

$$V_{mn}(x,y,z) = \left(E_{mn} e^{Lx} + F_{mn} e^{-Lx}\right) \sin(m\pi y/b) \sin(n\pi z/c), \tag{15.147}$$

where

$$L = \pi \sqrt{m^2/b^2 + n^2/c^2}. \tag{15.148}$$

It is to be noted that our separated solutions are labeled by *two* indices, m and n, where our previous examples were labeled by a single index n. How many summations will the solution to a partial differential equation require in general? We note that an arbitrary function in N-dimensional space can be represented by an N-variable Fourier series, with N summations. Our partial differential equation represents a constraint on the form of the solutions, which will typically reduce the number of summations by one. We typically expect $N-1$ summations in the solution of a PDE in N variables.

We now apply the final homogeneous boundary condition, $V(a,y,z) = 0$. This constraint leads to the relationship

$$F_{mn} = -E_{mn} e^{2La}, \tag{15.149}$$

so our general solution is of the form

$$V(x,y,z) = \sum_{m=1}^{\infty} \sum_{n=1}^{\infty} E_{mn} \left(e^{Lx} - e^{2La} e^{-Lx}\right) \sin(m\pi y/b) \sin(n\pi z/c). \tag{15.150}$$

With the requirement that this must equal $V_0(y,z)$ on the boundary $x = 0$, we have the requirement

$$V_0(y,z) = \sum_{m=1}^{\infty} \sum_{n=1}^{\infty} E_{mn} \left(1 - e^{2La}\right) \sin(m\pi y/b) \sin(n\pi z/c). \tag{15.151}$$

We can use techniques for a two-variable Fourier series to solve the problem, and arrive at the expression for E_{mn},

$$E_{mn} = \frac{4}{bc} \frac{1}{1 - e^{2la}} \int_0^b \int_0^c V_0(y,z) \sin(m\pi y/b) \sin(n\pi z/c) dy dz. \tag{15.152}$$

We have arrived at the complete solution to the problem.

The problem becomes slightly more difficult if more than one face of the cuboid has a nonzero potential. We now consider the case where $V(0,y,z) = V_0(y,z)$ and $V(a,y,z) = V_1(y,z)$. We return to Eq. (15.147), which represents the separated solution before the boundary conditions for $x = 0$ and $x = a$ are applied. The general solution can be written as

$$V(x,y,z) = \sum_{m=1}^{\infty} \sum_{n=1}^{\infty} \left(E_{mn} e^{Lx} + F_{mn} e^{-Lx} \right) \sin(m\pi y/b) \sin(n\pi z/c). \tag{15.153}$$

Considering our final two boundary conditions together at the same time, we have the expressions

$$V_0(y,z) = \sum_{m=1}^{\infty} \sum_{n=1}^{\infty} (E_{mn} + F_{mn}) \sin(m\pi y/b) \sin(n\pi z/c), \tag{15.154}$$

$$V_1(y,z) = \sum_{m=1}^{\infty} \sum_{n=1}^{\infty} \left(E_{mn} e^{La} + F_{mn} e^{-La} \right) \sin(m\pi y/b) \sin(n\pi z/c). \tag{15.155}$$

We may again use Fourier series tricks to extract the coefficents, but in this case we only extract linear combinations of the coefficients, i.e.

$$E_{mn} + F_{mn} = \frac{4}{bc} \int_0^b \int_0^c V_0(y,z) \sin(m\pi y/b) \sin(n\pi z/c) dy dz, \tag{15.156}$$

$$E_{mn} e^{La} + F_{mn} e^{-La} = \frac{4}{bc} \int_0^b \int_0^c V_1(y,z) \sin(m\pi y/b) \sin(n\pi z/c) dy dz. \tag{15.157}$$

By taking appropriate linear combinations of these expressions, we find that

$$E_{mn} = \frac{1}{e^{-La} - e^{La}} \frac{4}{bc} \int_0^b \int_0^c \left[e^{-La} V_0(y,z) - V_1(y,z) \right] \sin(m\pi y/b) \sin(n\pi z/c) dy dz, \tag{15.158}$$

$$F_{mn} = \frac{1}{e^{La} - e^{-La}} \frac{4}{bc} \int_0^b \int_0^c \left[e^{La} V_0(y,z) - V_1(y,z) \right] \sin(m\pi y/b) \sin(n\pi z/c) dy dz. \tag{15.159}$$

◇

15.6 Parabolic equations

The archetypical parabolic equation is the diffusion equation, which in a single spatial variable x takes on the form

$$u_{xx} - \alpha^2 u_t = 0, \tag{15.160}$$

where α is a constant which characterizes the rate of diffusion. Boundary conditions specified can be of Dirichlet, Neumann, or Robin type. For initial conditions, one only need specify the function $u(x,t)$ at $t = 0$.

Example 15.5 Heat diffusion in a finite length metal rod

We consider the solution of the diffusion equation,

$$u_{xx} - \alpha^2 u_t = 0, \tag{15.161}$$

on a metal rod which extends from $0 \leq x \leq 1$, subject to the boundary conditions

$$u(0,t) = 0,$$
$$u(1,t) = 0 \tag{15.162}$$

and the initial condition

$$u(x,0) = \phi(x), \quad 0 \leq x \leq 1. \tag{15.163}$$

To solve this by separation of variables, we assume a solution of the form

$$u(x,t) = X(x)T(t). \tag{15.164}$$

On substituting this into the diffusion equation, we have the result

$$X(x)T'(t) = \alpha^2 X''(x)T(t), \tag{15.165}$$

which we can simplify further by dividing each side of the equation by XT,

$$\frac{T'}{\alpha^2 T} = \frac{X''}{X}. \tag{15.166}$$

Each side of this equation must be equal to a constant, say $-\lambda^2$; we then have two differential equations,

$$T' + \lambda^2 \alpha^2 T = 0, \tag{15.167}$$

$$X'' + \lambda^2 X = 0. \tag{15.168}$$

Each equation is now a standard ODE which can be readily solved,

$$T(t) = Ae^{-\lambda^2\alpha^2 t},$$ (15.169)

$$X(x) = B\sin(\lambda x) + C\cos(\lambda x).$$ (15.170)

All functions of the form

$$u_\lambda(x,t) = e^{-\lambda^2\alpha^2 t}[A\sin(\lambda x) + B\cos(\lambda x)]$$ (15.171)

therefore satisfy the PDE.

The next step is to use this set of solutions to attempt to match the boundary conditions, Eqs. (15.162). The requirement that $u(0,t) = 0$ can be automatically satisfied by letting $B = 0$. To satisfy the requirement that $u(1,t) = 0$, we find that

$$u(1,t) = Ae^{-\lambda^2\alpha^2 t}\sin\lambda = 0, \quad \text{or} \quad \sin\lambda = 0.$$ (15.172)

This condition can be satisfied only by choosing $\lambda = \lambda_n$, where

$$\lambda_n = n\pi.$$ (15.173)

Now we have an infinite set of solutions which satisfy the PDE and the boundary conditions,

$$u_n(x,t) = A_n e^{-(n\pi\alpha)^2 t}\sin(n\pi x).$$ (15.174)

Our last step is to attempt to match the initial condition. We require that

$$u(x,0) = \sum_{n=1}^{\infty} A_n \sin(n\pi x) = \phi(x).$$ (15.175)

The left-hand side of this equation is just a Fourier series. We may determine the coefficients A_n in the usual manner,

$$A_n = 2\int_0^1 \phi(x)\sin(n\pi x)\mathrm{d}x.$$ (15.176)

With the determination of A_n, we have completely solved our PDE. The differential equation is solved by

$$u(x,t) = \sum_{n=1}^{\infty} A_n e^{-(n\pi\alpha)^2 t}\sin(n\pi x).$$ (15.177)

A few words on interpretation may be useful. Looking at the general solution above, we can see that it consists of terms of increasingly higher spatial frequency, $n\pi$. Each distinct spatial frequency is damped by an exponential in time, the higher frequency terms being damped more quickly. The solution to the diffusion equation therefore quickly loses the higher frequency terms, and "smooths" itself out.

It is to be noted that, being a parabolic equation, we only needed to specify the value of the function at $t = 0$, and not its time derivative as well, as we did for the wave equation.

◇

Example 15.6 Diffusion in a plane, with inhomogeneous boundary condition

We consider diffusion in a square region, i.e. $0 \leq x \leq a$, $0 \leq y \leq a$, where the amplitude $u(x,y,t)$ satisfies the two-dimensional diffusion equation,

$$u_{xx} + u_{yy} - \alpha^2 u_t = 0. \tag{15.178}$$

The initial condition is taken to be $u(x,y,0) = u_i(x,y)$. The amplitude is taken to be zero on the boundaries $x = a$, $y = 0$, and $y = a$, but is assumed to be *nonzero* on the boundary $x = 0$, i.e. $u(0,y,t) = u_0(y)$.

This inhomogeneous boundary condition makes the problem difficult to solve – it cannot be matched independently for each separable solution, as could be done with homogeneous conditions. We can approach the problem another way, however, by subtracting the steady-state solution of the problem from the general solution to convert it into a problem with homogeneous boundary conditions.

As $t \to \infty$, we expect the amplitude $u(x,y,t)$ to approach a steady-state value $u_s(x,y)$ which is independent of t. For this steady-state value, $\partial u_s / \partial t = 0$, so the solution satisfies Laplace's equation,

$$\frac{\partial^2 u_s}{\partial x^2} + \frac{\partial^2 u_s}{\partial y^2} = 0, \tag{15.179}$$

and must satisfy the same boundary conditions that the general solution $u(x,y,t)$ must satisfy. The solution to this problem is readily found to be

$$u_s(x,y) = \sum_{n=1}^{\infty} 2U_n \sin(n\pi y/a) e^{n\pi} \sinh(n\pi x/a), \tag{15.180}$$

with

$$U_n = \frac{2}{a} \frac{1}{1 - e^{2n\pi}} \int_0^a u_0(y) \sin(n\pi y/a) dy. \tag{15.181}$$

We now define a new function $v(x,y,t) = u(x,y,t) - u_s(x,y)$. This new function also satisfies the diffusion equation,

$$v_{xx}^2 + v_{yy}^2 - \alpha^2 v_t = 0, \tag{15.182}$$

and $v = 0$ on all four boundaries. We may now solve this problem, with *homogeneous* boundary conditions, in the usual manner. We assume a separable solution of the form

$$v(x,y,t) = X(x)Y(y)T(t), \tag{15.183}$$

which results in a separable differential equation

$$\frac{X''}{X} + \frac{Y''}{Y} - \alpha^2 \frac{T'}{T} = 0. \tag{15.184}$$

It is again most convenient to choose constants so that X and Y satisfy the harmonic oscillator equation. We let the first term of Eq. (15.184) be equal to $-M^2$, the second be equal to $-N^2$, and the third equal to λ^2, with the condition that

$$\lambda^2 = M^2 + N^2. \tag{15.185}$$

The solutions to each of the equations are as follows,

$$X_M(x) = A_M \sin(Mx) + B_M \cos(Mx), \tag{15.186}$$

$$Y_N(y) = C_N \sin(Ny) + D_N \cos(Ny), \tag{15.187}$$

$$T_\lambda(t) = E_\lambda e^{-\alpha^2 \lambda^2 t}. \tag{15.188}$$

Applying the homogeneous boundary conditions on the sides of the square, we have $B_M = 0$, $D_N = 0$, and

$$M = m\pi/a, \quad m = 1,2,\ldots, \tag{15.189}$$

$$N = n\pi/a, \quad n = 1,2,\ldots. \tag{15.190}$$

Our general solution may be written in the form

$$v(x,y,t) = \sum_{m=1}^{\infty} \sum_{n=1}^{\infty} F_{mn} \sin(m\pi x/a) \sin(n\pi y/a) e^{-\alpha^2 \lambda^2 t}. \tag{15.191}$$

Here we have an additional challenge: we need to match the initial conditions for the function $u(x,y,t)$, not for the function $v(x,y,t)$. However, comparing Eqs. (15.180) and (15.191), we see that the equation for u_s consists of a single sum, while the equation for v consists of a double sum. We can remedy this by Fourier expanding the x-component of u_s,

$$2e^{n\pi} \sinh(n\pi x/a) = \sum_{m=1}^{\infty} Q_{mn} \sin(m\pi x/a), \tag{15.192}$$

where

$$Q_{mn} = \frac{2}{a} \int_0^a 2e^{n\pi} \sinh(n\pi x/a) \sin(m\pi x/a) dx. \tag{15.193}$$

We may then write the solution for $u(x,y,t)$ in the form

$$u(x,y,t) = \sum_{m=1}^{\infty}\sum_{n=1}^{\infty}(F_{mn}e^{-\alpha^2\lambda^2 t} + U_n Q_{mn})\sin(m\pi x/a)\sin(n\pi y/a). \qquad (15.194)$$

We may now use Fourier techniques to match the initial conditions,

$$F_{mn} = -U_n Q_{mn} + \frac{4}{a^2}\int_0^a\int_0^a u_i(x,y)\sin(m\pi x/a)\sin(n\pi y/a)dxdy. \qquad (15.195)$$

◇

15.7 Solutions by integral transforms

The strategy behind separation of variables is to reduce the partial differential equation to a set of ordinary differential equations by looking for factorized solutions. An alternative strategy, particularly useful when the domain of one of the variables is infinite or semi-infinite, is to reduce a problem to a set of ordinary differential equations by the use of integral transforms.

We consider the use of the Laplace transform, discussed in Section 12.3, as an illustration. The Laplace transform $F(s)$ of a function $f(t)$ is given by

$$F(s) = \int_0^{\infty} f(t)e^{-st}dt. \qquad (15.196)$$

One may also perform an inverse transformation, i.e. determine the function $f(t)$ from the function $F(s)$. The inverse Laplace transform is nontrivial and involves an integral in the complex domain,

$$f(t) = \frac{1}{2\pi}\int_{-\infty}^{\infty} F(s)e^{st}d\omega, \qquad (15.197)$$

where $s = \sigma + i\omega$. The inverse transform involves extending the function $F(s)$ into the complex plane, and integrating it along a contour parallel to the imaginary axis, as discussed in Section 12.3.6. The inverse Laplace transform can typically be evaluated by using the residue theorem, with a contour enclosing the left-hand half-space.

How can we use this transform to solve a differential equation? As noted in Section 14.10, the Laplace transform can be used to reduce a differential equation to an algebraic equation. It can also be applied to a partial differential equation to eliminate one or more derivatives from a multi-variable equation. However, this simplification is often traded for difficulties in the solution further along.

As an example, we solve again the diffusion problem of an earlier subsection using integral transforms.

Example 15.7 Heat diffusion in a finite length metal rod. Integral transforms
We again consider the solution of the diffusion equation,

$$u_{xx} - \alpha^2 u_t = 0, \qquad (15.198)$$

on a metal rod which extends from $0 \leq x \leq 1$, subject to the boundary conditions

$$u(0,t) = 0,$$

$$u(1,t) = 0 \tag{15.199}$$

and the initial condition

$$u(x,0) = \phi(x), \quad 0 \leq x \leq 1. \tag{15.200}$$

We take the Laplace transform of the differential equation with respect to the time variable; the transformed equation is

$$U_{xx} - sU + u_0 = 0, \tag{15.201}$$

where $u_0 = u(x,0) = \phi(x)$, and for simplicity we have chosen $\alpha = 1$ (this problem will be hard enough, in the end). This is an inhomogeneous, second-order *ordinary* differential equation, which we can solve! The solution to the homogeneous equation is

$$U_h(x,s) = Ae^{-\sqrt{s}x} + Be^{\sqrt{s}x}. \tag{15.202}$$

The solution to the inhomogeneous equation can be found by the use of the method of *variation of constants*, discussed in Section 14.7; we assume a solution of the form

$$U_i(x,s) = v_1(x,s)e^{-\sqrt{s}x} + v_2(x,s)e^{\sqrt{s}x}, \tag{15.203}$$

and the Wronskian of the homogeneous solutions is

$$W(x,s) = \begin{vmatrix} e^{-\sqrt{s}x} & e^{\sqrt{s}x} \\ -\sqrt{s}e^{-\sqrt{s}x} & \sqrt{s}e^{\sqrt{s}x} \end{vmatrix} = 2\sqrt{s}. \tag{15.204}$$

The solutions for v_1 and v_2 are, from Eq. (14.110),

$$v_1(x,s) = -\int_{x_1}^{x} \frac{e^{\sqrt{s}x'}\phi(x')}{2\sqrt{s}}dx', \quad v_2(x,s) = \int_{x_2}^{x} \frac{e^{-\sqrt{s}x'}\phi(x')}{2\sqrt{s}}dx'. \tag{15.205}$$

The inhomogeneous solution takes on the form

$$U_i(x,s) = -\int_{0}^{x} \frac{e^{\sqrt{s}(x'-x)}\phi(x')}{2\sqrt{s}}dx' + \int_{0}^{x} \frac{e^{-\sqrt{s}(x'-x)}\phi(x')}{2\sqrt{s}}dx', \tag{15.206}$$

where we have taken $x_1 = 0$, $x_2 = 0$. We need to match our boundary conditions, however; this suggests that the complete solution must include both the inhomogeneous part and a homogeneous part,

$$U(x,s) = U_i(x,s) + Ae^{-\sqrt{s}x} + Be^{\sqrt{s}x}. \tag{15.207}$$

We find the coefficients A and B by requiring that the boundary conditions are satisfied, which in "Laplace transform space" become

$$U(0,s) = 0, \qquad (15.208)$$

$$U(1,s) = 0. \qquad (15.209)$$

If we use these boundary conditions, we get two equations and two unknowns, A and B, and we may solve them, to find

$$A = \frac{1}{2}\frac{1}{\sinh\sqrt{s}}\int_0^1 \frac{\sinh[\sqrt{s}(1-x')]}{\sqrt{s}}\phi(x')\mathrm{d}x', \qquad (15.210)$$

$$B = -\frac{1}{2}\frac{1}{\sinh\sqrt{s}}\int_0^1 \frac{\sinh[\sqrt{s}(1-x')]}{\sqrt{s}}\phi(x')\mathrm{d}x' = -A, \qquad (15.211)$$

where we simplified this equation using the fact that $\sinh x = (e^x - e^{-x})/2$. Grouping terms a bit further, we find that the overall solution takes on the messy form

$$U(x,s) = \int_0^x \frac{\sinh[\sqrt{s}(1-x')]}{\sqrt{s}}\phi(x')\mathrm{d}x' - \frac{\sinh[\sqrt{s}x]}{\sinh\sqrt{s}}\int_0^x \frac{\sinh[\sqrt{s}(1-x')]}{\sqrt{s}}\phi(x')\mathrm{d}x'. \qquad (15.212)$$

Our boundary conditions are satisfied, and our initial condition is satisfied; the only step left to solve the problem is to take the inverse Laplace transform, which can easily be written using Eq. (15.197) as

$$u(x,t) = \frac{1}{2\pi}\int_{-\infty}^{\infty} e^{zt}\left\{ \int_0^x \frac{\sinh[\sqrt{z}(1-x')]}{\sqrt{z}}\phi(x')\mathrm{d}x' \right.$$
$$\left. -\frac{\sinh[\sqrt{z}x]}{\sinh\sqrt{z}}\int_0^x \frac{\sinh[\sqrt{z}(1-x')]}{\sqrt{z}}\phi(x')\mathrm{d}x' \right\}\mathrm{d}\omega, \qquad (15.213)$$

where $z = \sigma + i\omega$. Okay, so it isn't particularly easy at all! We have a complex integral of functions which evidently involve a branch point, since all terms contain $\sqrt{z}$. At this point most people would either give up or flee in desperation to the nearest table of integrals. *This is more or less the problem one always encounters in solving PDEs by integral equation techniques*: one can find the solution in the transformed space quite easily, but returning to the original, "real" space is extremely challenging. When an analytical solution is needed, and we are faced with the choice between separation of variables or integral transforms, separation of variables is almost always easier. Integral transform techniques are often used, however, in finding numerical solutions, for Eq. (15.213) is an exact solution for which the integral can be taken numerically.

It turns out that we can in fact evaluate Eq. (15.213), using the integration contour of Fig. 12.2 and the residue theorem. First of all, we note that each term in the integral is a ratio of functions which contain $\sqrt{z}$; if we look at the simplest of these, and perform a power series expansion,

$$\frac{\sinh[\sqrt{z}x]}{\sqrt{z}} = \frac{1}{\sqrt{z}} \sum_{n=0}^{\infty} \frac{(\sqrt{z}x)^{2n+1}}{(2n+1)!} = \sum_{n=0}^{\infty} \frac{x^{2n+1}}{(2n+1)!} z^n, \tag{15.214}$$

which is a function which contains no branch point at all! A similar result is found for the other complex terms of the series. Evidently the point $z = 0$ is a *removable singularity*, and the function is not multivalued. In fact, the only singularities which occur are at the points where $\sinh \sqrt{z} = 0$, i.e. points where $z = -n^2\pi^2$, with $n \geq 1$.

We may use the residue theorem to calculate the integral, which may be expressed formally as

$$u(x,t) = 2\pi i \sum_{n=1}^{\infty} \left[\text{Res}\{z = -n^2\pi^2\} \right]. \tag{15.215}$$

What are the residues at these points? Looking at Eq. (15.213), the first half of the integral has no singularities at all, so we need only consider the residues of the function

$$f(z) = -\frac{1}{2\pi} e^{zt} \frac{\sinh[\sqrt{z}x]}{\sinh \sqrt{z}} \int_0^x \frac{\sinh[\sqrt{z}(1-x')]}{\sqrt{z}} \phi(x') dx'. \tag{15.216}$$

The singular part of this integral is the $1/(\sinh[\sqrt{z}])$, so we need to consider this function, expanded about the points $z_0 = -n^2\pi^2$,

$$\sinh[\sqrt{z}] \approx \sinh[\sqrt{z_0}] - \frac{1}{2} \frac{\cosh[\sqrt{z_0}]}{\sqrt{z_0}} (z - z_0). \tag{15.217}$$

Now, noting that the residue of a function $f(z)$ at a simple pole $z = z_0$ is given by

$$\text{Res}(z_0) = \lim_{z \to z_0} (z - z_0) f(z), \tag{15.218}$$

we may write that

$$\text{Res}(z = -n^2\pi^2) = \frac{1}{\pi i} e^{-(n\pi)^2 t} (-1)^n \sin(n\pi x) \int_0^x \sin(n\pi(1-x')) \phi(x') dx'. \tag{15.219}$$

Using the sum rule for sine functions, $\sin(A - B) = \sin A \cos B - \cos A \sin B$, we find that

$$\text{Res}(z = -n^2\pi^2) = \frac{1}{\pi i} e^{-(n\pi)^2 t} \sin(n\pi x) \int_0^x \sin(n\pi x') \phi(x') dx'. \tag{15.220}$$

Plugging this result into Eq. (15.215), we find that

$$u(x,t) = \sum_{n=1}^{\infty} 2e^{-(n\pi)^2 t} \sin(n\pi x) \int_0^x \sin(n\pi x')\phi(x')dx'. \qquad (15.221)$$

On comparison with the solution of the same problem using separation of variables, Eq. (15.177), we find that they are in fact the same solution!

In solving this problem, we have truly begun to use a variety of mathematical techniques. In the above, we used ODE solutions, integral transforms, Taylor series methods, and complex variables methods to solve a seemingly simple PDE. This example should emphasize that none of the techniques we are learning in this book exists in a vacuum and that each should be considered as one piece of a much larger mathematical puzzle.

◇

15.8 Inhomogeneous problems and eigenfunction solutions

The problems we have considered so far are *homogeneous* partial differential equations, which are satisfied by the trivial solution $u = 0$. It is quite common, however, to encounter inhomogeneous partial differential equations; for instance, the electric potential $V(\mathbf{r})$ in the presence of a charge density $\rho(\mathbf{r})$ satisfies Poisson's equation,

$$\nabla^2 V(\mathbf{r}) = -4\pi\rho(\mathbf{r}). \qquad (15.222)$$

The most elegant technique for solving such problems is through the use of Green's functions, to be discussed in Chapter 19. However, when one is interested in solving an inhomogeneous problem in a finite domain, it is also possible to use a Fourier series expansion to derive the result; we demonstrate this possibility with an example.

Example 15.8 (Diffusion with a source) We consider the solution of the inhomogeneous diffusion equation,

$$u_{xx} - \alpha^2 u_t = f(x,t), \qquad (15.223)$$

subject to the boundary conditions $u(0,t) = 0$, $u(L,t) = 0$ and the initial condition $u(x,0) = \phi(x)$. If we solved the homogeneous equation, we would have a solution of the form

$$u(x,t) = \sum_{n=1}^{\infty} A_n e^{-(n\pi\alpha)^2 t} \sin(n\pi x), \qquad (15.224)$$

with A_n determined by the initial conditions. The boundary conditions are completely encapsulated in the functions $\sin(n\pi x)$ and, furthermore, we note that these sine functions form a *complete basis* of functions on the domain $0 \le x \le L$. This mean we may try a solution to the differential equation of the form

$$u(x,t) = \sum_{n=1}^{\infty} T_n(t) \sin(n\pi x), \tag{15.225}$$

and may also expand our function $f(x,t)$ in a Fourier series,

$$f(x,t) = \sum_{n=1}^{\infty} f_n(t) \sin(n\pi x). \tag{15.226}$$

We may substitute these series into our original differential equation; because the equation must be satisfied term-by-term, we find that

$$\alpha^2 T_n'(t) + (n\pi)^2 T_n(t) = f_n(t). \tag{15.227}$$

We now only have to satisfy an inhomogeneous first-order ordinary differential equation! If we mulitply this equation by the integrating factor $\exp[(n\pi/\alpha)^2 t]$, we may rewrite it as

$$\alpha^2 \left[\exp[(n\pi/\alpha)^2 t] T_n(t) \right]' = f_n(t) \exp[(n\pi/\alpha)^2 t]. \tag{15.228}$$

The solution is immediately found to be

$$T_n(t) = \frac{1}{\alpha^2} \exp[-(n\pi/\alpha)^2 t] \int_0^t f_n(t') \exp[(n\pi/\alpha)^2 t'] dt' + A_n \exp[(n\pi/\alpha)^2 t]. \tag{15.229}$$

The constants A_n may be found by Fourier analysis as well: we expand the initial condition in the form

$$\phi(x) = \sum_{n=1}^{\infty} \phi_n \sin(n\pi x). \tag{15.230}$$

Matching this with Eq. (15.225) with $t = 0$, and Eq. (15.229) for the form of $T_n(t)$, we find that

$$\phi_n = A_n. \tag{15.231}$$

◇

15.9 Infinite domains; the d'Alembert solution

Solutions of the wave equation and the diffusion equation on the infinite line $-\infty < x < \infty$ are not typically needed in modern applications, but they have a certain elegance and conceptual clarity that makes them worth deriving. In these cases, a closed-form solution can be found.

We first consider the solution of the wave equation,

$$\frac{\partial^2 u}{\partial x^2} - \frac{1}{c^2}\frac{\partial^2 u}{\partial t^2} = 0. \tag{15.232}$$

In Section 15.2, we saw that this equation could be reduced to the canonical form

$$\frac{\partial^2 u}{\partial \xi \partial \eta} = 0, \tag{15.233}$$

where $\xi = x - ct$ and $\eta = x + ct$. Direct integration of this equation shows that the general solution to the differential equation is of the form

$$u(x,t) = F(x - ct) + G(x + ct). \tag{15.234}$$

The functions F and G depend on the initial conditions. Let us assume

$$u(x,0) = f(x), \tag{15.235}$$

$$u_t(x,0) = g(x). \tag{15.236}$$

These conditions translate to the equations

$$f(x) = F(x) + G(x), \tag{15.237}$$

$$g(x) = -cF'(x) + cG'(x), \tag{15.238}$$

where $F'(x) = \mathrm{d}F/\mathrm{d}x$. We may take the derivative of the first of these equations and then solve for F', G',

$$F'(x) = \frac{1}{2c}[cf'(x) - g(x)], \tag{15.239}$$

$$G'(x) = \frac{1}{2c}[cf'(x) + g(x)]. \tag{15.240}$$

We may integrate the function $F'(x)$ from 0 to the value $x - ct$, and the function $G'(x)$ from 0 to the value $x + ct$. This gives

$$F(x - ct) = \frac{1}{2}f(x - ct) - \frac{1}{2c}\int_0^{x-ct} g(x')\mathrm{d}x' + A, \tag{15.241}$$

$$G(x + ct) = \frac{1}{2}f(x + ct) + \frac{1}{2c}\int_0^{x+ct} g(x')\mathrm{d}x' + B. \tag{15.242}$$

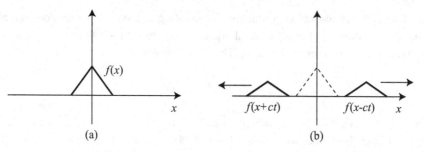

Figure 15.4 The d'Alembert solution. (a) Initial triangular displacement. (b) Displacement at later time.

The constants of integration must satisfy $A + B = 0$, which can be determined from Eq. (15.237). We may then determine the final result,

$$u(x,t) = \frac{1}{2}[f(x-ct)+f(x+ct)] + \frac{1}{2c}\int_{x+ct}^{x-ct} g(x')dx'. \tag{15.243}$$

This is the d'Alembert solution to the wave equation, first derived in 1746. It demonstrates that, in general, an initial disturbance at $t = 0$ will produce two waves, a right-going and a left-going wave. This is illustrated in Fig. 15.4 for an initial disturbance of the form

$$f(x) = \begin{cases} 1 - |x|, & |x| \le 1, \\ 0, & |x| > 1, \end{cases} \tag{15.244}$$

$$g(x) = 0. \tag{15.245}$$

We now turn to the solution of the diffusion equation,

$$u_{xx} - \alpha^2 u_t = 0, \tag{15.246}$$

subject to an initial disturbance $u(x,0) = \phi(x)$. The most straightforward technique for deriving the result is the use of Laplace transforms, as done in Example 15.7. We instead attempt a derivation through a careful study of the properties of the PDE.

The most important observation is that Eq. (15.246) is unchanged by a particular scaling of the coordinates. Let us suppose that $u(x,t)$ is a solution of the PDE, and ask under what conditions the function $u(X,T)$, with $X \equiv ax$, $T \equiv bT$, is also a solution. By use of the chain rule, we find Eq. (15.246) takes the form

$$a^2 u_{XX} - b\alpha^2 u_T = 0. \tag{15.247}$$

Provided $a = \sqrt{b}$, the differential equation is still satisfied. If $u(x,t)$ is a solution, so is any function of the form $u(\sqrt{b}x, bt)$.

We now look for a solution with the initial condition

$$\phi(x) = S(x), \tag{15.248}$$

where $S(x)$ is the Heaviside step function. The scaling of the coordinates does not affect this initial condition nor the differential equation, so the solution $u(x,t)$ itself should be unchanged under the process of scaling. This only happens if

$$u(x,t) = F(x/\sqrt{t}).$$ (15.249)

It can be easily seen that the function $u(\sqrt{b}x, bt) = u(x,t)$.

By use of the chain rule, we may use Eq. (15.249) to convert our PDE into an ODE in variable $y \equiv x/\sqrt{t}$,

$$u_{xx} = \frac{d^2 F}{dy^2} \left(\frac{dy}{dx} \right)^2 = \frac{d^2 F}{dy^2} \frac{1}{t},$$ (15.250)

$$u_t = \frac{dF}{dy} \frac{dy}{dt} = -\frac{1}{2t} \frac{x}{\sqrt{t}} \frac{dF}{dy}.$$ (15.251)

Substituting into our PDE, and canceling a factor of t, we have

$$F'' + \frac{\alpha^2}{2} y F' = 0.$$ (15.252)

This equation is readily solved for F', so that

$$F'(y) = c_1 \exp[-\alpha^2 y^2/4].$$ (15.253)

Integrating a second time, we have

$$F(x/\sqrt{t}) = c_1 \int_0^{x/\sqrt{t}} \exp[-\alpha^2 y'^2/4] dy' + c_2.$$ (15.254)

To determine the constants c_1 and c_2, we match this solution to the initial condition, Eq. (15.248). This gives us two equations in the limit $t \to 0$, for $x > 0$ and $x < 0$,

$$1 = c_1 \int_0^\infty \exp[-\alpha^2 y'^2/4] dy' + c_2 = c_1 \sqrt{\frac{\pi}{\alpha^2}} + c_2,$$ (15.255)

$$0 = c_1 \int_0^{-\infty} \exp[-\alpha^2 y'^2/4] dy' + c_2 = -c_1 \sqrt{\frac{\pi}{\alpha^2}} + c_2.$$ (15.256)

This results in $c_2 = 1/2$ and $c_1 = \alpha/(2\sqrt{\pi})$. The solution to the diffusion equation for the Heaviside step function is therefore

$$u_h(x,t) = \frac{1}{2} + \frac{\alpha}{2\sqrt{\pi}} \int_0^{x/\sqrt{t}} e^{-\alpha^2 y^2/4} dy.$$ (15.257)

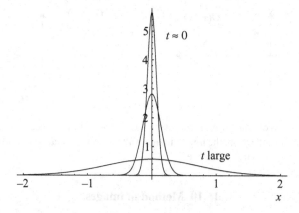

Figure 15.5 The behavior of solutions to the diffusion equation. Narrowly concentrated initial distributions spread rapidly.

If we are willing to consider initial conditions which are distributions, we recall that the derivative of the Heaviside step function is the Dirac delta function. This suggests that the derivative of the solution $u_h(x,t)$ is the solution to the diffusion equation for $\phi(x) = \delta(x)$,

$$\frac{du_h(x,t)}{dx} = u_\delta(x,t) = \frac{\alpha}{2\sqrt{\pi t}} e^{-\alpha^2 x^2/4t}. \tag{15.258}$$

It is to be noted that this function behaves as a delta sequence in the limit $t \to 0$. This solution should remain unchanged if we shift the origin of the coordinate system. The solution to the diffusion equation for a delta-like excitation at a point x', then, is

$$u_\delta(x,t) = \frac{\alpha}{2\sqrt{\pi t}} e^{-\alpha^2 (x-x')^2/4t}. \tag{15.259}$$

Finally, we note that an arbitrary initial condition $\phi(x)$ may be written in a delta representation using the sifting property,

$$\phi(x) = \int_{-\infty}^{\infty} \delta(x - x')\phi(x')dx'. \tag{15.260}$$

The corresponding solution to the diffusion equation satisfying $u(x,0) = \phi(x)$ should be

$$u(x,t) = \frac{\alpha}{2\sqrt{\pi t}} \int_{-\infty}^{\infty} \phi(x')e^{-\alpha^2 (x-x')^2/4t}dx'. \tag{15.261}$$

This solution can be interpreted quite nicely by the use of delta sequences. If $\phi(x) = \delta(x-x')$ represents a concentration of heat at a point at the initial time $t = 0$, the solution above indicates that the heat spreads (diffuses) from the point x' as a Gaussian function. This is illustrated in Fig. 15.5.

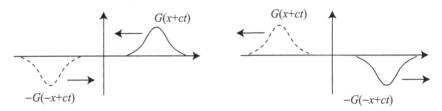

Figure 15.6 The method of images. A mirror image $-G(-x+ct)$ is introduced to match the boundary condition at $x = 0$. After reflection, the mirror image wave is now the real wave and the original wave has propagated into the virtual $x < 0$ space.

15.10 Method of images

We have seen that in a bounded domain, separation of variables can be used to derive solutions to partial differential equations, while in unbounded domains solutions can often be found by other analytic means. When the domain of interest is *semi-infinite*, for instance the line $0 \leq x < \infty$, one can often use a technique known as the *method of images*.

We illustrate this with the simple example of the wave equation on the semi-infinite line $x \geq 0$, subject to the boundary condition $u(0,t) = 0$. Let us suppose we have found a solution $u(x,t) = F(x - ct) + G(x + ct)$ for the infinite line that satisfies a given set of initial conditions; the final step is then to incorporate the boundary condition into the solution.

We can do this by considering the domain of the entire line, and introducing a "mirror image" of the solution into the $x < 0$ domain of the form

$$u_m(x,t) = -F(-x - ct) - G(-x + ct). \tag{15.262}$$

This solution has the opposite amplitude of the infinite line solution, and opposing position. We readily see that

$$u(0,t) + u_m(0,t) = F(-ct) + G(ct) - F(-ct) - G(ct) = 0, \tag{15.263}$$

automatically satisfying our boundary condition! Because any solution that satisfies the wave equation and matches the boundary conditions and initial conditions is unique, we have found the unique solution to the wave equation on the semi-infinite line satisfying a Dirichlet boundary condition. The solution via the method of images is illustrated in Fig. 15.6.

This technique may also be applied for the case of a Neumann boundary condition, $u_x(0,t) = 0$. In this case, the appropriate mirror image is

$$u_m(x,t) = F(-x - ct) + G(-x + ct). \tag{15.264}$$

For a Dirichlet boundary condition, the reflected wave is inverted, while for a Neumann condition, the reflected wave has the same amplitude.

The method of images will be applied again when we discuss Green's functions in Chapter 19.

15.11 Additional reading

- S. J. Farlow, *Partial Differential Equations for Scientists and Engineers* (New York, Wiley, 1982). An easy introduction to the theory and solution of PDEs. Also in Dover.
- H. F. Weinberger, *A First Course in Partial Differential Equations* (New York, Blaisdell Publishing Company, 1965). Another introductory book, which includes complex variables and integral transforms almost as a "bonus". Also in Dover.
- W. A. Strauss, *Partial Differential Equations: an Introduction* (New York, Wiley, 1992). Another introductory text, this one is much more oriented towards mathematicians. (This was my undergraduate PDE text.)
- I. G. Petrovsky, *Lectures on Partial Differential Equations* (New York, Interscience, 1954). This one is a translation of a series of lectures by a Russian mathematician, and is not for the faint of heart. Contains a very rigorous discussion of the classification of PDEs. Also in Dover.

15.12 Exercises

1. Determine the locations in (x, t) for which the following PDEs are hyperbolic, elliptic, and parabolic:
   1. $x \frac{\partial^2 u}{\partial t^2} + 2t \frac{\partial^2 u}{\partial x \partial t} + \frac{\partial^2 u}{\partial x^2} = 0,$
   2. $x^2 \frac{\partial^2 u}{\partial t^2} + \frac{\partial^2 u}{\partial x \partial t} + t^2 \frac{\partial^2 u}{\partial x^2} + x^2 \frac{\partial u}{\partial x} = 0,$
   3. $\frac{\partial^2 u}{\partial t^2} + 2t^3 \frac{\partial^2 u}{\partial x \partial t} + \frac{\partial^2 u}{\partial x^2} + xtu = 0.$

2. Determine the locations in (x, t) for which the following PDEs are hyperbolic, elliptic, and parabolic:
   1. $\frac{\partial^2 u}{\partial t^2} + 2 \frac{\partial^2 u}{\partial x \partial t} + x^3 \frac{\partial^2 u}{\partial x^2} + u = 0,$
   2. $t^2 \frac{\partial^2 u}{\partial t^2} + \frac{\partial^2 u}{\partial x \partial t} + t^2 \frac{\partial^2 u}{\partial x^2} + x^2 \frac{\partial u}{\partial t} = 0,$
   3. $x \frac{\partial^2 u}{\partial t^2} + 2t \frac{\partial^2 u}{\partial x \partial t} + x \frac{\partial^2 u}{\partial x^2} = 0.$

3. We consider waves of amplitude $u(x, t)$ on a string of length L, which satisfy the wave equation

$$\frac{\partial^2 u}{\partial t^2} - \frac{1}{c^2} \frac{\partial^2 u}{\partial x^2} = 0.$$

   Assuming the initial conditions $u(x, 0) = \phi(x)$, $u_t(x, 0) = \eta(x)$, find the solution for the boundary conditions
   1. $u(0, t) = 0$, $u_x(L, t) = 0$,
   2. $u_x(0, t) = 0$, $u_x(L, t) = 0$.
   Explain qualitatively how the difference in boundary conditions affects the allowed wavelengths of excitation.

4. Consider a string of length L which is fixed at both ends ($x = 0$ and $x = L$). If air resistance is taken into account, the string displacement $u(x, t)$ satisfies the PDE

$$\frac{\partial^2 u}{\partial t^2} + 2\Gamma \frac{\partial u}{\partial t} - \frac{1}{c^2} \frac{\partial^2 u}{\partial x^2} = 0,$$

where Γ and c are positive constants. Suppose that the displacement and velocity on the string at $t = 0$ are

$$u(x,0) = f(x),$$

$$\frac{\partial u(x,0)}{\partial t} = 0,$$

and that $\Gamma < \pi c/L$.

1. Find the solution $u(x,t)$ for $t > 0$.
2. Using the answer to part (a), show that if $\Gamma \ll \pi c/L$, then

$$u(x,t) \approx \exp[-\Gamma t]u_{nr}(x,t),$$

where $u_{nr}(x,t)$ is the solution in the absence of air resistance, $\Gamma = 0$.

5. Consider a string of length L which is fixed at both ends ($x = 0$ and $x = L$). If the medium surrounding the string is rubber (instead of air), the string displacement $u(x,t)$ satisfies the PDE

$$\frac{\partial^2 u(x,t)}{\partial x^2} - \frac{1}{c^2}\frac{\partial^2 u(x,t)}{\partial t^2} - \mu^2 u(x,t) = 0,$$

where μ and c are positive constants. Suppose that the displacement and velocity on the string at $t = 0$ are

$$u(x,0) = f(x),$$

$$\frac{\partial u(x,0)}{\partial t} = 0.$$

Find $u(x,t)$ for $t > 0$.

6. The telegrapher's equation arises in the transmission of electrical signals in a wire with distributed capacitance, resistance, and inductance. It is given by

$$\frac{\partial^2 u}{\partial t^2} + 2\Gamma\frac{\partial u}{\partial t} + \Delta u - \frac{1}{c^2}\frac{\partial^2 u}{\partial x^2} = 0,$$

where Γ, Δ, and c are positive constants. Solve this equation on the domain $0 \le x \le L$ with boundary conditions $u(0,t) = 0$, $u(L,t) = 0$, with the initial conditions $u(x,0) = 0$, $u_t(x,0) = g(x)$.

7. Waves propagating on a rectangular drumhead ($-L \le x \le L$, $-L \le y \le L$) satisfy the wave equation

$$u_{xx} + u_{yy} - \frac{1}{c^2}u_{tt} = 0.$$

The ends of the drumhead are assumed to be fixed ($u(-L,t) = 0$ and so forth). Assuming initial conditions $u(x,y,0) = f(x,y)$ and $u_t(x,y,0) = 0$, find the solution of the wave equation using separation of variables.

8. Find the solution of Laplace's equation in polar coordinates,

$$\frac{1}{\rho}\frac{\partial}{\partial \rho}\left(\rho\frac{\partial V}{\partial \rho}\right) + \frac{1}{\rho^2}\frac{\partial^2 V}{\partial \phi^2} = 0,$$

for the two systems displayed below.

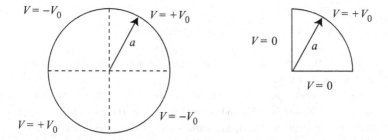

Explain why the two solutions should be identical in the first quadrant of the coordinate system.

9. Find the solution to Laplace's equation valid for the *exterior* of a circle of radius a, with boundary value $V(a,\phi) = V_0(\phi)$.

10. Find the solution to Laplace's equation for the inside of a rectangle $0 \le x \le a, 0 \le y \le b$, satisfying the boundary conditions $V(0,y) = 0$, $V(x,0) = 0$, $V(a,y) = V_a(y)$, $V(x,b) = V_b(x)$.

11. Find the solution to Laplace's equation for the "slotted" region illustrated below.

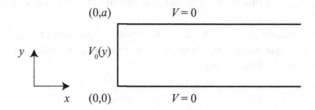

Explain clearly any "hidden" boundary conditions that are applied.

12. Find the solution to Laplace's equation,

$$\frac{1}{\rho}\frac{\partial}{\partial\rho}\left(\rho\frac{\partial V}{\partial\rho}\right) + \frac{1}{\rho^2}\frac{\partial^2 V}{\partial\phi^2} = 0,$$

in an annular domain of inner radius a and outer radius b, subject to the boundary conditions $V(a,\phi) = V_a(\phi)$, $V(b,\phi) = V_b(\phi)$. (Recall that you will need to keep all possible solutions to match the boundary conditions, including the logarithmic solution.)

13. We consider the diffusion equation,

$$u_{xx} - \alpha^2 u_t = 0,$$

on a finite domain $0 \le x \le L$. The boundary conditions are taken to be $u(0,t) = u_0$, $u(L,t) = u_L$, and the initial condition is $u(x,0) = \phi(x)$. Find the solution $u(x,t)$.

14. Find the solution to the diffusion equation,

$$u_{xx} - \alpha^2 u_t = 0,$$

on the finite domain $0 \le x \le L$, subject to the boundary conditions $u(0,t) = 0$, $u(L,t) + \beta u_x(L,t) = 0$ and the initial condition $u(x,0) = \phi(x)$.

15. Heat-flow in a finite-length metal rod with lateral loss through the sides of the rod can be described by the equation

$$u_{xx} - \alpha^2 u_t + \beta u = 0.$$

We assume that the ends of the rod are heat sinks, i.e. $u(0,t) = 0$, $u(L,t) = 0$, and that the initial condition is $u(x,0) = \phi(x)$.

   1. Simplify the differential equation by first looking for a solution $u(t)$ to the equation in the absence of diffusion, $u_{xx} = 0$, and assuming $u(x,t) = u(t)w(x,t)$, find an equation for $w(x,t)$.

   2. Solve the equation for $w(x,t)$, and subsequently find the solution $u(x,t)$.

16. Heat-flow in a finite-length metal rod with convection can be described by the equation

$$u_{xx} - \alpha^2 u_t - \beta u_x = 0.$$

We assume that the ends of the rod are heat sinks, i.e. $u(0,t) = 0$, $u(L,t) = 0$, and that the initial condition is $u(x,0) = \phi(x)$.

   1. Simplify the differential equation by first looking for a solution $v(x,t)$ to the equation in the absence of diffusion, $u_{xx} = 0$, and assuming $u(x,t) = v(x,t)w(x,t)$, find an equation for $w(x,t)$.

   2. Solve the equation for $w(x,t)$, and subsequently find the solution $u(x,t)$.

17. The boundary condition $\frac{\partial u}{\partial n} = 0$, where n is the normal to the boundary, represents an insulating boundary condition. Solve the diffusion equation,

$$u_{xx} + u_{yy} - \alpha^2 u_t = 0,$$

on the square domain $0 \le x \le L$, $0 \le y \le L$, where all boundaries are insulating and the initial condition is $u(x,y,0) = \phi(x,y)$. Interpret physically the result as $t \to \infty$.

18. We have seen that the solution to the diffusion equation on the infinite line is given by

$$u(x,t) = \frac{\alpha}{2\sqrt{\pi t}} \int_{-\infty}^{\infty} \phi(x') e^{-\alpha^2 (x-x')^2/4t} dx'.$$

Now suppose that the diffusion only occurs on the half-line, $0 \le x < \infty$. Use the method of images to find the solution on this half-line (integrals now range from 0 to ∞) given that the boundary condition at $x = 0$ is given by

   1. $u(0,t) = 0$,

   2. $u_x(0,t) = 0$.

19. Consider a square membrane which occupies an area $0 \le x \le L$, $0 \le y \le L$, and the four edges of the membrane are clamped down. The displacement of the membrane satisfies

$$\nabla^2 u(x,y) = f(x,y),$$

where $f(x,y)$ is the external force per unit area applied to the membrane. Find a series solution for $u(x,y)$.

20. We consider waves on a finite length string $0 \le x \le L$ subject to an applied force $f(x,t)$, and fixed at both ends. The waves satisfy the equation

$$u_{xx} - \frac{1}{c^2} u_{tt} = f(x,t),$$

and have initial conditions $u(x,0) = F(x)$, $u_t(x,0) = 0$. Find a series solution for $u(x,t)$.

21. We consider an electric potential $V(x,y)$ satisfying Laplace's equation in the half-space $x \ge 0$, with a boundary condition $V(0,y) = V_0(y)$, subject to the condition $\int_{-\infty}^{\infty} |V_0(y)| dy < \infty$. Use Fourier transformations to find an integral representation for the potential in $x > 0$.

16

Bessel functions

16.1 Introduction: propagation in a circular waveguide

We return to the problem of propagation in a waveguide, but now consider a circular waveguide of radius a with extension along the z-direction. The wave still satisfies the Helmholtz equation, but it is now natural to look for solutions in cylindrical coordinates,

$$\frac{1}{\rho}\frac{\partial}{\partial\rho}\left(\rho\frac{\partial U(\rho,\phi,z)}{\partial\rho}\right) + \frac{1}{\rho^2}\frac{\partial^2 U(\rho,\phi,z)}{\partial\phi^2} + \frac{\partial^2 U(\rho,\phi,z)}{\partial z^2} + k^2 U(\rho,\phi,z) = 0. \quad (16.1)$$

The wave is assumed to vanish on the boundary of the waveguide, i.e. $U(a,\phi,z) = 0$. We are primarily interested in solutions which behave as traveling waves along the length of the waveguide (the $+z$-direction); we therefore try a solution of the form

$$U(\rho,\phi,z) = u(\rho,\phi)e^{ik_L z}, \quad (16.2)$$

where k_L is an as yet undetermined constant, to be referred to as the longitudinal wavenumber. It is to be noted that we have, in essence, assumed a solution to the Helmholtz equation which is partially separable. On substitution, we readily find that

$$\frac{1}{\rho}\frac{\partial}{\partial\rho}\left(\rho\frac{\partial u(\rho,\phi)}{\partial\rho}\right) + \frac{1}{\rho^2}\frac{\partial^2 u(\rho,\phi)}{\partial\phi^2} + (k^2 - k_L^2)u(\rho,\phi) = 0. \quad (16.3)$$

Making a further assumption that

$$u(\rho,\phi) = R(\rho)\Phi(\phi), \quad (16.4)$$

i.e. that the solution is separable into distinct ρ and ϕ components, we can transform our differential equation into the form

$$\frac{\rho}{R(\rho)}\frac{d}{d\rho}\left(\rho\frac{dR(\rho)}{d\rho}\right) + (k^2 - k_L^2)\rho^2 = -\frac{1}{\Phi(\phi)}\frac{d\Phi(\phi)}{d\phi^2}. \quad (16.5)$$

The left-hand side of this equation is dependent only upon ρ, while the right-hand side is dependent only upon ϕ. Each side of this equation must therefore be equal to a constant, defined for convenience as m^2, which results in the following equation for Φ

$$\frac{d\Phi(\phi)}{d\phi^2} = -m^2\Phi(\phi). \tag{16.6}$$

This is a simple harmonic oscillator equation and, as noted in Section 15.5, a "hidden" condition of the solution is that the function be single-valued, which restricts m to an integer value. The equation for $R(\rho)$ may then be written as

$$\rho\frac{d}{d\rho}\left(\rho\frac{dR(\rho)}{d\rho}\right) + (\hat{k}^2\rho^2 - m^2)R(\rho) = 0, \tag{16.7}$$

where we have now written $\hat{k}^2 \equiv k^2 - k_L^2$. Defining a new variable $x \equiv \hat{k}\rho$, we may write this equation in the simplified form

$$x^2R''(x) + xR'(x) + (x^2 - m^2)R(x) = 0. \tag{16.8}$$

This ordinary differential equation is nontrivial and its solutions cannot be expressed as a simple combination of elementary functions. Equation (16.8) is known as *Bessel's equation* and the solutions to it are known as *Bessel functions*. Like any second-order ODE, there are two linearly independent solutions to Bessel's equation for integer m, which are generally written as $J_m(x)$ and $N_m(x)$, the latter being referred to as a *Neumann function*. Functions such as the Bessel functions, which have a rich set of mathematical properties and arise naturally in a variety of physical and mathematical problems, are typically referred to as *special functions*.[1]

What can the properties of Bessel functions tell us about the behavior of waves in circular waveguides? First, we will see that the Neumann functions are singular (infinite) at the origin, so they represent unphysical solutions. For this problem, we may focus our attention on the behavior of the Bessel functions $J_m(x)$.

Several of the low-order Bessel functions are shown in Fig. 16.1. It is to be noted that each Bessel function characterized by order m has an infinite set of zeros appearing roughly periodically in x. We label the nth zero of the mth Bessel function as x_{nm}, so that $J_m(x_{nm}) = 0$. Applying our boundary condition $U(a,\phi,z) = 0$, we find that we require

$$a\sqrt{k^2 - k_L^2} = x_{nm}. \tag{16.9}$$

Rewriting this in terms of k_L, we find that

$$k_L^2 = \frac{k^2a^2 - x_{nm}^2}{a}. \tag{16.10}$$

If $x_{nm}^2 > k^2a^2$, we find that $k_L^2 < 0$ and k_L must be imaginary. This implies that the wave is exponentially damped as it travels in the positive z-direction, i.e. it is a nonpropagating

---

[1] Michael Berry has written a nice essay about the need for special functions, "Why are special functions special?" [Ber01]

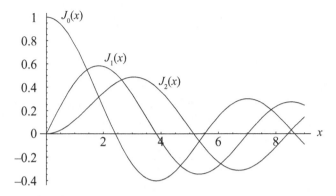

Figure 16.1 The first few Bessel functions of integer order.

wave. Just like in the case of the rectangular waveguide discussed in Section 15.1, we find that the size of a waveguide limits the total number of modes which can propagate within it. A single-mode waveguide in this case is one for which the $m = 0, n = 1$ mode is propagating, but no others.

A general solution for a wave propagating in our waveguide would be of the form

$$U(\rho,\phi,z) = \sum_{m=0}^{\infty}\sum_{n=0}^{\infty} J_m(x_{nm}\rho/a)\left[A_{nm}e^{im\phi} + B_{nm}e^{-im\phi}\right]e^{ik_L z}, \tag{16.11}$$

with A_{mn} and B_{mn} to be determined by boundary conditions.

Bessel functions are ubiquitous in the field of optics, and in this chapter we delve into their mathematical properties in significant detail.

16.2 Bessel's equation and series solutions

Bessel's differential equation of order m is defined by the expression

$$x^2 R''(x) + x R'(x) + (x^2 - m^2)R(x) = 0. \tag{16.12}$$

We consider for the moment integer values of m, and discuss fractional orders in Section 16.11. We attempt to find solutions to Bessel's equation using the Frobenius method of Section 14.9; we assume a solution $J_m(x)$ such that

$$J_m(x) = \sum_{n=0}^{\infty} c_n x^{n+s}, \tag{16.13}$$

with derivatives

$$J_m'(x) = \sum_{n=0}^{\infty} (s+n)c_n x^{n+s-1}, \tag{16.14}$$

$$J_m''(x) = \sum_{n=0}^{\infty} (s+n)(s+n-1)c_n x^{n+s-2}. \tag{16.15}$$

On substituting into Eq. (16.12), and simplifying, we have the equation

$$\sum_{n=0}^{\infty} \left[(n+s)(n+s-1) + (n+s) - m^2 \right] c_n x^{n+s} + \sum_{n=0}^{\infty} c_n x^{n+s+2} = 0. \tag{16.16}$$

The x^0 term of the series gives us the indicial equation,

$$s(s-1) + s - m^2 = 0, \tag{16.17}$$

or $s = \pm m$. We try the positive root, $s = m$, and consider the x^1 term of the series,

$$(2m+1)c_1 = 0, \tag{16.18}$$

or $c_1 = 0$. The recursion relation for the higher-order terms of the series takes on the form

$$c_n = \frac{(-1)c_{n-2}}{n(n+2m)}. \tag{16.19}$$

Reorganizing terms, one can write a series expression for the Bessel function of order m,

$$J_m(x) = \sum_{n=0}^{\infty} \frac{(-1)^n}{n!(n+m)!} \left(\frac{x}{2}\right)^{2n+m}. \tag{16.20}$$

This equation will also turn out to be valid for non-integer m, provided the factorial functions are replaced with the appropriate gamma functions (see Appendix A). The first few Bessel functions were shown in Fig. 16.1. It is to be noted that, for small x, $J_m(x) \sim x^m$.

The Bessel functions are oscillatory but nonperiodic, and the zeros of the Bessel functions are regularly spaced but not equally spaced. We will discuss the zeros in more detail in Section 16.9.

As a well-behaved second-order linear ordinary differential equation, Bessel's equation should have a second solution which is linearly independent of $J_m(x)$. For non-integer order m, the second solution can be readily shown to be $J_{-m}(x)$. For integer order, however, a second solution of Bessel's equation is not as easy to come by: we can show that $J_{-m}(x)$ is not linearly independent of $J_m(x)$. This can be seen by substituting $-m$ directly into the equation (16.20),

$$J_{-m}(x) = \sum_{s=0}^{\infty} \frac{(-1)^s}{s!(s-m)!} \left(\frac{x}{2}\right)^{2s-m}. \tag{16.21}$$

For $s - m < 0$, $(s - m)! \to \infty$, and so these terms of the series vanish; the series effectively starts with $s = m$. Defining a new summation index $t = s - m$, Eq. (16.21) may be rewritten as

$$J_{-m}(x) = \sum_{t=0}^{\infty} \frac{(-1)^{t+m}}{t!(t+m)!} \left(\frac{x}{2}\right)^{2t+m}. \tag{16.22}$$

This is simply related to the positive order by

$$J_{-m}(x) = (-1)^m J_m(x). \tag{16.23}$$

Therefore, for integer order, the two series solutions of Bessel's equation are not independent. We could have predicted this from the discussion of Section 14.9: it was noted there that one of the solutions may be of logarithmic form when the roots of the indicial equation differ by an integer.

There is some freedom in choosing this second solution, and historically a number of different choices for the second solution were proposed, each with its own advantages; we examine the form which is now considered standard.

Let us define a new function $N_m(x)$ by the expression

$$N_m(x) \equiv \frac{\cos(m\pi)J_m(x) - J_{-m}(x)}{\sin(m\pi)}, \tag{16.24}$$

and this function is known[2] as a *Neumann function of order m*. For noninteger m, this function is clearly linearly independent of J_m, because it is a linear combination of J_m and J_{-m}. For integer m, however, this expression reduces to the indeterminate form $0/0$. We may use L'Hôpital's rule to evaluate what happens, and may write the outcome in differential form as

$$N_m(x) = \lim_{\nu \to m} \frac{1}{\pi} \left[\frac{\partial J_\nu(x)}{\partial \nu} - (-1)^\nu \frac{\partial J_{-\nu}(x)}{\partial \nu} \right]. \tag{16.25}$$

Using the power series expansion for J_m [Eq. (16.20)] and J_{-m} [Eq. (16.21)], one may also write an explicit power series for $N_m(x)$. It is not pretty! It can be written as

$$N_m(x) = \frac{1}{\pi} \left[2\log\frac{x}{2} + 2\gamma - \sum_{k=1}^{m} \frac{1}{k} \right] J_m(x) - \frac{1}{\pi} \sum_{n=0}^{m-1} \frac{(m-n-1)!}{n!} \left(\frac{x}{2}\right)^{-m+2n}$$

$$- \frac{1}{\pi} \sum_{n=1}^{\infty} \frac{(-1)^n}{(n+m)!n!} \left[\sum_{k=1}^{n} \left(\frac{1}{k} + \frac{1}{m+k}\right) \right] \left(\frac{x}{2}\right)^{m+2n}, \tag{16.26}$$

---

[2] The function is named after Neumann but it is actually due to Weber and Schläfli.

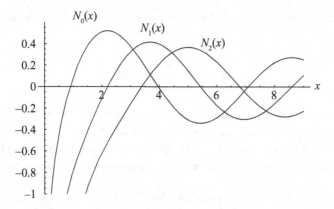

Figure 16.2 The first few Neumann functions of integer order.

where γ is the so-called Euler–Mascheroni constant, defined as

$$\gamma \equiv \lim_{N\to\infty}\left(\sum_{k=1}^{N}\frac{1}{k}-\log N\right). \tag{16.27}$$

The first few integer Neumann functions are plotted in Fig. 16.2. It is to be noted that the solution has logarithmic terms, as we would expect. Furthermore, the mth Neumann function diverges in the limit as $x \to 0$ as x^{-m}, with the exception that N_0 diverges as $\log x$.

Why define the second solution to Bessel's equation in this manner? First of all, this form of the second solution is linearly independent of $J_m(x)$ for both integer and non-integer m. Second, it can be shown that the Neumann functions satisfy the same recurrence relations (to be discussed in Section 16.4) that the Bessel functions satisfy.

The Neumann functions are extremely important in the solution of differential equations, and actually appear in some electromagnetic Green's functions, as we will see in Chapter 19. It should be noted that one rarely, if ever, needs to use the series representation of the Neumann functions given in Eq. (16.26). A major goal of this chapter is to derive enough properties of the Bessel and Neumann functions that an explicit calculation of the series can be avoided.

It is to be noted that the Bessel functions have a close relationship with the gamma function $\Gamma(z)$. We will use some results relating to gamma functions in proving some general properties of Bessel functions; gamma functions are discussed in Appendix A.

16.3 The generating function

We will see that there is a strong connection between the various Bessel functions of integer order, and that those connections can be used to derive useful formulas relating the functions. Both the connections, and the formulas demonstrating them, are elegantly encapsulated in a single function known as the *generating function*.

The generating function is derived by postulating that there exists a function $g(x,t)$ such that

$$g(x,t) = \sum_{n=-\infty}^{\infty} J_n(x)t^n. \qquad (16.28)$$

If we treat t as a complex variable, this equation suggests that we are looking for a complex function such that its Laurent series coefficients are given by the Bessel functions. Generating functions exist for all of the special functions we will consider in the following chapters; for the Bessel functions, the generating function has the closed form

$$g(x,t) = e^{(x/2)(t-1/t)}. \qquad (16.29)$$

This may be demonstrated by considering the Taylor series representations of the exponential function,

$$\exp[xt/2] = \sum_{k=0}^{\infty} \frac{(xt)^k}{k!2^k}, \qquad (16.30)$$

$$\exp[-x/2t] = \sum_{l=0}^{\infty} \frac{(-1)^l x^l}{l!(2t)^l}. \qquad (16.31)$$

These Taylor series were noted in Section 9.7 to be absolutely and uniformly convergent in t everywhere in the complex plane except the point at infinity and zero, respectively, and therefore their product is also well-behaved in that region. The product of these two series may be written as

$$e^{(x/2)(t-1/t)} = \sum_{k=0}^{\infty} \sum_{l=0}^{\infty} \frac{(-1)^l}{k!l!} 2^{-k-l} x^{k+l} t^{k-l}. \qquad (16.32)$$

Comparing this expression term by term in powers of t with Eq. (16.28), we have the constraint that $k - l = n$, or $l = k - n$. This leads to the relation

$$J_n(x) = \sum_{k=0}^{\infty} \frac{(-1)^{k-n}}{k!(k-n)!} \left(\frac{x}{2}\right)^{2k-n}. \qquad (16.33)$$

On comparison with Eq. (16.20), we find that Eq. (16.29) for the generating function is valid. We may therefore write

$$e^{(x/2)(t-1/t)} = \sum_{n=-\infty}^{\infty} J_n(x)t^n. \qquad (16.34)$$

This expression can be used to derive a number of important formulas and relationships amongst the Bessel functions, though one subtlety should be noted: the generating function only applies to Bessel functions of integer order, while many of the results we will describe apply equally well to non-integer order. One can derive the results described in later sections for general order by the use of Bessel's equation, Eq. (16.12).

Example 16.1 (Jacobi–Anger expansion) In optics, one regularly analyzes the effect of an optical system on monochromatic plane waves with spatial dependence

$$U(\mathbf{r}) = U_0 \exp[i k \mathbf{n} \cdot \mathbf{r}], \tag{16.35}$$

where $k = \omega/c$ is the wavenumber of the light, ω is the angular frequency of the wave, and $\mathbf{n}$ is the direction of propagation. When considering two-dimensional wave propagation problems (propagation in the $x - y$ plane, $\mathbf{r} = (x, y)$), this plane wave can be readily expanded into a series of cylindrical waves using the Bessel generating function. From Eq. (16.34), we choose $t = i \exp[i\theta]$, where θ represents the angle between $\mathbf{n}$ and $\mathbf{r}$. We may then write the generating function in the form

$$\exp\left[iu\frac{\exp[i\theta] + \exp[-i\theta]}{2}\right] = \exp[iu\cos\theta] = \sum_{n=-\infty}^{\infty} i^n J_n(u) \exp[in\theta]. \tag{16.36}$$

Let us choose $\mathbf{n} = \hat{\mathbf{x}}$, and we may then write the spatially dependent part of our plane wave in the form

$$U(\mathbf{r}) = U_0 \exp[i k r \cos\theta], \tag{16.37}$$

with $r = \sqrt{x^2 + y^2}$. Making the identification $u \equiv kr$, we may write

$$\exp[i k r \cos\theta] = \sum_{n=-\infty}^{\infty} i^n J_n(kr) \exp[in\theta]. \tag{16.38}$$

This is the Jacobi–Anger expansion, which represents a two-dimensional plane wave as a coherent sum of cylindrical waves. This can be particularly useful if one is analyzing a problem in which a plane wave is propagating in a system with cylindrical symmetry.
◇

16.4 Recurrence relations

If we take the derivative of the generating function with respect to t, we find that

$$\frac{\partial g(x,t)}{\partial t} = \frac{1}{2}x\left(1 + \frac{1}{t^2}\right)e^{(x/2)(t-1/t)} = \frac{1}{2}x \sum_{n=-\infty}^{\infty} J_n(x)t^n + \frac{1}{2}x \sum_{n=-\infty}^{\infty} J_n(x)t^{n-2}, \tag{16.39}$$

where we have resubstituted the definition of the generating function into the expression in the last step. The derivative of the other side of the generating function formula (16.29), however, shows that

$$\frac{\partial g(x,t)}{\partial t} = \sum_{n=-\infty}^{\infty} n J_n(x) t^{n-1}. \tag{16.40}$$

Equating these two formulas, we have

$$\frac{1}{2} x \sum_{n=-\infty}^{\infty} J_n(x) t^n + \frac{1}{2} x \sum_{n=-\infty}^{\infty} J_n(x) t^{n-2} = \sum_{n=-\infty}^{\infty} n J_n(x) t^{n-1}. \tag{16.41}$$

We may equate like powers of t in this expression, and are immediately led to the expression

$$J_{n-1}(x) + J_{n+1}(x) = \frac{2n}{x} J_n(x). \tag{16.42}$$

This is a *recurrence relation* (or recursion relation) for the Bessel functions. Given $J_{n-1}(x)$ and $J_n(x)$, we can immediately find $J_{n+1}(x)$.

Alternatively, we can take the derivative of the generating function with respect to x, and find that

$$\frac{\partial g(x,t)}{\partial x} = \frac{1}{2}\left(t - \frac{1}{t}\right) e^{(x/2)(t-1/t)} = \sum_{n=-\infty}^{\infty} J_n'(x) t^n. \tag{16.43}$$

Again, if we substitute into this equation the series definition of the generating function, we have

$$\frac{1}{2}\left(t - \frac{1}{t}\right) e^{(x/2)(t-1/t)} = \frac{1}{2}\left(t - \frac{1}{t}\right) \sum_{n=-\infty}^{\infty} J_n(x) t^n = \sum_{n=-\infty}^{\infty} J_n'(x) t^n. \tag{16.44}$$

By equating like powers of t again, we get a different recurrence relation,

$$J_{n-1}(x) - J_{n+1}(x) = 2J_n'(x). \tag{16.45}$$

A special case of this can be found for $n = 0$. Using $J_{-1}(x) = -J_1(x)$, we have

$$J_0'(x) = -J_1(x). \tag{16.46}$$

One more pair of recurrence relations can be found by combining the two developed previously. If we add Eqs. (16.42) and (16.45) and divide by 2, we have

$$J_{n-1} = \frac{n}{x} J_n(x) + J_n'(x). \tag{16.47}$$

If we multiply this equation by x^n and rearrange terms appropriately, we have

$$\frac{d}{dx}\left[x^n J_n(x)\right] = x^n J_{n-1}(x). \tag{16.48}$$

Similarly, if we *subtract* Eq. (16.45) from Eq. (16.42), we can group terms in a different way to find that

$$\frac{d}{dx}\left[x^{-n}J_n(x)\right] = -x^{-n}J_{n+1}(x).$$

(16.49)

These recurrence relations can be used to derive one other useful relation. Dividing Eq. (16.49) by x, it may be written in the form

$$\frac{J_{n+1}(x)}{x^{n+1}} = \frac{1}{x}\frac{d}{dx}\left[\frac{J_n(x)}{x^n}\right].$$

(16.50)

If we define a new set of functions

$$U_n(x) \equiv \frac{J_n(x)}{x^n},$$

(16.51)

we find that

$$U_{n+1}(x) = -\frac{1}{x}\frac{d}{dx}\left[U_n(x)\right].$$

(16.52)

Working up from $n = 0$, we can derive an expression for all higher-order Bessel functions from $J_0(x)$,

$$J_n(x) = (-1)^n x^n \left(\frac{1}{x}\frac{d}{dx}\right)^n J_0(x).$$

(16.53)

This expression is analogous to Rodrigues' formula for Legendre polynomials, to be discussed in Section 17.4.

As noted in Section 16.2, the definition of the Neumann function has been chosen in part to make it satisfy exactly the same recurrence relations the Bessel function $J_n(x)$ satisfies. With the Neumann function defined by Eq. (16.24), we may in general write

$$Z_{n-1}(x) + Z_{n+1}(x) = \frac{2n}{x}Z_n(x),$$

(16.54)

where Z_n is either J_n or N_n, as well as

$$Z_{n-1}(x) - Z_{n+1}(x) = 2Z_n'(x).$$

(16.55)

Also, recursion relations between the Bessel and Neumann functions can be found, such as

$$J_n N_{n+1} - J_{n+1} N_n = -\frac{2}{\pi x}.$$

(16.56)

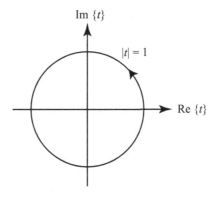

Figure 16.3 The contour of integration used to derive an integral representation of a Bessel function.

16.5 Integral representations

In addition to the series representations of Bessel functions, we may also construct integral representations using the generating function and the tools of complex analysis. Letting t be a complex variable, we integrate the function $t^{-m-1}g(x,t)$, where m is an integer, about the unit circle $|t| = 1$ in the complex plane, as illustrated in Fig. 16.3.

This integral can be written as

$$\oint_{|t|=1} t^{-m-1}g(x,t)\mathrm{d}t = \oint_{|t|=1} \sum_{n=-\infty}^{\infty} J_n(x)t^{n-m-1}\mathrm{d}t$$

$$= 2\pi\,\mathrm{i}\mathrm{Res}(t=0) = 2\pi\,\mathrm{i}J_m(x). \qquad (16.57)$$

The last step follows from the residue theorem; the only power of the Laurent series expansion which contributes to the residue is t^{-1}. We may rewrite this integral as

$$J_m(x) = \frac{1}{2\pi\,\mathrm{i}} \oint_{|t|=1} t^{-m-1}g(x,t)\mathrm{d}t. \qquad (16.58)$$

This integral may be put in a more useful form by letting $t = \mathrm{e}^{\mathrm{i}\theta}$, $\mathrm{d}t = \mathrm{i}\mathrm{e}^{\mathrm{i}\theta}$, so that

$$J_m(x) = \frac{1}{2\pi} \int_0^{2\pi} \mathrm{e}^{-\mathrm{i}m\theta}\mathrm{e}^{\mathrm{i}x\sin\theta}\mathrm{d}\theta. \qquad (16.59)$$

The Bessel function is real-valued, so the right-hand side of this equation must also be real-valued; we can take the real part of both sides of this equation to find

$$J_m(x) = \frac{1}{2\pi} \int_0^{2\pi} \cos[m\theta - x\sin\theta]\mathrm{d}\theta. \qquad (16.60)$$

Provided m is an integer, it is to be noted that the integrand is periodic and the limits of integration may be replaced by any 2π range, for instance $-\pi < \theta \leq \pi$.

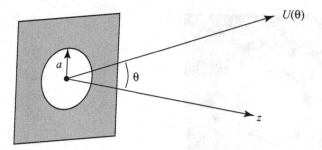

Figure 16.4 Notation related to aperture diffraction.

There are many other ways to write Eq. (16.59). Perhaps the most useful of these involves the special case $m = 0$, for which we may write

$$J_0(x) = \frac{1}{2\pi} \int_0^{2\pi} e^{ix\sin\theta}d\theta. \tag{16.61}$$

We may transform coordinates to $\theta' = \theta - \pi/2$, to get the new formula

$$J_0(x) = \frac{1}{2\pi} \int_0^{2\pi} e^{ix\cos\theta'}d\theta'. \tag{16.62}$$

This formula appears in diffraction theory, which we now consider as an example.

Example 16.2 (Diffraction by a circular aperture) In the theory of light diffraction through a circular aperture in a black screen, the field U far from the screen is roughly proportional to the Fourier transform of the aperture field,

$$U(\theta) \sim \int_0^a \int_0^{2\pi} e^{ikr\sin\theta\cos\phi}d\phi rdr, \tag{16.63}$$

where a is the radius of the aperture, θ is the angle at which the field is observed relative to the normal z of the screen, and $k = 2\pi/\lambda$ is the wavenumber of the light, where λ is the wavelength. The geometry is illustrated in Fig. 16.4.

The integral over ϕ is given immediately by Eq. (16.62), so that

$$U(\theta) \sim 2\pi \int_0^a J_0(kr\sin\theta)rdr. \tag{16.64}$$

Letting $x \equiv kr\sin\theta$ this integral may be rewritten as

$$U(\theta) \sim \frac{2\pi}{(k\sin\theta)^2} \int_0^{x_a} J_0(x)xdx, \tag{16.65}$$

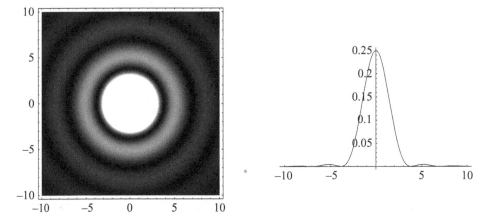

Figure 16.5 The intensity pattern resulting from diffraction by a circular aperture, and a cross-section of that pattern.

where now $x_a = ka \sin\theta$. From Eq. (16.48), we may rewrite this equation as

$$U(\theta) \sim \frac{2\pi}{(k \sin\theta)^2} \int_0^{x_a} \frac{d}{dx}[xJ_1(x)]dx = \frac{2\pi}{k \sin\theta} J_1(ka \sin\theta). \qquad (16.66)$$

The intensity of the field, given by the squared modulus of the field, is therefore

$$I(\theta) \sim a^2 \left(\frac{J_1(ka \sin\theta)}{ka \sin\theta} \right)^2, \qquad (16.67)$$

which is the famous *Airy pattern* of diffraction theory.[3] The intensity of the diffracted field observed on a screen beyond the aperture will appear as shown in Fig. 16.5, where we have plotted the total intensity as observed on the screen and a plot of the cross-section through it.
◊

The formula (16.59) cannot be directly extended to fractional values of m, because such a transformation introduces a branch point at the origin. With some identities related to the gamma function (discussed in Appendix A), we may develop a general integral formula for the Bessel functions, and from this we may develop a general integral formula for the Neumann functions.

Our starting point is the following complex integral representation of the gamma function,

$$\frac{1}{\Gamma(n+m+1)} = \frac{1}{2\pi i} \int_C e^s s^{-(n+m+1)} ds, \qquad (16.68)$$

---

[3] It should be noted that I dropped some unimportant factors of 2π in the final expression.

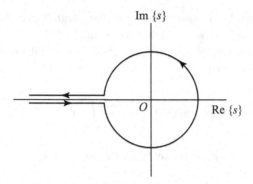

Figure 16.6 The contour used to evaluate Eq. (16.68).

where the contour C is the path shown in Fig. 16.6. In general, m is non-integer, which leaves a branch point at the origin of the integrand and the contour is chosen such that a cut line lies along the negative real axis.

Writing our series solution for the Bessel functions in its general, gamma, form (using $\Gamma(z+1) = z!$, for z integer), we have

$$J_m(x) = \sum_{n=0}^{\infty} \frac{(-1)^n}{\Gamma(n+1)\Gamma(n+m+1)} \left(\frac{x}{2}\right)^{2n+m}. \tag{16.69}$$

We may substitute from Eq. (16.68) into this equation; rearranging terms, we may write

$$J_m(x) = \left(\frac{x}{2}\right)^m \frac{1}{2\pi i} \int_C \sum_{n=0}^{\infty} \frac{(-1)^n}{\Gamma(n+1)} e^s s^{-m-1} \left(\frac{x^2}{4s}\right)^n ds, \tag{16.70}$$

where we have interchanged the orders of integration and summation. The summation, however, is simply an exponential function,

$$\sum_{n=0}^{\infty} \frac{(-1)^n}{\Gamma(n+1)} \left(\frac{x^2}{4s}\right)^n = \exp\left[-x^2/4s\right]. \tag{16.71}$$

We may therefore write Eq. (16.70) in the compact form

$$J_m(x) = \left(\frac{x}{2}\right)^m \frac{1}{2\pi i} \int_C e^{-x^2/4s} s^{-m-1} e^s ds. \tag{16.72}$$

Making a change of variables to $t = s/2x$, we find that

$$J_m(x) = \frac{1}{2\pi i} \int_{C'} e^{\frac{x}{2}(t-t^{-1})} t^{-m-1} dt, \tag{16.73}$$

where C' is a contour similar to C. It is to be noted that this integral is of the exact structural form of Eq. (16.57), which was determined with the generating function; the only difference is the choice of contour. For integer values of m, the integrals over the two infinite legs of the contour are equal and opposite and therefore cancel.

We write our complex variable $t = \rho e^{i\theta}$, and choose the circular part of the contour to be of radius 1. We may then write

$$J_m(x) = \frac{1}{\pi}\int_0^\pi \cos(x\sin\theta - m\theta)d\theta - \frac{\sin m\pi}{\pi}\int_1^\infty e^{-\frac{1}{2}x(\rho-\rho^{-1})}\rho^{-m-1}d\rho. \tag{16.74}$$

with the further choice $\rho = e^\alpha$, this may be written as

$$J_m(x) = \frac{1}{\pi}\int_0^\pi \cos(x\sin\theta - m\theta)d\theta - \frac{\sin m\pi}{\pi}\int_0^\infty e^{-x\sinh\alpha - m\alpha}d\alpha. \tag{16.75}$$

This equation is an integral formula for a Bessel function of general order m. It is to be noted that the second term vanishes when m is integer, due to the factor $\sin m\pi$, and we then reproduce Eq. (16.57).

By use of Eq. (16.24) for the Neumann function and Eq. (16.75), we may determine an integral form for the Neumann function,

$$N_m(x) = \frac{1}{\pi}\int_0^\pi \sin(x\sin\theta - m\theta)d\theta - \frac{1}{\pi}\int_0^\infty e^{-x\sinh\alpha}\left[e^{m\alpha} + e^{-m\alpha}\cos m\pi\right]d\alpha. \tag{16.76}$$

16.6 Hankel functions

In wave problems, it is convenient to introduce a second linearly independent set of solutions to Bessel's equation which are complex valued,

$$H_m^{(1)}(x) = J_m(x) + iN_m(x), \tag{16.77}$$

$$H_m^{(2)}(x) = J_m(x) - iN_m(x). \tag{16.78}$$

The functions $H_m^{(1)}(x)$ and $H_m^{(2)}(x)$ are called *Hankel functions of the first and second kinds*, respectively. The definition is analogous to the expression of a complex exponential in terms of sines and cosines, i.e.

$$e^{\pm ix} = \cos x \pm i \sin x. \tag{16.79}$$

In Chapter 19, we will see that Hankel functions appear quite naturally as the Green's functions for two-dimensional monochromatic wave systems. In such systems, the Hankel functions of the first and second kind represent cylindrical waves emanating from or converging to the point $x = 0$.

16.7 Modified Bessel functions

Occasionally, problems arise in which a modified form of Bessel's equation appears,

$$x^2 R''(x) + x R'(x) - (x^2 + m^2) R(x) = 0. \tag{16.80}$$

The only difference between this equation and Bessel's equation is the sign of x^2 in the last term. It can, however, be readily converted into Bessel's equation by making a coordinate transformation $x = -\mathrm{i}z$. It follows that the solutions to this equation are the Bessel and Neumann functions but with imaginary argument, i.e. $J_m(\mathrm{i}x)$ and $N_m(\mathrm{i}x)$. The expression for $J_m(\mathrm{i}x)$ follows immediately from Eq. (16.20),

$$J_m(\mathrm{i}x) = \mathrm{i}^m \sum_{n=0}^{\infty} \frac{1}{n!(n+m)!} \left(\frac{x}{2}\right)^{2n+m}. \tag{16.81}$$

To make the function real-valued, we choose a normalization such that

$$I_m(x) \equiv \frac{1}{\mathrm{i}^m} J_m(\mathrm{i}x). \tag{16.82}$$

The function $I_m(x)$ is referred to as a *modified Bessel function of the first kind*. It can be seen from Eq. (16.81) that $I_{-m}(x) = I_m(x)$.

Recurrence relations for I_m may be readily found by applying the formulas of Section 16.4 with $J_m(x) = \mathrm{i}^m I_m(-\mathrm{i}x)$. One finds that

$$I_{m-1}(x) - I_{m+1}(x) = \frac{2m}{x} I_m(x). \tag{16.83}$$

Similarly, one can show that

$$I_{m-1}(x) + I_{m+1}(x) = 2 I'_m(x). \tag{16.84}$$

Further relationships can be developed by a modification of the generating function,

$$\mathrm{e}^{(\mathrm{i}x/2)(t-1/t)} = \sum_{n=-\infty}^{\infty} I_n(x)(\mathrm{i}t)^n. \tag{16.85}$$

A second modified Bessel function $K_m(x)$ can be introduced which is valid for both integer and non-integer m in the form

$$K_m(x) = \frac{\pi}{2} \frac{I_{-m}(x) - I_m(x)}{\sin m\pi}. \tag{16.86}$$

This expression is chosen to give $K_m(x)$ an asymptotic form complementary to $I_m(x)$, as discussed in the next section.

16.8 Asymptotic behavior of Bessel functions

We have calculated Taylor (and Laurent) series expansions for the Bessel and Neumann functions; this means that, for small x, we can approximate the function by the lowest-order terms of the series. It is often of use to consider the opposite extreme, i.e. what happens to the Bessel functions for large x.

This is much more difficult than it might sound, and new theoretical tools must be applied to the problem. We consider the integral form of the Bessel function, given by Eq. (16.59),

$$J_m(x) = \frac{1}{2\pi} \int_0^{2\pi} e^{-im\theta} e^{ix\sin\theta} d\theta. \tag{16.87}$$

This integral is of a form for which the *method of stationary phase* can be applied, to be discussed in Section 21.4; in short, if we have an integral of the form

$$F(k) = \int_a^b f(z) e^{ikg(z)} dz, \tag{16.88}$$

it can be shown in the limit of large k to have the approximate form

$$F(k) \approx f(x_0) \exp[ikg(z_0)] \exp[i\pi/4] \sqrt{\frac{2\pi}{g''(z_0)k}}, \tag{16.89}$$

where z_0 are points for which $g'(z) = 0$. We may associate $f(z) \leftrightarrow \exp[-im\theta]$, $k \leftrightarrow x$, $\theta \leftrightarrow z$, and $g(x) \leftrightarrow \sin\theta$.

Applying stationary phase to the integral for the Bessel function, it can be shown that for values of x much larger than the integer order n of a Bessel function, it may be approximated by the simple form

$$J_m(x) \approx \sqrt{\frac{2}{\pi x}} [\cos(x - m\pi/2 - \pi/4)]. \tag{16.90}$$

With a bit more effort, the asymptotic form of the Neumann function of order m can be determined using Eq. (16.76) to be

$$N_m(x) \approx \sqrt{\frac{2}{\pi x}} [\sin(x - m\pi/2 - \pi/4)]. \tag{16.91}$$

In the asymptotic regime, then, the Bessel and Neumann functions become approximately periodic, with their zeros occurring at regular intervals. The Bessel and Neumann functions can be seen to have a "complementary" nature in this limit: they are essentially phase-shifted sine and cosine functions. The order m determines the nature of the phase shift. Both functions decrease in amplitude as $x^{-1/2}$.

The asymptotic forms of the Hankel functions may be readily derived from the Bessel and Neumann functions. We have

$$H_m^{(1)}(x) \approx \sqrt{\frac{2}{\pi x}} \exp[i(x - m\pi/2 - \pi/4)], \tag{16.92}$$

$$H_m^{(2)}(x) \approx \sqrt{\frac{2}{\pi x}} \exp[-i(x - m\pi/2 - \pi/4)]. \tag{16.93}$$

The asymptotic form of the modified Bessel functions may also be determined using integral representations, for instance by replacing x by ix in Eq. (16.59). The asymptotic forms are given by

$$I_m(x) \sim \frac{1}{\sqrt{2\pi x}} e^x, \tag{16.94}$$

$$K_m(x) \sim \sqrt{\frac{\pi}{2x}} e^{-x}. \tag{16.95}$$

Again we have a complementary nature between the modified Bessel functions of the first and second kinds. It is to be noted that the modified Bessel functions of any order have the same asymptotic form.

16.9 Zeros of Bessel functions

We noted in the previous section that the zeros of the Bessel and Neumann functions are asymptotically periodic; that is, they become periodic in the limit of large x. For small values of x, however, the best we can say is that the zeros appear regularly. However, specific knowledge of the locations of the zeros is needed to solve problems with homogeneous boundary conditions such as the waveguide problem of Section 16.1.

Table 16.1 lists the lowest zeros for the first few Bessel functions; Table 16.2 provides the lowest zeros of the first few Neumann functions. Further values are tabulated in [AS65].

We label the zeros of the Bessel function as α_{mn}, which refers to the nth zero of the Bessel function of order m; we similarly label β_{mn} as the zeros of the Neumann function. Several theorems concerning the zeros of the Bessel functions are worth mentioning.

Theorem 16.1 *The zeros of the Bessel and Neumann functions are all of first order, with the possible exception of a zero at the origin.*

In other words, the leading term of the Taylor expansion of $J_m(x)$ about any zero α_{mn} is $(x - \alpha_{mn})$. This is easy to prove by contradiction: if the zero were of second order, i.e. if the leading term of the Taylor series was of the form $(x - \alpha_{mn})^2$, then $Z_m(\alpha_{mn}) = 0$ and $Z_m'(\alpha_{mn}) = 0$. By Bessel's differential equation (16.12), we must therefore have $Z_m''(\alpha_{mn}) = 0$ as well. By differentiating Bessel's equation, we can show that all of the derivatives of $Z_m(x)$ must vanish at $x = \alpha_{mn}$, which implies that the function is identically zero, a

Table 16.1 *Zeros of the low-order Bessel functions.*

Zero order α_{mn}	$J_0(x)$	$J_1(x)$	$J_2(x)$	$J_3(x)$	$J_4(x)$
0	2.4048	3.8317	5.1356	6.3802	7.5883
1	5.5201	7.0156	8.4172	9.7610	11.0647
2	8.6537	10.1735	11.6198	13.0152	14.3725
3	11.7915	13.3237	14.7960	16.2235	17.6160

Table 16.2 *Zeros of the low-order Neumann functions.*

Zero order β_{mn}	$N_0(x)$	$N_1(x)$	$N_2(x)$	$N_3(x)$	$N_4(x)$
0	0.8936	2.1971	3.3842	4.5270	5.6452
1	3.9577	5.4297	6.7938	8.0976	9.3616
2	7.0861	8.5960	10.0235	11.3965	12.7301
3	10.2223	11.7492	13.2100	14.6231	15.9996

contradiction. The origin is an exceptional point because the differential equation has a singularity at $x = 0$.

Theorem 16.2 *For $m > -1$, the zeros of the Bessel functions $J_m(x)$ are interlaced; that is,*

$$0 < \alpha_{m,1} < \alpha_{m+1,1} < \alpha_{m,2} < \alpha_{m+1,2} < \alpha_{m,3} < \cdots \qquad (16.96)$$

Simply put, the zeros of $J_m(x)$ and $J_{m+1}(x)$ lie between one another. This theorem can be seen to hold for the data of Table 16.1; a general proof follows from the recurrence formulas (16.48) and (16.49),

$$\frac{d}{dx}\left[x^n J_n(x)\right] = x^n J_{n-1}(x), \qquad (16.97)$$

$$\frac{d}{dx}\left[x^{-n} J_n(x)\right] = -x^{-n} J_{n+1}(x). \qquad (16.98)$$

Between zeros of a Bessel function $J_n(x)$, the slope of the function, and consequently the slope of $x^n J_n(x)$, must change sign at least once. Equation (16.97) demonstrates that each of these points is the location of a zero of $J_{n-1}(x)$, and therefore a zero of $J_{n-1}(x)$ lies between each zero of $J_n(x)$. Likewise, Eq. (16.98) shows that between each zero of $J_n(x)$ lies a zero of $J_{n+1}(x)$. This theorem can be shown to hold for both Bessel and Neumann functions of arbitrary real order, provided $m \geq 1$.

More details about the zeros of Bessel functions may be found in [Wat44].

16.10 Orthogonality relations

We have seen that the Bessel functions show up naturally in the solution of the Helmholtz equation via separation of variables in a cylindrical geometry. As one might expect, we can show that the Bessel functions satisfy orthogonality relations comparable to those of the sine and cosine functions which appear in Fourier series analysis.

With our cylindrical waveguide in mind, we consider the set of all Bessel functions which vanish on the surface of the waveguide, i.e. for $\rho = a$. Such functions may be written as $J_m(\alpha_{mn}\rho/a)$, where α_{mn} is the nth zero of the mth Bessel function. We postulate that, for a fixed value of m, these functions are orthogonal to one another.

To prove this, we write Bessel's equation for the functions $X_{mn}(\rho) \equiv J_m(\alpha_{mn}\rho/a)$ and $X_{mk}(\rho) \equiv J_m(\alpha_{mk}\rho/a)$,

$$\rho\frac{d^2}{d\rho^2}X_{mn}(\rho) + \frac{d}{d\rho}X_{mn}(\rho) + \left(\frac{\alpha_{mn}^2\rho}{a^2} - \frac{m^2}{\rho}\right)X_{mn}(\rho) = 0, \tag{16.99}$$

$$\rho\frac{d^2}{d\rho^2}X_{mk}(\rho) + \frac{d}{d\rho}X_{mk}(\rho) + \left(\frac{\alpha_{mk}^2\rho}{a^2} - \frac{m^2}{\rho}\right)X_{mk}(\rho) = 0. \tag{16.100}$$

We multiply the first of these equations by $X_{mk}(\rho)$ and the second by $X_{mn}(\rho)$ and subtract, which results in

$$X_{mn}(\rho)\frac{d}{d\rho}\left[\rho\frac{d}{d\rho}X_{mk}(\rho)\right] - X_{mk}(\rho)\frac{d}{d\rho}\left[\rho\frac{d}{d\rho}X_{mn}(\rho)\right]$$

$$= \frac{\alpha_{mn}^2 - \alpha_{mk}^2}{a^2}\rho X_{mk}(\rho)X_{mn}(\rho). \tag{16.101}$$

We may integrate this function from $\rho = 0$ to $\rho = a$. Performing an integration by parts on the left-hand side of the equation, we find that

$$\int_0^a \frac{\alpha_{mn}^2 - \alpha_{mk}^2}{a^2}\rho X_{mk}(\rho)X_{mn}(\rho)d\rho = \left|\rho X_{mn}(\rho)\frac{d}{d\rho}X_{mk}(\rho)\right|_0^a - \left|\rho X_{mk}(\rho)\frac{d}{d\rho}X_{mn}(\rho)\right|_0^a. \tag{16.102}$$

As long as $m \geq 0$, the lower limits at $\rho = 0$ will automatically be zero. At the upper limit, by definition $X_{mn}(a) = 0$. We therefore may write

$$\int_0^a J_m(\alpha_{mn}\rho/a)J_m(\alpha_{mk}\rho/a)\rho d\rho = 0, \quad n \neq k. \tag{16.103}$$

The functions $X_{mn}(\rho)$ are therefore orthogonal over the domain $0 \leq \rho \leq a$.

We may make these functions orthonormal by using the following integral relation:

$$\int_0^a [J_m(\alpha_{mn}\rho/a)]^2 \rho d\rho = \frac{a^2}{2}[J_{m+1}(\alpha_{mn})]^2. \tag{16.104}$$

It is to be noted that the right-hand side of this equation is necessarily nonzero because of the interlacing of zeros discussed in the previous section. It may be derived in a manner similar to Eq. (16.103), except that we start with

$$\rho\frac{d^2}{d\rho^2}J_m(\alpha_{mn}\rho/a) + \frac{d}{d\rho}J_m(\alpha_{mn}\rho/a) + \left(\frac{\alpha_{mn}^2\rho}{a^2} - \frac{m^2}{\rho^2}\right)J_m(\alpha_{mn}\rho/a) = 0, \quad (16.105)$$

$$\rho\frac{d^2}{d\rho^2}J_m(\alpha\rho/a) + \frac{d}{d\rho}J_m(\alpha\rho/a) + \left(\frac{\alpha^2\rho}{a^2} - \frac{m^2}{\rho^2}\right)J_m(\alpha\rho/a) = 0, \quad (16.106)$$

where α is a free parameter. We multiply the first equation by $J_m(\alpha\rho/a)$ and the second by $J_m(\alpha_{mn}\rho/a)$, subtract, and integrate from $\rho = 0$ to $\rho = a$, resulting in the formula

$$a^2 J_m(\alpha)\alpha_{mn}J_m'(\alpha_{mn}) = (\alpha^2 - \alpha_{mn}^2)\int_0^a \rho J_m(\alpha_{mn}\rho)J_m(\alpha\rho)d\rho. \quad (16.107)$$

We may take the derivative of this with respect to α,

$$a^2 J_m'(\alpha)\alpha_{mn}J_m'(\alpha_{mn}) = 2\alpha\int_0^a \rho J_m(\alpha_{mn}\rho)J_m(\alpha\rho)d\rho$$

$$+ (\alpha^2 - \alpha_{mn}^2)\int_0^a \rho J_m(\alpha_{mn}\rho)J_m'(\alpha\rho)d\rho. \quad (16.108)$$

In the limit $\alpha \to \alpha_{mn}$, the last term vanishes and we are left with

$$\int_0^a \rho J_m(\alpha_{mn}\rho)J_m(\alpha\rho)d\rho = \frac{a^2}{2}\left[J_m'(\alpha_{mn})\right]^2. \quad (16.109)$$

An application of the recursion relations gives Eq. (16.104).

We may construct an orthonormal set of Bessel functions by use of Eq. (16.104); introducing functions $Y_{mn}(\rho)$ such that

$$Y_{mn}(\rho) \equiv \sqrt{\frac{2}{a^2}}\frac{1}{|J_{m+1}(\alpha_{mn})|}J_m(\alpha_{mn}\rho), \quad (16.110)$$

we may expand any function $f(\rho)$ defined for $0 \le \rho \le a$ in a Bessel series of the form

$$f(\rho) = \sum_{n=0}^{\infty}c_{mn}Y_{mn}(\rho), \quad (16.111)$$

where

$$c_{mn} = \int_0^a f(\rho)Y_{mn}(\rho)\rho\,d\rho. \quad (16.112)$$

Example 16.3 (Propagation in a circular waveguide) We briefly return to the problem of propagation in a circular waveguide to see how a complete solution could be formulated using the orthogonality of the Bessel functions. The length of the waveguide is assumed to run in the z-direction, and its width is taken to be a.

Following the results of Section 16.1, the general solution must be of the form

$$U(\rho,\phi,z) = \sum_{m=0}^{\infty}\sum_{n=0}^{\infty} J_m(x_{mn}\rho/a)\left[A_{mn}e^{im\phi} + B_{mn}e^{-im\phi}\right]e^{ik_Lz}. \tag{16.113}$$

Let us assume that the field in the plane $z=0$ has the value $U_0(\rho,\phi)$; because the Helmholtz equation is an elliptic differential equation, a unique solution may be found by specifying the field alone on the boundary of the system. We now wish to write expressions for the coefficients A_{mn}, B_{mn} in terms of $U_0(\rho,\phi)$. We may write

$$U_0(\rho,\phi) = \sum_{m=0}^{\infty}\sum_{n=0}^{\infty} J_m(x_{mn}\rho/a)\left[A_{mn}e^{im\phi} + B_{mn}e^{-im\phi}\right]. \tag{16.114}$$

To solve for the coefficients, we simultaneously apply the orthogonality relations for the functions $\exp[\pm im\phi]$ and for the functions $J_m(x_{mn}\rho/a)$. Noting that

$$\int_0^a [J_m(x_{mn}\rho/a)]^2\,\rho\mathrm{d}\rho = \frac{a^2}{2}[J_{m+1}(x_{mn})]^2, \tag{16.115}$$

$$\int_0^{2\pi} e^{il\phi}e^{-il'\phi}\mathrm{d}\phi = 2\pi\delta_{ll'}, \tag{16.116}$$

we may multiply Eq. (16.114) by $J_{m'}(x_{m'n'}\rho/a)e^{-im'\phi}$ and integrate to find

$$A_{mn} = \frac{2}{2\pi a^2}\frac{1}{[J_{m+1}(x_{mn})]^2}\int_0^a J_m(x_{mn}\rho/a)e^{-im\phi}U_0(\rho,\phi)\rho\mathrm{d}\rho\mathrm{d}\phi. \tag{16.117}$$

Similarly, multiplying by $J_{m'}(x_{m'n'}\rho/a)e^{im'\phi}$ and integrating results in

$$B_{mn} = \frac{2}{2\pi a^2}\frac{1}{[J_{m+1}(x_{mn})]^2}\int_0^a J_m(x_{mn}\rho/a)e^{im\phi}U_0(\rho,\phi)\rho\mathrm{d}\rho\mathrm{d}\phi. \tag{16.118}$$

Equations (16.113), (16.117), and (16.118) represent a complete solution to the waveguide problem with boundary value $U(\rho,\phi,0) = U_0(\rho,\phi)$.

Several observations are in order. First, it is to be noted that the orthogonality over m is enforced by the exponential functions $\exp[im\phi]$, while the orthogonality over n is enforced by the Bessel functions $J_m(x_{mn}\rho/a)$. Orthogonality in the plane $z=0$ is therefore enforced by the two-dimensional basis functions

$$U_{mn}(\rho,\phi) = \sqrt{\frac{2}{2\pi a^2}}\frac{1}{|J_{m+1}(\alpha_{mn})|}J_m(x_{mn}\rho/a)e^{\pm im\phi}. \tag{16.119}$$

Second, an astute observer may note that it seems that we have not specified boundary conditions on a *closed* surface, as is required for an elliptic partial differential equation –

what happened to the "open" end of the waveguide at $z \to \infty$? We have in fact enforced a "hidden" boundary condition by assuming that the z-behavior of the wave is $\exp[ik_L z]$. Physically, this corresponds to assuming that the wave is propagating in the $+z$-direction, and this condition specifies the behavior of the wave as $z \to \infty$.

◇

One additional orthogonality relationship between Bessel functions should be mentioned. Over a semi-infinite domain, we can demonstrate

$$\int_0^\infty J_m(\alpha\rho)J_m(\alpha'\rho)\rho\,d\rho = \frac{1}{\alpha}\delta(\alpha-\alpha'), \tag{16.120}$$

provided that $m > -1/2$. This is another example of a *closure relationship* such as discussed in Section 6.4. Equation (16.120) demonstrates that an arbitrary function $f(\rho)$ on the semi-infinite line may be expanded as a sum of Bessel functions over a continuous parameter α.

16.11 Bessel functions of fractional order

We first derived Bessel's equation from one of the separated terms of the Helmholtz equation in cylindrical coordinates. In the solution of the Helmholtz equation by separation of variables in spherical coordinates, we will show in Section 17.10 that the differential equation for $R(r)$ takes on the form

$$r^2\frac{d^2R}{dr^2} + 2r\frac{dR}{dr} + [k^2r^2 - l(l+1)]R = 0. \tag{16.121}$$

This equation looks *very* similar to the Bessel equation,

$$\rho^2\frac{d^2P}{d\rho^2} + \rho\frac{dP}{d\rho} + (k^2\rho^2 - m^2)P = 0, \tag{16.122}$$

but is functionally different. In particular, the factor of 2 on the first derivative makes this equation distinct – the weights of the first and second derivatives are different. However, let us simplify this equation by first letting $kr \equiv x$, so that it becomes

$$x^2\frac{d^2R}{dx^2} + 2x\frac{dR}{dx} + [x^2 - l(l+1)]R = 0. \tag{16.123}$$

Now we define a new function, $Z(x)$, related to $R(x)$ as follows,

$$R(x) = \frac{Z(x)}{x^{1/2}}. \tag{16.124}$$

From this we have

$$R'(x) = \frac{Z'(x)}{x^{1/2}} - \frac{1}{2}\frac{Z(x)}{x^{3/2}}, \tag{16.125}$$

$$R''(x) = \frac{Z''(x)}{x^{1/2}} - \frac{Z'(x)}{x^{3/2}} + \frac{3}{4}\frac{Z(x)}{x^{5/2}}. \tag{16.126}$$

On substituting into Eq. (16.123), we find that

$$x^2 Z'' + xZ' + \left[x^2 - (l+1/2)^2\right]Z = 0, \tag{16.127}$$

which is just Bessel's equation, with $v = l + 1/2$. Evidently the solution of what we will call the spherical Bessel equation, Eq. (16.123), is related to Bessel and Neumann functions of half-integer order.

We define the *spherical Bessel functions* and *spherical Neumann functions* by the formulas

$$j_n(x) \equiv \sqrt{\frac{\pi}{2x}}J_{n+1/2}(x), \tag{16.128}$$

$$n_n(x) \equiv \sqrt{\frac{\pi}{2x}}N_{n+1/2}(x). \tag{16.129}$$

The normalization has been introduced for reasons which will become clear in a moment. In Section 16.2, we defined the series solution of the Bessel functions, which we noted was valid for integer or fractional order. We can use this to evaluate the behavior of the spherical Bessel functions; first, we note that

$$J_{n+1/2}(x) = \sum_{n=0}^{\infty} \frac{(-1)^s}{s!(s+n+1/2)!}\left(\frac{x}{2}\right)^{2s+n+1/2}. \tag{16.130}$$

This can be simplified using a formula called the *Legendre duplication formula*, proven in Appendix A,

$$z!(z+1/2)! = 2^{-2z-1}\pi^{1/2}(2z+1)!. \tag{16.131}$$

We find that the spherical Bessel function takes on the form

$$j_n(x) = 2^n x^n \sum_{s=0}^{\infty} \frac{(-1)^s(s+n)!}{s!(2s+2n+1)!}x^{2s}. \tag{16.132}$$

This probably does not seem like much of an improvement. However, let us consider the $n = 0$ spherical Bessel function,

$$j_0(x) = \sum_{s=0}^{\infty} \frac{(-1)^s}{(2s+1)!}x^{2s} = \frac{\sin x}{x}. \tag{16.133}$$

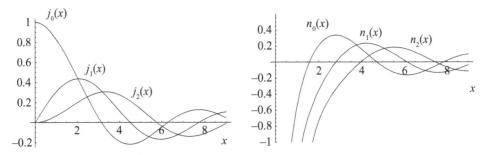

Figure 16.7 The first few spherical Bessel functions $j_m(x)$ and spherical Neumann functions $n_m(x)$.

The zeroth-order spherical Bessel function is simply related to the sine function! We may do a similar analysis for the zeroth Neumann function. First, we note from the definition (16.24) of the Neumann function that

$$N_{n+1/2}(x) = \frac{\cos(n\pi + \pi/2)J_{n+1/2}(x) - J_{-(n+1/2)}(x)}{\sin(n\pi + \pi/2)} = (-1)^{n+1}J_{-n-1/2}(x). \quad (16.134)$$

It is to be recalled that for non-integer order, the solution $J_{-n-1/2}$ is independent of $J_{n+1/2}$. The series expansion of the spherical Neumann function is

$$n_n(x) = \frac{(-1)^{n+1}}{2^n x^{n+1}} \sum_{s=0}^{\infty} \frac{(-1)^s (s-n)!}{s!(2s-2n)!} x^{2s}. \quad (16.135)$$

For $n = 0$, we find that

$$n_0(x) = -\frac{\cos x}{x}. \quad (16.136)$$

The first few spherical Bessel and Neumann functions are plotted in Fig. 16.7.

One can see that the spherical Bessel and Neumann functions are qualitatively very similar to their ordinary counterparts. The higher-order spherical Bessel and Neumann functions satisfy recursion relations very similar to those of the ordinary Bessel functions; this can be shown explicitly by careful manipulation of the usual Bessel function recursion relations. For instance, Eq. (16.42) can be multiplied by $\sqrt{\pi/2x}$, with $n \to n + 1/2$, to find

$$j_{n-1}(x) + j_{n+1}(x) = \frac{2n+1}{x} j_n(x), \quad (16.137)$$

with an identical formula for $n_n(x)$. This formula can then be used, in conjunction with Eq. (16.45), to show that

$$n j_{n-1}(x) - (n+1)j_{n+1}(x) = (2n+1)j'_n(x), \quad (16.138)$$

again with an identical formula for the Neumann functions. The derivative relations may also be derived,

$$\frac{d}{dx}\left[x^{n+1}j_n(x)\right] = x^{n+1}j_{n-1}(x), \tag{16.139}$$

$$\frac{d}{dx}\left[x^{-n}j_n(x)\right] = -x^{-n}j_{n+1}(x). \tag{16.140}$$

Furthermore, one can use the above expressions combined with mathematical induction to derive the Rayleigh formulas,

$$j_n(x) = (-1)^n x^n \left(\frac{1}{x}\frac{d}{dx}\right)^n \left(\frac{\sin x}{x}\right), \tag{16.141}$$

$$n_n(x) = -(-1)^n x^n \left(\frac{1}{x}\frac{d}{dx}\right)^n \left(\frac{\cos x}{x}\right). \tag{16.142}$$

From these Rayleigh formulas one can see that all the spherical Bessel functions are in fact expressible in terms of trigonometric functions. For instance,

$$j_1(x) = \frac{\sin x}{x^2} - \frac{\cos x}{x}, \tag{16.143}$$

$$n_1(x) = -\frac{\cos x}{x^2} - \frac{\sin x}{x}. \tag{16.144}$$

The spherical Bessel functions are quite common in problems of radiation and scattering, as well as in boundary value problems in spherical coordinates.

The spherical Bessel functions satisfy an orthogonality relation that can be derived directly from the corresponding one for the ordinary Bessel functions. Recalling Eqs. (16.103) and (16.104), we may write

$$\int_0^a J_{m+1/2}(\alpha_{(m+1/2)n}r/a)J_{m+1/2}(\alpha_{(m+1/2)k}r/a)r\,dr$$

$$= \delta_{nk}\frac{a^2}{2}[J_{m+3/2}(\alpha_{(m+1/2)n})]^2. \tag{16.145}$$

If we multiply and divide the integrand by $\pi a/r$, and the right-hand side by π, and use the definition of the spherical Bessel functions, we may write

$$\int_0^a j_m(\beta_{mn}r/a)j_m(\beta_{mk}r/a)r^2\,dr = \delta_{nk}\frac{a^3}{2}[j_{m+1}(\beta_{mn})]^2, \tag{16.146}$$

where $\beta_{mn} \equiv \alpha_{(m+1/2)k}$ is the nth zero of the mth spherical Bessel function. This in turn suggests that we may expand a function over the domain $0 \le r \le a$ in terms of spherical

Bessel functions in the form

$$f(r) = \sum_{n=0}^{\infty} a_n j_m(\beta_{mn} r/a),$$ (16.147)

where

$$a_n = \frac{2}{a^3 [j_{m+1}(\beta_{mn})]^2} \int_0^a f(r) j_m(\beta_{mn} r/a) r^2 dr.$$ (16.148)

16.12 Addition theorems, sum theorems, and product relations

There is a large collection of identities for which infinite sums of Bessel functions or their products can be shown to have a simple, closed form. In this section we describe some of the most important ones.

Looking at the definition of the generating function, Eq. (16.34), and setting $t = 1$, we may write

$$1 = J_0(x) + 2 \sum_{m=1}^{\infty} J_{2m}(x).$$ (16.149)

This formula introduces a normalization condition for the Bessel functions, and is particularly useful in their numeric evaluation. The standard technique for calculating a numeric value for $J_m(x_0)$ is assume, for a value $n \gg m$, that

$$J_{n+1}(x_0) = 0, \quad J_n(x_0) = a,$$ (16.150)

where a is a small number. The recurrence relation (16.42) may then be used to determine $J_p(x)$ for all $p < n$, down to $p = 0$. The undetermined constant a is chosen by requiring the derived Bessel values to satisfy the normalization condition, Eq. (16.149).

This process is required because the Neumann functions satisfy the same recurrence relations as the Bessel functions, and the Neumann functions are "dominant": any error introduced into the values of $J_{n+1}(x_0)$ and $J_{n+1}(x_0)$ inevitably allow a small amount of Neumann solution to enter and swamp the Bessel solution. The problem is the same one discussed in Section 14.12.4 in connection with "stiff" ODEs. Working the recursion relation backwards makes the Bessel functions the "dominant" ones.

Also from the generating function, we may show that

$$J_n(x+y) = \sum_{s=-\infty}^{\infty} J_s(x) J_{n-s}(y).$$ (16.151)

We note from Eq. (16.34) that $g(x+y,t) = g(x,t)g(y,t)$. From the series representation of the generating function, we may write

$$\sum_{n=-\infty}^{\infty} J_n(x+y)t^n = \sum_{s=-\infty}^{\infty} J_s(x)t^s \sum_{r=-\infty}^{\infty} J_r(y)t^r. \tag{16.152}$$

Equating like powers of t on either side of this expression, i.e. $r = n - s$, we arrive at the result.

An expression that is very important in statistical optics and two-dimensional wave problems is the addition formula,

$$J_0(k|\rho_1 - \rho_2|) = \sum_{m=-\infty}^{\infty} J_m(k\rho_1)J_m(k\rho_2)e^{im(\phi_2-\phi_1)}, \tag{16.153}$$

where $\rho = (\rho,\phi)$ is a two-dimensional vector in polar coordinates. We may demonstrate this by first writing J_0 using Eq. (16.62),

$$J_0(k|\rho_1 - \rho_2|) = \frac{1}{2\pi} \int_0^{2\pi} \exp[ik|\rho_1 - \rho_2|\cos\theta]d\theta. \tag{16.154}$$

Because the integrand is periodic with period 2π, we may shift the origin of integration and write the equation in the form

$$J_0(k|\rho_1 - \rho_2|) = \frac{1}{2\pi} \int_0^{2\pi} \exp[ik|\rho_1 - \rho_2|\cos(\theta - \alpha)]d\theta, \tag{16.155}$$

where α is defined in Fig. 16.8. By elementary geometry, we may readily show that

$$|\rho_1 - \rho_2|\cos\alpha = \rho_1 \sin\Delta\phi, \tag{16.156}$$
$$|\rho_1 - \rho_2|\sin\alpha = \rho_2 - \rho_1 \cos\Delta\phi, \tag{16.157}$$

where $\Delta\phi \equiv \phi_2 - \phi_1$. Applying these formulas to Eq. (16.155), we have

$$J_0(k|\rho_1 - \rho_2|) = \frac{1}{2\pi} \int_0^{2\pi} \exp[ik\rho_2 \sin\theta]\exp[ik\rho_1 \sin(\Delta\phi - \theta)]d\theta. \tag{16.158}$$

Using the generating function, we may write the first exponential as

$$\exp[ik\rho_2 \sin\theta] = \exp\left[\frac{k}{2}\rho_2(e^{i\theta} - e^{-i\theta})\right] = \sum_{m=-\infty}^{\infty} J_m(k\rho_2)e^{im\theta}. \tag{16.159}$$

Our expression (16.158) then becomes

$$J_0(k|\rho_1 - \rho_2|) = \frac{1}{2\pi} \int_0^{2\pi} \sum_{m=-\infty}^{\infty} J_m(k\rho_2)e^{im\theta} \exp[ik\rho_1 \sin(\Delta\phi - \theta)]d\theta. \tag{16.160}$$

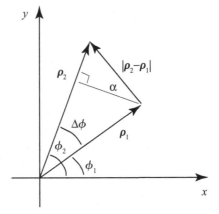

Figure 16.8 The geometry related to the derivation of the addition theorem, in particular Eq. (16.155).

Changing the variable of integration to $\theta' = \theta - \phi$,

$$J_0(k|\boldsymbol{\rho}_1 - \boldsymbol{\rho}_2|) = \frac{1}{2\pi} \sum_{m=-\infty}^{\infty} J_m(k\rho_2)e^{im\phi} \int_0^{2\pi} e^{ik\rho_1 \sin\theta' - im\theta'} d\theta'. \tag{16.161}$$

The integral is simply 2π times $J_m(k\rho_1)$. On substitution, we arrive at the addition theorem.
An analogous formula exists for the spherical Bessel functions, of the form

$$j_0(k|\mathbf{r}_1 - \mathbf{r}_2|) = \sum_{n=0}^{\infty}(2n+1)j_n(kr_1)j_n(kr_2)P_n(\cos\Theta), \tag{16.162}$$

where $P_n(x)$ is a Legendre polynomial, and Θ is the angle between vectors $\mathbf{r}_1$ and $\mathbf{r}_2$. This formula may also be written in terms of spherical harmonics Y_n^m, to be discussed in the next chapter,

$$j_0(k|\mathbf{r}_1 - \mathbf{r}_2|) = 4\pi \sum_{n=0}^{\infty} \sum_{m=-n}^{n} j_n(kr_1)j_n(kr_2)Y_n^{m*}(\theta_1,\phi_1)Y_n^m(\theta_2,\phi_2), \tag{16.163}$$

where $\mathbf{r} = (r,\theta,\phi)$ in spherical coordinates.

We may also write a more general addition theorem, known as *Gegenbauer's addition theorem*, which includes Eq. (16.162) as a special case, of the form

$$\frac{J_\nu(k|\mathbf{r}_1 - \mathbf{r}_2|)}{(k|\mathbf{r}_1 - \mathbf{r}_2|)^\nu} = 2^\nu \Gamma(\nu) \sum_{n=0}^{\infty}(\nu+n)\frac{J_{\nu+n}(kr_1)}{(kr_1)^\nu}\frac{J_{\nu+n}(kr_2)}{(kr_2)^\nu}\mathcal{G}_n^\nu(\cos\Theta), \tag{16.164}$$

where $\mathcal{G}_n^\mu$ is a Gegenbauer polynomial. Legendre polynomials, Gegenbauer polynomials, and spherical harmonics will be discussed in Chapter 17. Equation (16.164) reduces to Eq. (16.162) for $\nu = 1/2$; it is valid for any value of $\nu \neq 0, -1, -2, \ldots$.

A similar addition theorem for the Neumann function may be derived,

$$\frac{N_\nu(k|\mathbf{r}_1 - \mathbf{r}_2|)}{(k|\mathbf{r}_1 - \mathbf{r}_2|)^\nu} = 2^\nu \Gamma(\nu) \sum_{n=0}^\infty (\nu + n) \frac{N_{\nu+n}(kr_1)}{(kr_1)^\nu} \frac{J_{\nu+n}(kr_2)}{(kr_2)^\nu} \mathcal{G}_n^\nu(\cos\Theta), \qquad (16.165)$$

and the addition formulas can be extended to modified Bessel functions by the substitution $r_1 \to ir_1, r_2 \to ir_2$, e.g.

$$\frac{I_\nu(k|\mathbf{r}_1 - \mathbf{r}_2|)}{(k|\mathbf{r}_1 - \mathbf{r}_2|)^\nu} = 2^\nu \Gamma(\nu) \sum_{n=0}^\infty (-1)^n (\nu + n) \frac{I_{\nu+n}(kr_1)}{(kr_1)^\nu} \frac{I_{\nu+n}(kr_2)}{(kr_2)^\nu} \mathcal{G}_n^\nu(\cos\Theta) \qquad (16.166)$$

For a proof of Gegenbauer's addition theorem and related ones, we refer the reader to [Wat44, Section 11.4].

16.13 Focus: nondiffracting beams

One of the consequences of the wave theory of light is that a light beam of finite width tends to spread as it propagates, a process broadly referred to as diffraction. The diffraction of light on passing through a small aperture is a special case, but diffraction also occurs in optical beams with finite cross-section such as produced by a laser. In 1987, however, it was pointed out, both theoretically [Dur87] and experimentally [DME87], that solutions to the Helmholtz equation exist which do not spread on propagation, and that such solutions could be approximately realized experimentally. The solutions are related to the Bessel functions and are referred to both as *Bessel beams* and *nondiffracting beams*.

Such solutions can be readily found from the Helmholtz equation in cylindrical coordinates,

$$\frac{1}{\rho}\frac{\partial}{\partial\rho}\left(\rho\frac{\partial U(\rho,\phi,z)}{\partial\rho}\right) + \frac{1}{\rho^2}\frac{\partial^2 U(\rho,\phi,z)}{\partial\phi^2} + \frac{\partial^2 U(\rho,\phi,z)}{\partial z^2} + k^2 U(\rho,\phi,z) = 0. \quad (16.167)$$

We assume a solution $U(\rho,\phi,z)$ of the form

$$U(\rho,\phi,z) = R(\rho)\Phi(\phi)\exp[ik_L z], \qquad (16.168)$$

which is simply a separable solution to the Helmholtz equation which propagates in the positive z-direction. The solution for $\Phi(\phi)$ is of the form $\exp[im\phi]$, with m any integer, and the solution for $R(\rho)$ which is well-behaved at the origin is simply a Bessel function of order m,

$$R(\rho) = J_m(k_T\rho), \qquad (16.169)$$

where $k_T = \sqrt{k^2 - k_L^2}$. We therefore have as a valid solution of the wave equation,

$$U(\rho,\phi,z) = J_m(k_T\rho)\exp[im\phi]\exp[ik_L z]. \qquad (16.170)$$

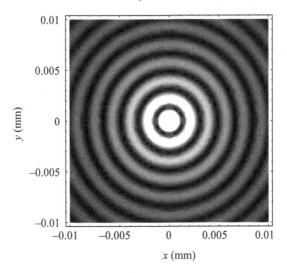

Figure 16.9 The transverse intensity profile of a nondiffracting beam. Here we have taken $\lambda = 500$ nm, $k = 2\pi/\lambda$, $k_L = 0.99k$.

The intensity of a scalar wave field is $I(\mathbf{r}) = |U(\mathbf{r})|^2$, and the intensity of our Bessel wave is

$$I(\rho,\phi,z) = |J_m(k_T \rho)|^2, \qquad (16.171)$$

which is independent of the propagation distance z! A wave of the form of Eq. (16.170) does not change its transverse shape on propagation, and is *nondiffracting*. An example of the transverse intensity of such a beam is shown in Fig. 16.9.

We may use the angular spectrum representation, discussed in Section 11.9, to better understand the origin of the nondiffracting effect. The angular spectrum $a(p,q)$ of a Bessel beam of order m may be written as

$$a(p,q) = \frac{k^2}{(2\pi)^2} \iint_{-\infty}^{\infty} U(x,y,0)e^{i(kpx+kqy)}\,dxdy. \qquad (16.172)$$

This integral is easier to evaluate in polar coordinates. We introduce a polar coordinate representation $(p,q) = (r,\theta)$, such that $r = \sqrt{p^2+q^2}$, and write our angular spectrum in the form

$$a(p,q) = \frac{k^2}{(2\pi)^2} \int_0^{\infty} \int_0^{2\pi} J_m(k_T\rho)\exp[im\phi]\exp[ikr\rho\cos(\theta-\phi)]\rho\,d\rho\,d\phi. \qquad (16.173)$$

Changing the variable of integration to $\phi' = \phi - \theta$, we may write

$$a(p,q) = \frac{k^2}{(2\pi)^2}\exp[im\theta]\int_0^{\infty}\int_0^{2\pi} J_m(k_T\rho)\exp[im\phi']\exp[ikr\rho\cos\phi']\rho\,d\rho\,d\phi'. \qquad (16.174)$$

By Eq. (16.59), the integral over ϕ' may be written as $2\pi i^m J_m(kr\rho)$, and we have

$$a(p,q) = \frac{i^m k^2}{2\pi} \exp[im\theta] \int_0^\infty J_m(k_T\rho)J_m(kr\rho)\rho d\rho. \tag{16.175}$$

We may now use the orthogonality relation, Eq. (16.120), to write

$$a(p,q) = \frac{i^m k^2}{2\pi k_T} \exp[im\theta]\delta(kr - k_T). \tag{16.176}$$

The angular spectrum of a Bessel beam is a "delta ring" of radius $r = k_T/k$. Each plane wave which contributes to the total field has the same value of k_T, and therefore the same value of k_L. The plane waves remain in phase as they propagate, and the shape of the wave does not change. This illustration suggests that diffraction may be interpreted as a dephasing of the constituent plane waves of a beam as it propagates. Only fields which consist entirely of plane waves with the same value of k_T will remain unchanged on propagation.

Equation (16.176) suggests that an approximate Bessel beam may be constructed using the technique of spatial filtering, described in Section 11.8. The field in the focal plane of a spatial filtering system is proportional to the Fourier transform of the field which emerges from the system. By using a screen with an annular aperture in it, the angular spectrum of light passing through the filtering system will be approximately a "delta ring" and the field which emerges from the system will be approximately a Bessel beam.

We may also use the properties of Bessel functions to illustrate a difficulty related to Bessel beams. The total integrated energy of a beam in a transverse plane may be defined as

$$E = \iint_{-\infty}^\infty |U(x,y,z)|^2 dxdy. \tag{16.177}$$

In polar coordinates, the energy of a Bessel beam may be written as

$$E = 2\pi \int_0^\infty |J_m(k_T\rho)|^2 \rho d\rho. \tag{16.178}$$

By the closure relation, Eq. (16.120), this integral is necessarily infinite. An ideal Bessel beam is therefore not realizable, as it would contain an infinite amount of energy.

Quasi-nondiffracting beams may be constructed using a Bessel beam with a Gaussian envelope, of the form

$$U(\rho,\phi,0) = J_m(k_T\rho)\exp[im\phi]\exp[-\rho^2/\sigma^2]. \tag{16.179}$$

If the Gaussian envelope is much wider than the characteristic length of the Bessel function, i.e. if $k_T\sigma \gg 1$, the field will be effectively nondiffracting over some extended but finite propagation distance.

More information about nondiffracting beams and their applications may be found in the review [Bou03].

16.14 Additional reading

- G. N. Watson, *A Treatise on the Theory of Bessel Functions* (Cambridge, Cambridge University Press, 1922). Simply *the* reference book on Bessel functions. Pretty much anything you might possibly want to know about the functions is contained in this volume.
- N. N. Lebedev, *Special Functions & Their Applications* (New York, Dover Publications, 1972). A relatively short and readable introduction to a variety of special functions, including the Bessel (cylinder) functions.
- M. Abramowitz and I. A. Stegun, *Handbook of Mathematical Functions* (New York, Dover Publications, 1965). This reference book lists many relations and integrals involving Bessel functions.

16.15 Exercises

1. Use the product of the generating functions $g(x,t)$ and $g(x,-t)$ to show that

$$1 = [J_0(x)]^2 + 2\sum_{n=1}^{\infty}[J_n(x)]^2.$$

2. By the use of the generating function, show that

$$\cos x = J_0(x) + 2\sum_{n=1}^{\infty}(-1)^n J_{2n}(x),$$

$$\sin x = 2\sum_{n=1}^{\infty}(-1)^{n+1} J_{2n+1}(x).$$

3. Use the power series representation of the Bessel function $J_m(x)$ to prove the recurrence relation

$$J_{m-1}(x) + J_{m+1}(x) = \frac{2m}{x}J_m(x).$$

4. Use the power series representation of the Bessel function $J_m(x)$ to prove the recurrence relation

$$J_{m-1}(x) - J_{m+1}(x) = 2J_m'(x).$$

5. Use the Bessel recurrence relations to show that

$$J_2(x) - J_0(x) = 2J_0''(x).$$

6. Use the Bessel recurrence relations to show that

$$J_2(x) = J_0''(x) - x^{-1}J_0'(x).$$

7. Prove the relation

$$\int_0^z J_\nu(x)\,\mathrm{d}x = 2\sum_{n=0}^{\infty} J_{\nu+2k+1}(z), \quad \mathrm{Re}(\nu) > -1.$$

8. Prove the following integral recurrence relation

$$\int_0^z x^\mu J_\nu(x)dx = z^\mu J_{\nu+1}(x) - (\mu - \nu - 1)\int_0^z x^{\mu-1}J_{\nu+1}(x)dx, \quad \mathrm{Re}(\mu+\nu) > -1.$$

9. Derive the recurrence relation

$$2J_n'(x) = J_{n-1}(x) - J_{n+1}(x)$$

by direct differentiation of

$$J_n(x) = \frac{1}{\pi}\int_0^\pi \cos(n\theta - x\sin\theta)d\theta.$$

10. Find a closed-form expression for the Laplace transform of $J_0(bx)$, i.e.

$$\int_0^\infty e^{-ax}J_0(bx)dx.$$

11. Find a closed-form expression for the Laplace transform of $J_\nu(bx)$, i.e.

$$\int_0^\infty e^{-ax}J_\nu(bx)dx,$$

with $\nu > -1$.

12. Let us suppose we define a function

$$Y_\nu(x) = \frac{1}{\pi}\int_0^\pi \cos(\nu\theta - x\sin\theta)d\theta,$$

which is comparable to the integral definition of the Neumann function. By direct differentiation, find the second-order ODE that $Y_\nu(x)$ satisfies, and show that it reduces to Bessel's equation for integer ν.

13. Derive the integral representation

$$J_n^2(x) = \frac{2}{\pi}\int_0^{\pi/2} J_{2n}(2x\cos\theta)d\theta = (-1)^n \frac{2}{\pi}\int_0^{\pi/2} J_0(2x\cos\theta)\cos(2n\theta)d\theta,$$

with n an integer.

14. Waves vibrating on a circular drumhead of radius a satisfy the two-dimensional wave equation

$$\nabla^2 u - \frac{1}{c^2}\frac{\partial^2 u}{\partial t^2} = 0,$$

subject to the boundary condition (in polar coordinates) that $u(a,\theta,t) = 0$. Use separation of variables to find the solution of the wave equation subject to the initial conditions $u(r,\theta,0) = f(x)$ and $u_t(r,\theta,0) = 0$.

15. A cylinder of radius a and length l has a steady-state heat distribution $u(r,\phi,z)$ distributed through its volume that satisfies Laplace's equation. Assuming that $u(a,\phi,z) = 0$ and that $u(\rho,\phi,0) = u(\rho,\phi,l) = u_0$, find the distribution.

16. A cylinder of radius a and length l has a steady-state heat distribution $u(\rho,\phi,z)$ distributed through its volume that satisfies Laplace's equation. Assuming that $u(a,\phi,z) = u_0(\phi)$ and that $u(\rho,\phi,0) = u(\rho,\phi,l) = 0$, find the distribution.

17. A cylinder of radius a and length l which has an insulating surface ($\partial u/\partial n = 0$, where n is the normal to the surface) has a heat distribution $u(\rho,\phi,z,t)$ that satisfies the three-dimensional diffusion equation

$$\nabla^2 u - \alpha^2 u_t = 0.$$

Assuming that the cylinder has an initial temperature distribution $u_0(\rho)$, find the distribution of temperature at time t.

18. Bessel's equation is a second-order ODE with two linearly independent solutions. Find expressions for the Wronskian of the two solutions when we choose as those solutions
    1. $J_\nu(x), N_\nu(x)$,
    2. $J_\nu(x), J_{-\nu}(x)$.

19. Bessel's equation is a second-order ODE with two linearly independent solutions. Find expressions for the Wronskian of the two solutions when we choose as those solutions
    1. $H_\nu^{(1)}(x), H_\nu^{(2)}(x)$,
    2. $J_\nu'(x), N_\nu'(x)$.

20. Prove the Bessel function relation,

$$J_1(x) + J_3(x) + J_5(x) + \cdots = \frac{1}{2}\left[J_0(x) + \int_0^x [J_0(t) + J_1(t)]\mathrm{d}t - 1\right].$$

21. The formula for the field at a point P in the neighborhood of focus of a general illuminating field U_i is

$$U(P) \propto \int_0^1 \int_0^{2\pi} U_i(\rho,\theta)\mathrm{e}^{-\mathrm{i}[v\rho\cos(\theta-\psi)+u\rho^2/2]}\rho\mathrm{d}\rho\mathrm{d}\theta,$$

where ψ is the azimuthal angle of the point P and u, v are dimensionless variables indicating the longitudinal and transverse distance from the focal point. Let us assume that we are focusing a vortex beam, i.e.

$$U_i(\rho,\theta) = U_0\rho^m\mathrm{e}^{\mathrm{i}m\theta},$$

where m is an integer. Evaluate the angular integral to express the field in the neighborhood of focus as an integral of a Bessel function. Derive a closed form expression for the field on the axis ($v = 0$) and in the geometric focal plane ($u = 0$).

17

Legendre functions and spherical harmonics

17.1 Introduction: Laplace's equation in spherical coordinates

Let us consider the solution of Laplace's equation, $\nabla^2 V = 0$, inside a spherical region of radius a, i.e. we wish to find the potential $V(r,\theta,\phi)$ which satisfies

$$\frac{1}{r^2 \sin\theta}\left[\sin\theta\frac{\partial}{\partial r}\left(r^2\frac{\partial V}{\partial r}\right) + \frac{\partial}{\partial\theta}\left(\sin\theta\frac{\partial V}{\partial\theta}\right) + \frac{1}{\sin\theta}\frac{\partial^2 V}{\partial\phi^2}\right] = 0. \tag{17.1}$$

We will consider two different boundary conditions: the general case where $V(a,\theta,\phi) = V_0(\theta,\phi)$, and the azimuthally symmetric case $V(a,\theta,\phi) = V_1(\theta)$.

Let us assume an intermediate separable solution of the form

$$V(r,\theta,\phi) = R(r)Y(\theta,\phi), \tag{17.2}$$

which on substitution into Eq. (17.1) results in the expression

$$\frac{1}{R}\frac{1}{r^2}\frac{d}{dr}\left(r^2\frac{dR}{dr}\right) + \frac{1}{r^2}\left\{\frac{1}{Y}\frac{1}{\sin\theta}\left[\frac{\partial}{\partial\theta}\left(\sin\theta\frac{\partial Y}{\partial\theta}\right) + \frac{1}{\sin\theta}\frac{\partial^2 Y}{\partial\phi^2}\right]\right\} = 0. \tag{17.3}$$

The term in the curved brackets is independent of r, and must therefore be equal to a constant, which we write for later convenience as $-l(l+1)$. We will later find that only $l > 0$ solutions are valid. Equation (17.3) results in a pair of separated equations of the form

$$\frac{1}{R}\frac{1}{r^2}\frac{d}{dr}\left(r^2\frac{dR}{dr}\right) - \frac{l(l+1)}{r^2} = 0, \tag{17.4}$$

$$\frac{1}{Y}\frac{1}{\sin\theta}\left[\frac{\partial}{\partial\theta}\left(\sin\theta\frac{\partial Y}{\partial\theta}\right) + \frac{1}{\sin\theta}\frac{\partial^2 Y}{\partial\phi^2}\right] = -l(l+1). \tag{17.5}$$

The solution for $R(r)$ can be readily found by assuming a solution of the form $R(r) = r^\alpha$; on substitution into the equation for $R(r)$, we are left with

$$(\alpha+1)\alpha - l(l+1) = 0, \tag{17.6}$$

585

with roots

$$\alpha = l, -l - 1. \tag{17.7}$$

For positive l, the only solution which is regular within the sphere is $R(r) = r^l$.

The angular expression, Eq. (17.5), is rather complicated. We investigate it by further assuming a separable solution of the form

$$Y(\theta, \phi) = P(\theta)\Phi(\phi). \tag{17.8}$$

On substitution, we may write

$$\frac{1}{P}\frac{1}{\sin\theta}\frac{d}{d\theta}\left(\sin\theta\frac{dP}{d\theta}\right) + \frac{1}{\sin^2\theta}\frac{1}{\Phi}\frac{d^2\Phi}{d\phi^2} + l(l+1) = 0. \tag{17.9}$$

The Φ-term may be set equal to a constant, say $-m^2$, which results in the separated equations,

$$\frac{1}{\sin\theta}\frac{d}{d\theta}\left(\sin\theta\frac{dP}{d\theta}\right) + \left(l(l+1) - \frac{m^2}{\sin^2\theta}\right)P = 0, \tag{17.10}$$

$$\frac{1}{\Phi}\frac{d^2\Phi}{d\phi^2} = -m^2. \tag{17.11}$$

The equation for Φ has the solution

$$\Phi_m(\phi) = A_m e^{im\phi} + B_m e^{-im\phi}. \tag{17.12}$$

Because the functions must be 2π-periodic, m is constrained to be an integer; to avoid redundancy, we assume $m \geq 0$.

The solution for P is not quite so trivial, but it can be written in a simpler form by making the coordinate transformation $x = \cos\theta$; we then get

$$\frac{d}{dx}\left[(1-x^2)\frac{dP}{dx}\right] + \left[l(l+1) - \frac{m^2}{1-x^2}\right]P = 0, \tag{17.13}$$

with $-1 \leq x \leq 1$. This second-order ODE presumably has two linearly independent solutions, which we label $P_l^m(x)$ and $Q_l^m(x)$. A general solution may be written as

$$P(x) = C_{lm}P_l^m(x) + D_{lm}Q_l^m(x). \tag{17.14}$$

We will see that only one solution exists which is finite for $-1 \leq x \leq 1$; we label this solution $P_l^m(x)$ and the functions which result are known as *associated Legendre functions*. Equation (17.13) is referred to as the *associated Legendre equation*.

In the special case of azimuthal symmetry, i.e. $V(a, \theta, \phi) = V_1(\theta)$, only solutions with $m = 0$ are possible. The equation for P takes on the simplified form

$$\frac{d}{dx}\left[(1-x^2)\frac{dP}{dx}\right] + l(l+1)P = 0. \tag{17.15}$$

The solutions to this equation will be shown to be polynomials of order l and are labeled $P_l(x)$; they are referred to as the *Legendre polynomials*, and Eq. (17.15) is the *Legendre equation*.

The Legendre functions are the natural basis for describing the solutions of partial differential equations in spherical coordinates. In this chapter we describe the properties of Legendre polynomials, associated Legendre functions, and spherical harmonics, which are the most commonly used functions in separation of variables in spherical coordinates. We later discuss several useful generalizations of these concepts, namely vector spherical harmonics and Gegenbauer polynomials.

17.2 Series solution of the Legendre equation

We begin by considering the solution of the Legendre equation, which may also be written in the form

$$(1-x^2)\frac{d^2P}{dx^2} - 2x\frac{dP}{dx} + l(l+1)P = 0. \tag{17.16}$$

This equation has regular singular points[1] at $x = \pm 1$, but no singularities at $x = 0$. We may therefore attempt to find a series solution about $x = 0$ of the form

$$P(x) = \sum_{n=0}^{\infty} c_n x^n. \tag{17.17}$$

On substitution into the differential equation, we are immediately led to the relation

$$\frac{c_{n+2}}{c_n} = \frac{n(n+1) - l(l+1)}{(n+2)(n+1)}. \tag{17.18}$$

A pair of linearly independent solutions may be found by an independent choice of c_0 and c_1. We may test for convergence by application of the ratio test of Section 7.3,

$$\lim_{n\to\infty} \frac{c_{n+2}x^{n+2}}{c_n x^n} = x^2 \lim_{n\to\infty} \frac{n(n+1) - l(l+1)}{(n+2)(n+1)} = x^2. \tag{17.19}$$

This shows that the series converges for all $|x| < 1$, but is ambiguous for $x = \pm 1$. By the application of Kummer's test, described in Section 7.3, with $b_n = 1/(n\log n)$, one can show that the series diverges for $x = 1$. However, $x = 1$ is part of the physical domain of interest in spherical coordinates, corresponding to $\theta = 0$, and we require a solution which is valid over the entire domain.

A non-divergent solution can be found in the special case that l is a positive integer. Then, for $n = l$, by Eq. (17.18), we find that $c_{l+2} = 0$. It follows that $c_m = 0$ for $m > l$, and the

---

[1] Recall Section 14.9.

Table 17.1 *The first few Legendre polynomials.*

$$P_0(x) = 1$$
$$P_1(x) = x$$
$$P_2(x) = \tfrac{1}{2}(3x^2 - 1)$$
$$P_3(x) = \tfrac{1}{2}(6x^3 - 3x)$$
$$P_4(x) = \tfrac{1}{8}(35x^4 - 30x^2 + 3)$$
$$P_5(x) = \tfrac{1}{8}(63x^5 - 70x^3 + 15x)$$
$$P_6(x) = \tfrac{1}{16}(231x^6 - 315x^4 + 105x^2 - 5)$$

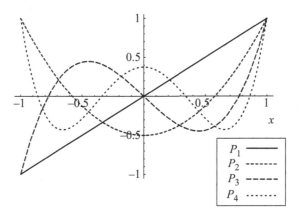

Figure 17.1 The Legendre polynomials P_1, P_2, P_3, and P_4.

solution is a polynomial of order l. If l is even, only the function with nonzero c_0 is a finite polynomial; if l is odd, only the function with nonzero c_1 is a finite polynomial. We label the finite (and therefore convergent) solution $P_l(x)$; by convention, the solution is chosen such that

$$P_l(1) = 1. \tag{17.20}$$

The functions $P_l(x)$ are the *Legendre polynomials*; they can be used as the separable solutions in θ in a problem in spherical coordinates with azimuthal symmetry. The function $P_l(x)$ is a polynomial with highest order x^l. Using Eq. (17.20) and the recursion relation (17.18), the Legendre polynomials of any order may be derived and properly normalized.

Table 17.1 lists the first few Legendre polynomials, and the several of them are plotted in Fig. 17.1. It is to be noted that the even-order polynomials have even symmetry about $x = 0$, while the odd-order polynomials have odd symmetry.

We may write a straightforward series expression for the Legendre functions by the use of Eq. (17.18), along with a change of summation variable to r such that $l - 2r = n$. With this change, we may write

$$P_l(x) = \sum_{r=0}^{R} a_r x^{l-2r},$$
(17.21)

where

$$R = \begin{cases} \frac{l}{2}, & l \text{ even}, \\ \frac{l-1}{2}, & l \text{ odd}. \end{cases}$$
(17.22)

The result of this change of summation results in a simpler expression for the coefficients a_r, which may be determined directly from Eq. (17.18),

$$\frac{a_r}{a_{r-1}} = \frac{c_{l-2r}}{c_{l-2r+2}} = -\frac{(l - 2r + 1)(l - 2r + 2)}{2r(2l - 2r + 1)}.$$
(17.23)

This results in the relation

$$P_l(x) = a_0 \sum_{r=0}^{R} (-1)^r \frac{l!}{(l-2r)!} \frac{1}{2^r r!} \frac{(2l-2r)! 2^l l!}{(2l)! 2^{l-r}(l-r)!} x^{l-2r}.$$
(17.24)

To satisfy the condition $P_l(1) = 1$, one can show that a_0 must have the form

$$a_0 = \frac{(2l)!}{2^l (l!)^2}.$$
(17.25)

We then have a series expression for the Legendre functions as

$$P_l(x) = \frac{1}{2^l} \sum_{r=0}^{R} \frac{(-1)^r}{r!} \frac{(2l - 2r)!}{(l-r)!(l-2r)!} x^{l-2r}.$$
(17.26)

This derivation of the Legendre functions, and many of the tricks we apply in the following sections, can feel a bit like "magic", in that the results appear seemingly from nowhere without justification. The reality is that many of these identities were originally found by a simple combination of luck and mathematical stubbornness. We will attempt to make the study of such orthogonal polynomials a little more systematic in Chapter 18.

17.3 Generating function

Analogous to the Bessel functions, the Legendre polynomials may be constructed from a generating function, and this function may be used to determine various useful properties of the polynomials. The generating function may be shown to be

$$w(x,t) = (1 - 2xt + t^2)^{-1/2} = \sum_{n=0}^{\infty} P_n(x)t^n. \tag{17.27}$$

For the moment, we simply assume this equation is correct; we will justify it in the next section.

Equation (17.27) has a very important physical interpretation in terms of the multipole expansion, which will be discussed in Section 17.13.

17.4 Recurrence relations

We may use Eq. (17.27) to construct recurrence relations amongst the Legendre functions, just as was done in Section 16.4 for the Bessel functions. The first of these may be found by taking the derivative of the generating function with respect to t,

$$\frac{\partial w(x,t)}{\partial t} = \frac{x-t}{(1-2xt+t^2)^{3/2}} = \sum_{n=0}^{\infty} nP_n(x)t^{n-1}. \tag{17.28}$$

We may multiply both side of this equation by $1 - 2xt + t^2$; then, using the definition of the generating function again, we may write

$$(x-t)\sum_{n=0}^{\infty} P_n(x)t^n = (1-2xt+t^2)\sum_{n=0}^{\infty} nP_n(x)t^{n-1}. \tag{17.29}$$

This equation must be satisfied in a continuous neighborhood of $t=0$, which means each power of t must vanish identically. We arrive at the relation

$$xP_n(x) - P_{n-1}(x) = (n+1)P_{n+1}(x) - 2xnP_n(x) + (n-1)P_{n-1}(x), \tag{17.30}$$

or

$$(2n+1)xP_n(x) = (n+1)P_{n+1}(x) + nP_{n-1}(x). \tag{17.31}$$

This recurrence relation connects the value of $P_{n+1}(x)$ to the values $P_n(x)$ and $P_{n-1}(x)$.

If we instead take the derivative of the generating function with respect to x, we have

$$\frac{t}{(1-2xt+t^2)^{3/2}} = \sum_{n=0}^{\infty} P_n'(x)t^n. \tag{17.32}$$

Again, we use the definition of the generating function to write

$$t\sum_{n=0}^{\infty} P_n(x)t^n = (1-2xt+t^2)\sum_{n=0}^{\infty} P_n'(x)t^n. \tag{17.33}$$

Equating like powers of t, we find that

$$P_n(x) = P'_{n+1}(x) - 2xP'_n(x) + P'_{n-1}(x).$$ (17.34)

A modified version of this formula may be found by taking the derivative of Eq. (17.31), multiplying it by 2, and adding the result to $(2n+1)$ times Eq. (17.34),

$$(2n+1)P_n(x) = P'_{n+1}(x) - P'_{n-1}(x).$$ (17.35)

Recurrence formulas such as these may be used to numerically determine the values of $P_n(x)$ based on the values $P_0(x) = 1$ and $P_1(x) = x$. There are many other recurrence relations which may be constructed from the two given and from other manipulations of the generating function. For instance, comparing the derivatives of the generating function with respect to x and t, we may readily see that

$$\frac{x-t}{t}\frac{\partial w}{\partial t}(x,t) = \frac{\partial w}{\partial x}(x,t).$$ (17.36)

Comparing like powers of t, we find that

$$xP'_n(x) - P'_{n-1}(x) = nP_n(x).$$ (17.37)

Our recurrence formulas can be used to demonstrate that the functions $P_n(x)$ in the generating function are indeed the Legendre polynomials. We may substitute $P'_{n-1}(x)$ from Eq. (17.37) into Eq. (17.35), such that

$$(n+1)P_n(x) = P'_{n+1}(x) - xP'_n(x).$$ (17.38)

This formula may be down-shifted from index n to $n-1$,

$$nP_{n-1}(x) = P'_n(x) - xP'_{n-1}(x).$$ (17.39)

We now add x times Eq. (17.37) to this formula,

$$x^2P'_n(x) + nP_{n-1}(x) = xnP_n(x) + P'_n(x),$$ (17.40)

or

$$(1-x^2)P'_n(x) = n[P_{n-1}(x) - xP_n(x)].$$ (17.41)

Let us now differentiate this equation with respect to x, and use Eq. (17.37) to eliminate the term $P'_{n-1}(x)$; we have

$$(1-x^2)P''_n(x) - 2xP'_n(x) = -n(n+1)P_n(x).$$ (17.42)

This is just the Legendre equation, Eq. (17.16); we may further show that the functions $P_n(x)$ defined by the generating function have the proper normalization by setting $x = 1$; we then have

$$w(1,t) = \frac{1}{\sqrt{1 - 2t + t^2}} = \frac{1}{1-t} = \sum_{n=0}^{\infty} t^n. \tag{17.43}$$

By definition, this is equal to

$$\sum_{n=0}^{\infty} t^n = \sum_{n=0}^{\infty} P_n(1) t^n, \tag{17.44}$$

and since a power series representation is unique, we have that $P_n(1) = 1$.

We may also derive an explicit formula for the Legendre polynomials which represents them as the derivatives of a fundamental function, comparable to Eqs. (16.48) and (16.49) for Bessel functions. This formula is known as *Rodrigues' formula* and is given as

$$P_l(x) = \frac{1}{2^l l!} \frac{d^l}{dx^l} (x^2 - 1)^l. \tag{17.45}$$

We may prove this expression by the use of the series form of the Legendre polynomials, Eq. (17.26). We first note that

$$\frac{d^l}{dx^l} x^{2l-2r} = \frac{(2l - 2r)!}{(l - 2r)!} x^{l-2r}. \tag{17.46}$$

On substitution into Eq. (17.26), we may write the Legendre polynomials as

$$P_l(x) = \frac{1}{2^l} \frac{d^l}{dx^l} \sum_{r=0}^{R} \frac{(-1)^r}{r!} \frac{1}{(2l - r)!} x^{2l-2r}. \tag{17.47}$$

We may now use the binomial theorem of Section 7.5 to identify

$$(1 - x^2)^l = \sum_{n=0}^{} \frac{l!}{n!(l-n)!} x^{2n} = \sum_{r=0}^{} \frac{l!}{r!(l-r)!} x^{2l-2r}. \tag{17.48}$$

On substitution from Eq. (17.48) into Eq. (17.47), we immediately find Rodrigues' formula.[2]

17.5 Integral formulas

A number of different integral representations of the Legendre polynomials may be developed. The most straightforward comes from the observation that the generating function

[2] Rodrigues' original paper is [Rod16]. It can be found online, if one searches for the journal.

$w(x,t)$ is an analytic function in the domain $|x| < 1$. Via the Cauchy formula (9.76), we may express the nth Legendre polynomial as a contour integral,

$$P_n(x) = \frac{1}{2\pi i} \oint_C (1 - 2xt + t^2)^{-1/2} t^{-n-1} dt, \tag{17.49}$$

where C is a closed contour surrounding the point $t = 0$ and with radius $r < 1$. A more elegant and useful form can be found by introducing a new variable u by the nonlinear coordinate transformation

$$1 - ut = (1 - 2xt + t^2)^{1/2}, \tag{17.50}$$

or

$$t = \frac{2(u - x)}{u^2 - 1}. \tag{17.51}$$

This nontrivial transformation moves the zero at $t = 0$ to the point $u = x$. The differential may be written as

$$dt = \frac{2}{u^2 - 1} - \frac{4(u - x)}{(u^2 - 1)^2}. \tag{17.52}$$

On substitution, our integral takes on the form

$$P_n(x) = \frac{1}{2\pi i} \oint_{C'} \frac{(u^2 - 1)^n}{2^n (u - x)^{n+1}} du, \tag{17.53}$$

where the curve C' encloses the point $u = x$. We may use Cauchy's formula (9.77) to evaluate the integral, which takes on the form

$$P_n(x) = \frac{1}{2^n n!} \frac{d^n}{dx^n} (x^2 - 1)^n. \tag{17.54}$$

This is just Rodrigues' formula, Eq. (17.45); this procedure has provided us with an alternative proof of the said formula. The integral expression (17.53) is known as the *Schlaefli integral*.

The Schlaefli integral can be put into the form of a real-valued integral by taking the path C' to be a circle of radius $\sqrt{|x^2 - 1|}$ in the complex plane with center at $u = x$. We may write

$$u = x + \sqrt{x^2 - 1} e^{i\phi}, \quad -\pi \leq \phi \leq \pi, \tag{17.55}$$

and with this choice of path the Schlaefli integral becomes

$$P_n(x) = \frac{1}{2\pi} \int_{-\pi}^{\pi} \left[\frac{x^2 + 2x\sqrt{x^2 - 1} e^{i\phi} + (x^2 - 1)e^{2i\phi} - 1}{2\sqrt{x^2 - 1} e^{i\phi}} \right]^n d\phi. \tag{17.56}$$

With a little manipulation, this integral can be shown to depend only upon $\cos\phi$, which is symmetric about $\phi = 0$. Equation (17.56) may then be written as

$$P_n(x) = \frac{1}{\pi} \int_0^\pi \left[x + \sqrt{x^2 - 1} \cos\phi \right]^n d\phi. \tag{17.57}$$

This integral is known as *Laplace's integral*.

At this point, we have developed integral [Eq. (17.53)], differential [Eq. (17.45)], and series [Eq. (17.26)] representations of the Legendre polynomials. Each of these representations may be used as a *definition* of the polynomials, though some can be used more conveniently than others in deriving other properties such as the recurrence relations.

17.6 Orthogonality

We can show that the Legendre polynomials are orthogonal over the interval $-1 \le x \le 1$. To do so, we consider the Legendre equation in the form

$$\frac{d}{dx} \left[(1 - x^2) \frac{dP_l}{dx} \right] + l(l+1)P_l = 0. \tag{17.58}$$

We multiply the Legendre equation for P_m by the function P_n, multiply the Legendre equation for P_n by P_m, and consider the difference,

$$[(1 - x^2)P_m']'P_n - [(1 - x^2)P_n']P_m + [m(m+1) - n(n+1)]P_mP_n = 0. \tag{17.59}$$

This expression may be rewritten as

$$\left\{ (1 - x^2)[P_m'P_n - P_n'P_m] \right\}' + (m-n)(m+n+1)P_mP_n = 0. \tag{17.60}$$

We now integrate this equation over the interval $-1 \le x \le 1$. The first term of the equation is a perfect differential, and its integral vanishes on account of the $(1 - x^2)$ term. We are left with

$$(m-n)(m+n+1) \int_{-1}^{1} P_mP_n dx = 0. \tag{17.61}$$

Provided $m \neq n$, this results in

$$\int_{-1}^{1} P_mP_n dx = 0, \quad m \neq n. \tag{17.62}$$

The Legendre polynomials are therefore orthogonal on the interval $-1 \le x \le 1$. It can be shown that they form an orthogonal basis, and a well-behaved (piecewise continuous) function can be expanded in a series of Legendre polynomials.

To perform such an expansion, however, we need to consider the normalization of the Legendre polynomials. This normalization can be found in a number of ways; we consider a trick using the recurrence relation, Eq. (17.31). Let us take this equation with $n = k - 1$ and multiply it by $(2k + 1)P_k$, and take Eq. (17.31) with $n = k$ and multiply it by $(2k - 1)P_{k-1}$, and take the difference of the two equations. The result is

$$k(2k + 1)P_k^2 + (k - 1)(2k + 1)P_{k-2}P_k - (k + 1)(2k - 1)P_{k-1}P_{k+1} - k(2k - 1)P_{k-1}^2 = 0.$$

(17.63)

Integrating this formula over $-1 \leq x \leq 1$, and taking into account the orthogonality of the functions P_k, we find that

$$\int_{-1}^{1} P_k^2(x)dx = \frac{2k - 1}{2k + 1} \int_{-1}^{1} P_{k-1}^2(x)dx.$$

(17.64)

This is a recurrence relation for the integral of P_k^2; repeated application of it results in

$$\int_{-1}^{1} P_k^2(x)dx = \frac{3}{2k + 1} \int_{-1}^{1} P_1^2(x)dx = \frac{2}{2k + 1}.$$

(17.65)

We can verify by direct integration that this result also holds for $m = 1$ and $m = 0$. An arbitrary function $f(x)$ on the interval $-1 \leq x \leq 1$ may be expanded in Legendre polynomials in the form

$$f(x) = \sum_{n=0}^{\infty} c_n P_n(x),$$

(17.66)

where

$$c_n = \frac{2n + 1}{2} \int_{-1}^{1} f(x)P_n(x)dx.$$

(17.67)

Example 17.1 (The Rayleigh equation) In Section 16.2, we introduced the Jacobi–Anger expansion, which can be used to expand a two-dimensional plane wave into an infinite sum of spherical waves. For three-dimensional problems, there is an analogous expansion of a plane wave into spherical waves known as the Rayleigh equation,

$$\exp[ikr\cos\Theta] = \sum_{l=0}^{\infty} i^n(2l + 1)j_l(kr)P_l(\cos\Theta).$$

(17.68)

We may demonstrate this formula as follows. We note that a plane wave is a solution of the Helmholtz equation, and independent of azimuthal angle ϕ. By separation of variables, we can show that all solutions of the Helmholtz equation that are well-behaved at the origin must be a sum of terms of the form $j_l(kr)P_l(\cos\Theta)$, i.e.

$$\exp[ikr\cos\Theta] = \sum_{l=0}^{\infty} c_l j_l(kr)P_l(\cos\Theta). \tag{17.69}$$

We may use the orthogonality properties of the Legendre polynomials to write an integral expression for c_n,

$$c_l j_l(kr) = \frac{2l+1}{2} \int_{-1}^{1} e^{ikrx} P_l(x)\mathrm{d}x. \tag{17.70}$$

We now use Rodrigues' formula, Eq. (17.45), to evaluate this integral. We have

$$c_l j_l(kr) = \frac{2l+1}{2^{l+1}l!} \int_{-1}^{1} e^{ikrx} \frac{\mathrm{d}^l}{\mathrm{d}x^l}(x^2-1)^l \mathrm{d}x. \tag{17.71}$$

Integrating by parts l times, we get

$$c_l j_l(kr) = \frac{2l+1}{2^{l+1}l!}(-1)^l(ikr)^l \int_{-1}^{1} e^{ikrx}(x^2-1)^l \mathrm{d}x. \tag{17.72}$$

We now play a little "trick" via integration by parts. We arrange our integral as follows,

$$c_l j_l(kr) = \frac{2l+1}{2^{l+1}l!}(-1)^l(ikr)^l \int_{-1}^{1} \frac{1}{ikr}\frac{\mathrm{d}}{\mathrm{d}x}\left(e^{ikrx}\right)(x^2-1)^l \mathrm{d}x. \tag{17.73}$$

Integration by parts results in the expression

$$c_l j_l(kr) = \frac{2l+1}{2^{l+1}l!}(-1)^l(ikr)^l \left\{ (-2l)\int_{-1}^{1} \frac{e^{ikrx}}{ikr}x(x^2-1)^{l-1}\mathrm{d}x \right\}. \tag{17.74}$$

We may use the relation

$$\frac{\mathrm{d}}{\mathrm{d}(kr)}e^{ikrx} = ixe^{ikrx} \tag{17.75}$$

to rewrite the integral in the form

$$c_l j_l(kr) = \frac{2l+1}{2^{l+1}l!}(ikr)^l(-1)^l \left\{ (2l)\frac{1}{kr}\frac{\mathrm{d}}{\mathrm{d}(kr)}\int_{-1}^{1} e^{ikrx}(x^2-1)^{l-1}\mathrm{d}x \right\}. \tag{17.76}$$

Repeating this process l times in total reduces our expression to the form

$$c_l j_l(kr) = \frac{2l+1}{2^{l+1}l!}(-1)^l(ikr)^l 2^l l! \left(\frac{1}{kr}\frac{\mathrm{d}}{\mathrm{d}(kr)}\right)^l \int_{-1}^{1} e^{ikrx}\mathrm{d}x. \tag{17.77}$$

The integral is simply $2j_0(kr)$; we then have

$$c_l j_l(kr) = \frac{2l+1}{2^{l+1}} (-1)^l (ikr)^l 2^{l+1} \left(\frac{1}{kr} \frac{d}{d(kr)} \right)^l j_0(kr). \tag{17.78}$$

Referring to Eq. (16.141), we find that

$$c_l j_l(kr) = (2l+1) i^l j_l(kr), \tag{17.79}$$

or $c_l = (2l+1) i^l$.

Rayleigh's formula is very useful in solving problems relating to the scattering of plane waves from spherically symmetric objects. By the use of the spherical harmonic addition theorem, Eq. (17.133), it may be written entirely in terms of spherical harmonics. For a plane wave propagating in the direction $\mathbf{n}$, the formula takes on the form

$$e^{i k \mathbf{n} \cdot \mathbf{r}} = 4\pi \sum_{l=0}^{\infty} \sum_{m=-l}^{l} i^l j_l(kr) Y_l^{m*}(\theta_n, \phi_n) Y_l^m(\theta, \phi), \tag{17.80}$$

where θ_n, ϕ_n are the angles associated with the propagation direction $\mathbf{n}$.
◇

17.7 Associated Legendre functions

The Legendre functions may be used to form solutions of Laplace's equation in spherical coordinates for problems of azimuthal symmetry, i.e. $m = 0$. For cases without azimuthal symmetry, the θ dependence of the separable solutions satisfies the *associated Legendre equation*, Eq. (17.13), i.e. with $x = \cos\theta$,

$$\frac{d}{dx} \left[(1-x^2) \frac{dP_l^m}{dx} \right] + \left[l(l+1) - \frac{m^2}{1-x^2} \right] P_l^m = 0. \tag{17.81}$$

Because this equation is (like the Legendre equation) regular at the origin, it can be solved by looking for a power series solution. This strategy follows closely the derivation of the Legendre polynomials, so we will instead jump ahead to the most useful form of the solution, and then justify it and derive its properties.

The solutions of the associated Legendre equation which are well-behaved in the domain $-1 \le x \le 1$ are the *associated Legendre functions*, which may be defined using a Rodrigues-type expansion,

$$P_l^m(x) = (1-x^2)^{m/2} \frac{d^m}{dx^m} P_l(x) = \frac{(1-x^2)^{m/2}}{2^l l!} \frac{d^{l+m}}{dx^{l+m}} (x^2-1)^l. \tag{17.82}$$

The quantity l must again be an integer for the functions to be well-behaved; however, because of the quantity $(1-x^2)^{m/2}$, the associated Legendre functions are not polynomials.

For $m = 0$, Eq. (17.82) reduces to the familiar Rodrigues formula, Eq. (17.45). It is to be noted that $P_l^m(x) = 0$ for $m > l$, on account of P_l being a polynomial of order l.

To justify Eq. (17.82), we assume that Eq. (17.81) has a solution of the form $P_l^m(x) = (1 - x^2)^{m/2} u(x)$, with $u(x)$ to be determined. On substitution into the ODE, we find that

$$(1 - x^2)u'' - 2(m+1)u' + [l(l+1) - m(m+1)]u = 0. \qquad (17.83)$$

We compare this differential equation with the mth derivative of the Legendre equation (17.15), which may be determined by applying Leibniz's formula, namely

$$\frac{d^m}{dx^m}[u(x)v(x)] = \sum_{k=0}^{m} \binom{m}{k} \frac{d^{m-k}}{dx^{m-k}} u(x) \frac{d^k}{dx^k} v(x), \qquad (17.84)$$

with

$$\binom{m}{n} = m!/[n!(m-n)!]. \qquad (17.85)$$

With a little work (noting that only derivatives of $(1 - x^2)$ up to order 2 are nonzero), we may show that

$$(1 - x^2)\frac{d^m}{dx^m}P_l'' - 2(m+1)\frac{d^m}{dx^m}P_l' + [l(l+1) - m(m+1)]\frac{d^m}{dx^m}P_l = 0. \qquad (17.86)$$

Comparing Eqs. (17.83) and (17.86), we find that the function $u(x)$ must be the mth derivative of the Legendre polynomial P_l.

We can use Eq. (17.82) to define $P_l^m(x)$ for negative values of m, as well; such values are still valid solutions of the associated Legendre equation, which only contains m^2. Using $-m$ in Eq. (17.82), we have

$$P_l^{-m}(x) = \frac{(1 - x^2)^{-m/2}}{2^l l!} \frac{d^{l-m}}{dx^{l-m}} (x^2 - 1)^l. \qquad (17.87)$$

We can demonstrate that this function is related in a straightforward way to $P_l^m(x)$. We begin by using Leibniz's formula on the derivatives of Eq. (17.87) by writing $(x^2 - 1)^l = (x-1)^l(x+1)^l$, such that

$$P_l^{-m}(x) = \frac{(1 - x^2)^{-m/2}}{2^l l!} \sum_{k=0}^{l+m} \frac{(l-m)!}{k!(l-m-k)!} \frac{d^{l-m-k}}{dx^{l-m-k}} (x+1)^l \frac{d^k}{dx^k} (x-1)^l. \qquad (17.88)$$

The derivatives may be evaluated explicitly, with the result

$$P_l^{-m}(x)$$

$$= (-1)^{-m/2} \frac{1}{2^l l!} \sum_{k=0}^{l-m} \frac{(l-m)!}{k!(l-m-k)!} \frac{l!}{(m+k)!} \frac{l!}{(l-k)!} (x+1)^{k+m/2}(x-1)^{l-k-m/2}.$$

$$(17.89)$$

We now change to the summation variable $k' = k + m$, which transforms the equation to the form

$$P_l^{-m}(x) = (-1)^{-m/2} \frac{1}{2^l l!}$$

$$\times \sum_{k'=m}^{l} \frac{(l-m)!(l!)^2}{(k'-m)!(l-k')!k'!(l-k'+m)!}(x+1)^{k'-m/2}(x-1)^{l-k'+m/2}. \quad (17.90)$$

Similarly, we apply Leibniz's formula to Eq. (17.82), with the result

$$P_l^m(x) = \frac{(1-x^2)^{m/2}}{2^l l!} \sum_{k=0}^{l+m} \frac{(l+m)!}{k!(l+m-k)!} \frac{d^{l+m-k}}{dx^{l+m-k}}(x+1)^l \frac{d^k}{dx^k}(x-1)^l. \quad (17.91)$$

Each derivative is of a polynomial of order l, which implies that the only nonzero terms of the sum are those such that $l + m - k \le l$, or $k \ge m$, and $k \le l$. We may therefore write

$$P_l^m(x) = \frac{(1-x^2)^{m/2}}{2^l l!} \sum_{k=m}^{l} \frac{(l+m)!}{k!(l+m-k)!} \frac{d^{l+m-k}}{dx^{l+m-k}}(x+1)^l \frac{d^k}{dx^k}(x-1)^l. \quad (17.92)$$

Evaluating the derivatives, we find that

$$P_l^{-m}(x) = (-1)^{m/2} \frac{1}{2^l l!} \sum_{k=m}^{l} \frac{(l+m)!(l!)^2}{k!(l+m-k)!(k-m)!(l-k)!}(x+1)^{k-m/2}(x-1)^{l-k+m/2}.$$

$$(17.93)$$

On comparison of Eqs. (17.90) and (17.93), we readily see that the associated Legendre functions of positive and negative m are related by the expression

$$P_l^{-m}(x) = (-1)^m \frac{(l-m)!}{(l+m)!} P_l^m(x). \quad (17.94)$$

Equation (17.82) can be used to determine many of the other useful properties of associated Legendre functions. For instance, we may define recurrence relations for the associated Legendre polynomials by taking the mth derivative of their Legendre polynomial counterparts. As an example, we recall Eq. (17.31),

$$(2n+1)xP_n(x) = (n+1)P_{n+1}(x) + nP_{n-1}(x). \quad (17.95)$$

We take the mth derivative of this formula and multiply by $(1-x^2)^{m/2}$, obtaining

$$(2l+1)xP_l^m + (2l+1)m\sqrt{1-x^2}P_l^{m-1} = (l+1)P_{l+1}^m + lP_{l-1}^m. \quad (17.96)$$

This expression relates functions of differing l and m values. If we consider the mth derivative of Eq. (17.37),

$$xP'_l(x) - P'_{l-1}(x) = lP_l(x), \qquad (17.97)$$

we may write

$$xP^m_l(x) - P^m_{l-1} = (l-m+1)\sqrt{1-x^2}P^{m-1}_l. \qquad (17.98)$$

We may eliminate $\sqrt{1-x^2}P^{m-1}_l$ from Eqs. (17.96) and (17.98) to obtain

$$(l-m+1)P^m_{l+1}(x) - (2l+1)xP^m_l(x) + (l+m)P^m_{l-1} = 0. \qquad (17.99)$$

This equation is noteworthy because it is a derivative-free relationship amongst the associated functions of the same m-value. An analogous relationship amongst associated function of the same l-value can be found,

$$\sqrt{1-x^2}P^{m+1}_l(x) - 2mxP^m_l(x) + (l+m)(l-m+1)\sqrt{1-x^2}P^{m-1}_l(x) = 0. \qquad (17.100)$$

Many other recurrence relations can be derived; we list some significant ones below including those which relate the derivatives of the associated functions.

$$(2l+1)xP^m_l = (l+m)P^m_{l-1} + (n-m+1)P^m_{l+1}, \qquad (17.101)$$

$$(2l+1)\sqrt{1-x^2}P^m_l = P^{m+1}_{l+1} - P^{m+1}_{l-1}, \qquad (17.102)$$

$$(2l+1)\sqrt{1-x^2}P^m_l = (l+m)(l+m-1)P^{m-1}_{l-1}$$
$$- (l-m+1)(l-m+2)P^{m-1}_{l+1}, \qquad (17.103)$$

$$(1-x^2)\frac{dP^m_l}{dx} = \sqrt{1-x^2}P^{m+1}_l - mxP^m_l, \qquad (17.104)$$

$$(1-x^2)\frac{dP^m_l}{dx} = mxP^m_l - (l+m)(l-m+1)\sqrt{1-x^2}P^{m-1}_l, \qquad (17.105)$$

$$(1-x^2)\frac{dP^m_l}{dx} = (l+m)P^m_{l-1} - lxP^m_l, \qquad (17.106)$$

$$(1-x^2)\frac{dP^m_l}{dx} = (l+1)xP^m_l - (l-m+1)P^m_{l+1}. \qquad (17.107)$$

Orthogonality of the associated Legendre functions of the same m value may be determined from Eq. (17.81). We look at the integral of two functions P^m_p and P^m_q, and assume that $p < q$. We have, neglecting normalizing constants for the moment,

$$\int_{-1}^{1} P^m_p(x)P^m_q(x)\,dx \propto \int_{-1}^{1} X^m \left[\frac{d^{p+m}}{dx^{p+m}}X^p\right]\left[\frac{d^{q+m}}{dx^{q+m}}X^q\right]dx, \qquad (17.108)$$

where we have taken $X(x) = (x^2 - 1)$ for convenience. We perform $q + m$ integrations by parts to eliminate the derivatives on X^q; the boundary terms will vanish in the presence of a term $(1 - x^2)$. We get

$$\int_{-1}^{1} P_p^m(x) P_q^m(x) dx \propto (-1)^{p+q} \int_{-1}^{1} X^q \frac{d^{q+m}}{dx^{q+m}} \left[X^m \frac{d^{p+m}}{dx^{p+m}} X^p \right] dx. \qquad (17.109)$$

We may use the Leibniz formula, Eq. (17.84), to expand the derivatives in the integrand in the form

$$\int_{-1}^{1} P_p^m(x) P_q^m(x) dx$$

$$\propto (-1)^{p+q} \int_{-1}^{1} X^q \sum_{k=0}^{q+m} \frac{(q+m)!}{k!(q+m-k)!} \frac{d^{q+m-k}}{dx^{q+m-k}} X^m \frac{d^{p+m+k}}{dx^{p+m+k}} X^p dx. \qquad (17.110)$$

Because X^m is a polynomial of order $2m$, all derivatives of X^m vanish save those for which

$$q + m - k \leq 2m. \qquad (17.111)$$

Likewise, because X^p is a polynomial of order $2p$, all derivatives of X^p vanish save those for which

$$p + m + k \leq 2p. \qquad (17.112)$$

We may add these two inequalities together, and all nonzero terms must satisfy

$$q \leq p. \qquad (17.113)$$

However, this contradicts our assumption that $p < q$. Therefore there are no nonzero terms in the integral, and it vanishes.

The normalization of the functions may also be done by integration by parts. In this case, a single term of the integration by parts survives, such that $p = q$, or $k = q - m$. It can be shown that

$$\int_{-1}^{1} [P_l^m(x)]^2 dx = \frac{2}{2l+1} \frac{(l+m)!}{(l-m)!}. \qquad (17.114)$$

This relationship is important in the next section, when we look for orthogonal separable solutions to Laplace's equation in spherical coordinates. In terms of the angular variable θ, we may write the orthogonality relation as

$$\int_{0}^{\pi} P_l^m(\cos\theta) P_{l'}^m(\cos\theta) \sin\theta d\theta = \frac{2}{2l+1} \frac{(l+m)!}{(l-m)!} \delta_{ll'}. \qquad (17.115)$$

17.8 Spherical harmonics

Returning to the solution of Laplace's equation within a spherical domain of radius a, we now see that a complete separated solution V_{lm} characterized by integers l and m is given by

$$V_{lm}(r,\theta,\phi) = C_{lm}r^l P_l^m(\cos\theta)e^{im\phi}, \tag{17.116}$$

where we now allow m to take on positive and negative values for convenience. For the angular functions, it is convenient to introduce a normalized combined function $Y_l^m(\theta,\phi)$ defined as

$$Y_l^m(\theta,\phi) = (-1)^m\sqrt{\frac{2l+1}{2}\frac{(l-m)!}{2\pi(l+m)!}}P_l^m(\cos\theta)e^{im\phi}. \tag{17.117}$$

These functions, called the *spherical harmonics*,[3] have been chosen to be orthonormal with respect to an integral over the solid angle of a sphere, i.e.

$$\int_\Omega Y_l^m(\theta,\phi)Y_{l'}^{m'*}(\theta,\phi)d\Omega = \int_0^\pi\int_0^{2\pi}Y_l^m(\theta,\phi)Y_{l'}^{m'*}(\theta,\phi)\sin\theta d\theta d\phi = \delta_{ll'}\delta_{mm'}. \tag{17.118}$$

It is to be noted that the orthogonality is between one spherical harmonic and the complex conjugate of another.

By the application of Eq. (17.94), it is straightforward to show that the spherical harmonics satisfy the relation

$$Y_l^{-m}(\theta,\phi) = (-1)^m Y_l^{m*}(\theta,\phi). \tag{17.119}$$

The first few spherical harmonics are given in Table 17.2. They are complete, in that any well-behaved function of θ and ϕ on the sphere may be represented by a series of spherical harmonics, i.e. a function $f(\theta,\phi)$ may be written as

$$f(\theta,\phi) = \sum_{l=0}^\infty\sum_{m=-l}^l c_{lm}Y_l^m(\theta,\phi), \tag{17.120}$$

where

$$c_{lm} = \int_\Omega Y_l^{m*}(\theta',\phi')f(\theta',\phi')d\Omega'. \tag{17.121}$$

Several recurrence relations for the spherical harmonics may be derived by use of the recurrence relations for the associated Legendre functions. For instance, from Eq. (17.101),

---

[3] The term "harmonic" in this case refers to the the solutions of Laplace's equation being harmonic functions, such as discussed in Section 9.4.

Table 17.2 *Low-order spherical harmonics.*

l	m	Y_l^m
0	0	$Y_0^0(\theta,\phi) = \frac{1}{\sqrt{4\pi}}$
1	1	$Y_1^1(\theta,\phi) = -\sqrt{\frac{3}{8\pi}}\sin\theta\, e^{i\phi}$
1	0	$Y_1^0(\theta,\phi) = -\sqrt{\frac{3}{4\pi}}\cos\theta$
1	-1	$Y_1^{-1}(\theta,\phi) = \sqrt{\frac{3}{8\pi}}\sin\theta\, e^{-i\phi}$
2	2	$Y_2^2(\theta,\phi) = \frac{1}{4}\sqrt{\frac{15}{2\pi}}\sin^2\theta\, e^{2i\phi}$
2	1	$Y_2^1(\theta,\phi) = -\sqrt{\frac{15}{8\pi}}\sin\theta\cos\theta\, e^{i\phi}$
2	0	$Y_2^0(\theta,\phi) = \frac{1}{2}\sqrt{\frac{5}{4\pi}}(3\cos^2\theta - 1)$
2	-1	$Y_2^{-1}(\theta,\phi) = \sqrt{\frac{15}{8\pi}}\sin\theta\cos\theta\, e^{-i\phi}$
2	-2	$Y_2^2(\theta,\phi) = \frac{1}{4}\sqrt{\frac{15}{2\pi}}\sin^2\theta\, e^{-2i\phi}$

we may readily write

$$\cos\theta Y_l^m(\theta,\phi) = \left[\frac{(l-m+1)(l+m+1)}{(2l+1)(2l+3)}\right]^{1/2} Y_{l+1}^m(\theta,\phi)$$

$$+ \left[\frac{(l-m)(l+m)}{(2l-1)(2l+1)}\right]^{1/2} Y_{l-1}^m, \tag{17.122}$$

and from Eqs. (17.102) and (17.102) we may write

$$e^{\pm i\phi}\sin\theta Y_l^{m\mp1} = \mp\left[\frac{(l\pm m)(l\pm m+1)}{(2l+1)(2l+3)}\right]^{1/2} Y_{l+1}^m$$

$$\pm \left[\frac{(l\mp m+1)(l\mp m)}{(2l-1)(2l+1)}\right]^{1/2} Y_{l-1}^m. \tag{17.123}$$

We may write a closure relation for the spherical harmonics by substituting from Eq. (17.121) into Eq. (17.120), and switching the order of integration and summation,

$$f(\theta,\phi) = \int_\Omega f(\theta',\phi')\left[\sum_{l=0}^{\infty}\sum_{m=-l}^{l} Y_l^{m*}(\theta',\phi')Y_l^m(\theta,\phi)\right]\sin\theta'd\theta'd\phi'. \tag{17.124}$$

This equation is a "sifting" equation for the function $f(\theta,\phi)$, which suggests that we may write

$$\sum_{l=0}^{\infty}\sum_{m=-l}^{l}Y_l^{m*}(\theta',\phi')Y_l^m(\theta,\phi)=\frac{\delta(\theta-\theta')\delta(\phi-\phi')}{\sin\theta}, \qquad (17.125)$$

in accordance with the results of Section 6.6.

It is convenient to write the delta functions for the spherical harmonics as a delta function of solid angle, in the form

$$\delta(\Omega-\Omega')\equiv\frac{\delta(\theta-\theta')\delta(\phi-\phi')}{\sin\theta}, \qquad (17.126)$$

where $\Omega=(\theta,\phi)$ is a short-hand notation for the complete angular description of a vector. The new delta function then satisfies

$$\int_{\Omega}\delta(\Omega-\Omega')d\Omega'=1. \qquad (17.127)$$

With this expression, the closure relation takes on the form

$$\delta(\Omega-\Omega')=\sum_{l=0}^{\infty}\sum_{m=-l}^{l}Y_l^{m*}(\Omega')Y_l^m(\Omega). \qquad (17.128)$$

It is to be noted that there are a number of different definitions of the spherical harmonics. Some references define them using $\sin m\phi$ and $\cos m\phi$ instead of complex exponentials, but even with the complex exponential definition there are a number of variations, based on the signs of the harmonics and the relationship between the harmonics for positive and negative values of m. We are using the so-called *Condon–Shortley* choice of phase for the spherical harmonics, which includes a factor of $(-1)^m$ in the definition of Y_l^m and relates positive and negative m functions by Eq. (17.119). Other possibilities include the *Darwin* choice of phase, for which

$$Y_{lm}^{\text{Darwin}}(\theta,\phi)=\sqrt{\frac{2l+1}{2}\frac{(l-|m|)!}{2\pi(l+|m|)!}}P_l^{|m|}(\cos\theta)e^{im\phi}. \qquad (17.129)$$

A third choice is the *Bethe* phase, which is related to the Darwin by

$$Y_{lm}^{\text{Bethe}}(\theta,\phi)=\begin{cases}Y_{lm}^{\text{Darwin}}(\theta,\phi), & m\geq 0,\\(-1)^mY_{lm}^{\text{Darwin}}(\theta,\phi), & m<0.\end{cases} \qquad (17.130)$$

The Darwin convention includes no changes of sign for different m values or for positive and negative m values. The Bethe convention is chosen so that the negative m harmonics have an overall $(-1)^m$ sign relation. The Condon–Shortley convention instead varies the signs

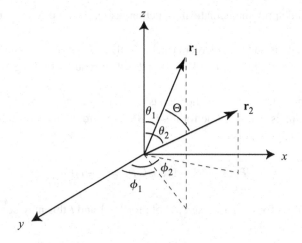

Figure 17.2 Illustrating the notation related to the addition theorem.

of the positive m harmonics by $(-1)^m$. In using results from other references discussing the spherical harmonics, one should always begin by checking which phase convention is used.

17.9 Spherical harmonic addition theorem

In this section, we prove an important result known as the *spherical harmonic addition theorem*. Let us suppose we are given two vectors $\mathbf{r}_1$ and $\mathbf{r}_2$, represented in polar coordinates by (r_1, θ_1, ϕ_1) and (r_2, θ_2, ϕ_2). We define the angle between these two vectors as Θ, such that

$$\mathbf{r}_1 \cdot \mathbf{r}_2 = r_1 r_2 \cos \Theta. \tag{17.131}$$

This is illustrated in Fig. 17.2.

The angles are related by the formula

$$\cos \theta_1 \cos \theta_2 + \sin \theta_1 \sin \theta_2 \cos(\phi_1 - \phi_2) = \cos \Theta. \tag{17.132}$$

The spherical harmonic addition theorem asserts that the following identity holds

$$P_l(\cos \Theta) = \frac{4\pi}{2l+1} \sum_{m=-l}^{l} Y_l^{m*}(\theta_1, \phi_1) Y_l^m(\theta_2, \phi_2). \tag{17.133}$$

It relates functions of the relative angle Θ to functions of the absolute angles (θ_1, ϕ_1) and (θ_2, ϕ_2).

There are a number of techniques that can be used to prove the addition theorem, including a proof using group theoretical arguments and a proof using complex integration. Here we follow [CH91] and apply mathematical induction, using the recurrence formulas we have previously derived.

We note first that the addition theorem can be easily proven for $l = 0$ and $l = 1$. For $l = 0$, it reduces to the trivial identity $1 = 1$, and for $l = 1$, it reduces to the geometric equation (17.132).

We now assume that the addition theorem is true for $l - 1$ and l, where $l = 1, 2, 3, \ldots$, and prove that this implies that it is true for $l + 1$. We return to the recurrence relation (17.31) for the Legendre polynomials,

$$-lP_{l-1} + (2l+1)\cos\Theta P_l = (l+1)P_{l+1}, \qquad (17.134)$$

and with the addition formula assumed to hold for $l - 1$ and l this may be written as

$$-\frac{l}{2l-1}\sum_{m=-l+1}^{l-1} Y_{l-1}^{m*}Y_{l-1}^{m} + \cos\Theta \sum_{m=-l}^{l} Y_{l}^{m*}Y_{l}^{m} = \frac{(l+1)}{4\pi}P_{l+1}. \qquad (17.135)$$

We have suppressed the functional dependence for brevity; the first spherical harmonic of any product is assumed to depend upon (θ_1, ϕ_1), while the second depends on (θ_2, ϕ_2). If we can prove that

$$-\frac{l}{2l-1}\sum_{m=-l+1}^{l-1} Y_{l-1}^{m*}Y_{l-1}^{m} + \cos\Theta \sum_{m=-l}^{l} Y_{l}^{m*}Y_{l}^{m} = \frac{l+1}{2l+3}\sum_{m=-l-1}^{l+1} Y_{l+1}^{m*}Y_{l+1}^{m}, \qquad (17.136)$$

we will have equated the right-hand sides of Eqs. (17.135) and (17.136) and completed our proof.

We now apply the spherical harmonic recursion formulas (17.122) and (17.123), which we write in the form

$$\cos\theta Y_l^m = A_{lm}Y_{l+1}^m + B_{lm}Y_{l-1}^m, \qquad (17.137)$$

$$\sin\theta e^{i\phi}Y_l^m = C_{lm}^+ Y_{l+1}^{m+1} + D_{lm}^+ Y_{l-1}^{m+1}, \qquad (17.138)$$

$$\sin\theta e^{-i\phi}Y_l^m = C_{lm}^- Y_{l+1}^{m-1} + D_{lm}^- Y_{l-1}^{m-1}, \qquad (17.139)$$

with

$$A_{lm} = \sqrt{\frac{(l-m+1)(l+m+1)}{(2l+1)(2l+3)}}, \qquad (17.140)$$

$$B_{lm} = \sqrt{\frac{(l-m)(l+m)}{(2l-1)(2l+1)}}, \qquad (17.141)$$

$$C_{lm}^{\pm} = \mp \sqrt{\frac{(l \pm m + 1)(l \pm m + 2)}{(2l+1)(2l+3)}}, \tag{17.142}$$

$$D_{lm}^{\pm} = \pm \sqrt{\frac{(l \mp m)(l \mp m - 1)}{(2l-1)(2l+1)}}. \tag{17.143}$$

Noting that $\cos\Theta$ satisfies Eq. (17.132), we may write

$$\cos\Theta Y_l^m = (A_{lm}Y_{l+1}^{m*} + B_{lm}Y_{l-1}^{m*})(A_{lm}Y_{l+1}^{m*} + B_{lm}Y_{l-1}^m)$$

$$+ \frac{1}{2}(C_{lm}^+ Y_{l+1}^{m+1*} + D_{lm}^+ Y_{l-1}^{m+1*})(C_{lm}^+ Y_{l+1}^{m+1} + D_{lm}^+ Y_{l-1}^{m+1})$$

$$+ \frac{1}{2}(C_{lm}^- Y_{l+1}^{m-1*} + D_{lm}^- Y_{l-1}^{m-1*})(C_{lm}^- Y_{l+1}^{m-1} + D_{lm}^- Y_{l-1}^{m-1}). \tag{17.144}$$

This mess may be simplified using the notation

$$\mathcal{X}_l^m \equiv Y_l^{m*} Y_l^m, \tag{17.145}$$

$$\mathcal{Y}_l^m \equiv Y_{l+1}^{m*} Y_{l-1}^m, \tag{17.146}$$

$$\mathcal{Z}_l^m \equiv Y_{l-1}^{m*} Y_{l+1}^m. \tag{17.147}$$

It is to be noted that Eq. (17.136) must be satisfied independently for every value of m due to the linear independence of the spherical harmonics. Also, it must be satisfied independently for $\mathcal{X}_l^m$, $\mathcal{Y}_l^m$, and $\mathcal{Z}_l^m$. Considering first the case $-l+1 \le m \le l-1$, we may write Eq. (17.136) as three separate conditions,

$$-\frac{l}{2l-1}\mathcal{X}_{l-1}^m + A_{lm}^2 \mathcal{X}_{l+1}^m + B_{lm}^2 \mathcal{X}_{l-1}^m + \frac{1}{2}C_{l,m-1}^{+2}\mathcal{X}_{l+1}^m$$

$$+ \frac{1}{2}C_{l,m+1}^{-2}\mathcal{X}_{l+1}^m + \frac{1}{2}D_{l,m-1}^{+2}\mathcal{X}_{l-1}^m + \frac{1}{2}D_{l,m+1}^{-2}\mathcal{X}_{l-1}^m = \frac{l+1}{2l+3}\mathcal{X}_{l+1}^m. \tag{17.148}$$

$$A_{lm}B_{lm}\mathcal{Y}_l^m + \frac{1}{2}C_{l,m-1}^+ D_{l,m-1}^+ \mathcal{Y}_l^m + \frac{1}{2}C_{l,m+1}^- D_{l,m+1}^- \mathcal{Y}_l^m = 0, \tag{17.149}$$

$$A_{lm}B_{lm}\mathcal{Z}_l^m + \frac{1}{2}C_{l,m-1}^+ D_{l,m-1}^+ \mathcal{Z}_l^m + \frac{1}{2}C_{l,m+1}^- D_{l,m+1}^- \mathcal{Z}_l^m = 0. \tag{17.150}$$

The conditions for $\mathcal{Y}$ and $\mathcal{Z}$ are redundant. We may further break down the condition for $\mathcal{X}$ into conditions for $l+1$ and $l-1$, such that

$$-\frac{l}{2l-1} + B_{lm}^2 + \frac{1}{2}D_{l,m-1}^{+2} + \frac{1}{2}D_{l,m+1}^{-2} = 0, \tag{17.151}$$

$$A_{lm}^2 + \frac{1}{2}C_{l,m-1}^{+2} + \frac{1}{2}C_{l,m+1}^{-2} = \frac{l+1}{2l+3}. \tag{17.152}$$

Each of these conditions can be verified by direct substitution. A similar set of conditions can be derived for the case $|m| = |l|$, and shown to be valid. With this demonstration, the proof of the addition theorem by induction is achieved.

This theorem can be used to decompose a rotationally symmetric function (one which depends only upon Θ) into a set of vectors orthogonal over solid angle. It is applied, for instance, in going from Eq. (16.162) to Eq. (16.163).

17.10 Solution of PDEs in spherical coordinates

The solution of PDEs using separation of variables in spherical coordinates typically requires the use of a number of special functions. We consider an example which highlights strategies for the solution of such problems.

Example 17.2 (Waves in a sphere) We consider the solution $u(r,\theta,\phi,t)$ of the wave equation,

$$\nabla^2 u - \frac{1}{c^2}\frac{\partial^2 u}{\partial t^2} = 0, \tag{17.153}$$

in a spherical domain of radius a, subject to the boundary condition $u(a,\theta,\phi,t) = 0$ and the initial conditions $u(r,\theta,\phi,0) = f(\mathbf{r})$ and $u_t(r,\theta,\phi,0) = g(\mathbf{r})$. This problem may represent the behavior of acoustic waves propagating in a spherical cavity. Writing the Laplacian operator out explicitly, we have

$$\frac{1}{r}\frac{\partial^2}{\partial r^2}(ru) + \frac{1}{r^2\sin\theta}\left[\frac{\partial}{\partial\theta}\left(\sin\theta\frac{\partial u}{\partial\theta}\right) + \frac{1}{\sin\theta}\frac{\partial^2 u}{\partial\phi^2}\right] - \frac{1}{c^2}\frac{\partial^2 u}{\partial t^2} = 0. \tag{17.154}$$

We try a separable solution of the form

$$u(r,\theta,\phi,t) = R(r)Y(\theta,\phi)T(t), \tag{17.155}$$

which results in the expression

$$\frac{1}{Rr}\frac{\partial^2}{\partial r^2}(rR) + \frac{1}{r^2}\left\{\frac{1}{Y\sin\theta}\left[\frac{\partial}{\partial\theta}\left(\sin\theta\frac{\partial Y}{\partial\theta}\right) + \frac{1}{\sin\theta}\frac{\partial^2 Y}{\partial\phi^2}\right]\right\} - \frac{1}{Tc^2}\frac{\partial^2 T}{\partial t^2} = 0. \tag{17.156}$$

Referring back to Eq. (17.5), we immediately see that the quantity in the curved brackets must be equal to $-l(l+1)$, and the function Y must be a spherical harmonic $Y_l^m(\theta,\phi)$. In fact, provided the only spatial derivatives in the PDE are in the Laplacian and the PDE has constant coefficients, *the angular dependence will always be characterized by the spherical harmonics.*

The two non-constant terms which remain must each be equal to a constant k^2 themselves, and this gives us a pair of equations for R and T,

$$\frac{1}{r}\frac{\partial^2}{\partial r^2}(rR) - \frac{l(l+1)}{r^2}R + k^2 R = 0, \tag{17.157}$$

$$\frac{\partial^2 T}{\partial t^2} = -k^2 c^2 T. \tag{17.158}$$

The latter equation is a harmonic oscillator equation with solution

$$T(t) = A_k \cos(kct) + B_k \sin(kct),$$ (17.159)

while the former equation is simply the equation for the spherical Bessel functions, discussed in Section 16.11, and the solution is

$$R(r) = C_k j_l(kr) + D_k n_l(kr).$$ (17.160)

We may immediately set $D_k = 0$, as we must satisfy the "hidden" boundary condition that the solution is well-behaved at the origin. We must then satisfy the explicit boundary condition $u(a, \theta, \phi, t) = 0$, which can only be satisfied if $k = k_{ln}$, where $k_{ln}a$ is the nth zero of the lth Bessel function. Our general solution therefore takes the form

$$u(r, \theta, \phi, t) = \sum_{n=0}^{\infty} \sum_{l=0}^{\infty} \sum_{m=-l}^{l} j_l(k_{ln}r) Y_l^m(\theta, \phi) [A_{lmn} \cos(\omega_{ln}t) + B_{lmn} \sin(\omega_{ln}t)], \quad (17.161)$$

where we have written $\omega_{ln} \equiv k_{ln}c$ for convenience. It is to be noted that the solution involves a triple summation; in general, a partial differential equation in N variables will result in a solution which has $N - 1$ summations.

In matching the boundary conditions, we arrive at the pair of series relations

$$f(\mathbf{r}) = \sum_{n=0}^{\infty} \sum_{l=0}^{\infty} \sum_{m=-l}^{l} A_{lmn} j_l(k_{ln}r) Y_l^m(\theta, \phi),$$ (17.162)

$$g(\mathbf{r}) = \sum_{n=0}^{\infty} \sum_{l=0}^{\infty} \sum_{m=-l}^{l} \omega_{ln} B_{lmn} j_l(k_{ln}r) Y_l^m(\theta, \phi).$$ (17.163)

We may now use the orthogonality relations for the Bessel functions and spherical harmonics to find our coefficients. The result is

$$A_{lmn} = \frac{2}{a^3 [j_{m+1}(\beta_{mn})]^2} \int_V f(\mathbf{r}) j_l(k_{ln}r) Y_l^{m*}(\theta, \phi) d^3r,$$ (17.164)

$$B_{lmn} = \frac{2}{\omega_{ln} a^3 [j_{m+1}(\beta_{mn})]^2} \int_V g(\mathbf{r}) j_l(k_{ln}r) Y_l^{m*}(\theta, \phi) d^3r.$$ (17.165)

This is as far as we can go without introducing a specific form of the initial conditions $f(\mathbf{r})$ and $g(\mathbf{r})$. Even with a specific form, the integrals in question (and consequently the solution) will likely need to be evaluated numerically.

◇

A similar strategy may be employed to solve the diffusion equation and Laplace's equation in spherical coordinates.

17.11 Gegenbauer polynomials

We have seen that the Legendre polynomials may be constructed from the generating function

$$\frac{1}{(1-2xt+t^2)^{1/2}} = \sum_{n=0}^{\infty} P_n(x)t^n, \tag{17.166}$$

where the Legendre polynomials are intimately connected with the angular behavior of functions in spherical coordinates. We may also introduce a more general generating function of the form

$$\frac{1}{(1-2xt+t^2)^{\nu}} = \sum_{n=0}^{\infty} \mathcal{G}_n^{\nu}(x)t^n, \tag{17.167}$$

where $\nu \neq 0, -1, -2, \ldots$ and $\mathcal{G}_n^{\nu}(x)$ are known as the *Gegenbauer polynomials* or *ultraspherical polynomials*. These functions are not as common as the Legendre polynomials, but do appear in some important problems such as Gegenbauer's addition theorem, discussed in Section 16.12. In this section, we work from this generating function and derive some useful properties of the Gegenbauer polynomials. It is to be noted that they are a special case of the orthogonal polynomials to be discussed in the next chapter; furthermore, the special cases $\nu = 0$ and $\nu = 1$ are referred to as *Chebyshev polynomials*, and will be discussed in the next chapter as well. $\nu = 0$ requires special consideration.

Using the generating function with $t = 0$, we readily find that $\mathcal{G}_0^{\nu}(x) = 1$. If we take the derivative of the generating function with respect to t, we find that

$$\frac{-\nu(-2x+2t)}{(1-2xt+t^2)^{\nu+1}} = \sum_{n=1}^{\infty} n\mathcal{G}_n^{\nu}t^{n-1}, \tag{17.168}$$

which immediately leads to the result $\mathcal{G}_1^{\nu}(x) = 2\nu x$. Substituting from Eq. (17.167) into Eq. (17.168) leads to the expression

$$(2x-2t)\nu \sum_{n=0}^{\infty} \mathcal{G}_n^{\nu}(x)t^n = (1-2xt+t^2) \sum_{n=1}^{\infty} n\mathcal{G}_n^{\nu}t^{n-1}. \tag{17.169}$$

From this we may derive the recurrence relation

$$(n+1)\mathcal{G}_{n+1}^{\nu}(x) = 2x(\nu+n)\mathcal{G}_n^{\nu}(x) - (2\nu+n-1)\mathcal{G}_{n-1}^{\nu}(x). \tag{17.170}$$

With this recurrence relation and the values of $\mathcal{G}_0^{\nu}$ and $\mathcal{G}_1^{\nu}$, the functions for all values of n may be determined. By working up from $n = 0$ and $n = 1$, it may also be seen from these expressions that $\mathcal{G}_n^{\nu}$ is a polynomial of order n.

We state a few additional results without proof. Further manipulation of the generating function and recurrence relation leads to the differential equation

$$(1-x^2)\frac{d^2\mathcal{G}_n^\nu}{dx^2} - (2\nu+1)x\frac{d\mathcal{G}_n^\nu}{dx} + n(n+2\nu)\mathcal{G}_n^\nu = 0. \tag{17.171}$$

The Gegenbauer polynomials are orthogonal on $-1 \le x \le 1$ with respect to a weighting function $(1-x^2)^{\nu-1/2}$, i.e.

$$\int_{-1}^{1} (1-x^2)^{\nu-1/2}\mathcal{G}_n^\nu(x)\mathcal{G}_m^\nu(x)dx = 0, \quad m \neq n. \tag{17.172}$$

The normalization may be found to be

$$\int_{-1}^{1} (1-x^2)^{\nu-1/2}\mathcal{G}_n^\nu(x)\mathcal{G}_n^\nu(x)dx = \frac{\pi 2^{1-2\nu}\Gamma(n+2\nu)}{n!(n+\nu)[\Gamma(\nu)]^2}. \tag{17.173}$$

17.12 Focus: multipole expansion for static electric fields

One of the most significant uses of Legendre functions and spherical harmonics is in the characterization of static distributions of charges and currents and the fields that they produce, in what is known as a *multipole expansion*.

We consider the behavior of the electric potential $\phi(\mathbf{r})$ produced by a charge density $\rho(\mathbf{r})$ that is confined to a spherical volume V of radius R centered on the origin. We are interested in determining the potential in the infinite region *outside* of the spherical volume; in this region, the potential satisfies Laplace's equation,

$$\nabla^2\phi(\mathbf{r}) = 0. \tag{17.174}$$

If we apply separation of variables to Laplace's equation in spherical coordinates, we may readily find that the general solution may be written in the form

$$\phi(\mathbf{r}) = \sum_{l=0}^{\infty} \sum_{m=-l}^{l} \frac{4\pi}{2l+1}\rho_{lm}\frac{Y_l^m(\theta,\phi)}{r^{l+1}}. \tag{17.175}$$

Here we have applied the "hidden boundary condition" that the potential must approach zero as $r \to \infty$; the factor of $4\pi/(2l+1)$ is for later convenience. This expression represents the multipole expansion of the electrostatic potential, an expansion of the potential into a series of functions which are orthogonal with respect to solid angle. Far from the source domain (large r), the behavior of the potential will be dominated by the lowest nonzero term in l, because terms decay as $1/r^{l+1}$. The $l = 0$ term is the *monopole* term, while the $l = 1$ terms are the *dipole* terms. The $l = 2$ terms are referred to as the *quadrupole* terms. We will explain the origin of this terminology momentarily.

The coefficients ρ_{lm} depend on the specific form of the charge distribution and are known as the *moments* of it. We now consider how they may be calculated.

It can be shown (for instance, in Chapter 19) that the potential due to a charge density $\rho(\mathbf{r})$ may be written as the integral,

$$\phi(\mathbf{r}) = \int_V \frac{\rho(\mathbf{r}')}{|\mathbf{r} - \mathbf{r}'|} d^3 r'. \tag{17.176}$$

The denominator may be written in terms of r, r', and the angle Θ between the vectors as

$$\frac{1}{|\mathbf{r} - \mathbf{r}'|} = \frac{1}{r} \frac{1}{\sqrt{1 + (r'/r)^2 - 2(r'/r)\cos\Theta}}. \tag{17.177}$$

This expression, however, may be compared with the generating function of the Legendre polynomials,

$$(1 - 2xt + t^2)^{-1/2} = \sum_{l=0}^{\infty} P_l(x) t^l. \tag{17.178}$$

Making the association $x \equiv \cos\Theta$ and $t \equiv r'/r$, we may expand the denominator of Eq. (17.176) as a series of Legendre polynomials,

$$\phi(\mathbf{r}) = \frac{1}{r^{l+1}} \sum_{l=0}^{\infty} \int_V \rho(\mathbf{r}') P_l(\cos\Theta) r'^l d^3 r'. \tag{17.179}$$

This expression is already in partial agreement with Eq. (17.175). Before finishing the comparison, we look at the physical interpretation of the first two terms of the series. For $l = 0$, we find the potential takes on the form

$$\phi_0(\mathbf{r}) = \frac{1}{r} \int_V \rho(\mathbf{r}') d^3 r' = \frac{Q_{\text{net}}}{r}, \tag{17.180}$$

where Q_{net} is the net charge of the distribution. The lowest-order term of the expansion has the same potential as a point charge at the origin, and therefore acts like a single "pole".

For $l = 1$, we may write the potential in the form

$$\phi_1(\mathbf{r}) = \frac{1}{r^2} \int_V \cos\Theta \, r' \rho(\mathbf{r}') d^3 r' = \frac{1}{r^2} \int_V \hat{\mathbf{r}} \cdot \mathbf{r}' \rho(\mathbf{r}') d^3 r', \tag{17.181}$$

where we have written $r' \cos\Theta = \hat{\mathbf{r}} \cdot \mathbf{r}'$. Pulling the $\hat{\mathbf{r}}$ out of the integral, we may write this as

$$\phi_1(\mathbf{r}) = \frac{1}{r^2} \hat{\mathbf{r}} \cdot \int_V \mathbf{r}' \rho(\mathbf{r}') d^3 r' = \frac{\mathbf{r} \cdot \mathbf{p}}{r^3}, \tag{17.182}$$

where

$$\mathbf{p} \equiv \int_V \mathbf{r}' \rho(\mathbf{r}') d^3 r' \tag{17.183}$$

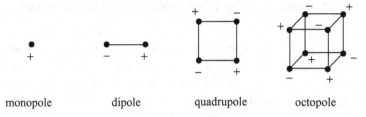

monopole   dipole   quadrupole   octopole

Figure 17.3 Illustration of some basic multipole distributions.

is the dipole moment of the distribution. An example of a collection of charges for which the dipole term is the leading term in the potential is a physical dipole, consisting of equal and opposite charges separated by a distance d. The net charge of a physical dipole is zero, meaning that the monopole contribution to the potential vanishes.

If the net dipole moment of the distribution vanishes, for example if the distribution consists of two oppositely oriented physical dipoles, the $l = 2$ (quadrupole) term of the potential is dominant. If the quadrupole term vanishes, the $l = 3$ (octopole) term is dominant, etc. A dipole is made of opposing monopoles, a quadrupole is made of opposing dipoles, and so on. Illustrations of the simplest monopole, dipole, quadrupole, and octopole distributions are shown in Fig. 17.3.

Equation (17.179) can be cumbersome to work with for higher-order multipoles because Θ depends on the directions of both $\mathbf{r}$ and $\mathbf{r}'$. It can be written in a more convenient form by use of the addition theorem, Eq. (17.133), replacing the Legendre function as a sum over spherical harmonics,

$$\phi(\mathbf{r}) = \sum_{l=0}^{\infty} \frac{4\pi}{2l+1} \sum_{m=-l}^{l} Y_l^m(\theta,\phi) \int_V r'^l \rho(\mathbf{r}') Y_l^{m*}(\theta',\phi') \mathrm{d}^3 r'. \tag{17.184}$$

Comparing with Eq. (17.175), we see that the coefficients ρ_{lm} may be written as

$$\rho_{lm} = \int_V r'^l \rho(\mathbf{r}') Y_l^{m*}(\theta',\phi') \mathrm{d}^3 r'. \tag{17.185}$$

A cursory glance at Eq. (17.185) suggests that the moments q_{lm} depend upon the choice of origin, which in turn suggests that they are somewhat unphysical. However, it can be shown that the lowest nonzero moments with respect to l are in fact independent of origin. The lowest-order multipole moments are a good way to characterize the potential of a charge distribution, and can be used to accurately determine the potential in a region far from the source.

17.13 Focus: vector spherical harmonics and radiation fields

As noted in Section 3.1, the radiation emitted from a localized source or scattered from a localized scattering object form outgoing spherical waves far from the source domain, making such problems ideally suited for using spherical coordinates and, presumably, the spherical harmonics associated with them. In this section we show that the solution of Maxwell's equations in spherical coordinates leads naturally to a *vector* form of the spherical harmonics, known as the *vector spherical harmonics*.

We restrict ourselves to time harmonic fields of frequency ω propagating in free space. The time-harmonic Maxwell's equations are of the form

$$\nabla \times \mathbf{E} = ik\mathbf{B}, \tag{17.186}$$

$$\nabla \times \mathbf{B} = -ik\mathbf{E}, \tag{17.187}$$

$$\nabla \cdot \mathbf{E} = 0, \tag{17.188}$$

$$\nabla \cdot \mathbf{B} = 0, \tag{17.189}$$

with $k = \omega/c$. By combining the curl equations, we may readily show that $\mathbf{E}$ and $\mathbf{B}$ each satisfy the Helmholtz equation, i.e.

$$(\nabla^2 + k^2)U = 0, \tag{17.190}$$

where U can represent any component of the electric or magnetic field. By separation of variables in spherical coordinates, we have seen that a general solution to the Helmholtz equation may be constructed of particular solutions of the form

$$U_{lm}(r,\theta,\phi) = \left[A_{lm}h_l^{(1)}(kr) + B_{lm}h_l^{(2)}(kr)\right]Y_l^m(\theta,\phi). \tag{17.191}$$

Constructing a particular solution of Maxwell's equations, however, is not so simple, because the fields must satisfy not just the Helmholtz equation but the full set of Maxwell's equations. It is of interest, then, to develop a new set of orthogonal spherical functions with the necessary vector properties "built in".

We begin by constructing a new function $\mathbf{M}(\mathbf{r})$, defined as

$$\mathbf{M}(\mathbf{r}) = \nabla \times [\mathbf{c}\psi(\mathbf{r})], \tag{17.192}$$

where $\psi(\mathbf{r})$ is a scalar function and $\mathbf{c}$ a vector to be determined. We note that the vector $\mathbf{M}$ is automatically transverse, i.e. $\nabla \cdot \mathbf{M} = 0$, as required of solutions of Maxwell's equations. We next ask under what conditions $\mathbf{M}$ satisfies the Helmholtz equation. The Laplacian operator applied to $\mathbf{M}$ has the form

$$\nabla^2\mathbf{M} = \nabla^2[\nabla \times (\mathbf{c}\psi)] = \nabla \times [\nabla^2(\mathbf{c}\psi)] = \nabla \times \left[\mathbf{c}\nabla^2\psi + 2(\nabla\psi \cdot \nabla)\mathbf{c} + \psi\nabla^2\mathbf{c}\right]. \tag{17.193}$$

This suggests that the result of acting on $\mathbf{M}$ with the Helmholtz operator is

$$(\nabla^2 + k^2)\mathbf{M} = \nabla \times \mathbf{c}[\nabla^2 \psi + k^2 \psi] + 2\nabla \times [(\nabla \psi \cdot \nabla)\mathbf{c}] + \nabla \times (\psi \nabla^2 \mathbf{c}). \quad (17.194)$$

We consider two possible possibilities for $\mathbf{c}$. In the event that $\mathbf{c}$ is a constant vector, the latter two terms on the right of the above equation vanish and we are left with

$$(\nabla^2 + k^2)\mathbf{M} = \nabla \times \mathbf{c}[\nabla^2 \psi + k^2 \psi]. \quad (17.195)$$

We find that $\mathbf{M}$ will satisfy the Helmholtz equation if $\mathbf{c}$ is a constant vector and $\psi(\mathbf{r})$ is a scalar function satisfying the Helmholtz equation.

In the event that we take $\mathbf{c} = \mathbf{r}$, we may readily show that

$$\nabla \times [(\nabla \psi \cdot \nabla)\mathbf{r}] = \nabla \times [\nabla \psi] = 0. \quad (17.196)$$

Also, because $\mathbf{r}$ is linear in x, y, and z, $\nabla^2 \mathbf{r} = 0$. We therefore find that $\mathbf{M}$ will also satisfy the Helmholtz equation if $\mathbf{c} = \mathbf{r}$ and $\psi(\mathbf{r})$ is a scalar function satisfying the Helmholtz equation.

The function $\mathbf{M}$ now satisfies two of three necessary conditions to be a solution of Maxwell's equations: it is transverse and it satisfies the Helmholtz equation. To satisfy Maxwell's equations entirely, however, we need to introduce an additional function $\mathbf{N}$ such that $\mathbf{M}$ and $\mathbf{N}$ satisfy the curl relations (17.186) and (17.187). We introduce the function

$$\mathbf{N} = \frac{\nabla \times \mathbf{M}}{k}. \quad (17.197)$$

Because $\mathbf{M}$ satisfies the Helmholtz equation, so does $\mathbf{N}$; $\mathbf{N}$ is automatically transverse because the divergence of a curl is zero. Furthermore, $\mathbf{N}$ and $\mathbf{M}$ satisfy curl relations comparable to the electromagnetic fields $\mathbf{E}$ and $\mathbf{B}$. In essence, we have "pre-solved" Maxwell's equations for the vector properties, and now we need only find appropriate solutions to the Helmholtz equation.

The natural choice is given by Eq. (17.191); if we generally refer to $z_l(kr)$ as a solution of the Bessel equation of the form

$$z_l(kr) = A_l h_1^{(1)}(kr) + B_l h_2^{(2)}(kr), \quad (17.198)$$

we may write

$$\psi_{lm}(\mathbf{r}) = z_l(kr) Y_l^m(\theta, \phi). \quad (17.199)$$

With this choice, we have

$$\mathbf{M}_{lm}(\mathbf{r}) = \nabla \times [\mathbf{r} z_l(kr) Y_l^m(\theta, \phi)], \quad (17.200)$$

$$\mathbf{N}_{lm}(\mathbf{r}) = \frac{1}{k}\nabla \times \{\nabla \times [\mathbf{r} z_l(kr) Y_l^m(\theta, \phi)]\} \quad (17.201)$$

A couple of further simplifications will help here. We note the following vector identity

$$\nabla \times [\mathbf{r}\psi(\mathbf{r})] = \psi \nabla \times \mathbf{r} + \nabla \psi \times \mathbf{r} = -\mathbf{r} \times \nabla \psi, \qquad (17.202)$$

where we have used $\nabla \times \mathbf{r} = 0$. Furthermore, the operator $\mathbf{r} \times \nabla$ contains only derivatives with respect to θ and ϕ. Finally, we use $\nabla \times (\nabla \times) = \nabla(\nabla \cdot) - \nabla^2$ on the expression for $\mathbf{N}$ to write

$$\mathbf{M}_{lm}(\mathbf{r}) = -z_l(kr)[\mathbf{r} \times \nabla Y_l^m(\theta, \phi)], \qquad (17.203)$$

$$\mathbf{N}_{lm}(\mathbf{r}) = -\frac{2}{k}\nabla[z_l(kr)Y_l^m(\theta, \phi)]. \qquad (17.204)$$

Equations (17.203) and (17.204) are one form of the functions referred to as *vector spherical harmonics*.

Because $\mathbf{M}$ and $\mathbf{N}$ satisfy Maxwell's equations for $\mathbf{E}$ and $\mathbf{B}$, or vice versa, we may introduce two different classes of radiation field. A *magnetic multipole* of order (l,m) is defined by the equations

$$\mathbf{E}_{lm}^{(M)}(\mathbf{r}) = -\frac{1}{i}\mathbf{M}_{lm}(\mathbf{r}), \qquad (17.205)$$

$$\mathbf{B}_{lm}^{(M)}(\mathbf{r}) = -i\mathbf{N}_{lm}(\mathbf{r}). \qquad (17.206)$$

Such a field has an electric field which is transverse to the radial direction, i.e. $\mathbf{r} \cdot \mathbf{E} = 0$, but a magnetic field with a radial component.

An *electric multipole* of order (l,m) is defined by the equations

$$\mathbf{B}_{lm}^{(E)}(\mathbf{r}) = -\frac{1}{i}\mathbf{M}_{lm}(\mathbf{r}), \qquad (17.207)$$

$$\mathbf{E}_{lm}^{(E)}(\mathbf{r}) = i\mathbf{N}_{lm}(\mathbf{r}). \qquad (17.208)$$

One can verify that these fields satisfy the curl equations of Maxwell's equations. A general solution to Maxwell's equations can be written as

$$\mathbf{B} = \sum_{lm}\left[A_{lm}^{(E)}\mathbf{B}_{lm}^{(E)}(\mathbf{r}) + A_{lm}^{(M)}\mathbf{B}_{lm}^{(M)}(\mathbf{r})\right], \qquad (17.209)$$

$$\mathbf{E} = \sum_{lm}\left[A_{lm}^{(E)}\mathbf{E}_{lm}^{(E)}(\mathbf{r}) + A_{lm}^{(M)}\mathbf{E}_{lm}^{(M)}(\mathbf{r})\right], \qquad (17.210)$$

where $A_{lm}^{(E)}$ and $A_{lm}^{(M)}$ are coefficients to be determined by the boundary conditions. It can be shown that the solution is uniquely specified in a region between two spheres of radii r_1 and r_2 by the normal components of the electric and magnetic fields on the spheres.

The pair of functions $\mathbf{M}_{lm}$ and $\mathbf{N}_{lm}$ are somewhat inconvenient because they depend in a nontrivial way upon the radial coordinate r, and are not in this sense true angular functions analogous to $Y_l^m(\theta, \phi)$. It is convenient to write the vector spherical harmonics in a slightly condensed, normalized form, using

$$\mathbf{X}_{lm}(\theta, \phi) = \frac{1}{i\sqrt{l(l+1)}} \mathbf{r} \times \nabla Y_l^m(\theta, \phi). \tag{17.211}$$

The function $\mathbf{X}_{lm}$ is what is referred to as a "vector spherical harmonic" (VSH) in [Jac75]. These functions can be shown to satisfy the following orthogonality relationships with respect to solid angle

$$\int_\Omega \mathbf{X}_{l'm'}^* \cdot \mathbf{X}_{lm} d\Omega = \delta_{ll'} \delta_{mm'}, \tag{17.212}$$

$$\int_\Omega \mathbf{X}_{l'm'}^* \cdot (\mathbf{r} \times \mathbf{X}_{lm}) d\Omega = 0. \tag{17.213}$$

With these definitions, we may write

$$\mathbf{B}_{lm}^{(E)}(\mathbf{r}) = f_l(kr) \mathbf{X}_{lm}(\mathbf{r}), \tag{17.214}$$

$$\mathbf{E}_{lm}^{(E)}(\mathbf{r}) = \frac{i}{k} \nabla \times \mathbf{B}_{lm}^{(E)}(\mathbf{r}), \tag{17.215}$$

$$\mathbf{E}_{lm}^{(M)}(\mathbf{r}) = g_l(kr) \mathbf{X}_{lm}(\mathbf{r}), \tag{17.216}$$

$$\mathbf{B}_{lm}^{(M)}(\mathbf{r}) = -\frac{i}{k} \nabla \times \mathbf{B}_{lm}^{(M)}(\mathbf{r}), \tag{17.217}$$

where f_l and g_l are functions of the form of Eq. (17.198). The advantage here is that all fields are expressed in terms of a single, "fundamental", harmonic function which has a pure angular dependence.[4]

Another method of writing the VSHs is described in [CELL91], in which an electric field multipole of order (l, m) is expressed as

$$\mathbf{E}(\mathbf{r}, t) = E_{lm}^r(r, t) \mathbf{Y}_{lm} + E_{lm}^{(1)}(r, t) \mathbf{\Psi}_{lm} + E_{lm}^{(2)}(r, t) \mathbf{\Phi}_{lm}, \tag{17.218}$$

with

$$\mathbf{\Psi}_{lm} = r \nabla Y_l^m, \tag{17.219}$$

$$\mathbf{\Phi}_{lm} = \hat{\mathbf{r}} \times \mathbf{\Psi}_{lm}, \tag{17.220}$$

$$\mathbf{Y}_{lm} = \hat{\mathbf{r}} Y_l^m. \tag{17.221}$$

These functions are a complete set of vector functions orthogonal with respect to an integration over solid angle. On comparison with Eqs. (17.203) and (17.204), we find that $\mathbf{N}$ is a combination of $\mathbf{Y}$ and $\mathbf{\Psi}$; $\mathbf{\Phi}$ is comparable to $\mathbf{M}$. $\mathbf{\Phi}$ is also proportional to $\mathbf{X}$. As written,

---

[4] The r in Eq. (17.211) is canceled by a $1/r$ in the definition of ∇ in spherical coordinates.

these VSHs (with a similar expression for **B**) do not necessarily represent an electromagnetic field; one must still impose additional constraints on the functions $E_{lm}^r(r,t)$, etc., to satisfy Maxwell's equations. With some effort, one can show that these VSHs satisfy the orthogonality relations

$$\int \mathbf{Y}_{lm}^* \cdot \mathbf{Y}_{l'm'} d\Omega = \delta_{ll'}\delta_{mm'}, \tag{17.222}$$

$$\int \mathbf{\Psi}_{lm}^* \cdot \mathbf{\Psi}_{l'm'} d\Omega = l(l+1)\delta_{ll'}\delta_{mm'}, \tag{17.223}$$

$$\int \mathbf{\Phi}_{lm}^* \cdot \mathbf{\Phi}_{l'm'} d\Omega = l(l+1)\delta_{ll'}\delta_{mm'}, \tag{17.224}$$

with all other other integrals of pairs equal to zero.

There are a number of uses for the vector spherical harmonics, the most obvious being the multipole expansion of the electromagnetic field produced by a radiation source; this is discussed in more detail in [Jac75]. These functions may also be used to determine the amount of electromagnetic radiation scattered by a homogeneous spherical particle, a problem now referred to as Mie scattering [BH83, Chapter 4]. Useful in this regard is an expansion of an electromagnetic plane wave into a series of vector spherical harmonics, analogous to the Rayleigh expansion derived in Section 17.6. If we consider a circularly polarized plane wave with electric field

$$\mathbf{E}(\mathbf{r}) = (\hat{\mathbf{x}} \pm i\hat{\mathbf{y}})e^{ikz}, \tag{17.225}$$

one can show that this may be written in terms of VSHs as

$$\mathbf{E}(\mathbf{r}) = \sum_{l=1}^{\infty} i^l \sqrt{4\pi(2l+1)} \left[j_l(kr)\mathbf{X}_{l,\pm1} \pm \frac{1}{k}\nabla \times j_l(kr)\mathbf{X}_{l,\pm1} \right], \tag{17.226}$$

$$\mathbf{B}(\mathbf{r}) = \sum_{l=1}^{\infty} i^l \sqrt{4\pi(2l+1)} \left[-\frac{i}{k}\nabla \times j_l(kr)\mathbf{X}_{l,\pm1} \mp ij_l(kr)\mathbf{X}_{l,\pm1} \right]. \tag{17.227}$$

Details of the derivation may be found in [Jac75, Section 16.8].

Other alternative definitions of the VSHs have been introduced; see, for instance, [Hil54] and [BW78]. There is no set of standard vector spherical harmonics; the proper choice will depend upon the application being considered.

17.14 Exercises

1. Using the recurrence relations of the Legendre polynomials, and that $P_0(x) = 1$ and $P_1(x) = x$, find the value of $P_k(0)$ for all $k > 1$.
2. Expand the function

$$f(x) = x^6 + 3x^5 + x^3 - 2x^2 + x$$

in terms of Legendre polynomials on $-1 \le |x| \le 1$.
3. Expand the function

$$f(x) = x^5 + 2x^5 + 5x^4 - x^3 - 2x^2 + x + 1$$

in terms of Legendre polynomials on $-1 \le |x| \le 1$.
4. Expand the function

$$f(x) = |x|$$

in terms of Legendre polynomials on $-1 \le |x| \le 1$.
5. Show that the function $P_n^{(n)}(\cos\theta) = A\sin^n\theta$ for $n = 0, 1, 2, \ldots$, where A is a constant.
6. Convert the function $f(x, y, z) = x^2 + y^2 + 4z$ to spherical coordinates and express as a sum of spherical harmonics.
7. Convert the function $f(x, y, z) = xy + 2z + z^2$ to spherical coordinates and express as a sum of spherical harmonics.
8. Write the function

$$f(\theta, \phi) = Y_1^1(\theta, \phi) + Y_2^2(\theta, \phi)$$

in terms of x, y, and z.
9. Write the function

$$f(\theta, \phi) = Y_2^1(\theta, \phi) + Y_1^0(\theta, \phi)$$

in terms of x, y, and z.
10. Verify the recurrence relations

$$(2l+1)xP_l^m = (l+m)P_{l-1}^m + (n-m+1)P_{l+1}^m,$$

$$(2l+1)\sqrt{1-x^2}P_l^m = P_{l+1}^{m+1} - P_{l-1}^{m+1}.$$

11. Verify the recurrence relations

$$(1-x^2)\frac{dP_l^m}{dx} = \sqrt{1-x^2}P_l^{m+1} - mxP_l^m,$$

$$(1-x^2)\frac{dP_l^m}{dx} = mxP_l^m - (l+m)(l-m+1)\sqrt{1-x^2}P_l^{m-1}.$$

12. Determine a solution to the diffusion equation in a spherical domain of radius a, subject to the boundary condition $\partial u(a,\theta,\phi,t)/\partial r = 0$, and with the initial condition $u(r,\theta,\phi,0) = f(\mathbf{r})$.
13. Determine an expression for the solution of Laplace's equation in a spherical domain when the boundary condition is of the form

$$V(a,\theta,\phi) = V_0(\theta)\cos\phi. \tag{17.228}$$

14. Find the solution of Laplace's equation in a *hemispherical* domain $r < a$, $0 \le \theta \le \pi/2$, satisfying the boundary conditions $V(a,\theta,\phi) = V_0(\theta)$ and $V(r,\pi/2,\phi) = 0$. (Hint: consider the solution to

the problem on the whole sphere, using the method of images to satisfy the boundary condition in θ.)

15. Determine the solution to Laplace's equation between an inner spherical surface of radius a and an outer surface of radius b, subject to the boundary conditions $V(a,\theta,\phi) = V_a(\theta)$, $V(b,\theta,\phi) = V_b(\theta)$.

16. Use the Rayleigh formula to show that

$$j_n(x) = \frac{1}{2i^n} \int_{-1}^{1} e^{ikx} P_n(k)dk.$$

This is a demonstration that the Fourier transform of the Legendre polynomial $P_n(k)$ is proportional to the spherical Bessel function $j_n(x)$.

17. Show that, for an arbitrary charge distribution $\rho(\mathbf{r})$, the values of the moments for the first non-vanishing multipole in l are independent of the choice of origin.

18. Find the multipole moments for the lowest two nonzero values of l of a distribution consisting of point charges q, $-2q$, $2q$, and q at positions $z = 2a$, a, $-a$, $-2a$ respectively.

19. A distribution consists of three point charges: $+q$ at $z = a$, $-2q$ at $z = 0$, and $+q$ at $z = -a$. Find the multipole moments for the lowest nonzero value of l.

20. A sphere of radius R centered on the origin carries a charge density

$$\rho(r,\theta) = k(R - 4r/3)\sin\theta,$$

where k is a constant. Find an approximate form for the potential on the z-axis.

21. A sphere of radius R centered on the origin carries a charge density

$$\rho(r,\theta) = k(r^2 - 4Rr/5)\cos(2\theta),$$

where k is a constant. Find an approximate form for the potential on the z-axis.

22. It is possible to determine a multipole expansion in *Cartesian* coordinates, by expanding $1/|\mathbf{r} - \mathbf{r}'|$ in a Taylor series in terms of x, y, and z. Determine this Cartesian multipole expansion.

23. Verify the vector spherical harmonic orthogonality relations

$$\int_\Omega \mathbf{X}^*_{l'm'} \cdot \mathbf{X}_{lm}d\Omega = \delta_{ll'}\delta_{mm'},$$

$$\int_\Omega \mathbf{X}^*_{l'm'} \cdot (\mathbf{r} \times \mathbf{X}_{lm})d\Omega = 0.$$

24. Verify the vector spherical harmonic orthogonality relations

$$\int \mathbf{Y}^*_{lm} \cdot \mathbf{Y}_{l'm'}d\Omega = \delta_{ll'}\delta_{mm'},$$

$$\int \mathbf{\Psi}^*_{lm} \cdot \mathbf{\Psi}_{l'm'}d\Omega = l(l+1)\delta_{ll'}\delta_{mm'},$$

$$\int \mathbf{\Phi}^*_{lm} \cdot \mathbf{\Phi}_{l'm'}d\Omega = l(l+1)\delta_{ll'}\delta_{mm'}.$$

25. Read the paper by R. G. Barrera, G. A. Estévez and J. Giraldo, Vector spherical harmonics and their application to magnetostatics, *Eur. J. Phys.* **6** (1985), 287–294. Describe the set of VSHs

employed in the paper, and what types of system they are being applied to. What specific examples are solved using these harmonics?

26. Read the paper by E. L. Hill, The theory of vector spherical harmonics, *Am. J. Phys.* **22** (1954), 211–214. Write the expressions for the VSHs defined by Hill. Is there any obvious connection between Hill's VSHs and those defined in this chapter, e.g. $\mathbf{Y}$, $\mathbf{\Psi}$, and $\mathbf{\Phi}$?

18

Orthogonal functions

18.1 Introduction: Sturm–Liouville equations

In the description of the Legendre functions and Bessel functions in the previous two chapters, it is immediately obvious that the two classes of functions share many common behaviors: the existence of a generating function, recurrence relations, orthogonality, and completeness. The Legendre and Bessel equations, with appropriate boundary conditions, are in fact special cases of the so-called *Sturm–Liouville equation*, solutions of which have the properties mentioned above. In this section, we discuss the general properties of Sturm–Liouville equations and their solutions.

A Sturm–Liouville equation is an ordinary differential equation of the form

$$\frac{d}{dx}\left[p(x)\frac{dy}{dx}\right] - q(x)y + \lambda w(x)y = 0, \tag{18.1}$$

where λ is an unspecified constant to be determined. The solution $y(x)$ is evaluated on a domain $a \leq x \leq b$, with homogeneous boundary conditions given at the ends of the domain; these conditions determine the allowed values of λ. Appropriate conditions will be discussed below, and they are essential ingredients in deriving the special properties of Sturm–Liouville solutions.

Example 18.1 (Legendre's equation) Legendre's equation (17.15), which may be written as

$$\frac{d}{dx}\left[(1-x^2)\frac{dP}{dx}\right] + l(l+1)P = 0, \tag{18.2}$$

is a Sturm–Liouville equation with $p = (1 - x^2)$, $q = 0$, $w = 1$ and $\lambda = l(l+1)$.
◇

Example 18.2 (Bessel's equation) Bessel's equation (16.12) may be written in the form

$$\frac{d}{dx}\left[x\frac{dy}{dx}\right] + \left[x - \frac{m^2}{x}\right]y(x) = 0. \tag{18.3}$$

622

At first glance, this equation is not exactly in Sturm–Liouville form: it is typically derived from separation of variables, and the constants m are constrained by the behavior of the azimuthal function, not by the boundary values of $y(x)$. From Section 16.10, however, we know that the solutions are typically of the form $J_m(x/\alpha)$, where α is a parameter chosen to satisfy the boundary conditions. We therefore define $x = \alpha u$ and $Y(u) = y(x)$. With these definitions, Bessel's equation becomes

$$\frac{\mathrm{d}}{\mathrm{d}u}\left[u\frac{\mathrm{d}Y}{\mathrm{d}u}\right] + \left[\alpha^2 u - \frac{m^2}{u}\right]Y(u) = 0, \tag{18.4}$$

which is a Sturm–Liouville equation with $p = u$, $q = m^2/u$, $w = 1$ and $\lambda = \alpha^2$.

◇

The allowed values of λ are typically determined by the boundary conditions of the problem, and it is convenient to write the equation as a generalized eigenvalue equation with $-\lambda$ as the eigenvalue, in the form

$$\mathcal{L}|y\rangle = -\lambda w|y\rangle, \tag{18.5}$$

where

$$\mathcal{L} = \frac{\mathrm{d}}{\mathrm{d}x}\left[p(x)\frac{\mathrm{d}}{\mathrm{d}x}\right] - q(x) \tag{18.6}$$

is a differential operator which acts on $|y\rangle$. We refer to Eq. (18.5) as a *generalized eigenvalue equation* because of the presence of the generally nontrivial function $w(x)$ on the right-hand side.

We can readily show that solutions $y_m(x)$ related to distinct eigenvalues λ_m are orthogonal to each other in a generalized sense. Let us assume $\lambda_m \neq \lambda_n$, and we consider the equations for y_m and y_n,

$$\frac{\mathrm{d}}{\mathrm{d}x}\left[p(x)\frac{\mathrm{d}y_m}{\mathrm{d}x}\right] - q(x)y_m + \lambda_m w(x)y_m = 0, \tag{18.7}$$

$$\frac{\mathrm{d}}{\mathrm{d}x}\left[p(x)\frac{\mathrm{d}y_n}{\mathrm{d}x}\right] - q(x)y_n + \lambda_n w(x)y_n = 0. \tag{18.8}$$

We multiply the first of these equations by y_n and the second by y_m, and take the difference,

$$y_n\frac{\mathrm{d}}{\mathrm{d}x}\left[p(x)\frac{\mathrm{d}y_m}{\mathrm{d}x}\right] - y_m\frac{\mathrm{d}}{\mathrm{d}x}\left[p(x)\frac{\mathrm{d}y_n}{\mathrm{d}x}\right] = w(x)(\lambda_n - \lambda_m)y_m y_n. \tag{18.9}$$

We now integrate this equation over the domain $a \leq x \leq b$. Each of the derivative terms may be simplified by the use of integration by parts, i.e.

$$\int_a^b y_n\frac{\mathrm{d}}{\mathrm{d}x}\left[p(x)\frac{\mathrm{d}y_m}{\mathrm{d}x}\right]\mathrm{d}x = \left[y_n p(x)\frac{\mathrm{d}y_m}{\mathrm{d}x}\right]_a^b - \int_a^b \frac{\mathrm{d}y_n}{\mathrm{d}x}\frac{\mathrm{d}y_m}{\mathrm{d}x}\mathrm{d}x. \tag{18.10}$$

Equation (18.9) takes on the form

$$\left[y_n p(x) \frac{dy_m}{dx} - y_m p(x) \frac{dy_n}{dx} \right]_a^b = \int_a^b w(x)(\lambda_n - \lambda_m) y_m y_n dx. \tag{18.11}$$

Provided the functions y_n and y_m have appropriate behaviors on the boundary of the domain, we may write

$$\int_a^b w(x) y_m(x) y_n(x) dx = 0, \quad n \neq m. \tag{18.12}$$

The eigenfunctions of a Sturm–Liouville operator are therefore orthogonal to each other with respect to a *weight function* $w(x)$.

A number of common boundary conditions will result in the vanishing of the boundary term of Eq. (18.11) and hence the orthogonality of the eigenfunctions. It is straightforward to show that linear homogeneous conditions of the form

$$y(a) = 0, \quad y(b) = 0 \qquad \text{(Dirichlet)} \tag{18.13}$$

$$\frac{dy(a)}{dx} = 0, \quad \frac{dy(b)}{dx} = 0 \qquad \text{(Neumann)} \tag{18.14}$$

$$y(a) + \alpha \frac{dy(a)}{dx} = 0, \quad y(b) + \beta \frac{dy(b)}{dx} = 0 \qquad \text{(Robin)} \tag{18.15}$$

will work. Also, periodic boundary conditions of the form

$$y_m(a) = y_m(b), \quad \frac{dy_m(a)}{dx} = \frac{dy_m(b)}{dx} \tag{18.16}$$

will work. A special case involves $p(a) = 0$ or $p(b) = 0$, which suggests that the boundary term of Eq. (18.11) will be automatically satisfied. Such a case, however, results in a singular point of the differential equation lying on the boundary, and the behavior of the eigenfunctions is constrained by the requirement that the solution of the equation be regular in the domain; an example of this is the Legendre equation. The eigenvalues which result from satisfying the boundary conditions usually, but not always, form a countable semi-infinite set.

It can be shown that the eigenfunctions of a Sturm–Liouville operator form a complete orthonormal basis on the domain $a \leq x \leq b$, i.e. given a piecewise continuous function $f(x)$, we may write

$$f(x) = \sum_{n=0}^{\infty} c_n y_n(x), \tag{18.17}$$

where $y_n(x)$ are the orthonormal eigenfunctions defined by our Sturm–Liouville problem and

$$c_n = \int_a^b w(x')f(x')y_n(x')dx'. \tag{18.18}$$

The series (18.17) converges in the mean to the function $f(x)$.

Sturm–Liouville operators are, in essence, an infinite-dimensional analogue to the real symmetric and Hermitian matrices of Chapter 4. To see this, let us define an inner product of real-valued functions $|u\rangle$, $|v\rangle$ which satisfy an appropriate set of Sturm–Liouville boundary conditions; we may write

$$\langle u| v \rangle \equiv \int_a^b u(x)v(x)dx. \tag{18.19}$$

We consider the behavior of the inner product

$$\langle u|\mathcal{L}|v\rangle = \int_a^b u(x)\frac{d}{dx}\left[p(x)\frac{dv(x)}{dx}\right]dx - \int_a^b u(x)v(x)q(x)dx. \tag{18.20}$$

Let us consider performing integration by parts twice on the first integral we find that the boundary terms cancel and we are left with

$$\langle u|\mathcal{L}|v\rangle = \int_a^b \int_a^b v(x)\frac{d}{dx}\left[p(x)\frac{du(x)}{dx}\right]dx - \int_a^b u(x)v(x)q(x)dx = \langle v|\mathcal{L}|u\rangle, \tag{18.21}$$

or

$$\langle u|\mathcal{L}|v\rangle = \langle v|\mathcal{L}|u\rangle. \tag{18.22}$$

The operator $\mathcal{L}$ is said to be *self-adjoint*, and is analogous to a real symmetric matrix for which $\tilde{\mathbf{A}} = \mathbf{A}$.

This result may be generalized to complex operators, by introducing the inner product

$$\langle u| v \rangle \equiv \int_a^b u^*(x)v(x)dx. \tag{18.23}$$

An operator $\mathcal{L}$ is said to be *Hermitian* if

$$\langle u|\mathcal{L}|v\rangle = \left(\langle v|\mathcal{L}|u\rangle\right)^*, \tag{18.24}$$

or equivalently

$$\int_a^b u^*(x)\mathcal{L}v(x)dx = \int_a^b v(x)\mathcal{L}^*u^*(x)dx. \tag{18.25}$$

Hermitian operators, of which self-adjoint operators are a special case, have three important properties.

1. Their eigenvalues are real-valued.
2. Their eigenfunctions are orthogonal.
3. The eigenfunctions form a complete set.

These properties are directly analogous to the properties of the eigenvalues and eigenvectors of Hermitian matrices, discussed in Section 4.6.1.

As one might expect, it is certainly possible to come across linear second-order ODEs in problems which are not self-adjoint or Hermitian. Such equations, however, can always be transformed into a self-adjoint form by multiplication. Suppose we have an operator $\mathcal{L}'$ such that

$$\mathcal{L}' = f(x)\frac{d^2}{dx^2} + g(x)\frac{d}{dx} + h(x). \tag{18.26}$$

If we multiply this equation by

$$F(x) = \frac{\exp[\int_x g(x')/f(x')dx']}{f(x)}, \tag{18.27}$$

it takes on the form

$$F(x)\mathcal{L}' = \exp\left[\int_x g(x')/f(x')dx'\right]\frac{d^2}{dx^2} + \exp\left[\int_x g(x')/f(x')dx'\right]$$
$$\times \frac{g(x)}{f(x)}\frac{d}{dx} + F(x)h(x). \tag{18.28}$$

This may be rewritten as

$$F(x)\mathcal{L}' = \frac{d}{dx}\left[\exp[\int_x g(x')/f(x')dx']\frac{d}{dx}\right] + F(x)h(x), \tag{18.29}$$

which is self-adjoint. Multiplication of a non-self-adjoint equation by a factor (18.27) converts it to a self-adjoint form.

In general, one finds that *most ordinary differential equations which arise in separation of variables are, or can be, converted to Hermitian eigenfunction equations.* From this observation, we can readily conclude that the derived eigenfunctions form a complete set of functions over the given domain.

The Bessel and Legendre equations are the two most commonly used Sturm–Liouville equations. In this chapter, we review a number of other sets of orthogonal functions which occur regularly in physical and optical problems. For each set of functions, we will introduce the defining ODE, the series solution, the generating function, recurrence relations, and the orthogonality condition. We conclude the discussion of each set with an illustration of its physical relevance.

One very general class of differential equations that are not discussed here are the so-called *hypergeometric functions*. They are discussed in Appendix B.

18.2 Hermite polynomials

Hermite's differential equation is of the form

$$H'' - 2xH' + 2mH = 0, \tag{18.30}$$

where $-\infty < x < \infty$ and m will be restricted to positive integers in most physical problems. This equation is *not* self-adjoint; we will see how to convert it to self-adjoint form momentarily.

18.2.1 Series solution

The point $x = 0$ is an ordinary point of the differential equation, which indicates that there are two linearly independent solutions of the form

$$H(x) = \sum_{n=0}^{\infty} c_n x^n. \tag{18.31}$$

On substitution into Hermite's equation and matching like powers of x, we derive the recursion relation

$$c_{n+2} = -\frac{2(m-n)}{(n+1)(n+2)} c_n. \tag{18.32}$$

Two independent solutions are provided by choosing c_0 or c_1 nonzero. With m an integer, it can be seen that one of the solutions will be a polynomial of order m; this solution is what we refer to as a *Hermite polynomial*.

We will see that the most convenient normalization for these functions is the choice $c_0 = (-1)^{m/2} m!/(m/2)!$ for m even and $c_1 = (-1)^{(m-1)/2} 2m!/[(m-1)/2]!$ for m odd. The Hermite polynomials $H_m(x)$ of order m may then be written as

$$H_m(x) = \sum_{n=0}^{M} (-1)^n \frac{m!}{n!(m-2n)!} (2x)^{m-2n}, \tag{18.33}$$

where

$$M = \begin{cases} \frac{m}{2}, & m \text{ even}, \\ \frac{m-1}{2}, & m \text{ odd}. \end{cases} \tag{18.34}$$

The first few Hermite polynomials are depicted in Fig. 18.1, and tabulated in Table 18.1. It is to be noted that the Hermite polynomials satisfy the parity relation

$$H_m(x) = (-1)^m H_m(-x). \tag{18.35}$$

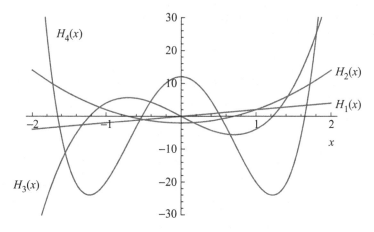

Figure 18.1 The Hermite polynomials H_1, H_2, H_3, and H_4.

Table 18.1 *The first few Hermite polynomials.*

$$H_0(x) = 1$$
$$H_1(x) = 2x$$
$$H_2(x) = 4x^2 - 2$$
$$H_3(x) = 8x^3 - 12x$$
$$H_4(x) = 16x^4 - 48x^2 + 12$$
$$H_5(x) = 32x^5 - 160x^3 + 120x$$
$$H_6(x) = 64x^6 - 480x^4 + 720x^2 - 120$$

In the case that m is not an integer, the solution to Hermite's equation is an infinite series. In the limit of large n, the recursion relation (18.32) approximately takes on the form

$$\lim_{n\to\infty} \frac{c_{n+2}}{c_n} = \lim_{n\to\infty} -\frac{2(n-m)}{(n+1)(n+2)} \approx \frac{2}{n+2}. \tag{18.36}$$

However, if we compare this to the series representation of $\exp[x^2]$,

$$\exp[x^2] = \sum_{n=0}^{\infty} \frac{x^{2n}}{n!} = \sum_{n=0}^{\infty} d_n x^{2n}, \tag{18.37}$$

we find that the ratio of coefficients d_{n+1}/d_n for this representation also has the form $d_{n+1}/d_n \approx 2/(n+2)$ for large n. This suggests that the function $H_\nu(x) \approx \exp[x^2]$ for large values of x; this observation will be important in justifying the use of integer m in physical problems.

18.2.2 Generating function

With the normalization chosen above, it can be shown that the Hermite polynomials possess the generating function

$$g(x,t) = e^{-t^2+2tx} = \sum_{n=0}^{\infty} H_n(x)\frac{t^n}{n!}.$$ (18.38)

This generating function, which we take as given, can be used to derive most of the other properties of the Hermite polynomials.

18.2.3 Recurrence relations

If we take the derivative of the generating function with respect to t, we can derive the recurrence relation

$$H_{n+1}(x) = 2xH_n(x) - 2nH_{n-1}(x).$$ (18.39)

Taking the derivative of the generating function instead with respect to x, we can derive the recurrence relation

$$H_n'(x) = 2nH_{n-1}(x).$$ (18.40)

We can show that a function which satisfies these recurrence relations satisfies the Hermite equation. We take the first derivatives of Eqs. (18.39) and (18.40), and use the former to eliminate $2nH_{n-1}'$ from the latter,

$$H_n'' = 2H_n + 2xH_n' - H_{n+1}'.$$ (18.41)

Applying Eq. (18.40) to H_{n+1}', we arrive at

$$H_n'' = 2xH_n' - 2nH_n,$$ (18.42)

which is Hermite's equation.

With $x = 0$ in the generating function, we can justify the normalization of the Hermite functions as well. By direct series expansion of $\exp[-t^2]$, the even Hermite functions satisfy

$$H_{2n}(0) = c_0 = (-1)^n \frac{(2n)!}{n!}.$$ (18.43)

We may also apply Eq. (18.40) to determine the value of c_1 for the odd Hermite functions,

$$H_{2n+1}'(0) = 2c_1 = (4n+2)H_{2n}(0) = (-1)^n \frac{(2n+1)!}{n!}.$$ (18.44)

Both of these results can be seen to be in agreement with our initial choices of the previous section.

We may rewrite the generating function in the form,

$$g(x,t) = e^{x^2}e^{-(x-t)^2}.\tag{18.45}$$

Noting that

$$\frac{\partial}{\partial t}e^{-(x-t)^2} = -\frac{\partial}{\partial x}e^{-(x-t)^2},\tag{18.46}$$

we can derive a Rodrigues formula for the Hermite polynomials by differentiating the generating function n times with respect to t and then setting $t = 0$,

$$H_n(x) = (-1)^n e^{x^2}\frac{d^n}{dx^n}\left(e^{-x^2}\right).\tag{18.47}$$

18.2.4 Orthogonality

We have noted that Hermite's equation is not self-adjoint. It can be made self-adjoint by multiplication by a factor $\exp[-x^2]$, so it takes on the form

$$\frac{d}{dx}\left[e^{-x^2}\frac{dH_n}{dx}\right] + 2ne^{-x^2}H_n = 0.\tag{18.48}$$

Comparing with the self-adjoint form of Eq. (18.1), we have $p(x) = \exp[-x^2]$, $q(x) = 0$ and $w(x) = \exp[-x^2]$. This suggests that the Hermite functions are orthogonal with respect to the weight function $w(x)$, i.e.

$$\int_{-\infty}^{\infty}e^{-x^2}H_n(x)H_m(x)dx = 0, \quad m \neq n.\tag{18.49}$$

To use the Hermite polynomials to represent an arbitrary function, however, we need to determine their normalization. We can do so using the Rodrigues formula for the Hermites, Eq. (18.47), and write

$$\int_{-\infty}^{\infty}e^{-x^2}[H_n(x)]^2dx = (-1)^n\int_{-\infty}^{\infty}H_n(x)\frac{d^n}{dx^n}\left(e^{-x^2}\right)dx.\tag{18.50}$$

We integrate by parts n times to get the formula

$$\int_{-\infty}^{\infty}e^{-x^2}[H_n(x)]^2dx = \int_{-\infty}^{\infty}e^{-x^2}\frac{d^n H_n(x)}{dx^n}dx.\tag{18.51}$$

Because the Hermite functions are polynomials of order n, only one term survives in the differentiation, namely

$$\frac{d^n H_n(x)}{dx^n} = 2^n n!,\tag{18.52}$$

as can be shown by direct differentiation of Eq. (18.33). The integral can be evaluated directly to find

$$\int_{-\infty}^{\infty} e^{-x^2}[H_n(x)]^2 dx = 2^n n! \sqrt{\pi}. \tag{18.53}$$

An arbitrary function $f(x)$ may therefore be expanded on the domain $-\infty < x < \infty$ using Hermite functions in the form

$$f(x) = \sum_{n=0}^{\infty} a_n H_n(x), \tag{18.54}$$

where

$$a_n = \frac{1}{2^n n! \sqrt{\pi}} \int_{-\infty}^{\infty} f(x) H_n(x) e^{-x^2} dx. \tag{18.55}$$

18.2.5 The quantum harmonic oscillator

We consider the motion of a quantum-mechanical particle of mass m moving in a one-dimensional, time-independent potential $V(x)$. The stationary (time-harmonic) behavior of the particle is described by the time-independent Schrödinger equation,

$$-\frac{\hbar^2}{2m} \frac{d^2}{dz^2} \psi(z) + V(z)\psi(z) = E\psi(z), \tag{18.56}$$

where $\psi(z)$ is the wavefunction of the particle and E is the energy of the stationary state, to be determined. This equation is self-adjoint, with eigenvalue E, and therefore in general possesses a complete set of orthogonal solutions. We consider the special case

$$V(z) = \frac{1}{2} m\omega^2 z^2, \tag{18.57}$$

where ω is a characteristic frequency; this is the potential of a harmonic oscillator, and Eq. (18.56) can be used to find the energies and wavefunctions of the stationary states of the harmonic oscillator. This equation can be written in a simple form by defining

$$x = \alpha z, \tag{18.58}$$

$$\alpha^2 = \frac{m\omega}{\hbar}, \tag{18.59}$$

$$\lambda = \frac{2E}{\hbar\omega}. \tag{18.60}$$

The Schrödinger equation then becomes

$$\frac{d^2}{dx^2} \psi + \lambda\psi - x^2\psi = 0. \tag{18.61}$$

This may be put into a more familiar form by making the substitution

$$\psi(x) = e^{-x^2/2}\eta(x),$$ (18.62)

so that

$$\eta'' - 2x\eta + (\lambda - 1)\eta = 0.$$ (18.63)

With the choice $\lambda = 2n + 1$, this equation is simply the Hermite equation, and the normalized solutions to Schrödinger's equation are of the form

$$\psi_n(x) = 2^{-n/2}\pi^{-1/4}(n!)^{-1/2}e^{-x^2/2}H_n(x).$$ (18.64)

The requirement that n be an integer arises from the "hidden" boundary condition that the quantum mechanical wavefunction must vanish for large values of x. In Section 18.2.1, we demonstrated that for non-integer n the Hermite function $H_n(x) \sim \exp[x^2]$ as $x \to \infty$, which implies that the function $\psi_n(x)$ diverges as $x \to \infty$ for non-integer n.

Because n is constrained to take integer values, the energy of the system is constrained to take discrete values

$$E_n = (n + 1/2)\hbar\omega.$$ (18.65)

The most elegant treatment of the quantum mechanical oscillator introduces *raising and lowering operators* to relate the wavefunctions of different energy states; these raising and lowering operators are denoted $\hat{a}^\dagger$ and $\hat{a}$, respectively, and are defined as

$$\hat{a}^\dagger \equiv \frac{1}{\sqrt{2}}\left(x - \frac{d}{dx}\right),$$ (18.66)

$$\hat{a} \equiv \frac{1}{\sqrt{2}}\left(x + \frac{d}{dx}\right).$$ (18.67)

The effect of the operators on the wavefunctions may be found by the use of the recurrence relations. Combining Eqs. (18.39) and (18.40), we may write

$$H_{n+1} = 2xH_n - H_n'.$$ (18.68)

Multiplying by $\exp[-x^2/2]$, this may be rewritten in terms of ψ_n as

$$2^{(n+1)/2}\sqrt{(n+1)!}\psi_{n+1} = 2^{n/2}x\sqrt{n!}\psi_n - 2^{n/2}\sqrt{n!}\psi_n'.$$ (18.69)

This in turn may be readily rewritten as

$$\frac{1}{\sqrt{2}}\left(x - \frac{d}{dx}\right)\psi_n = \hat{a}^\dagger\psi_n = \sqrt{(n+1)}\psi_{n+1}.$$ (18.70)

Similarly, if we downshift Eqs. (18.39) and (18.40) by one index, and eliminate H_{n-2} between them, we can find that

$$\frac{1}{\sqrt{2}}\left(x+\frac{d}{dx}\right)\psi_n = \hat{a}\psi_n = \sqrt{n}\psi_{n-1}. \tag{18.71}$$

18.2.6 Hermite–Gauss laser beams

In the study of the propagation of optical waves, it is highly convenient to work with planar and spherical waves, for instance in the angular spectrum representation (Section 11.9) and the Rayleigh–Sommerfeld diffraction formula (Section 11.1). Although plane waves and spherical waves are extremely useful models for many applications, they represent extreme ends of what could be called a "spectrum" of model wavefields. A plane wave is a highly directional field, with rays which are all parallel to a unit vector **s**. A spherical wave is a highly nondirectional field, with rays that propagate equally in all directions. There is obviously a lot of room for wavefields which fall between these two extremes.

Turning to physical observation for inspiration, the output of many lasers is a highly directional beam with Gaussian intensity profile, i.e.

$$I(x,y,z_0) \sim I_0 e^{-2(x^2+y^2)/w^2}, \tag{18.72}$$

where the beam is propagating in the z-direction and the effective width of the beam is denoted by w. Furthermore, it has been observed that the shape of such Gaussian beams does not change as the beam propagates; only the width of the Gaussian and its brightness change. Additional Gaussian-like "shape-invariant" beams may be derived; in this section we show that a complete set of such beams can be represented in Cartesian coordinates using the Hermite–Gauss functions. We begin by deriving the propagation characteristics of Gaussian beams.

We are interested in finding solutions to the Helmholtz equation,

$$\nabla^2 U(\mathbf{r}) + k^2 U(\mathbf{r}) = 0, \tag{18.73}$$

which have a Gaussian intensity of the form of Eq. (18.72). Because a laser is known to produce a highly directional output, we restrict ourselves to solutions which look more like beams,

$$U(\mathbf{r}) = u(\mathbf{r})e^{ikz}, \tag{18.74}$$

where the function $u(\mathbf{r})$ is assumed to be such that its variations in the z-direction are negligible over the distance of a wavelength, i.e.

$$\lambda\left|\frac{\partial u}{\partial z}\right| \ll |u|, \tag{18.75}$$

$$\lambda\left|\frac{\partial^2 u}{\partial z^2}\right| \ll \left|\frac{\partial u}{\partial z}\right|. \tag{18.76}$$

These assumptions basically enforce the requirement that the beam does not change its size and shape appreciably as it propagates in the z-direction, i.e. that it is directional. If we substitute from Eq. (18.74) into the Helmholtz equation, we may expand the z-derivative,

$$\frac{\partial^2}{\partial z^2} u e^{ikz} = \left(\frac{\partial^2}{\partial z^2} u + 2ik \frac{\partial}{\partial z} u - k^2 u \right) e^{ikz} \approx \left(2ik \frac{\partial}{\partial z} u - k^2 u \right) e^{ikz}, \tag{18.77}$$

where in the last step we have used our assumption of directionality. On substitution into the Helmholtz equation, that equation takes on the form

$$\nabla_T^2 u + 2ik \frac{\partial}{\partial z} u = 0, \tag{18.78}$$

where

$$\nabla_T^2 = \frac{\partial^2}{\partial x^2} + \frac{\partial^2}{\partial y^2} \tag{18.79}$$

is referred to as the *transverse Laplacian*. Equation (18.78) is known as the *paraxial wave equation*.

We now try to construct a solution of the paraxial wave equation of Gaussian form whose shape is invariant on propagation, i.e.

$$u(\mathbf{r}) = A_0 e^{ik(x^2+y^2)/2q(z)} e^{ip(z)}, \tag{18.80}$$

where $q(z)$ and $p(z)$ are z-dependent, possibly complex, functions to be determined. On substitution of this form into the paraxial equation, we find that

$$A_0 \left[\frac{k^2}{q^2} (x^2 + y^2) \left(\frac{dq}{dz} - 1 \right) - 2k \left(\frac{dp}{dz} - \frac{i}{q} \right) \right] = 0. \tag{18.81}$$

Since p and q only depend on z, this equation will only be satisfied if

$$\frac{dq}{dz} = 1, \tag{18.82}$$

$$\frac{dp}{dz} = \frac{i}{q}. \tag{18.83}$$

We can solve these equations quite readily, first integrating the q equation and then substituting this result into the p equation. The results are

$$q(z) = q_0 + z, \tag{18.84}$$

$$p(z) = i \log \frac{q_0 + z}{q_0}, \tag{18.85}$$

where $q_0 = q(0)$ and we have assumed that $p(0) = 0$. The quantity $q(z)$ is in general a complex number, and it is convenient to write it in the form

$$\frac{1}{q(z)} = \frac{1}{R(z)} + \frac{i\lambda}{\pi w^2(z)}, \tag{18.86}$$

where R and w are real. The latter term was chosen to match the intensity to the "observed" intensity profile, given by Eq. (18.72). With this choice of $q(z)$, we may write

$$e^{ip(z)} = \exp\left(-\log\frac{q_0 + z}{q_0}\right) = \frac{1}{1 + z/R_0 + i\lambda z/\pi w_0^2}, \tag{18.87}$$

where R_0 and w_0 are the values of R and w at $z = 0$. If we match the real parts of Eqs. (18.84) and (18.86), we can readily find that

$$\frac{1}{R(z)} = \frac{\text{Re}(q_0) + z}{|q_0|^2 + 2z\text{Re}(q_0) + z^2}. \tag{18.88}$$

From this we note that there exists some value of z for which $1/R = 0$, or $R = \infty$. Because q_0 is unspecified, let us choose $R_0 = \infty$, so that

$$\frac{1}{q_0} = \frac{i\lambda}{\pi w_0^2}. \tag{18.89}$$

If we define a new parameter z_0 such that

$$z_0 = \frac{\pi w_0^2}{\lambda}, \tag{18.90}$$

we may then write

$$R(z) = z + \frac{z_0^2}{z}, \tag{18.91}$$

$$w(z) = w_0\sqrt{1 + z^2/z_0^2}. \tag{18.92}$$

The parameter z_0 is known as the *Rayleigh range* and is the distance at which the beam width is $\sqrt{2}$ times the waist width. With its definition, and the specification of $R_0 = \infty$, we further find that

$$e^{ip(z)} = \frac{1}{1 + iz/z_0} = \frac{1}{1 + z^2/z_0^2}e^{-i\Phi(z)}, \tag{18.93}$$

where

$$\Phi(z) = \tan^{-1}(z/z_0). \tag{18.94}$$

636   *Orthogonal functions*

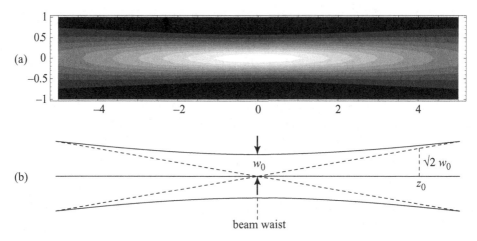

Figure 18.2 (a) Intensity $|u|^2$ and (b) beam width $w(z)$ of a Gaussian beam, as a function of position z. For illustration purposes, $z_0 = 4$ and $w_0 = 0.5$.

Our solution for a Gaussian beam may be written completely in the form

$$u(\mathbf{r}) = A_0 e^{-i\Phi(z)} \left[\frac{1}{\sqrt{1+z^2/z_0^2}} e^{ikz} e^{ik(x^2+y^2)/2R(z)} \right] e^{-(x^2+y^2)/w^2(z)}. \qquad (18.95)$$

Each of these terms has a clear physical meaning. The term A_0 is the amplitude of the beam. The last term, dependent on $w(z)$, represents the amplitude profile of the beam as a function of z; it is a Gaussian profile which decreases in width as z increases towards the plane $z = 0$, where it is minimum, and increases again afterwards. The plane $z = 0$ is known as the *beam waist*, and represents the focal plane of a Gaussian beam. The intensity profile of a Gaussian beam is illustrated in Fig. 18.2.

The term which depends on $R(z)$ is a phase-only term. Its meaning may be deduced by looking at an outgoing spherical wave for $z \gg \sqrt{x^2+y^2}$,

$$\frac{e^{ikr}}{r} = \frac{e^{ik\sqrt{x^2+y^2+z^2}}}{\sqrt{x^2+y^2+z^2}} = \frac{e^{ikz\sqrt{1+(x^2+y^2)/z^2}}}{z\sqrt{1+(x^2+y^2)/z^2}} \approx \frac{1}{z} e^{ikz} e^{ik(x^2+y^2)/2z}. \qquad (18.96)$$

For large z, the term in square brackets in Eq. (18.95) has an identical form to a spherical wave. Evidently $R(z)$ represents a radius, the *wavefront curvature* of the beam, which approaches a spherical wave as $z \to \infty$. In the waist plane, the curvature is infinite and the wavefronts are planar. The wavefronts of a Gaussian beam are illustrated in Fig. 18.3.

The term which includes $\Phi(z)$ is a peculiar one; $\Phi(z)$ approaches $\pi/2$ as $z \to \infty$ and $-\pi/2$ as $z \to -\infty$; otherwise it is nearly constant, except at the beam waist, where it rapidly changes from negative to positive. This is illustrated in Fig. 18.4. This π phase shift is referred to as the *phase anomaly at focus* or the *Gouy phase shift*. In geometrical

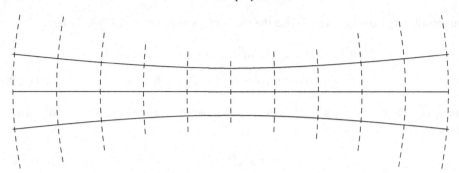

Figure 18.3 Illustrating the surfaces of constant phase (wavefronts) in a Gaussian beam.

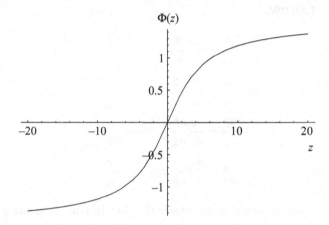

Figure 18.4 A plot of the Gouy phase shift, as a function of z. Here $z_0 = 4$.

optics this anomaly is interpreted as a light ray picking up an additional, "mysterious", π phase shift as it passes through focus. A number of explanations have been proposed for this effect, some quite absurd! Perhaps the best statement which can be made is that the phase anomaly is only an anomaly if one makes inappropriate assumptions about the wave propagation to begin with, namely the rectilinear propagation of light of geometrical optics. A nice description of the Gouy effect, and an intuitive explanation of it, is given in [Boy80].

With the Gaussian solution of Eq. (18.95) developed, we now look for other shape-invariant beams, of the form

$$v(\mathbf{r}) = f[\sqrt{2}x/w(z)]g[\sqrt{2}y/w(z)]u(\mathbf{r})\exp[i\Phi(z)], \qquad (18.97)$$

where $f[\sqrt{2}x/w]$ and $g[\sqrt{2}y/w]$ are the propagation-dependent transverse profiles to be determined; the factor $\sqrt{2}$ is for later convenience. The function $\Phi(z)$ is a propagation-dependent phase shift, assumed to be more general than the Gouy shift mentioned above. The transverse intensity of this solution scales in size by a factor $w(z)/w_0$ on propagation.

On substituting from Eq. (18.97) into the paraxial wave equation (18.78), we have

$$gu\partial_x^2 f + 2g\,\partial_x u\partial_x f + fu\partial_y^2 g + 2f\,\partial_y u\partial_y g + 2ik[gu\partial_z f$$
$$+fu\partial_z g] + fg\{\partial_x^2 u + \partial_y^2 u + 2ik\partial_z u\} - 2kfgu\partial_z\Phi = 0. \tag{18.98}$$

The term in the curly braces satisfies Eq. (18.78), and is therefore equal to zero. We change to coordinates

$$\xi \equiv \sqrt{2}x/w, \tag{18.99}$$

$$\eta \equiv \sqrt{2}y/w, \tag{18.100}$$

and note by the chain rule that

$$\frac{\partial}{\partial x} = \frac{\sqrt{2}}{w}\frac{\partial}{\partial\xi}, \tag{18.101}$$

$$\frac{\partial}{\partial y} = \frac{\sqrt{2}}{w}\frac{\partial}{\partial\eta}, \tag{18.102}$$

$$\frac{\partial f}{\partial z} = \frac{\partial\xi}{\partial z}\frac{\partial f}{\partial\xi} = -\frac{\xi w'}{w}\frac{\partial f}{\partial\xi}, \tag{18.103}$$

$$\frac{\partial g}{\partial z} = \frac{\partial\eta}{\partial z}\frac{\partial g}{\partial\eta} = -\frac{\eta w'}{w}\frac{\partial g}{\partial\eta}, \tag{18.104}$$

and $w' = \partial w/\partial z$. Using these transformations in Eq. (18.98), and then dividing by $2fgu/w^2$, we have

$$\frac{\partial_\xi^2 f}{f} + \left(\frac{ikw^2}{R} - 2\right)\xi\frac{\partial_\xi f}{f} + \frac{\partial_\eta^2 g}{g} + \left(\frac{ikw^2}{R} - 2\right)\eta\frac{\partial_\eta g}{g}$$
$$- ikww'\left(\frac{\xi\partial_\xi f}{f} + \frac{\eta\partial_\eta g}{g}\right) - kw^2\partial_z\Phi = 0. \tag{18.105}$$

We note that

$$ww' = w_0^2\frac{z}{z_0^2} = \frac{w^2}{R}. \tag{18.106}$$

The imaginary terms in the above equation cancel, and we are left with

$$\frac{\partial_\xi^2 f}{f} - 2\xi\frac{\partial_\xi f}{f} + \frac{\partial_\eta^2 g}{g} - 2\eta\frac{\partial_\eta g}{g} - kw^2\partial_z\Phi = 0. \tag{18.107}$$

This equation may be grouped into terms which depend only upon a single variable. As in separation of variables, each grouping must therefore be equal to a constant: $-2m$ for the first, $-2n$ for the second, and C for the third. The constants satisfy the equation

$$2m + 2n = C. \tag{18.108}$$

We get the following separated set of solutions

$$\partial_\xi^2 f - 2\xi f + 2mf = 0, \tag{18.109}$$

$$\partial_\eta^2 g - 2\eta g + 2ng = 0, \tag{18.110}$$

$$\partial_z \Phi = \frac{C}{kw_0^2} \frac{1}{1 + z^2/z_0^2}. \tag{18.111}$$

The equation for ϕ can be directly integrated, to find

$$\Phi(z) = \frac{C}{2} \arctan(z/z_0). \tag{18.112}$$

The equations for f and g are both in the form of the Hermite equation. For the functions to converge for large values of ξ and η, m and n are constrained to integer values. We find that there exist an infinite number of solutions to the paraxial wave equation of the form

$$v(\mathbf{r}) = H_m\left[\sqrt{2}\frac{x}{w}\right] H_n\left[\sqrt{2}\frac{y}{w}\right] u(\mathbf{r}) \exp[i\Phi(z)], \tag{18.113}$$

with

$$\Phi(z) = (m+n)\arctan(z/z_0). \tag{18.114}$$

In the plane $z = 0$, these solutions appear as

$$v(x,y,0) = H_m\left[\sqrt{2}\frac{x}{w_0}\right] \exp[-x^2/w_0^2] H_n\left[\sqrt{2}\frac{y}{w_0}\right] \exp[-y^2/w_0^2]. \tag{18.115}$$

The intensities $|v(x,y,0)|^2$ of several of these modes are illustrated in Fig. 18.5.

Just as an arbitrary function of x may be expanded in terms of a Hermite series of functions $H_m(x)\exp[-x^2/2]$, an arbitrary paraxial wave in the plane $z = 0$ may be expanded in terms of functions of the form of Eq. (18.115). We may introduce normalized versions of these functions,

$$v_{mn}(x,y,z)$$

$$= \sqrt{\frac{2}{\pi}} 2^{-(m+n)/2} \frac{1}{\sqrt{n!m!w^2}} H_m\left[\sqrt{2}\frac{x}{w_0}\right] \exp[-x^2/w_0^2] H_n\left[\sqrt{2}\frac{y}{w_0}\right] \exp[-y^2/w_0^2], \tag{18.116}$$

which satisfy the orthogonality condition

$$\int_{-\infty}^{\infty} \int_{-\infty}^{\infty} v_{mn}(x,y,0) v_{m'n'}(x,y,0) \mathrm{d}x\mathrm{d}y = \delta_{mm'}\delta_{nn'}. \tag{18.117}$$

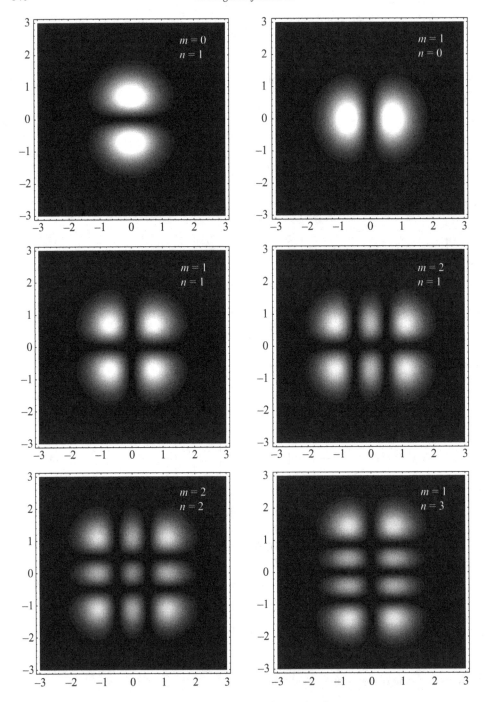

Figure 18.5 The intensities of Hermite–Gauss beams of several orders.

This suggests a quite remarkable and useful result. Assuming the set v_{mn} is complete, an arbitrary wavefield $U(x,y,z)$ may be expanded in terms of the functions $v_{mn}(x,y,z)$ in the form

$$U(x,y,z) = \sum_{m=0}^{\infty} \sum_{n=0}^{\infty} c_{mn} v_{mn}(x,y,z), \qquad (18.118)$$

where

$$c_{mn} = \int_{-\infty}^{\infty} \int_{-\infty}^{\infty} U(x,y,0) v_{mn}(x,y,0) dx dy. \qquad (18.119)$$

We have, in essence, developed another method of propagating a wavefield from a plane $z = 0$, akin to the angular spectrum method of Section 11.9 and the Rayleigh–Sommerfeld diffraction formula of Section 11.1. Given a field distribution in the plane $z = 0$, the coefficients c_{mn} may be calculated. The field at any plane $z > 0$ may be determined by substituting these coefficients into Eq. (18.118).

This method is only practical, however, when the field of interest is very close in shape to one of the modes v_{mn}; otherwise, a large number of coefficients c_{mn} will need to be calculated to accurately propagate the field.

18.3 Laguerre functions

The Laguerre differential equation is of the form

$$xL'' + (1-x)L' + nL = 0, \qquad (18.120)$$

where $0 \le x < \infty$ and n will typically be restricted to be a positive integer. Like Hermite's equation, Laguerre's equation is also not self-adjoint, and will need to be converted to a self-adjoint form to derive orthogonality relations.

18.3.1 Series solution

In Laguerre's equation, the point $x = 0$ is a regular singular point. Following Section 14.9, we expect that there exists at least one solution of the form

$$L(x) = \sum_{l=0}^{\infty} c_l x^{l+s}. \qquad (18.121)$$

The indicial equation for this solution has the form $s^2 = 0$, so that the Frobenius-type solution is an ordinary power series,

$$L(x) = \sum_{l=0}^{\infty} c_l x^l. \qquad (18.122)$$

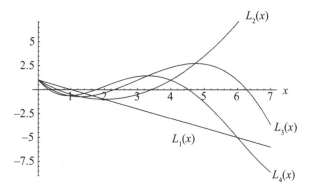

Figure 18.6 The Laguerre polynomials L_1, L_2, L_3, and L_4.

Table 18.2 *The first few Laguerre polynomials.*

$L_0(x) = 1$
$L_1(x) = -x + 1$
$2!L_2(x) = x^2 - 4x + 2$
$3!L_3(x) = -x^3 + 9x^2 - 18x + 6$
$4!L_4(x) = x^4 - 16x^3 + 72x^2 - 96x + 24$
$5!L_5(x) = -x^5 + 25x^4 - 200x^3 + 600x^2 - 600x + 120$

The recursion relation for this series may be written as

$$c_{l+1} = \frac{l-n}{(l+1)^2}c_l. \tag{18.123}$$

For integer n, the series terminates after $l = n$. We choose $a_0 = 1$ as our normalization, and the series solution may be written as

$$L_n(x) = \sum_{l=0}^{n}(-1)^l \frac{n!}{(l!)^2(n-l)!}x^l. \tag{18.124}$$

We refer to $L_n(x)$ as the nth *Laguerre polynomial*; the first few polynomials are depicted in Fig. 18.6 and tabulated in Table 18.2.

18.3.2 Generating function

The generating function for the Laguerre polynomials can be shown to be of the form

$$g(x,t) = \frac{e^{-xt/(1-t)}}{1-t} = \sum_{n=0}^{\infty}L_n(x)t^n. \tag{18.125}$$

As in the Hermite case, we take this function as given and later show that it reproduces the Laguerre functions.

18.3.3 Recurrence relations

Recurrence relations can be readily found through the generating function. Differentiation with respect to x results in the derivative formula

$$L'_{n-1}(x) - L'_n(x) = L_{n-1}(x).\tag{18.126}$$

Differentiating the generating function with respect to t gives the formula

$$(n+1)L_{n+1}(x) = (2n+1-x)L_n(x) - nL_{n-1}(x).\tag{18.127}$$

Finally, these two expressions may be combined to get

$$xL'_n(x) = nL_n(x) - nL_{n-1}(x).\tag{18.128}$$

We may use these relations to show that the coefficients of the generating function satisfy Laguerre's equation. Taking the derivative of Eq. (18.128), we have

$$xL''_n + L'_n = n(L'_n - L'_{n-1}).\tag{18.129}$$

We may use Eq. (18.126) to replace the right-hand side of this equation,

$$xL''_n + L'_n = -nL_{n-1}.\tag{18.130}$$

Finally, we apply Eq. (18.128) again to find that

$$xL''_n + L'_n = xL'_n - nL_n.\tag{18.131}$$

A simple rearrangement of terms results in Laguerre's equation.

A Rodrigues formula for the Laguerre polynomials may be shown to take the form

$$L_n(x) = \frac{e^x}{n!}\frac{d^n}{dx^n}\left(x^n e^{-x}\right).\tag{18.132}$$

As noted, Laguerre's equation is not self-adjoint. It may be put into self-adjoint form by multiplication by $\exp[-x]$, so that

$$\frac{d}{dx}\left[e^{-x}\frac{dL_n}{dx}\right] + e^{-x}nL_n = 0.\tag{18.133}$$

The weight function is $w(x) = \exp[-x]$, which suggests that the Laguerre polynomials satisfy the following orthogonality relation:

$$\int_0^\infty e^{-x}L_m(x)L_n(x)dx = 0, \quad m \neq n.\tag{18.134}$$

In fact, by following a procedure similar to that of Section 18.2.4, we can show that

$$\int_0^\infty e^{-x} L_m(x) L_n(x) dx = \delta_{mn}. \tag{18.135}$$

18.3.4 Associated Laguerre functions

We will show momentarily that the Laguerre functions appear in problems of separation of variables in curved coordinate systems. The associated Legendre functions were introduced in order to construct a complete basis of functions in curved coordinates; we will see that an analogous set of associated Laguerre functions are needed for certain problems in curved systems. The associated Laguerre functions are defined as

$$L_n^m(x) = (-1)^m \frac{d^m}{dx^m} L_{n+m}(x), \tag{18.136}$$

where L_m are the ordinary Laguerre polynomials, and n and m are taken to be non-negative integers. These functions satisfy the associated Laguerre equation,

$$xy'' + (m + 1 - x)y' + ny = 0. \tag{18.137}$$

The associated Laguerre functions are also polynomials of order n, as can be seen from their definition and the definition of the ordinary Laguerre functions. A generating function may be derived for them by straightforward differentiation of the generating function (18.125),

$$g_m(x,t) = \frac{e^{-xt/(1-t)}}{(1-t)^{m+1}} = \sum_{n=0}^\infty L_n^m(x) t^n. \tag{18.138}$$

Recurrence relations may be derived from these generating functions; several examples are

$$(n+1)L_{n+1}^m(x) = (2n + m + 1 - x)L_n^m(x) - (n + m)L_{n-1}^m(x), \tag{18.139}$$

$$x(L_n^m)'(x) = nL_n^m(x) - (n + m)L_{n-1}^m(x). \tag{18.140}$$

A Rodrigues formula may be found for the associated Laguerre functions of the form

$$L_n^m(x) = \frac{e^x x^{-m}}{n!} \frac{d^n}{dx^n} (x^{n+m} e^{-x}). \tag{18.141}$$

The associated Laguerre equation may be put into self-adjoint form by multiplying by $x^m e^{-x}$, such that

$$\frac{d}{dx}\left[x^{m+1} e^{-x} \frac{dL_n^m}{dx} \right] + nx^m e^{-x} L_n^m = 0. \tag{18.142}$$

The polynomials are thus orthogonal with respect to the integer n such that

$$\int_0^\infty x^m e^{-x} L_n^m(x) L_{n'}^m(x) \mathrm{d}x = 0, \quad n \neq n'. \tag{18.143}$$

The proper normalization may be found by application of the Rodrigues formula and integration by parts. One finds that

$$\int_0^\infty x^m e^{-x} L_{n'}^m(x) L_n^m(x) \mathrm{d}x = \frac{(n+m)!}{n!} \delta_{nn'}. \tag{18.144}$$

Finally, it is to be noted that the associated Laguerre functions have a different normalization than the ordinary Laguerre functions. This normalization may be found using the generating function with $x = 0$, such that

$$\sum_{n=0}^\infty L_n^m(0) t^n = \frac{1}{(1-t)^{m+1}} = \sum_{n=0}^\infty \frac{(m+1)!}{n!(m-n+1)!} t^n. \tag{18.145}$$

The last equality was made using the binomial theorem. Equating powers of t, and using the appropriate definition for the factorial of a negative number, we readily find that

$$L_n^m(0) = \frac{(n+m)!}{n!m!}. \tag{18.146}$$

18.3.5 Wavefunction of the hydrogen atom

Associated Laguerre polynomials appear in the solution of Schrödinger's equation for the hydrogen atom. The wavefunction $\psi(\mathbf{r})$ of the electron in a stationary state satisfies the time-independent equation

$$-\frac{\hbar^2}{2m} \nabla^2 \psi(\mathbf{r}) - \frac{e^2}{r} \psi(\mathbf{r}) = E\psi(\mathbf{r}), \tag{18.147}$$

where E is the as-yet unspecified energy of the electron and e is the fundamental unit of charge.

We may use separation of variables on this equation, in the form

$$\psi(\mathbf{r}) = R(r) Y(\theta, \phi). \tag{18.148}$$

The function Y may be expressed as a spherical harmonic, i.e. $Y \equiv Y_l^m(\theta, \phi)$; the equation for $R(r)$ reduces to

$$\frac{\mathrm{d}^2 R}{\mathrm{d}r^2} + \frac{2}{r} \frac{\mathrm{d}R}{\mathrm{d}r} + \frac{2m}{\hbar^2} [E + \frac{e^2}{r}] R - \frac{l(l+1)R}{r^2} = 0. \tag{18.149}$$

We may introduce the dimensionless variable

$$\rho = \left(-8mE/\hbar^2\right)^{1/2} r, \qquad (18.150)$$

where E will be assumed to be negative. We may then write

$$\frac{\mathrm{d}^2 R}{\mathrm{d}\rho^2} + \frac{2}{\rho}\frac{\mathrm{d}R}{\mathrm{d}\rho} - \frac{l(l+1)}{\rho^2}R + \left(\frac{\lambda}{\rho} - \frac{1}{4}\right)R = 0, \qquad (18.151)$$

where

$$\lambda = \frac{e^2}{\hbar}\left(\frac{m}{2|E|}\right)^{1/2} = \alpha\left(\frac{mc^2}{2|E|}\right)^{1/2}, \qquad (18.152)$$

and $\alpha = e^2/\hbar c \approx 1/137$ is the fine structure constant.

Equation (18.151) is not immediately familiar; we attempt to simplify it by looking for the asymptotic form of the solution as $\rho \to \infty$; the differential equation becomes

$$\frac{\mathrm{d}^2 R}{\mathrm{d}r^2} - \frac{1}{4}R \approx 0. \qquad (18.153)$$

This has a solution well-behaved at infinity of the form $R \sim \exp[-\rho/2]$; we may therefore assume a solution of the form

$$R(\rho) = \exp[-\rho/2]u(\rho). \qquad (18.154)$$

With this, our differential equation may be written as

$$\frac{\mathrm{d}^2 u}{\mathrm{d}\rho^2} - \left(1 - \frac{2}{\rho}\right)\frac{\mathrm{d}u}{\mathrm{d}\rho} + \left[\frac{\lambda - 1}{\rho} - \frac{l(l+1)}{\rho^2}\right]u = 0. \qquad (18.155)$$

This equation has a regular singular point, and we expect that it will have at least one Frobenius solution

$$u(\rho) = \rho^k \sum_{n=0}^{\infty} c_n \rho^n. \qquad (18.156)$$

On substitution, we find that the indicial equation gives us $k = l$; we therefore simplify the equation further with the solution

$$u(\rho) = \rho^l v(\rho). \qquad (18.157)$$

With this, our differential equation takes the form

$$\frac{\mathrm{d}^2 v}{\mathrm{d}\rho^2} + \left(\frac{2l+2}{\rho} - 1\right)\frac{\mathrm{d}v}{\mathrm{d}\rho} + \frac{\lambda - 1 - l}{\rho}v = 0. \qquad (18.158)$$

This is the associated Laguerre equation (18.137), with

$$m = 2l + 1, \tag{18.159}$$

$$\lambda = n + l + 1. \tag{18.160}$$

The eigenvalue λ is therefore constrained to be an integer, in order for the Laguerre polynomials to be well-behaved. The integer n must be non-negative, and so the integer λ will also be non-negative, and is referred to as the *principal quantum number*. The motivation for this label is given by Eq. (18.152), which shows that the energy of the electron is given by

$$E_n = -\frac{1}{2} mc^2 \frac{\alpha^2}{n^2}. \tag{18.161}$$

The energy levels are discrete, i.e. quantized, and larger n values correspond to higher energies.

The wavefunction of the hydrogen atom in one of its stationary states may be conveniently written as

$$\psi_{nlm}(\mathbf{r}) = \left(-8mE_n/\hbar^2\right)^{-1/4} \sqrt{\frac{(n-l-1)!}{(2l+1)!}} Y_l^m(\theta,\phi) e^{-\rho/2} \rho^l L_{n-l-1}^{2l+1}(\rho), \tag{18.162}$$

where the normalization follows directly from the normalization of the associated Laguerre functions.

18.3.6 Laguerre–Gauss laser beams

In Section 18.2.6 we derived the set of Hermite–Gauss laser modes by looking for Gaussian-like solutions to the paraxial wave equation in Cartesian coordinates. We may also look for Gaussian-like solutions in cylindrical coordinates, which results in the set of so-called Laguerre–Gauss laser modes. We now consider their derivation.

In cylindrical coordinates (ρ, ϕ, z), the wavefunction of a Gaussian beam may be written as

$$u(\mathbf{r}) = A_0 e^{-i\Phi(z)} \left[\frac{1}{\sqrt{1+z^2/z_0^2}} e^{ikz} e^{ik\rho^2/2R(z)} \right] e^{-\rho^2/w^2(z)}. \tag{18.163}$$

We assume a more general solution of the paraxial wave equation of the form

$$v(\mathbf{r}) = F[\sqrt{2}\rho/w(z)] G(\phi) u(\mathbf{r}) \exp[i\alpha(z)]. \tag{18.164}$$

We may substitute this solution into the paraxial wave equation, which in cylindrical coordinates may be written as

$$\frac{\partial^2 v}{\partial \rho^2} + \frac{1}{\rho}\frac{\partial v}{\partial \rho} + \frac{1}{\rho^2}\frac{\partial^2 v}{\partial \phi^2} + 2ik\frac{\partial v}{\partial z} = 0. \tag{18.165}$$

On substitution, the terms depending solely upon u satisfy the paraxial wave equation and are equal to zero. The remaining terms are

$$\frac{1}{F}\frac{\partial^2 F}{\partial \rho^2} + \frac{2}{Fu}\frac{\partial u}{\partial \rho}\frac{\partial F}{\partial \rho} + \frac{1}{\rho F}\frac{\partial F}{\partial \rho} + \frac{1}{\rho^2 G}\frac{\partial^2 G}{\partial \phi^2} + \frac{2ik}{F}\frac{\partial F}{\partial z} - 2k\frac{\partial \alpha}{\partial z} = 0. \tag{18.166}$$

We modify the remaining terms by the coordinate transformation

$$\zeta \equiv \sqrt{2}\rho/w, \tag{18.167}$$

and note by the chain rule that

$$\frac{\partial}{\partial \rho} = \frac{\sqrt{2}}{w}\frac{\partial}{\partial \zeta}, \tag{18.168}$$

$$\frac{\partial F}{\partial z} = \frac{\partial \zeta}{\partial z}\frac{\partial F}{\partial \zeta} = -\frac{\zeta}{R}\frac{\partial F}{\partial \zeta}. \tag{18.169}$$

We may write the paraxial wave equation in the form

$$\frac{1}{F}\frac{\partial^2 F}{\partial \zeta^2} + \frac{1}{F}\left(\frac{ikw^2}{R} - 2\right)\zeta\frac{\partial F}{\partial \zeta} + \frac{1}{\zeta F}\frac{\partial F}{\partial \zeta} - \frac{ikw^2}{R}\frac{\zeta}{F}\frac{\partial F}{\partial \zeta} + \frac{1}{\zeta^2}\frac{1}{G}\frac{\partial^2 G}{\partial \phi^2} - kw^2\frac{\partial \alpha}{\partial z} = 0. \tag{18.170}$$

As in the case of the Hermite–Gauss beams, we let the last term be equal to a constant C, and the function $\alpha(z)$ may be written as

$$\alpha(z) = \frac{C}{2}\arctan(z/z_0). \tag{18.171}$$

The second to last term of the paraxial equation we set equal to a constant $-m^2$, as it is clear that the function G has the solution

$$G(\phi) = A_m e^{im\phi} + B_m e^{-im\phi}, \tag{18.172}$$

with m an integer. We may then write

$$\frac{d^2 F}{d\zeta^2} + \left(\frac{1}{\zeta} - 2\zeta\right)\frac{dF}{d\zeta} + \left[C - \frac{m^2}{\zeta^2}\right]F = 0. \tag{18.173}$$

Now we must rely on a little luck and a little intuition guided by the example of the Hermite–Gauss modes. For Hermite functions, the weight function is e^{-x^2}, while for Laguerre functions, the weight function is e^{-x}. This suggests that we make a coordinate transformation $y = \zeta^2$, for which our equation becomes

$$y\frac{d^2 F}{dy^2} + (1-y)\frac{dF}{dy} + \frac{1}{4}\left(C - \frac{m^2}{y}\right)F = 0. \tag{18.174}$$

The singularity at $y = 0$ is a regular singularity, which suggests that a Frobenius solution exists of the form

$$F(y) = \sum_{k=0}^{\infty} c_k y^{k+s}. \tag{18.175}$$

The indicial equation gives $s = \pm m/2$; choosing the positive root as the only solution regular at the origin, we define

$$H(y) = y^{m/2} F(y). \tag{18.176}$$

Our differential equation becomes

$$yH'' + (m+1-y)H' + \left(\frac{1}{4}C - \frac{m}{2}\right)H = 0. \tag{18.177}$$

This is exactly the associated Laguerre equation, Eq. (18.137), with $n = C/4 - m/2$ constrained to be an integer, or $C = 4n + 2m$. Consolidating our results, we may write

$$v_n^m(\mathbf{r}) = \sqrt{\frac{2n!}{\pi w_0^2(n+m)!}} \left(\frac{\sqrt{2}\rho}{w(z)}\right)^m L_n^m\left(\frac{2\rho^2}{w^2(z)}\right)e^{im\Phi}u(\mathbf{r})e^{i\alpha(z)}, \tag{18.178}$$

where

$$\alpha(z) = (2n+m)\arctan(z/z_0), \tag{18.179}$$

and the functions have been normalized such that

$$\int_0^{\infty}\int_0^{2\pi}|v(\rho,\phi,0)|^2\rho d\rho d\phi = 1. \tag{18.180}$$

The intensity and phase of several of these beams in the plane $z = 0$ are plotted in Fig. 18.7. Beams with $m > 0$ have a phase which circulates around the central axis, and are therefore vortex beams, as discussed in Section 9.15.

Just as in the case of the Hermite–Gauss beams, the Laguerre–Gauss beams form a complete set of orthonormal paraxial fields. The propagation of any field in the plane $z = 0$ can be found by decomposing into a collection of Laguerre–Gauss fields in that plane.

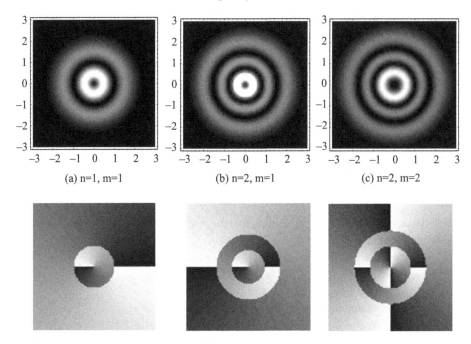

Figure 18.7 The intensity (top) and phase (bottom) of several Laguerre–Gauss beams in the plane $z = 0$. The phase runs from 0 (black) to 2π (white).

18.4 Chebyshev polynomials

The Chebyshev functions (also transliterated as "Tchebyshef") are special cases of the Gegenbauer polynomials $\mathcal{G}_n^\nu(x)$, discussed in Section 17.11. Those functions were defined by a generating function of the form

$$\frac{1}{(1 - 2xt + t^2)^\nu} = \sum_{n=0}^{\infty} \mathcal{G}_n^\nu(x)t^n. \tag{18.181}$$

For $\nu = 1/2$, this reduces to the expression for the Legendre polynomials.

After $\nu = 1/2$, the cases $\nu = 0$ and $\nu = 1$ are the next most commonly occurring and are referred to as *Chebyshev polynomials of the first and second kind*, respectively denoted $T_n(x)$ and $U_n(x)$. Polynomials of the second kind follow directly from the recurrence relations of the Gegenbauer functions, while polynomials of the first kind require a little more subtlety.

18.4.1 Polynomials of the second kind

From the Gegenbauer results, we know that the functions $U_n(x)$ satisfy a recurrence relation of the form

$$U_{n+1}(x) = 2xU_n(x) - U_{n-1}(x). \tag{18.182}$$

Table 18.3 *The first few Chebyshev polynomials of the second kind.*

$$U_0(x) = 1$$
$$U_1(x) = 2x$$
$$U_2(x) = 4x^2 - 1$$
$$U_3(x) = 8x^3 - 4x$$
$$U_4(x) = 16x^4 - 12x^2 + 1$$

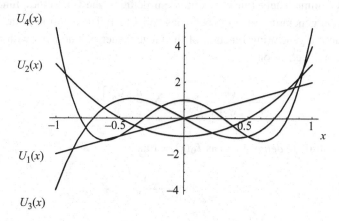

Figure 18.8 The Chebyshev polynomials U_1, U_2, U_3, and U_4.

Given that $U_0(x) = 1$ and $U_1(x) = 2x$, we can find the polynomials of higher order, some of which are listed in Table 18.3 and plotted in Fig. 18.8.

From Eq. (17.171), we may readily write the differential equation that the functions $U_n(x)$ satisfy,

$$(1 - x^2)U_n'' - 3xU_n' + n(n+2)U_n = 0. \tag{18.183}$$

This equation may be put into self-adjoint form by multiplying by $(1 - x^2)^{1/2}$. This suggests that the functions are orthogonal with respect to the weighting function $(1 - x^2)^{1/2}$, i.e.

$$\int_{-1}^{1} (1 - x^2)^{1/2} U_n(x) U_m(x) dx = 0, \quad n \neq m. \tag{18.184}$$

It can be shown that the functions U_n have a Rodrigues representation

$$U_n(x) = \frac{(-1)^n (n+1)\pi^{1/2}}{2^{n+1}(n+1/2)!(1 - x^2)^{1/2}} \frac{d^n}{dx^n}(1 - x^2)^{n+1/2}. \tag{18.185}$$

From this representation, one can show that the normalization of the functions is of the form

$$\int_{-1}^{1} (1-x^2)^{1/2} U_n(x) U_m(x) dx = \delta_{mn} \frac{\pi}{2}. \tag{18.186}$$

This can be shown by writing $x = \cos\theta$ and rewriting the Rodrigues representation and the orthogonality integral in terms of an integral over $0 \le \theta \le \pi$.

18.4.2 Polynomials of the first kind

For $\nu = 0$, we immediately run into a problem defining the Chebyshev function of the first kind, as the generating function becomes equal to 1. The solution to this problem is to differentiate the generating function of the Gegenbauer polynomials with respect to t, which results in the equation

$$\frac{x-t}{(1-2xt+t^2)^{\nu+1}} = \sum_{n=1}^{\infty} \frac{n}{2} \left[\frac{G_n^{\nu}}{\nu} \right] t^{n-1}. \tag{18.187}$$

It is then convention to *define* $T_n(x)$ as $T_0(x) = 1$ and

$$T_n(x) = \frac{n}{2} \lim_{\nu \to 0} \frac{G_n^{\nu}(x)}{\nu}, \quad n > 1. \tag{18.188}$$

A generating function for this new set may be determined by multiplying Eq. (18.187) by $2t$ and adding 1 to both sides. A rearrangement of terms gives

$$\frac{1-t^2}{1-2xt+t^2} = T_0(x) + 2 \sum_{n=1}^{\infty} T_n(x) t^n. \tag{18.189}$$

A recurrence relation can be derived from this generating function, or from the recurrence relation (17.170) for the Gegenbauer polynomials in the appropriate limit,

$$T_{n+1} = 2x T_n - T_{n-1}. \tag{18.190}$$

Using $G_1^{\nu}(x) = 2\nu x$, we may readily determine that $T_1(x) = x$. With the recurrence relation, we may find values of all the polynomials, some of which are listed in Table 18.4 and plotted in Fig. 18.9.

From Eq. (17.171), we may readily write the differential equation that the functions $T_n(x)$ satisfy,

$$(1-x^2) T_n'' - x T_n' + n^2 T_n = 0. \tag{18.191}$$

Table 18.4 *The first few Chebyshev polynomials of the first kind.*

$$T_0(x) = 1$$
$$T_1(x) = x$$
$$T_2(x) = 2x^2 - 1$$
$$T_3(x) = 4x^3 - 3x$$
$$T_4(x) = 8x^4 - 8x^2 + 1$$

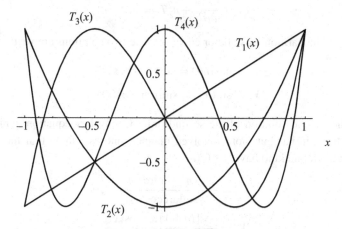

Figure 18.9 The Chebyshev polynomials T_1, T_2, T_3, and T_4.

This equation may be put into self-adjoint form by multiplying by $(1-x^2)^{-1/2}$. This suggests that the functions are orthogonal with respect to the weighting function $(1-x^2)^{-1/2}$, i.e.

$$\int_{-1}^{1} (1-x^2)^{-1/2} T_n(x) T_m(x) dx = 0, \quad n \neq m. \tag{18.192}$$

It also can be shown that the functions T_n have a Rodrigues representation

$$T_n(x) = \frac{(-1)^n \pi^{1/2} (1-x^2)^{1/2}}{2^n (n-1/2)!} \frac{d^n}{dx^n} (1-x^2)^{n-1/2}. \tag{18.193}$$

From this representation, one can show that the normalization of the functions has the form

$$\int_{-1}^{1} (1-x^2)^{-1/2} T_n(x) T_m(x) dx = G_n \delta_{mn} \frac{\pi}{2}, \tag{18.194}$$

where

$$G_n \equiv \begin{cases} 2, & n = 0, \\ 1, & n > 0. \end{cases} \tag{18.195}$$

This can again be shown by writing $x = \cos\theta$ and rewriting the Rodrigues representation and the orthogonality integral in term of an integral over $0 \le \theta \le \pi$.

The Chebyshev polynomials of the first kind take on an exceedingly simple form under the transformation $x = \cos\theta$. One can show by making this coordinate transformation that the differential equation reduces to

$$\frac{d^2 T_n}{d\theta^2} + n^2 T_n = 0, \tag{18.196}$$

which is simply the harmonic oscillator equation. The two solutions are written as

$$T_n = \cos(n\theta) = \cos[n\arccos(x)], \tag{18.197}$$

$$V_n = \sin(n\theta) = \sin[n\arccos(x)]. \tag{18.198}$$

Here V_n is the second solution of the second-order Chebyshev equation, which is not a polynomial. In a similar but more involved manner, one can show that the Chebyshev functions of the second kind have the form

$$U_n = \frac{\sin[(n+1)\arccos(x)]}{\sin[\arccos(x)]}, \tag{18.199}$$

$$W_n = \frac{\cos[(n+1)\arccos(x)]}{\sin[\arccos(x)]}, \tag{18.200}$$

The Chebyshev functions rarely appear directly in physical problems. However, they can be extremely useful in numeric calculations, for instance in Gaussian quadrature formulas. Quite a bit of discussion of their uses is given in [PTVF92].

18.5 Jacobi polynomials

One other set of polynomials can be defined that are very closely related to the Legendre and Gegenbauer polynomials and also may be considered a generalization of them. These polynomials, called the *Jacobi polynomials*, satisfy a differential equation on the domain $-1 \le x \le 1$ of the form

$$(1 - x^2)\frac{d^2 P_n^{(\nu,\mu)}}{dx^2} + [\mu - \nu - (\nu + \mu + 2)x]\frac{dP_n^{(\nu,\mu)}}{dx} + n(n + \nu + \mu + 1)P_n^{(\nu,\mu)} = 0. \tag{18.201}$$

For $\nu = \mu = 0$, this equation reduces to the Legendre equation, with the Legendre polynomials P_n as solutions. For $\mu = \nu = m - 1/2$, this equation reduces to the Gegenbauer equation, with the Gegenbauer polynomials $\mathcal{G}_n^m$ as solutions.

As one might expect, these polynomials share similar properties to their simpler brethren. For instance, the Rodrigues formula for the polynomials is of the form

$$P_n^{(\nu,\mu)}(x) = \frac{(-1)^n}{2^n n!}(1-x)^{-\nu}(1+x)^{-\mu}\frac{d^n}{dx^n}\left[(1-x)^{\nu+n}(1+x)^{\mu+n}\right].\qquad(18.202)$$

Equation (18.201) can be made self-adjoint by multiplying it by $(1-x)^\nu(1+x)^\mu$; this suggests that the functions satisfy the orthogonality relationship,

$$\int_{-1}^{1}(1-x)^\nu(1+x)^\mu P_n^{(\nu,\mu)}(x)P_m^{(\nu,\mu)}(x)dx = 0, \quad m \neq n.\qquad(18.203)$$

A normalized set of polynomials $p_n^{(\nu,\mu)}$ can be constructed as

$$p_n^{(\nu,\mu)} = P_n^{(\nu,\mu)}/\sqrt{A_n},\qquad(18.204)$$

where

$$A_n = \frac{2^{\nu+\mu+1}}{\nu+\mu+2n+1}\frac{\Gamma(\nu+n+1)\Gamma(\mu+n+1)}{n!\Gamma(\nu+\mu+n+1)}.\qquad(18.205)$$

Various recurrence relations between the Jacobi polynomials may be derived; we only note one of them

$$2(n+1)(n+\nu+\mu+1)(2n+\nu+\mu)P_{n+1}^{(\nu,\mu)}$$
$$= (2n+\nu+\mu+1)[(2n+\nu+\mu)(2n+\nu+\mu+2)x+\nu^2-\mu^2]P_n^{(\nu,\mu)}$$
$$- 2(n+\nu)(n+\mu)(2n+\nu+\mu+2)P_{n-1}^{(\nu,\mu)}.\qquad(18.206)$$

The generating function for the Jacobi polynomials is truly a frightening thing,

$$R^{-1}(1-z+R)^{-\nu}(1+z+R)^{-\mu} = \sum_{n=0}^{\infty}2^{-\nu-\mu}P_n^{(\nu,\mu)}(x)z^n,\qquad(18.207)$$

where $R = \sqrt{1-2xz+z^2}$.

These polynomials, though quite general, are not commonly used. However, they are quite helpful in deriving properties of the Zernike polynomials discussed in the next section.

18.6 Focus: Zernike polynomials

In Section 11.8, we discussed the focusing of light in the context of spatial filtering. Under the assumption of Fresnel diffraction and a thin lens approximation, we showed that the field focused through an aperture of area $\mathcal{A}$ may be written as

$$
\begin{aligned}
U(x,y,z) \\
= \frac{e^{ikz}}{z} \frac{1}{i\lambda} \int_{\mathcal{A}} U^{(i)}(x',y') T(x',y') \exp\left\{ i\frac{k}{2z}[(x-x')^2 + (y-y')^2] \right\} dx' dy',
\end{aligned} \tag{18.208}
$$

where $U^{(i)}(x,y)$ is the field illuminating the lens, and $T(x,y)$ is the transmission function of the lens. An ideal lens introduces a quadratic curvature in the wavefront, i.e.

$$
T(x,y) = \exp\left[-i\frac{k}{2f}(x^2 + y^2) \right], \tag{18.209}
$$

where f is the focal length of the lens, such that the field in the focal plane of the lens simplifies to the form

$$
U(x,y;f) = e^{ikf} \frac{e^{ik(x^2+y^2)/2f}}{i\lambda f} (2\pi)^2 \tilde{U}^{(i)}\left[\frac{kx}{f}, \frac{ky}{f} \right], \tag{18.210}
$$

where $\tilde{U}^{(i)}$ is the two-dimensional spatial Fourier transform of the illuminating field. Such an optical system is referred to as *diffraction-limited*, in that the size of the focal spot produced by a point source is limited only by aperture diffraction.

In general, however, optical systems possess *aberrations*, distortions of the wavefront emerging from the aperture deviating from the ideal diffraction-limited case. These aberrations arise in a number of ways: imperfections in the manufacture of the lens, inherent higher-order effects of perfectly spherical lenses (especially for large apertures), and deviations from the thin lens approximation. We restrict ourselves to lens imperfections that are rotationally symmetric about a central axis.

To account for such effects, we introduce an aberration function $\Phi(x,y)$ into Eq. (18.208), such that

$$
\begin{aligned}
U(x,y,z) = \frac{e^{ikz}}{z} \frac{1}{i\lambda} \int_{\mathcal{A}} U^{(i)}(x',y') T(x',y') \exp[ik\Phi(x',y')] \\
\times \exp\left\{ i\frac{k}{2z}[(x-x')^2 + (y-y')^2] \right\} dx' dy'.
\end{aligned} \tag{18.211}
$$

In general, the aberration function $\Phi(x,y)$ depends not only on the position within the aperture but also on the form of the incident wavefield: a plane wave normally incident upon a lens with aberrations experiences a different distortion than a plane wave incident upon the lens from an angle.

In the limit that aberrations are weak, we may expand the aberration function $\Phi(x',y')$ in a series of polynomials in x' and y' or, for a circular aperture, in ρ and θ. We may then ask what effect the lowest terms in the series have on the properties of the focused wave.

Because of the symmetry of the system, the aberration function must be rotationally invariant in form, depending only upon the position of the object to be imaged, designated

(x_0, y_0), and the position in the aperture, (x', y'). The rotationally invariant combinations of these variables are $l_0^2 = x_0^2 + y_0^2$, $\rho^2 = x'^2 + y'^2$, and $x_0 x' + y_0 y' = l_0 \rho \cos\theta$. The lowest-order polynomial in x' and y' is a constant, referred to as the *piston* term. The next lowest term is $x_0 x' + y_0 y'$ or, alternatively, $\rho \cos\theta$, with θ measured from the direction of (x_0, y_0). This linear phase term results in a transverse shift of the origin and is referred to as *tilt*. The next possible term is quadratic, $\rho^2 = x'^2 + y'^2$, which results in a lateral shift in the focus, referred to as *defocus*.

The next lowest-order polynomials are fourth-order in the variables, and are referred to as the *Seidel aberrations*. Neglecting an overall constant, these are ρ^4, which is known as *spherical aberration*, $l_0^2 \rho^2 \cos^2\theta$, known as *astigmatism*, $l_0^2 \rho^2$, known as *curvature of field*, $l_0^3 \rho \cos\theta$, known as *distortion*, and $l_0 \rho^3 \cos\theta$, known as *coma*. Each of these aberrations has a distinct effect on the field in the focal region.

It is especially convenient to choose a set of polynomials that are orthogonal over the unit circle (aperture). There are in principle an infinite number of polynomial sets that possess this orthogonality, but a particularly useful and now standard set is referred to as the *Zernike polynomials*. We derive the polynomials by requiring certain symmetry conditions, following [BW52, BW54]; we then discuss the mathematical properties of the Zernike polynomials and finally discuss their usefulness in certain applications.

Noting again that the lens imperfections are assumed to be rotationally symmetric, with no preferred direction of orientation, it is natural to look for a set of polynomials $Z(x, y)$ that are invariant under rotations in the xy-plane; that is, if we introduce a coordinate rotation as discussed in Section 1.2, i.e.

$$x' = x\cos\phi + y\sin\phi, \tag{18.212}$$

$$y' = -x\sin\phi + y\cos\phi, \tag{18.213}$$

the transformed polynomial may differ only by a function dependent on ϕ,

$$Z(x, y) = F(\phi)Z(x', y'). \tag{18.214}$$

It is not difficult to see that the only functions which satisfy this condition are of the form

$$Z(\rho, \theta) = R(\rho)e^{il\theta}, \tag{18.215}$$

where l must be an integer for the function to be single-valued.

With the requirement that the function Z must be a polynomial, Eq. (18.215) puts strict conditions on the form of $R(\rho)$. Because $x + iy = \rho e^{i\theta}$, we see that

$$\rho^\alpha e^{il\theta} = \rho^{\alpha - |l|}\rho^{|l|}e^{il\theta} = (x^2 + y^2)^{\alpha/2 - |l|/2}(x \pm iy)^{|l|}. \tag{18.216}$$

We see that the quantity $(x^2 + y^2)^{\alpha/2 - |l|/2}$ will be in polynomial form if and only if $\alpha/2 - |l|/2$ is a non-negative integer value m, i.e.

$$\alpha/2 - |l|/2 = m \geq 0. \tag{18.217}$$

This implies that all terms in the polynomial must be of the form

$$\alpha = |l| + 2m, \quad m \geq 0. \tag{18.218}$$

We state the result as a theorem.

Theorem 18.1 *A rotationally-invariant polynomial in x and y must be of the form*

$$Z(\rho, \theta) = \left[\sum_{m=0}^{p} a_m \rho^{|l|+2m} \right] e^{il\theta}. \tag{18.219}$$

The function $Z(\rho, \theta)$ is an even or an odd polynomial in ρ according as l is even or odd, and contains no powers of r lower than $|l|$. The order of the polynomial is $|l| + 2p$.

This observation, even coupled with the requirement that polynomials with different l and p values be orthogonal, does not produce a unique set. An additional condition that provides a unique set is provided in the following theorem.

Theorem 18.2 *There is one and only one set of polynomials $Z_n^m(x, y)$ that:*

1. *is orthogonal over the unit circle,*
2. *consists entirely of polynomials that are rotationally invariant in form,*
3. *contains a polynomial for every integer value of m and n such that $n \geq |m|$ and $n - |m|$ is even.*

The proof of this theorem is given in [BW54]; we will see that these conditions are satisfied by the specific set of Zernike polynomials.

Let us write our polynomials in the form

$$Z_n^m(\rho, \theta) = R_n^m(\rho) e^{im\theta}. \tag{18.220}$$

These functions must be orthogonal with respect to integrations over the unit circle,

$$\int_0^1 \int_0^{2\pi} Z_n^m(\rho, \theta) Z_{n'}^{m'*}(\rho, \theta) \rho \, d\rho \, d\theta = \alpha_n^m \delta_{mm'} \delta_{nn'}, \tag{18.221}$$

where α_n^m is a constant to be determined. The functions R_n^m must themselves satisfy an orthogonality condition

$$\int_0^1 R_n^m(\rho) R_{n'}^m(\rho) \rho \, d\rho = \beta_n^m \delta_{nn'}, \tag{18.222}$$

with β_n^m to be determined.

Surprisingly, Theorem 18.2 gives us enough information to derive the form of the polynomials $R_n^m(\rho)$. Because it has a degree n which has no power lower than $|m|$ and is even or odd according to the evenness or oddness of n, it may be written as

$$R_n^m(\rho) = t^{|m|/2} P_{(n-|m|)/2}(t), \tag{18.223}$$

where $t = \rho^2$ and $P_l(t)$ is a polynomial in t of degree l. The orthogonality relation (18.222) takes on the form

$$\frac{1}{2}\int_0^1 t^m P_k(t)P_{k'}(t)\mathrm{d}t = \beta_n^m \delta_{kk'}, \tag{18.224}$$

where $k = (n - |m|)/2$, $k' = (n' - |m|)/2$. What is remarkable is that this orthogonality relation can be directly connected to the Jacobi polynomials, for which we already have tabulated important relations.

The orthogonality relation for the Jacobi polynomials was given by Eqs. (18.203) and (18.205) as

$$\int_{-1}^1 (1-x)^\nu (1+x)^\mu P_n^{(\nu,\mu)}(x)P_{n'}^{(\nu,\mu)}(x)\mathrm{d}x = \frac{2^{\nu+\mu+1}}{\nu+\mu+2n+1}\frac{\Gamma(\nu+n+1)\Gamma(\mu+n+1)}{n!\Gamma(\nu+\mu+n+1)}\delta_{nn'}. \tag{18.225}$$

We make the coordinate transformation $x = 2t - 1$ to find

$$\int_0^1 (1-t)^\nu t^\mu Q_n^{(\nu,\mu)}(t)Q_{n'}^{(\nu,\mu)}(t)\mathrm{d}t = \frac{1}{\nu+\mu+2n+1}\frac{\Gamma(\nu+n+1)\Gamma(\mu+n+1)}{n!\Gamma(\nu+\mu+n+1)}\delta_{nn'}, \tag{18.226}$$

where we have defined $Q_n^{(\nu,\mu)}(t) \equiv P_n^{(\nu,\mu)}(2t - 1)$. With $\nu = 0$, $\mu = m$, this becomes

$$\int_0^1 t^m P_n^{(0,m)}(2t-1)P_{n'}^{(0,m)}(2t-1)\mathrm{d}t = \frac{1}{2n+m+1}\delta_{nn'}, \tag{18.227}$$

which is in agreement with Eq. (18.224). The orthogonal functions $P_k(t)$ may be directly identified with $P_k^{(0,m)}(2t - 1)$. We may use the Jacobi polynomial Rodrigues formula, Eq. (18.202), to write

$$P_k(t) = P_k^{(0,m)}(2t-1) = \frac{1}{k!t^m}\frac{\mathrm{d}^k}{\mathrm{d}t^k}\left[(t-1)^k t^{m+k}\right]. \tag{18.228}$$

We may determine an expression for $R_n^m(\rho)$ by substitution into Eq. (18.223),

$$R_n^m(\rho) = \frac{1}{2^k k!}\frac{1}{\rho^m}\left[\frac{1}{\rho}\frac{\mathrm{d}}{\mathrm{d}\rho}\right]^k\left[(\rho^2-1)^k \rho^{n+m}\right], \tag{18.229}$$

where again $k = (n - |m|)/2$; recall that this will always be an integer value. We have

$$\int_0^1 R_n^m(\rho)R_{n'}^m(\rho)\rho\,\mathrm{d}\rho = \frac{1}{2(n+1)}\delta_{nn'}. \tag{18.230}$$

If we define a set of circle functions $Z_n^m(\rho,\theta)$ by the expression

$$Z_n^m(\rho,\theta) = R_n^m(\rho)\mathrm{e}^{\mathrm{i}m\theta}, \tag{18.231}$$

Table 18.5 *The first few radial polynomials* $R_n^m(\rho)$.

n/m	0	1	2	3	4	5
0	1					
1		ρ				
2	$2\rho^2 - 1$		ρ^2			
3		$3\rho^3 - 2\rho$		ρ^3		
4	$6\rho^4 - 6\rho^2 + 1$		$4\rho^4 - 3\rho^2$		ρ^4	
5		$10\rho^5 - 12\rho^3 + 3\rho$		$5\rho^5 - 4\rho^3$		ρ^5

these functions satisfy the orthogonality relation over the circle,

$$\int_{\mathcal{A}} Z_n^m(\rho,\theta) Z_{n'}^{m'}(\rho,\theta)\rho \mathrm{d}\rho = \frac{\pi}{n+1}\delta_{nn'}\delta_{mm'}. \tag{18.232}$$

Because we are interested in using these polynomials to represent phase aberrations, it is more appropriate to use the set of functions

$$U_n^m(\rho,\theta) = R_n^m(\rho)\cos(m\theta), \tag{18.233}$$

$$U_n^{-m}(\rho,\theta) = R_n^m(\rho)\sin(m\theta). \tag{18.234}$$

For aberration theory, the sine functions are typically not necessary as aberrations are symmetric around a central azimuthal axis. For other applications, however, both sets of functions are necessary.

Some of the low-order polynomials are shown in Table 18.5.

The aberration function can be expanded in a series of Zernike polynomials, of the form

$$\Phi(r,\theta) = A_{00} + \frac{1}{\sqrt{2}}\sum_{n=2}^{\infty} A_{n0}R_n^0(\rho) + \sum_{n=1}^{\infty}\sum_{m=1}^{n} A_{nm}R_n^m(\rho)\cos(m\theta). \tag{18.235}$$

The factor $\sqrt{2}$ is introduced for later convenience. It should again be noted that the aberration function depends upon the nature of the illuminating wavefield; for simplicity, we assume the lens is illuminated by a normally-incident plane wave;[1] a more detailed discussion of the relationship between aberrations and illumination can be found in [BW99], Chapters 5 and 9.

To illustrate the usefulness of the Zernike polynomials, we first consider the intensity $I_0 \equiv |U(0,0,f)|^2$ of the wavefield at the geometrical focus, located at $(x,y,z) = (0,0,f)$. We

---

[1] We're cheating somewhat, as a lens with rotational symmetry and normally-incident plane illumination would have no aberrations with $m > 0$! We can pretend for illustration purposes that the illumination is incident from slightly off the normal axis, which breaks the symmetry.

assume the illuminating plane wave is of unit amplitude and that the aperture has a radius a. We may evaluate Eq. (18.210) to find that

$$I_0 = \left[\frac{\pi a^2}{\lambda f} \right]^2 . \tag{18.236}$$

In the presence of aberrations, the normalized intensity at focus may be written as

$$\frac{I}{I_0} = \frac{1}{\pi^2} \int_{\mathcal{A}} e^{ik\Phi(\rho,\theta)} \rho \, d\rho \, d\theta , \tag{18.237}$$

where ρ is a variable normalized to unity on the circumference of the aperture. Let us further assume that the aberration function has an amplitude small compared to the wavelength; in such a case, the exponential may be approximated by its lowest-order Taylor series terms,

$$e^{ik\Phi(\rho,\theta)} \approx 1 + ik\Phi(\rho,\theta) + \frac{1}{2} [ik\Phi(\rho,\theta)]^2 . \tag{18.238}$$

If we only keep terms up to second-order in the aberration function, Eq. (18.237) takes the form

$$\frac{I}{I_0} = 1 - k^2 \left[\overline{\Phi^2} - (\overline{\Phi})^2 \right] , \tag{18.239}$$

where

$$\overline{\Phi^n} \equiv \frac{1}{\pi} \int_0^1 \int_0^{2\pi} [\Phi(\rho,\theta)]^n \rho \, d\rho \, d\theta . \tag{18.240}$$

The bracketed quantity in Eq. (18.239) represents the mean square deformation of the wavefront. The equation as a whole suggests that the intensity of the wavefield at focus with aberrations is always lower than the intensity of an aberration-free wavefront. However, aberrations of different orders can be played against one another, in a technique known as the *balancing of aberrations*. For instance, one can balance a fourth-order spherical aberration (with aberration function $A\rho^4$) with an appropriate amount of defocus (with aberration function $B\rho^2$). An appropriate choice of B relative to A will increase the intensity in the focal plane relative to the intensity with A alone.

Remarkably, the Zernike polynomials are optimally balanced to provide a minimum variance of the aberration function and, consequently, a maximum of intensity at focus. This can be most readily observed by noting that the phase function defined by Eq. (18.235) results in an intensity of the form

$$\frac{I}{I_0} = 1 - \frac{k^2}{2} \sum_{n=1}^{\infty} \sum_{m=0}^{\infty} \frac{A_{nm}^2}{n+1} . \tag{18.241}$$

Suppose that one aberration is fixed, i.e. A_{NM} is a fixed aberration of the system. It is clear from Eq. (18.241) that the addition of any further aberrations will decrease the intensity at focus. An aberration based on a Zernike polynomial is already optimally balanced.

Zernike polynomials have found use beyond the classification and balancing of aberrations in focusing. They have also been used to characterize the effects of atmospheric turbulence [Nol76], and for correcting those effects by the use of adaptive optics [RW96].

18.7 Additional reading

Most mathematical methods books cover at least some of the theory of orthogonal polynomials; two books which specialize in them are:

- D. Jackson, *Fourier Series and Orthogonal Polynomials* (Ohio, Mathematical Association of America, 1941). Also in Dover.
- H. F. Davis, *Fourier Series and Orthogonal Functions* (Boston, Allyn and Bacon, 1963). Also in Dover.

18.8 Exercises

1. Express the function $f(x) = x^4 + 2x^3 - 2x^2 + 4$ as a series of Hermite polynomials and as a series of Legendre polynomials.
2. Express the function $f(x) = x^5 - 3x^3 + 2x^2 + x$ as a series of Chebyshev polynomials of the first kind and of the second kind.
3. Demonstrate the orthogonality relation for the Chebyshev polynomials,

$$\int_{-1}^{1} (1-x^2)^{1/2} U_n(x) U_m(x) dx = \delta_{mn} \frac{\pi}{2},$$

   by writing $x = \cos\theta$ and using the Rodrigues representation of U_n.
4. Show that the Laguerre–Gauss beam $v_0^1(\mathbf{r})$ may be written as a sum of Hermite–Gauss modes in the form

$$v_0^1(\mathbf{r}) = e^{i\phi_1} H_0^1(\mathbf{r}) + e^{i\phi_2} H_1^0(\mathbf{r}).$$

   Determine the constants ϕ_1 and ϕ_2.
5. Suppose that in the plane $z = 0$ a laser beam has the form

$$u(x,y,0) = x^2 e^{-(x^2+y^2)/w_0^2}. \tag{18.242}$$

   Write this beam both as a sum of Hermite–Gauss beams and as a sum of Laguerre–Gauss beams.
6. Suppose that in the plane $z = 0$ a laser beam has the form

$$u(x,y,0) = (x^2 + xy)e^{-(x^2+y^2)/w_0^2}. \tag{18.243}$$

   Write this beam both as a sum of Hermite–Gauss beams and as a sum of Laguerre–Gauss beams.

7. Derive the following integral representations of the Hermite polynomials:

$$H_{2n}(x) = \frac{2^{2n+1}(-1)^n e^{x^2}}{\sqrt{\pi}} \int_0^\infty e^{-t^2} t^{2n} \cos(2xt)dt,$$

$$H_{2n+1}(x) = \frac{2^{2n+2}(-1)^n e^{x^2}}{\sqrt{\pi}} \int_0^\infty e^{-t^2} t^{2n+1} \sin(2xt)dt.$$

(Hint: write e^{-x^2} in terms of its Fourier transform, symmetrize the integral, and apply the Rodrigues formula to both sides of the expression.)

8. Show that the results of the preceding problem can be combined into a single expression of the form

$$H_n(x) = \frac{2^{2n+2}(-i)^n e^{x^2}}{\sqrt{\pi}} \int_{-\infty}^\infty e^{-t^2 + 2itx} t^n dt.$$

9. Demonstrate that the Hermite functions are eigenfunctions of the Fourier transform, i.e.

$$e^{-x^2/2} H_n(x) = \frac{1}{i^n \sqrt{2\pi}} \int_{-\infty}^\infty e^{ixy} e^{-y^2/2} H_n(y)dy.$$

(The generating function will help here.)

10. Prove the orthogonality of the Laguerre polynomials, i.e.

$$\int_0^\infty L_m(x)L_n(x)dx = \delta_{mn}.$$

11. Determine the differential equation for the Chebyshev polynomials of the second kind,

$$(1-x^2)U_n'' - 3xU_n' + n(n+2)U_n = 0,$$

directly from the generating function

$$\frac{1}{1-2xt+t^2} = \sum_{n=0}^\infty U_n(x)t^n.$$

12. Determine the differential equation for the Chebyshev polynomials of the first kind,

$$(1-x^2)T_n'' - xT_n' + n^2 T_n = 0,$$

directly from the generating function

$$\frac{1-t^2}{1-2xt+t^2} = T_0(x) + 2\sum_{n=1}^\infty T_n(x)t^n.$$

13. Find an expression for the Wronskian of the two solutions to the Chebyshev equation of the first kind, $T_n(x)$ and $V_n(x)$.

14. Find an expression for the Wronskian of the two solutions to the Chebyshev equation of the second kind, $U_n(x)$ and $W_n(x)$.

15. Use the generating function to prove the following addition theorem for the Hermite polynomials

$$H_n(x\cos\alpha + y\sin\alpha) = n! \sum_{m=0}^{n} \frac{H_m(x)H_{n-m}(y)}{m!(n-m)!} \cos^m \alpha \sin^{n-m} \alpha.$$

16. Read the paper by V. N. Majahan, Zernike annular polynomials for imaging systems with annular pupils, *J. Opt. Soc. Am.* **71** (1981), 75–85. Describe how the derivation of the Zernike polynomials changes for an annular aperture, and directly compare the first few polynomials for a circular and annular aperture. For what applications might an annular aperture be a better choice than a circular one?

17. Read the paper by R. J. Noll, Zernike polynomials and atmospheric turbulence, *J. Opt. Soc. Am.* **66** (1976), 207–211. What aspect of an optical system are the Zernike polynomials being used to characterize? How does the form of Noll's polynomials differ from those described in this chapter? Explain Noll's numbering system for ordering of the modes.

18. We have said that the Zernike polynomials are balanced to have maximum intensity at the focal point. Let us suppose we have a system with a fixed amount of sixth-order coma, $\Phi^{(6)} = A\rho^5 \cos\theta$, and we can introduce a tunable amount of fourth-order coma $\Phi^{(4)} = B\rho^3 \cos\theta$ and shift $\Phi^{(2)} = C\rho\cos\theta$, where B and C are tunable. Determine the value of B and C which produces the largest normalized intensity by Eq. (18.239). (Look for the value of B such that $\partial I/\partial B = 0$; substitute back in and search for the value of C such that $\partial I/\partial C = 0$.) Compare with the polynomial $R_6^1(\rho)$.

19. We have said that the Zernike polynomials are balanced to have maximum intensity at the focal point. Let us suppose we have a system with a fixed amount of sixth-order spherical aberration, $\Phi^{(6)} = A\rho^6$, and we can introduce a tunable amount of fourth-order spherical aberration $\Phi^{(4)} = B\rho^4$, defocus $\Phi^{(2)} = C\rho^2$ and piston $\Phi^{(0)} = D$, where B, C, and D are tunable. Determine the values of B, C, and D which produce the largest normalized intensity by Eq. (18.239). (Look for the value of B such that $\partial I/\partial B = 0$; substitute back in and search for the value of C such that $\partial I/\partial C = 0$, and so on. Assume that $\overline{\Phi} = 0$.) Compare with the polynomial $R_6^0(\rho)$.

19

Green's functions

19.1 Introduction: the Huygens–Fresnel integral

We return once more to the diffraction of light by an aperture in a planar screen. In Sections 11.1 and 12.1, we derived the formulas of Fraunhofer diffraction and Fresnel diffraction by first assuming the Rayleigh–Sommerfeld solution to the diffraction problem; we now take the first step towards deriving that R-S solution.

Let us consider a monochromatic wave $U(\mathbf{r})$, with time dependence $\exp[-i\omega t]$. We have already seen that in free space the field $U(\mathbf{r})$ satisfies the Helmholtz equation,

$$(\nabla^2 + k^2)U = 0, \tag{19.1}$$

where $k = \omega/c$. We consider a volume V_0 bounded by a closed surface S_0, with *inward* unit normal $\mathbf{n}$, and a point P which lies within it. This is illustrated in Fig. 19.1 (a). It is assumed that U possesses continuous first and second-order partial derivatives within the volume. If U' is another function which satisfies the same continuity requirements, we have the following relation from Green's theorem, Eq. (2.82)

$$\int_{V_0} \left(U\nabla^2 U' - U'\nabla^2 U \right) d\tau = -\int_{S_0} \left(U\frac{\partial U'}{\partial n} - U'\frac{\partial U}{\partial n} \right) da, \tag{19.2}$$

where $\partial/\partial n$ denotes the derivative along the *inward* normal to S_0. If U' is also a solution of the Helmholtz equation, then the two terms on the left-hand size of Eq. (19.2) cancel and we are left with

$$\int_S \left(U\frac{\partial U'}{\partial n} - U'\frac{\partial U}{\partial n} \right) da = 0. \tag{19.3}$$

Let us choose U' to be a spherical wave, i.e.

$$U'(\mathbf{r}) = \frac{e^{ikR}}{R}, \tag{19.4}$$

where R denotes the distance from the point P to the point of integration $\mathbf{r}$. This provides us with a significant problem: the spherical wave is singular at the point P, and we have

665

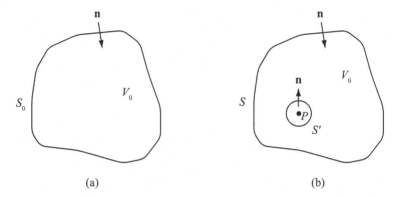

Figure 19.1 Illustrating (a) the initial volume of consideration, and (b) the volume used when a singularity exists at point P.

assumed in using Green's theorem that the fields are continuous within the volume. We resolve this by excluding a small sphere of radius ϵ and surface S' around the point P from the volume; this is illustrated in Fig. 19.1 (b). With this exclusion, the total surface S_0 of the volume consists of two disconnected surfaces, the exterior surface S and the interior surface S'. The integral over S_0 then becomes

$$\int_{S+S'} \left[U \frac{\partial}{\partial n}\left(\frac{e^{ikR}}{R} \right) - \frac{e^{ikR}}{R} \frac{\partial U}{\partial n} \right] da = 0. \tag{19.5}$$

Separating out the two contributions, we may write

$$\int_{S} \left[U \frac{\partial}{\partial n}\left(\frac{e^{ikR}}{R} \right) - \frac{e^{ikR}}{R} \frac{\partial U}{\partial n} \right] da = -\int_{S'} \left[U \frac{\partial}{\partial n}\left(\frac{e^{ikR}}{R} \right) - \frac{e^{ikR}}{R} \frac{\partial U}{\partial n} \right] da. \tag{19.6}$$

The integral on the right-hand side is now an integral over a spherical surface of constant $R = \epsilon$; we may rewrite it in spherical coordinates centered on the point P as

$$-\int_{S'} \left[U \frac{\partial}{\partial n}\left(\frac{e^{ikR}}{R} \right) - \frac{e^{ikR}}{R} \frac{\partial U}{\partial n} \right] da = -\int_{\Omega} \left[U \frac{e^{ik\epsilon}}{\epsilon} \left(ik - \frac{1}{\epsilon} \right) - \frac{e^{ik\epsilon}}{\epsilon} \frac{\partial U}{\partial s} \right] \epsilon^2 d\Omega, \tag{19.7}$$

where $d\Omega$ is an element of solid angle. The integral is in fact independent of ϵ, and since ϵ is an arbitrary radius, we may take the limit as $\epsilon \to 0$. The only term in the square brackets which does not vanish is the term proportional to $1/\epsilon^2$, and we may write

$$\int_{S} \left[U \frac{\partial}{\partial n}\left(\frac{e^{ikR}}{R} \right) - \frac{e^{ikR}}{R} \frac{\partial U}{\partial n} \right] da = 4\pi U(P). \tag{19.8}$$

We therefore have the result

$$U(P) = \frac{1}{4\pi} \int_S \left[U \frac{\partial}{\partial n} \left(\frac{e^{ikR}}{R} \right) - \frac{e^{ikR}}{R} \frac{\partial U}{\partial n} \right] da. \tag{19.9}$$

This equation allows us to determine the field at any point within a volume V from the value of the field on the boundary of that volume. It is one form of what is known as the *integral theorem of Helmholtz and Kirchoff* (H-K).

If the point P lies outside the volume V, we may still choose U' in the form of Eq. (19.4), but now there is no need to introduce the second surface S' into the calculation, which means the integral over S is identically zero. We may combine the two results into the form

$$\frac{1}{4\pi} \int_S \left[U \frac{\partial}{\partial n} \left(\frac{e^{ikR}}{R} \right) - \frac{e^{ikR}}{R} \frac{\partial U}{\partial n} \right] da = \begin{cases} U(P), & P \in V, \\ 0, & P \notin V. \end{cases} \tag{19.10}$$

We now attempt to use the H-K integral theorem to solve the explicit problem of diffraction by a plane screen. The formula (19.10) is valid for any field which satisfies the Helmholtz equation and presumably encompasses the problem of diffraction, although further assumptions are required to apply it to such a problem. It was Kirchoff who put the Helmholtz–Kirchoff theorem into a form which allows the calculation of diffraction effects at an aperture.

We consider a monochromatic wave impinging on an aperture in a plane opaque screen, and we let P be the point at which the field is to be determined. We choose the surface S to consist of three parts: (1) the aperture $\mathcal{A}$, (2) a large part of the unilluminated part of the screen $\mathcal{B}$, and (3) a portion $\mathcal{C}$ of a large sphere of radius Δ centered at P which, together with $\mathcal{A}$ and $\mathcal{B}$, form a closed surface. This is illustrated in Fig. 19.2. The Helmholtz–Kirchoff theorem therefore may be written as

$$U(P) = \frac{1}{4\pi} \int_{\mathcal{A}+\mathcal{B}+\mathcal{C}} \left[U \frac{\partial}{\partial n} \left(\frac{e^{ikR}}{R} \right) - \frac{e^{ikR}}{R} \frac{\partial U}{\partial n} \right] da. \tag{19.11}$$

This equation is in a more appropriate form for evaluating the field diffracted by an aperture. However, a significant problem remains, in that we do not know the values of the field on the surfaces $\mathcal{A}$, $\mathcal{B}$, and $\mathcal{C}$. Working from physical intuition, it is reasonable to *assume*, however, that the field and its derivative must be zero on the opaque screen $\mathcal{B}$. Furthermore, it is reasonable to assume that within the aperture, the field is roughly equal to the field $U^{(i)}$ illuminating the aperture. In mathematical terms, we write

$$\begin{aligned} \text{on } \mathcal{A}: \quad & U = U^{(i)}, \quad \frac{\partial U}{\partial n} = \frac{\partial U^{(i)}}{\partial n}, \\ \text{on } \mathcal{B}: \quad & U = 0, \quad \frac{\partial U}{\partial n} = 0. \end{aligned} \tag{19.12}$$

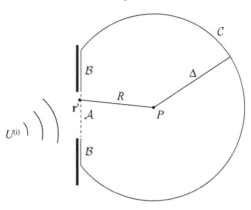

Figure 19.2 Illustrating the surface of choice for solving diffraction problems.

All that remains is to consider the effect of the spherical portion C on the integral. It should be clear that the field amplitude will approach zero as $\Delta \to \infty$, but this is not sufficient to neglect the contribution of the field on C, because the area of the surface also increases with Δ. There are two ways of handling this: one is to require that the solution satisfy the condition

$$\lim_{\Delta \to \infty} \Delta \left(\frac{\partial U}{\partial n} - ikU \right) = 0, \tag{19.13}$$

a requirement known as the *Sommerfeld radiation condition*. To understand this condition, we note that on the surface C the spherical wave in Eq. (19.11) takes on the simple form

$$\frac{e^{ikR}}{R} = \frac{e^{ik\Delta}}{\Delta}. \tag{19.14}$$

We also note that $\mathbf{n}$ is in the radial direction, so the normal derivatives in Eq. (19.11) are with respect to Δ, i.e.

$$\frac{\partial}{\partial \Delta} \frac{e^{ik\Delta}}{\Delta} = \left(ik - \frac{1}{\Delta} \right) \frac{e^{ik\Delta}}{\Delta} \approx ik \frac{e^{ik\Delta}}{\Delta}, \tag{19.15}$$

where in the last step we assume that $k\Delta \gg 1$. On substitution, the contribution for the surface C is given by

$$\frac{1}{4\pi} \int_C \left[U \frac{\partial}{\partial n} \left(\frac{e^{ikR}}{R} \right) - \frac{e^{ikR}}{R} \frac{\partial U}{\partial n} \right] da \approx \int_C \frac{e^{ik\Delta}}{\Delta} \left(ikU - \frac{\partial U}{\partial n} \right) \Delta^2 d\Omega. \tag{19.16}$$

The term in parenthesis, together with the Δ term, vanishes if the Sommerfeld radiation condition is satisfied. This condition therefore results in no contribution of C to the integral.

What does it mean? Let us assume that the field U on the spherical surface is well-approximated by the form

$$U(\mathbf{r}) = \frac{e^{\pm ik\Delta}}{\Delta}. \tag{19.17}$$

This is effectively assuming that the field far from the aperture has either the form of an outgoing spherical wave (+ sign) or the form of an incoming spherical wave (− sign). On substituting this form of the field into the radiation condition, we find

$$\frac{e^{\pm ik\Delta}}{\Delta}\Delta\left(\pm ik - \frac{1}{\Delta} - ik\right) = -e^{\pm ik\Delta}\left(ik(\mp 1 + 1) + \frac{1}{\Delta}\right). \tag{19.18}$$

In the limit $\Delta \to \infty$, the condition is only satisfied if the field takes on the form of an outgoing spherical wave. Essentially, the Sommerfeld radiation condition is the requirement that the field is behaving causally, i.e. propagating away from, not towards, the aperture.

There is another, less rigorous but more physical, way to justify the neglect of the spherical surface. Realistically, no field is truly monochromatic, and all fields have a finite starting point and take time to propagate any distance. We can imagine choosing the distance R large enough such that the field has not yet reached the surface, but has been effectively monochromatic over a large number of cycles near the aperture.

With the preceding conditions satisfied, our formula for the field at a point P takes on the form

$$U(P) = \frac{1}{4\pi}\int_A\left[U^{(i)}\frac{\partial}{\partial n}\left(\frac{e^{ikR}}{R}\right) - \frac{e^{ikR}}{R}\frac{\partial U^{(i)}}{\partial n}\right]da. \tag{19.19}$$

This formula is known as the *Fresnel–Kirchoff diffraction formula*. It describes how the field at a point P beyond the aperture depends on the field and its normal derivative which are incident on the aperture.

Equation (19.19) is significant because it allows one to express the diffracted wavefield as the superposition of a number of "elementary" sources, expressed by the behavior of the function given by Eq. (19.4), which is referred to as a *Green's function*. This is a special case of the method of Green's functions, which we discuss in detail in this chapter. A better understanding of the properties of Green's functions will allow us to derive the Rayleigh–Sommerfeld diffraction integrals from the Fresnel–Kirchoff integral derived here.

19.2 Inhomogeneous Sturm–Liouville equations

Green's functions can be derived for inhomogeneous ordinary differential equations as well as partial differential equations. The ODE case has the advantage of relative mathematical simplicity and highlights most of the relevant properties of all Green's functions.

We therefore begin by considering an inhomogeneous ordinary differential equation of the form

$$\frac{d}{dx}\left[p(x)\frac{dy}{dx}\right] - q(x)y = f(x),$$ (19.20)

with boundary conditions

$$\alpha y(a) + \beta y'(a) = 0,$$ (19.21)

$$\gamma y(b) + \delta y'(b) = 0$$ (19.22)

at $x = a$ and $x = b$. This differential equation is an inhomogeneous version of the *Sturm–Liouville equation*, discussed in Section 18.1. Such inhomogeneous equations appear quite regularly in mathematical physics; for example, in working with the Laplace transform in Section 15.7, we came across the following equation

$$U_{xx} - sU = -u(x,0),$$ (19.23)

which was Eq. (15.201), written in slightly different form. This equation is of the Sturm–Liouville form, with $p(x) = 1$, $q(x) = s$, and $f(x) = -u(x,0)$. We originally found the solution to this problem by finding the homogeneous solutions of the equation, calculating the Wronskian, using variation of constants to determine the result, and then matching the boundary conditions.

It would be nice, however, if the solution to Eq. (19.20) could be written as a single integral of the inhomogeneous term with a function which depends only on the properties of the differential equation, i.e.

$$y(x) = \int_a^b G(x,t)f(t)dt,$$ (19.24)

and the solution automatically satisfies the boundary conditions at $x = a, x = b$. The function $G(x,t)$ is called the *Green's function* of the differential equation. What properties must this function have? If we define an *operator* $\mathcal{L}$ by

$$\mathcal{L} \equiv \frac{d}{dx}\left[p(x)\frac{d}{dx}\right] - q(x),$$ (19.25)

such that

$$\mathcal{L}y(x) = \frac{d}{dx}\left[p(x)\frac{dy}{dx}\right] - q(x)y,$$ (19.26)

we may let this operator act on both sides of Equation (19.24),

$$\mathcal{L}y(x) = \frac{d}{dx}\left[p(x)\frac{dy}{dx}\right] - q(x)y = f(x) = \mathcal{L}\int_a^b G(x,t)f(t)dt = \int_a^b [\mathcal{L}G(x,t)]f(t)dt.$$

(19.27)

In the last step we interchanged the orders of integration and $\mathcal{L}$; it is assumed, though not proven here, that such an interchange is valid. From the above equation, we have the requirement that

$$f(x) = \int_a^b [\mathcal{L}G(x,t)]f(t)dt, \tag{19.28}$$

which can evidently only be satisfied if

$$\mathcal{L}G(x,t) = \frac{d}{dx}\left[p(x)\frac{dG(x,t)}{dx}\right] - q(x)G(x,t) = \delta(x-t), \tag{19.29}$$

where $\delta(x-t)$ is the Dirac delta function! Furthermore, to have the solution match the boundary conditions, it is clear that we require

$$\alpha G(a,t) + \beta G'(a,t) = 0,$$
$$\gamma G(b,t) + \delta G'(b,t) = 0. \tag{19.30}$$

The question which arises, of course, is whether or not such a function $G(x,t)$ exists. We may attempt to find it, however, by deducing properties of the function from Eqs. (19.29) and (19.30). First, we have already noted that $G(x,t)$ must satisfy the boundary conditions. Second, we note that for $x \neq t$, $G(x,t)$ satisfies the *homogeneous* differential equation. Apparently for $x \neq t$ the Green's function can be constructed from solutions to the homogeneous equation. Third, it is reasonable to assume that $G(x,t)$ must be continuous everywhere in the interval $a \leq x \leq b$; in particular, it should be continuous across the point $x = t$:

$$\lim_{x \to t_-} G(x,t) = \lim_{x \to t_+} G(x,t). \tag{19.31}$$

Finally, let us integrate Eq. (19.29) from $x = t - \epsilon$ to $x = t + \epsilon$, and let $\epsilon \to 0$. From the property of the delta function, we find that

$$\lim_{x \to t_+}\frac{dG(x,t)}{dx} - \lim_{x \to t_-}\frac{dG(x,t)}{dx} = -\frac{1}{p(t)}. \tag{19.32}$$

Therefore the derivative of the Green's function is discontinuous at the point t.

From these observations, we come up with the following rules for constructing $G(x,t)$.

1. We let $G(x,t) = G_1(x)$ for $a \leq x < t$ and $G(x,t) = G_2(x)$ for $t < x \leq b$.
2. The functions $G_1(x)$ and $G_2(x)$ must be solutions to the homogeneous differential equation, i.e. $\mathcal{L}G_1(x) = 0$, $\mathcal{L}G_2(x) = 0$.
3. The function $G_1(x)$ satisfies the boundary condition $\alpha G_1(a) + \beta G_1'(a) = 0$, and the function $G_2(x)$ satisfies the boundary condition $\gamma G_2(b) + \delta G_2'(b) = 0$.
4. The function $G(x,t)$ must be continuous, or $\lim_{x \to t_-} G_1(x) = \lim_{x \to t_+} G_2(x)$.

5. The derivative of $G(x,t)$ must be discontinuous, such that

$$\lim_{x \to t_+} \frac{dG_2(x)}{dx} - \lim_{x \to t_-} \frac{dG_1(x)}{dx} = -\frac{1}{p(t)}. \tag{19.33}$$

If we apply the above rules (in order), we should be able to find the Green's function for the problem.

The best way to understand what we are doing here is to consider an example.

Example 19.1 We want to find the Green's function for the equation

$$y'' - k^2 y = f(x), \tag{19.34}$$

in the region $0 \le x \le L$, subject to the boundary conditions $y(0) = 0$, $y(L) = 0$. To do so, we follow the steps described above for constructing the Green's function.

Steps 1 and 2. The Green's function is constructed from solutions to the homogeneous differential equation

$$y'' - k^2 y = 0. \tag{19.35}$$

The solutions to this equation are $y_1(x) = e^{-kx}$, $y_2(x) = e^{+kx}$. We therefore let

$$G_1(x) = A_1 e^{-kx} + B_1 e^{kx}, \tag{19.36}$$

$$G_2(x) = A_2 e^{-kx} + B_2 e^{kx}, \tag{19.37}$$

where the constants A_i and B_i will be determined by the next steps. Our Green's function can be written, so far, as

$$G(x,t) = \begin{cases} A_1 e^{-kx} + B_1 e^{kx}, & \text{for } 0 \le x < t, \\ A_2 e^{-kx} + B_2 e^{kx}, & \text{for } t < x \le L. \end{cases} \tag{19.38}$$

Step 3. The functions $G_1(x)$ and $G_2(x)$ must satisfy the boundary conditions, i.e. $G_1(0) = 0$, $G_2(L) = 0$. The first condition gives us

$$A_1 + B_1 = 0, \tag{19.39}$$

or let $A_1 \equiv A$, $B_1 = -A$. The second boundary condition gives us

$$A_2 e^{-kL} + B_2 e^{kL} = 0, \tag{19.40}$$

or, letting $A_2 \equiv B$, $B_2 = -Be^{-2kL}$. We may rewrite our Green's function in the form

$$G(x,t) = \begin{cases} A e^{-kx} - A e^{kx} = -2A\sinh(kx), & \text{for } 0 \le x < t, \\ Be^{-kx} - Be^{kx-2kL} = -2Be^{-kL}\sinh[kx - kL], & \text{for } t < x \le L. \end{cases} \tag{19.41}$$

Step 4. Now we require that at $x = t$, the Green's function is continuous. This is the requirement that $G_1(t) = G_2(t)$, or that

$$-2A \sinh kt = -2Be^{-kL} \sinh[kt - kL]. \tag{19.42}$$

We therefore have the requirement that

$$B = Ae^{kL} \frac{\sinh kt}{\sinh[kt - kL]}. \tag{19.43}$$

The Green's function may now be written as

$$G(x,t) = \begin{cases} -2A \sinh(kx), & \text{for } 0 \le x < t, \\ -2A \sinh kt \frac{\sinh[kx-kL]}{\sinh[kt-kL]}, & \text{for } t < x \le L. \end{cases} \tag{19.44}$$

Step 5. Finally, we attempt to match the discontinuity condition, Eq. (19.33). In this case, $p(x) = 1$, so our condition becomes

$$\lim_{x \to t_+} \frac{dG_2(x)}{dx} - \lim_{x \to t_-} \frac{dG_1(x)}{dx} = -2Ak \sinh kt \frac{\cosh[kt - kL]}{\sinh[kt - kL]} + 2Ak \cosh(kt) = -1. \tag{19.45}$$

Solving for A, we find that

$$A = \frac{1}{2k} \frac{1}{\frac{\sinh(kt)\cosh(kt-kL)}{\sinh(kt-kL)} - \cosh(kt)}. \tag{19.46}$$

If we substitute this into our expression for the Green's function, and use the identity

$$\sinh u \cosh v - \sinh v \cosh u = \sinh(u - v), \tag{19.47}$$

we find that

$$G(x,t) = \begin{cases} -\frac{1}{k} \frac{\sinh(kx)\sinh(kt-kL)}{\sinh(kL)}, & \text{for } 0 \le x < t, \\ -\frac{1}{k} \frac{\sinh(kt)\sinh(kx-kL)}{\sinh(kL)}, & \text{for } t < x \le L. \end{cases} \tag{19.48}$$

We therefore have our Green's function! To remind the reader, that means that the solution to the inhomogeneous equation $y'' - k^2y = f(x)$, for any $f(x)$, is given by

$$\int_0^L G(x,t)f(t)dt, \tag{19.49}$$

with $G(x,t)$ given by Eq. (19.48). It is obvious from this equation that the Green's function satisfies the boundary conditions, is continuous, and satisfies the homogeneous equation everywhere except $x = t$.

◇

A few comments about such solutions are worth mentioning. First, it is to be noted that the Green's function may be written in the form

$$G(x,t) = \begin{cases} u(x)v(t), & \text{for } 0 \le x < t, \\ v(x)u(t), & \text{for } t < x \le L. \end{cases} \tag{19.50}$$

There is therefore a reciprocity in the Green's function such that $G(x,t) = G(t,x)$. Physically this can be interpreted as the fact that a "cause" at t produces the same "effect" at x as a cause at x produces at t. This is a general property of Green's functions and does not just hold for our particular example. It is also to be noted that there exist some cases for which the Green's function does not exist: if one solution of the homogeneous equation satisfies *both* boundary conditions simultaneously, we cannot find the Green's function. It is possible to define a "generalized" Green's function in such circumstances, but we will not discuss this here. Finally, it is worth noting that the Green's function depends on the differential equation and the boundary conditions, but not the inhomogeneous term. It is a fundamental property of the background system, without source terms.

19.3 Properties of Green's functions

We may extend the concept of a Green's function to linear inhomogeneous partial differential equations in N variables of the form[1]

$$\mathcal{L}u(\mathbf{r}') = -4\pi\rho(\mathbf{r}'), \tag{19.51}$$

where $\mathcal{L}$ is a differential operator and $\mathbf{r}'$ represents the N variables of the PDE; for the moment we restrict ourselves to elliptic equations, for which all variables are spatial variables. This equation is defined on an N-dimensional domain D with boundary B; an appropriate set of boundary conditions must be specified, to be discussed below.

Examples of the operator $\mathcal{L}$ include the Helmholtz operator,

$$\mathcal{L} = \nabla^2 + k^2, \tag{19.52}$$

and the Laplacian,

$$\mathcal{L} = \nabla^2. \tag{19.53}$$

We assume that a Green's function $G(\mathbf{r},\mathbf{r}')$ exists for this operator which satisfies the equation

$$\mathcal{L}G(\mathbf{r},\mathbf{r}') = -4\pi\delta(\mathbf{r}-\mathbf{r}'), \tag{19.54}$$

---

[1] It is to be noted that the factor of 4π (representing total solid angle) is a convenient one for three-dimensional problems that eliminates such factors in the formula for the Green's function itself. For consistency, we keep this factor for lower-dimensional problems, though it is common to use a factor of 2π for 2-D problems and a factor of unity for 1-D problems.

where δ is a delta function in N variables, and the operator $\mathcal{L}$ is understood to act on the variable $\mathbf{r}'$. The factor of 4π is chosen to simplify the form of the Green's function in three-dimensional problems, and only changes the Green's function by a constant. Provided this function exists, we derive an integral equation from it by multiplying Eq. (19.51) by $G(\mathbf{r},\mathbf{r}')$, Eq. (19.54) by $u(\mathbf{r}')$, subtracting the resulting equations, and integrating $\mathbf{r}'$ over the domain D,

$$\int_D [G(\mathbf{r},\mathbf{r}')\mathcal{L}u(\mathbf{r}') - u(\mathbf{r}')\mathcal{L}G(\mathbf{r},\mathbf{r}')] \mathrm{d}^N r'$$

$$= -4\pi \int_D G(\mathbf{r},\mathbf{r}')\rho(\mathbf{r}') \mathrm{d}^N r' + 4\pi \int_D \delta(\mathbf{r}-\mathbf{r}')u(\mathbf{r}') \mathrm{d}^N r'. \qquad (19.55)$$

Using the sifting property of the delta function, we may write this as

$$u(\mathbf{r}) = \int_D G(\mathbf{r},\mathbf{r}')\rho(\mathbf{r}') \mathrm{d}^N r' + \frac{1}{4\pi} \int_D [G(\mathbf{r},\mathbf{r}')\mathcal{L}u(\mathbf{r}') - u(\mathbf{r}')\mathcal{L}G(\mathbf{r}',\mathbf{r})] \mathrm{d}^N r'. \qquad (19.56)$$

This equation is comparable to the integral equation of Helmholtz and Kirchoff, Eq. (19.19), except for the absence of a source term $\rho(\mathbf{r}')$ in the Helmholtz–Kirchoff equation.

It is difficult to make statements about Eq. (19.56) in general; we consider as a special case an operator analogous to the Sturm–Liouville operators of ordinary differential equations,

$$\mathcal{L} = \nabla \cdot [p(\mathbf{r})\nabla] + q(\mathbf{r}). \qquad (19.57)$$

With this operator, the second integral of Eq. (19.56) becomes

$$\frac{1}{4\pi} \int_D [G(\mathbf{r},\mathbf{r}')\mathcal{L}u(\mathbf{r}') - u(\mathbf{r}')\mathcal{L}G(\mathbf{r}',\mathbf{r})] \mathrm{d}^N r' = \frac{1}{4\pi} \int_D \{G\nabla \cdot [p\nabla u] - u\nabla \cdot [p\nabla G]\} \mathrm{d}^N r'. \qquad (19.58)$$

Noting that

$$\nabla \cdot [Gp\nabla u] = p\nabla G \cdot \nabla u + G\nabla \cdot (p\nabla u), \qquad (19.59)$$

$$\nabla \cdot [up\nabla G] = p\nabla u \cdot \nabla G + u\nabla \cdot (p\nabla G), \qquad (19.60)$$

We may use this to simplify the second integral to the form

$$\frac{1}{4\pi} \int_D [G(\mathbf{r},\mathbf{r}')\mathcal{L}u(\mathbf{r}') - u(\mathbf{r}')\mathcal{L}G(\mathbf{r}',\mathbf{r})] \mathrm{d}^N r' = \frac{1}{4\pi} \int_D \nabla \cdot [Gp\nabla u - up\nabla G] \mathrm{d}^N r'. \qquad (19.61)$$

With this expression, we may use an N-dimensional generalization of the divergence theorem, Eq. (2.77), which may be written as

$$\int_D \nabla \cdot \mathbf{F} \mathrm{d}^N r' = \oint_B \mathbf{F} \cdot \mathrm{d}\mathbf{S}, \qquad (19.62)$$

where d**S** is an infinitesimal boundary element. We may finally write

$$u(\mathbf{r}) = \int_D G(\mathbf{r}, \mathbf{r}') \rho(\mathbf{r}') \mathrm{d}^N r' + \frac{1}{4\pi} \oint_B p[G\nabla u - u\nabla G] \cdot \mathrm{d}\mathbf{S}. \tag{19.63}$$

This expression suggests that a unique solution to Eq. (19.51) depends not only upon the value of the source term, but also upon the value of u and $\frac{\partial u}{\partial n}$ upon the boundary of the domain. However, this seems to be in conflict with our observation in Section 15.2.3 that an elliptic PDE is uniquely specified by either the value of u or the value of $\frac{\partial u}{\partial n}$ on the boundary, and is overspecified with both conditions.

This problem is resolved by observing that the Green's function is itself not uniquely specified by Eq. (19.54); if $G(\mathbf{r}, \mathbf{r}')$ is a solution of Eq. (19.54), and $H(\mathbf{r}')$ is a solution of the homogeneous equation

$$\mathcal{L}H(\mathbf{r}') = 0, \tag{19.64}$$

then $G(\mathbf{r}, \mathbf{r}') + H(\mathbf{r}')$ is also a valid solution of Eq. (19.54). It is possible, then, to choose the Green's function in such a manner that only one boundary condition of u is needed to specify the solution. We will see examples of this momentarily.

19.4 Green's functions of second-order PDEs

Unfortunately, simple expressions for Green's functions are only derivable for certain special cases, typically systems with a high degree of symmetry. In this section we derive expressions for some of these, and discuss their properties.

19.4.1 Elliptic equation: Poisson's equation

We consider first the Green's function for Poisson's equation in unbounded three-dimensional space, which must satisfy

$$\nabla^2 G(\mathbf{r}, \mathbf{r}') = -4\pi \delta(\mathbf{r} - \mathbf{r}'), \tag{19.65}$$

where ∇ is assumed to act on $\mathbf{r}'$.

Because the delta function is rotationally symmetric around the central point $\mathbf{r}$ and the domain unbounded, it follows that the Green's function itself must be rotationally symmetric about this point, i.e.

$$G(\mathbf{r}, \mathbf{r}') = G(|\mathbf{r} - \mathbf{r}'|). \tag{19.66}$$

Without loss of generality, we may take $\mathbf{r} = \mathbf{0}$. In spherical coordinates, the Green's function will depend only upon r, and we may write Eq. (19.65) in the form

$$\frac{1}{r'^2} \frac{\mathrm{d}}{\mathrm{d}r'} \left(r'^2 \frac{\mathrm{d}G(r')}{\mathrm{d}r'} \right) = -4\pi \delta(\mathbf{r}'). \tag{19.67}$$

Let us integrate this equation over a spherical volume of radius r. We readily find that

$$r^2 \frac{dG}{dr} = -1. \tag{19.68}$$

By direct integration, we find that

$$G_{P,3D}(\mathbf{r} - \mathbf{r}') = \frac{1}{|\mathbf{r} - \mathbf{r}'|}. \tag{19.69}$$

This is exactly what we would expect from the discussion of Gauss' law in Section 6.1; our solution automatically satisfies the boundary condition at infinity. In electrostatics, the Green's function represents the electric potential produced by a point charge at position $\mathbf{r}'$.

We may make a similar analysis of Poisson's equation in unbounded two-dimensional space. Assuming polar symmetry, we may write

$$\frac{1}{\rho'} \frac{d}{d\rho'} \left(\rho' \frac{dG(\rho')}{d\rho'} \right) = -4\pi \delta(\rho'). \tag{19.70}$$

Integrating over a circle of radius ρ, we find that

$$\rho \frac{dG}{d\rho} = -2. \tag{19.71}$$

On integration, we readily find that

$$G_{P,2D}(|\rho - \rho'|) = -2\log(|\rho - \rho'|). \tag{19.72}$$

19.4.2 Elliptic equation: Helmholtz equation

We now consider the Green's function for the Helmholtz equation in unbounded three-dimensional space, which satisfies the equation

$$(\nabla^2 + k^2)G(\mathbf{r}, \mathbf{r}') = -4\pi \delta(\mathbf{r} - \mathbf{r}'), \tag{19.73}$$

and ∇ is again assumed to act on $\mathbf{r}'$. We again assume that the solution is rotationally symmetric and consider the case $\mathbf{r} = 0$. We may then write

$$\frac{1}{r'^2} \frac{d}{dr'} \left(r'^2 \frac{dG(r')}{dr'} \right) + k^2 G(r') = -4\pi \delta(\mathbf{r}'). \tag{19.74}$$

This formula may be written in the alternative form

$$\frac{1}{r'} \frac{d^2}{dr'^2} [r'G(r')] + k^2 G(r') = -4\pi \delta(\mathbf{r}'). \tag{19.75}$$

We may solve this in a manner analogous to that used to solve inhomogeneous Sturm–Liouville problems. First, we note that, for points $r' \neq 0$, we have

$$\frac{d^2}{dr'^2}[r'G(r')] + k^2[r'G(r')] = 0. \tag{19.76}$$

From this, we see that the function $r'G(r')$ satisfies the harmonic oscillator equation, so that

$$G(r) = A_+ \frac{e^{ikr}}{r} + A_- \frac{e^{-ikr}}{r}. \tag{19.77}$$

We may remove one of the constants by physical considerations. Supposing that the solution to the Helmholtz equation represents monochromatic waves with time dependence, $e^{-i\omega t}$, we find that the A_+ term represents a wave outgoing from the origin with phase $kr - \omega t$, while the A_- term represents a wave incoming to the origin from infinity with phase $kr + \omega t$. For radiation problems, only the first term represents a physical wave emanating from the origin, so we restrict ourselves to a Green's function of the form

$$G(r) = A_+ \frac{e^{ikr}}{r}. \tag{19.78}$$

The constant A_+ may be deduced by integrating Eq. (19.74) over a very small volume of radius r. In this limit, the term with k^2 is negligible, and we may write

$$\lim_{r \to 0} r^2 \frac{dG}{dr} = -1. \tag{19.79}$$

On substitution, we find that $A_+ = 1$, and we then have

$$G(|\mathbf{r} - \mathbf{r}'|) = \frac{e^{ik|\mathbf{r}-\mathbf{r}'|}}{|\mathbf{r}-\mathbf{r}'|}. \tag{19.80}$$

The Green's function for the two-dimensional Helmholtz equation may be found by similar reasoning. It must satisfy the equation

$$\frac{1}{\rho'}\frac{d}{d\rho'}\left(\rho'\frac{dG(\rho')}{d\rho'}\right) + k^2 G(\rho') = -4\pi\delta(\rho'). \tag{19.81}$$

For $\rho' \neq 0$, the Green's function satisfies the homogeneous equation

$$G'' + \frac{1}{\rho}G' + k^2 G = 0. \tag{19.82}$$

From Eq. (16.12), we see that this is simply Bessel's equation of order 0, whose solutions may be written in terms of the Hankel functions as

$$G(\rho) = A_+ H_0^{(1)}(k\rho) + A_- H_0^{(2)}(k\rho). \tag{19.83}$$

We may again eliminate one of the terms by physical considerations. From Section 16.8, we know that for large $k\rho$ the Hankel functions have the asymptotic forms

$$H_0^{(1)}(k\rho) \sim \sqrt{\frac{2}{\pi k\rho}} \exp[i(k\rho - \pi/4)], \tag{19.84}$$

$$H_0^{(2)}(k\rho) \sim \sqrt{\frac{2}{\pi k\rho}} \exp[-i(k\rho - \pi/4)]. \tag{19.85}$$

Assuming these represent monochromatic waves with time dependence $\exp[-i\omega t]$, only the $H_0^{(1)}$ term has the behavior of a circular wave outgoing from the origin. The Green's function therefore has the form

$$G(\rho) = A_+ H_0^{(1)}(k\rho). \tag{19.86}$$

The constant may be determined by integrating Eq. (19.81) over a circular region of small radius ρ. For sufficiently small ρ, the k^2 term is negligible and we may write

$$\lim_{\rho \to 0} \rho \frac{dG(\rho)}{d\rho} = -2. \tag{19.87}$$

From the series expansions of the Bessel and Neumann functions, Eqs. (16.20) and (16.26), we see that we may approximate

$$N_0(x) \approx \frac{2}{\pi} \log(x/2) J_0(x). \tag{19.88}$$

Only the derivative of the logarithmic term gives a nonzero value in the limit $\rho \to 0$; we therefore find that $A_+ = i\pi$ and

$$G(\rho) = i\pi H_0^{(1)}(k\rho). \tag{19.89}$$

Several comments are in order. First, it is to be noted that our requirement that the Green's function represent an outgoing wave is equivalent to imposing the Sommerfeld radiation condition on the solution. Second, it is to be noted that the dimensional units of the Green's function depend upon the dimension of the space it is introduced in and the second-order nature of the operator. A three-dimensional Green's function must therefore have dimensions of inverse length, while a two-dimensional Green's function is dimensionless.

A Green's function may also be found for the one-dimensional Helmholtz equation on an infinite domain, which satisfies

$$\frac{d^2 G}{dx^2} + k^2 G = -4\pi \delta(x - x'). \tag{19.90}$$

This equation may be solved in the usual Sturm–Liouville manner, applying the boundary condition that the waves must be outgoing from the source point x. The result is

$$G(x - x') = -\frac{2\pi i}{k} \exp[ik|x - x'|].$$ (19.91)

19.4.3 Hyperbolic equation: wave equation

In this section we determine the Green's function for the wave equation in unbounded, three-dimensional space, of the form

$$\nabla^2 G(\mathbf{r}, \mathbf{r}', t - t') - \frac{1}{c^2} \frac{\partial^2}{\partial t'^2} G(\mathbf{r}, \mathbf{r}', t - t') = -4\pi \delta(\mathbf{r} - \mathbf{r}')\delta(t - t'),$$ (19.92)

where again the derivatives are with respect to the primed coordinates. We have assumed that the Green's function depends only upon the time delay $t - t'$, which is equivalent to assuming that the Green's function is independent of the origin of time.

We may take the Fourier transform of this equation with respect to time, i.e.

$$\tilde{G}(\mathbf{r}, \mathbf{r}', \omega) = \frac{1}{2\pi} \int_{-\infty}^{\infty} G(\mathbf{r}, \mathbf{r}', t) e^{i\omega t} dt,$$ (19.93)

and find that

$$\nabla^2 \tilde{G}(\mathbf{r}, \mathbf{r}', \omega) + k^2 \tilde{G}(\mathbf{r}, \mathbf{r}', \omega) = -2\delta(\mathbf{r} - \mathbf{r}'),$$ (19.94)

with $k = \omega/c$. This equation is directly related to the defining equation for the Green's function of the Helmholtz equation, Eq. (19.73); the solution is of the form

$$\tilde{G}(\mathbf{r}, \mathbf{r}', \omega) = \frac{1}{2\pi} \frac{e^{ik|\mathbf{r} - \mathbf{r}'|}}{|\mathbf{r} - \mathbf{r}'|}.$$ (19.95)

We may then take the inverse Fourier transform of this function, i.e.

$$G(\mathbf{r}, \mathbf{r}', \tau) = \int_{-\infty}^{\infty} \tilde{G}(\mathbf{r}, \mathbf{r}', \omega) e^{-i\omega \tau} d\omega,$$ (19.96)

and find that

$$G(\mathbf{r}, \mathbf{r}', t - t') = \frac{\delta(t - t' - |\mathbf{r} - \mathbf{r}'|/c)}{|\mathbf{r} - \mathbf{r}'|}.$$ (19.97)

This is the Green's function for the wave equation. The presence of the delta function suggests that a disturbance felt at time t and position $\mathbf{r}$ is the result of a source at position $\mathbf{r}'$ radiating at an earlier time $t' = t - |\mathbf{r} - \mathbf{r}'|/c$. Equation (19.97) is known as the *retarded*

Green's function, and is the one which exhibits causal behavior. We may also define an *advanced* Green's function of the form

$$G(\mathbf{r},\mathbf{r}',t-t') = \frac{\delta(t-t'+|\mathbf{r}-\mathbf{r}'|/c)}{|\mathbf{r}-\mathbf{r}'|},\tag{19.98}$$

which comes from the advanced form of the Helmholtz equation Green's function $\exp[-ikR]/R$.

A similar strategy may be used to derive the Green's function for the two-dimensional wave equation, but the math is much more difficult and the results somewhat counter-intuitive. We start with the Green's function for the Helmholtz equation in two dimensions,

$$G(\boldsymbol{\rho},\boldsymbol{\rho}',\omega) = i\pi H_0^{(1)}(k|\boldsymbol{\rho}-\boldsymbol{\rho}'|),\tag{19.99}$$

where $H_0^{(1)}$ is the Hankel function of the first kind and zeroth order, defined in Section 16.6. To get the Green's function of the two-dimensional wave equation, we consider an inverse Fourier transform of this Green's function with respect to time, i.e.

$$G(\boldsymbol{\rho},\boldsymbol{\rho}',\tau) = \int_{-\infty}^{\infty} \tilde{G}(\boldsymbol{\rho},\boldsymbol{\rho}',\omega)e^{-i\omega\tau}\,d\omega.\tag{19.100}$$

This Fourier transform is not easily determined. We may evaluate it by using an integral representation of the Hankel function of the form

$$H_0^{(1)}(x) = -\frac{2i}{\pi}\int_0^{\infty} e^{ix\cosh\alpha}\,d\alpha,\tag{19.101}$$

which is valid for $x > 0$. Letting $|\boldsymbol{\rho}-\boldsymbol{\rho}'| = \rho$, we may readily find that

$$G(\rho,\omega) = 2\int_0^{\infty} e^{i\omega\rho\cosh[\alpha]/c}\,d\alpha.\tag{19.102}$$

The Fourier transform of this function results in the expression

$$G(\rho,\tau) = 4\pi\int_0^{\infty}\delta\left(\tau-\frac{\rho}{c}\cosh\alpha\right)d\alpha.\tag{19.103}$$

This integral may be evaluated using the composition property of the delta function, dis-cussed in Section 6.3.3. Before doing so, however, it is to be noted that the integrand is only nonzero over the range $0 \le x < \infty$ if $\tau - \frac{\rho}{c}\cosh\alpha > 0$ for some value of alpha. Since $\cosh\alpha \ge 1$, this produces the result that

$$G(\rho,\tau) = 0 \text{ for } \rho > ct.\tag{19.104}$$

We may multiply our Green's function by the Heaviside step function $S(ct-\rho)$ without modifying its value,

$$G(\rho,\tau) = 4\pi S(ct-\rho)\int_0^{\infty}\delta\left(\tau-\frac{\rho}{c}\cosh\alpha\right)d\alpha.\tag{19.105}$$

We now use the composition property of the delta function, i.e.

$$\delta[g(x)] = \sum_{a=\text{zeros of } g} \frac{\delta(x-a)}{|g'(a)|}, \tag{19.106}$$

where the only zero of the function is at the point $\alpha_0 = \cosh^{-1}[c\tau/\rho]$. We may write

$$\delta\left(\tau - \frac{\rho}{c}\cosh\alpha\right) = \frac{c}{\rho\sinh[\alpha_0]}\delta(\alpha - \alpha_0). \tag{19.107}$$

Equation (19.105) simplifies to the form

$$G(\rho,\tau) = \frac{4\pi c}{\rho}\frac{1}{\sinh[\cosh^{-1}(c\tau/\rho)]}S(ct - \rho). \tag{19.108}$$

Furthermore, using

$$\sinh[\cosh^{-1}(x)] = \sqrt{x^2 - 1}, \tag{19.109}$$

we may write

$$G(\rho,\tau) = \frac{4\pi c}{\sqrt{[ct]^2 - \rho^2}}S(ct - \rho). \tag{19.110}$$

Writing in terms of the appropriate difference variables, we have

$$G(\boldsymbol{\rho} - \boldsymbol{\rho}', t - t') = \frac{4\pi c}{\sqrt{c^2(t-t')^2 - |\boldsymbol{\rho} - \boldsymbol{\rho}'|^2}}S\left[c(t - t') - \sqrt{|\boldsymbol{\rho} - \boldsymbol{\rho}'|^2}\right]. \tag{19.111}$$

This result is rather surprising. The function $G(\rho,\tau)$ may be interpreted as the wave produced by a delta function impulse occurring at time t' and position $\boldsymbol{\rho}'$, but unlike the three-dimensional case the response is not delta-like in time. This is illustrated in Fig. 19.3, where it is assumed that an excitation of duration Δt occurs at the origin. The three-dimensional excitation produces a localized spherical wave which spreads from the origin, while the two-dimensional excitation produces a "tail" which gradually decays as time progresses. The two-dimensional excitation is still causal, however, as no wave amplitude travels faster than c from the origin.

19.4.4 Parabolic equation: diffusion equation

Here we determine the Green's function for the diffusion equation in unbounded three-dimensional space, which satisfies

$$\nabla^2 G(\mathbf{r} - \mathbf{r}', t - t') - \alpha^2\frac{\partial}{\partial t'}G(\mathbf{r} - \mathbf{r}', t - t') = -4\pi\delta(\mathbf{r} - \mathbf{r}')\delta(t - t'). \tag{19.112}$$

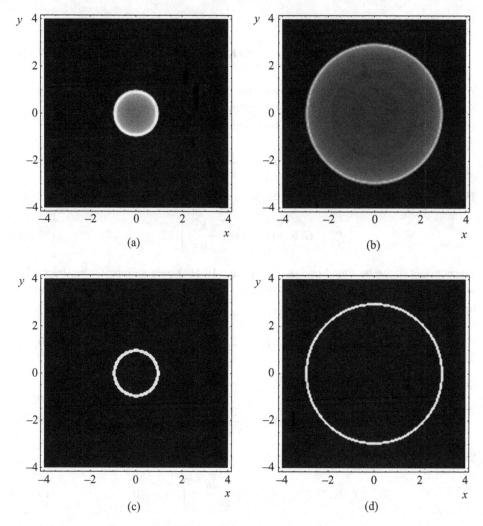

Figure 19.3 The amplitude of the Green's function of the wave equation in two dimensions and three dimensions, with $c = 1$, $\Delta t = 0.1$. (a) Two-dimensional, with $t = 1$, (b) two-dimensional, with $t = 3$, (c) three-dimensional (in the plane $z = 0$), with $t = 1$, (d) three-dimensional, with $t = 3$.

In principle, we can solve for the Green's function by taking a Fourier or Laplace transform with respect to time; however, this method results in complicated mathematics similar to that used in the example of Section 15.7. An easier approach is to take the *spatial* Fourier transform of the equation,

$$\tilde{G}(\mathbf{K}, \tau) = \frac{1}{(2\pi)^3} \int \int \int_{-\infty}^{\infty} G(\mathbf{R}, \tau) e^{-i\mathbf{K}\cdot\mathbf{R}} d^3 R. \qquad (19.113)$$

This results in the first-order ODE

$$-K^2\tilde{G}(\mathbf{K},\tau) - \alpha^2\frac{\partial}{\partial\tau}\tilde{G}(\mathbf{K},\tau) = -\frac{2}{(2\pi)^2}\delta(\tau). \qquad (19.114)$$

We now have a one-variable inhomogeneous problem that can be solved similarly to a Sturm–Liouville problem. Assuming the solution is causal, we expect $\tilde{G}(\mathbf{K},\tau) = 0$ for $\tau < 0$; for $\tau > 0$, we need to solve the inhomogeneous equation

$$-K^2\tilde{G}(\mathbf{K},\tau) - \alpha^2\frac{\partial}{\partial\tau}\tilde{G}(\mathbf{K},\tau) = 0. \qquad (19.115)$$

This first-order equation is simply the equation of exponential decay, with solution

$$\tilde{G}(\mathbf{K},\tau) = A\exp\left[-\frac{K^2}{\alpha^2}t\right]. \qquad (19.116)$$

The constant A can be found by integrating Eq. (19.114) over an infinitesmal interval of width 2ϵ centered on the point $\tau = 0$. Assuming that the function $\tilde{G}$ itself is continuous, we have the requirement

$$\tilde{G}(\mathbf{K},+\epsilon) - \tilde{G}(\mathbf{K},-\epsilon) = \frac{2}{(2\pi)^2\alpha^2}, \qquad (19.117)$$

Since $\tilde{G}(\mathbf{K},-\epsilon) = 0$, this implies that

$$A = \frac{2}{(2\pi)^2\alpha^2}. \qquad (19.118)$$

We therefore have

$$\tilde{G}(\mathbf{K},\tau) = \frac{2}{(2\pi)^2\alpha^2}\exp\left[-\frac{K^2}{\alpha^2}\tau\right], \quad \tau > 0. \qquad (19.119)$$

We may now take the inverse spatial Fourier transform,

$$G(\mathbf{R},\tau) = \int\int\int_{-\infty}^{\infty}\tilde{G}(\mathbf{K},\tau)e^{i\mathbf{K}\cdot\mathbf{R}}d^3K. \qquad (19.120)$$

Using Eq. (11.25), we may finally write

$$G(\mathbf{R},\tau) = \frac{\alpha}{2\pi^{1/2}\tau^{3/2}}\exp[-\alpha^2R^2/\tau], \quad \tau > 0. \qquad (19.121)$$

This expression has the form of a three-dimensional Gaussian delta sequence in t, comparable to Eq. (6.18). The Green's function has all the expected behaviors of a diffusive process: as t increases from 0, an excitation localized to $R = 0$ gradually broadens throughout space, as illustrated in Fig. 19.4.

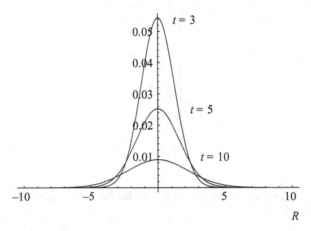

Figure 19.4 The behavior of the diffusion Green's function with $\alpha = 1$, for $t = 3$, $t = 5$, and $t = 10$.

The diffusion Green's function in unbounded one-dimensional space can be found in a strictly similar manner to be

$$G(x,\tau) = \frac{2\sqrt{\pi}}{\alpha\tau^{1/2}} \exp[-\alpha^2 x^2/\tau], \quad \tau > 0. \tag{19.122}$$

As the examples of this section show, even determining the Green's function of a differential equation in unbounded space can be a difficult exercise. When one considers problems with nontrivial boundary conditions, it is not in general possible to find a closed form expression for the Green's function, or even any expression at all. In the next sections, we consider a few special cases for which boundary conditions can be readily incorporated into Green's function solutions.

19.5 Method of images

In uncommon but important cases of high symmetry, homogeneous boundary conditions can be incorporated into Green's functions by introducing a "mirror image" of the physical source in the problem, in a technique referred to as the *method of images*. A similar strategy was used in Section 15.10 to solve certain homogeneous partial differential equations. We introduce this method by a simple example.

Example 19.2 (Laplace's equation with grounded conducting plane) Let us suppose we are interested in determining the Green's function for Poisson's equation in the presence of an infinite grounded conducting plane, on which we have the boundary condition $G = 0$. The geometry is illustrated in Fig. 19.5; the Green's function satisfies the equation

$$\nabla^2 G(\mathbf{r}, \mathbf{r}') = -4\pi \delta(\mathbf{r} - \mathbf{r}'), \tag{19.123}$$

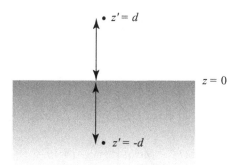

Figure 19.5 The geometry for the method of images for Poisson's equation in the presence of a grounded conducting plane.

subject to the Dirichlet boundary condition

$$G(\mathbf{r}, \mathbf{r}')|_{z=0} = 0. \tag{19.124}$$

The method of images takes advantage of the symmetry of the problem and the knowledge we already have of the Green's function for unbounded space, and the observation that we may add any solution of Laplace's equation to the Green's function without affecting the form of Eq. (19.123). Let us suppose that our source is located at a point above the plane such that $z' = d$. In unbounded space, the Green's function would be of the form

$$G_u(\mathbf{r}, \mathbf{r}') = \frac{1}{\sqrt{(x - x')^2 + (y - y')^2 + (z - d)^2}}. \tag{19.125}$$

We introduce a mirror image to this function which represents a source in unbounded space located at $z' = -d$ and which has the opposite sign,

$$F(\mathbf{r}, \mathbf{r}') = -\frac{1}{\sqrt{(x - x')^2 + (y - y')^2 + (z + d)^2}}. \tag{19.126}$$

In the half-space $z > 0$, $F(\mathbf{r}, \mathbf{r}')$ is a solution of Laplace's equation, and $G_u + F$ is a solution of Eq. (19.123) in that same half-space. We define

$$G(\mathbf{r}, \mathbf{r}') = G_u(\mathbf{r}, \mathbf{r}') + F(\mathbf{r}, \mathbf{r}'). \tag{19.127}$$

On the plane $z = 0$,

$$G_u(\mathbf{r}, \mathbf{r}')|_{z=0} = -F(\mathbf{r}, \mathbf{r}')|_{z=0}, \tag{19.128}$$

which shows that $G(\mathbf{r}, \mathbf{r}')|_{z=0} = 0$. Our function G satisfies the boundary condition and Eq. (19.123), and is therefore the unique Green's function for the inhomogeneous problem.

In physical terms, it is to be noted that the boundary condition is enforced by free charges moving on the surface of the conductor. Those charges, however, arrange themselves in such a way that they produce a field in the space $z > 0$ equivalent to a negative point charge located a distance d below the surface of the conductor.

◇

The method of images can also be applied to hyperbolic or parabolic problems, as the next example shows.

Example 19.3 (The diffusion equation on the semi-infinite line) We wish to determine the Green's function for the diffusion equation on the semi-infinite line $x > 0$, which satisfies

$$\frac{d^2}{dx'^2} G(x - x', t - t') - \alpha^2 \frac{\partial}{\partial t'} G(x - x', t - t') = -4\pi \delta(x - x')\delta(t - t'), \qquad (19.129)$$

subject to the condition

$$G(x - x', t - t') = 0, \quad x' = 0. \qquad (19.130)$$

The Green's function for the infinite line has been shown to be

$$G_u(x - x', t - t') = \frac{2\sqrt{\pi}}{\alpha \tau^{1/2}} \exp[-\alpha^2 (x - x')^2/(t - t')], \quad t > t'. \qquad (19.131)$$

We introduce a homogeneous solution to the diffusion equation of the form

$$F(x - x', t - t') = -\frac{2\sqrt{\pi}}{\alpha \tau^{1/2}} \exp[-\alpha^2 (x + x')^2/(t - t')], \quad t > t'. \qquad (19.132)$$

This solution represents the diffusion from a source located at $x = -x'$, with an opposite sign. For $x = 0$, we readily find that $G_u + F = 0$, which satisfies our boundary condition, so that we may write

$$G(x - x', t - t') = G_u(x - x', t - t') + F(x - x', t - t'). \qquad (19.133)$$

◇

The method of images typically only works for problems with planar geometry; however, a Green's function for Laplace's equation may be determined for a spherical boundary, as well.

Example 19.4 (Laplace's equation with grounded conducting sphere) We consider the Green's function for Poisson's equation in the presence of a grounded conducting sphere of radius a, as illustrated in Fig. 19.6. It is assumed that

$$G(\mathbf{r}, \mathbf{r}')|_{r=a} = 0. \qquad (19.134)$$

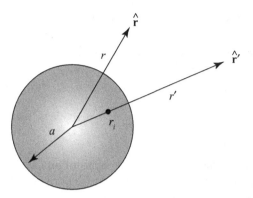

Figure 19.6 The geometry for the method of images for Poisson's equation in the presence of a grounded conducting sphere.

By symmetry, it appears that an image charge which is present must lie within the sphere, and on a line connecting the origin to the position $\mathbf{r}'$ of the external charge. The relative magnitude of the image charge, and its position, are not immediately obvious, however; we designate the radial position of the image r_i and the magnitude of the image charge $-\alpha_i$. The total Green's function in the presence of the conducting sphere is assumed to be the sum of the unbounded Green's function and the Green's function of the image charge, written as

$$G(\mathbf{r},\mathbf{r}') = \frac{1}{|\mathbf{r}-\mathbf{r}'|} - \frac{\alpha_i}{|\mathbf{r}-\mathbf{r}_i|}. \tag{19.135}$$

We may write this in a more useful form by expressing the position vectors in terms of unit vectors, i.e. $\mathbf{r} = r\hat{\mathbf{r}}$, such that

$$G(\mathbf{r},\mathbf{r}') = \frac{1}{|r\hat{\mathbf{r}}-r'\hat{\mathbf{r}}'|} - \frac{\alpha_i}{|r\hat{\mathbf{r}}-r_i\hat{\mathbf{r}}'|}. \tag{19.136}$$

It is to be noted that $\mathbf{r}'$ and $\mathbf{r}_i$ share a unit vector. When the observation point $\mathbf{r}$ lies on the surface of the sphere, we may write

$$G(\mathbf{r},\mathbf{r}')|_{r=a} = \frac{1}{|a\hat{\mathbf{r}}-r'\hat{\mathbf{r}}'|} - \frac{\alpha_i}{|a\hat{\mathbf{r}}-r_i\hat{\mathbf{r}}'|} = 0. \tag{19.137}$$

This expression must hold for all values of $\hat{\mathbf{r}}$; defining θ as the angle between $\hat{\mathbf{r}}$ and $\hat{\mathbf{r}}'$, we may write

$$\frac{1}{\sqrt{[a^2-r'^2]-2ar'\cos\theta}} - \frac{1}{\sqrt{[a^2-r_i^2]/\alpha_i^2 - 2ar_i\cos\theta/\alpha_i^2}} = 0. \tag{19.138}$$

This equation is satisfied if

$$a^2 - r'^2 = (a^2 - r_i^2)/\alpha_i^2, \tag{19.139}$$

$$2ar' = 2ar_i/\alpha_i^2. \tag{19.140}$$

With a little bit of work, these equations may be solved for α_i and r_i,

$$\alpha_i = \frac{a}{r'}, \tag{19.141}$$

$$r_i = \frac{a^2}{r'}. \tag{19.142}$$

The magnitude of the image charge decreases as the source point is moved further away from the sphere. Also, the image charge moves to the center of the sphere in the limit $r' \to \infty$ and to the surface of the sphere as $r' \to a$.

◇

19.6 Modal expansion of Green's functions

Let us return to solutions of an inhomogeneous partial differential equation in N variables of the form

$$\mathcal{L}u(\mathbf{r}') + \lambda u(\mathbf{r}') = -4\pi\rho(\mathbf{r}'), \tag{19.143}$$

where λ is a constant. We are interested in writing an expression for the Green's function of this equation,

$$\mathcal{L}G(\mathbf{r}, \mathbf{r}') + \lambda G(\mathbf{r}, \mathbf{r}') = -4\pi\delta(\mathbf{r} - \mathbf{r}'), \tag{19.144}$$

which is also subject to a set of boundary conditions. If the operator $\mathcal{L}$ is of Sturm–Liouville form, we expect that it has a complete set of orthonormal eigenfunctions ψ_n which can be chosen to satisfy the boundary conditions and corresponding eigenvalues λ_n such that

$$\mathcal{L}\psi_n(\mathbf{r}') + \lambda_n\psi_n(\mathbf{r}') = 0. \tag{19.145}$$

The index n in general represents $N - 1$ integer indices. We now distinguish two cases: one for which the eigenfunctions are orthonormal with respect to a real inner product, i.e.

$$\int_D \psi_n(\mathbf{r}')\psi_{\mathrm{m}}(\mathbf{r}')\mathrm{d}^N r' = \delta_{mn}, \tag{19.146}$$

and one for which the eigenfunctions are orthonormal with respect to a complex inner product, i.e.

$$\int_D \psi_n^*(\mathbf{r}')\psi_{\mathrm{m}}(\mathbf{r}')\mathrm{d}^N r' = \delta_{mn}. \tag{19.147}$$

We consider the real inner product first. Because the functions ψ_n are complete and orthonormal, we may expand the Green's function in the form

$$G(\mathbf{r}, \mathbf{r}') = \sum_n \alpha_n(\mathbf{r}) \psi_n(\mathbf{r}'),$$ (19.148)

with the functions $\alpha_n(\mathbf{r})$ to be determined. Furthermore, from Section 6.4, we know that we can in general write the Dirac delta function as

$$\delta(\mathbf{r} - \mathbf{r}') = \sum_n \psi_n(\mathbf{r}) \psi_n(\mathbf{r}').$$ (19.149)

On substituting from Eqs. (19.148) and (19.149) into Eq. (19.144), and using Eq. (19.145), we have

$$-\sum_n \alpha_n(\mathbf{r}) \lambda_n \psi_n(\mathbf{r}') + \sum_n \alpha_n(\mathbf{r}) \lambda \psi_n(\mathbf{r}') = -4\pi \sum_n \psi_n(\mathbf{r}) \psi_n(\mathbf{r}').$$ (19.150)

We may multiply this equation by $\psi_n(\mathbf{r}')$ and integrate over the domain D; using Eq. (19.146), we have

$$\alpha_n(\mathbf{r}) = 4\pi \frac{\psi_n(\mathbf{r})}{\lambda_n - \lambda}.$$ (19.151)

We therefore have an expression for the Green's function of the form

$$G(\mathbf{r}, \mathbf{r}') = 4\pi \frac{\psi_n(\mathbf{r}) \psi_n(\mathbf{r}')}{\lambda_n - \lambda}.$$ (19.152)

This equation is an expansion of the Green's function into a complete basis of orthonormal modes. Since the modes were assumed to satisfy the boundary conditions, the Green's function automatically satisfies the same boundary conditions.

From this expression we may immediately see an important symmetry property of the Green's function, namely

$$G(\mathbf{r}, \mathbf{r}') = G(\mathbf{r}', \mathbf{r}).$$ (19.153)

Physically, this equation suggests that the field produced at $\mathbf{r}$ by a source at $\mathbf{r}'$ is equivalent to the field produced at $\mathbf{r}'$ by a source at $\mathbf{r}$.

If Eq. (19.147) is instead satisfied, i.e. the eigenfunctions are complex, we may expand the delta function in the form

$$\delta(\mathbf{r} - \mathbf{r}') = \sum_n \psi_n^*(\mathbf{r}) \psi_n(\mathbf{r}'),$$ (19.154)

which leads instead to the expression

$$G(\mathbf{r}, \mathbf{r}') = 4\pi \sum_n \frac{\psi_n^*(\mathbf{r}) \psi_n(\mathbf{r}')}{\lambda_n - \lambda}.$$ (19.155)

In such a case, we may write

$$G(\mathbf{r},\mathbf{r}') = G^*(\mathbf{r}',\mathbf{r}). \tag{19.156}$$

There is some subtlety involved in these symmetry relationships. As we will see, it is possible to have a complex Green's function, with complex eigenfunctions, which still satisfies Eq. (19.153).

Modal expansions of Green's functions are also in general quite challenging to find. The use of addition theorems, as described in Sections 16.12 and 17.9, is often key in determining an expansion.

Example 19.5 (The Poisson equation) As a first example, we consider the modal expansion of the Green's function for the Poisson equation,

$$G(\mathbf{r},\mathbf{r}') = \frac{1}{|\mathbf{r} - \mathbf{r}'|}. \tag{19.157}$$

We may rewrite this function as

$$G(\mathbf{r},\mathbf{r}') = \frac{1}{\sqrt{r^2 + r'^2 - 2rr'\cos\Theta}}, \tag{19.158}$$

where Θ is the angle between $\mathbf{r}$ and $\mathbf{r}'$.

We consider two cases: $r > r'$ and $r < r'$. For $r > r'$, we may rewrite Eq. (19.158) as

$$G(\mathbf{r},\mathbf{r}') = \frac{1}{r}\frac{1}{\sqrt{1 - \cos\Theta r'/r + (r'/r)^2}}. \tag{19.159}$$

With the identification $\cos\Theta \equiv x$ and $r'/r \equiv t$, the latter term is simply the generating function of the Legendre polynomials, Eq. (17.27). We may therefore write

$$G(\mathbf{r},\mathbf{r}') = \frac{1}{r}\sum_{n=0}^{\infty} P_n(\cos\Theta) \left(\frac{r'}{r}\right)^n, \quad r > r'. \tag{19.160}$$

Similarly, for $r < r'$ we may remove a factor of r' from the square root in Eq. (19.158) and write

$$G(\mathbf{r},\mathbf{r}') = \frac{1}{r'}\sum_{n=0}^{\infty} P_n(\cos\Theta) \left(\frac{r}{r'}\right)^n, \quad r < r'. \tag{19.161}$$

The $P_n(\cos\Theta)$ may be further expanded by use of the spherical harmonic addition theorem of Section 17.9, which states that

$$P_n(\cos\Theta) = \frac{4\pi}{2n+1}\sum_{m=-n}^{n} Y_n^{m*}(\theta,\phi)Y_n^m(\theta',\phi'). \tag{19.162}$$

We may then write a complete modal expansion for the Green's function of the form

$$G(\mathbf{r},\mathbf{r}') = \sum_{n=0}^{\infty} \frac{4\pi}{2n+1} g_n(r,r') \sum_{m=-n}^{n} Y_n^{m*}(\theta,\phi) Y_n^m(\theta',\phi'), \tag{19.163}$$

where

$$g_n(r,r') = \begin{cases} \dfrac{1}{r}\left(\dfrac{r'}{r}\right)^n, & r > r', \\[2mm] \dfrac{1}{r'}\left(\dfrac{r}{r'}\right)^n, & r < r'. \end{cases} \tag{19.164}$$

◇

Example 19.6 (The Helmholtz equation) We now look for a modal expansion of the Green's function of the Helmholtz equation,

$$G(\mathbf{r},\mathbf{r}') = \frac{\exp[ik|\mathbf{r}-\mathbf{r}'|]}{|\mathbf{r}-\mathbf{r}'|}. \tag{19.165}$$

This Green's function, like the Poisson one, only depends upon the variables r, r', and Θ. It is reasonable then to assume a modal expansion of the form

$$G(\mathbf{r},\mathbf{r}') = \sum_{n=0}^{\infty} g_n(r,r') P_n(\cos\Theta), \tag{19.166}$$

where $P_n(\cos\Theta)$ may again be expanded in terms of spherical harmonics using the addition theorem. The Green's function satisfies the differential equation

$$(\nabla^2 + k^2) G(\mathbf{r},\mathbf{r}') = -4\pi\delta(\mathbf{r}-\mathbf{r}'). \tag{19.167}$$

From Section 6.6, we know that we may expand the delta function in spherical coordinates as

$$\delta(\mathbf{r}-\mathbf{r}') = \delta(r-r')\delta(\Omega-\Omega'). \tag{19.168}$$

Because the spherical harmonics form a complete orthonormal set of functions with respect to solid angle, the latter delta function may be written in the form

$$\delta(\Omega-\Omega') = \sum_{n=0}^{\infty}\sum_{m=-n}^{n} Y_n^{m*}(\theta,\phi) Y_n^m(\theta',\phi'). \tag{19.169}$$

On substitution from Eqs. (19.166), (19.168), and (19.169) into Eq. (19.167), we find that the functions $g_n(r,r')$ must satisfy

$$\frac{1}{r^2}\frac{d}{dr}\left(r^2\frac{dg_n}{dr}\right) - \frac{l(l+1)}{r^2}g_n + k^2 g_n = -4\pi\delta(r-r'). \tag{19.170}$$

This is now an equation of Sturm–Liouville form, which can be solved using the methods of Section 19.2. The solutions to the homogeneous equation are the spherical Bessel functions of Section 16.11, and our Green's function must be finite at the origin and behave as a spherical wave $h_l^{(1)}(r)$ for large values of r. The solution may be written as

$$g_n(r,r') = 4\pi i k \begin{cases} j_n(kr')h_n^{(1)}(kr), & r > r', \\ j_n(kr)h_n^{(1)}(kr'), & r < r'. \end{cases} \tag{19.171}$$

The complete modal expansion of the Green's function has the form

$$G(\mathbf{r},\mathbf{r}') = 4\pi i k \sum_{n=0}^{\infty} \begin{cases} j_n(kr')h_n^{(1)}(kr), & r > r' \\ j_n(kr)h_n^{(1)}(kr'), & r < r' \end{cases} \sum_{m=-n}^{n} Y_n^{m*}(\theta,\phi)Y_n^m(\theta',\phi'). \tag{19.172}$$

◇

19.7 Integral equations

As we have seen, Green's functions can be used to convert a differential equation into an integral equation. In many cases, however, the derived integral equation is also difficult to solve. Some strategies exist for solution, however, and the theory of integral equations and their solutions has been extensively studied; in this section we describe some of the most important results of that theory.

19.7.1 Types of linear integral equation

A linear integral equation is an equation in which the unknown function $\phi(x)$ lies within an integral of the form

$$\int_a^b K(x,x')\phi(x')dx'. \tag{19.173}$$

The function $K(x,x')$ is referred to as the *kernel* of the integral. There are four basic types of linear integral equation.

1. The *Fredholm equation of the first kind*, which has the form

$$f(x) = \int_a^b K(x,x')\phi(x')dx'. \tag{19.174}$$

2. The *Fredholm equation of the second kind*, of the form

$$\phi(x) = f(x) + \int_a^b K(x,x')\phi(x')dx'. \tag{19.175}$$

A homogeneous form of this equation has $f(x) = 0$.

3. The *Volterra equation of the first kind*, of the form

$$f(x) = \int_a^x K(x,x')\phi(x')dx'. \tag{19.176}$$

It is to be noted that the only difference from the Fredholm equation of the first kind is the upper limit of integration.

4. The *Volterra equation of the second kind*, of the form

$$\phi(x) = f(x) + \int_a^x K(x,x')\phi(x')dx'. \tag{19.177}$$

It is to be noted that the Volterra equations may be considered a special case of the Fredholm equations, with a kernel satisfying the condition

$$K(x,x') = 0, \quad x' > x. \tag{19.178}$$

These integral equations may be considered as a generalization of the matrix equations solved in Chapters 4 and 5. For instance, the Fredholm equation of the first kind is analogous to the solution of the matrix equation

$$\mathbf{K}\,|\phi\rangle = |f\rangle, \tag{19.179}$$

while the Fredholm equation of the second kind is analogous to the solution of the matrix equation

$$\mathbf{K}\,|\phi\rangle = |\phi\rangle - |f\rangle. \tag{19.180}$$

The homogeneous Fredholm equation of the second kind is then analogous to the eigenvector equation $\mathbf{K}\,|\phi\rangle = |\phi\rangle$. The Volterra equations are analogous to matrix equations with a lower triangular matrix $\mathbf{K}$. These analogies are useful in the numerical solution of integral equations, as noted in Section 5.1.

The strategy for solving a given integral equation typically depends upon the form of the kernel. A kernel is *separable* if it may be written in the form

$$K(x,x') = \lambda u(x)v^*(x'), \tag{19.181}$$

where λ is referred to as a *characteristic value* of the equation.

A kernel is *Hermitian* if it satisfies the relation

$$K^*(x',x) = K(x,x'). \tag{19.182}$$

A *homogeneous* kernel is one that may be written in the form

$$K(x,x') = K(x-x'). \tag{19.183}$$

A kernel is *square integrable* if it satisfies the condition

$$\int_a^b \int_a^b |K(x,x')|^2 \mathrm{d}x \mathrm{d}x' < \infty. \tag{19.184}$$

An integral equation that does not have a square integrable kernel or has an infinite range of integration is said to be a *singular* equation.

19.7.2 Solution of basic linear integral equations

It is relatively rare to be able to find exact solutions to integral equations. In this section, we note a few cases where a solution may be found.

The Fredholm equation of the second kind may be solved in closed form when the kernel is separable. The equation to solve is of the form

$$\phi(x) = f(x) + \lambda u(x) \int_a^b v^*(x')\phi(x')\mathrm{d}x'. \tag{19.185}$$

We multiply both sides of this equation by $v^*(x)$ and integrate over x from a to b, to find

$$\int_a^b v^*(x)\phi(x)\mathrm{d}x = \int_a^b f(x)v^*(x)\mathrm{d}x + \lambda \int_a^b u(x)v^*(x)\mathrm{d}x \int_a^b v^*(x')\phi(x')\mathrm{d}x'. \tag{19.186}$$

We may solve for the integral with respect to $\phi(x)$,

$$\int_a^b v^*(x')\phi(x')\mathrm{d}x' = \frac{\int_a^b v^*(x)f(x)\mathrm{d}x}{1 - \lambda \int_a^b v^*(x)u(x)\mathrm{d}x}. \tag{19.187}$$

We may substitute from this expression into Eq. (19.185), to find that

$$\phi(x) = f(x) + \lambda u(x) \frac{\int_a^b v^*(x')f(x')\mathrm{d}x'}{1 - \lambda \int_a^b v^*(x')u(x')\mathrm{d}x'}. \tag{19.188}$$

The homogeneous form of Eq. (19.185) only has a solution for one characteristic value; this can be seen by starting with the equation

$$\phi(x) = \lambda u(x) \int_a^b v^*(x')\phi(x')\mathrm{d}x', \tag{19.189}$$

again multiplying both sides by $v^*(x)$ and integrating, and dividing out the integral over ϕ. We are left with the condition

$$1 = \lambda \int_a^b v^*(x)u(x)\mathrm{d}x. \tag{19.190}$$

Provided λ satisfies this equation, the solution to Eq. (19.189) is $\phi(x) = cu(x)$, with c an arbitrary constant.

When the kernel of an integral equation is homogeneous, it can often be solved by the use of integral transforms such as the Laplace or Fourier transform. For instance, a Fredholm equation of the second kind with homogeneous kernel and infinite domain may be written as

$$\phi(t) = f(t) + \lambda \int_{-\infty}^{\infty} K(t-t')\phi(t')dt'. \tag{19.191}$$

The integral on the right is in the form of a convolution; by use of the convolution theorem of Section 11.5.1, we know that

$$\tilde{\phi}(\omega) = \tilde{f}(\omega) + 2\pi\lambda\tilde{K}(\omega)\tilde{\phi}(\omega). \tag{19.192}$$

We may rearrange this algebraic equation to write

$$\tilde{\phi}(\omega) = \frac{\tilde{f}(\omega)}{1 - 2\pi\tilde{K}(\omega)}, \tag{19.193}$$

and the solution to the equation may be written in integral form as

$$\phi(t) = \int_{-\infty}^{\infty} \frac{\tilde{f}(\omega)e^{-i\omega t}}{1 - 2\pi\tilde{K}(\omega)}d\omega. \tag{19.194}$$

This integral can potentially be evaluated by the use of contour integration.

19.7.3 Liouville–Neumann series

We consider again the Fredholm equation of the form

$$\phi(x) = f(x) + \int_a^b \lambda K(x,x')\phi(x')dx'. \tag{19.195}$$

It is not possible in general to find a closed-form solution to such an equation; however, provided the parameter λ is in some sense a "small" quantity, it is possible to find an infinite series of approximations to the solution known as the *Liouville–Neumann series*.

We look for a solution which is a power series in λ, i.e.

$$\phi(x) = \sum_{n=0}^{\infty} \lambda^n \phi_n(x), \tag{19.196}$$

where the functions $\phi_n(x)$ are to be determined. On substitution from this equation into Eq. (19.195), we have

$$\sum_{n=0}^{\infty} \lambda^n \phi_n(x) = f(x) + \sum_{n=0}^{\infty} \lambda^{n+1} \int_a^b K(x,x')\phi_n(x')dx'. \tag{19.197}$$

Treating λ as a free parameter, and requiring this equation to be satisfied for a continuous range of λ, we are led to the series of approximations

$$\phi_0(x) = f(x),\tag{19.198}$$

$$\phi_1(x) = \int_a^b K(x,x')\phi_0(x')dx',\tag{19.199}$$

$$\cdots$$

$$\phi_n(x) = \int_a^b K(x,x')\phi_{n-1}(x')dx'.\tag{19.200}$$

This solution to the integral equation is the Liouville–Neumann series. We have already seen it applied specifically to the case of scattering theory in Section 7.8, in which it is typically referred to as the Born series; the Born series is further discussed in Section 19.10. As noted in Section 7.8, this series is not guaranteed to converge, and may therefore not represent a valid solution in all cases.

A simple criterion for convergence may be introduced when the functions $f(x)$ and $K(x,x')$ are bounded, such that

$$|f(x)| \leq \mathcal{F}, \quad a \leq x \leq b,\tag{19.201}$$

$$|K(x,x')| \leq \mathcal{K}, \quad a \leq x \leq b.\tag{19.202}$$

The terms of the Liouville–Neumann series are then bounded, with $|\phi_0| \leq \mathcal{F}$,

$$|\phi_1| \leq \int_a^b |\phi_0(x')||K(x,x')|dx' \leq \mathcal{F}\mathcal{K}(b-a),\tag{19.203}$$

and, in general,

$$|\phi_n| \leq \mathcal{F}\mathcal{K}^n(b-a)^n.\tag{19.204}$$

The Liouville–Neumann series is subject to the upper bound

$$\left| \sum_{n=0}^{\infty} \lambda^n \phi_n(x) \right| \leq \mathcal{F} \sum_{n=0}^{\infty} |\lambda|^n \mathcal{K}^n (b-a)^n.\tag{19.205}$$

The sum on the right-hand side is comparable to the geometric series, of the form

$$\mathcal{F} \sum_{n=0}^{\infty} u^n = \frac{\mathcal{F}}{1-u}.\tag{19.206}$$

This suggests that the Liouville–Neumann series is absolutely and uniformly convergent, provided that

$$|\lambda| < \frac{1}{\mathcal{K}(b-a)}.\tag{19.207}$$

This condition is a sufficiency condition for convergence, though not a necessary one, as an investigation of the comparable Volterra equation of the second kind demonstrates. We assume the functions $f(x)$ and $K(x,x')$ are bounded as before, but the equation to be evaluated is now

$$\phi(x) = f(x) + \int_a^x \lambda K(x,x')\phi(x')dx'. \tag{19.208}$$

As noted previously, this may be considered a special case of the Fredholm equation with a lower triangular kernel. The first-order term of the Liouville-Neumann series satisfies the inequality

$$|\phi_1| \le \int_a^x |\phi_0(x')||K(x,x')|dx' \le \mathcal{F}\mathcal{K}(x-a), \tag{19.209}$$

which implies via integration that the nth-order term satisfies

$$|\phi_n| \le \mathcal{F}\frac{\mathcal{K}^n(b-a)^n}{n!}. \tag{19.210}$$

The Liouville–Neumann series of the Volterra equation is subject to the upper bound

$$\left|\sum_{n=0}^{\infty} \lambda^n \phi_n(x)\right| \le \mathcal{F}\sum_{n=0}^{\infty} \frac{|\lambda|^n \mathcal{K}^n(b-a)^n}{n!} = \mathcal{F}\exp[|\lambda|\mathcal{K}(b-a)]. \tag{19.211}$$

This function is always bounded, and the Liouville–Neumann series for the Volterra equation of the second kind (with bounded kernel and inhomogeneous term) always converges.

Returning to the general Liouville–Neumann series, it is to be noted that the nth order approximation may be written as

$$\phi_n(x) = \int_a^b K_n(x,x')f(x')dx', \tag{19.212}$$

where

$$K_n(x,x') = \int_a^b \cdots \int_a^b K(x,x_1)K(x_1,x_2)\cdots K(x_{n-2},x_{n-1})K(x_{n-1},x')dx_1\cdots dx_{n-1}. \tag{19.213}$$

It follows that the solution to the integral equation may also be written as

$$\phi(x) = f(x) + \lambda \int_a^b R(x,x';\lambda)f(x')dx', \tag{19.214}$$

where

$$R(x,x';\lambda) = \sum_{n=0}^{\infty} \lambda^n K_{n+1}(x,x') \tag{19.215}$$

is referred to as the *resolvent kernel*. In some cases, this resolvent kernel can be summed to a closed-form solution.

19.7.4 Hermitian kernels and Mercer's theorem

Particularly elegant and powerful results can be derived in the special case when the kernel of the integral equation is Hermitian, i.e.

$$K^*(x',x) = K(x,x'). \tag{19.216}$$

A Hermitian kernel which is square integrable is known as a *Hilbert–Schmidt kernel*, and we will focus our attention primarily on such kernels. We define the characteristic values of the kernel as those values of λ such that

$$\lambda \int_a^b K(x,x')\phi(x')dx' = \phi(x). \tag{19.217}$$

This is, in essence, an eigenvalue/eigenfunction equation for the kernel $K(x,x')$. In analogy with a Hermitian matrix, we can demonstrate many useful properties of the characteristic values and characteristic functions of a Hilbert-Schmidt kernel.

Theorem 19.1 *A non-null Hilbert-Schmidt kernel possesses at least one nonzero characteristic value.*

We refer the reader to [Moi05] for the proof.

Theorem 19.2 *The characteristic values of an H-S kernel are real-valued, and the characteristic functions corresponding to distinct values are orthogonal.*

This result is analogous to Theorem 4.1 for Hermitian matrices, and can be proven in much the same way. We introduce two equations for the characteristic values λ_i and λ_j,

$$\lambda_i \int_a^b K(x,x')\phi_i(x')dx' = \phi_i(x), \tag{19.218}$$

$$\lambda_j \int_a^b K(x,x')\phi_j(x')dx' = \phi_j(x). \tag{19.219}$$

We multiply the first equation by $\phi_j^*(x)$, the second by $\phi_i^*(x)$, and integrate each equation with respect to x. We then have

$$\int_a^b \int_a^b K(x,x')\phi_i(x')\phi_j^*(x)dx'dx = \int_a^b \phi_i(x)\phi_j^*(x)dx/\lambda_i, \tag{19.220}$$

$$\int_a^b \int_a^b K(x,x')\phi_i^*(x)\phi_j(x')dx'dx = \int_a^b \phi_i^*(x)\phi_j(x)dx/\lambda_j. \tag{19.221}$$

We take the complex conjugate of the second equation, apply the Hermitian property of the kernel, and then subtract the two equations, resulting in

$$(\lambda_i - \lambda_j^*) \int_a^b \phi_i^*(x)\phi_j(x)\mathrm{d}x = 0. \tag{19.222}$$

If $i = j$, the integral is necessarily greater than zero and we find that $\lambda_i^* = \lambda_i$. If $i \neq j$, we find that

$$\int_a^b \phi_i^*(x)\phi_j(x)\mathrm{d}x = 0. \tag{19.223}$$

The theorem is essentially proven, provided the double integrals used in the proof exist.

As with Hermitian matrices, it is possible to have degenerate characteristic values, and correspondingly non-orthogonal characteristic functions. These degenerate functions can always be made orthogonal by a Gram–Schmidt procedure.

Hilbert–Schmidt kernels may be shown to be expandable in a "diagonal" representation, of the form

$$K(x,x') = \sum_{n=1}^{\infty} \frac{\phi_n^*(x')\phi_n(x)}{\lambda_n}, \tag{19.224}$$

where the summation is over all characteristic values and is potentially infinite. This series is generally only convergent in the mean to the kernel $K(x,x')$. A stronger theorem may be applied to continuous kernels which are *non-negative definite*, i.e.

$$\int_a^b \int_a^b K(x,x')\phi^*(x)\phi(x')\mathrm{d}x\mathrm{d}x' \geq 0 \tag{19.225}$$

for all square-integrable functions $\phi(x)$.

Theorem 19.3 (Mercer's theorem) *A continuous, non-negative definite square integrable kernel $K(x,x')$ may be expanded in series form as*

$$K(x,x') = \sum_{n=1}^{\infty} \frac{\phi_n^*(x')\phi_n(x)}{\lambda_n}, \tag{19.226}$$

where $\lambda_n \geq 0$ are the characteristic values of the kernel and $\phi_n(x)$ are the characteristic functions. This series converges uniformly and absolutely to the kernel $K(x,x')$.

We again refer the reader to [Moi05] for the proof. When Mercer's theorem holds, the $\phi_n(x)$ are solutions of the Fredholm eigenvalue equation of the form

$$\phi_n(x) = \lambda_n \int_a^b K(x,x')\phi_n(x')\mathrm{d}x'. \tag{19.227}$$

19.8 Focus: Rayleigh–Sommerfeld diffraction

The Fresnel–Kirchoff diffraction theory discussed in Section 19.1 is perfectly adequate for most diffraction problems, but suffers from two related difficulties. First, we have noted that the formalism is problematic, since only the field or its derivative on the boundary is necessary to uniquely specify a solution to the elliptic Helmholtz equation, and the Fresnel–Kirchoff (FK) diffraction integral requires both. In fact, this leads to a significant inconsistency in the solution: the FK diffraction integral does not reproduce the assumed field in the aperture plane. Second, this difficulty suggests that the formula is in a sense wasteful, since it uses both the field and its derivative; in principle, a diffraction formula should only require one of these. In this section we derive a new pair of diffraction formulas, called the Rayleigh–Sommerfeld (RS) formulas, each of which involves only a single boundary condition.

We return to Eq. (19.10) of Section 19.1,

$$\frac{1}{4\pi} \int_S \left[U \frac{\partial}{\partial n} \left(\frac{e^{ikR}}{R} \right) - \frac{e^{ikR}}{R} \frac{\partial U}{\partial n} \right] da = \begin{cases} U(P), & P \in V, \\ 0, & P \notin V. \end{cases} \tag{19.228}$$

This formula was derived by the use of the Helmholtz equation and Green's theorem, and is generally valid. We derived the Kirchoff formula by assuming P lies within the volume, assumed a surface of the form of Fig. 19.2, and assumed boundary conditions on the screen given by

$$\begin{array}{llll} \text{on } \mathcal{A}: & U = U^{(i)}, & \frac{\partial U}{\partial n} = \frac{\partial U^{(i)}}{\partial n}, \\ \text{on } \mathcal{B}: & U = 0, & \frac{\partial U}{\partial n} = 0. \end{array} \tag{19.229}$$

It is worth reiterating these boundary conditions because they will also be used for the RS diffraction formulas; the same physical assumptions and limitations will also apply.

We use the same surface that we used for the Kirchoff diffraction formula, and furthermore assume that the Sommerfeld radiation condition is valid. The surface integral in Eq. (19.228) just reduces to an integral over the plane $z = 0$, i.e. the plane of the screen and aperture. If the point $P(x, y, z)$ lies within the half-space $z > 0$, the integral becomes

$$\frac{1}{4\pi} \int_{z'=0} \left[U \frac{\partial}{\partial z'} \left(\frac{e^{ikR^+}}{R^+} \right) - \frac{e^{ikR^+}}{R^+} \frac{\partial U}{\partial z'} \right] da = U(P), \tag{19.230}$$

where

$$R^+ = \sqrt{(x-x')^2 + (y-y')^2 + (z-z')^2}. \tag{19.231}$$

The point $P(x, y, -z)$ will then be located in the negative half-space, $z < 0$, outside the surface, and the surface integral becomes

$$\frac{1}{4\pi} \int_{z'=0} \left[U \frac{\partial}{\partial z'} \left(\frac{e^{ikR^-}}{R^-} \right) - \frac{e^{ikR^-}}{R^-} \frac{\partial U}{\partial z'} \right] da = 0. \tag{19.232}$$

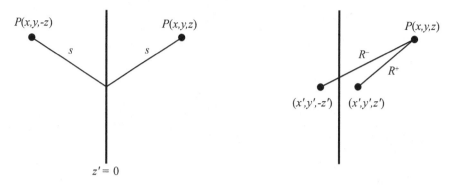

Figure 19.7 Illustrating the location of the two points used in calculating the R-S diffraction integrals.

where

$$R^- = \sqrt{(x-x')^2 + (y-y')^2 + (z+z')^2}.$$ (19.233)

These two points are illustrated in Fig. 19.7. It is to be noted that the distances R^+ and R^- are mathematically equivalent to the primed coordinates having their z-component flipped. The importance of these points is the observation that

$$\left.\frac{e^{ikR^+}}{R^+}\right|_{z'=0} = \left.\frac{e^{ikR^-}}{R^-}\right|_{z'=0},$$ (19.234)

$$\left.\frac{\partial}{\partial z'}\left(\frac{e^{ikR^+}}{R^+}\right)\right|_{z'=0} = -\left.\frac{\partial}{\partial z'}\left(\frac{e^{ikR^-}}{R^-}\right)\right|_{z'=0}.$$ (19.235)

We may rewrite Eq. (19.232) in terms of R^+ by using the above, and find that

$$\frac{1}{4\pi}\int_{z'=0}\left[U\frac{\partial}{\partial z'}\left(\frac{e^{ikR^+}}{R^+}\right) + \frac{e^{ikR^+}}{R^+}\frac{\partial U}{\partial z'}\right]da = 0.$$ (19.236)

If we add Eq. (19.236) from Eq. (19.230), and let $R^+ = s$, we get the equation

$$U(P) = \frac{1}{2\pi}\int_{z'=0}U\frac{\partial}{\partial z'}\left(\frac{e^{iks}}{s}\right)da.$$ (19.237)

Similarly, if we subtract Eq. (19.236) from Eq. (19.230), we find that

$$U(P) = -\frac{1}{2\pi}\int_{z'=0}\frac{\partial U}{\partial z'}\frac{e^{iks}}{s}da.$$ (19.238)

Equations (19.237) and (19.238) are known as the Rayleigh diffraction formulas of the first and second kind, respectively. These formulas relate the field on the plane $z' = 0$ to the field in the positive half-space. It is to be noted that we have essentially used the *method of images* to derive these formulas. The nice feature about these expressions is that each formula only uses knowledge of the field or its derivative on the surface to calculate the diffracted field, which is mathematically consistent. These formulas will reproduce the assumed conditions on the boundary, and are seemingly more rigorous than the FK formulas, although we will discuss below why this is not quite true.

To actually apply these formulas, of course, we need to specify what the field (or its derivative) looks like on the boundary, and we can again apply the Kirchoff boundary conditions (19.229). If we do so, the Rayleigh diffraction formulas take on the form

$$U(P) = \frac{1}{2\pi} \int_A U^{(i)} \frac{\partial}{\partial z'} \left(\frac{e^{iks}}{s} \right) da. \tag{19.239}$$

$$U(P) = -\frac{1}{2\pi} \int_A \frac{\partial U^{(i)}}{\partial z'} \frac{e^{iks}}{s} da. \tag{19.240}$$

With the assumed boundary conditions, these formulas are referred to as the *Rayleigh–Sommerfeld diffraction formulas of the first and second kind*.

It is worthwhile to compare these formulas to the FK formula of the earlier section. In its most general form at a planar screen, the FK formula has the form

$$U(P) = \frac{1}{4\pi} \int_A \left[U^{(i)} \frac{\partial}{\partial z'} \left(\frac{e^{iks}}{s} \right) - \frac{e^{iks}}{s} \frac{\partial U^{(i)}}{\partial z'} \right] da. \tag{19.241}$$

If we refer to the RS solutions as U_{RS}^I and U_{RS}^{II}, and the FK solution as U_{FK}, we san see immediately that we have the relation

$$U_{FK} = \frac{1}{2} \left(U_{RS}^I + U_{RS}^{II} \right). \tag{19.242}$$

In other words, the FK diffraction formula is simply the average of the first and second R-S formulas!

It has been mentioned that all three formulas give essentially the same results away from the immediate vicinity of the aperture (more than a few wavelengths away), and all three disagree with experiment in the aperture region. The choice of which formula to use is, in the end, a matter of taste. The R-S formulas are appealing because of their relative simplicity, but it is to be noted that they are always limited to planar geometries, due to the use of the method of images for derivation, while the FK formula is in principle valid for any surface of integration.

All three formulas run into difficulties due to the assumption that the field in the aperture is simply the incident wave, and the field on the screen is zero. If we could calculate exactly the field in the aperture and on the screen, we would find that all three formulas would give exactly the same results. In other words, each formula is perfectly rigorous, except for the assumed boundary conditions.

19.9 Focus: dyadic Green's function for Maxwell's equations

The method of Green's functions can also be applied to vector differential equations such as Maxwell's equations. The Green's function itself then generally has a tensor/matrix character, and is generally referred to as a *dyadic Green's function*. Such a dyad can be used, for instance, to calculate the electromagnetic radiation pattern produced or received by antennas, and can be used as the basis of computational techniques in scattering theory.

We start with Maxwell's equations, of the form

$$\nabla \cdot \mathbf{D} = 4\pi \rho_f, \tag{19.243}$$

$$\nabla \cdot \mathbf{B} = 0, \tag{19.244}$$

$$\nabla \times \mathbf{E} = -\frac{1}{c}\frac{d\mathbf{B}}{dt}, \tag{19.245}$$

$$\nabla \times \mathbf{H} = \frac{4\pi}{c}\mathbf{J}_f + \frac{1}{c}\frac{d\mathbf{D}}{dt}, \tag{19.246}$$

where $\mathbf{E}$ and $\mathbf{B}$ are the electric and magnetic fields, respectively, and $\mathbf{D}$ and $\mathbf{H}$ are the electric displacement and auxillary magnetic field. The quantities ρ_f and $\mathbf{J}_f$ are the *free* charge and current densities, those not due to material polarization and magnetization.

Let us now assume that the fields and sources are monochromatic; that is, they may be represented in the form

$$\mathbf{E}(\mathbf{r},t) = \mathbf{E}(\mathbf{r})e^{-i\omega t}, \tag{19.247}$$

and so on. With this assumption, Maxwell's equations reduce to their monochromatic form,

$$\nabla \cdot \mathbf{D} = 4\pi \rho_f, \tag{19.248}$$

$$\nabla \cdot \mathbf{B} = 0, \tag{19.249}$$

$$\nabla \times \mathbf{E} = ik\mathbf{B}, \tag{19.250}$$

$$\nabla \times \mathbf{H} = \frac{4\pi}{c}\mathbf{J} - ik\mathbf{D}, \tag{19.251}$$

where $k = \omega/c$. It is convenient to write these expressions in terms of a single pair of field quantities, $\mathbf{E}$ and $\mathbf{H}$, using the material relations

$$\mathbf{D} = \mathbf{E} + 4\pi \mathbf{P}, \tag{19.252}$$

$$\mathbf{B} = \mathbf{H} + 4\pi \mathbf{M}, \tag{19.253}$$

where $\mathbf{P}$ is the polarization density of matter and $\mathbf{M}$ is the magnetization density. With these relations, we may write

$$\nabla \cdot \mathbf{E} = 4\pi[\rho_f + \nabla \cdot \mathbf{P}], \tag{19.254}$$

$$\nabla \cdot \mathbf{H} = -4\pi \nabla \cdot \mathbf{M}, \tag{19.255}$$

$$\nabla \times \mathbf{E} - ik\mathbf{H} = 4\pi ik\mathbf{M}, \tag{19.256}$$

$$\nabla \times \mathbf{H} + ik\mathbf{E} = \frac{4\pi}{c}[\mathbf{J}_f - ikc\mathbf{P}]. \tag{19.257}$$

We may define an *electric current density* and a *magnetic current density* as

$$\mathbf{J}_e = \mathbf{J}_f - ikc\mathbf{P}, \tag{19.258}$$

$$\mathbf{J}_m = ikc\mathbf{M}, \tag{19.259}$$

and an *electric charge density* and a *magnetic charge density*[2] as

$$\rho_e = \rho_f + \nabla \cdot \mathbf{P}, \tag{19.260}$$

$$\rho_m = -\nabla \cdot \mathbf{M}. \tag{19.261}$$

These definitions allow us to write Maxwell's equations in the particularly symmetric form

$$\nabla \cdot \mathbf{E} = 4\pi\rho_e, \tag{19.262}$$

$$\nabla \cdot \mathbf{H} = 4\pi\rho_m, \tag{19.263}$$

$$\nabla \times \mathbf{E} - ik\mathbf{H} = \frac{4\pi}{c}\mathbf{J}_m, \tag{19.264}$$

$$\nabla \times \mathbf{H} + ik\mathbf{E} = \frac{4\pi}{c}\mathbf{J}_e. \tag{19.265}$$

This decomposition allows us to separate the radiation fields produced by an arbitrary radiation source into two classes: fields from electric sources and fields from magnetic sources. Because of the linearity of Maxwell's equations, these two sets of fields can be solved for independently. The electric source fields, for instance, can be determined by setting $\mathbf{J}_m = 0$ and $\rho_m = 0$. We consider first the calculation of the dyadic Green's function for electric sources, and return to magnetic sources at the end of the section.

For monochromatic fields, the charge density ρ_e is directly related to the current density by the continuity equation,

$$\rho_e = \frac{1}{i\omega}\nabla \cdot \mathbf{J}_e, \tag{19.266}$$

---

[2] It should be noted that these definitions are for convenience, and there is no physical distinction between "electric" and "magnetic" currents – a current is a current! If we wrote Maxwell's equations in terms of $\mathbf{E}$ and $\mathbf{B}$, the only current term would appear in the curl equation for $\mathbf{B}$, and the only charge term would appear in the divergence equation for $\mathbf{E}$.

as can be determined by taking the divergence of Eq. (19.265). We therefore can neglect the ρ term and focus entirely on the Maxwell curl equations. We begin by taking the curl of Eq. (19.264), with $\mathbf{J}_m = 0$, such that

$$\nabla \times \mathbf{H} = \frac{1}{ik} \nabla \times (\nabla \times \mathbf{E}). \tag{19.267}$$

We may substitute from this equation into Eq. (19.265), to find that

$$\nabla \times (\nabla \times \mathbf{E}) - k^2 \mathbf{E} = \frac{4\pi ik}{c} \mathbf{J}_e. \tag{19.268}$$

This linear relationship between the electric field and the current density suggests that we should be able to write an integral relation between them of the form

$$\mathbf{E}(\mathbf{r}) = \frac{ik}{c} \int_V \boldsymbol{\Gamma}_e(\mathbf{r}, \mathbf{r}') \cdot \mathbf{J}_e(\mathbf{r}') d\tau', \tag{19.269}$$

where $\boldsymbol{\Gamma}_e$ is the electric dyadic Green's function. The term "dyadic" refers to a second-rank tensor (i.e. a matrix) written as a sum of ordered pairs of vectors. For instance, we may decompose a matrix as shown below,

$$\mathbf{A} = \begin{bmatrix} 1 & 2 \\ 0 & 1 \end{bmatrix} = \begin{bmatrix} 1 \\ 0 \end{bmatrix} [\ 1 \quad 0 \ ] + \begin{bmatrix} 0 \\ 1 \end{bmatrix} [\ 0 \quad 1 \ ] + 2 \begin{bmatrix} 1 \\ 0 \end{bmatrix} [\ 0 \quad 1 \ ]. \tag{19.270}$$

If we make the following vector associations:

$$\hat{\mathbf{x}} = \begin{bmatrix} 1 \\ 0 \end{bmatrix}, \tag{19.271}$$

$$\hat{\mathbf{y}} = \begin{bmatrix} 0 \\ 1 \end{bmatrix}, \tag{19.272}$$

We may write our matrix in shorthand as

$$\mathbf{A} = \hat{\mathbf{x}}\hat{\mathbf{x}} + \hat{\mathbf{y}}\hat{\mathbf{y}} + 2\hat{\mathbf{x}}\hat{\mathbf{y}}. \tag{19.273}$$

This dyad may be premultiplied by the dot product of a vector or postmultiplied by the dot product of a vector; these actions are equivalent to premultiplying the matrix by a row vector or postmultiplying it by a column vector, respectively.

Returning to the dyadic Green's function, we may substitute from Eq. (19.269) into Eq. (19.268), so that

$$\nabla \times \nabla \times \frac{ik}{c} \int_V \boldsymbol{\Gamma}_e(\mathbf{r}, \mathbf{r}') \cdot \mathbf{J}_e(\mathbf{r}') d\tau' - k^2 \frac{ik}{c} \int_V \boldsymbol{\Gamma}_e(\mathbf{r}, \mathbf{r}') \cdot \mathbf{J}_e(\mathbf{r}') d\tau' = \frac{4\pi ik}{c} \mathbf{J}_e. \tag{19.274}$$

The latter term may be written as a volume integral as well by writing

$$\mathbf{J}_e(\mathbf{r}) = \int_V \mathbf{I} \cdot \mathbf{J}_e(\mathbf{r}')\delta(\mathbf{r} - \mathbf{r}')d\tau', \tag{19.275}$$

where $\mathbf{I}$ is the identity dyadic (identity matrix).

Here we must be a little careful. As in the case of the Green's function for the Helmholtz equation, Eq. (19.80), we expect that the Green's dyad may be singular for points $\mathbf{r} = \mathbf{r}'$. Provided $\mathbf{r} \neq \mathbf{r}'$ for all values of $\mathbf{r}'$ in V, however, the integrand is well-behaved and the order of integration and differentiation may be interchanged in Eq. (19.274). We then have

$$\int_V \nabla \times \nabla \times \boldsymbol{\Gamma}_e(\mathbf{r},\mathbf{r}') \cdot \mathbf{J}_e(\mathbf{r}')d\tau' - k^2 \int_V \boldsymbol{\Gamma}_e(\mathbf{r},\mathbf{r}') \cdot \mathbf{J}_e(\mathbf{r}')d\tau'$$

$$= 4\pi \int_V \mathbf{I} \cdot \mathbf{J}_e(\mathbf{r}')\delta(\mathbf{r} - \mathbf{r}')d\tau'. \tag{19.276}$$

Because this must hold generally for any current density $\mathbf{J}_e$ and any volume V which does not include $\mathbf{r}$, we may write

$$\nabla \times \nabla \times \boldsymbol{\Gamma}_e(\mathbf{r},\mathbf{r}') - k^2\boldsymbol{\Gamma}_e(\mathbf{r},\mathbf{r}') = 4\pi \mathbf{I}\delta(\mathbf{r} - \mathbf{r}'). \tag{19.277}$$

This equation is comparable to the general differential equation of a Green's function, Eq. (19.54). We may simplify it by using Eq. (2.76),

$$\nabla \times (\nabla \times \mathbf{A}) = -\nabla^2\mathbf{A} + \nabla\nabla \cdot \mathbf{A}. \tag{19.278}$$

The Green's dyad therefore satisfies the equation

$$(\nabla^2 + k^2)\boldsymbol{\Gamma}_e(\mathbf{r},\mathbf{r}') = -4\pi \mathbf{I}\delta(\mathbf{r} - \mathbf{r}') + \nabla\nabla \cdot \boldsymbol{\Gamma}_e(\mathbf{r},\mathbf{r}'). \tag{19.279}$$

We may simplify this further by noting from Eq. (19.277) that

$$\nabla \cdot \boldsymbol{\Gamma}_e(\mathbf{r},\mathbf{r}') = -\frac{4\pi}{k^2}\nabla\delta(\mathbf{r} - \mathbf{r}'). \tag{19.280}$$

The equation for the Green's dyad becomes

$$(\nabla^2 + k^2)\boldsymbol{\Gamma}_e(\mathbf{r},\mathbf{r}') = -4\pi \left(\mathbf{I} + \frac{1}{k^2}\nabla\nabla\right)\delta(\mathbf{r} - \mathbf{r}'). \tag{19.281}$$

Let us suppose the Green's dyad may be written in the form

$$\boldsymbol{\Gamma}_e(\mathbf{r},\mathbf{r}') = \left(\mathbf{I} + \frac{1}{k^2}\nabla\nabla\right)G(\mathbf{r} - \mathbf{r}'). \tag{19.282}$$

With this assumption, the function $G(\mathbf{r} - \mathbf{r}')$ satisfies the equation

$$(\nabla^2 + k^2)G(\mathbf{r},\mathbf{r}') = -4\pi\delta(\mathbf{r} - \mathbf{r}'). \tag{19.283}$$

This is simply the equation for the Green's function of the Helmholtz equation, which is satisfied by

$$G(\mathbf{r} - \mathbf{r}') = \frac{e^{ik|\mathbf{r} - \mathbf{r}'|}}{|\mathbf{r} - \mathbf{r}'|}. \tag{19.284}$$

We may therefore write the electric dyadic Green's function in the form

$$\boldsymbol{\Gamma}_e(\mathbf{r}, \mathbf{r}') = \left(\mathbf{I} + \frac{1}{k^2}\nabla\nabla\right)\frac{e^{ik|\mathbf{r} - \mathbf{r}'|}}{|\mathbf{r} - \mathbf{r}'|}. \tag{19.285}$$

This function characterizes the electric field produced by an infinitesmal electric point current; that is, if the current at the point $\mathbf{r}'$ is $\mathbf{j}$, the field produced by that current is

$$\mathbf{E}(\mathbf{r}) = \boldsymbol{\Gamma}_e(\mathbf{r}, \mathbf{r}') \cdot \mathbf{j}. \tag{19.286}$$

The Green's dyad can be used to calculate the electric field everywhere external to a volume containing an electric current density $\mathbf{J}_e$. Inside the volume, however, is a different story: because of the $\nabla\nabla$ derivative in the dyad, it diverges as $1/r^3$ for small values of $r = |\mathbf{r} - \mathbf{r}'|$, which is singular even when integrated over a volume. This singular behavior is an artifact of our interchange of differentiation and integration in Eq. (19.276), which is unjustified in the interior of the current source. This is especially a problem when using the Green's dyad to determine numerical solutions of Maxwell's equations; techniques for "fixing" the dyad can be found in [Bla91].

An expression may also be found for the magnetic field outside the source domain by applying the Maxwell equation (19.250) directly to the dyadic expression (19.269) for the electric field. One has

$$\mathbf{B}(\mathbf{r}) = \frac{1}{ik}\nabla \times \frac{ik}{c}\int_V \boldsymbol{\Gamma}_e(\mathbf{r}, \mathbf{r}') \cdot \mathbf{J}_e(\mathbf{r}')d\tau'. \tag{19.287}$$

Again assuming that the order of differentiation and integration may be interchanged, we may us Eq. (19.285) to write

$$\nabla \times \boldsymbol{\Gamma}_e(\mathbf{r}, \mathbf{r}') \cdot \mathbf{J}(\mathbf{r}') = \nabla G(\mathbf{r}, \mathbf{r}') \times \mathbf{J}_e(\mathbf{r}'). \tag{19.288}$$

The magnetic field is therefore of the form

$$\mathbf{B}(\mathbf{r}) = \frac{1}{c}\int_V \nabla G(\mathbf{r}, \mathbf{r}') \times \mathbf{J}_e(\mathbf{r}')d\tau'. \tag{19.289}$$

The electric dyadic Green's function is typically all that is needed for optics applications, as most sources and materials are nonmagnetic at optical frequencies. We may also construct a magnetic dyadic Green's function for the situation where $\mathbf{J}_m \neq 0$, by returning to

Eqs. (19.262)–(19.265) and letting $\mathbf{J}_e = 0$. The curl equations can be combined to find an expression for $\mathbf{H}$ of the form

$$\nabla \times (\nabla \times \mathbf{H}) = -\frac{4\pi i k}{c} \mathbf{J}_m. \tag{19.290}$$

This expression is almost the same as Eq. (19.268), with the exception of a minus sign. It follows that we may immediately write

$$\mathbf{H}(\mathbf{r}) = -\frac{ik}{c} \int_V \mathbf{\Gamma}_m(\mathbf{r}, \mathbf{r}') \cdot \mathbf{J}_m(\mathbf{r}') d\tau', \tag{19.291}$$

with

$$\mathbf{\Gamma}_m(\mathbf{r}, \mathbf{r}') = \left(\mathbf{I} + \frac{1}{k^2}\nabla\nabla\right) \frac{e^{ik|\mathbf{r}-\mathbf{r}'|}}{|\mathbf{r} - \mathbf{r}'|}. \tag{19.292}$$

We may also use Eq. (19.265) to write

$$\mathbf{E}(\mathbf{r}) = \frac{1}{c} \int_V \nabla G(\mathbf{r}, \mathbf{r}') \times \mathbf{J}_m(\mathbf{r}') d\tau'. \tag{19.293}$$

19.10 Focus: Scattering theory and the Born series

One important application of Green's functions is their use in converting partial differential equations into integral equations, which may be discretized and solved by matrix methods such as discussed in Section 5.1. This may be demonstrated by considering the problem of scattering from an inhomogeneous object, as touched upon in Section 7.8.

A scalar wave $U_0(\mathbf{r})$ traveling through free space is incident upon an object confined to a volume V of refractive index $n(\mathbf{r})$. We would like to determine the form of the scattered field $U_s(\mathbf{r})$, and formally define the total field in the region of the scattering object as $U(\mathbf{r}) = U_0(\mathbf{r}) + U_s(\mathbf{r})$. The geometry is illustrated in Fig. 19.8.

The total field will satisfy the Helmholtz equation with a nonuniform refractive index and hence a nonuniform wavenumber k,

$$[\nabla^2 + k^2 n^2(\mathbf{r})]U(\mathbf{r}) = 0. \tag{19.294}$$

Because the form of $n(\mathbf{r})$ may be quite complicated, it is not in general possible to solve the differential equation as written. However, solutions to the Helmholtz equation with $n(\mathbf{r}) = 1$ are much more tractable, so we add $k^2 U$ to each side of Eq. (19.294) and subtract the quantity $k^2 n^2 U$,

$$[\nabla^2 + k^2]U(\mathbf{r}) = k^2[1 - n^2(\mathbf{r})]U(\mathbf{r}). \tag{19.295}$$

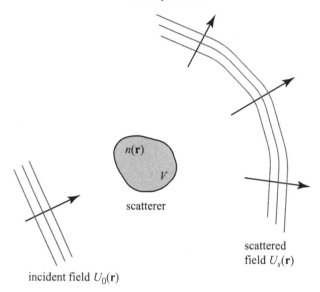

Figure 19.8 Illustration of the notation related to the scattering of a scalar wave $U_0(\mathbf{r})$ from an inhomogeneous object of refractive index $n(\mathbf{r})$.

This equation is formally the Helmholtz equation with a source term of the form $k^2[1-n^2]U$; we now define a quantity

$$F(\mathbf{r}) \equiv \frac{k^2}{4\pi}[n^2(\mathbf{r}) - 1], \tag{19.296}$$

to be referred to as the *scattering potential*, and our PDE may be written as

$$[\nabla^2 + k^2]U(\mathbf{r}) = -4\pi F(\mathbf{r})U(\mathbf{r}). \tag{19.297}$$

Referring back to Eq. (19.73), we see that the Green's function for the Helmholtz equation satisfies the expression

$$[\nabla^2 + k^2]U(\mathbf{r}) = -4\pi \delta(\mathbf{r} - \mathbf{r}'). \tag{19.298}$$

By the use of Eq. (19.63), and assuming that the boundary contributions vanish, we may write

$$U(\mathbf{r}) = H(\mathbf{r}) + \int_V F(\mathbf{r}')U(\mathbf{r}')\frac{e^{ikR}}{R}d^3r', \tag{19.299}$$

where $H(\mathbf{r})$ is a solution of the homogeneous Helmholtz equation. In the limit that $F(\mathbf{r}) \to 0$, the solution should reduce to $U(\mathbf{r}) = U_0(\mathbf{r})$, which implies that we may write

$$U(\mathbf{r}) = U_0(\mathbf{r}) + \int_V F(\mathbf{r}')U(\mathbf{r}')\frac{e^{ikR}}{R}d^3r', \tag{19.300}$$

or, for the scattered field,

$$U_s(\mathbf{r}) = \int_V F(\mathbf{r}')U(\mathbf{r}')\frac{e^{ikR}}{R}\mathrm{d}^3 r'. \tag{19.301}$$

By the use of Green's functions techniques, we have converted our differential equation for the scattered field into an integral equation, specifically a multi-variable Fredholm equation of the second kind. In general, however, this equation cannot be solved analytically, due to the presence of the scattered field on both sides of the equation. We may discretize the integral and solve it by numerical techniques as in Section 5.1; alternatively, we may develop a Liouville–Neumann series solution to the equation, namely the Born series.

The Born series is typically derived by assuming that the scattering potential is very weak, and may be written as $F \to \lambda F$, where λ is a dimensionless parameter. We then seek a series solution for the total field of the form

$$U(\mathbf{r}) = \sum_{n=0}^{\infty} \lambda^n V_n(\mathbf{r}). \tag{19.302}$$

On substituting from this expression into Eq. (19.300), we find the series

$$V_0(\mathbf{r}) = U_0(\mathbf{r}), \tag{19.303}$$

$$V_1(\mathbf{r}) = \int_V F(\mathbf{r}')U_0(\mathbf{r}')\frac{e^{ikR}}{R}\mathrm{d}^3 r', \tag{19.304}$$

$$\cdots$$

$$V_n(\mathbf{r}) = \int_V F(\mathbf{r}')V_{n-1}(\mathbf{r}')\frac{e^{ikR}}{R}\mathrm{d}^3 r'. \tag{19.305}$$

Introducing the operator $K[g](\mathbf{r})$ as

$$K[g] \equiv \int_V g(\mathbf{r}')\frac{e^{ikR}}{R}\mathrm{d}^3 r', \tag{19.306}$$

we may then write the Born series formally as

$$U(\mathbf{r}) = \sum_{n=0}^{\infty}(KF)^n U_0. \tag{19.307}$$

As noted in Section 7.8, the Born series does not necessarily converge and therefore does not necessarily represent a solution to the integral equation. The $n = 1$ term of the Born series, $V_1(\mathbf{r})$, is typically referred to as the *Born approximation* and forms the basis of the simplest inverse scattering algorithms, in which the scattering potential $F(\mathbf{r})$ is deduced from measurements of the scattered field $U_s(\mathbf{r})$. Under the assumption that the Born approximation is valid, the inverse problem is straightforward because there is a linear relationship between the scattered field and the scattering potential.

19.11 Exercises

1. Find a Green's function for the inhomogeneous equation

$$y'' - 2y' + 4y = f(x)$$

on the domain $0 \leq x \leq 1$, subject to the conditions $y(0) = 1$, $y'(1) = 0$.

2. Find a Green's function for the inhomogeneous equation

$$y'' - y' - 2y = f(x)$$

on the domain $0 \leq x \leq 1$, subject to the conditions $y(0) = 1$, $y(1) = 0$.

3. Find a Green's function for the inhomogeneous equation,

$$2y'' + y' + 4y = f(x)$$

on the domain $0 \leq x \leq 1$, subject to the conditions $y(0) = 0$, $y(1) + 2y'(1) = 0$.

4. Find a Green's function for the inhomogeneous equation

$$y'' - k^2 y = f(x)$$

on the *infinite* domain $-\infty \leq x \leq \infty$, subject to the condition that the Green's function must be finite at infinity.

5. Find a Green's function for the inhomogeneous equation

$$y'' + k^2 y = f(x)$$

on the *infinite* domain $-\infty \leq x \leq \infty$, subject to the condition that the Green's function $G(x,x')$ must produce waves that are outgoing from the source point x' (assuming time dependence $\exp[-i\omega t]$).

6. Find a Green's function for the inhomogeneous equation

$$y'' + 4y = f(x)$$

on the domain $0 \leq x \leq 1$, subject to the conditions $y(0) = 0$, $y(1) = 1$. (Hint: find a solution to the homogeneous equation that satisfies the boundary conditions and convert the inhomogeneous equation to one with homogeneous boundary conditions.)

7. Use the method of images to find the Green's function for Poisson's equation $\nabla^2 U = -4\pi\rho$ in three-dimensional space in the presence of a rectangular boundary, as shown in Fig. 19.9(a).

8. Use the method of images to find the Green's function for Poisson's equation $\nabla^2 U = -4\pi\rho$ in three-dimensional space in the presence of a wedge of opening angle α, as shown in Fig. 19.9(b).

9. Use the method of images to find the Green's function for Poisson's equation $\nabla^2 U = -4\pi\rho$ in three-dimensional space in between two grounded planes separated by a distance d, as shown in Fig. 19.9(c).

10. Determine the Green's function for the equation

$$y'' + 4y = f(x),$$

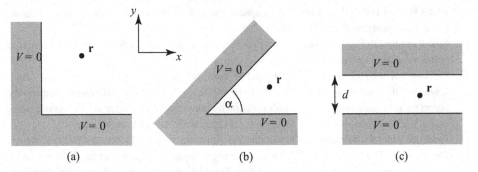

Figure 19.9 Figures relating to homework problems.

subject to the boundary conditions $y(0) = 0$ and $y(2\pi) = 0$. Expand the Green's function in a Fourier series and confirm that it has a modal expansion of the form of Eq. (19.152).

11. Integral transforms can be used to find an integral representation for many Green's functions. By use of the three-dimensional Fourier transform, show that the Green's function of the Helmholtz equation, satisfying

$$[\nabla^2 + k_0^2]G(\mathbf{r} - \mathbf{r}') = -4\pi \delta^{(3)}(\mathbf{r} - \mathbf{r}'),$$

can be written in the form

$$G(\mathbf{r} - \mathbf{r}') = \frac{4\pi}{(2\pi)^3} \int \frac{e^{i\mathbf{k}\cdot(\mathbf{r}-\mathbf{r}')}}{k^2 - k_0^2} d^3k.$$

12. Integral transforms can be used to find an integral representation for many Green's functions. By use of the three-dimensional Fourier transform, show that the Green's function of Laplace's equation, satisfying

$$\nabla^2 G(\mathbf{r} - \mathbf{r}') = -4\pi \delta^{(3)}(\mathbf{r} - \mathbf{r}'),$$

can be written in the form

$$G(\mathbf{r} - \mathbf{r}') = \frac{4\pi}{(2\pi)^3} \int \frac{e^{i\mathbf{k}\cdot(\mathbf{r}-\mathbf{r}')}}{k^2} d^3k.$$

13. Read the paper by M. R. Teague, Deterministic phase retrieval: a Green's function solution, *J. Opt. Soc. Am.* **73** (1983), 1434–1441. Describe the physical problem that is being solved, and explain the role of Green's functions in the solution.

14. Use the method of images to find the Green's function for a point charge outside of an infinite grounded conducting cylinder of radius $\rho = a$.

15. By the techniques of this chapter, find the retarded Green's function for the inhomogeneous Schrödinger equation

$$-\frac{\hbar^2}{2m}\nabla^2 \psi(\mathbf{r},t) - i\hbar \frac{\partial \psi(\mathbf{r},t)}{\partial t} = \frac{\hbar^2}{2m} 4\pi \rho(\mathbf{r},t)$$

which is valid for all three-dimensional space. Use this function to write an expression for the initial value problem $\psi(\mathbf{r}, 0) = f(\mathbf{r})$.

16. Find a solution to Poisson's equation in two dimensions in a circular domain of radius a subject to the boundary condition $V(a, \theta) = f(\theta)$.

17. Read the paper by G. Gbur, J. T. Foley and E. Wolf, Nonpropagating string excitations – finite length and damped strings, *Wave Motion* **30** (1999), 125. Describe the differential equation and boundary value problem that a Green's function has been applied to. What is a "nonpropagating excitation", and what conditions must the force distribution satisfy to create one?

18. Read the paper by S. Morgan, On the integral equations of laser theory, *IEEE Trans. Micro. Theory Tech.* **11** (1963), 191–193. How are integral equations applied to laser theory? What types of kernel arise in the theory, and why are the solutions of the integral equations thought to be problematic?

19. Read the paper by E. Wolf, New theory of partial coherence in the space-frequency domain. Part I: spectra and cross spectra of steady-state sources, *J. Opt. Soc. Am.* **72** (1982), 343–351. Describe what physical property of a light field is described by a Hermitian kernel. What important theorem of integral equations is applied, and what physical interpretation can be given to the characteristic functions of the theory?

20. Determine the Liouville–Neumann series for the integral equation

$$\phi(x) = 1 + \lambda \int_0^x \phi(x') \, dx'$$

and express this solution in closed form.

21. Determine the Liouville–Neumann series for the integral equation

$$\phi(x) = 1 + \lambda \int_0^x xx' \phi(x') \, dx'$$

and find the resolvent kernel of the equation. Can the solution of the equation be expressed in a closed form?

22. We consider the solution of a Volterra equation of the second kind of the form

$$\phi(x) = e^{-x} + \int_0^x \sin[\alpha(x - x')] \phi(x') \, dx'.$$

Use Laplace transforms to determine the solution of this equation. (Hint: convert the Volterra equation to a Fredholm equation with upper limit ∞ by redefining the kernel.)

23. We consider the solution of a Fredholm equation of the second kind of the form

$$\phi(x) = e^{-|x|} + \int_{-\infty}^{\infty} T(x - x') \phi(x') \, dx',$$

where $T(x - x')$ is the top hat function,

$$T(x) = \begin{cases} 1, & |x| \leq 1, \\ 0, & |x| > 1. \end{cases}$$

Use Fourier transform techniques to find a solution to the equation.

20

The calculus of variations

20.1 Introduction: principle of Fermat

In 1657 the French mathematician and physicist Pierre de Fermat enunciated the principle which forms the basis of all geometrical optics, now known as *Fermat's principle* (also as the *principle of shortest optical path* or the *principle of least time*). If we define an *optical length* between two points P_1 and P_2 by the formula

$$\int_{P_1}^{P_2} n(\mathbf{r})\mathrm{d}l, \tag{20.1}$$

where $n(\mathbf{r})$ is the index of refraction at spatial position $\mathbf{r}$ and $\mathrm{d}l$ is an infinitesimal path length, Fermat's principle states that *the optical length of an actual ray which passes between the two points P_1 and P_2 is shorter than any other curve which joins these points and lies within a certain regular neighborhood of it.*

The geometry relating to Fermat's principle is illustrated in Fig. 20.1. It is a brilliant combination of mathematical insight and philosophical thought, giving in some sense a reason for rays to travel the path they do – to minimize the optical path.

It only takes a little work to reproduce the usual laws of geometrical optics from Fermat's principle. First, if a ray is traveling in a medium of constant refractive index, the shortest optical path between two points is a straight line. Rays therefore propagate in straight lines through uniform media.

With this knowledge, we can now derive the law of refraction. We consider a ray propagating in the x, z plane from a region of index n_1 to a region of index n_2, and let its initial position be $(0, -z_1)$ and final position be (x_2, z_2). The geometry is illustrated in Fig. 20.2. The only place a ray can change direction is at the interface, and the position of intersection is denoted x. The distance the ray travels in region 1 is denoted l_1, and the distance the ray travels in region 2 is denoted l_2. By elementary geometry,

$$l_1 = \sqrt{x^2 + z_1^2}, \tag{20.2}$$

$$l_2 = \sqrt{(x_2 - x)^2 + z_2^2}, \tag{20.3}$$

The calculus of variations

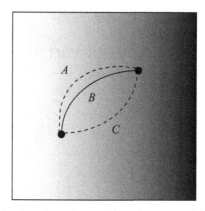

Figure 20.1 Illustrating Fermat's principle. Ray B minimizes the optical path, with the ray spending more time in the rarer medium. Ray A spends even more time in the rarer medium, but has a longer optical path. Although the geometrical length of ray C is the same as that of B, it spends more time in the denser medium, and consequently also has a longer optical path.

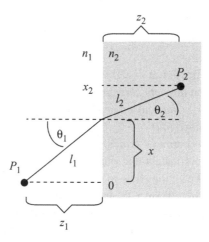

Figure 20.2 Notation relating to refraction with Fermat's principle.

and the optical path, which we denote by $S(x)$, is given by

$$S(x) = n_1 l_1 + n_2 l_2 = n_1 \sqrt{x^2 + z_1^2} + n_2 \sqrt{(x_2 - x)^2 + z_2^2}. \tag{20.4}$$

Using elementary calculus, the optical path is given by the minimum of S, or the point where the derivative of S with respect to x vanishes,

$$\frac{dS(x)}{dx} = \frac{n_1 x}{\sqrt{x^2 + z_1^2}} - \frac{n_2 (x_2 - x)}{\sqrt{(x_2 - x)^2 + z_2^2}} = 0. \tag{20.5}$$

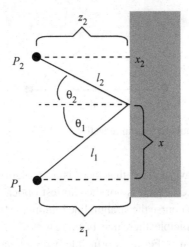

Figure 20.3 Notation relating to reflection with Fermat's principle.

Rearranging terms, and using

$$\sin \theta_1 = \frac{x}{\sqrt{x^2 + z_1^2}},$$ (20.6)

$$\sin \theta_2 = \frac{(x_2 - x)}{\sqrt{(x_2 - x)^2 + z_2^2}},$$ (20.7)

requiring the optical path to be a minimum results in

$$n_1 \sin \theta_1 = n_2 \sin \theta_2,$$ (20.8)

which is the law of refraction.

We may similarly attempt to derive the law of reflection. The geometry is illustrated in Fig. 20.3. The initial position of the ray is $(0, -z_1)$ and the final position is $(x_2, -z_2)$. The optical path in this case is given by

$$S(x) = n \left(\sqrt{x^2 + z_1^2} + \sqrt{(x_2 - x)^2 + z_2^2} \right).$$ (20.9)

Again requiring this optical path to be a minimum, we have the equation

$$\frac{dS(x)}{dx} = \frac{nx}{\sqrt{x^2 + z_1^2}} - \frac{n(x_2 - x)}{\sqrt{(x_2 - x)^2 + z_2^2}} = 0.$$ (20.10)

Referring to the figure, we immediately find the result

$$\sin \theta_1 = \sin \theta_2,$$ (20.11)

which is the law of reflection.

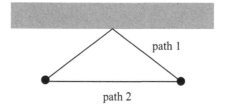

Figure 20.4 The problem of multiple rays in reflection.

The astute reader will have noticed that there is a problem with our derivation of the law of reflection: our choice of path is not the shortest optical path! The shortest path is shown in Fig. 20.4, and is of course the straight line joining the two points. This illustrates a complication of Fermat's principle: there may be more than one physical ray which connects two points. In our derivation of Fermat's principle, we have taken this into account with the statement that the ray "is shorter than any other curve which joins these points and *lies within a certain regular neighborhood of it*". In other words, the physical rays between two points are *local* minima of the optical path function. There may be one, two, or an infinite number of such rays (an ideal lens will put all rays emanating from a point source to the same focal point).

Fermat's principle is an example of what is called a *variational principle*. The Lagrangian formalism of mechanics is another well-known example. The application of Fermat's principle is straightforward in the simple cases of reflection and refraction at an interface, but how can we apply it to a gradient index medium, such as the one illustrated in Fig. 20.1? In such a case, we need to develop more sophisticated techniques, collectively referred to as *the calculus of variations*, and this is the topic of this chapter. From a variational principle, one can always derive a differential equation of motion; in the case of Fermat's principle,[1] this results in the *Eikonal equation*. Variational calculus can also be used in problems of *optimization*, when one needs to minimize or maximize a particular property of a system. We will discuss the problem of aperture apodization at the end of the chapter.

20.2 Extrema of functions and functionals

The search for maxima and minima of functions (collectively referred to as *extrema*) is a fundamental problem in calculus, and is closely related to problems of optimization. We begin our discussion by reviewing the mathematics of extrema of functions, as many of the concepts will find their analogues in the calculus of variations.

We begin by considering a function of one variable $f(x)$; an example of such a function is illustrated in Fig. 20.5. A number of observations concerning this curve will be relevant for our later analysis. First, it is well known that *local* extrema can be found by determining

---

[1] The derivation of the Eikonal equation from Fermat's principle is rather involved; a description of it can be found in [BW99, Appendix I].

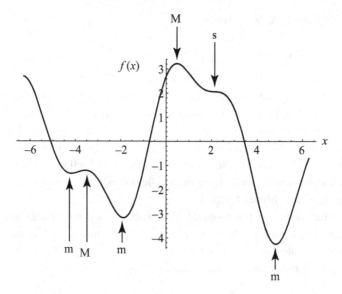

Figure 20.5 A function $f(x)$, which possesses many local maxima and minima over its range of x values. M refers to maxima, m refers to minima, and s refers to a saddle point.

those locations at which the slope of the function vanishes, i.e.

$$\text{extrema at } \frac{df}{dx} = 0. \tag{20.12}$$

By solving the resulting equation, we can identify all the points labeled maxima ('M'), minima ('m'), and saddle points ('s') in the figure. However, it is to be noted that the solution of this equation only identifies *stationary points* of the function, which may be maxima or minima, but also may be saddle points, at which the function is not maximum or minimum; we must determine the second derivative of the function at the stationary point to determine the type. Also, the solution of this equation does not tell us the *global* maximum or minimum of the function (the largest or smallest value of the function). To determine this, we would need to determine the value of the function at each stationary point (and also at the endpoints, where the slope is not necessarily zero but the function might have an extreme value). This latter point is especially important in problems of optimization. Finally, it is to be noted that for open domains (domains of a function which do not include the endpoints), there may be no well-defined stationary points. For instance, if we consider the function $f(x) = x$ on the entire real line, $x \in (-\infty, \infty)$, no stationary points exist, nor even a well-defined maximum or minimum: for any x we choose we can always find another value of x for which $f(x)$ is larger or smaller.

We can extend these observations to the problem of finding extrema of a function of N variables, $f(x_1, x_2, \ldots, x_N)$. In this case the extrema are located at positions where the

N-dimensional gradient, $\nabla_N f$, vanishes,

$$\nabla_N f(x_1,\dots,x_N) \equiv \frac{\partial f}{\partial x_1}\hat{\mathbf{x}}_1 + \cdots + \frac{\partial f}{\partial x_N}\hat{\mathbf{x}}_N = 0. \tag{20.13}$$

All of the comments regarding one-variable functions also apply here. The solutions of this equation only identify stationary points and do not specify the type, which may be a maximum, minimum, or saddle, and also do not specify which point is a global maximum or minimum. For example, a two-dimensional function $f(x,y)$ might represent the heights of mountains in the Himalayas. Solutions of Eq. (20.13) will tell us the location of all the peaks and valleys (and ridges, which are the saddle points) of the mountain range, but will not tell us which peak is Mount Everest.

An ordinary function is an input–output relationship between a (finite) number of input variables (x_1, x_2, etc.) and a single output $f(x_1,x_2,\dots)$. We now introduce the concept of a *functional*, which relates an input *function* $y(x)$ to a single output $F[y]$. This relationship typically takes the form of an integral, of the form

$$F[y] = \int_a^b f(x;y,y')\,dx. \tag{20.14}$$

The function $f(x;y,y')$ is generally a function of the independent variable x, as well as the function $y(x)$ (referred to as the *dependent variable*) and its first derivative $y'(x)$. The integral is taken from the limits a to b, which are usually dictated by the physics of the problem. We use square brackets on the functional $F[y]$ to indicate that the argument is a function, and not a variable.

One example of a functional is the optical path which Fermat's principle is based upon, which we rewrite here to make the dependence on the functions explicit,

$$F[\mathbf{r}] = \int_0^{l_0} n[\mathbf{r}(l)]\,dl. \tag{20.15}$$

Here the path through the medium is parameterized by the vector function $\mathbf{r}(l)$, where l is a parameter ranging from 0 to l_0, and $\mathbf{r}(0) \equiv \mathbf{r}_0$ and $\mathbf{r}(l_0) = \mathbf{r}_1$. This functional therefore depends on the properties of three functions, $x(l)$, $y(l)$, and $z(l)$. Fermat's principle states that a ray follows the path from $\mathbf{r}_0$ to $\mathbf{r}_1$ which minimizes the functional $F[\mathbf{r}]$.

Another well-known example of a functional is the *action* of a system, which for a mechanical system based on one coordinate $x(t)$ may be written as

$$A[x] = \int_{t_1}^{t_2} L[x(t),\dot{x}(t)]\,dt, \tag{20.16}$$

where $L(x,\dot{x}) = T(\dot{x}) - U(x)$ is the difference between the kinetic energy T of the system and the potential energy U, and is known as the *Lagrangian*. Hamilton's principle (developed in 1834 and 1835) may then be stated that *over a time interval from t_1 to t_2, a dynamical*

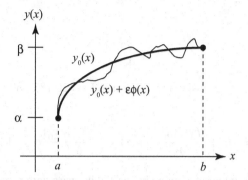

Figure 20.6 Illustrating an example of a stationary solution $y_0(x)$ and a "variation" $\tilde{y}(x)$ on that solution. It is to be noted that the variation starts and ends at the same locations as the stationary solution.

system follows the path which minimizes the quantity A[x]. This principle may be extended to a mechanical system of any number of coordinates.

Our general goal, then, is as follows. Given a functional $F[y]$, find the function $y(x)$ which produces a stationary point (typically a minimum is hoped for) of the functional. We will see, as in the case of locating extrema of functions, that (a) our efforts do not necessarily distinguish between the types of stationary points, (b) we may find multiple stationary points (recall Fig. 20.4), and (c) no proper solution may exist.

20.3 Euler's equation

We now consider finding a function $y(x)$ which results in a stationary value of the functional

$$F[y] = \int_a^b f(x;y,y')\,dx, \qquad (20.17)$$

with the constraint that $y(a) = \alpha$ and $y(b) = \beta$. Posing this problem, however, immediately raises the question of how one characterizes a stationary point of a functional. We resolve this as follows: first, *assume* that $y_0(x)$ is a stationary solution. Now, let us imagine slight *variations* of the solution from the stationary value, which we characterize by the function $\tilde{y}(x) = y_0(x) + \epsilon\phi(x)$. The number ϵ is taken to be extremely small, and the function $\phi(x)$ is arbitrary, save for assumptions of continuity and $\phi(a) = 0$ and $\phi(b) = 0$. Different functions $\phi(x)$ result in different "paths" through the integration region; this is illustrated in Fig. 20.6.

The functional relating to the variation solution can therefore be written as

$$F[y_0, \epsilon] = \int_a^b f(x;y_0 + \epsilon\phi, y_0' + \epsilon\phi')\,dx. \qquad (20.18)$$

This quantity is now an ordinary function of the variable ϵ. If $F[y]$ is truly an extremum when $y(x) = y_0(x)$, it should have a stationary point with respect to the ordinary variable ϵ

when $\epsilon = 0$, i.e.

$$\lim_{\epsilon \to 0} \frac{dF[y_0, \epsilon]}{d\epsilon} = 0. \tag{20.19}$$

This derivative can be taken inside the integral in Eq. (20.18), and we find by use of the chain rule that

$$\lim_{\epsilon \to 0} \frac{dF[y_0, \epsilon]}{d\epsilon} = \int_a^b \left[\frac{\partial f(x; y_0, y_0')}{\partial y_0} \phi(x) + \frac{\partial f(x; y_0, y_0')}{\partial y_0'} \phi'(x) \right] dx = 0. \tag{20.20}$$

The second term in the integrand can be integrated by parts with respect to x; we have

$$\int_a^b \frac{\partial f(x; y_0, y_0')}{\partial y_0'} \phi'(x) dx = \left[\frac{\partial f(x; y_0, y_0')}{\partial y_0'} \phi(x) \right]_a^b - \int_a^b \frac{d}{dx} \left[\frac{\partial f(x; y_0, y_0')}{\partial y_0'} \right] \phi(x) dx. \tag{20.21}$$

The first term on the right-hand side vanishes because $\phi(a) = \phi(b) = 0$. Our condition (20.20) for a stationary point of the functional F then takes on the form

$$\int_a^b \left\{ \frac{\partial f(x; y_0, y_0')}{\partial y_0} - \frac{d}{dx} \left[\frac{\partial f(x; y_0, y_0')}{\partial y_0'} \right] \right\} \phi(x) dx = 0. \tag{20.22}$$

Our function $\phi(x)$ was taken to be arbitrary, so for $y_0(x)$ to be a true stationary point this relation should hold regardless of the choice of variation. The only way for this to happen is for the quantity in curved brackets to vanish identically, i.e.

$$\frac{\partial f(x; y_0, y_0')}{\partial y_0} - \frac{d}{dx} \left[\frac{\partial f(x; y_0, y_0')}{\partial y_0'} \right] = 0. \tag{20.23}$$

This equation is known as *Euler's equation*. It defines a differential equation for the stationary solution y_0, which must then in turn be solved to determine the actual form of the solution. It is to be emphasized that Euler's equation only presents us with a stationary solution: additional work must be done to identify it as a maximum, minimum, or saddle.

Example 20.1 (Brachistocrone problem) Brachistocrone is taken from the Greek *brachistos* which means "shortest" and *chronos* which means "time". It refers to the problem of finding the path between two points by which a particle released at rest at the first point arrives at the second point in the least possible time. The solution of this problem (and more challenging variants) led directly to the calculus of variations, and it is referred to quite often in this context.

The problem is illustrated in Fig. 20.7. A particle of mass m starts at rest at position (x_1, y_1) and travels down a curved path $y(x)$ under the influence of gravity (acceleration g) to a position (x_2, y_2). The infinitesimal length of an element of the curve is specified by dl. It is to be noted that the positive y direction is taken downward for later convenience. We

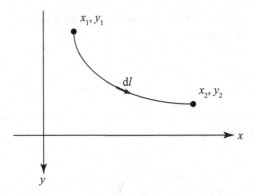

Figure 20.7 Illustrating the setup of the brachistochrone problem. A particle of mass m is released from rest at the point (x_1,y_1) and travels under the influence of gravity to the point (x_2,y_2). We wish to determine the path which minimizes the travel time.

wish to find the path which results in the shortest transit time between the start and end positions.

It is worth taking a moment to consider the nontriviality of the solution. A straight line connecting the two points is the shortest spatial distance but not a path of least time, because the particle will be traveling slowly during the early part of the journey, and wasting time. We can make the particle move faster initially by making the path steeper at the beginning (similar to Fig. 20.7), but this increases the overall length of the path. Evidently there is some path which represents a balance between the need for large initial speed and small overall distance.

We start by noting that the infinitesimal path length is related to x and y by the formula

$$dl = \sqrt{dx^2 + dy^2} = dx\sqrt{1 + (dy/dx)^2}. \qquad (20.24)$$

The time dt the particle takes to traverse this infinitesimal distance depends on its velocity v on that segment of path, i.e.

$$dt = \frac{dl}{v}. \qquad (20.25)$$

But what is this velocity? Here we rely on conservation of energy and note that the kinetic energy of the particle is equal to the amount of potential energy it has released on moving from its initial position,

$$\frac{1}{2}mv^2 = mgy, \qquad (20.26)$$

or that

$$v(y) = \sqrt{2gy}. \qquad (20.27)$$

We now have all the pieces to write a formula for the time it takes a particle to traverse the curve. This time may be written as

$$T \equiv \int_{t_1}^{t_2} dt = \int_{x_1,y_1}^{x_2,y_2} \frac{dl}{v} = \frac{1}{\sqrt{2g}} \int_{x_1}^{x_2} \sqrt{\frac{1+y'^2}{y}}\, dx. \tag{20.28}$$

We have parameterized the time in terms of an integral over the horizontal position x.
    This equation has exactly the form of the functional (20.14), with

$$f(x;y,y') = \sqrt{\frac{1+y'^2}{y}}. \tag{20.29}$$

We may immediately substitute this function into the Euler equation (20.23) to find an equation for the curve of least time, which after taking the partial derivatives takes the form

$$-\frac{1}{2}\frac{(1+y'^2)^{1/2}}{y^{3/2}} - \frac{d}{dx}\left[\frac{y'}{y^{1/2}(1+y'^2)^{1/2}}\right] = 0. \tag{20.30}$$

On taking the total derivative with respect to x, a number of terms cancel and we are left with the relatively simple formula

$$y'^2 + 2yy'' = -1. \tag{20.31}$$

This can be rewritten in terms of a total derivative with respect to x as

$$\frac{1}{y'}\frac{d}{dx}(yy'^2) = -1. \tag{20.32}$$

On integration, we find

$$y'^2 = \frac{C}{y} - 1, \tag{20.33}$$

where C is a constant of integration which in the end will be chosen such that the particle passes through the end point (x_2,y_2). This equation is a differential equation for the path of least time $y(x)$.
    Equation (20.33) illustrates one of the common challenges in applying the calculus of variations; it is often straightforward to find a differential equation for the stationary solution, but not so easy to solve this equation itself. Fortunately, this equation is well known to represent a *cycloid*, the curve traced out by a single point on a circle as it rolls along a horizontal plane; the cycloid is illustrated in Fig. 20.8. Parameterizing the cycloid by an angle θ, we may write

$$x(\theta) = a(\theta - \sin\theta), \tag{20.34}$$

$$y(\theta) = a(1 - \cos\theta). \tag{20.35}$$

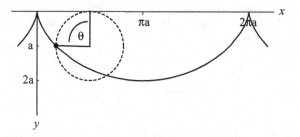

Figure 20.8 Illustration of a cycloid and the method of generating one.

One can readily show that Eqs. (20.34) and (20.35) satisfy Eq. (20.33). Using the chain rule, we may write

$$\frac{dy}{dx} = \frac{dy}{d\theta}\frac{d\theta}{dx}. \tag{20.36}$$

Then, noting that

$$\frac{dx}{d\theta} = a(1 - \cos\theta), \tag{20.37}$$

$$\frac{dy}{d\theta} = a\sin\theta, \tag{20.38}$$

we can rewrite the differential equation (20.33) in the form

$$a\sin\theta = a(1 - \cos\theta)\sqrt{\frac{C}{a(1 - \cos\theta)} - 1}. \tag{20.39}$$

This equation can be simplified by expanding the terms in the square root, so that

$$a\sin\theta = a\sqrt{(Ca - 1) + (2 - Ca)\cos\theta - \cos^2\theta}. \tag{20.40}$$

With the choice $C = 2/a$, this equation will be satisfied.

To match our end point, we need to solve our cycloid equations for a (the "radius" of the cycloid) and θ (the position at which the cycloid intersects the end point). It is clear from Eqs. (20.34) and (20.35) that this requires the solution of a nonlinear equation for θ. Several solutions are shown in Fig. 20.9. If the particle must travel long horizontal distances, the optimal path takes it *below* its final destination, to further increase speed.

◇

Example 20.2 (GRIN medium) Optical fibers and lenses are often made from so-called GRadient INdex materials, in which the refractive index is a function of radial position. We

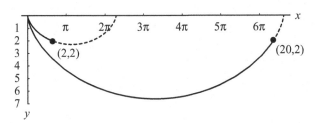

Figure 20.9 Solutions of the brachistochrone problem, for several values of the end points.

wish to study the propagation of rays in a material with a refractive index specified by

$$n(y) = n_0 - \alpha y^2, \tag{20.41}$$

and search for solutions which begin at $(0,0)$ and return to the axis at $y = 0$. The functional of Fermat's principle may be written in the form

$$F[\mathbf{r}] = \int_0^{l_0} n(\mathbf{r}(l))\mathrm{d}l, \tag{20.42}$$

and we choose to parameterize the path by $y(x)$. The infinitesimal path length may be written as

$$\mathrm{d}l = \sqrt{\mathrm{d}x^2 + \mathrm{d}y^2} = \sqrt{1 + y'^2}\,\mathrm{d}x, \tag{20.43}$$

so that our functional to be minimized may be written as

$$F[y] = \int_0^{x_0} (n_0 - \alpha y^2)\sqrt{1 + y'^2}\,\mathrm{d}x. \tag{20.44}$$

We may immediately apply Euler's equation, with the function

$$f(y, y') = (n_0 - \alpha y^2)\sqrt{1 + y'^2}. \tag{20.45}$$

The derivatives are complicated and require a little bit of care. Applying the partial derivatives first, we have

$$-2\alpha y\sqrt{1 + y'^2} - \frac{\mathrm{d}}{\mathrm{d}x}\left[\frac{(n_0 - \alpha y^2)y'}{\sqrt{1 + y'^2}}\right] = 0. \tag{20.46}$$

We may then take the derivative with respect to the variable x. Applying the product rule, we get a complicated quantity, which may be simplified somewhat by multiplying the equation through by $(1 + y'^2)^{3/2}$. We arrive at the formula

$$-2\alpha y(1 + y'^2)^2 - (n_0 - \alpha y^2)y''\left[(1 + y'^2) - y'^2\right] + 2\alpha yy'^2(1 + y'^2) = 0. \tag{20.47}$$

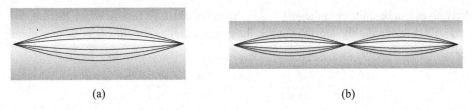

(a)   (b)

Figure 20.10 Rays propagating in a GRIN medium, used as (a) a lens, and (b) an optical fiber.

Grouping terms, and canceling some, we finally have the expression

$$2\alpha y(1+y'^2) + (n_0 - \alpha y^2)y'' = 0. \tag{20.48}$$

Although the variational calculus has allowed us to quite readily find a differential equation which the solution satisfies, finding the solution to that equation is another matter entirely, as it is highly nonlinear. We will settle for an approximate solution, by requiring one for which

$$y' \ll 1, \quad y \ll \sqrt{n_0/\alpha}. \tag{20.49}$$

With this approximation, we may neglect the y and y' parts in the parenthesis of Eq. (20.48), and immediately write the simple formula

$$2\alpha y + n_0 y'' = 0, \tag{20.50}$$

which is simply a harmonic oscillator! Solutions which start at $(0,0)$ may be simply written as

$$y(x) = A\sin(\sqrt{2\alpha/n_0}x). \tag{20.51}$$

As long as A is small (to satisfy inequalities (20.49)), any ray which passes through the origin will return to the y-axis at a distance $x_0 = \pi\sqrt{2\alpha/n_0}$. Short lengths of such a GRIN material may be used as a lens, or long lengths will serve as an optical fiber; both applications are illustrated in Fig. 20.10.

It is worth noting that although this makes a nice example of an application of Fermat's principle, a proper analysis of a GRIN-type medium would involve a solution of Maxwell's equations, which would take into account the full vector wave nature of fields confined to the medium.

◇

20.4 Second form of Euler's equation

Euler's equation (20.23) contains partial derivatives with respect to the dependent variables $y(x)$ and $y'(x)$, and a total derivative with respect to the independent variable x. These derivatives can be quite complicated, but in the rather common case in which the function $f(x; y, y')$ is not explicitly dependent on x, the problem can be simplified, as we now show.

It is to be noted that

$$\frac{df}{dx} = \frac{\partial f}{\partial x} + \frac{\partial f}{\partial y}\frac{dy}{dx} + \frac{\partial f}{\partial y'}\frac{dy'}{dx} = \frac{\partial f}{\partial x} + y'\frac{\partial f}{\partial y} + y''\frac{\partial f}{\partial y'}. \tag{20.52}$$

We may also note, however, that we may write

$$\frac{d}{dx}\left(y'\frac{\partial f}{\partial y'}\right) = y''\frac{\partial f}{\partial y'} + y'\frac{d}{dx}\frac{\partial f}{\partial y'}. \tag{20.53}$$

The first term on the right-hand side is equal to the last term of Eq. (20.52), and on substitution we find that

$$\frac{d}{dx}\left(y'\frac{\partial f}{\partial y'}\right) = \frac{df}{dx} - \frac{\partial f}{\partial x} - y'\frac{\partial f}{\partial y} + y'\frac{d}{dx}\frac{\partial f}{\partial y'}. \tag{20.54}$$

The rightmost two terms of this equation vanish if the Euler equation (20.23) is satisfied, and we are left with the expression

$$\frac{\partial f}{\partial x} - \frac{d}{dx}\left(f - y'\frac{\partial f}{\partial y'}\right) = 0. \tag{20.55}$$

This is a second form of the Euler equation. If the function f does not explicitly depend on the independent variable x, the partial derivative vanishes and we may integrate the term that remains. We are left with

$$f - y'\frac{\partial f}{\partial y'} = C, \tag{20.56}$$

where C is a constant of integration.

Example 20.3 (Fermat's principle and mirages) We consider as an example the propagation of an optical ray through a height-dependent refractive index given by

$$n(y) = n_0 + \alpha y. \tag{20.57}$$

This is a reasonable approximation to the inversion of refractive index which happens above blacktop (tarmac) on a hot, sunny day. On such days, one can often see mirages in which the surface of the road at a distance appears to be wet – what one is actually seeing is an image of the sky. We wish to use Fermat's principle to determine the path of such a "mirage ray". The initial position of the ray is taken to be $(x_1, y_1) = (-x_0, 0)$ and the final position is $(x_2, y_2) = (x_0, 0)$. The quantity n_0 is the refractive index of the atmosphere at $y = 0$, and α is a parameter which represents the gradient of the refractive index. The system is illustrated in Fig. 20.11.

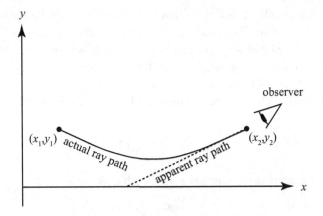

Figure 20.11 Illustrating the notation and principle behind mirage formation.

This problem is quite similar to the GRIN medium of the previous section, and could be solved in essentially the same manner; however, here we use the second form of the Euler equation to determine a solution.

The functional of Fermat's principle may again be written in the form

$$F[\mathbf{r}] = \int_0^{l_0} n(\mathbf{r}(l))\,dl, \tag{20.58}$$

where, if we parameterize the path by $y(x)$, we may write

$$dl = \sqrt{dx^2 + dy^2} = \sqrt{1 + y'^2}\,dx. \tag{20.59}$$

On substitution from Eqs. (20.57) and (20.59) into Fermat's principle, we have the following functional to minimize

$$\int_{-x_0}^{x_0} (n_0 + \alpha y)\sqrt{1 + y'^2}\,dx. \tag{20.60}$$

Because this functional does not depend explicitly on the independent variable x, we may use the second form of Euler's equation, and its solution (20.56). On taking the appropriate derivative and substituting, we have

$$(n_0 + \alpha y)\sqrt{1 + y'^2} - \frac{(n_0 + \alpha y)y'^2}{\sqrt{1 + y'^2}} = C. \tag{20.61}$$

We may multiply both sides of this equation by $\sqrt{1 + y'^2}$, and then square both sides of the equation; we are left with the formula

$$1 + y'^2 = \frac{1}{C^2}(n_0 + \alpha y)^2. \tag{20.62}$$

We have once again readily found the differential equation which the stationary solution satisfies; the challenge now lies in finding an actual solution to this equation.

First, it is to be noted that with the choice $y = 0$, $y' = 0$, we can satisfy the differential equation with $C = n_0$. This solution represents the straight-line path between the start and end points; recall Fig. 20.4 regarding reflection.

To find a nontrivial solution, we need to be more clever. It is to be noted that the hyperbolic identity

$$1 + \sinh^2 x = \cosh^2 x \tag{20.63}$$

may be written, by defining the function $g(x) = \cosh x$, as

$$1 + g'^2 = g^2. \tag{20.64}$$

This is structurally similar to Eq. (20.62), and suggests that we may try a solution of the form

$$y(x) = A\cosh(Bx) + D, \tag{20.65}$$

where A, B, and D are constants to be determined. On substitution into Eq. (20.62), we arrive at the equation

$$1 - (AB)^2 + (AB)^2 \cosh^2(Bx)$$
$$= \frac{1}{C^2}\left[(n_0 + \alpha D)^2 + 2(n_0 + \alpha D)\alpha A \cosh(Bx) + (\alpha A)^2 \cosh^2(Bx)\right]. \tag{20.66}$$

To satisfy this equation, we require $D = -n_0/\alpha$, $A = 1/B$, and $B = \alpha/C$. Our solution takes on the form

$$y(x) = -\frac{n_0}{\alpha} + \frac{C}{\alpha}\cosh(\alpha C x). \tag{20.67}$$

This solution has the form of a *catenary*, the shape of a chain or cable which is fixed at both ends and allowed to hang under its own weight. The constant C must be chosen to match the boundary condition that $y(x_0) = y(-x_0) = 0$. This leads to the condition

$$\cosh(\alpha C x_0) = n_0/C. \tag{20.68}$$

This is a nonlinear equation for C which cannot be solved analytically. An example of a solution is shown in Fig. 20.12.

◇

20.5 Calculus of variations with several dependent variables

In the examples considered so far, we had a functional which depends on a single dependent variable y (and its derivative, y'). Many problems, however, will involve a functional which

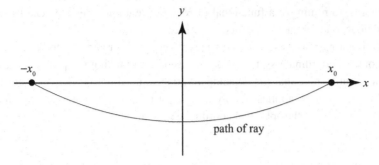

Figure 20.12 Illustrating a catenary solution to the mirage problem, with $\alpha/n_0 = 0.005$ m^{-1} and $x_0 = 100$ m.

depends on multiple dependent variables. For instance, the propagation of a ray through a three-dimensional system must satisfy Fermat's principle based on the functional

$$F[x(l),y(l),z(l)] = \int_0^{l_0} n(x(l),y(l),z(l))dl. \qquad (20.69)$$

The functional depends on $x(l)$, $y(l)$, and $z(l)$, which characterize the path of the ray through the medium of refractive index $n(x,y,z)$.

It is not difficult to determine the nature of the stationary solution in such cases. We consider a general function of N dependent variables,

$$F[\{y_n(x)\}] = \int_a^b f(x;y_1,y_1',\ldots,y_N,y_N')dx. \qquad (20.70)$$

We simultaneously introduce a variation into each of the dependent variables,

$$\tilde{y}_n(x) = y_{n,0}(x) + \epsilon\phi_n(x), \qquad (20.71)$$

where the set $\{y_{n,0}\}$ represents the optimal set of dependent variables. The epsilon is the same for each of the variations. When we determine the stationary point with respect to ϵ, by use of the product rule we find that

$$\int_a^b \sum_{n=1}^N \left\{ \frac{\partial f}{\partial y_n} - \frac{d}{dx}\left[\frac{\partial f}{\partial y_n'} \right] \right\} \phi_n(x)dx = 0. \qquad (20.72)$$

Since this equation must hold regardless of the choice of variations ϕ_n, we find that we must satisfy N independent equations,

$$\frac{\partial f}{\partial y_n} - \frac{d}{dx}\left[\frac{\partial f}{\partial y_n'} \right] = 0 \quad \text{for all } n = 1,2,\ldots,N. \qquad (20.73)$$

In other words, to minimize a functional of N dependent variables we need to satisfy N Euler equations, one for each dependent variable.

It is worth noting that these equations may be coupled, in addition to being nonlinear; a good choice of coordinate system is often essential in making the problem tractable. It is also important to note that the dependent variables need not be Cartesian coordinates: the calculus of variations applies equally well if, for instance, we choose as our dependent variables the (r,θ,ϕ) of the spherical coordinate system.

20.6 Calculus of variations with several independent variables

We may also consider the solution of variational problems where the functional depends on more than one *independent* variable. For instance, in two independent variables, we might have a functional of the form

$$F[z] = \int_D f(x,y;z,z_x,z_y)dxdy, \qquad (20.74)$$

where the integral is now over an area D and the functional depends on $z(x,y)$, as well as the partial derivatives z_x and z_y; we may visualize $z(x,y)$ as the height of a surface above the xy-plane. We assume that all possible surfaces have the same boundary curve C.

To find an extremum of this functional, we introduce a single-parameter family of surfaces $z(x,y,\epsilon)$,

$$z(x,y,\epsilon) = z(x,y) + \epsilon\phi(x,y), \qquad (20.75)$$

where it is assumed that $z(x,y,0) = z(x,y)$ is an extreme solution. This being the case, it should satisfy the equation

$$\lim_{\epsilon\to 0}\frac{dF[z,\epsilon]}{d\epsilon} = 0. \qquad (20.76)$$

Let us define $\phi_x(x,y) \equiv \chi(x,y)$ and $\phi_y(x,y) \equiv \psi(x,y)$. Our variational equation therefore takes the form

$$\lim_{\epsilon\to 0}\frac{dF[z,\epsilon]}{d\epsilon} = \int_D (f_\phi\phi + f_\chi\chi + f_\psi\psi)dxdy = 0. \qquad (20.77)$$

This may be simplified by observing that

$$\frac{\partial[f_\chi\phi]}{\partial x} = \frac{\partial f_\chi}{\partial x}\phi + f_\chi\chi, \qquad (20.78)$$

$$\frac{\partial[f_\psi\phi]}{\partial y} = \frac{\partial f_\psi}{\partial y}\phi + f_\psi\psi. \qquad (20.79)$$

The latter two integrations of Eq. (20.77) may therefore be written as

$$\int_D (f_\chi \chi + f_\psi \psi) dx dy = \int_D \left[\frac{\partial[f_\chi \phi]}{\partial x} + \frac{\partial[f_\psi \phi]}{\partial y} \right] dx dy - \int_D \left[\frac{\partial f_\chi}{\partial x} + \frac{\partial f_\psi}{\partial y} \right] \phi dx dy. \quad (20.80)$$

Here we run into a situation where the existing notation is somewhat inadequate; the derivative $\partial[f_\chi]/\partial x$ is what may be called a "total partial derivative", in which ϕ, χ, and ψ are still considered functions of x and only y is considered constant. The result of this is that, for instance,

$$\frac{\partial f_\chi}{\partial x} = f_{\chi x} + f_{\chi \phi} \frac{\partial \phi}{\partial x} + f_{\chi \chi} \frac{\partial \chi}{\partial x} + f_{\chi \psi} \frac{\partial \psi}{\partial x}. \quad (20.81)$$

We may apply Stokes' theorem to the first integral of Eq. (20.80), so that

$$\int_D \left[\frac{\partial[f_\chi \phi]}{\partial x} + \frac{\partial[f_\psi \phi]}{\partial y} \right] dx dy = \oint_C [f_\chi dy - f_\psi dx] \phi. \quad (20.82)$$

This integral is identically zero, because of the requirement that all admissible surfaces have the same height on the boundary, i.e. $\phi = 0$. We therefore find that

$$\int_D (f_\chi \chi + f_\psi \psi) dx dy = - \int_D \left[\frac{\partial f_\chi}{\partial x} + \frac{\partial f_\psi}{\partial y} \right] \phi dx dy. \quad (20.83)$$

Substituting back into Eq. (20.77), our necessary condition for an extremum is

$$\int_D \left(f_\phi - \frac{\partial f_\chi}{\partial x} - \frac{\partial f_\psi}{\partial y} \right) \phi dx dy = 0. \quad (20.84)$$

Since this equation should be independent of ϕ, we find the condition

$$f_\phi - \frac{\partial f_\chi}{\partial x} - \frac{\partial f_\psi}{\partial y} = 0. \quad (20.85)$$

In terms of the original partial derivatives, we may write

$$\frac{\partial f}{\partial z} - \frac{\partial}{\partial x} \left[\frac{\partial f}{\partial z_x} \right] - \frac{\partial}{\partial y} \left[\frac{\partial f}{\partial z_y} \right] = 0. \quad (20.86)$$

This equation is known as the *Ostrogradski equation*, after Mikhail Vasilievich Ostrogradski who first derived it in 1834. It is analogous to Euler's equation (20.23) for a single independent variable.

Equation (20.86) can be generalized to N independent variables in a straightforward manner. The minimum solution for the functional

$$F[z] = \int_D f(x_1, \ldots, x_N; z, z_{x_1}, \ldots, z_{x_N}) dx dy \quad (20.87)$$

satisfies the equation

$$\frac{\partial f}{\partial z} - \sum_{n=1}^{N} \frac{\partial}{\partial x_n} \left[\frac{\partial f}{\partial z_{x_n}} \right] = 0. \tag{20.88}$$

The Ostrogradski equation can be used to derive partial differential equations of physical systems from a corresponding variational principle.

Example 20.4 (Laplace's equation) We search for extreme solutions of the functional

$$F[U(x,y)] = \int_D \left[\left(\frac{\partial U}{\partial x} \right)^2 + \left(\frac{\partial U}{\partial y} \right)^2 \right] dxdy, \tag{20.89}$$

subject to the condition that $U(x,y) = U_0(x,y)$ on a boundary curve C of the domain D. The function $f(x,y;z,z_x,z_y)$ of Eq. (20.86) is

$$f(x,y;U,U_x,U_y) = \left(\frac{\partial U}{\partial x} \right)^2 + \left(\frac{\partial U}{\partial y} \right)^2. \tag{20.90}$$

On substitution into Eq. (20.86), we readily find that

$$\frac{\partial^2 U}{\partial x^2} + \frac{\partial^2 U}{\partial y^2} = 0. \tag{20.91}$$

This is, of course, simply Laplace's equation in two dimensions. Given that $f(x,y;U,U_x,U_y) = (\nabla U)^2$, we may observe that solutions to Laplace's equation minimize the mean square value of the gradient of U on the domain of interest.

◇

20.7 Euler's equation with auxiliary conditions: Lagrange multipliers

In solving variational problems, particularly when applying Hamilton's principle and studying mechanical systems, there are often one or more *constraints* on the behavior of the system,[2] usually often implicitly stated. This is illustrated in Fig. 20.13. In the case of the weight-and-pulley system, the positions of the weights are not independent: they are related by the free-hanging length of the rope L such that $x_1 + x_2 = L$. Similarly, for a particle constrained to a sphere the coordinates are related by the equation $x^2 + y^2 + z^2 = R^2$.

Often we can apply our constraint directly to eliminate one of the dependent variables in the problem. In the weight-and-pulley system, we can substitute $x_2 = L - x_1$ into our expressions for the potential and kinetic energy and solve the problem entirely in terms

---

[2] Rigid constraints on the direction of a light ray are not encountered in practice, so this discussion applies more to mechanical systems and optimization problems then to Fermat's principle.

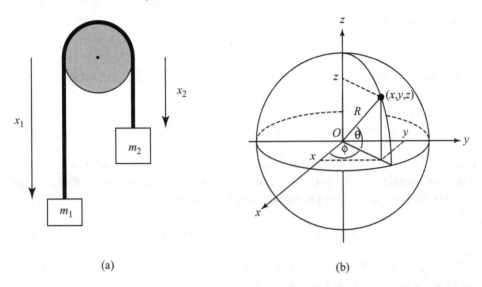

(a)   (b)

Figure 20.13 (a) A simple weight-and-pulley system, where the positions of the weights are specified by x_1 and x_2. These positions are constrained by the free-hanging length of the rope L, such that $x_1 + x_2 = L$. (b) Motion on the surface of a sphere. The particle position is constrained by the condition $x^2 + y^2 + z^2 = R^2$. In spherical coordinates, this constraint takes on the simpler form $r = R$.

of x_1. In other circumstances, an appropriate choice of coordinate system will separate the constraint from the rest of the problem. In the particle constrained to a sphere, by choosing spherical coordinates we separate the r, θ, and ϕ dependencies and can solve readily for the θ and ϕ motion. In other cases, however, the application of the constraint is not so straightforward. If we consider the motion of a particle constrained to move on the surface of a cycloid (Eqs. (20.34) and (20.34)), for instance, there is no simple formula relating x and y. Furthermore, we might have a constraint which depends on the derivative of a coordinate, rather than the coordinate itself. For instance, a particle may be constrained to move in a rotating or accelerating system. We would now like to develop a technique for explicitly dealing with such constraints.

Let us consider for simplicity a system which depends on two dependent variables $x_1(t)$ and $x_2(t)$, and one independent variable t. A constraint is referred to as *holonomic* if it depends only on the dependent variables and not their derivatives, i.e. it may be written in the form

$$g(x_1, x_2; t) = 0. \tag{20.92}$$

It is to be noted that this includes constraints which seemingly depend on the derivatives of variables but may be integrated, i.e. $x_1'(t) = q$ can be integrated to $x_1(t) = qt + C$. If a constraint cannot be written in the form of Eq. (20.92), which typically means that it depends on $x_1'(t)$ and/or $x_2'(t)$, it is referred to as *non-holonomic*, in which case it can be written formally as

$$g(x_1, x_2, x_1', x_2'; t) = 0. \tag{20.93}$$

We restrict ourselves to the holonomic case from now on. Let us return to our functional $F[x_1, x_2]$,

$$F[x_1, x_2] = \int_0^1 f(x_1, x_2, x_1', y_1'; t) \mathrm{d}t. \tag{20.94}$$

Because $g = 0$, we can add it to the function f without changing the value of the overall functional, introducing an as yet unspecified multiplier $\lambda(t)$, which is known as the *Lagrange undetermined multiplier*. Our functional then takes on the form

$$F[x_1, x_2] = \int_0^1 \left\{ f(x_1, x_2, x_1', x_2'; t) + \lambda(t) g(x_1, x_2; t) \right\} \mathrm{d}t. \tag{20.95}$$

Euler's equation (20.23) must be satisfied for this new form of the functional as well, and we are left with the following pair of equations:

$$\frac{\partial f}{\partial x_n} - \frac{\mathrm{d}}{\mathrm{d}t} \left[\frac{\partial f}{\partial x_n'} \right] + \lambda(t) \frac{\partial g}{\partial x_n} = 0 \quad \text{for } n = 1, 2. \tag{20.96}$$

To see what we have gained by this approach, it is helpful to consider what we would be left with by using Euler's equation without undetermined multipliers. In such a case, we would have reduced our two dependent variables to one dependent variable using our constraint equation, and then solve for the differential equation defining that single dependent variable. By the use of Lagrange undetermined multipliers, we keep our two independent variables, introduce an undetermined multiplier $\lambda(t)$, and find a pair of differential equations for the dependent variables. We also still have the equation of constraint; in total, we have three unknowns ($x_1(t)$, $x_2(t)$, and $\lambda(t)$) and three equations, so in principle we should be able to solve this system. The undetermined multipliers allow us to apply the constraints *after* applying the calculus of variations, instead of before.

In mechanics problems, the use of Lagrange undetermined multipliers provides an additional piece of physical information about the system: the forces required to satisfy the constraints on the mechanical system. We may see this in a straightforward way by considering the Euler equations for a particle of mass m traveling in a holonomic Cartesian coordinate system; we may write them as

$$\frac{\mathrm{d}}{\mathrm{d}t} \left(\frac{\partial f}{\partial \dot{x}_n} \right) = -\frac{\partial f}{\partial x_n} + \lambda \frac{\partial g}{\partial x_n}. \tag{20.97}$$

The dot indicates a time derivative. Because the system is Cartesian and holonomic, we may immediately write

$$\frac{\partial f}{\partial \dot{x}_n} = \frac{\partial}{\partial \dot{x}_n} \left[\frac{1}{2} m \dot{x}_n^2 \right] = m \ddot{x}_n. \tag{20.98}$$

If the potential energy of the system depends only on the position of the particle, we may also write

$$\frac{\partial f}{\partial x_n} = \frac{\partial U}{\partial x_n}.$$
(20.99)

In vector form, Eq. (20.97) becomes

$$m\ddot{\mathbf{x}} = -\nabla U + \lambda \nabla g.$$
(20.100)

This equation is simply Newton's second law. The first term on the right-hand side is the force arising from the potential energy of the system; the second term is associated with the forces required to satisfy the constraints on the system. In the case of non-Cartesian systems, the λs are still associated with the forces of constraint, but the relationship is not as straightforward as in Eq. (20.100).

Example 20.5 (Bead on a rotating wire) To illustrate the application of the undetermined multipliers, we consider a problem for which the forces of constraint can be calculated readily both by Newton's laws and by the use of variational calculus.

We consider a bead constrained to move along a straight wire which is rotating about the z-axis with angular frequency ω. The wire makes an angle α with the z-axis. The geometry is illustrated in Fig. 20.14 (a). Because of the axial symmetry, the cylindrical coordinates (r, ϕ, z) are most appropriate here. Our constraints on the motion of the bead may be broken down into the constraint of the bead along the length of the wire, and the constraint of the rotation of the wire, i.e.

$$\phi(t) = \phi_0 + \omega t,$$
(20.101)

$$r = z \tan \alpha.$$
(20.102)

We are interested in determining the force the wire must apply to the bead to constrain it to the wire. We can evaluate this in a straightforward way using Newtonian mechanics, using the force diagram illustrated in Fig. 20.14 (b). Two forces act upon the bead: the force of gravity mg and the centrifugal "force" $mr\omega^2$. Decomposing each of these forces into a component parallel to the wire and a component perpendicular, we find that the constraint force F_c the wire must apply is given by

$$F_c = mr\omega^2 \cos\alpha + mg \sin\alpha = (mz\omega^2 + mg)\sin\alpha.$$
(20.103)

We now wish to determine this force by using Hamilton's principle. The Lagrangian for this system is given by

$$L = \frac{1}{2}m\dot{r}^2 + \frac{1}{2}mr^2\dot{\phi}^2 + \frac{1}{2}m\dot{z}^2 - mgz,$$
(20.104)

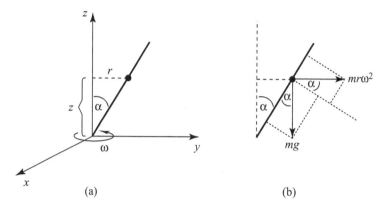

Figure 20.14 (a) A bead constrained to slide without friction along a rotating wire. (b) The force diagram.

subject to the constraints given above. We will explicitly include the rotational constraint, and deal with the other via an undetermined multiplier. Applying $\dot{\phi} = \omega$, we have

$$L = \frac{1}{2}m\dot{r}^2 + \frac{1}{2}mr^2\omega^2 + \frac{1}{2}m\dot{z}^2 - mgz. \tag{20.105}$$

Finally, we include our r, z-constraint by the use of an undetermined multiplier, creating the modified Lagrangian

$$\hat{L} = \frac{1}{2}m\dot{r}^2 + \frac{1}{2}mr^2\omega^2 + \frac{1}{2}m\dot{z}^2 - mgz + \lambda(r - z\tan\alpha). \tag{20.106}$$

We now apply Euler's formula for the two coordinates r and z. The formula for r takes on the form

$$mr\omega^2 + \lambda + m\ddot{r} = 0, \tag{20.107}$$

while the formula for z takes on the form

$$-mg - \lambda\tan\alpha - m\ddot{z} = 0. \tag{20.108}$$

These two formulas, combined with the constraint formula $r = z\tan\alpha$, constitute a set of equations for r, z, and λ, which it is to be noted may depend on the coordinates. To solve this set of equations for λ, which is directly related to the constraint force, we first take the second time derivative of our constraint equation, i.e.

$$\ddot{r} = \ddot{z}\tan\alpha. \tag{20.109}$$

Equation (20.107) may then be written entirely in terms of z as

$$m\ddot{z} = m\omega^2 z + \frac{\lambda}{\tan\alpha}. \tag{20.110}$$

We may substitute this expression for $\ddot{z}$ into Eq. (20.108). We are left with the formula

$$-mg - \lambda \left(\tan\alpha + \frac{1}{\tan\alpha} \right) - m\omega^2 z = 0. \tag{20.111}$$

We can immediately solve this for λ to find that

$$\lambda = -\frac{mg + m\omega^2 z}{\tan\alpha + 1/\tan\alpha} = -(mg + m\omega^2 z)\cos\alpha\sin\alpha. \tag{20.112}$$

From Eq. (20.107), we can see that this is the r-component of the constraint force. To get the z-component, we multiply λ by $\tan\alpha$. The magnitude of the force is then given by

$$F_c = (mz\omega^2 + mg)\sin\alpha\sqrt{\cos^2\alpha + \sin^2\alpha} = (mz\omega^2 + mg)\sin\alpha, \tag{20.113}$$

which is exactly the result derived by using the force diagram.

◇

20.8 Hamiltonian dynamics

We consider again the case of Section 20.5 in which we seek to minimize a functional of N dependent variables of the form

$$F[\{q_n(x)\}] = \int_a^b f(t; q_1, q_1', \ldots, q_N, q_N')dt, \tag{20.114}$$

where we label the dependent variables q_i and the independent variable t. Let us suppose, for the moment, that the function f is not explicitly dependent on t, such that

$$\frac{\partial f}{\partial t} = 0. \tag{20.115}$$

We now ask what consequences arise from this seemingly simple and very common condition. The total derivative of f with respect to t may be written in the form

$$\frac{df}{dt} = \sum_{j=1}^N \frac{\partial f}{\partial q_j} q_j' + \sum_{j=1}^N \frac{\partial f}{\partial q_j'} q_j'' + \frac{\partial f}{\partial t}. \tag{20.116}$$

We may use Euler's equation (20.73), of the form

$$\frac{\partial f}{\partial q_j} = \frac{d}{dt}\left[\frac{\partial f}{\partial q_j'}\right], \tag{20.117}$$

to modify the first term in Eq. (20.116); also applying Eq. (20.115), we then have

$$\frac{df}{dt} = \sum_{j=1}^{N} q_j' \frac{d}{dt}\left[\frac{\partial f}{\partial q_j'}\right] + \sum_{j=0}^{N} \frac{\partial f}{\partial q_j'} q_j''$$

$$= \frac{d}{dt}\left[\sum_{j=1}^{N} q_j' \frac{\partial f}{\partial q_j'}\right]. \tag{20.118}$$

Grouping the total derivatives together, we find that

$$\frac{d}{dt}\left[f - \sum_{j=0}^{N} q_j' \frac{\partial f}{\partial q_j'}\right] = 0. \tag{20.119}$$

Apparently the quantity H, defined as

$$H \equiv \sum_{j=1}^{N} q_j' p_j - f, \tag{20.120}$$

with

$$p_j \equiv \frac{\partial f}{\partial q_j'}, \tag{20.121}$$

is *constant* for all values of t. Borrowing terminology from mechanics, the quantity H will be referred to as the *Hamiltonian* of the system; the quantities p_j will be referred to as *generalized momenta*. The Hamiltonian is considered to be a function of q_j and p_j rather than q_j and q_j'. If t represents time, H represents a quantity of the system that is conserved in time; if t represents a parameter of a curve, H represents a quantity that is preserved along the path of the curve. It is to be noted that the Hamiltonian was constructed using Euler's equation, and therefore implicitly assumes an extremal solution to the variational problem. We may use the Hamiltonian to construct an alternative system of equations describing this extremal solution.

First, it is to be noted that, in general, the evolution of the Hamiltonian (referring to t as time for the moment) is given by

$$\frac{dH}{dt} = \frac{\partial H}{\partial t}. \tag{20.122}$$

That is, the Hamiltonian changes in time if and only if the Hamiltonian is explicitly time-dependent; if it is not, it is constant in time.

We may derive evolution equations by first noting that the total differential of H is

$$dH = \sum_{j=1}^{N} \left(\frac{\partial H}{\partial q_j} dq_j + \frac{\partial H}{\partial p_j} dp_j\right) + \frac{\partial H}{\partial t} dt. \tag{20.123}$$

From Eq. (20.120), we may also write

$$dH = \sum_{j=1}^{N} \left[q_j' dp_j + p_j dq_j' - \frac{\partial f}{\partial q_j} dq_j - \frac{\partial f}{\partial q_j'} dq_j' \right] - \frac{\partial f}{\partial t} dt. \tag{20.124}$$

However, we have seen that $\partial f / \partial q_j' = p_j$ and, from Euler's equation, we have $\partial f / \partial q_j = p_j'$; on substitution into Eq. (20.124), we find that the terms dependent on dq_j' cancel and we are left with

$$dH = \sum_{j=1}^{N} \left(q_j' dp_j - p_j' dq_j \right) - \frac{\partial f}{\partial t} dt. \tag{20.125}$$

On comparison with Eq. (20.123), we find that

$$q_j' = \frac{\partial H}{\partial p_j}, \tag{20.126}$$

$$-p_j' = \frac{\partial H}{\partial q_j}. \tag{20.127}$$

These equations are referred to in mechanics as *Hamilton's equations of motion*.

Example 20.6 (Mechanics and Hamilton's principle) In Section 20.2, we described Hamilton's principle, which states that a dynamical mechanical system follows a path that minimizes the quantity

$$A[x] = \int_{t_1}^{t_2} L[x(t), \dot{x}(t)] dt, \tag{20.128}$$

where L is the Lagrangian of the system, denoted

$$L(x, \dot{x}) = T(\dot{x}) - U(x), \tag{20.129}$$

with T the kinetic energy and U the potential energy of the system. Let us consider a simple system such that $T(x) = \frac{1}{2} m \dot{x}^2$ and $U(x) = \frac{1}{2} k x^2$, which describe the one-dimensional motion of a mass m attached to a spring with spring constant k. We then have

$$L(x, \dot{x}) = \frac{1}{2} m \dot{x}^2 - \frac{1}{2} k x^2. \tag{20.130}$$

The generalized momentum in this case is

$$p = \frac{\partial L}{\partial \dot{x}} = m \dot{x}. \tag{20.131}$$

The Hamiltonian of the system is then given by

$$H(x,\dot{x}) = \dot{x}p - L = m\dot{x}^2 - \frac{1}{2}\dot{x}^2 + \frac{1}{2}kx^2 = \frac{1}{2m}p^2 + \frac{1}{2}kx^2. \qquad (20.132)$$

The Hamiltonian in this case represents the total energy of the system. It can also be seen from Eq. (20.131) that p represents the momentum of the mass. The equations of motion for the system follow from Eqs. (20.127) and (20.126),

$$\dot{x} = \frac{p}{m}, \qquad (20.133)$$

$$\dot{p} = -kx. \qquad (20.134)$$

The former equation is simply the definition of momentum for the system, while the second expression represents Newton's second law for a mass–spring system.

◇

It is to be noted that the Hamiltonian does not always represent the total energy of the system in mechanical problems. Additional conditions need to be met for this to be true: the generalized coordinates used to describe the system must be independent of time, and the potential energy must be velocity independent.

20.9 Focus: aperture apodization

It is well known that diffraction effects can seriously impact the performance of imaging systems. In Fig. 20.15 we illustrate the familiar Fraunhofer diffraction pattern produced by a plane wave passing through a square aperture. The pattern is characterized by a central lobe surrounded by numerous lower-intensity sidelobes. In imaging applications these sidelobes can adversely affect the system resolution. In astronomy, for instance, there exist binary star systems where one star is much brighter than the other, and the dimmer star can be masked by the sidelobes of the brighter star in the image. In other applications which require a well-collimated beam, these sidelobes represent energy lost from the central, useful, part of the beam.

The strength of the sidelobes can be directly traced by the use of Fourier analysis to the fact that the field is cut off sharply at the edges of the aperture. We can reduce these sidelobes by a technique known as *apodization* (literally: "removing the foot"), in which a phase or amplitude mask is used to modify the field passing through the aperture and smooth the sharp cutoff at the edges. To determine the ideal mask, we can formulate the problem in a way amenable to the calculus of variations.

Such a variational approach to apodization was originally discussed by McCutchen [McC69]. Asakura and Ueno [AU76] later solved a more general class of problems, as did Gbur and Carney [GC98, CG99]. We closely follow here the approach of the latter authors.

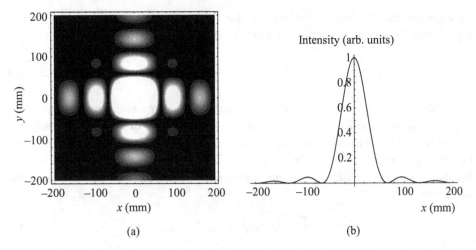

Figure 20.15 Illustrating (a) the Fraunhofer diffraction pattern for a rectangular aperture, and (b) the horizontal cross-section of this pattern. Here the horizontal width of the aperture is 7 mm, the vertical width is 8 mm, the wavelength is $\lambda = 5790\,\text{Å}$, and the distance from the screen is $z = 800$ m.

We consider for simplicity only a one-dimensional aperture (a "slit"). We assume the aperture extends from $-x_0$ to x_0, and that an amplitude mask covers it with the functional form $u(x)$. We require as boundary conditions that $u(-x_0) = u(x_0) = 0$. The radiation pattern produced by a plane wave passing through the aperture is directly related to the Fourier transform of the aperture function, i.e.

$$\tilde{u}(s) = \int_{-\infty}^{\infty} u(x)\mathrm{e}^{-2\pi\mathrm{i}sx}\mathrm{d}x. \tag{20.135}$$

This same formula may be used to describe the far-zone field in Fraunhofer diffraction or the field in the focal plane of a lens, depending on the interpretation of s. In Fraunhofer diffraction, $s \equiv kx'/2z$, where k is the wavenumber of the light, x' is the transverse coordinate of the point of observation, and z is the propagation distance. In focusing, $s \equiv kx'/f$, where f is the focal length of the lens.

We will seek to minimize the second moment of the intensity in the far field, defined as

$$\langle s^2 \rangle \equiv \frac{\int_{-\infty}^{\infty} s^2 |\tilde{u}(s)|^2 \mathrm{d}s}{\int_{-\infty}^{\infty} |\tilde{u}(s)|^2 \mathrm{d}s}. \tag{20.136}$$

It is to be noted that we choose our coordinate system such that

$$\langle s \rangle \equiv \frac{\int_{-\infty}^{\infty} s |\tilde{u}(s)|^2 \mathrm{d}s}{\int_{-\infty}^{\infty} |\tilde{u}(s)|^2 \mathrm{d}s} = 0. \tag{20.137}$$

Here we immediately have a challenge, as this functional is not in the standard form of Eq. (20.14). We can get around this by imposing a constraint on our solution of the form

$$\int_{-\infty}^{\infty} |\tilde{u}(s)|^2 ds = N_0. \tag{20.138}$$

Physically, this represents the requirement that the total energy in the aperture, and consequently in the radiation pattern, is fixed. This constraint is of a form previously not considered, and is referred to as an *isoperimetric constraint* (isoperimetric refers to the historical problem of finding the maximum area bounded by a perimeter of given length). We incorporate this constraint into our problem by considering the minimization of the functional

$$F[u] = \int_{-\infty}^{\infty} s^2 |\tilde{u}(s)|^2 ds - \lambda \int_{-\infty}^{\infty} |\tilde{u}(s)|^2 ds, \tag{20.139}$$

where $-\lambda$ is a Lagrange multiplier, and the minus sign is for later convenience. Minimizing this functional is equivalent to minimizing our original functional (20.136), for the energy constraint means that (a) the denominator of Eq. (20.136) is constrained to be constant, and can be omitted in our new functional, and (b) we can add our energy constraint to the functional without changing the functional's stationary values. Furthermore, using differentiation properties of Fourier transforms and Plancherel's identity (recall Sections 11.4.9 and 11.5.2), we can rewrite our functional in the form

$$F[u] = \int_{-\infty}^{\infty} \left[\frac{1}{(2\pi)^2} |u'(x)|^2 - \lambda |u(x)|^2 \right] dx. \tag{20.140}$$

This is now in the form of Eq. (20.14), with an integrand

$$f[u, u'; x] = \frac{1}{(2\pi)^2} |u'(x)|^2 - \lambda |u(x)|^2, \tag{20.141}$$

and assuming that the function $u(x)$ is real-valued we can immediately apply the Euler equation (20.23) to seek an equation for the minimum solution. We readily find that

$$u'' + (2\pi)^2 \lambda u = 0, \tag{20.142}$$

which is just the harmonic oscillator equation! A symmetric solution to this equation will have the form

$$u(x) = A\cos(Bx), \tag{20.143}$$

taking $B = 2\pi/\sqrt{\lambda}$. To match our boundary conditions $u(-x_0) = u(x_0) = 0$, we require $B = (2n+1)\pi/2x_0$. With $n = 0$, we have

$$u(x) = A\cos[\pi x/(2x_0)]. \tag{20.144}$$

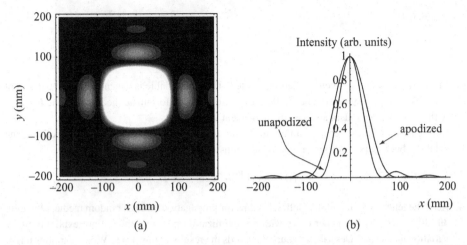

Figure 20.16 Illustrating (a) the Fraunhofer diffraction pattern for a rectangular aperture apodized according to Eq. (20.145), and (b) a comparison between the apodized and unapodized solutions. All parameters are as in Fig. 20.15.

Finally, applying the constraint (20.138) results in the solution

$$u(x) = \sqrt{N_0/x_0}\cos\left[\frac{\pi x}{2x_0}\right].\tag{20.145}$$

The radiation pattern for an aperture apodized according to Eq. (20.145) is shown in Fig. 20.16. One can see that the sidelobes are significantly reduced, although the width of the central peak has increased somewhat. The broadening of the central peak is a consequence of the Fourier nature of Fraunhofer diffraction – a Fourier uncertainty relationship must still be satisfied.

20.10 Additional reading

- L. D. Elsgolc, *Calculus of Variations* (New York, Pergamon Press, 1961). A short and clear book on the calculus of variations. Also in Dover.
- C. Lanczos, *The Variational Principles of Mechanics* (Toronto, University of Toronto Press, 1970, 4th edn.). This book gives a thorough discussion of variational calculus, with an emphasis on its application to mechanics.

20.11 Exercises

1. Occasionally a functional may depend on more than the first derivative of the dependent variable $y(x)$. Following the steps that lead to Euler's equation, find the comparable formula for the

functional

$$F[y] = \int_a^b f(x;y,y',y'') \, dx.$$

2. In two-dimensional problems, often the choice of which variable is dependent and independent can make a significant difference in the ease of solution. Reformulate the brachistrone problem with y as the independent and x as the dependent variable.

3. In discussion of the catenary solution to the mirage problem, we assumed a hyperbolic cosine solution, but others may exist. Look for solutions of the form

$$y(x) = A_+ e^{Bx} + A_- e^{-Bx} + D.$$

4. Read the letter by T. J. Schulz, Optimal beams for propagation through random media, *Opt. Lett.* **30** (2005), 1093. This paper deals with an optimization problem, though not explicitly using variational calculus. Describe in your own words the results of this paper. What quantities might be optimized using the calculus of variations?

5. Read the papers by G. Gbur and P. S. Carney, Convergence criteria and optimization techniques for beam moments, *Pure Appl. Opt.* **7** (1998), 1221, and by P. S. Carney and G. Gbur, Optimal apodizations for finite apertures, *J. Opt. Soc. Am. A* **16** (1999), 1638. Discuss how these results build upon previous apodization results.

6. Equation (20.145), which represents a solution to the apodization problem, is based on the choice $n = 0$. However, $n = 1, 2, \ldots$ must also represent *locally* minimum solutions. Consider the apodization function

$$u(x) = (1 - q)\cos[\pi x / 2x_0] + q\cos[3\pi x / 2x_0]$$

and calculate the value of $\langle s^2 \rangle$ as a function of q in the range $0 \le q \le 1$.

7. A particle lies on the top of a sphere of radius R. It slides without friction down the side of the sphere. Use the method of Lagrange multipliers to determine the position at which the particle leaves the surface of the sphere.

8. Given that Laplace's equation can be derived from the minimization of a functional with kernel

$$f(x,y;U,U_x,U_y) = \left(\frac{\partial U}{\partial x}\right)^2 + \left(\frac{\partial U}{\partial x}\right)^2,$$

find a kernel that will result in an extremum solution that satisfies Poisson's equation,

$$\nabla^2 U = -4\pi\rho(x,y).$$

9. Given that Laplace's equation can be derived from the minimization of a functional with kernel

$$f(x,y;U,U_x,U_y) = \left(\frac{\partial U}{\partial x}\right)^2 + \left(\frac{\partial U}{\partial x}\right)^2,$$

find a kernel that will result in an extremal solution that satisfies the Helmholtz equation,

$$\nabla^2 U + k^2 U = 0.$$

10. We consider the vibrations of a string with mass density ρ and tension T. Write expressions for the total kinetic and potential energy of the string and, minimizing the Lagrangian of the system, determine the wave equation for the string amplitude $u(x,t)$. (Hint: a length of string dx has a length $ds = \sqrt{1+u_x^2}\,dx$ in a deformed state; assume small displacements and Hooke's law, $F = T(ds - dx)^2$, to determine the potential energy per unit length.)

11. Find the extremal of the functional

$$F[y] = \int_{x_0}^{x_1} \frac{1+y^2}{y'^2}\,dx.$$

12. Find the extremal of the functional

$$F[y] = \int_{x_0}^{x_1} (y^2 + y'^2 - 2y\sin x)\,dx.$$

21

Asymptotic techniques

21.1 Introduction: foundations of geometrical optics

Light has wave properties, but the geometrical theory of light propagation was pre-eminent for hundreds if not thousands of years before being supplanted. One reason for this is the smallness of the wavelength of visible light: wavelengths are of the order of 10^{-5} cm, which to a good approximation is negligible compared with other length scales in traditional optical problems.

Geometrical rays behave very differently from waves, which raises a natural question: can one derive the law of geometrical optics from Maxwell's equations, and under what conditions can this derivation be made? The solution is suggested by the above observation of the smallness of optical wavelengths: geometrical optics should be derivable from Maxwell's equations in the limit $\lambda \to 0$ or, equivalently, $k \to \infty$.

In this section we consider a slightly simpler problem: the derivation of the Eikonal equation of geometrical optics from the Helmholtz equation for a monochromatic scalar wave $U(\mathbf{r})$ of frequency ω propagating in a medium of spatially-varying refractive index $n(\mathbf{r})$, i.e.

$$\nabla^2 U(\mathbf{r}) + n^2(\mathbf{r})k^2 U(\mathbf{r}) = 0, \tag{21.1}$$

where

$$k = \frac{\omega}{c} = \frac{2\pi}{\lambda_0} \tag{21.2}$$

is the *wavenumber* of the light, λ being its wavelength and c being the speed of light in vacuum.

It is not immediately obvious how one looks at the behavior of the wavefield in the geometrical optics limit $k \to \infty$, since we expect that the solution $U(\mathbf{r})$ is itself generally dependent on wavenumber k. It would be helpful if we had a simple example of a wavefield whose geometrical optics representation and wave optics representation were both known. Fortunately, we have two such examples: plane waves and spherical waves.

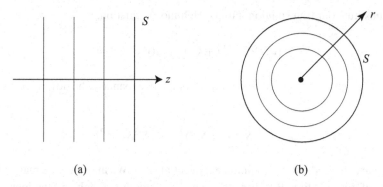

Figure 21.1 Illustration of the wavefronts and ray directions of (a) a plane wave, (b) a spherical wave.

A plane wave propagating in free space in the positive z-direction can be written in the form

$$U(\mathbf{r}) = U_0 e^{ikz}, \tag{21.3}$$

where U_0 is the amplitude of the wavefield. The geometric waves are simply straight lines propagating in the z-direction, and the wavefronts are the planes of constant z. The optical path, starting from the plane $z = 0$, is simply given by $S = z$.

A spherical wave propagating in free space from the origin can be written using the Green's function for the Helmholtz equation, i.e.

$$U(\mathbf{r}) = U_0 \frac{e^{ikr}}{kr}, \tag{21.4}$$

Because the point source is in free space, the optical path is simply given by $S = r$, the rays are straight lines emanating from the origin, and the wavefronts are just spherical surfaces centered on the point source. The geometry and notation are illustrated in Fig. 21.1.

In each of these examples the field may be expressed in the general form

$$U(\mathbf{r}) \sim a(\mathbf{r}) e^{ikS(\mathbf{r})}, \tag{21.5}$$

where $S(\mathbf{r})$ and $a(\mathbf{r})$ are taken to be *real* functions of position, which we assume for simplicity to be essentially independent of k_0. It seems reasonable to expect that any wavefield which satisfies the conditions of geometrical optics may be written in this way, and we take this as our trial solution of the Helmholtz equation.

Equation (21.5) is an *assumption* which we will use in investigating the foundations of geometrical optics. It is not expected to be valid in general, but may be used to represent a field which has a well-defined direction (ray) at any given point.

If we substitute this trial solution into the Helmholtz equation, we have

$$\nabla \cdot \nabla \left[a(\mathbf{r}) e^{ikS(\mathbf{r})} \right] + k^2 n^2(\mathbf{r}) a(\mathbf{r}) e^{ikS(\mathbf{r})} = 0. \tag{21.6}$$

If we substitute our trial solution into the Helmholtz equation, we have

$$\nabla \cdot \nabla \left[a(\mathbf{r}) e^{ikS(\mathbf{r})} \right] + k^2 n^2(\mathbf{r}) a(\mathbf{r}) e^{ikS(\mathbf{r})} = 0. \tag{21.7}$$

We may use the chain rule on the first term; completely expanded, the differential equation becomes

$$\nabla^2 a + 2ik \nabla a \cdot \nabla S - k^2 a (\nabla S)^2 + ika \nabla^2 S + k^2 n^2 a = 0, \tag{21.8}$$

where we have divided out the common exponential term. We may take the real and imaginary parts of this equation. Recalling that a and S are both real-valued functions, the real part of the above equation becomes

$$\nabla^2 a - k^2 a (\nabla S)^2 + k^2 n^2 a = 0. \tag{21.9}$$

If we let $\lambda \to 0$, then $k \to \infty$ and the first term becomes negligible compared to the second two. We then have, after dividing by $k^2 a$,

$$(\nabla S)^2 = n^2. \tag{21.10}$$

This equation is the Eikonal equation of geometrical optics, and is in essence a differential form of Fermat's principle. We briefly demonstrate this relationship.

We consider all rays emanating from the point P_1, and allow the point P_2 to vary. We may define the optical path of the ray which reaches point P_2 as

$$S(\mathbf{r}) = \int_{P_1}^{P_2(\mathbf{r})} n(\mathbf{r}') dl', \tag{21.11}$$

where the primed variable is now integrated over, and the path of integration is the path of the ray from P_1 to P_2. Solutions of this equation where

$$S(\mathbf{r}) = L = \text{constant} \tag{21.12}$$

define surfaces of constant optical path length (and hence phase), and are referred to as the *wavefronts* of geometrical optics. The direction of a particular ray at any point on the surface is defined by the normal to the surface, with unit vector

$$\mathbf{s}(\mathbf{r}) = \frac{\nabla S(\mathbf{r})}{|\nabla S(\mathbf{r})|}. \tag{21.13}$$

By elementary calculus, the infinitesimal change in the optical path is given by

$$dS(\mathbf{r}) = d\mathbf{r} \cdot \nabla S(\mathbf{r}). \tag{21.14}$$

If we evaluate the change along a ray, we have

$$dS(\mathbf{r})|_{\text{ray}} = dl\mathbf{s} \cdot \nabla S(\mathbf{r}) = n dl. \tag{21.15}$$

Canceling out dl, and substituting in for $\mathbf{s}$, we have the result

$$|\nabla S(\mathbf{r})| = n(\mathbf{r}), \tag{21.16}$$

or writing this in its standard form by squaring it, we get the Eikonal equation

$$|\nabla S(\mathbf{r})|^2 = n^2(\mathbf{r}). \tag{21.17}$$

We have so far only considered the requirement that the real part of Eq. (21.8) vanish. If we consider the imaginary part of the equation, we have (after dividing by k),

$$2\nabla a \cdot \nabla S + a\nabla^2 S = 0. \tag{21.18}$$

Let us multiply this equation by a; then it becomes

$$2a\nabla a \cdot \nabla S + a^2\nabla^2 S = 0. \tag{21.19}$$

This has the form of a total derivative, which may then be written as

$$\nabla \cdot (a^2\nabla S) = 0. \tag{21.20}$$

If we define $\mathbf{s} = \nabla S/|\nabla S|$, we may also use $|\nabla S| = n$ to write

$$\nabla \cdot (a^2 n\mathbf{s}) = 0. \tag{21.21}$$

If we define the intensity of the wavefield as $I \equiv a^2 n$, we may rewrite this equation in the form

$$\nabla \cdot (I\mathbf{s}) = 0. \tag{21.22}$$

At any point in space, the unit vector $\mathbf{s}$ represents the local direction of the ray. Let us consider a surface Σ which encloses a bundle of rays. This surface in general will be roughly cylindrical. The endcaps of the cylinder are of infinitesimal area, $d\sigma_1$ and $d\sigma_2$, and the remaining part of the cylinder form fits to the extreme rays of the bundle. This surface is illustrated in Fig. 21.2. If we integrate Eq. (21.22) through the volume of this surface, we may use Gauss' theorem to convert the volume integral to a surface integral:

$$\int_V \nabla \cdot (I\mathbf{s}) d^3 r = \int_\Sigma I\mathbf{s} \cdot \mathbf{n} d\Sigma = 0. \tag{21.23}$$

The vector $\mathbf{n}$ is the unit outward normal to the surface, and $d\Sigma$ is the infinitesimal area element of the surface. It is to be noted that the outward normal points in the direction of $\mathbf{s}$

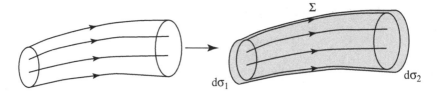

Figure 21.2 Illustrating the enclosure of a ray bundle.

on $d\sigma_2$, and points opposite to $\mathbf{s}$ on $d\sigma_1$. Furthermore, the outward normal is perpendicular to $\mathbf{s}$ on the remaining part of the surface. The surface integral reduces to the simple form

$$I_2 d\sigma_2 - I_1 d\sigma_1 = 0. \tag{21.24}$$

This is the *intensity law of geometrical optics*: the product of the intensity and the area of a bundle of rays is constant along the propagation direction. It was originally an empirically-determined formula, but can be justified, as shown, through an analysis of the wave equation in the asymptotic limit $k \to \infty$.

Equation (21.5) is the simplest approximation one can make for waves which propagate in a more or less geometric manner. We may attempt to make a more sophisticated model of a geometric wave by assuming a solution of the form

$$U(\mathbf{r}) = \sum_{n=0}^{\infty} \frac{a_n(\mathbf{r})}{(ik)^n} e^{ikS(\mathbf{r})}. \tag{21.25}$$

This approximation reduces to Eq. (21.5) in the limit of large k. We may substitute from Eq. (21.25) into the Helmholtz equation to find differential equations for $S(\mathbf{r})$ and the amplitudes $a_n(\mathbf{r})$,

$$[\nabla^2 + k^2 n^2(\mathbf{r})] \sum_{n=0}^{\infty} \frac{a_n(\mathbf{r})}{(ik)^n} e^{ikS(\mathbf{r})} = 0. \tag{21.26}$$

This results in the expression

$$\sum_{n=0}^{\infty} \frac{1}{(ik)^n} \left[\nabla^2 a_n + 2ik\nabla a_n \cdot \nabla S - k^2 a_n (\nabla S)^2 + ika_n \nabla^2 S + k^2 n^2 a_n \right] = 0. \tag{21.27}$$

If we treat k as a variable of the equation, it can only be satisfied term-by-term in $1/k$. The lowest-order terms have a k^2-dependence, and must satisfy

$$-k^2 a_0 (\nabla S)^2 + k^2 n^2 a_0 = 0. \tag{21.28}$$

This again results in the Eikonal equation. The next lowest-order terms have a k-dependence, and must satisfy

$$2ik\nabla a_0 \cdot \nabla S + ika_0 \nabla^2 S - a_1 (\nabla S)^2 + a_1 n^2 = 0. \tag{21.29}$$

The latter two terms of this expression satisfy the Eikonal equation and vanish, leaving us with the constraint that the coefficient a_0 must satisfy the intensity law of geometrical optics. Term-by-term, the rest of the series results in the set of differential equations

$$\nabla^2 a_n + 2\nabla a_{n+1} \cdot \nabla S + a_{n+1}\nabla^2 S + a_{n+2}(\nabla S)^2 - n^2 a_{n+2} = 0, \quad n \geq 0. \tag{21.30}$$

The last two terms satisfy the Eikonal equation and may be neglected, leaving

$$\nabla^2 a_n + 2\nabla a_{n+1} \cdot \nabla S + a_{n+1}\nabla^2 S = 0. \tag{21.31}$$

The result is a recursive set of differential equations for a_{n+1} in terms of a_n and S. Since S is determined by the Eikonal equation and a_0 is determined by the intensity law of geometrical optics, we can in principle determine a_n for any n. These represent higher-order wave corrections to the geometric theory of light propagation.

A few observations should be made. The series given by Eq. (21.25) might seem at first glance like a valid description of a wavefield for *all* values of k, but we have still made very strict assumptions about the form of the wavefield that do not apply in many, if not most, cases of wave propagation. Furthermore, it can be shown that the complete infinite series does not necessarily converge, even if a finite number of terms of the series give a good approximation to the exact wavefunction. A similar, though more involved, analysis can be done for the full set of Maxwell's equations, and is described in [BW99, Section 3.1]; such an analysis can be used to characterize polarization effects in geometrical optics.

The derivation described here is an example of an *asymptotic analysis*, in which one characterizes the behavior of a differential or integral equation in the limit that one of the parameters in the equation is large. Asymptotic techniques are frequently useful in optics, where the wavelength is sufficiently small that it may be taken to be effectively zero, with the wavenumber effectively infinite. In this chapter we discuss common techniques of asymptotic analysis.

21.2 Definition of an asymptotic series

Let us consider the function $f(k)$, defined by the real-valued integral

$$f(k) = \int_k^\infty t^{-1} e^{k-t} dt, \tag{21.32}$$

where k is a positive number. Noting that the integrand always satisfies the inequality

$$t^{-1} e^{k-t} \leq k^{-1} e^{k-t}, \tag{21.33}$$

we can determine that the integral satisfies the inequality

$$f(k) \le k^{-1} \int_k^\infty e^{k-t} dt = k^{-1},$$
(21.34)

and therefore approaches zero as $k \to \infty$. This inequality is a crude estimation of the behavior of the function, however, and we can develop a precise series expansion of $f(k)$ in terms of powers of k^{-1} by using integration by parts

$$f(k) = \frac{1}{k} - \frac{1}{k^2} + \frac{2!}{k^2} - \cdots + \frac{(-1)^{N-1}(N-1)!}{k^N} + (-1)^N N! \int_k^\infty \frac{e^{k-t}}{t^{N+1}} dt.$$
(21.35)

There seems to be no reason why we could not continue this process of integration by parts indefinitely, and determine an infinite series for $f(k)$:

$$f(k) = \sum_{n=1}^\infty u_n(k) = \sum_{n=1}^\infty \frac{(-1)^{n-1}(n-1)!}{k^n}.$$
(21.36)

However, a simple application of the ratio test of Section 7.3 shows that this series is not absolutely convergent for *any* value of k, i.e. $|u_{n+1}(k)/u_n(k)| = n/k$, and furthermore the terms of the series do not even approach zero as $n \to \infty$. This infinite series representation of $f(k)$ is divergent for all values of k.

If we consider ending the integration by parts at the Nth term, however, we may write an expression for $f(k)$ as

$$f(k) = \sum_{n=1}^N \frac{(-1)^{n-1}(n-1)!}{k^n} + R_N(k),$$
(21.37)

where we define a "remainder" term of the series as

$$R_N(k) = (-1)^N N! \int_k^\infty \frac{e^{k-t}}{t^{N+1}} dt.$$
(21.38)

We again note that the integrand satisfies a simple inequality,

$$\frac{e^{k-t}}{t^{N+1}} \le \frac{e^{k-t}}{k^{N+1}},$$
(21.39)

and that the remainder may therefore be written as

$$|R_N(k)| \le \frac{N!}{k^{N+1}}.$$
(21.40)

The remainder of the series approaches zero as $1/k^{N+1}$ as $k \to \infty$, while the fastest decaying term of the series itself approaches zero as $1/k^N$. For sufficiently large values of k, then, the series represents a very good approximation to the behavior of the function. Equation (21.37) is referred to as the *asymptotic series* of the function $f(k)$, and represents the behavior of the function as $k \to \infty$. The simplest class of asymptotic series is the asymptotic power series, defined below.

Definition 21.1 (Asymptotic power series) *An asymptotic power series of order N is a series of the form*

$$f(k) \sim \sum_{n=1}^{N} \frac{a_n}{k^n} + o(k^{-N}), \qquad (21.41)$$

where o is known as an order *symbol and* $o(k^{-N})$ *represents a function which decays at a rate* faster *than the function* k^{-N}.

The order symbols are defined formally as follows.

Definition 21.2 *A function* $f(k)$ *is of order* $o[g(k)]$ *if, in the limit* $k \to \infty$, $f(k)/g(k) \to 0$.

For example, the remainder R_N of the series given in Eq. (21.37) decays as $k^{-(N+1)}$ and may be written as $o(k^{-N})$.

It will also be useful to define the "big-O" order symbol O such that $O(k^{-N})$ represents a function which decays *at the rate* k^{-N}.

Definition 21.3 *A function* $f(k)$ *is of order* $O[g(k)]$ *in a neighborhood of a point* k_0 *(typically* $k = \infty$*) if* $|f(k)| \le M |g(k)|$ *in that neighborhood, i.e.* $f(k)$ *is bounded by* $Mg(k)$.

For example, the nth term of the asymptotic series, a_n/k^n, is $O(k^{-n})$.

The distinction between $f(k) = o[g(k)]$ and $f(k) = O[g(k)]$ is a subtle but important one: $O[g(k)]$ indicates that the function $f(k)$ is essentially proportional to $g(k)$, while $o[g(k)]$ indicates that the function decays *faster* than $g(k)$. If $f(k) = o[g(k)]$, then it automatically follows that $f(k) = O[g(k)]$; however, the converse is not necessarily true.

We will restrict our attention primarily to asymptotic power series, though in general other types of asymptotic series are possible. One relevant extension is an asymptotic series which may be written in the form

$$f(k) \sim g(k) \left[\sum_{n=1}^{N} \frac{a_n}{k^n} + o(k^{-N}) \right], \qquad (21.42)$$

where $g(k)$ encompasses the non-power series asymptotic behavior of the function, which is the same for all terms of the expansion. Even more generally, we may write

$$f(k) \sim \sum_{n=1}^{N} a_n \phi_n(k) + o[\phi_N(k)], \qquad (21.43)$$

where $\phi_{n+1}(k) = o[\phi_n(k)]$.

We note a few general observations about the behavior of asymptotic series. Two asymptotic series with the same order N and same asymptotic sequence ϕ_n may be added together, i.e. if $f(k) \sim \sum_{n=1}^{N} a_n \phi_n(k)$ and $g(k) \sim \sum_{n=1}^{N} b_n \phi_n(k)$, then

$$\alpha f(k) + \beta g(k) \sim \sum_{n=1}^{N} (\alpha a_n + \beta b_n) \phi_n(k). \tag{21.44}$$

An asymptotic series does not uniquely define a function. Two functions $f(k)$ and $g(k)$ can be said to be *asymptotically equal to order N* if their difference is of order $o(\phi_N)$.

An asymptotic series can depend on one or more parameters; in the geometrical optics example of the previous section, the spatial variables (x, y, z) represent the parameters of the series. We may therefore write

$$f(x, k) \sim \sum_{n=1}^{N} a_n(x) \phi_n(k), \tag{21.45}$$

where x represents one or more parameters. We may consider, as in Section 7.4, whether the asymptotic series is uniformly convergent in its parameters. It is said to be uniformly convergent if the series is $o(\phi_N(k))$ independent of the parameters x.

21.3 Asymptotic behavior of integrals

One of the simplest asymptotic series which can be developed is for the Fourier transform of a function of finite support. We consider a function $F(k)$ such that

$$F(k) = \int_a^b f(x) e^{ikx} dx, \tag{21.46}$$

where we assume the function $f(x)$ is N-times differentiable. We may then ask the question: what is the behavior of this function for large values of k? Assuming differentiability of $f(x)$, we may again perform an integration by parts, i.e.

$$F(k) = \left[f(x) \frac{e^{ikx}}{ik} \right]_a^b - \frac{1}{ik} \int_a^b f'(x) e^{ikx} dx. \tag{21.47}$$

This may then be written as

$$F(k) = \frac{1}{ik} \left[f(b) e^{ikb} - f(a) e^{ika} \right] - \frac{1}{ik} \int_a^b f'(x) e^{ikx} dx. \tag{21.48}$$

This process can be continued, provided higher-order derivatives exist. To the Nth derivative, we may write

$$F(k) = \sum_{n=0}^{N-1} \frac{(-1)^n}{(ik)^{n+1}} \left[f^{(n)}(b)e^{ikb} - f^{(n)}(a)e^{ika} \right] + \frac{(-1)^N}{(ik)^N} \int_a^b f^{(N)}(x)e^{ikx}dx. \quad (21.49)$$

The integral in the remainder term must approach zero as $k \to \infty$, according to the Riemann–Lebesque lemma, which may be stated as follows.

Lemma 21.1 (Riemann–Lebesgue lemma) *For $f(x)$ absolutely integrable and piecewise continuous,*

$$\lim_{k \to \infty} \int_{-\infty}^{\infty} f(x)e^{-ikx}dx = 0. \quad (21.50)$$

Intuitively, this lemma can be understood by a simple argument. Suppose $f(x)$ has a Fourier transform $F(k)$. The inverse Fourier transform of $F(k)$ is $f(x)$ and clearly must exist and be well-behaved (and not a distribution), and this is only possible if $F(k)$ tends to zero as $k \to \infty$.

Applying the lemma to Eq. (21.49) implies that the remainder term approaches zero at a rate *faster* than k^{-N}. We may therefore write the asymptotic behavior of the Fourier integral as

$$F(k) \sim \sum_{n=0}^{N-1} \frac{(-1)^n}{(ik)^{n+1}} \left[f^{(n)}(b)e^{ikb} - f^{(n)}(a)e^{ika} \right] + o(k^{-N}). \quad (21.51)$$

It is interesting to note that the asymptotic behavior of the function depends only upon the endpoints of the interval, $x = a$ and $x = b$. We can roughly understand this as follows. For sufficiently large values of k, the function $f(x)$ will be approximately constant over a period of the complex exponential $\exp[ikx]$. The integral over every complete period of the function will be effectively zero, and the main contributions to the integral will therefore come from the ends of the interval, where periods are "cut off". This is illustrated in Fig. 21.3.

In the next section, we consider *the method of stationary phase*, which may be considered a generalization of the asymptotic expansion of Fourier integrals. In the method of stationary phase, endpoint contributions will potentially play an important role in the asymptotic expansion.

Another straightforward asymptotic series is that of the Laplace transform. We consider the behavior of the function

$$F(\alpha) = \int_0^{\infty} e^{-\alpha x} f(x)dx, \quad (21.52)$$

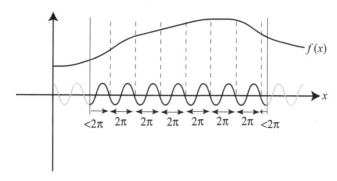

Figure 21.3 The endpoint contribution to the asymptotic behavior of a Fourier integral.

in the limit of large α. Assuming the function $f(x)$ is differentiable, we can perform an integration by parts as in the Fourier case to find that

$$F(\alpha) = \frac{1}{\alpha}f(0) + \frac{1}{\alpha}\int_0^\infty e^{-\alpha x}f'(x)dx. \tag{21.53}$$

If the function $f(x)$ is N times differentiable, we may continue the integration by parts to find the relation

$$F(\alpha) = \sum_{n=0}^{N-1}\frac{1}{\alpha^{n+1}}f^{(n)}(0) + \frac{1}{\alpha^N}\int_0^\infty e^{-\alpha x}f^{(N)}(x)dx. \tag{21.54}$$

Given that $f^{(N)}(x)$ is finite, the area under the function $\exp[-\alpha x]$ goes rapidly to zero, which implies that the latter integral in the above equation vanishes at a rate faster than α^{-N}. We may therefore express the asymptotic series of the function $F(\alpha)$ in the form

$$F(\alpha) \sim \sum_{n=0}^{N-1}\frac{1}{\alpha^{n+1}}f^{(n)}(0) + o(\alpha^{-N}). \tag{21.55}$$

This series, like the Fourier case, depends only upon the behavior of the function at the endpoint $x = 0$. It is quite clear why this is the case: as $\alpha \to \infty$, the exponential function $\exp[-\alpha x]$ becomes more sharply peaked and localized about the origin, as illustrated in Fig. 21.4.

Laplace himself applied similar reasoning to determine the asymptotic behavior of integrals with a more complicated exponential behavior, and his method is now referred to as *Laplace's method*. We consider an integral of the form

$$F(k) = \int_a^b f(x)e^{kg(x)}dx, \tag{21.56}$$

where the function $f(x)$ is assumed to be continuous and the function $g(x)$ is taken to be real-valued and twice differentiable. As $k \to \infty$, the exponential function will grow without

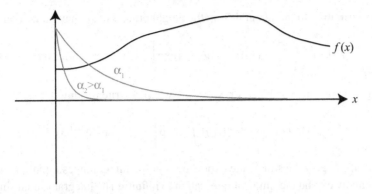

Figure 21.4 The endpoint contribution to the asymptotic behavior of a Laplace integral.

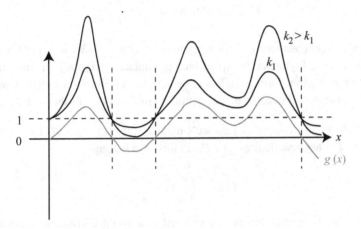

Figure 21.5 The behavior of $\exp[kg(x)]$ as $k \to \infty$.

bound for points such that $g(x) > 0$ and will decrease rapidly to zero for $g(x) < 0$. The peaks of the function $g(x)$ will become tall and narrow, very much like delta functions, and the contributions of the integral in the immediate neighborhood of the peaks will dominate. This is illustrated in Fig. 21.5.

Let us focus on the neighborhood of a single maximum of $g(x)$, say a distance ϵ around the maximum x_0. The primary contribution is from the immediate neighborhood of x_0, and we may expand $f(x)$ and $g(x)$ in their lowest significant Taylor series terms,

$$f(x) \approx f(x_0), \tag{21.57}$$

$$g(x) \approx g(x_0) + \frac{1}{2}g''(x_0)(x - x_0)^2. \tag{21.58}$$

Then the contribution to the integral from the neighborhood of x_0 may be written as

$$F(k) \approx \int_{x_0-\epsilon}^{x_0+\epsilon} f(x_0)\exp[kg(x_0)]\exp\left[\frac{k}{2}g''(x_0)(x-x_0)^2\right]dx. \qquad (21.59)$$

On making a change of variables to $x' = x - x_0$, we may write this integral as

$$F(k) \approx f(x_0)\exp[kg(x_0)]\int_{-\epsilon}^{\epsilon}\exp\left[-\frac{k}{2}|g''(x_0)|x'^2\right]dx'. \qquad (21.60)$$

For sufficiently large values of k, the exponential in the integral goes rapidly to zero, and we may formally extend the limits of integration to infinity. The integral is then simply that of a Gaussian, and the result is

$$F(k) \approx f(x_0)\exp[kg(x_0)]\sqrt{\frac{2\pi}{|g''(x_0)|k}}. \qquad (21.61)$$

This is the most significant term of the asymptotic series for $F(k)$. We have derived this result by reasoning from a Taylor series expansion, but Laplace's original derivation involved a nonlinear coordinate transformation. It is to be noted that if the point x_0 lies at an endpoint of integration, the contribution to the asymptotic series will be half that of Eq. (21.61).

Example 21.1 As an example of Laplace's method, let $f(x) = x^2$, $g(x) = \sin x - 1$, and $a = 0$, $b = \pi$. The asymptotic form of $F(k)$ in this case is simply

$$F(k) \sim \left(\frac{\pi}{2}\right)^2\sqrt{\frac{2\pi}{k}}. \qquad (21.62)$$

In Fig. 21.6, the asymptotic result is compared to a result derived by exact numerical integration. It can be seen that, for sufficiently large values of k, the asymptotic result agrees well with the exact result.

The heuristic method used to derive Eq. (21.61) suffers from two related limitations. First, we have only derived the leading-order term of the asymptotic expansion, and there is no clear strategy for developing the higher-order terms. Second, without the higher-order terms or, more importantly, an estimate of the error, we cannot rigorously show that we have actually derived the true asymptotic form of the integral.

An extension of Laplace's method to higher asymptotic orders was first undertaken by Erdélyi [Erd56]. We state the result as a theorem and outline the ideas behind the proof.

Theorem 21.1 (Erdélyi's theorem) *Let $f(x)$ and $g(x)$ be functions on the interval $a \leq x \leq b$, such that the integral*

$$F(k) = \int_a^b f(x)e^{kg(x)}dx \qquad (21.63)$$

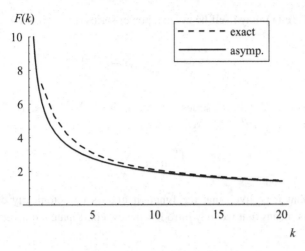

Figure 21.6 An example of Laplace's method, with $f(x) = x^2$, $g(x) = \sin x - 1$, $a = 0$, $b = \pi$. The exact result (dashed) is compared to the asymptotic result (solid).

exists. The function $g(x)$ is taken to be real and continuous at $x = a$, continuously differentiable and such that $g'(x) < 0$ for $a \leq x \leq a + \delta$ with $\delta > 0$, and $g(x) \leq g(a) + \epsilon$, with $\epsilon > 0$, for $a + \delta < x \leq b$. Furthermore, $f(x)$ and $g(x)$ have the following approximate power series representations at $x \to a$:

$$-g'(x) \sim \sum_{n=0}^{N-1} a_n(x-a)^{\nu+n-1},$$
(21.64)

$$f(x) \sim \sum_{n=0}^{N-1} b_n(x-a)^{\lambda+n-1}.$$
(21.65)

We may then write an asymptotic series for Eq. (21.63) which has the form

$$F(k) \sim e^{kg(a)} \sum_{n=0}^{N-1} \gamma_n \Gamma\left(\frac{\lambda+n}{\nu}\right) k^{-(\lambda+n)/\nu},$$
(21.66)

where the constants γ_n depend in a nontrivial way on the coefficients a_n and b_n.

The proof of this begins by writing the integral (21.63) in the form

$$F(k) = \int_a^\delta \frac{f(x)}{g'(x)} e^{kg(x)} g'(x) dx + \int_\delta^b f(x) e^{kg(x)} dx.$$
(21.67)

Because it is given that $g(x) \leq g(a) + \epsilon$ in the second integral, one can show that it is asymptotically negligible compared to the first integral. For the first integral, we first note that the ratio $f(x)/g'(x)$ may itself be given a power series representation,

$$h(x) = \frac{f(x)}{g'(x)} = \sum_{n=0}^{N-1} c_n (x-a)^{\lambda - \nu + n}. \tag{21.68}$$

Since the lowest-order term of the series for $g'(x)$ is $(x-a)^{\nu-1}$, the series for $h(x)$ consists of terms of the form $(x-a)^{\lambda - \nu + n}$. We now have

$$F(k) \sim \int_a^\delta h(x) e^{kg(x)} g'(x) dx \tag{21.69}$$

We now recall that it is given that the function $g(x)$ is monotonically decreasing with increasing x. This means that we may introduce a new coordinate u defined by

$$u(x) = -\int_a^x g'(x') dx' = g(a) - g(x) = g(a) + \sum_{n=0}^{N-1} \frac{a_n}{\nu + n} (x-a)^{\nu + n}. \tag{21.70}$$

On substitution into the first integral, we have

$$F(k) = e^{kg(a)} \int_{u(a)}^{u(\delta)} h[x(u)] e^{-ku} du. \tag{21.71}$$

This integral is very nearly in the form of a Laplace-type integral with an asymptotic series given by Eq. (21.55). However, there are two subtleties that make the calculation nontrivial.

The first complication is the presence of the function $h[x(u)]$, which means that we need to invert Eq. (21.70) to find an expression for $(x-a)$ in terms of u. No general expression exists for this function, but we may reason that

$$x - a = \sum d_n u^{n/\nu}. \tag{21.72}$$

The second complication is more serious: the function $h[x(u)]$ may not be well-behaved at the origin, making a straightforward integration by parts inappropriate. Fortunately, a more general method for evaluating the asymptotic behavior of Laplace-type integrals exists, known as Watson's lemma.

Lemma 21.2 (Watson's lemma) *We consider an integral of the form*

$$F(k) = \int_a^b f(x) e^{-kx} dx, \tag{21.73}$$

where $f(t)$ is an integrable function and may be written in the limit $x \to a$ as

$$f(t) \sim t^\alpha \sum_{n=0}^\infty a_n t^{\beta n}. \tag{21.74}$$

Then the asymptotic form of $F(k)$ is

$$F(k) \sim \sum_{n=0}^{\infty} a_n \frac{\Gamma(\alpha + \beta n + 1)}{k^{\alpha + \beta n + 1}}. \tag{21.75}$$

Watson's lemma can be proven as follows: divide the integration range into two separated by a value R which is very close to a. The upper integration range is asymptotically negligible, by arguments similar to those stated earlier. The lower integration range may be written as

$$\int_0^R t^{\alpha + \beta n} e^{-kt} \mathrm{d}t = \int_0^{\infty} t^{\alpha + \beta n} e^{-kt} \mathrm{d}t - \int_R^{\infty} t^{\alpha + \beta n} e^{-kt} \mathrm{d}t. \tag{21.76}$$

Because $t^{\alpha + \beta n}$ is finite at $t = R$, the second integral can be shown to be asymptotically negligible again. The first integral is simply related to the gamma function, so that

$$\int_0^R t^{\alpha + \beta n} e^{-kt} \mathrm{d}t \sim \frac{\Gamma(\alpha + \beta n + 1)}{k^{\alpha + \beta n + 1}}. \tag{21.77}$$

Watson's lemma immediately follows.

Combining Eqs. (21.68), (21.72), and Watson's lemma, we find that

$$F(k) = e^{kg(a)} \sum_{n=0}^{N-1} \gamma_n \Gamma\left(\frac{\lambda + n}{\nu}\right) k^{-(\lambda + n)/\nu}, \tag{21.78}$$

which is the result of Erdélyi's theorem. The coefficients γ_n depend upon the coefficients c_n of $h(x)$ and the coefficients d_n of $x(u)$ in a nontrivial way.

Fortunately, one rarely needs to explicitly determine an asymptotic series beyond the leading term in physical applications. Erdélyi's theorem also, however, gives us a description of the behavior of the next term in the series or the order of the remainder. An alternative method for determining the asymptotic series and, more importantly, the coefficients γ_n, is given in [Woj06].

21.4 Method of stationary phase

Laplace's method is, broadly speaking, an generalization of the arguments used to derive the asymptotic series of the Laplace transform. Laplace's method can be used to determine the lowest-order term of the asymptotic series when the asymptotic parameter k appears in a *real-valued* exponential.

We may also generalize the arguments made in evaluating the Fourier integral and determine the asymptotic series of an integral when the asymptotic parameter k appears in a *purely imaginary* exponential. The technique, which we derive using nonrigorous Taylor

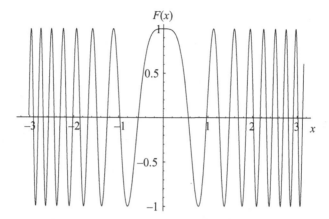

Figure 21.7 The function $F(x) = \cos[kg(x)]$, with $k = 5$ and $g(x) = x^2$, evaluated from $x = -\pi$ to $x = \pi$.

series reasoning, is known as the *method of stationary phase*. We consider an integral of the form

$$F(k) = \int_a^b f(x)e^{ikg(x)}\,dx, \tag{21.79}$$

where the function $f(x)$ is assumed to be continuous and the function $g(x)$ is taken to be real-valued and twice differentiable. The only difference between this integral and the integral of the Laplace method is the presence of "i" in the exponential. If $g(x)$ were a linear function, i.e. $g(x) = \alpha x$, this integral would be a Fourier integral. The only contributions to the asymptotic series would be at the endpoints, because the rapid oscillations of the exponential would cancel any contribution in the interior. For more general functions $g(x)$, however, there also potentially exist points within the domain $a \le x \le b$ where $g'(x) = 0$, known as *stationary points*. An example is $g(x) = x^2$, for which $g'(0) = 0$. The real part of the complex exponential is shown in Fig. 21.7. At the point $x = 0$ the function is locally flat and non-oscillatory. We would therefore expect that this region will provide the bulk of the contribution to the integral when $k \to \infty$.

We evaluate the contribution using a nonrigorous Taylor series argument as we did for Laplace's method. We assume that a stationary point exists at $x = x_0$, and that the major contribution to the asymptotic series occurs in the neighborhood of this point, $x_0 - \epsilon < x < x_0 + \epsilon$. We again expand $f(x)$ and $g(x)$ in their lowest significant Taylor series terms,

$$f(x) \approx f(x_0), \tag{21.80}$$

$$g(x) \approx g(x_0) + \frac{1}{2}g''(x_0)(x - x_0)^2. \tag{21.81}$$

Our integral then takes on the form

$$F(k) \approx f(x_0) \exp[ikg(x_0)] \int_{-\epsilon}^{\epsilon} \exp\left[i\frac{k}{2}g''(x_0)x'^2\right] dx'.$$ (21.82)

For sufficiently large k, we can formally extend the limits of integration to infinity. The integral is then a Fresnel integral which may be readily evaluated[1] to find that

$$F(k) \approx f(x_0) \exp[ikg(x_0)] \exp[i\pi/4]\sqrt{\frac{2\pi}{g''(x_0)k}}.$$ (21.83)

It is to be noted that $g''(x_0)$ may be positive or negative, so that an additional complex term may arise from within the square root.

Stationary points such as those we have just discussed are referred to as *stationary points of the first kind*; however, we have also seen that the endpoints of the domain can contribute to the asymptotic expansion of an oscillatory integral. Such points are referred to as *stationary points of the second kind*, and we can evaluate their contribution by generalizing the integration by parts we performed for Fourier integrals. In particular, starting with the integral given by Eq. (21.79), we perform an integration by parts by letting

$$u \equiv \frac{f(x)}{ikg'(x)},$$ (21.84)

$$v' \equiv ikg'(x)\exp[ikg(x)].$$ (21.85)

Integrating over the entire domain, we have

$$F(k) = \frac{1}{ik}\left[\frac{f(b)}{g'(b)}e^{ikg(b)} - \frac{f(a)}{g'(a)}e^{ikg(a)}\right] - \frac{1}{ik}\int_a^b \frac{d}{dx}\left[\frac{f(x)}{g'(x)}\right]e^{ikg(x)}dx.$$ (21.86)

The integral on the right, by similar arguments as described previously, is $o(k^{-1})$ and is negligible compared to the first two terms. We find that stationary points of the second kind are of order $O(k^{-1})$, while stationary points of the first kind are of order $O(k^{-1/2})$. The asymptotic series of an integral of the form (21.79) will be dominated by stationary points of the first kind.

The higher-order terms of the asymptotic series in the method of stationary phase may also be determined, though the process is potentially more complicated due to the presence of two types of stationary point. We refer the interested reader to [Erd56] and [Sta86, Chapter 8].

---

[1] See Example 9.14.

21.5 Method of steepest descents

We have now developed a pair of tools for evaluating the asymptotic behavior of integrals
of the form

$$F(k) = \int_a^b f(x) e^{kg(x)} dx. \tag{21.87}$$

When $g(x)$ is a purely real-valued function, we may determine the leading asymptotic
term by Laplace's method; when $g(x)$ is pure imaginary, we may determine the leading
asymptotic term by the method of stationary phase. If $g(x)$ is a generally complex function,
however, neither method can be directly applied. If the functions $f(z)$ and $g(z)$ are analytic,
however, it is possible to convert the integral into a Laplace-type form by distorting the
contour; this is known as the *method of steepest descents*.

To illustrate this method, let us consider the general behavior of a stationary point z_0 of
an analytic function at which $g'(z) = 0$. The Taylor series representation of $g(z)$ at point
z_0 is of the form

$$g(z) = a_0 + a_{m+1}(z - z_0)^{m+1} + \sum_{n=m+2}^{\infty} a_n(z - z_0)^n, \tag{21.88}$$

with m an integer greater than zero. In the immediate neighborhood of z_0, the function is
well-approximated by

$$g(z) \approx a_0 + a_{m+1}(z - z_0)^{m+1}. \tag{21.89}$$

Let us assume for the moment $a_0 = 0$; we may write $g(z)$ in polar coordinates $z - z_0 = r e^{i\theta}$
as

$$g(r,\theta) \approx |a_{m+1}| r^{m+1} e^{i(m+1)\theta + \phi_{m+1}}, \tag{21.90}$$

where ϕ_{m+1} is the phase of a_{m+1}. A straight-line path through the origin is a path of constant
θ. If we make the choice

$$\theta = -\phi_{m+1}/(m+1) + n\pi/(m+1), \tag{21.91}$$

with $n = 0, 1, 2, \ldots, m$, the function $g(r,\theta)$ will be real-valued; if we make the choice

$$\theta = -\phi_{m+1}/(m+1) + (2n+1)\pi/2(m+1), \tag{21.92}$$

the function $g(r,\theta)$ will be pure imaginary.

If we now consider the function

$$h(z) = e^{kg(z)}, \tag{21.93}$$

we readily see that a path satisfying Eq. (21.91) will make the $h(z)$ act like a Laplace-type exponential, for which we can apply Laplace's method. A path satisfying Eq. (21.92) will make the exponential act like a Fourier-type integral, for which we can apply the method of stationary phase. Because Cauchy's integral theorem, Theorem 9.3, demonstrates that we may distort the path of integration of an analytic function without changing its value, we may take a general integral of the form of Eq. (21.87) and convert it to a form that can be readily asymptotically evaluated.

It can be seen that there are $m+1$ straight-line paths satisfying Eq. (21.92) and $\mathrm{Re}[g(z)] = 0$, which are referred to as *level curves*. There are also $m+1$ straight-line paths satisfying Eq. (21.91) along which $\mathrm{Im}[g(z)] = 0$ and the modulus of the function changes as rapidly as possible; these are called the *steepest paths*. The point x_0 is referred to as a *saddle of order m*.

More generally, if we consider functions such that $a_0 \neq 0$, a saddle of order m has the local form given by Eq. (21.89). Paths of constant $\mathrm{Re}[g(z)]$ are referred to as level curves and paths of constant $\mathrm{Im}[g(z)] = 0$ are referred to as steepest paths. A saddle of order m has $m+1$ level curves and $m+1$ steepest paths; these curves are no longer in general straight lines.

Example 21.2 (Level curves and steepest paths) We consider as an example the properties of the function

$$h(x) = e^{kg(x)} = e^{k(1+\mathrm{i})x^2}, \qquad (21.94)$$

which clearly has a saddle point at $x = 0$. Writing the exponential in polar coordinates, we have

$$g(r,\theta) = \sqrt{2}r^2 e^{\mathrm{i}\pi/4} e^{2\mathrm{i}\theta}. \qquad (21.95)$$

The level curves satisfy Eq. (21.92), and are lines along the angles $\theta = \pi/8, 5\pi/8$, which the steepest paths satisfy Eq. (21.91) and are lines along the angles $\theta = -\pi/8, 3\pi/8$. The geometry is illustrated in Fig. 21.8(a). If we add a constant to the function $g(x)$, e.g. $g(x) = \mathrm{i} + (1+\mathrm{i})x^2$, the position of the curves does not change.

We now consider as an example the properties of the function

$$h(x) = e^{kg(x)} = e^{k[x^2 + \mathrm{i}x^3]}. \qquad (21.96)$$

The function now has two saddle points for which $g'(z) = 0$, at $z = 0$ and $z = 2\mathrm{i}/3$. Noting that $\mathrm{Im}[g(0)] = 0$ and $\mathrm{Im}[g(2\mathrm{i}/3)] = 0$, the equation for the steepest paths is

$$x[2y + x^2 - y^2 - 2y^2] = 0. \qquad (21.97)$$

This equation has three solutions, $x = 0$ and two solutions to the quadratic equation

$$y = \frac{1}{3} \pm \frac{1}{3}\sqrt{1 + 3x^2}. \qquad (21.98)$$

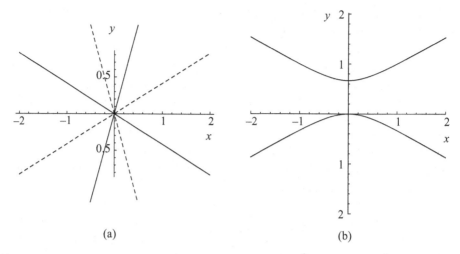

Figure 21.8 (a) Steepest paths and level curves for $g(z) = (1+i)z^2$; the level curves are dashed lines. (b) Steepest paths for $g(z) = z^2 + iz^3$.

Only the negative root passes through the origin, while the positive root passes through $z = 2i/3$. We have the following steepest paths for our function

$$x = 0, \quad z = 0, z = 2i/3, \tag{21.99}$$

$$y = \frac{1}{3} + \frac{1}{3}\sqrt{1+3x^2}, \quad z = 2i/3, \tag{21.100}$$

$$y = \frac{1}{3} - \frac{1}{3}\sqrt{1+3x^2}, \quad z = 0. \tag{21.101}$$

These paths are illustrated in Fig. 21.8(b). To find the level curves, we note that $\text{Re}[g(0)] = 0$, $\text{Re}[g(2i/3)] = -4/27$. The level curves for $z = 0$ must therefore satisfy the equation

$$(1-y)(x^2 - y^2) - 2x^2 y = 0, \tag{21.102}$$

while the level curves for $z = 2i/3$ must satisfy the equation

$$(1-y)(x^2 - y^2) - 2x^2 y = -4/27. \tag{21.103}$$

◇

The method of steepest descents involves distorting the path of integration to lie as close as possible along one of the steepest paths. The integral then behaves like the integral of Laplace's method, and can be evaluated by that technique. It is to be noted, however, that the contour of integration can only be distorted if both $f(z)$ and $g(z)$ are analytic functions everywhere within the region of distortion. The path cannot be pushed past a singularity, for instance, without changing the value of the integral and making the technique invalid.

Laplace's method and the method of stationary phase both result in closed-form expressions for the leading asymptotic term that can be directly applied. If these techniques are to be referred to as "methods", it is perhaps best to refer to steepest descent as a "strategy" for evaluating the integral! The appropriate contour of steepest descent must be determined anew for each different integral, and the determination is often nontrivial. Nevertheless, the technique allows the asymptotic evaluation of integrals whose asymptotic behavior would otherwise be indeterminate.

Example 21.3 (Airy's integral) Airy's integral is of the form

$$Ai(k) = \frac{1}{\pi} \int_0^\infty \cos\left[\frac{1}{3}x^3 + kx\right] dx, \tag{21.104}$$

and we wish to determine the asymptotic form of this integral for large values of k.[2] We may make a coordinate transformation

$$x = k^{1/2}z, \quad K = k^{3/2}, \tag{21.105}$$

to write the integral in the form

$$Ai(K^{2/3}) = \frac{K^{1/3}}{2\pi} \int_{-\infty}^\infty \exp\left[iK\left(\frac{1}{3}z^3 + z\right)\right] dz. \tag{21.106}$$

We let

$$g(z) = i\left(\frac{1}{3}z^3 + z\right). \tag{21.107}$$

The saddle points such that $g'(z) = 0$ are readily found to be at $z = \pm i$, and $\text{Im}[g(i)] = \text{Im}[g(-i)] = 0$. The steepest paths are therefore those paths such that $\text{Im}[g(z)] = 0$; letting $z = x + iy$ we find that they satisfy

$$\text{Im}[g(z)] = \frac{1}{3}x^3 - xy^2 + x = 0. \tag{21.108}$$

One root of this equation is $x = 0$; the other roots satisfy the quadratic equation

$$-3y^2 + x^2 + 3 = 0, \tag{21.109}$$

or

$$y = \pm\sqrt{\frac{1}{3}x^2 + 1}. \tag{21.110}$$

[2] This example comes from Erdélyi [Erd56], who took it originally from Copson [Cop46].

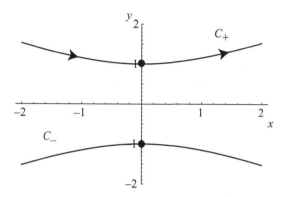

Figure 21.9 Steepest paths for the Airy function.

These paths are illustrated in Fig. 21.9. The function $g(z)$ has no singularities in the complex plane, so we may readily distort the path of integration to the upper hyperbolic path. The integral therefore has the form

$$Ai(K^{2/3}) = \frac{K^{1/3}}{2\pi} \int_{C_+} \exp\left[iK\left(\frac{1}{3}z^3 + z\right)\right]dz. \qquad (21.111)$$

The integrals along the contours is evaluated by using x as a parameter and letting

$$y(x) = \sqrt{\frac{1}{3}x^2 + 1}, \qquad (21.112)$$

$$dz = [1 + y'(x)]dx. \qquad (21.113)$$

With this parameterization, we may apply Laplace's method, Eq. (21.61), and find, with some effort, that

$$Ai(K^{2/3}) = \frac{K^{1/3}}{2\pi}e^{-2K/3}\sqrt{\pi/K}. \qquad (21.114)$$

Transforming back to the original parameter k, we have

$$Ai(k) = \frac{\sqrt{\pi}}{2\pi}e^{-\frac{2k^{3/2}}{3}}k^{-1/4}. \qquad (21.115)$$

◇

21.6 Method of stationary phase for double integrals

The method of stationary phase can be extended to the evaluation of two-dimensional integrals of the form

$$F(k) = \int_D f(x,y)e^{ikg(x,y)}\,dxdy, \qquad (21.116)$$

where f and g are real-valued functions and D represents a two-dimensional, closed, simply connected domain with boundary C. This sort of integral arises, for instance, in diffraction theory in the calculation of the far-zone behavior of diffracted wavefields.

There are in general three types of critical point that may contribute to the asymptotic behavior of the integral. *Critical points of the first kind* are those at which $\nabla g(x,y) = 0$; their contribution to the asymptotic form of the integral is of order $1/k$. *Critical points of the second kind* are those points on the boundary curve for which $\partial g/\partial t = 0$, where t represents the path of the boundary; their contribution is of order $1/k^{3/2}$. *Critical points of the third kind* are points where the boundary curve has sharp edges (such as the corners of a square), and their contribution is of order $1/k^2$. From the form of the contributions, it is clear that boundary points of the first kind generally dominate the asymptotic behavior of the integral.

A rigorous derivation of the asymptotic behavior is quite complicated; we follow the heuristic description given in [MW95, Chapter 3] for critical points of the first kind.

Following the discussion of Section 21.4, we assume that in the neighborhood of a critical point of the first kind (x_0,y_0) the function $g(x,y)$ may be well-approximated by the lowest terms of its Taylor series expansion,

$$g(x,y) \approx g(x_0,y_0) + \frac{1}{2}\left[g_{xx}(x-x_0)^2 + 2g_{xy}(x-x_0)(y-y_0) + g_{yy}(y-y_0)^2\right], \qquad (21.117)$$

where

$$g_{xx} \equiv \frac{\partial^2 g(x_0,y_0)}{\partial x^2}, \qquad (21.118)$$

and so on. It is further assumed that $g_{xx}g_{yy} - g_{xy}^2 \neq 0$ and that $f(x,y)$ can be approximated by its lowest-order Taylor series term, $f(x_0,y_0)$.

For sufficiently large k, we may extend the domain of integration in Eq. (21.116) to infinity; introducing $x' \equiv (x-x_0)$, $y' \equiv (y-y_0)$, we may write

$$F(k) \sim f(x_0,y_0)\exp[ikg(x_0,y_0)]\int_{-\infty}^{\infty}\int_{-\infty}^{\infty}\exp\left[\frac{1}{2}ik(g_{xx}x'^2\right.$$

$$\left. + 2g_{xy}x'y' + g_{yy}y'^2)\right]dx'dy'. \qquad (21.119)$$

This integral can be converted to a simpler form by seeking a coordinate rotation

$$x' = \cos\theta\xi + \sin\theta\eta, \tag{21.120}$$

$$y' = -\sin\theta\xi + \cos\theta\eta, \tag{21.121}$$

such that the exponent depends only upon ξ^2 and η^2, in the form

$$F(k) \sim f(x_0, y_0) \int_{-\infty}^{\infty}\int_{-\infty}^{\infty} \exp\left[\frac{1}{2}ik(\alpha\xi^2 + \beta\eta^2)\right]d\xi\,d\eta. \tag{21.122}$$

In this case, our two-dimensional stationary phase integral breaks into the product of two one-dimensional integrals, such that

$$F(k) \sim 4f(x_0, y_0)\exp[ikg(x_0, y_0)]\int_0^{\infty}\exp[ik\alpha\xi^2]d\xi\int_0^{\infty}\exp[ik\beta\eta^2]d\eta. \tag{21.123}$$

Each integral is of the form of a Fresnel integral of Example 9.14,

$$\int_0^{\infty} e^{itx^2}dx = \frac{1}{2}\sqrt{\frac{\pi}{t}}e^{i\pi/4} = \frac{1}{2}\sqrt{\frac{\pi}{|t|}}e^{i\,\mathrm{sgn}[t]\pi/4}, \tag{21.124}$$

where $\mathrm{sgn}[t]$ is the sign of the variable t. We may then write

$$F(k) \sim 4f(x_0, y_0)\exp[ikg(x_0, y_0)]\frac{2\pi e^{i\,\mathrm{sgn}[\alpha]\pi/4}e^{i\,\mathrm{sgn}[\beta]\pi/4}}{k\sqrt{|\alpha||\beta|}}. \tag{21.125}$$

With some effort, one can explicitly determine the values of α and β. They satisfy the equations

$$\mathrm{Tr}[\mathbf{g}] = \alpha + \beta, \tag{21.126}$$

$$\det[\mathbf{g}] = \alpha\beta, \tag{21.127}$$

where $\mathbf{g}$ is the matrix formed by the derivatives g_{ij}.

21.7 Additional reading

There are a number of excellent books on asymptotic techniques:

- A. Erdélyi, *Asymptotic Expansions* (New York, Dover, 1956).
- N. Bleistein and R. A. Handelsman, *Asymptotic Expansions of Integrals* (New York, Holt, Reinhart and Winston, 1975).
- R. Wong, *Asymptotic Approximations of Integrals* (Boston, Academic Press, 1989).

21.8 Exercises

1. Show that the following asymptotic expansion holds

$$\int_0^\infty \frac{e^{-zt}}{1+t^2}\,dt \sim \frac{1}{z} - \frac{2!}{z^3} + \frac{4!}{z^5} - \cdots,$$

for positive real z.

2. Show, for large values of z, that

$$e^z z^{-a} \int_z^\infty e^{-x} x^{a-1}\,dx \sim \frac{1}{z} + \frac{a-1}{z^2} + \frac{(a-1)(a-2)}{z^3} + \cdots.$$

3. We may introduce a complementary error function, erfc(x), of the form

$$\operatorname{erfc}(x) = \frac{2}{\sqrt{\pi}} \int_x^\infty e^{-t^2}\,dt.$$

Show that this function has an asymptotic series for large x of the form

$$\operatorname{erfc}(x) \sim \frac{e^{-x^2}}{\sqrt{\pi}x} \left[1 + \sum_{k=1}^{n}(-1)^k \frac{(2k-1)!!}{(2x^2)^k} \right].$$

(Try an inductive argument, using integration by parts to get the first terms.) See Appendix A for an explanation of the gamma function.

4. We may introduce a complementary incomplete gamma function $\Gamma(x,\alpha)$ of the form

$$\Gamma(x,\alpha) = \int_x^\infty e^{-t} t^{\alpha-1}\,dt.$$

Show that, for large x,

$$\Gamma(x,\alpha) = e^{-x} x^{\alpha-1} \sum_{k=0}^{n} \frac{\Gamma(\alpha)}{\Gamma(\alpha-k-1)x^k} + O(x^{-n-1}).$$

5. Read the paper by K. E. Oughstun and G. C. Sherman, Propagation of electromagnetic pulses in a linear dispersive medium with absorption (the Lorentz medium), *J. Opt. Soc. Am.* B **5** (1988), 817–849. Describe the physical problem being considered, and explain qualitatively what a precursor is. How is asymptotic analysis applied to the study of such precursors?

6. The asymptotic technique used to relate rays and waves in the beginning of this chapter is in fact the most primitive technique. Read the paper by G. W. Forbes and M. A. Alonso, Using rays better. I. Theory for smoothly-varying media, *J. Opt. Soc. Am.* A **18** (2001), 1132–1145, and describe the technique used within to apply wave effects to geometrical optics. Describe the asymptotic methods applied. What mathematical model do the authors use to describe a geometrical ray?

7. Starting from the integral forms of the Bessel and Neumann functions, Eqs. (16.59) and (16.76), determine the leading asymptotic terms of these functions as $x \to \infty$ using the method of stationary phase.

8. Evaluate the leading asymptotic term of the following integral:

$$I(k) = \int_{-\pi/8}^{\pi/8} x^2 \exp[ik(\cos(2x))^2]dx,$$

using the method of stationary phase.

9. Evaluate the leading asymptotic term of the following integral:

$$I(k) = \int_{-\pi/4}^{\pi/4} (x^2 + 1)\exp[ik(\cos x)^3]dx,$$

using the method of stationary phase.

10. Following a method similar to Watson's lemma for Laplace-type integrals, determine the leading term of the asymptotic expansion for

$$F(k) = \int_a^b f(x)e^{ikg(x)}dx,$$

when the function $f(x) \sim (x - x_0)^\alpha$ in the neighborhood of the stationary point x_0, with $\alpha > -1$.

11. We consider the function

$$h(x) = \exp[k(x^2 + ix + 1)]. \tag{21.128}$$

Determine the saddle points, the level curves and the steepest paths of this function.

12. We consider the function

$$h(x) = \exp[k(x^3 + 3ix + 2)]. \tag{21.129}$$

Determine the saddle points, the level curves and the steepest paths of this function.

13. Determine the leading asymptotic term of the integral

$$I(k) = \int_{-\infty}^{\infty} \frac{e^{k(x^2 + ix + 1)}}{1 + x^2}dx, \tag{21.130}$$

using the method of steepest descents.

14. Determine the leading asymptotic term of the integral

$$I(k) = \int_{-\infty}^{\infty} \frac{e^{k(x^5 + i)}}{1 + x^4}dx, \tag{21.131}$$

using the method of steepest descents.

15. Complete the solution for Airy's integral by proceeding from Eq. (21.111) to (21.114).

Appendix A

The gamma function

The gamma function, to be denoted $\Gamma(z)$, is a bit of a curious beast. It rarely appears directly in physical problems, and unlike most of the other special functions such as the Bessel functions and Legendre polynomials, it does not satisfy a second-order ODE and cannot be used to construct a set of orthogonal basis functions. However, it is extremely useful in deriving advanced and general formulas relating to many of those more common functions.

A.1 Definition

The gamma function is a continuous and analytic interpolation of the discrete factorial function $n! = n(n-1)\cdots 2 \cdot 1$. We initially define it by the formula

$$\Gamma(z) = \int_0^\infty e^{-t} t^{z-1} \, dt, \tag{A.1}$$

with the requirement that $\mathrm{Re}(z) > 0$ for convergence. We may decompose this integral into two parts, in the form

$$\Gamma(z) = \int_0^1 e^{-t} t^{z-1} \, dt + \int_1^\infty e^{-t} t^{z-1} \, dt. \tag{A.2}$$

It can be shown that the first integral is analytic in the half-plane $\mathrm{Re}(z) > 0$, while the second integral is entire analytic; this expression of the gamma function is therefore analytic in the right half-plane. We may analytically continue to the entire complex plane by using the power series representation of the exponential,

$$\int_0^1 e^{-t} t^{z-1} \, dt = \int_0^1 t^{z-1} \sum_{n=0}^\infty \frac{(-1)^n t^n}{n!} \, dt. \tag{A.3}$$

It can be shown that integral is convergent and that the order of integration and summation may be reversed. We may then write

$$\int_0^1 t^{z-1} \sum_{n=0}^{\infty} \frac{(-1)^n t^n}{n!} dt = \sum_{n=0}^{\infty} \frac{(-1)^n}{n!} \int_0^1 t^{n+z-1} dt = \sum_{n=0}^{\infty} \frac{(-1)^n}{n!} \frac{1}{z+n}. \qquad (A.4)$$

For any value of z not zero or a negative integer, this series is convergent, as can be shown via the comparison test. We may therefore analytically continue the function $\Gamma(z)$ to all values of the complex plane except the points $z = 0, -1, -2, \ldots$, in the form

$$\Gamma(z) = \sum_{n=0}^{\infty} \frac{(-1)^n}{n!} \frac{1}{z+n} + \int_1^{\infty} e^{-t} t^{z-1} dt. \qquad (A.5)$$

The points $z = 0$ and the negative integers are clearly simple poles.

A.2 Basic properties

Perhaps the most fundamental property of the gamma function is its identification with the factorial function for positive integers; this may be characterized by the expression

$$\Gamma(z+1) = z\Gamma(z). \qquad (A.6)$$

Assuming $\mathrm{Re}(z) > 0$, we may prove this relation using integration by parts on Eq. (A.1). We therefore have

$$\Gamma(z+1) = \int_0^{\infty} e^{-t} t^z dt = -\left[e^{-t} t^z\right]_0^{\infty} + z \int_0^{\infty} e^{-t} t^{z-1} dt = z\Gamma(z). \qquad (A.7)$$

Since it is clear that $\Gamma(1) = 1$, we find that for integer z, $\Gamma(z+1) = z!$.

For fractional values of z, the evaluation of $\Gamma(z)$ is much more complicated. The most commonly occurring case is $\Gamma(1/2)$; Eq. (A.1) takes on the form

$$\Gamma(1/2) = \int_0^{\infty} e^{-t} t^{-1/2} dt. \qquad (A.8)$$

Introducing a new variable $w^2 = t$, we may write

$$\Gamma(1/2) = 2 \int_0^{\infty} e^{-w^2} dw = \sqrt{\pi}. \qquad (A.9)$$

From this, we may apply Eq. (A.6) to find that

$$\Gamma(n+1/2) = \frac{1 \cdot 3 \cdot 5 \cdots (2n-1)}{2^n} \sqrt{\pi}, \qquad (A.10)$$

with n an integer. The numerator of the fraction is often abbreviated using the double factorial notation, i.e.

$$(2n+1)!! \equiv 1 \cdot 3 \cdot 5 \cdots (2n+1), \tag{A.11}$$

$$(2n)!! \equiv 2 \cdot 4 \cdot 6 \cdots (2n). \tag{A.12}$$

The gamma function may be directly related to the trigonometric functions through the formula

$$\Gamma(z)\Gamma(1-z) = \frac{\pi}{\sin \pi z}. \tag{A.13}$$

To prove this, let us assume first that $0 < \mathrm{Re}(z) < 1$. We may use the integral relation (A.1) to write

$$\Gamma(z)\Gamma(1-z) = \int_0^\infty \int_0^\infty e^{-(s+t)} t^{z-1} s^{-z} ds dt. \tag{A.14}$$

Let us introduce new variables $x = s+t$ and $y = t/s$. The above integral may then be written as

$$\Gamma(z)\Gamma(1-z) = \int_0^\infty \int_0^\infty e^{-x} y^{z-1} \frac{dxdy}{1+y} = \int_0^\infty y^{z-1} \frac{dy}{1+y}. \tag{A.15}$$

The latter integral can be evaluated using contour integration as done in Example 9.15, and the result is

$$\Gamma(z)\Gamma(1-z) = \frac{\pi}{\sin \pi z}. \tag{A.16}$$

This function can be used to prove that the function $\Gamma(z)$ has no zeros in the complex plane.
Another important formula is the so-called *Legendre duplication formula*,

$$2^{2z-1}\Gamma(z)\Gamma(z+1/2) = \sqrt{\pi}\,\Gamma(2z). \tag{A.17}$$

To prove it, we begin by writing the left-hand side of the equation with the integral definition of the gamma function,

$$2^{2z-1}\Gamma(z)\Gamma(z+1/2) = \int_0^\infty \int_0^\infty e^{-(s+t)} (2\sqrt{st})^{2z-1} t^{-1/2} ds dt. \tag{A.18}$$

We introduce new variables $p = \sqrt{s}$, $q = \sqrt{t}$, so that

$$2^{2z-1}\Gamma(z)\Gamma(z+1/2) = \int_0^\infty \int_0^\infty e^{-(p^2+q^2)} (2pq)^{2z-1} p dp dq. \tag{A.19}$$

To make a symmetric form of this equation, we also use the complementary transformation $p = \sqrt{t}$, $q = \sqrt{s}$, and add the resulting equations together,

$$2^{2z-1}\Gamma(z)\Gamma(z+1/2) = \int_0^\infty \int_0^\infty e^{-(p^2+q^2)}(2pq)^{2z-1}(p+q)\mathrm{d}p\mathrm{d}q. \tag{A.20}$$

We make a third set of transformations, $u = p^2 + q^2$, $v = 2pq$; the terms of the resulting formula may be rearranged to the form

$$2^{2z-1}\Gamma(z)\Gamma(z+1/2) = \int_0^\infty v^{2z-1}\mathrm{d}v \int_0^\infty \frac{e^{-u}}{\sqrt{u-v}}\mathrm{d}u. \tag{A.21}$$

One final coordinate transformation, $w^2 = u - v$, brings us to the expression

$$2^{2z-1}\Gamma(z)\Gamma(z+1/2) = 2\int_0^\infty e^{-v}v^{2z-1}\mathrm{d}v \int_0^\infty e^{-w^2}\mathrm{d}w. \tag{A.22}$$

The integral over w has value $\sqrt{\pi}/2$, while the integral over v is the gamma function $\Gamma(2z)$; the Legendre duplication formula follows.

A.3 Stirling's formula

For large values of n, is it well known that the factorial function takes on the asymptotic form known as *Stirling's formula*,

$$n! \approx \sqrt{2\pi n}(n/e)^n. \tag{A.23}$$

This result may be derived by looking at the asymptotic form of the gamma function using the techniques of Chapter 21, in particular Laplace's method. We begin with the definition of the gamma function,

$$\Gamma(z+1) = \int_0^\infty e^{-t}t^z\mathrm{d}t, \tag{A.24}$$

and rewrite it in a form suitable for asymptotic analysis by defining $t = z\tau$,

$$\Gamma(z+1) = z^{z+1}\int_0^\infty e^{-z\tau}\tau^z\mathrm{d}\tau. \tag{A.25}$$

We may write $\tau^z = \exp[z\log\tau]$, so that we have

$$\Gamma(z+1) = z^{z+1}\int_0^\infty \exp[-z(\tau - \log\tau)]\mathrm{d}\tau. \tag{A.26}$$

We now apply Laplace's method, Eq. (21.61), to the integral. Laplace's method approximates the integral asymptotically as

$$\int_a^b f(x) e^{kg(x)} dx \approx f(x_0) \exp[kg(x_0)] \sqrt{\frac{2\pi}{|g''(x_0)|k}}. \tag{A.27}$$

For our integral, $a = 0$, $b = \infty$, $f(\tau) = 1$, $g(\tau) = \tau - \log \tau$, and the stationary point of the exponent is $x_0 = 1$. On substitution, we readily find that

$$\Gamma(z+1) \sim z^{z+1} e^{-z} \left[\frac{\pi}{2z}\right]^{1/2}, \tag{A.28}$$

or

$$\Gamma(z+1) \sim z^{z+1/2} e^{-z} \sqrt{2\pi}. \tag{A.29}$$

In the special case $z = n$, we reproduce Stirling's formula. It is often derived and written using $\log[\Gamma(z+1)]$, such that

$$\log[\Gamma(z+1)] = (z+1/2) \log z - z + \frac{1}{2} \log(2\pi). \tag{A.30}$$

A.4 Beta function

The *beta function* is defined by the expression

$$\beta(m,n) = \int_0^1 x^{m-1} (1-x)^{n-1} dx, \tag{A.31}$$

and the integral converges for $m, n > 0$. A simple coordinate transformation to $y = 1 - x$ can be used to show that $\beta(m,n) = \beta(n,m)$. The coordinate transformation $x = \sin^2 \theta$ results in an alternative expression,

$$\beta(m,n) = 2 \int_0^{\pi/2} \sin^{2m-1} \theta \cos^{2n-1} \theta d\theta. \tag{A.32}$$

The beta function may be readily expressed in terms of the gamma function, in the form

$$\beta(m,n) = \frac{\Gamma(m)\Gamma(n)}{\Gamma(m+n)}. \tag{A.33}$$

This may be proven by first transforming the integral representation of $\Gamma(z)$ using $t = x^2$, so that

$$\Gamma(z) = \int_0^\infty e^{-t} t^{z-1} dt = 2 \int_0^\infty e^{-x^2} x^{2z-1} dx. \tag{A.34}$$

We now consider the product $\Gamma(m)\Gamma(n)$, which takes on the form

$$\Gamma(m)\Gamma(n) = 4 \int_0^\infty e^{-x^2} x^{2z-1} dx \int_0^\infty e^{-y^2} y^{2z-1} dy. \tag{A.35}$$

This may be interpreted as an integral over the first quadrant of the xy-plane. Introducing polar coordinates $x = r\cos\theta$ and $y = r\sin\theta$, we may write

$$\Gamma(m)\Gamma(n) = 4 \int_0^{\pi/2} \int_0^\infty r^{2(m+n-1)} e^{-r^2} \sin^{2m-1}\theta \cos^{2n-1}\theta \, r dr d\theta. \tag{A.36}$$

Rearranging terms, we may write

$$\Gamma(m)\Gamma(n) = 4 \int_0^{\pi/2} \sin^{2m-1}\theta \cos^{2n-1}\theta d\theta \int_0^\infty r^{2(m+n)-1} e^{-r^2} dr. \tag{A.37}$$

The r-integral is just $\Gamma(m+n)/2$, while the integral over θ gives $\beta(m,n)/2$; the result is therefore proven.

This result can be used to prove Eq. (A.16), provided $0 \le z \le 1$, as well as the Legendre duplication formula, Eq. (A.17).

A.5 Useful integrals

There are a large number of integral formulas for the gamma function, some quite surprising in form. Here we derive a number of integral relations involving the gamma function that appear elsewhere in the text.

We begin by considering the integral

$$\int_D (-w)^{z-1} e^{-w} dw, \tag{A.38}$$

where D is the complex domain contour shown in Fig. A.1(a). Unlike most integrals in the complex domain, we cannot evaluate this one using the residue theorem: it is not a closed contour because it encloses a branch point. We can, however, evaluate each term explicitly, and it is important to note that we may distort the path without changing the value of the integral via Cauchy's integral formula. The circular part of the contour is taken to be radius δ, and the phase of $-w$ is taken to be $-\pi$ on the inward infinite line and $+\pi$ on the outward line. We may write

$$\int_D (-w)^{z-1} e^{-w} dw = e^{-i\pi(z-1)} \int_\infty^\delta w^{z-1} e^{-w} dw + \oint_\circ (-w)^{z-1} e^{-w} dw$$
$$+ e^{i\pi(z-1)} \int_\delta^\infty w^{z-1} e^{-w} dw, \tag{A.39}$$

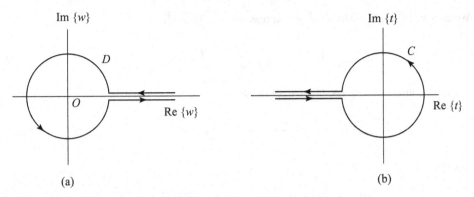

Figure A.1 The infinite contours used for integral representations of the gamma function.

where the central integral is over the circular part of the contour. Setting $w = \delta e^{i\theta}$ for this circular part, we readily find that in the limit $\delta \to 0$ this contribution vanishes. We are left with

$$\int_D (-w)^{z-1} e^w \, dw = e^{-i\pi(z-1)} \int_\infty^0 w^{z-1} e^{-w} \, dw + e^{i\pi(z-1)} \int_0^\infty w^{z-1} e^{-w} \, dw$$

$$= -2i \sin(\pi z) \int_0^\infty w^{z-1} e^{-w} \, dw. \tag{A.40}$$

Identifying the integral on the right with the gamma function, this leads to the result

$$\Gamma(z) = -\frac{1}{2i \sin(\pi z)} \int_D (-w)^{z-1} e^{-w} \, dw. \tag{A.41}$$

This result is known as Hankel's formula for the gamma function.

Because the function $(-w)^{z-1} e^{-w}$ has no singularities in the finite complex plane, we may rotate our contour by π without changing its value, i.e. $w = -t$. We then find, for the contour of Fig. A.1(b), that

$$\Gamma(z) = \frac{1}{2i \sin(\pi z)} \int_C t^{z-1} e^t \, dt. \tag{A.42}$$

We may use this result, together with Eq. (A.16), to prove the formula

$$\frac{1}{\Gamma(n)} = \frac{1}{2\pi i} \int_C e^t t^{-n} \, dt. \tag{A.43}$$

We first write

$$\frac{1}{\Gamma(z)} = \frac{\sin(\pi z)\Gamma(1-z)}{\pi}. \tag{A.44}$$

We now use Eq. (A.42) to write an expression for $\Gamma(1-z)$,

$$\Gamma(1-z) = \frac{1}{2i\sin(\pi z)} \int_C t^{-z} e^t dt. \tag{A.45}$$

On substitution into Eq. (A.44), we find that

$$\frac{1}{\Gamma(z)} = \frac{1}{2\pi i} \int_C e^t t^{-z} dt, \tag{A.46}$$

as expected.

Appendix B
Hypergeometric functions

A *hypergeometric function* is broadly defined as a function $f(z)$ that may be represented as a power series,

$$f(z) = \sum_{k=0}^{\infty} \alpha_k z^k, \tag{B.1}$$

where the ratio of successive coefficients is a rational power of n, i.e.

$$\frac{a_{k+1}}{a_k} = \frac{A(n)}{B(n)}, \tag{B.2}$$

where $A(n)$ and $B(n)$ are polynomials in n. The most general form of such a function may be written as

$$_pF_q(a_1,\ldots,a_p;b_1,\ldots,b_q;z) = \sum_{k=0}^{\infty} \frac{(a_1)_k \cdots (a_p)_k}{k!(b_1)_k \cdots (b_q)_k} z^k, \tag{B.3}$$

where $(a)_k$ is called a *Pochhammer symbol*, defined as

$$(a)_k = a(a+1)(a+2)\cdots(a+k-1). \tag{B.4}$$

It acts as a "reverse factorial" of sorts, and for integer a has the form $(a)_k = (a+k-1)!/(a-1)!$. It can be readily demonstrated that the function defined by Eq. (B.3) satisfies Eq. (B.2). The leading subscript p of $_pF_q$ indicates the number Pochhammer symbols in the numerator while the following subscript q indicates the number of Pochhammer symbols in the denominator. By convention, the denominator of Eq. (B.3) possesses an extra factor of $k!$.

Many elementary functions may be expressed as hypergeometric functions. For instance, the exponential function has the hypergeometric representation

$$e^z = {}_0F_0(;;z), \tag{B.5}$$

which possesses no Pochhammer symbols. The natural logarithm may be expressed as

$$\log(1+z) = z\,{}_2F_1(1,1;2;-z). \tag{B.6}$$

783

The hypergeometric functions are helpful for a number of reasons. Many of the special functions that we consider in this text may be treated as special cases of the hypergeometric functions, and they therefore serve as a unifying principle. Also, the hypergeometric function can be used to quantify a number of otherwise analytically intractable integrals.

Two special cases of the general hypergeometric function are of special interest, the first referred to simply as the *hypergeometric function* and the second referred to as the *confluent hypergeometric function*.

B.1 Hypergeometric function

The *hypergeometric equation* is defined as the equation

$$z(1-z)y'' + [c - (a+b+1)z]y' - aby = 0. \tag{B.7}$$

The regular solution to this expression is of the form

$$_2F_1(a,b;c;z) = \sum_{k=0}^{\infty} \frac{(a)_k(b)_k}{k!(c)_k} z^k, \tag{B.8}$$

This function is what is commonly referred to as the hypergeometric function. Many of the special functions discussed in this book can be represented by this function with an appropriate choice of a, b, and c. For instance, the Legendre functions and associated Legendre functions can be expressed as

$$P_n(x) = {_2F_1}\left(-n, n+1; 1; \frac{1-x}{2}\right) \tag{B.9}$$

and

$$P_n^m(x) = \frac{(n+m)!}{(n-m)!} \frac{(1-x^2)^{m/2}}{2^m m!} {_2F_1}\left(m-n, m+n+1; m+1; \frac{1-x}{2}\right). \tag{B.10}$$

The hypergeometric functions have many properties that are reminiscent of the properties of other special functions, namely recurrence and derivative relations. The recurrence relations, which relate functions which differ by integer values of a, b, and c, are referred to as *contiguous function relations*. For instance, relations between functions of different a values include

$$(c-a){_2F_1}(a-1,b;c;z) + (2a-c-az+bz){_2F_1}(a,b;c;z) + a(z-1){_2F_1}(a+1,b;c;z), \tag{B.11}$$

while different b values satisfy

$$(c-b){_2F_1}(a,b-1;c;z) + (2b-c-bz+az){_2F_1}(a,b;c;z) + b(z-1){_2F_1}(a,b+1;c;z). \tag{B.12}$$

Derivative relations amongst contiguous functions also exist; for instance,

$$\frac{d}{dz}{}_2F_1(a,b;c;z) = \frac{ab}{c}{}_2F_1(a+1,b+1;c+1;z).$$ (B.13)

Many, many other relations exist amongst the hypergeometric functions; see [AS65] for additional details.

B.2 Confluent hypergeometric function

The *confluent hypergeometric equation* is given by

$$zy'' + (b-z)y' - ay.$$ (B.14)

A regular solution of this equation is another hypergeometric function, called the *confluent hypergeometric function*, of the form

$$_1F_1(a;b;z) \equiv M(a;b;z) = \sum_{k=0}^{\infty} \frac{(a)_k}{k!(b)_k} z^k.$$ (B.15)

A second solution of this equation may be written in the form

$$U(a;b;z) = \frac{\pi}{\sin(\pi b)}\left[\frac{M(a;b;z)}{\Gamma(1+a-b)\Gamma(b)} - z^{1-b}\frac{M(a+1-b;2-b;z)}{\Gamma(a)\Gamma(2-b)}\right].$$ (B.16)

This definition, like the definition of the Neumann function, Eq. (16.24), allows the function to have meaning in the limiting case that b takes an integer value.

The confluent functions also can be used to represent a number of special functions; for example, the Bessel functions can be written as

$$J_\nu(z) = \frac{e^{-iz}}{\Gamma(\nu+1)}\left(\frac{z}{2}\right)^\nu M(\nu+1/2;2\nu+1;2iz).$$ (B.17)

B.3 Integral representations

There are a number of integral representations of the hypergeometric functions, and also a number of important integrals whose solutions can be expressed in terms of them. We list a few representative examples here.

A standard integral form of the hypergeometric function is

$$_2F_1(a,b;c;z) = \frac{\Gamma(c)}{\Gamma(b)\Gamma(c-b)}\int_0^1 t^{b-1}(1-t)^{c-b-1}(1-tz)^{-a}dt,$$ (B.18)

with the requirement that $\text{Re}\{c\} > \text{Re}\{b\} > 0$. This function is single-valued and analytic in the complex plane with a branch cut running from $z=1$ to $z=\infty$.

The confluent hypergeometric function may be written in integral form as

$$_1F_1(a;b;z) = \frac{\Gamma(b-a)}{\Gamma(a)\Gamma(b)} \int_0^1 e^{zt} t^{a-1}(1-t)^{b-a-1} dt,$$ (B.19)

with the condition that $\mathrm{Re}\{b\} > \mathrm{Re}\{a\} > 0$.

The error function $\mathrm{erf}(x)$ may be written in terms of the confluent hypergeometric function as

$$\mathrm{erf}(x) = \frac{2}{\sqrt{\pi}} \int_0^x \exp^{-t^2} dt = \frac{2x}{\sqrt{\pi}} M(1/2; 3/2; -x^2).$$ (B.20)

Elliptic integrals, useful in the solution of rigid pendulum problems, can be written in terms of hypergeometric functions. The complete elliptic integral of the first kind may be written as

$$K(m) = \int_0^{\pi/2} (1 - m\sin^2\theta)^{-1/2} d\theta = \frac{\pi}{2} {}_2F_1(1/2, 1/2; 1; m),$$ (B.21)

while the complete elliptic integral of the second kind may be written as

$$E(m) = \int_0^{\pi/2} (1 - m\sin^2\theta)^{1/2} d\theta = \frac{\pi}{2} {}_2F_1(-1/2, 1/2; 1; m).$$ (B.22)

The application of elliptic integrals to the rigid pendulum can be found in mechanics texts such as [MT88]; a list of the mathematical properties of such integrals can be found in [AS65].

References

[ABP03]   L. Allen, S. M. Barnett, and M. J. Padgett. *Optical Angular Momentum*. Bristol and Philadelphia, IOP Publishing, 2003.

[AE09]   A. Alù and N. Engheta. Cloaking a sensor. *Phys. Rev. Lett.*, **102**:233901, 2009.

[AF97]   M. A. Alonso and G. W. Forbes. Uniform asymptotic expansions for wave propagators via fractional transformations. *J. Opt. Soc. Am. A*, **14**:1279–1292, 1997.

[AFN$^+$01] L. J. Allen, H. M. L. Faulkner, K. A. Nugent, M. P. Oxley, and D. Paganin. Phase retrieval from images in the presence of first-order vortices. *Phys. Rev. E*, **63**:037602, 2001.

[Ara92]   P. K. Aravind. A simple proof of Pancharatnam's theorem. *Opt. Commun.*, **94**:191–196, 1992.

[AS65]   M. Abramowitz and I. A. Stegun. *Handbook of Mathematical Functions*. New York, Dover Publications, 1965.

[Atr75]   S. R. Atre. An alternate derivation of final and initial value theorems in Z domain. *Proc. IEEE*, **63**:537–538, 1975.

[AU76]   T. Asakura and T. Ueno. Apodization for minimizing the second moment of the intensity distribution in the fraunhofer diffraction pattern. *Nouv. Rev. Opt.*, **7**:199–203, 1976.

[Bas78]   M. J. Bastiaans. The Wigner distribution function applied to optical signals and systems. *Opt. Commun.*, **25**:26–30, 1978.

[Ber01]   M. V. Berry. Why are special functions special? *Phys. Today*, **54**:11–12, 2001.

[BH75]   N. Bleistein and R. A. Handelsman. *Asymptotic Expansions of Integrals*. New York, Holt, Rinehart and Winston, 1975.

[BH83]   C. F. Bohren and D. R. Huffman. *Absorption and Scattering of Light by Small Particles*. New York, Wiley, 1983.

[BK96]   M. V. Berry and S. Klein. Integer, fractional and fractal talbot effects. *J. Mod. Opt.*, **43**:2139–2164, 1996.

[Bla91]   J. Van Bladel. *Singular Electromagnetic Fields and Sources*. New York, IEEE Press, 1991.

[BMS01]   M. Berry, I. Marzoli, and W. Schleich. Quantum carpets, carpets of light. *Physics World*, 39–44, June 2001.

[Bou03]   Z. Bouchal. Nondiffracting optical beams: physical properties, experiments, and applications. *Czech. J. Phys.*, **53**:537–578, 2003.

[Boy80]   R. W. Boyd. Intuitive explanation of the phase anomaly of focused light-beams. *J. Opt. Soc. Am. A*, **70**:877–880, 1980.

[Bra14]   W. L. Bragg. The diffraction of short electromagnetic waves by a crystal. *Proc. Cam. Philos. Soc.*, **17**:43, 1914.

[Bro98]   C. Brosseau. *Fundamentals of Polarized Light*. New York, Wiley, 1998.

[But08]   J. C. Butcher. *Numerical Methods for Ordinary Differential Equations*. New York, Wiley, second edition, 2008.

[BW52]   A. B. Bhatia and E. Wolf. The Zernike circle polynomials occurring in diffraction theory. *Proc. Phys. Soc. B*, **65**:909–910, 1952.

[BW54]   A. B. Bhatia and E. Wolf. On the circle polynomials of zernike and related orthogonal sets. *Proc. Camb. Phil. Soc.*, **50**:40–48, 1954.

[BW78]   J. M. Blatt and V. F. Weisskopf. *Theoretical Nuclear Physics*. New York, Wiley, second edition, 1978.

[BW99]   M. Born and E. Wolf. *Principles of Optics*. Cambridge, Cambridge University Press, seventh edition, 1999.

[CELL91] B. Carrascal, G. A. Estévez, P. Lee, and V. Lorenzo. Vector spherical harmonics and their application to classical electrodynamics. *Eur. J. Phys.*, **12**:184–191, 1991.

[CG99]   P. S. Carney and G. Gbur. Optimal apodizations for finite apertures. *J. Opt. Soc. Am. A*, **16**:1638–1640, 1999.

[CH91]   J. Coster and H. B. Hart. A simple proof of the addition theorem for spherical harmonics. *Am. J. Phys.*, **59**:371–373, 1991.

[CK98]   D. Colton and R. Kress. *Inverse Acoustic and Electromagnetic Scattering Theory*. Berlin, Springer, second edition, 1998.

[Cop35] E. T. Copson. *An Introduction to the Theory of Functions of a Complex Variable*. Oxford, Clarendon Press, 1935.

[Cop46] E. T. Copson. *The Asymptotic Expansion of a Function Defined by a Definite Integral or a Contour Integral*. London, Admiralty Computing Service, 1946.

[Cor97] J. F. Cornwell. *Group Theory in Physics: an Introduction*. San Diego, CA, Academic Press, 1997.

[CTDL77] C. Cohen-Tannoudji, B. Diu, and F. Laloë. *Quantum Mechanics*, volume 1. New York, Wiley, 1977.

[CWZK07] H. Chen, B. Wu, B. Zhang, and J. A. Kong. Electromagnetic wave interactions with a metamaterial cloak. *Phys. Rev. Lett.*, **99**:063903, 2007.

[Dau88] I. Daubechies. Orthonormal bases of compactly supported wavelets. *Comm. Pure Appl. Math.*, **41**:909–996, 1988.

[Dav60] H. Davies. On the convergence of the Born approximation. *Nuclear Physics*, **14**:465–471, 1960.

[Des54] R. Descartes. *The Geometry of Rene Descartes*. New York, Dover, 1954.

[Det65] J. W. Dettman. *Applied Complex Variables*. New York, Macmillan Company, 1965.

[Dev78] A. J. Devaney. Nonuniqueness in the inverse scattering problem. *J. Math. Phys.*, **19**:1526–1531, 1978.

[Dir30] P. A. M. Dirac. *The Principles of Quantum Mechanics*. Oxford, Clarendon Press, 1930.

[DME87] J. Durnin, J. M. Miceli, and J. H. Eberly. Diffraction-free beams. *Phys. Rev. Lett.*, **58**:1499–1501, 1987.

[DOP09] M. R. Dennis, K. O'Holleran, and M. J. Padgett. Singular optics: optical vortices and polarization singularities. In E. Wolf, ed., *Progress in Optics*, volume 53, p. 293–363, Amsterdam, Elsevier, 2009.

[Dur87] J. Durnin. Exact solutions for nondiffracting beams. I. The scalar theory. *J. Opt. Soc. Am. A*, **4**:651–654, 1987.

[Erd56] A. Erdélyi. *Asymptotic Expansions*. New York, Dover Publications, 1956.

[Eul48] L. Euler. *Introductio in Analysin Infinitorum*, volume 1. Lausanne, M. M. Bousquet, 1748.

[EW77]   J. H. Eberly and K. Wódkiewicz. The time-dependent physical spectrum of light. *J. Opt. Soc. Am.*, **67**:1252–1261, 1977.

[Far46] M. Faraday. Experimental researches in electricity. Nineteenth series. *Phil. Trans. Roy. Soc. Lond.*, **136**:1–20, 1846.

[Far51] M. Faraday. On the possible relation of gravity to electricity. *Phil. Trans. Roy. Soc. Lond.*, **141**:1–6, 1851.

[Fol45] L. L. Foldy. The multiple scattering of waves. *Phys. Rev.*, **67**:107–119, 1945.

[FP97] C. Fabry and A. Perot. Sur les franges des lames minces argentées et leur application a la mesure de petites épaisseurs d'air. *Ann. Chim. Phys.*, **12**:459, 1897.

[FP99] C. Fabry and A. Perot. Théorie et applications d'une nouvelle méthode de spectroscopie interférentielle. *Ann. Chim. Phys.*, **16**(7):115, 1899.

[Fre98] I. Freund. '1001' correlations in random wave fields. *Waves Rand. Media*, **8**:119–158, 1998.

[Gab46] D. Gabor. Theory of communication. *J. IEE*, **93**:429–457, 1946.

[GC98] G. Gbur and P. S. Carney. Convergence criteria and optimization techniques for beam moments. *Pure Appl. Opt.*, **7**:1221–1230, 1998.

[GCP+04] G. Gibson, J. Courtial, M. J. Padgett, M. Vasnetsov, V. Pas'ko, S. M. Barnett, and S. Franke-Arnold. Free-space information transfer using light beams carrying orbital angular momentum. *Opt. Exp.*, **12**:5448–5456, 2004.

[Gib99] J. W. Gibbs. Fourier's series. *Nature*, **59**:200, 1898–1899.

[GL96] G. H. Golub and C. F. Van Loan. *Matrix Computations*. Baltimore, Johns Hopkins University Press, third edition, 1996.

[Gor94] F. Gori. Why is the Fresnel transform so little known? In J. C. Dainty, ed., *Current Trends in Optics*, New York, Academic Press, 1994.

[GT08] G. Gbur and R. K. Tyson. Vortex beam propagation through atmospheric turbulence and topological charge conservation. *J. Opt. Soc. Am. A*, **25**:225–230, 2008.

[GW01] G. Gbur and E. Wolf. Relation between computed tomography and diffraction tomography. *J. Opt. Soc. Am. A*, **18**:2132–2137, 2001.

[Ham64] M. Hamermesh. *Group Theory and its Application to Physical Problems*. Reading, MA, Addison-Wesley, 1964.

[Her75] R. Hermann. *Ricci and Levi-Civita's tensor analysis paper*. Brookline, MA, Math Sci Press, 1975.

[Hil54] E. L. Hill. The theory of vector spherical harmonics. *Am. J. Phys.*, **22**:211–214, 1954.

[Hou73] G. N. Hounsfield. Computerized transverse axial scanning (tomography): Part I. description of system. *British J. Radiol.*, **46**:1016–1022, 1973.

[Ise96] A. Iserles. *A First Course in the Numerical Analysis of Differential Equations*. Cambridge, Cambridge University Press, 1996.

[Jac75] J. D. Jackson. *Classical Electrodynamics*. New York, Wiley, second edition, 1975.

[Jac08] J. D. Jackson. Examples of the zeroth theorem of the history of science. *Am. J. Phys.*, **76**:704–719, 2008.

[Jer54] H. G. Jerrard. Transmission of light through birefringent and optically active media: the Poincaré sphere. *J. Opt. Soc. Am.*, **44**:634–640, 1954.

[JMW95] J. D. Joannopoulos, R. D. Meade, and J. N. Winn. *Photonic Crystals*. Princeton, NJ, Princeton University Press, 1995.

[Jon41] R. C. Jones. New calculus for the treatment of optical systems. *J. Opt. Soc. Am.*, **31**:488–493, 1941.

[Kno51] K. Knopp. *Theory and Application of Infinite Series*. Glasgow, Blackie & Son, Ltd., second edition, 1951.

[Kow75] S. Kowalevski. Zur Theorie der partiellen Differentialgleichung. *J. reine angewandte Math.*, **80**:1–32, 1875.

[Kre78] E. Kreyszig. *Introductory Functional Analysis with Applications*. New York, Wiley, 1978.

[Lax52] M. Lax. The multiple scattering of waves. II. The effective field in dense systems. *Phys. Rev.*, **85**:621–629, 1952.

[Leo06]   U. Leonhardt. Optical conformal mapping. *Science*, **312**:1777–1780, 2006.

[LG04]   K. Ladavac and D. G. Grier. Microoptomechanical pumps assembled and driven by holographic optical vortex arrays. *Opt. Exp.*, **12**:1144–1149, 2004.

[Loh88]   A. W. Lohmann. An array illuminator based on the Talbot-effect. *Optik*, **79**:41–45, 1988.

[Loh93]   A. W. Lohmann. Image rotation, Wigner rotation, and the fractional Fourier transform. *J. Opt. Soc. Am. A*, **10**:2181–2186, 1993.

[Lor67]   L. Lorenz. On the identity of the vibrations of light with electrical currents. *Phil. Mag.*, **34**:287–301, 1867.

[LP09]   U. Leonhardt and T. G. Philbin. Transformation optics and the geometry of light. In E. Wolf, ed., *Progress in Optics*, volume 53, p. 69–152, Amsterdam, Elsevier, 2009.

[Mal98]   S. Mallat. *A Wavelet Tour of Signal Processing*. San Diego, Academic Press, 1998.

[McC69]   C. W. McCutchen. Two families of apodization problems. *J. Opt. Soc. Am.*, **59**:1163–1171, 1969.

[MO93]   D. Mendlovic and H. M. Ozaktas. Fractional Fourier transforms and their optical implementation. I. *J. Opt. Soc. Am. A*, **10**:1875–1881, 1993.

[Moi05]   B. L. Moiseiwitsch. *Integral Equations*. New York, Dover, 2005.

[MP98]   O. J. F. Martin and N. B. Piller. Electromagnetic scattering in polarizable backgrounds. *Phys. Rev. E*, **58**:3909–3915, 1998.

[MT88]   J. B. Marion and S. T. Thornton. *Classical Dynamics of Particles and Systems*. San Diego, Harcourt Brace Jovanovich, third edition, 1988.

[MW95]   L. Mandel and E. Wolf. *Optical Coherence and Quantum Optics*. Cambridge, Cambridge University Press, 1995.

[Nac88]   A. I. Nachman. Reconstructions from boundary measurements. *Annals Math.*, **128**:531–576, 1988.

[Nah98]   P. J. Nahin. *An Imaginary Tale*. Princeton, NJ, Princeton University Press, 1998.

[Nam80]   V. Namias. The fractional order Fourier transform and its application to quantum mechanics. *J. Inst. Maths. Applics.*, **25**:241–265, 1980.

[NB74]   J. F. Nye and M. V. Berry. Dislocations in wave trains. *Proc. Roy. Soc. Lond. A*, **336**:165–190, 1974.

[Nol76]   R. J. Noll. Zernike polynomials and atmospheric turbulence. *J. Opt. Soc. Am.*, **66**:207–211, 1976.

[OM93]   H. M. Ozaktas and D. Mendlovic. Fractional Fourier transforms and their optical implementation. II. *J. Opt. Soc. Am. A*, **10**:2522–2531, 1993.

[OM95]   H. M. Ozaktas and D. Mendlovic. Fractional Fourier optics. *J. Opt. Soc. Am. A*, **12**:743–751, 1995.

[Pan56]   S. Pancharatnam. Generalized theory of interference, and its applications. *Proc. Indian. Acad. Sci.*, A**44**:247–262, 1956.

[Pel11]   A. J. Pell. *Biorthogonal Systems of Functions*. PhD thesis, University of Chicago, 1911.

[Pen00]   J. B. Pendry. Negative refraction makes a perfect lens. *Phys. Rev. Lett.*, **85**:3966–3969, 2000.

[PF94]   P. Pellat-Finet. Fresnel diffraction and the fractional-order Fourier transform. *Opt. Lett.*, **19**:1388–1390, 1994.

[Pic79]   C. E. Picard. Sur une propriété des fonctions entières. *Comptes Rendus*, **88**:1024–1027, 1879.

[Poh92]   K. C. Pohlmann. *The Compact Disc Handbook*. Oxford, Oxford University Press, second edition, 1992.

[Poi92]   H. Poincaré. *Théorie Mathématique de la Lumiere*, volume 2. Paris, Gauthiers-Villars, 1892.

[PP37]   M. Plancherel and G. Pólya. Fonctions entieres et integrales de Fourier multiples. *Comm. Math. Helv.*, **9**:224–248, 1936–1937.

[PSS06]   J. B. Pendry, D. Schurig, and D. R. Smith. Controlling electromagnetic fields. *Science*, **312**:1780–1782, 2006.

[PTVF92] W. H. Press, S. A. Teukolsky, W. T. Vetterling, and B. P. Flannery. *Numerical Recipes in C++*. Cambridge, Cambridge University Press, second edition, 1992.

[Rad17]   J. Radon. Über die Bestimmung von Funktionen durch ihre Integralwerte längs gewisser Mannigfaltigkeiten. *Berichte Sächsische Akademie der Wissenschaften, Leipzig, Math. Phys. Kl*, **69**:262–277, 1917.

[Ram85]   A. G. Ramm. Some inverse scattering problems of geophysics. *Inverse Problems*, **1**:133–172, 1985.

[Ray81]   Lord Rayleigh. On copying diffraction-gratings, and on some phenomena connected therewith. *Phil. Mag.*, **11**:196–205, 1881.

[Red53]   R. M. Redheffer. On a theorem of Plancherel and Pólya. *Pacific J. Math.*, **3**:823–835, 1953.

[Rie68]   B. Riemann. Ueber die Darstellbarkeit einer Function durch eine trigonometrische Reihe. *Abh. d. Ges. d. Wiss. z. Göttingen*, **13**:87–131, 1868.

[Rod16]   O. Rodrigues. Mémoire sur l'attraction des sphéroïdes. *Corresp. sur l'École polytechnique*, **3**:361–385, 1814–1816.

[RW96]   M. C. Roggemann and B. M. Welsh. *Imaging Through Turbulence*. Boca Raton, FL, CRC Press, 1996.

[SD92]   A. Schatzberg and A. J. Devaney. Super-resolution in diffraction tomography. *Inverse Problems*, **8**:149–164, 1992.

[Sha80]   R. Shankar. *Principles of Quantum Mechanics*. New York, Plenum Press, 1980.

[SMJ⁺06] D. Schurig, J. J. Mock, B. J. Justice, S. A. Cummer, J. B. Pendry, A. F. Starr, and D. R. Smith. Metamaterial electromagnetic cloak at microwave frequencies. *Science*, **314**:977–980, 2006.

[Som64]   A. Sommerfeld. *Optics*. New York, Academic Press, 1964.

[Spr81]   G. Springer. *Introduction to Riemann Surfaces*. New York, Chelsea Publishing, 1981.

[Sta86]   J. J. Stamnes. *Waves in Focal Regions*. Bristol, Adam Hilger, 1986.

[Sul92]   D. M. Sullivan. Frequency-dependent FDTD methods using Z transforms. *IEEE Trans. Ant. Prop.*, **40**:1223–1230, 1992.

[Sul97]   T. J. Suleski. Generation of Lohmann images from binary-phase Talbot array illuminators. *Appl. Opt.*, **36**:4686–4691, 1997.

[Sul00]   D. M. Sullivan. *Electromagnetic Simulation Using the FDTD Method*. New York, IEEE Press, 2000.

[SV01]   M. S. Soskin and M. V. Vasnetsov. Singular optics. In E. Wolf, ed., *Progress in Optics*, volume 42, 219–276, Amsterdam, Elsevier, 2001.

[Tal36]   H. F. Talbot. Facts relating to optical science. No. IV. *Phil. Mag.*, **9**:401–440, 1836.

[TH05]   A. Taflove and S. C. Hagness. *Computational Electrodynamics: The Finite-Difference Time Domain Method*. Norwood, MA, Artech House, Inc., 2005.

[TSC06]   M. Testorf, T. J. Suleski, and Y.-C. Chuang. Design of Talbot array illuminators for three-dimensional intensity distributions. *Opt. Exp.*, **14**:7623–7629, 2006.

[VBL99]   T. D. Visser, H. Blok, and D. Lenstra. Theory of polarization-dependent amplification in a slab waveguide with anisotropic gain and losses. *IEEE J. Quant. Elect.*, **35**:240–249, 1999.

[Ves68]   V. G. Vesalago. The electrodynamics of substances with simultaneously negative values of ϵ and μ. *Soviet Physics Uspekhi*, **10**:509–514, 1968.

[vL13]   M. von Laue. Röntgenstrahlinterferenzen. *Physikalische Zeitschrift*, **14**:1075–1079, 1913.

[Wat44]   G. N. Watson. *A Treatise on the Theory of Bessel Functions*. Cambridge, Cambridge University Press, 1944.

[Wei76]   K. Weierstrass. Zur Theorie der eindeutigen analytischen Functionen. *Abh. der Preuss. Akad. Wiss. zu Berlin (Math. Klasse)*, **11**:11–60, 1876.

[WH93]   E. Wolf and T. Habashy. Invisible bodies and uniqueness of the inverse scattering problem. *J. Mod. Opt.*, **40**:785–792, 1993.

[Wig32]   E. P. Wigner. On the quantum correction for thermodynamic equilibrium. *Phys. Rev.*, **40**:749–759, 1932.

[WNV85]   E. Wolf and M. Nieto-Vesperinas. Analyticity of the angular spectrum amplitude of scattered fields and some of its consequences. *J. Opt. Soc. Am. A*, **2**:886–890, 1985.

[Woj06]   J. Wojdylo. On the coefficients that arise from Laplace's method. *J. Comp. Appl. Math.*, **196**:241–266, 2006.

[Wol54]   E. Wolf. Optics in terms of observable quantities. *Nuovo Cimento*, **12**:884–888, 1954.

[Wol78]   E. Wolf. Coherence and radiometry. *J. Opt. Soc. Am.*, **68**:6–17, 1978.

[Wol79]   K. B. Wolf. *Integral Transforms in Science and Engineering*. New York, Plenum Press, 1979.

[Wol83]   E. Wolf. Recollections of Max Born. *Optics News*, **9**:10–16, 1983.

[Wol01]   E. Wolf. *Selected Works of Emil Wolf with Commentary*. Singapore, World Scientific, 2001.

[Wol07]   E. Wolf. *Introduction to the Theory of Coherence and Polarization of Light*. Cambridge, Cambridge University Press, 2007.

[Yee66]   K. S. Yee. Numerical solution of initial boundary value problems involving Maxwell's equations in isotropic media. *IEEE Trans. Ant. Prop.*, **AP-14**:302–307, 1966.

Index

Italics refer to the primary entry for a topic.

aberrations, 656
  astigmatism, 657
  balancing of, 661
  coma, 657
  curvature, 657
  defocus, 657
  distortion, 657
  piston, 657
  spherical, 657
  tilt, 657
absolute convergence, 203
absolutely integrable, 359
adaptive optics, 662
addition formula
  Bessel, 577
  Gegenbauer, 578
  spherical harmonic, 605
adjoint matrix, 105
Airy pattern, 562
Airy's integral, 769
aliasing, 426
almost linear system, 466
alternating harmonic series, 201
Ampère's law, 54
analytic continuation, 312
  and Laplace transform, 397
analyticity, 263
angular spectrum representation, 378
apodization, 742
argument principle, 337
asymptotic series, 755
asymptotics
  Erdélyi's theorem, 760
  Fourier transform, 756
  Laplace transform, 757
  Laplace's method, 758
  Watson's lemma, 762

BAC-CAB rule, 13
balancing of aberrations, 661
bandlimited, 419
basis, 15, 118, 242

Beer's law, 411
Bessel function, 551
  addition formula, 577
  asymptotic form, 566
  generating function of, 555
  Hankel, 564
  integral representations, 560
  modified, 565
  orthogonality, 569
  recurrence relation, 558
  Rodrigues' formula, 559
  series form, 553
  spherical, 573
  zeros of, 567
Bessel's equation, 551, *552*
Bethe phase, 604
bilinear transformation, 329
binomial series, 216
biorthogonal system, 145, 160
blazed grating, 233
Born approximation, 141, 223, 711
Born series, 222, 697, 711
  convergence of, 221
boundary conditions, 460, 475
  Cauchy, 515
  Dirichlet, 515, 519
  hidden, 525, 526, 611, 632
  Neumann, 515, 519
  Robin, 519
bra-ket notation, 14, 89
brachistochrone, 722
Bragg condition, 20
branch point, 279
  algebraic, 282
  logarithmic, 282
  transcendental, 282
Bravais lattice, 17

calculus of variations, 718
canonical form, 510, 514
catenary, 730
Cauchy conditions, 515

Cauchy principal value, 298
Cauchy's integral formula, 270
Cauchy's theorem, 267
Cauchy–Goursat theorem, 267
Cauchy–Riemann conditions, 262
Cauchy-Kowalevski theorem, 516
causality, 299
chaotic behavior, 468
characteristic value, 694
Chebyshev function, 610, *650*
  first kind, 650, 652
    generating function, 652
    orthogonality, 653
    recurrence relation, 652
    Rodrigues' formula, 653
  generating function, 650
  second kind, 650
    orthogonality, 652
    recurrence relation, 650
    Rodrigues' formula, 651
Christoffel symbol, 173
cloaking device, 345
closure relation, 118, 186, 242
cofactor, 97
column vector, 89
comb function, 424, 453
comparison test, 203
complex conjugate, 257
complex derivatives, 261
complex number, 252, *254*
  argument of, 256
  imaginary part, 254, 257
  modulus of, 256
  real part, 254, 257
  roots of, 256
complex plane, 254
  compactified, 317
computed tomography, 410
  filtered back-projection algorithm, 415
  projection-slice theorem, 414
condition number, 153
conditional convergence, 203
Condon–Shortley phase, 604
confluent hypergeometric equation, 785
confluent hypergeometric function, 785
conformal mapping, 325, 327
conjugate matrix, 128
conservative force, 48
  and complex functions, 265
constraint
  holonomic, 735
  isoperimetric, 744
convergence
  absolute, 203
  in the mean, 246
  tests
    comparison, 203
    Gauss', 209
    guidelines for, 209
    integral, 205

Kummer's, 207
Raabe's, 208
ratio, 205
root, 204
uniform, 211
convolution theorem, 368
  for DFTs, 432
  for z-transforms, 443
cosines
  direction, 4
  law of, 10
Coulomb's law, 177
covariant derivative, 173
curl, 41
  curvilinear, 71
current
  displacement, 54
curvilinear coordinate system, 66
cut line, 279
cycloid, 724
cylindrical coordinates, *73*

d'Alembert solution, 541
Darwin phase, 604
delta
  Dirac, *see* Dirac delta function, 181
  Kronecker, *see* Kronecker delta, 8
  sequence, 182
dense periodic orbits, 469
determinant, 94
  and inverse matrix, 94
  and system solution, 95
diffraction
  circular aperture, 561
  Fraunhofer, 352, 386, 743
    and lenses, 376
  Fresnel, 351, 386
  Fresnel–Kirchoff, 669
  Rayleigh–Sommerfeld integral, 350, 386, 701
diffraction grating, 230
diffraction order, 230
diffraction-limited, 656
diffusion equation, 508, 682
dimension, 15
Dirac delta function, 181
  calculus of, 184
  cylindrical coordinates, 190
  multi-variable, 188
  sequence, 182
  sifting property of, 181
  spherical coordinates, 189
direct product, 118, 162
direction cosines, 4, 7
Dirichlet conditions, 515, 519
Dirichlet series, 204
discrete Fourier transform, 428, 430
displacement vector, 5
distribution, *182*
divergence, 37
  curvilinear, 70

divergence theorem, 45
dyadic, 165, 706
dyadic Green's function, 704

eigenvalue equation
    generalized, 623
eigenvalues, 110
    degenerate, 113
    left-right, 143
    of Fourier transform, 372
    of normal matrix, 114
eigenvectors, 110
Eikonal equation, 718, 748, 750
Einstein summation convention, 159
elliptic equation, 511, 525
entire analytic, 263
equilibrium points, 462
    asymptotically stable, 468
Erdélyi's theorem, 760
Euler's equation, 722
    second form, 728
    several dependent variables, 733
    several independent variables, 731
    with undetermined multipliers, 736
Euler's formula, 216, 255, 259
Euler's method, 490
    improved, 494
Euler-Mascheroni constant, 555
Ewald sphere, 344
    limiting, 344
exact equation, *see* ordinary differential equation,
        exact, 471
exponential decay, 458
exponential order, 397

Fabry-Perot interferometer, 195
fan-beam tomography, 416
Faraday rotator, 85
Faraday's law, 53
fast Fourier transform, 433
Fermat's principle, 715, 750
filtered back-projection algorithm, 415
finesse, 197
finite difference time domain method, 445
focal planes, 126
Foldy-Lax theory, 139
Fourier series
    complex-valued, 240
    cosine, 240
    fundamental frequency, 234
    real-valued, 233
    sine, 241
Fourier transform, *352*
    discrete, 428, 430
    fast, 433
    fractional, 401
        and Wigner functions, 410
    inverse, 353
    multi-variable, 374
    of derivatives, 365

windowed, 403
fractional Fourier transform, *see* Fourier transform,
        fractional, 401
Fraunhofer diffraction, *see* diffraction,
        Fraunhofer, 352
Fredholm equation, 693, 711
free spectral range, 197
Fresnel diffraction, *see* diffraction, Fresnel, 351
Fresnel integrals, 294
Fresnel transform, 386
    inverse of, 388
Fresnel–Kirchoff diffraction formula, 669
Frobenius method, 482
functional, 720
fundamental theorem of algebra, 339

gauge
    Coulomb, 59
    freedom, 57
    Lorenz, 58
Gauss' law, 52, 177
Gauss' test, 209
Gauss' theorem, 45
Gauss–Jordan matrix inversion, 100
Gaussian elimination, 99
Gaussian quadratures, 494
Gegenbauer polynomial, 610
    addition formula, 578
    generating function, 610
    orthogonality, 611
generalized momenta, 740
generating function
    Bessel, 555
    Chebyshev, 650
    Chebyshev first kind, 652
    Gegenbauer, 610
    Jacobi, 655
    Laguerre, 642
    Laguerre function, 644
    Legendre, 589
geometric phase, 325
geometric series, 200
geometrical optics, 715
    and matrices, 120
    foundations of, 748
    intensity law, 752
Gibbs phenomenon, 212, 243, 360
Gouy phase shift, 636
gradient, 35
    curvilinear, 69
gradient index, 725
Gram–Schmidt orthonormalization,
        116, 700
Green's function, 669
    advanced, 681
    diffusion, 682
    dyadic, 704
    modal expansion of, 689
    Helmholtz, 677
    Poisson, 676

Green's function (*cont.*)
  retarded, 681
  wave, 680
Green's functions
Green's theorem, 46
group theory, 14, 107

Haar basis, 408
half-wave plate, 85
Hamilton's principle, 720
Hamiltonian, 740
Hankel function, 564
harmonic conjugates, 264
harmonic function, 264
harmonic oscillator, 458
  quantum, 631
harmonic series, *see* series, harmonic, 200
Heaviside function, 187
Helmholtz equation, 505, 665, 677
Helmholtz-Kirchoff integral theorem, 667
Hermite equation, 627
Hermite polynomial, 627
  generating function, 629
  orthogonality, 630
  recurrence relation, 629
  Rodrigues' formula, 630
Hermite–Gauss beams, 633
Hermite-Gauss beams, 639
Hermitian matrix, 105
  eigenvalues of, 112
Hermitian operator, 625
Hessian matrix, 220
Heun formula, 494
Hilbert-Schmidt kernel, 699
holonomic constraint, 735
hydrogen atom, 645
hyperbolic equation, 510, 519
hypergeometric equation, 784
  confluent, 785
hypergeometric function, 783
  confluent, 785
  contiguous function relations, 784

imaginary number, 252, 253
improper rotations, 168
infinite sequence, *198*
infinite series, *see* series, infinite, 200
initial conditions, 515
inner product, 15, 162
inner product space, 15
integral equation, 693
  Fredholm, 693
  kernel, 693
  singular, 695
  Volterra, 694
integral test, 205
integrating factor, 472
integration
  complex, 265
  path, 29, 265

surface, 31
vector, 29
volume, 35
intensity law of geometrical optics, 752
interferometer
  Fabry-Perot, 195
inverse problems, 342, 411
irrotational, 43
isoperimetric constraint, 744

Jacobi polynomial, 654
  generating function, 655
  orthogonality, 655
  recurrence relation, 655
  Rodrigues' formula, 655
Jacobi-Anger expansion, 557
Jacobian vector, 220
Jones vector, *83*, 322
Jordan's lemma, 291

kernel, 693
  Hilbert–Schmidt, 699
  homogeneous, 694
  separable, 694
Kramers–Kronig relations, 300
Kronecker delta, 8
Kummer's test, 207

Lagrange undetermined multiplier,
      736
Lagrangian, 720
Laguerre equation, 641
Laguerre function, 644
  generating function, 644
  Rodrigues' formula, 644
Laguerre polynomial, 642
  generating function, 642
  recurrence relations, 643
Laguerre–Gauss beams, 647
Laplace transform, 395
  and ODEs, 485
  inverse of, 399
Laplace's equation, 44, 180, 508
  in spherical coordinates, 585
Laplace's integral, 594
Laplace's method, 758
Laplacian, 44
  transverse, 634
lattice planes, 19
Laurent series, 273
  and z-transform, 437
law of cosines, 10
Lebesgue integration
  and distributions, 181
Legendre duplication formula, 573
Legendre equation, 587, *587*
  associated, 586, 597
Legendre function, 587, 588
  associated, 586, 597
    orthogonality, 601

recurrence, 599
  Rodrigues' formula, 597
generating function, 589
integral representations, 592
Laplace's integral, 594
orthogonality, 594
recurrence relations, 590
Rodrigues' formula, 592
Schlaefli integral, 593
series form, 589
lensmaker's formula, 121
level curves, 767
Levi-Civita tensor, 11, 169
generalized, 96
limit
of sequence, 198
linear canonical transform, 391
inverse of, 393
linear independence, 14
of functions, 475
linear optical system, 366
linear vector space
basis, 15
definition, 13
dimension, 15
Liouville's theorem, 333
Liouville–Neumann series, 696, 711
lowering operator, 632
LU decomposition, 155

Möbius transformation, 329
Maclaurin series, 215
mapping, 316, *325*
inverse of, 326
one-to-one, 326
onto, 326
matrix
addition, 89
adjoint, 105
definition, 88
diagonalization of, 107
functions of, 120
Hermitian, 105
inverse, *93*
lower triangular, 155
multiplication, 90
normal, 106
null, 89
null space of, 143
orthogonal, 102
Pauli, 106
range of, 143
rank of, 143
transpose, 102
unit, 92
unitary, 106
upper triangular, 155
Maximum modulus principle, 335
Maxwell's equation, 704
Maxwell's equations, 51, 614

and the z-transform, 445
Mercer's theorem, 700
metamaterials, 345
method of images, 544
Green's functions, 685
Rayleigh-Sommerfeld diffraction, 703
method of stationary phase, 764
double integrals, 771
method of steepest descent, 766
metric, 68, 164
chordal, 320
metric space, 16, 257
Mexican hat wavelet, 406
mirage, 728
Morera's theorem, 334
multipole
electric, 616
magnetic, 616
multipole expansion, 611
multivalued, 256, 259, 279

neighborhood, 263
Neumann conditions, 515, 519
Neumann function, 551, *554*
asymptotic form, 566
series form, 554
spherical, 573
nodes
improper, 467
proper, 467
non-Euclidean space, 171
nondiffracting beam, 579
normal matrix, 106
null space, 143
Nyquist frequency, 421
and compact discs, 422
and the DFT, 430

optical vortices, 305
order parameter, 38, 215, 491, 755
ordinary differential equation, 459
autonomous, 460
boundary conditions, 460, 475
exact, 471
homogeneous, 460
inhomogeneous, 460
integrating factor, 472
numerical methods
Euler's method, 490
explicit and implicit, 492
order, 489
Runge–Kutta, 497
stiffness, 500
order of, 460
power series solution, 480
separable, 470
systems of, 487
variation of parameters, 478
Wronskian, 476
orthogonal matrix, 102

orthogonality
  of vectors, 16
orthogonality condition, 8
Ostrogradski equation, 733
outer product, 118, 162

Pancharatnam phase, 325
parabolic equation, 511, 530
paraxial wave equation, 634, 648
Parseval's theorem, 369
  for DFTs, 432
partial differential equation, 505
  canonical form, 510, 514
  elliptic, 511, 525
  hyperbolic, 510, 519
  integral transforms, 534
  method of images, 544
  parabolic, 511, 530
Pauli matrices, 106
phase anomaly, 636
phase portrait, 462
phase singularities, 304
phase space, 460, *462*
  equilibrium points, 462
  trajectories, 462
  vector field, 462
photonic crystals, 17
Picard's theorem, 277
pivoting, 154
Plancherel's identity, 369
  for Fresnel transform, 390
  for LCTs, 393
  for wavelets, 408
Plancherel-Polya theorem, 341
Pochhammer symbol, 783
Poincaré sphere, 322
point at infinity, 317
point conditions, 475
Poisson equation, 676
Poisson sum formula, 424
Poisson's equation, 180
polarization, 83
  circular, 83
  elliptical, 83
  linear, 83
polarizer, 84
position vector, 3
potential theory, 48
power series, 220
  and ODEs, 480
primitive vectors, 17
principal planes, 127
principal value, 298
projection-slice theorem, 414
proper rotations, 168
pseudotensor, 169
pseudovector, 168

quadratures, 494
quantum number, 647

quarter-wave plate, 84
quotient theorem, 163

Raabe's test, 208
radiation condition
  Sommerfeld, 668
Radon transform, 413
raising operator, 632
range, 143
rank
  of matrix, 143
  of tensor, 159
ratio test, 205
ray, 715
Rayleigh equation, 595
Rayleigh range, 635
Rayleigh–Sommerfeld diffraction, 350, 386, 701, 703
rays
  meridional, 122
reciprocal lattice, 22
recurrence relation
  associated Legendre, 599
  Bessel, 558
  Chebyshev first kind, 652
  Chebyshev second kind, 650
  Hermite, 629
  Jacobi, 655
  Laguerre, 643
  Legendre, 590
  spherical harmonics, 602
residue, 285
residue theorem, 286
  and definite integrals, 288
  extended, 298
Riemann mapping theorem, 329
Riemann sphere, 317
Riemann surface, 280
  and stereographic projection, 320
Riemann–Lebesgue lemma, 757
Robin conditions, 519
Rodrigues' formula
  associated Legendre, 597
  Bessel, 559
  Chebyshev first kind, 653
  Chebyshev second kind, 651
  Hermite, 630
  Jacobi, 655
  Laguerre function, 644
  Legendre, 592
  Zernike, 659
root test, 204
Rouché's theorem, 338
round-off error, 489
row vector, 89
Runge–Kutta method, 497

saddle point
  in ODEs, 467
sampling theorem, 419
  aliasing, 426

sawtooth wave, 236
scale factors, 68
scattering amplitude, 224
scattering potential, 710
scattering theory, 77, 221, 709
 Foldy-Lax, 139
Schlaefli integral, 593
Schwarz-Christoffel transformation, 331
secular equation, 110
Seidel aberrations, 657
self-adjoint, 625
separable equation (ODE), 470
separable systems, 64
separation of variables, 505, *517*
sequence
 infinite, 198
 of partial sums, 199
series
 alternating harmonic, 201
 binomial, 216
 Dirichlet, 204
 geometric, 200
 harmonic, 200, 207
 infinite, 200
 of functions, 210
 power, 220
 Taylor, *see* Taylor series, 213
shadowgram, 411
similarity transformation, 105
singular optics, 302
singular points, 263
 of ODEs, 482
singular value decomposition, 147
singularities
 essential singularity, 277
 higher-order pole, 277
 removable, 276
 simple pole, 276
skew coordinates, 160
solenoidal, 43
solid angle, *77*
Sommerfeld radiation condition, 668, 679
span, 15
spatial filtering, 375
spatial frequencies, 380
special function, 551
spherical coordinates, *76*
spherical harmonic
 addition theorem, 605
 vector, 614, 616
spherical harmonics, 602
 recurrence relations, 602
spiral point, 467
square wave, 237
stationary phase, 566, 764
 double integrals, 771
stationary points, 719, 764
steepest descents, 766
steepest paths, 767
step function, *see* Heaviside function, 187

stereographic projection, 317
stiff equations, 500
 and Bessel functions, 576
Stokes parameters, 323
Stokes' theorem, 48, 266
Sturm–Liouville equation, 622
 Green's function, 670
 inhomogeneous, 670
summation convention, 159
system matrix, 125

Talbot carpet, 452
Talbot distance, 451
Talbot effect, 449
 fractional, 455
tangent line method, 490
Taylor series, 213
 complex, 271
 multi-variable, 218
 shortcuts, 217
Tchebyshef function, *see* Chebyshev function, 650
telescope, 132
tensor, *159*
 Christoffel symbol, 173
 contraction, 162
 covariant derivative, 173
 Levi-Civita, 11, 169
 metric, 164
 outer product, 162
 quotient theorem, 163
 summation convention, 159
time–frequency atoms, 404
topological charge, 306
topological mixing, 468
topology, 320
trajectory, 462
transformation optics, 345
triangle inequality, 257
truncation error, 489

ultraspherical polynomial, *see* Gegenbauer
 polynomial, 610
uncertainty relation, 371
uniform convergence, 211
unitary matrix, 106

variation of parameters, 478
variational principle, 718
vector
 addition, 2
 coordinate representation, 3
 coordinate system invariance of, 4
 definition, 1
 derivative
  curl, 41
  divergence, 37
  gradient, 35
 displacement, 5
 integration, 29
 irrotational, 43

vector (*cont.*)
  Jones, 83
  multiplication
    cross product, 10
    dot product, 9
    scalar, 9
    triple scalar, 12
    triple vector, 13
  position, 3
  primitive, 17
  solenoidal, 43
vector field, 28
  of phase space, *see* phase space, vector field, 462
vector potential, 44, 50
vector space, 1
vector spherical harmonic, 614, 616
Volterra equation, 694
von Laue condition, 22

Watson's lemma, 762
wave equation, 508, 680
wavefront curvature, 636
waveguides, 505, 550, 570

wavelet, 406
  admissibility condition, 408
  Haar basis, 408
  Mexican hat, 406
  transform, 407
Weierstrass' theorem, 277
weight function, 624
Weyl representation, 382
Whittaker–Shannon sampling formula, 419
Wigner transform, 409
winding number, 306
windowed Fourier transform, 403
Wronskian, 476

X-ray diffraction, 17, 246

z-transform, 437
  final value theorem, 444
  initial value theorem, 444
  inverse of, 438
Zernike polynomial, 657
  orthogonality, 660
  Rodrigues' formula, 659

Printed in the United States
by Baker & Taylor Publisher Services

Printed in the United States
by Baker & Taylor Publisher Services